Plant Speed Breeding and High-Throughput Technologies

The research field of "plant speed breeding" is promising to meet the future demand for crop improvement in an efficient and effective manner. The process accelerates breeding and research programs through the shortening of the generation time of crops. *Plant Speed Breeding and High-Throughput Technologies* explores and analyzes a range of innovations from CRISPR/Cas9-based genome editing to state-of-the-art controlled-environment growth chambers.

This book integrates achievements of plant speed breeding technologies with an emphasis on the fundamental and advanced systems for genome editing, functional genomics, controlled-environment growth chambers, genomic and phenotypic selection methods, and high-throughput technologies. It examines the applications of technologies across major model plants and key crops such as cereals, vegetables, and fruit trees. This book explores high-throughput genotyping using next-generation sequencing systems, enabling precise single nucleotide polymorphism markers, quantitative trait loci mapping, and genome-wide association studies. These tools are used in identifying candidate genes that will revolutionize crop breeding, as they are capable of tolerating complex abiotic and biotic stresses induced by global climate change.

Serving as an ideal reference on plant speed breeding technologies, this book is for graduate students, teachers, researchers, and experts in the fields of plant sciences and plant breeding as well as agronomy and agriculture.

Plant Speed Breeding and High-Throughput Technologies

Edited by

Jen-Tsung Chen

CRC Press
Taylor & Francis Group
Boca Raton London New York

CRC Press is an imprint of the
Taylor & Francis Group, an **informa** business

Designed cover image: Shutterstock

First edition published 2025
by CRC Press
2385 NW Executive Center Drive, Suite 320, Boca Raton FL 33431

and by CRC Press
4 Park Square, Milton Park, Abingdon, Oxon, OX14 4RN

CRC Press is an imprint of Taylor & Francis Group, LLC

ISBN: 978-1-032-55327-6 (hbk)
ISBN: 978-1-032-56268-1 (pbk)
ISBN: 978-1-003-43473-3 (ebk)

DOI: 10.1201/b23372

Typeset in Times
by MPS Limited, Dehradun

Contents

Preface....vii
About the Editor....ix
Contributors....x

Chapter 1 Technological Development and Prospects of Crop Speed Breeding....1

Marcelo Picanço de Farias, Charles Hobi Zimmer, João Kafer, Mayla Daiane Correa Molinari, Fabricio Barbosa Monteiro Arraes, Renata Fuganti-Pagliarini, Isabela Yamauchi Torres, Samantha Vieira Abbad, Liliane Mertz-Henning, Alexandre Nepomuceno, and Hugo Molinari

Chapter 2 Speed Breeding in the Era of High-Throughput OMICS: Recent Developments and Future Outlook....17

Furqan Tayyab, Momna Mehmood, Aqsa Parvaiz, Ghulam Mustafa, Muhammad Sarwar Khan, and Faiz Ahmad Joyia

Chapter 3 Accelerate Crop Breeding Utilizing Modern Genomics and Phenomics Tools....37

Amooru Harika, Sivapragasam Ezhumalai, Biswabiplab Singh, Elangovan Allimuthu, Sukumar Taria, Dhandapani Raju, Sudhir Kumar, Mahesh Kumar, Madhurima Das, and Viswanathan Chinnusamy

Chapter 4 Implementation of Novel Genomic and Biotechnological Interventions for Accelerated Breeding of Crops....53

Muhammad Ramzan Khan, Nazia Rehman, Safeena Inam, Muhammad Kashif Naeem, Amna Muhammad, Muhammad Uzair, Aamir Riaz, Obaid Ur Rehman, Fara Muqaddas, Maryam Murtaza, Bilal Saleem, Muhammad Aqeel, and Muhammad Zahid

Chapter 5 Recent Advances in Plant Phenomics and Speed Breeding for Climate-Smart Agriculture....82

Rubab Zahra Naqvi, Muhammad Farooq, Rida Zahid, Ifrah Imran, Muhammad Arslan Mahmood, Muhammad Jawad Akbar Awan, Shahid Mansoor, Imran Amin, and Muhammad Asif

Chapter 6 Genome Editing Technologies for Accelerating Crop Improvement: An Updated Overview....104

Anil Namdev Kale, Raj Kumar, Rushikesh Sanjay Mane, Sagar Krushnaji Rangari, Praful Mukund Damdar, Zeba Quaiyum, Karma L. Bhutia, and Akbar Hossain

Chapter 7 The CRISPR/Cas System for Plant Speed Breeding: Advances and Prospects....122

Muhammad Jawad Akbar Awan, Muhammad Ismail Buzdar, Asma Zia, Aiman Raza, Rubab Zahra Naqvi, Imran Amin, and Shahid Mansoor

Chapter 8 Genomic Selection Tools for Plant Speed Breeding 149

Sumi Sarkar, Marium Khatun, F. M. Era, and A. K. M. Aminul Islam

Chapter 9 Network of Low-Cost RGB-Depth Cameras and Mini-Computers for High-Throughput Monitoring of Plant Stress Response 168

Cordier Mathis, Rasti Pejman, Torres Cindy, and Rousseau David

Chapter 10 Speed Breeding and High-Throughout Phenotypic Platforms to Accelerate Improvements in Cereal Crops 187

Indra Rautela, Sonika Kalia, Megha Katoch, Shanvi Nautiyal, Natasha Charaya, Monika Kalia, Himanshu Sharma, and Vinay Sharma

Chapter 11 Proteomics: Aims at Stress Mitigation in Soybean under Abiotic Stress 207

Farhat Yasmeen, Rizwana Parveen, and Setsuko Komatsu

Chapter 12 Genome Editing Systems for Speed Breeding of Tomato (*Solanum lycopersicon* L.) 242

Sonupriya Sahu, Madhusmita Barik, Jatindra Nath Mohanty, Animesh Pattnaik, Rukmini Mishra, and Raj Kumar Joshi

Chapter 13 Speed Germplasm Generation and Precision-Designed Breeding in Tomato Based on the CRISPR/Cas9 System 254

Jiao Wang, Shuting Ding, and Kai Shi

Chapter 14 Speed Breeding in Fruit Trees: Current Achievements and Future Perspectives 264

Vadim Lebedev

Chapter 15 Molecular, Morpho-Functional, and Agronomic Phenotyping Tools for Grapevine Selection and Breeding 294

J. Miguel Costa, Luísa Carvalho, Rui Figueira, Hiroyuki Shimono, and Jorge Marques da Silva

Chapter 16 Rice Submergence Tolerance: Emerging Molecular Breeding Strategies in the Post-Genomic Era 318

Ashish Gautam, Anamika Thakur, Charupriya Chauhan, Vinay Sharma, Karthick Babu Sivakumar, Rakesh Kumar, and Sanjay Kumar Chetia

Chapter 17 Speed Breeding in Vegetable Crops: Implications and Challenges 343

Ruma Devi and Shivani Chauhan

Chapter 18 Strategies for Speed Breeding in Crops to Accelerate Plant Improvement 354

Naglaa A. Abdallah, Aladdin Hamwieh, Khaled Radwan, and Michael Baum

Index 372

Preface

In modern times, global food security is a critical issue due to the increasing human population and changing climate, and one of the greatest challenges is to organize strategies for improving time-consuming plant breeding programs. In current years, the emerging research field of "plant speed breeding" promises to meet the future demand for crop improvement more efficiently. It chiefly accelerates breeding and research programs through the shortening of the generation time of crops. Importantly, upgraded plant speed breeding needs the involvement and coordination of some modern molecular tools and high-throughput technologies.

This book summarizes the recent achievements of plant speed breeding technologies through refined literature with an emphasis on the fundamental and advanced systems for genome editing (GE), functional genomics, controlled-environment growth chambers, genomic and phenotypic selection methods, and high-throughput technologies and the applications in significant model plants and crops.

To shorten the generation period and reduce the selection cycle in plant breeding programs, some methods have been developed and applied chiefly for early flowering, including accelerated single-seed descent, efficient in vitro culturing, fast generation cycling, rapid generation cycling, and rapid generation turnover. Another breakthrough in plant speed breeding is the introduction of high-throughput phenotyping technologies which involves a series of components including automated machinery, bioinformatics tools of artificial intelligence and machine learning, robotics, remote sensing technology, imaging technology, and computational/network platforms. Importantly, high-throughput phenotyping efficiently accumulates performances of interest traits and leads to the integration of functional phenomics to advance crop breeding.

This book highlights the constructions and applications of functional phenomics in some significant cereals, vegetables, and fruit trees. In recent years, scientists have started to integrate the part of phenomics with plant functional genomics, particularly acquiring valuable data that explore multiple levels of phenotypes under abiotic stress and the attacks by pests and pathogens. In the meantime, the tools used to enrich plant functional genomics upgrade fast through the innovations of multiple omics and bioinformatics/systems biology to achieve results in a high-throughput way that leads to an explosive increase in literature which gradually takes shape to organize an emerging topic of "high-throughput omics".

The game-changing technologies based on the Clustered Regularly Interspaced Short Palindromic Repeats (CRISPR)/Cas GE system are promising to accelerate crop breeding and are comprehensively summarized for their technical advancements and current applications in this book. Using the CRISPR/Cas system, directed genetic diversity at multiple sites of the genome can be generated, and therefore, plant breeders are able to precisely deliver target mutagenesis and thereby develop "CRISPR crops" with capacities to combat stressors. Ways to organize future crops to be "climate resilient" or "climate smart" that are designed to face more serious and complicated abiotic as well as biotic stresses under global climate change are increasingly proposed by researchers. This book provides summaries of innovative approaches using CRISPR-enabled speed breeding, including examples for the domestication of wild cultivars, GE-accelerated re-domestication, CRISPR screening, GE-organized haploid inducer lines, and GE-mediated development of male sterile lines that lead to the acceleration of breeding programs.

One of the essential tools in the next generation of crop breeding is "genomic selection" and over the past years, it has been upgraded by modern molecular technologies to evolve into "integrated genomic selection" in a more powerful and efficient way to obtain genetic gains and trait improvements. This book discusses plant speed breeding strategies assisted by the system of genomic selection when coordinated with genetic engineering, genetic mapping, marker-assisted selection, and phenotyping, particularly based on the introduction of high-throughput technologies from different available levels. The advantages and limitations of genomic selection were proposed briefly by evaluating the approaches in barley, cassava, maize, rapeseed, rice, sorghum, and wheat. For high-throughput genotyping, genetic marker platforms are upgraded based on the system of next-generation sequencing to perform approaches using single nucleotide polymorphism markers, quantitative trait loci mapping, and genome-wide association study for identifying candidate genes benefiting crop breeding. Nowadays, with the fast development of high-throughput sequencing systems together with integrated genomic selection, plant breeders have more powerful tools to enable the "design" of future crops. This book provides detailed descriptions of various genetic marker platforms and discussions on their emerging applications for crop improvement and the development of stress tolerance.

In the way to develop stress-resilient agriculture against global changing climate, undoubtedly, plant speed breeding with high-throughput technologies plays a crucial role through the coordinated breeding strategies based on CRISPR/Cas and functional genomics, and the knowledge explored in this book supports such goal that fits the vision of sustainable agriculture that expects to enable global food and nutritional security when facing increasingly human population. Stress-resilient crops are expected to produce better yield and higher quality food in the changing climate, and therefore, this book also supports the Sustainable Development Goals by the United Nations to achieve zero hunger in the world. This book is an ideal reference to provide updated information on diverse aspects of plant speed breeding technologies for graduate students, teachers, researchers, and experts in plant sciences, plant breeding, agronomy, and agriculture.

As the book editor, I'd like to thank all contributors for their insightful chapters. The instructions and help of the publisher are extremely grateful.

About the Editor

Jen-Tsung Chen is a professor of cell biology at the National University of Kaohsiung in Taiwan. He also teaches genomics, proteomics, plant physiology, and plant biotechnology. Dr. Chen's research interests include bioactive compounds, chromatography techniques, plant molecular biology, plant biotechnology, bioinformatics, and systems pharmacology. He is an active editor of academic books and journals to advance the exploration of multidisciplinary knowledge involving plant physiology, plant biotechnology, nanotechnology, ethnopharmacology, and systems biology. He serves as an editorial board member in reputed journals, including *Plant Methods*, *GM Crops & Food*, *Plant Nano Biology*, *Biomolecules*, *International Journal of Molecular Sciences*, and a guest editor in *Frontiers in Plant Science*, *Frontiers in Pharmacology*, *Journal of Fungi*, and *Current Pharmaceutical Design*. Dr. Chen has published books in collaboration with Springer Nature, CRC Press/Taylor & Francis Group, and In-TechOpen and he is handling book projects for some more international publishers on diverse topics such as drug discovery, herbal medicine, medicinal biotechnology, nanotechnology, bioengineering, plant functional genomics, plant speed breeding, and CRISPR-based plant genome editing.

Contributors

Samantha Vieira Abbad
Sempre AgTech/WIN
Brazil

Naglaa A. Abdallah
Department of Genetics Faculty of Agriculture
Cairo University
Cairo, Egypt
and
National Biotechnology Network of Expertise
ASRT Egypt

Elangovan Allimuthu
Division of Plant Physiology and Nanaji Deshmukh Plant Phenomics Center (NDPPC), Division of Plant Physiology
ICAR-Indian Agricultural Research Institute (IARI)
New Delhi, India

Imran Amin
Agricultural Biotechnology Division
National Institute for Biotechnology and Genetic Engineering (NIBGE), Constituent College of Pakistan Institute of Engineering and Applied Sciences
Faisalabad, Pakistan

Muhammad Aqeel
Functional Genomics and Bioinformatics Program
National Institute for Genomics and Advanced Biotechnology, National Agricultural Research Centre
Islamabad, Pakistan

Fabricio Barbosa Monteiro Arraes
Sempre AgTech/WIN
Brazil

Muhammad Asif
Agricultural Biotechnology Division
National Institute for Biotechnology and Genetic Engineering (NIBGE), Constituent College of Pakistan Institute of Engineering and Applied Sciences
Faisalabad, Pakistan

Muhammad Jawad Akbar Awan
Agricultural Biotechnology Division
National Institute for Biotechnology and Genetic Engineering (NIBGE), Constituent College of Pakistan Institute of Engineering and Applied Sciences
Faisalabad, Pakistan

Madhusmita Barik
Department of Botany
School of Applied Sciences, Centurion University of Technology and Management
Jatni, Bhubaneswar, India

Michael Baum
Department of Biotechnology
International Centre for Agricultural Research in the Dry Areas (ICARDA)
Giza, Egypt

Karma L. Bhutia
Department of Agricultural Biotechnology and Molecular Biology
Dr. Rajendra Prasad Central Agricultural University
Pusa, Bihar, India

Muhammad Ismail Buzdar
Agricultural Biotechnology Division
National Institute for Biotechnology and Genetic Engineering (NIBGE), Constituent College of Pakistan Institute of Engineering and Applied Sciences
Faisalabad, Pakistan

Luísa Carvalho
LEAF – Linking Landscape, Environment, Agriculture and Food – Research Center, Associated Laboratory TERRA, Instituto Superior de Agronomia
Universidade de Lisboa
Tapada da Ajuda, Lisboa, Portugal

Natasha Charaya
Department of Biotechnology
School of Applied and Life Sciences, Uttaranchal University
Dehradun, India

Charupriya Chauhan
Department of Genetics and Plant Breeding
School of Agriculture, Lovely Professional University
Phagwara, Punjab, India

Sanjay Kumar Chetia
Department of Plant Breeding and Genetics
Assam Agricultural University
Jorhat, Assam, India
and
Regional Agricultural Research Station
Assam Agricultural University
Titabar, Assam, India

Viswanathan Chinnusamy
Division of Plant Physiology and Nanaji Deshmukh Plant Phenomics Center (NDPPC), Division of Plant Physiology
ICAR-Indian Agricultural Research Institute (IARI)
New Delhi, India

Torres Cindy
R&D Artificial Vision and Automation
Vilmorin-Mikado
La Ménitré, France

J. Miguel Costa
LEAF – Linking Landscape, Environment, Agriculture and Food – Research Center, Associated Laboratory TERRA, Instituto Superior de Agronomia
Universidade de Lisboa
Tapada da Ajuda, Lisboa, Portugal

Jorge Marques da Silva
BioISI – Biosystems and Integrative Sciences Institute
Universidade de Lisboa, Faculdade de Ciências
Lisboa, Portugal

Praful Mukund Damdar
Department of Agricultural Biotechnology and Molecular Biology
Dr. Rajendra Prasad Central Agricultural University
Pusa, Bihar, India

Madhurima Das
Division of Plant Physiology and Nanaji Deshmukh Plant Phenomics Center (NDPPC), Division of Plant Physiology
ICAR-Indian Agricultural Research Institute (IARI)
New Delhi, India

Rousseau David
Laboratoire Angevin de Recherche en Ingénierie des Systèmes (LARIS)
UMR INRAe-IRHS, Université d'Angers
Angers, France

Marcelo Picanço de Farias
Sempre AgTech/WIN
Brazil

Shuting Ding
Department of Horticulture
Zhejiang University
Hangzhou, China

F. M. Era
Department of Genetics and Plant Breeding
Bangabandhu Sheikh Mujibur Rahman Agricultural University
Gazipur, Bangladesh

Sivapragasam Ezhumalai
Division of Plant Physiology and Nanaji Deshmukh Plant Phenomics Center (NDPPC), Division of Plant Physiology
ICAR-Indian Agricultural Research Institute (IARI)
New Delhi, India

Muhammad Farooq
Agricultural Biotechnology Division
National Institute for Biotechnology and Genetic Engineering (NIBGE), Constituent College of Pakistan Institute of Engineering and Applied Sciences
Faisalabad, Pakistan

Rui Figueira
CIBIO/InBio, Centro de Investigação em Biodiversidade e Recursos Genéticos, Laboratório Associado, Instituto Superior de Agronomia
Universidade de Lisboa
Tapada da Ajuda, Lisboa, Portugal
and
CIBIO/InBio, Centro de Investigação em Biodiversidade e Recursos Genéticos, Laboratório Associado
Universidade do Porto
Campus Agrário de Vairão, Vairão, Portugal

Renata Fuganti-Pagliarini
Piccola Consulting
Canada

Ashish Gautam
Department of Life Science, School of Life Sciences
Central University of Karnataka
Kadaganchi, Kalaburagi, Karnataka, India
and
Department of Plant Breeding and Genetics
Assam Agricultural University
Jorhat, Assam, India

Aladdin Hamwieh
Department of Biotechnology
International Centre for Agricultural Research in the Dry Areas (ICARDA)
Giza, Egypt

Amooru Harika
Division of Plant Physiology and Nanaji Deshmukh Plant Phenomics Center (NDPPC), Division of Plant Physiology
ICAR-Indian Agricultural Research Institute (IARI)
New Delhi, India

Akbar Hossain
Division of Soil Science
Bangladesh Wheat and Maize Research Institute
Dinajpur, Bangladesh

Ifrah Imran
Agricultural Biotechnology Division
National Institute for Biotechnology and Genetic Engineering (NIBGE), Constituent College of Pakistan Institute of Engineering and Applied Sciences
Faisalabad, Pakistan

Safeena Inam
Functional Genomics and Bioinformatics Program
National Institute for Genomics and Advanced Biotechnology, National Agricultural Research Centre
Park Road, Islamabad, Pakistan

A. K. M. Aminul Islam
Department of Genetics and Plant Breeding
Bangabandhu Sheikh Mujibur Rahman Agricultural University
Gazipur, Bangladesh

Raj Kumar Joshi
Department of Biotechnology
Rama Devi Women's University
Bhubaneswar, Odisha, India

Faiz Ahmad Joyia
Centre of Agricultural Biochemistry and Biotechnology (CABB)
University of Agriculture
Faisalabad, Pakistan

João Kafer
Embrapa Soybean
Brazil

Anil Namdev Kale
Department of Agricultural Biotechnology and Molecular Biology
Dr. Rajendra Prasad Central Agricultural University
Pusa, Bihar, India

Monika Kalia
Department of Entomology
CSK HPKV
Palampur, Himachal Pradesh, India

Sonika Kalia
Department of Biotechnology, School of Applied and Life Sciences
Uttaranchal University
Dehradun, India
and
Department of Agricultural Biotechnology
CSK HPKV
Palampur, Himachal Pradesh, India

Megha Katoch
Department of Agricultural Biotechnology
CSK HPKV
Palampur, Himachal Pradesh, India

Muhammad Ramzan Khan
Functional Genomics and Bioinformatics Program
National Institute for Genomics and Advanced Biotechnology, National Agricultural Research Centre
Park Road, Islamabad, Pakistan

Muhammad Sarwar Khan
Centre of Agricultural Biochemistry and Biotechnology (CABB)
University of Agriculture
Faisalabad, Pakistan

Marium Khatun
Department of Genetics and Plant Breeding
Bangabandhu Sheikh Mujibur Rahman Agricultural University
Gazipur, Bangladesh

Setsuko Komatsu
Faculty of Life and Environmental Sciences
Fukui University of Technology
Fukui, Japan

Mahesh Kumar
Division of Plant Physiology and Nanaji Deshmukh Plant Phenomics Center (NDPPC), Division of Plant Physiology
ICAR-Indian Agricultural Research Institute (IARI)
New Delhi, India

Raj Kumar
Department of Agricultural Biotechnology and Molecular Biology
Dr. Rajendra Prasad Central Agricultural University
Pusa, Bihar, India

Rakesh Kumar
Department of Life Science
School of Life Sciences, Central University of Karnataka
Kadaganchi, Kalaburagi, Karnataka, India

Sudhir Kumar
Division of Plant Physiology and Nanaji Deshmukh Plant Phenomics Center (NDPPC), Division of Plant Physiology
ICAR-Indian Agricultural Research Institute (IARI)
New Delhi, India

Vadim Lebedev
Branch of the Shemyakin-Ovchinnikov Institute of Bioorganic Chemistry of the Russian Academy of Sciences
Prospekt Nauki 6
Pushchino, Moscow Region, Russia

Muhammad Arslan Mahmood
Agricultural Biotechnology Division
National Institute for Biotechnology and Genetic Engineering (NIBGE), Constituent College of Pakistan Institute of Engineering and Applied Sciences
Jhang Road, Faisalabad, Pakistan
and
Plant Sciences Division
Research School of Biology, The Australian National University
Canberra, Australia

Rushikesh Sanjay Mane
Department of Agricultural Biotechnology and Molecular Biology
Dr. Rajendra Prasad Central Agricultural University
Pusa, Bihar, India

Shahid Mansoor
Agricultural Biotechnology Division
National Institute for Biotechnology and Genetic Engineering (NIBGE), Constituent College of Pakistan Institute of Engineering and Applied Sciences
Jhang Road, Faisalabad, Pakistan
and
International Center for Chemical and Biological Sciences
University of Karachi
Karachi, Pakistan

Cordier Mathis
Laboratoire Angevin de Recherche en Ingénierie des Systèmes (LARIS), UMR INRAe-IRHS
Université d'Angers
Angers, France
and
R&D Artificial Vision and automation
Vilmorin-Mikado
La Ménitré, France

Momna Mehmood
Department of Agricultural Biotechnology, Faculty of Agriculture
Ondokuz Mayis University
Samsun, Turkey

Liliane Mertz-Henning
Embrapa Soybean
Brazil

Rukmini Mishra
Department of Botany
School of Applied Sciences, Centurion University of Technology and Management
Jatni, Bhubaneswar, India

Jatindra Nath Mohanty
Department of Botany
School of Applied Sciences, Centurion University of Technology and Management
Jatni, Bhubaneswar, India

Hugo Molinari
Sempre AgTech/WIN
Brazil

Mayla Daiane Correa Molinari
Sempre AgTech/WIN
Brazil

Amna Muhammad
Functional Genomics and Bioinformatics Program
National Institute for Genomics and Advanced Biotechnology, National Agricultural Research Centre
Islamabad, Pakistan

Fara Muqaddas
Functional Genomics and Bioinformatics Program
National Institute for Genomics and Advanced Biotechnology, National Agricultural Research Centre
Islamabad, Pakistan

Maryam Murtaza
Functional Genomics and Bioinformatics Program
National Institute for Genomics and Advanced Biotechnology, National Agricultural Research Centre
Islamabad, Pakistan

Ghulam Mustafa
Centre of Agricultural Biochemistry and Biotechnology (CABB)
University of Agriculture
Faisalabad, Pakistan

Muhammad Kashif Naeem
Functional Genomics and Bioinformatics Program
National Institute for Genomics and Advanced Biotechnology, National Agricultural Research Centre
Islamabad, Pakistan

Rubab Zahra Naqvi
Agricultural Biotechnology Division, National Institute for Biotechnology and Genetic Engineering (NIBGE)
Constituent College of Pakistan Institute of Engineering and Applied Sciences
Faisalabad, Pakistan

Shanvi Nautiyal
Department of Biotechnology
School of Applied and Life Sciences, Uttaranchal University
Dehradun, India

Alexandre Nepomuceno
Embrapa Soybean
Brazil

Aqsa Parvaiz
Department of Biochemistry and Biotechnology
The Women University Multan
Multan, Pakistan

Rizwana Parveen
Department of Biosciences
University of Wah
Wah Cantt, Pakistan

Animesh Pattnaik
Department of Botany
School of Applied Sciences, Centurion University of Technology and Management
Jatni, Bhubaneswar, India

Rasti Pejman
Centre d'Études et de Recherche pour l'Aide à la Décision (CERADE)
ESAIP
Saint-Barthélemy-d'Anjou, France
and
Laboratoire Angevin de Recherche en Ingénierie des Systèmes (LARIS)
UMR INRAe-IRHS, Université d'Angers
Angers, France

Zeba Quaiyum
Department of Agricultural Biotechnology and Molecular Biology
Dr. Rajendra Prasad Central Agricultural University
Pusa, Bihar, India

Khaled Radwan
National Biotechnology Network of Expertise
ASRT Egypt
and
Department of Biotechnology
Agricultural Genetic Engineering Research Institute (AGERI), ARC
Giza, Egypt

Dhandapani Raju
Division of Plant Physiology and Nanaji Deshmukh Plant Phenomics Center (NDPPC), Division of Plant Physiology
ICAR-Indian Agricultural Research Institute (IARI)
New Delhi, India

Sagar Krushnaji Rangari
Department of Agricultural Biotechnology and Molecular Biology
Dr. Rajendra Prasad Central Agricultural University
Pusa, Bihar, India

Indra Rautela
Department of Biotechnology
School of Applied and Life Sciences, Uttaranchal University
Dehradun, India

Aiman Raza
Agricultural Biotechnology Division
National Institute for Biotechnology and Genetic Engineering (NIBGE), Constituent College of Pakistan Institute of Engineering and Applied Sciences
Faisalabad, Pakistan

Nazia Rehman
Functional Genomics and Bioinformatics Program
National Institute for Genomics and Advanced Biotechnology, National Agricultural Research Centre
Park Road, Islamabad, Pakistan

Obaid Ur Rehman
Functional Genomics and Bioinformatics Program
National Institute for Genomics and Advanced Biotechnology, National Agricultural Research Centre
Park Road, Islamabad, Pakistan

Aamir Riaz
Functional Genomics and Bioinformatics Program
National Institute for Genomics and Advanced Biotechnology, National Agricultural Research Centre
Park Road, Islamabad, Pakistan

Sonupriya Sahu
Department of Botany
School of Applied Sciences
Centurion University of Technology and Management
Bhubaneswar, Odisha, India

Bilal Saleem
Functional Genomics and Bioinformatics Program
National Institute for Genomics and Advanced Biotechnology, National Agricultural Research Centre
Islamabad, Pakistan

Sumi Sarkar
Department of Genetics and Plant Breeding
Bangabandhu Sheikh Mujibur Rahman Agricultural University
Gazipur, Bangladesh

Himanshu Sharma
Agri-Biotechnology Division
National Agri-Food Biotechnology Institute
Knowledge City Sector-81, Mohali, Punjab, India

Vinay Sharma
International Crops Research Institute for the Semi-Arid Tropics (ICRISAT)
Hyderabad, India
and
Department of Genetics and Plant Breeding
Ch. Charan Singh University
Meerut, India

Kai Shi
Department of Horticulture
Zhejiang University
Hangzhou, China

Hiroyuki Shimono
Crop Science Laboratory, Faculty of Agriculture
Iwate University, Agri-Innovation Center, Iwate University
Morioka, Iwate, Japan

Biswabiplab Singh
Division of Plant Physiology and Nanaji Deshmukh Plant Phenomics Center (NDPPC), Division of Plant Physiology
ICAR-Indian Agricultural Research Institute (IARI)
New Delhi, India

Karthick Babu Sivakumar
Department of Plant Breeding and Genetics
Punjab Agricultural University
Ludhiana, Punjab, India

Sukumar Taria
Division of Plant Physiology and Nanaji Deshmukh Plant Phenomics Center (NDPPC), Division of Plant Physiology
ICAR-Indian Agricultural Research Institute (IARI)
New Delhi, India

Furqan Tayyab
Centre of Agricultural Biochemistry and Biotechnology (CABB)
University of Agriculture
Faisalabad, Pakistan

Anamika Thakur
Department of Biotechnology
Dr. Y.S. Parmar University of Horticulture and Forestry
Nauni, Solan, Himachal Pradesh, India

Isabela Yamauchi Torres
Sempre AgTech/WIN
Brazil

Muhammad Uzair
Functional Genomics and Bioinformatics Program, National Institute for Genomics and Advanced Biotechnology
National Agricultural Research Centre
Islamabad, Pakistan

Jiao Wang
Department of Horticulture
Zhejiang University
Hangzhou, China

Farhat Yasmeen
Department of Biosciences
University of Wah
Wah Cantt, Pakistan

Muhammad Zahid
Wheat Program, Crop Sciences Institute
National Agricultural Research Centre
Islamabad, Pakistan

Rida Zahid
Agricultural Biotechnology Division
National Institute for Biotechnology and Genetic Engineering (NIBGE), Constituent College of Pakistan Institute of Engineering and Applied Sciences
Faisalabad, Pakistan

Asma Zia
Agricultural Biotechnology Division
National Institute for Biotechnology and Genetic Engineering (NIBGE), Constituent College of Pakistan Institute of Engineering and Applied Sciences
Faisalabad, Pakistan

Charles Hobi Zimmer
Sempre AgTech/WIN
Brazil

1 Technological Development and Prospects of Crop Speed Breeding

Marcelo Picanço de Farias and Charles Hobi Zimmer
Sempre AgTech/WIN, Brazil

João Kafer
Embrapa Soybean, Brazil

Mayla Daiane Correa Molinari and Fabricio Barbosa Monteiro Arraes
Sempre AgTech/WIN, Brazil

Renata Fuganti-Pagliarini
Piccola Consulting, Canada

Isabela Yamauchi Torres and Samantha Vieira Abbad
Sempre AgTech/WIN, Brazil

Liliane Mertz-Henning and Alexandre Nepomuceno
Embrapa Soybean, Brazil

Hugo Molinari
Sempre AgTech/WIN, Brazil

1.1 INTRODUCTION

Traditional plant breeding plays an indispensable role in generating cultivars adapted to diverse climatic conditions. However, conventional methodologies, which involve prolonged cycles of crossbreeding and inbreeding, require a lengthy developmental process, often spanning 10–15 years (Watson et al., 2018; Gudi et al., 2022). This extensive timeframe challenges the acceptance and effective incorporation of methodological and technological innovations into crops of agronomic importance.

To address this challenge, techniques known as Speed Breeding have been developed to speed up the vegetative cycle of cultivated plants. These protocols optimize various environmental parameters such as light, photoperiod, temperature, and humidity, to shorten the growth cycle and benefit from high planting density and early seed harvesting. Since their introduction, these techniques have been applied to a range of plant species including rice, cassava, canola, barley, peas, chickpeas, cowpeas, maize, soybeans, and wheat, among others (Watson et al., 2018; Figure 1.1).

Furthermore, the Speed Breeding approach integrates seamlessly with other emerging technological tools in the field of genetic improvement. Notable among these is CRISPR/Cas

DOI: 10.1201/b23372-1

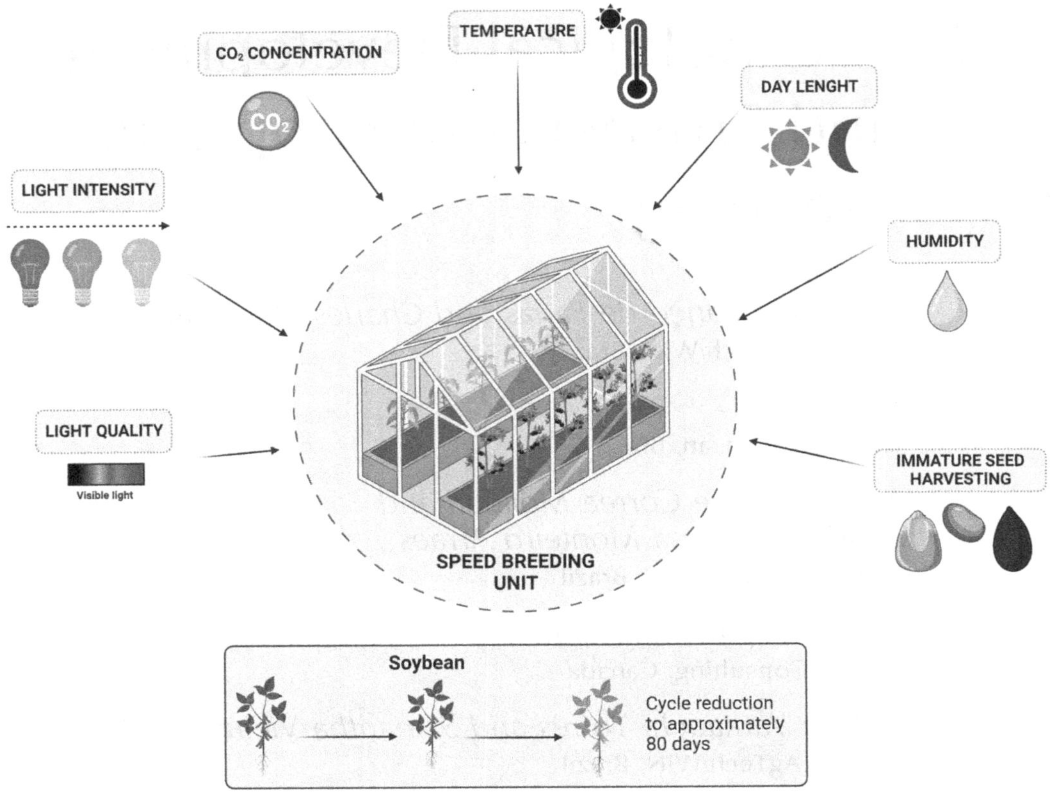

FIGURE 1.1 Schematic representation of a Speed Breeding unit, highlighting its primary operational requirements and advancements achieved in soybeans, with an emphasis on necessary adaptions for corn and cotton based on Xu et al. (2023) (Created with http://BioRender.com).

(Clustered Regularly Interspaced Short Palindromic Repeats) technology, Next-Generation Sequencing (NGS), Marker-Assisted Selection (MAS), Genome-Wide Selection, and Genome-Wide Association Studies (GWAS) (Hickey et al., 2019; Bhatta et al., 2021; Molinari, 2021c; Molinari et. al., 2024).

In this context, Speed Breeding technologies accelerate research in plant breeding and catalyze interdisciplinary and international scientific collaborations. Such advancement fosters partnerships among academic institutions, governmental sectors, and industrial players, propelling innovation, disseminating knowledge, and ultimately promoting investment in research and development.

This chapter aims to explore the role and impact of Speed Breeding technologies in expediting plant breeding, identify emerging trends and technological innovations, and discuss challenges and opportunities in the global landscape.

1.2 EMERGING TRENDS AND TECHNOLOGICAL INNOVATION

Speed Breeding emerges as a pool of transformative techniques within the context of modern agriculture. Initially developed by NASA (National Aeronautics and Space Administration) for plant cultivation in space, its potential was recognized for terrestrial applications, leading researchers to adapt it for use in greenhouses and growth chambers (Watson et al., 2018).

The underlying process of Speed Breeding involves meticulous manipulation of the growing environment. adapting some protocols for the development of various crops. In wheat, for example, the protocol was established considering an extended photoperiod of 22 h of LED lights at 22°C, followed by 2 h of darkness at 17°C, the technique seeks to replicate optimal dawn and dusk conditions to speed up the cycle. This simulation, combined with controlled humidity, aims to stimulate the expression of circadian cycle-related genes. This manipulation results in an accelerated growth cycle, allowing plants to reach maturity in a significantly reduced timeframe. Research has validated the effectiveness of these methodologies, highlighting their potential to increase the number of yearly generations of crops such as wheat, barley, and chickpeas under greenhouse conditions (Watson et al., 2018).

Furthermore, the technique is not limited to these crops. Encouraged by the promising results, scientists have crafted Speed Breeding protocols for a variety of other plants, including rice, subterranean clover, pigeon-pea, and oat (Collard et al., 2017; Pazos-Navarro et al., 2017; Saxena et al., 2019; González-Barrios et al., 2020).

1.2.1 How Does Speed Breeding Improve Cotton, Maize, and Soybean Breeding?

Considering the benefits of applying Speed Breeding technology, it is anticipated that protocols for globally significant crops—such as cotton, soybeans, and maize—will be developed. These are crops with substantial economic value, and the time required for each generation is a limiting factor in their breeding. Such breeding is lucrative for plant breeding companies, as it can expedite their programs, thereby providing more cultivars to farmers in less time.

Soybean (*Glycine Max* Merril., 2n=40) is a short-day crop with a total cycle ranging from 100 to 160 days until harvest. The primary method for its breeding is through crossbreeding, a time-consuming process requiring eight generations to achieve stable lines for selection and evaluation, and an additional four generations to develop pure lines. Therefore, the large number of generations required for soybean breeding is a limiting factor. Speed Breeding techniques can accelerate this process (Fang & Han, 2019; Fang et al., 2021). Protocols have already been developed in soybean breeding that reduce the crop cycle time and allow for a higher number of cycles per year. Harrison et al. (2020) demonstrated that under Speed Breeding conditions, the soybean cycle could be reduced from 120 days to 63 and 81 days when using blue and red light (RB) and full-spectrum white light (FB) conditions, respectively, under a 12-hour light/12-hour dark regime. Similar findings were reported by Jähne et al. (2020), who achieved soybean maturity in 77 days by employing a photoperiod of 10 hours with blue light.

These results showed the potential for increasing the number of soybean generations from three to five in a single year. Similarly, Fang et al. (2021) observed that fresh seeding and MAS allowed for four generations per year. Using a different approach, Nagatoshi and Fujita (2019) accelerated soybean growth by applying CO_2 (400 ppm) and white light under a 14-hour light, 30°C/10-hour dark, 25°C regime, enabling the cultivation of five soybean generations within a year.

Maize (*Zea Mays* L., 2n = 2x = 20) is an economically significant outcrossing species with a crop cycle ranging from 115 to 160 days, depending on the variety. Although maize is categorized as a short-day plant, photoperiod is not a prerequisite for completing its life cycle. However, temperature plays a significant role in crop development, with growth and development considerably reduced below 10°C compared to higher temperatures (Hatfield et al., 2015). Given these conditions, maize breeding programs stand to benefit from Speed Breeding techniques (Singh et al., 2021; Farooqi et al., 2022). To date, there are no established protocols specifically for Speed Breeding in maize. However, technologies exist that can reduce the generation time of the crop. In this context, Liotino et al. (2019) developed a protocol capable of increasing maize generations through embryo rescue techniques. While not classified as Speed Breeding per se, this methodology involves the removal of immature embryos at the R3 developmental stage and cultivating them in tissue culture under controlled light and temperature conditions. After the tissue culture phase, the embryos are transferred to pots (3:1 ratio) for acclimatization and subsequently moved to a greenhouse under a 16-hour light/8-hour dark photoperiod with a light intensity of 650 $\mu E/m^2$, and a 30/20°C day/night temperature cycle.

Cotton (*Gossypium* spp.) is an annual crop with a cycle ranging from 120 to 220 days, depending on the variety. The most widely cultivated species is *Gossypium hirsutum* (2n = 4x = 52), an allotetraploid, and its genetic breeding may take up to 8 years, offering the possibility of two cycles per year. The process primarily involves hybridization followed by biparental selection. In more complex programs, the breeding time until cultivar release can extend to 15 years (Li et al., 2020; Mubarik et al., 2020; Wang et al., 2011). Given this timeline, cotton would greatly benefit from Speed Breeding protocols to expedite cultivar release to farmers. However, there is limited information available in the literature on technologies that increase the number of generations of cotton. Li et al. (2010) investigated the impact of various light sources on explant growth, finding that the highest growth rates were observed when explants were exposed to a combination of blue and red light (1:1 ratio), followed by explants under the effect of blue light alone. In this context, the development of Speed Breeding protocols for cotton is critically important and demands the optimization of various factors such as temperature and CO_2 release, among others.

The value of Speed Breeding is amplified when integrated with precision agriculture. The latter, which relies on meticulous data collection from various sources like sensors, drones, and satellite imagery, can complement the Speed Breeding approach, offering insights on when and how best to deploy Speed Breeding protocols for optimizing agricultural production (Cherubin et al., 2022).

Over the years, the initial concept of Speed Breeding has emerged as a groundbreaking tool in agrobiotechnology. When combined with advanced technologies like genomics, it promises rapid advancements in plant breeding, since genomics delivers intricate details about genes, genetic regulators, and DNA sequences governing specific plant traits. By applying this information, molecular markers are identified, and linked to desirable traits like disease resistance and yield, for example. This approach has found increasing applications in agribusiness to enhance traits in vegetable crops. However, its potential in predicting responses to stresses, especially abiotic ones, remains not fully understood and explored (Derbyshire et al., 2022).

In another aspect, Genome sequencing has been made more accessible by NGS, enabling rapid analysis of multiple genomes. One of the pioneering approaches within genomics is pangenomics. Instead of focusing on a single reference genome, pangenomics sequences and analyzes genomes from multiple varieties of the same plant. This unveils genetic diversity, crucial for identifying genes and genetic variants associated with desirable traits (Tao et al., 2019). Such an approach has already led to notable findings. For instance, a USDA (United States Department of Agriculture) project sequenced a pangenome from about 1,000 soybean accessions, uncovering 3,765 genes missing in the reference genome. According to the authors, domestication may have resulted in this genetic diversity loss and a significant reduction in the average number of protein-coding genes per individual (Bayer et al., 2021). Similar pangenomic projects were conducted for maize (Hufford et al., 2021; https://maize-pangenome.gramene.org/) and cotton (Li et al., 2020), leading to analogous conclusions.

In conclusion, the integration of Speed Breeding with detailed genomic analyses, especially the pangenomic approach, stands as a promise to accelerate plant breeding, benefiting from a rich genetic diversity that's often overlooked.

Building upon the foundation of pangenomics, Speed Breeding enters a new chapter with the integration of the GWAS into its arsenal. This genomic tool, which focuses on single nucleotide polymorphisms (SNPs) and is supported by advanced statistical analysis, accelerates the pace at which researchers can identify crucial genes and genomic regions (Brachi et al., 2011).

A deeper dimension to GWAS in the Speed Breeding landscape is the Meta-QTL (MQTL) approach. Essentially an extension of GWAS, MQTL connects data from multiple studies, pinpointing chromosome regions that consistently relate to specific traits across diverse populations. Such holistic perspectives enhance the potential of Speed Breeding, a notion echoed by the studies of Khahani et al. (2021).

The union of GWAS and Speed Breeding tools is reshaping the timelines of plant breeding. The traditionally drawn-out processes, dependent on phenotype-based selections across generations, are now being expedited. With GWAS, breeders can rapidly identify and zero in on superior plant candidates early in the process, reflecting the paradigm shifts heralded by researchers like Brachi and coworkers back in 2011.

For instance, in *Gossypium hirsutum*, a myriad of SNPs within Quantitative Trait Loci (QTLs) crucial for essential agronomic traits have been unveiled, thanks to Speed Breeding driven by GWAS insights, as seen in studies by Yasir et al. (2022). In *Z. mays*, pivotal genomic regions governing vital traits have been identified, drawing upon studies like those by Zila et al. (2013) and Tao et al. (2013).

The synergy of Speed Breeding and GWAS was also accessed in soybean (*Glycine max*). An expansive study by Fang et al. (2017), involving 809 soybean accessions, identified vital genes for an array of plant traits, such as growth period, plant architecture, seed color, seed development, as well as oil and protein content in the seeds (Fang et al., 2017). Among all these characteristics evaluated, an important highlight of the study was the identification of several genes related to the biosynthesis of fatty acids. The identified genes showed substantial differences in total fatty acid content, demonstrating the relevance of these findings in the context of agronomic research (Fang et al., 2017). Follow-up research employed GWAS to detect SNPs linked to the highest levels of yield, maturity, plant height, and seed weight characteristics (Ravelombola et al., 2019).

In summary, when Speed Breeding is applied together with tools like GWAS and MQTL, plant breeding is not only sped up but also enhanced, resulting in a vibrant and accelerated era in agronomy.

The advancement in the field of plant breeding has been boosted by a combination of "omic" technologies and Speed Breeding. While transcriptomics, proteomics, and metabolomics provide an in-depth view of the molecular biology of crops, Speed Breeding offers the ability to apply these insights more rapidly and effectively (Eldakak et al., 2013; Komatsu et al., 2013; Fernandez et al., 2021; Litvinov et al., 2021; Mustafa & Komatsu, 2021; Sakurai, 2022; Tyagi et al., 2022; Abdul and Masmoudi, 2023).

Transcriptomics, for instance, has offered an unparalleled view of gene expression in plants under various conditions. By 2021, over 3,000 soybean-related transcriptomic datasets had been made available in public repositories, unveiling the intricate patterns of gene expression under various conditions such as biotic and abiotic stresses (Marcolino-Gomes et al., 2015; Almeida-Silva et al., 2021; Molinari et al., 2021a; 2021b; 2022; Kafer et al., 2023). With the implementation of Speed Breeding, the desired traits identified by these transcriptomic studies can be quickly introduced into new varieties, significantly speeding up breeding programs.

While metabolomics aids in identifying metabolites produced in response to different stresses, Speed Breeding ensures that these findings are rapidly translated into practical traits, whether in soybean, maize, or cotton (Chatham & Juvik, 2021; Liang et al., 2021; Owusu et al., 2023).

On the proteomics side, it is possible to understand the regulation at the protein level in various crops (Florentino et al., 2022; Molinari et al., 2022; 2023). Speed Breeding, combined with these proteomic insights, can accelerate the introduction of desired variations in crops.

In summary, the "omic" tools provide a detailed map of a plant's complex biological network, and Speed Breeding acts as the catalyst that allows breeders to navigate this map more efficiently, promising to revolutionize the field of plant breeding (Eldakak et al., 2013; Komatsu et al., 2013; Mustafa & Komatsu, 2021; Arraes et al., 2022; Tyagi et al., 2022; Abdul & Masmoudi, 2023).

In addition to the gene and protein expression, the epigenetic mechanisms triggered in responses to stress have been widely studied in plants. Epigenetic regulation is an intricate control system that incorporates diverse regulatory elements such as small non-coding RNAs, DNA methylation, histone modifications, and chromatin remodeling. This intricate interaction between these components enables precise control of gene expression, giving plants the remarkable ability to resist infections and adapt to significant environmental changes in nature (Huang & Jin, 2022). Associated with this concept, the priming response in plants refers to a phenomenon where exposure to certain environmental stresses or stimuli can enhance plants' defense mechanisms and prepare them for more effective responses to future stressors. This concept is often associated with the plant's ability to remember previous stress events and prepare a quicker and more robust response when faced with similar challenges in the future, which include biotic and abiotic stresses (Hilker & Schmülling, 2019).

In this way, "omic" technologies that aim to elucidate priming responses have become the focus of epigenomics, which can be defined as a study involving epigenetic modifications in an organism. Among them, the massive sequencing of small RNAs (sRNA-Seq) stands out. Using NGS technology, the sRNA-Seq pipeline allows the fraction of small RNAs (18–34 nucleotides) expressed in each plant tissue to be evaluated after a given treatment (Garg & Varshney, 2022).

Another molecular biology technique used to elucidate epigenetic and regulatory mechanisms in plants is "Chromatin Immunoprecipitation Sequencing" (ChIP-Seq), which maps, in a genome, interactions between specific proteins and DNA. This technique is especially useful for identifying binding sites for proteins, such as transcription factors or modified histones, throughout the plant genome, which helps to understand how genes are regulated and how changes in chromatin structure can affect gene expression. (Chen et al., 2018). Finally, the methylome is an "omic" technology that evaluates the DNA methylation profile in a plant species. DNA methylation is an epigenetic modification that involves the addition of methyl groups (CH_3) to cytosines in the CpG regions (dinucleotides composed of a cytosine followed by a guanine) of DNA. This modification is a crucial part of controlling gene expression and regulating biological processes (Kumar & Mohapatra, 2021). Methylome analysis allows us to identify which parts of the genome are methylated (or demethylated) at a given time. This is fundamental to understanding how genes are regulated and how epigenetic modifications can affect gene expression and, consequently, the phenotypes observed in plants. Several studies using such technologies have provided a better understanding of the epigenetic mechanisms associated with priming responses in maize, cotton, and soybean, which has allowed the implementation of practices such as the prior treatment of seeds of these crops with compounds that mimic stress situations observed on the field. This procedure is capable of increasing drought tolerance, reducing susceptibility to high salt levels, as well as increasing seed vigor and resistance to pathogens (Bakht et al., 2011; Syaiful et al., 2015; Pallaoro et al., 2016; Fang et al., 2018; Zhao et al., 2020; Michereff et al., 2021; Yung et al., 2021; Zhang et al., 2021; Sintaha et al., 2022).

All examples presented so far illustrate the large volume of data and techniques developed over the years and the considerably increased number of parameters to be monitored for the implementation of Speed Breeding techniques. This fact makes automation procedures and intelligent monitoring increasingly necessary. Therefore, Speed Breeding systems based on artificial intelligence (AI), as well as machine learning (ML) and deep learning (DL) have increasingly gained ground in the breeding of maize, soybean, cotton, and other crops. An important aspect to be considered is the automation of the phenotypic characterization process (phenomics) from image processing by ML and/or DL. For example, Singh and coworkers (2023) developed the image-based traits (i-Traits) system capable of predicting different characteristics related to biomass in wheat, such as fresh weight, dry weight, and aerial area, from image processing. In a similar initiative, Rolland and coworkers (2022) developed a very curated DL model, HairNet, capable of scoring leaf hairiness, an important parameter to evaluate fiber yield and insect resistance.

In parallel, various other platforms similarly generate phenomics data for several other plant species, but an important problem faced in this field is inefficient data storage, retrieval, analysis, and reuse (Marsh et al., 2021). In response to this challenge, software such as PhytoOracle (PO) has emerged. PO comprises a set of modular, scalable pipelines created to manage substantial volumes of field phenomics data (Gonzalez et al., 2023). This software facilitated the extraction of phenotypic traits for lettuce and sorghum, intending to expand the scope to encompass a broader spectrum of supported species in the future (Gonzalez et al., 2023). A further challenge is to integrate genomic data, obtained mostly by GWAS and pan-genomic, with phenomics data. The AI plays a fundamental role in analyzing and associating two vast sets of phenotypic and genomic data, contributing to significant advances in improvement strategies. Through DL and ML models, AI can correlate high-yielding crops, such as soybeans, maize, and cotton, as well as predict the performance of crops‘ main characteristics. Thus, the effective integration of AI with phenomics and genomic tools facilitates the rapid identification of genes associated with specific phenotypes, promoting a significant boost in crop breeding programs (Khan et al., 2022).

Finally, one of the two major challenges today is integrating data from all "omic" technologies (genomics, epigenomics, transcriptomics, proteomics, and metabolomics) (Figure 1.2).

It is very difficult to integrate these data given that they differ in several parameters, which can lead to a large accumulation of errors in biological mechanisms. To solve this problem, Cembrowska-Lech et al. (2023) proposed a pipeline that can efficiently integrate multi-omics data, where the key steps are: (i) data collection; (ii) data pre-processing and normalization; (iii) data integration; (iv) ML modeling; (v) model evaluation and interpretation; (vi) application to agricultural research and practice. According to Cembrowska-Lech, this pipeline allows genomic selection and phenotypic prediction, high-throughput phenotyping, and disease detection, which, in general, encourages predictability and sensitivity in the field.

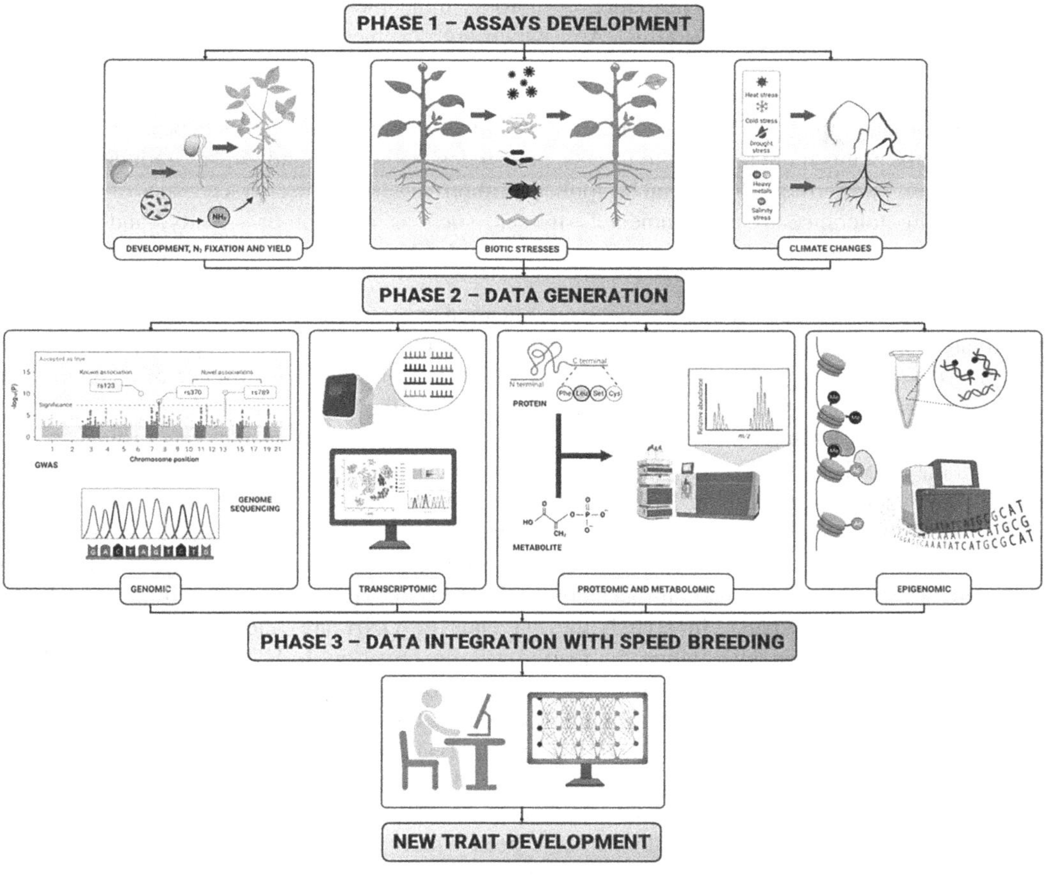

FIGURE 1.2 Integrated approach in agricultural and biotechnological research: from plant responses to biotic stresses and climatic changes, through to genomic, transcriptomic, proteomic, and epigenomic analyses. When combined with Speed Breeding methodologies, this allows for the expedited development of novel plant traits tailored to environmental challenges (Created with BioRender.com).

Despite difficulties, the large volume of data associated with "omics" technologies has provided a better understanding of countless molecular mechanisms associated with development and stress responses in maize, soybean, and cotton, which allows the identification of interesting new traits and reduces the time needed for genetic breeding. Furthermore, biotechnology can use this data to develop new crop genotypes, or by inserting these traits into new cultivars through genetic transformation (gene knock-in), gene silencing (gene knockdown), or even gene depletion (gene knockout). The gene knock-in, or insertion of a heterologous DNA into the plant genome, is generally performed by two main genetic transformation methodologies: transformation mediated by *Agrobacterium tumefaciens* or biolistic methods. These technologies allow the introduction of a foreign DNA molecule into both the nuclear genome and the genome of plant organelles such as chloroplasts (Verma & Daniell, 2007; Jin & Daniell, 2015; Basso et al., 2020). Current genetic engineering tools make it possible to insert any DNA sequence from any organism such as genes of external interest and regulatory elements to stimulate the expression of native genes (Basso et al., 2020). Gene knockdown, also known as gene silencing, is a process in molecular biology that involves temporarily reducing the expression of a specific gene in a cell or organism. This process is based on RNA interference (RNAi) technology, which, in plants, is a fundamental biological process initially described as a mechanism for regulating gene expression and epigenetics through the formation of double-stranded RNAs (dsRNAs) and subsequent cleavage of a target RNA. RNAi is achieved through the controlled introduction of RNA molecules, such as short interfering RNAs (siRNAs) or short hairpin RNAs (shRNAs), which are designed to bind to the messenger RNA (mRNA) of the target gene, leading to its degradation or inhibiting its translation into protein (Muhammad et al., 2019). RNAi can be designed for topical use or to create genetically modified plants that produce small dsRNA molecules capable of silencing specific genes in pests or pathogens. This can make plants more resistant to insects, nematodes, pathogens, and viruses, reducing the need for pesticides. Furthermore, RNAi can be applied to improve agronomic characteristics of crops, such as increasing tolerance to environmental stresses, such as drought or salinity, and increasing nutritional quality and productivity (Jagtap et al., 2011; Das & Sherif, 2020; Rajput et al., 2021; Távora et al., 2022).

Finally, gene knockout refers to a genetic modification process where a specific gene is completely inactivated or "turned off" in an organism. This is usually achieved by completely removing or stopping the functioning of the target gene, which results in the complete absence of its expression. Among the emerging methodologies used for this process, the CRISPR/Cas (Clustered Regularly Interspaced Short Palindromic Repeats/CRISPR-associated proteins) system has proven to be a revolutionary tool in agriculture and agricultural biotechnology. It is an extremely specific technology where a single guide RNA molecule (sgRNA), which is complementary to a specific DNA sequence, directs the cleavage of the target gene that is desired to be "turned off" by the Cas endonuclease. After cleavage, the DNA repair system itself repairs the damage caused to the molecule. There are two main repair mechanisms: NHEJ (Non-Homologous End Joining) and HDR (Homology-Directed Repair). In NHEJ, the cut ends of DNA are rejoined, often resulting in insertions or deletions (indels), which can lead to gene inactivation. In HDR, a segment of DNA is inserted to replace the original sequence, allowing for more precise editing. However, HDR is less efficient than NHEJ in many cases (Impens et al., 2022; Molinari et al., 2024). This technology can be widely used to create new non-GM (genetically modified) varieties of maize, soybean, and cotton that are more productive and less susceptible to pests and diseases (Gong et al., 2021).

Even though genetic engineering provides several tools to reduce the generation time of new plant varieties, several other technologies have been widely used for this purpose. Among them, the double haploid (DH) technology stands out, commonly used in maize, sorghum, and wheat breeding. This methodology is based on the initiation of haploid cell generation from plant somatic cells, followed by the subsequent duplication of chromosomes. This process produces cells possessing a singular complete set of chromosomes (haploid), which are subsequently replicated to two complete sets of chromosomes (diploid) in a state of homozygosity (Maqbool et al., 2020).

The preferential haploid induction in maize is based on the use of inductor lines. These genotypes can generate androgenetic or gynogenetic haploids. In the case of the androgenetic system, after fertilization, one of the reproductive nuclei of the pollen grain develops into a "pseudo-embryo," while in the gynogenetic system, it is the oosphere that develops. To identify haploids generated using these inductor genotypes, a phenotypic marker system based on anthocyanin pigmentation was developed, which, unfortunately, is not the most efficient, even though it is the most used (Chase & Nanda, 1965). The dominant allele causes anthocyanin pigmentation in both the endosperm and the embryo of diploid seeds, while haploid seeds present pigmentation only in the endosperm, keeping the embryo white. After haploid induction, the development of an effective chromosome duplication protocol is essential to obtain DH plants, as the frequency of spontaneous chromosome duplication in maize haploids is generally low and highly variable (Wan et al., 1991). Colchicine is one of the most used doubling agents. This chemical is known for its ability to arrest cell division at specific stages of the cell cycle by binding to proteins involved in the formation of the mitotic spindle, a structure critical for the proper segregation of chromosomes during cell division (Barnabás et al., 1999). The resulting diploid cells are then grown into complete plants. These plants are genetically pure (homozygous) for all their genes, meaning all alleles are identical, ideal for generating new maize hybrids. In addition to reducing the development time of new cultivars, DH technology reduces costs and generates new plants with greater genetic homogeneity, which has led multinationals from agribusiness to introduce DHs into their breeding programs (Rajcan et al., 2011).

1.3 CHALLENGES AND OPPORTUNITIES IN THE GLOBAL SCENARIO

In the current global landscape marked by climate change and ongoing population growth, the agricultural sector confronts the escalating demand for cultivars that produce high yields, optimized nutritional content, and enhanced resistance to pests and diseases. This demand accentuates the urgency for efficient agricultural production that observes environmental sustainability while ensuring economic viability (Cobb et al., 2019).

Modern plant biotechnology presents itself as a critical solution to these constraints. Advanced techniques such as Speed Breeding, genomic editing, and trait selection are promoting significant shifts in breeding programs, leading to the accelerated and precise development of commercial cultivars. In the context of soybean cultivation, traditional breeding methods, which typically require up to eight generations for the development of a new cultivar, are now being surpassed. Speed Breeding techniques, as documented by Jahne et al. (2020), considerably shorten the generation advancement, enabling up to four cycles annually. Moreover, innovations in photoperiod management, as outlined by Fang et al. (2021) and Harrison et al. (2020), induce early flowering, thus accelerating soybean maturation. Their studies suggested a cycle reduction from the standard 120 days under natural conditions to 63–81 days in controlled settings.

One of the methodologies applied to reduce growth cycles is the Rapid Generation Advancement (RGA) method which emphasizes the manipulation of external growth influencers, ranging from light quality and intensity to nutritional aspects. The primary objective of this method is to reduce the timeline between sowing and physiological maturity. Despite its efficacy, the RGA's implementation requires specialized infrastructure and highly skilled personnel (Gieco et al., 2007). Simultaneously, genomic editing, especially leveraging the CRISPR/Cas technology, has emerged as crucial in introducing desired traits in plants.

In the maize cultivation context, the culture of premature embryos stands out among the techniques adopted to accelerate the breeding cycle. This technique, detailed by McCaw et al. (2021), involves cultivating isolated embryonic cells on sterile media to plantlets. Additionally, the DH method is recognized for its ability to rapidly generate 100% homozygous lines, cutting down the conventional time frame of six to seven generations to a mere two.

However, the progression of plant biotechnology, despite its noteworthy advancements, confronts considerable challenges. From the prerequisites for advanced infrastructure and intensive professional training to the pursuit of sustainable financing, the path ahead is complex. With the ongoing evolution of technological innovation, producing new tools and methodologies, the agricultural community must adapt to effectively integrate these emerging methodologies.

1.4 COLLABORATION, INVESTMENT, AND INNOVATION IN SPEED BREEDING TECHNOLOGY

1.4.1 Collaboration in Speed Breeding Research

With the rise of innovative techniques such as Speed Breeding, research institutions have expanded their knowledge output, formulated new protocols, and implemented them across a diverse range of crops. A bibliometric analysis carried out on the Web of Science database from 2018 to 2022 recorded the publication of 128 scientific papers related to Speed Breeding, marking a significant increase of 59.23% over this period. This substantial growth in publications highlights the extensive research and application of these techniques across different crop species.

Two articles have gained global attention. Watson et al. (2018) published a comprehensive paper in *Nature Plants* that compiled several protocols adapted for both short-day and long-day crops. This work summarizes the scientific community's commitment to implementing these methodologies for a broad spectrum of crops. However, Hickey et al. (2019) presented a holistic review in *Nature Biotechnology*, delineating both the historical trajectory of genetic breeding and the details of Speed Breeding protocols. Notably, their discussion accentuates how Speed Breeding, synergized with methodologies like CRISPR-Cas, holds the potential to profoundly advance plant breeding.

Beyond the outbreak in publication volume, international collaboration among researchers and institutions offers pivotal insights into the adoption and assimilation of this technology. India, Australia, China, and the United States emerge as frontrunners in collaborative publications on Speed Breeding. Noteworthy is India's predominance in papers outcoming from internal collaborations, while Australia excels in co-authorship with international counterparts.

From this context, it is evident that Speed Breeding techniques are gaining prominence within the scientific community. A series of customized protocols for diverse crops are being unveiled, meeting diverse objectives in the public and private sectors.

1.4.2 Speed Breeding and Regulatory Challenges in Crop Varieties Release

Speed Breeding stands as a transformative innovation in plant biotechnology, expediting the developmental cycle of crops. Yet, its adoption triggers questions about the regulatory landscape for the varieties it produces.

Contrary to certain expectations, plants derived via Speed Breeding are not inherently demanding alterations in the existing protocols for variety registration and release. These guidelines, predominantly established by the International Union for the Protection of New Varieties of Plants (UPOV), aim to safeguard the investments of plant breeders. Under this regulatory umbrella, new varieties are mandated to undergo rigorous Distinctness, Uniformity, and Stability (DUS) tests, as well as Value for Cultivation and Use (VCU) assessments (Gudi et al., 2022).

Speed Breeding techniques offer an opportunity to streamline this evaluative process. The Speed DUS testing method, in particular, truncates cultivation periods without infringing upon established DUS protocols, thereby facilitating swifter generational advancements (Jamali et al., 2020). As underscored in studies on wheat and barley, this approach permits a meticulous assessment of certain UPOV-pertinent traits such as awn presence, plant height, and flowering time (Hickey et al., 2017; Watson et al., 2018; Bhatta et al., 2021).

However, it is pivotal to acknowledge that not all traits assessed by conventional DUS tests are amenable to the speed DUS technique. Furthermore, wider adoption of this technique faces challenges, including a shortage of specialized facilities.

In summary, while "Speed Breeding" holds the potential to revolutionize plant breeding methodologies, the industry and regulatory bodies must harmonize the industry, and regulatory bodies must harmonize their approaches to ensure the efficacious and safe release of produced varieties.

1.4.3 Investments in Speed Breeding Technologies

Speed Breeding techniques have gained attention within the scientific community due to their potential to significantly rush genetic breeding processes. Yet, despite this growing interest, the lack of economic evaluations and the substantial costs associated with establishing and maintaining these units (estimated at $3,930 annually or $720.72 per annum) act as constraints to their widespread adoption (Gudi et al., 2022).

Considering the infrastructural aspect of Speed Breeding, the initiation of a novel unit without pre-existing facilities might reach tens of thousands of dollars. Conversely, several institutions already possess infrastructure that can be repurposed for this objective, with component costs approximated at $758 (Watsonet al., 2018). These units are crucial for reducing crop generation time by taking advantage of ideal growing conditions such as precise controls of light, temperature, and growing medium (Watson et al., 2018; Chiurugwi et al., 2019; Gudi et al., 2022).

Comparative analysis by Lenaerts et al. (2018) highlighted the economic advantages when genetic breeding procedures are speeded. In alignment with this principle, Gudi et al. (2022) determined that deploying Speed Breeding could culminate in a timesaving of up to 4 years compared to traditional breeding methodologies. This efficiency escalates when integrated with other methodologies, potentially leading to a time reduction of up to 6 years in cultivating novel varieties.

Considering the global investment in Research and Development (R&D), the significant capital injections led by the United States, China, and others in 2022 deserve attention (https://www.statista.com/statistics/732247/worldwide-research-and-development-gross-expenditure-top-countries/). The appropriation of Speed Breeding techniques is predominantly advantageous in the private sector, where genetic breeding programs attempt to speedily produce novel cultivars. Within the public sphere, nations with robust R&D investments, such as the United States and China, stand to reap the most benefits from these techniques.

To democratize access to Speed Breeding, Chiurugwi et al. (2019) suggested the establishment of Speed Breeding capsules and dedicated centers, employing cost-effective components and offering adaptability across a spectrum of crops. Respected institutes, including ICRISAT and IITA, have already signaled their predisposition to develop such facilities, underscoring the global potential of this approach (Ceballos et al., 2017).

1.5 CONCLUSION

In conclusion, reinvigorating investment in R&D is indispensable to optimize reproduction speed, especially when integrated with other sophisticated methodologies such as CRISPR-Cas9 and MAS. This strategic merger could accelerate the development of cultivars that are naturally attuned to contemporary and future agricultural demands. However, challenges encompassing high operational expenses and the need for specialized knowledge highlight the importance of inter-institutional collaboration. The immediate development of specific protocols for economically important crops is an urgent need, poised to induce substantial advances in global agriculture.

REFERENCES

Abdul Aziz, M. and Masmoudi, K. (2023). Insights into the transcriptomics of crop wild relatives to unravel the salinity stress adaptive mechanisms. *International Journal of Molecular Sciences*, *24* (12), 9813.

Almeida-Silva, F., Moharana, K. C., and Venancio, T. M. (2021). The state of the art in soybean transcriptomics resources and gene coexpression networks. *in silico Plants*, *3* (1), 1–7.

Arraes, F. B. M., Vasquez, D. D. N., Tahir, M., Pinheiro, D. H., Muhammed, F., Freitas-Alves, N. S., Pinto, C. E. M., Moreira, V. J. V., Paes-De-Melo, B., Lisei-De-Sá, M. E., Morgante, C. V., Mota, A. P. Z., Lourenço-Tessutti, I. T., Togawa, R. C., Grynberg, P., Fragoso, R. R., De Almeida-Engler, J., Larsen, M. J., and Grossi-de-Sá, M. F. (2022). Integrated omic approaches reveal molecular mechanisms of tolerance during soybean and meloidogyne incognita interactions. *Plants*, *2744* (20), 1–38.

Bakht, J., Shafi, M., Jamal, Y., and Sher, H. (2011). Response of maize (Zea mays L.) to seed priming with NaCl and salinity stress. *Spanish Journal of Agricultural Research*, *252* (1), 252–261.

Barnabás, B., Obert, B., and Kovács, G. (1999). Colchicine, an efficient genome-doubling agent for maize (Zea mays L.) microspores cultured in anthero. *Plant Cell Reports*, (10), 858–862.

Basso, M. F., Arraes, F. B. M., Grossi-De-Sá, M. F., Moreira, V. J. V., Alves-Ferreira, M., and Grossi-De-Sá, M. F. (2020). Insights into genetic and molecular elements for transgenic crop development. *Frontiers in Plant Science*, *11*, 498560.

Bayer, P. E., Valliyodan, B., Hu, H., Marsh, J. I., Yuan, Y., Vuong, T. D., Patil, G., Song, Q., Batley, J., Varshney, R. K., Lam, H., Edwards, D., and Nguyen, H. T. (2021). Sequencing the USDA core soybean collection reveals gene loss during domestication and breeding. *The Plant Genome*, *15* (1), 1–12.

Bhatta, M., Sandro, P., Smith, M. R., Delaney, O., Voss-Fels, K. P., Gutierrez, L., and Hickey, L. T. (2021). Need for speed: manipulating plant growth to accelerate breeding cycles. *Current Opinion in Plant Biology*, *60*, 101986.

Brachi, B., Morris, G. P., and Borevitz, J. O. (2011). Genome-wide association studies in plants: the missing heritability is in the field. *Genome Biology*, *12* (10), 232.

Ceballos, H., Jaramillo, J. J., Salazar, E., Pineda, L. M., Calle, F., and Setter, T. L. (2017). Induction of flowering in cassava through grafting. *Journal of Plant Breeding and Crop Science*, *9* (2), 19–29.

Cembrowska-Lech, D., Krzemińska, A., Miller, T., Nowakowska, A., Adamski, C. J., Radaczyńska, M., Mikiciuk, G., and Mikiciuk, M. (2023). An integrated multi-omics and artificial intelligence framework for advance plant phenotyping in horticulture. *Biology*, *1298* (10), 1–34.

Chase, S. S. and Nanda, D. K. (1965). Comparison of variability in inbred lines and monoploid-derived lines of maize (Zea mays L.). *Crop Science*, *5* (3), 275–276.

Chatham, L. A. and Juvik, J. A. (2021). Linking anthocyanin diversity, hue, and genetics in purple corn. *G3: Genes, Genomes, Genetics*, *11* (2), 1–18.

Chen, X., Bhadauria, V., and Ma, B. (2018). ChIP-seq: A powerful tool for studying protein-DNA interactions in plants. *Current Issues in Molecular Biology*, *27*, 171–180.

Cherubin, M. R., Damian, J. M., Tavares, T. R., Trevisan, R. G., Colaço, A. F., Eitelwein, M. T., Martello, M., Inamasu, R. Y., Pias, O.H. de C., and Molin, J. P. (2022). Precision agriculture in Brazil: The trajectory of 25 years of scientific research. *Agriculture*, *12* (11), 1882.

Chiurugwi, T., Kemp, S., Powell, W., and Hickey, L. T. (2019). Speed breeding orphan crops. *Theoretical and Applied Genetics*, (3), 607–616.

Cobb, J. N., Juma, R. U., Biswas, P., Arbelaez, J. D., Rutkoski, J., Atlin, G. N., Hagen, T., Quinn, M. J., and Ng, E. H. (2019). Enhancing the rate of genetic gain in public-sector plant breeding programs: Lessons from the breeder's equation. *Theoretical and Applied Genetics*, *132* (3), 627–645.

Collard, B. C. Y., Beredo, J. C., Lenaerts, B., Mendoza, R., Santelices, R., Lopena, V., Verdeprado, H., Raghavan, C., Gregorio, G. B., Vial, L., Demont, M., Biswas, P. S., Iftekharuddaula, K. M., Rahman, M. A., Cobb, J. N., and Islam, M. R. (2017). Revisiting rice breeding methods – evaluating the use of rapid generation advance (RGA) for routine rice breeding. *Plant Production Science*, *20* (4), 337–352.

Das, P. R. and Sherif, S. M. (2020). Application of exogenous dsRNAs-induced RNAi in agriculture: Challenges and triumphs [online]. *Frontiers in Plant Science*. Available from: 10.3389/fpls.2020.00946.

Derbyshire, M. C., Batley, J., and Edwards, D. (2022). Use of multiple 'omics techniques to accelerate the breeding of abiotic stress-tolerant crops. *Current Plant Biology*, *32*, 100262.

Eldakak, M., Milad, S. I. M., Nawar, A. I., and Rohila, J. S. (2013). Proteomics: A biotechnology tool for crop improvement. *Frontiers in Plant Science*, *4*.

Fang, C., Ma, Y., Wu, S., Liu, Z., Wang, Z., Yang, R., Hu, G., Zhou, Z., Yu, H., Zhang, M., Pan, Y., Zhou, G., Ren, H., Du, W., Yan, H., Wang, Y., Han, D., Shen, Y., Liu, S., Liu, T., Zhang, J., Qin, H., Yuan, J., Yuan, X., Kong, F., Liu, B., Li, J., Zhang, Z., Wang, G., Zhu, B., and Tian, Z. (2017). Genome-wide association studies dissect the genetic networks underlying agronomical traits in soybean. *Genome Biology*, *18* (1), 1–14.

Fang, S., Gao, K., Hu, W., Snider, J. L., Wang, S., Chen, B., and Zhou, Z. (2018). Chemical priming of seed alters cotton floral bud differentiation by inducing changes in hormones, metabolites, and gene expression. *Plant Physiology and Biochemistry*, *130*, 633–640.

Fang, Y. and Han, T. (2019). Research progress in speed breeding of crops. *Crops*, *35* (2), 1–7.

Fang, Y., Wang, L., Sapey, E., Fu, S., Wu, T., Zeng, H., Sun, X., Qian, S., Khan, M. A. A., Yuan, S., Wu, C., Hou, W., Sun, S., and Han, T. (2021). Speed-breeding system in soybean: Integrating off-site generation advancement, fresh seeding, and marker-assisted selection. *Frontiers in Plant Science*, *12*.

Farooqi, M. Q. U., Nawaz, G., Wani, S. H., Choudhary, J. R., Rana, M., Sah, R. P., Afzal, M., Zahra, Z., Ganie, S. A., Razzaq, A., Reyes, V. P., Mahmoud, E. A., Elansary, H. O., El-Abedin, T. K. Z., and Siddique, K. H. M. (2022). Recent developments in multi-omics and breeding strategies for abiotic stress tolerance in maize (Zea mays L.). *Frontiers in Plant Science*, *13*, 965878.

Fernandez, O., Millet, E. J., Rincent, R., Prigent, S., Pétriacq, P., and Gibon, Y. (2021). Plant metabolomics and breeding. *Advances in Botanical Research*, *98*, 207–235.

Florentino, L. H., Lima, R. N., and Molinari, M. D. C. (2022). Soybean functional proteins and the synthetic biology. In: *Soybean – recent advances and application*. IntechOpen. https://www.intechopen.com/chapters/82209

Garg, V. and Varshney, R. K. (2022). Analysis of small RNA sequencing data in plants. *Methods in Molecular Biology*, *2443*, 497–509.

Gieco, L. C., Caviglia, O. P., and Lassaga, S. L. (2007). Rescate in vitro de embriones inmaduros de soja como técnica para programas de mejoramiento genético. *Revista Cientfica Agropecuaria*, *11* (2), 87–93.

Gong, Z., Cheng, M., and Botella, J. R. (2021). Non-GM genome editing approaches in crops. *Frontiers in Genome Editing*, *3*, 817279.

Gonzalez, E., Zarei, A., Hendler, N., Simmons, T., Zarei, A., Demieville, J., Strand, R. J., Rozzi, B., Calleja, S., Ellingson, H., Cosi, M., Davey, S., Lavelle, D., Truco, M. J., Swetnam, T. L., Merchant, N., Michelmore, R. W., Lyons, E., and Pauli, D. (2023). PhytoOracle: Scalable, modular phenomics data processing pipelines. *Frontiers in Plant Science*, *14*, 1112973.

González-Barrios, P., Bhatta, M., Halley, M., Sandro, P., and Gutiérrez, L. (2020). Speed breeding and early panicle harvest accelerate oat (Avena sativa L.) breeding cycles. *Crop Science*, *61* (1), 320–330.

Gudi, S., Kumar, P., Singh, S., Tanin, M. J., and Sharma, A. (2022). Strategies for accelerating genetic gains in crop plants: Special focus on speed breeding. *Physiology and Molecular Biology of Plants*, *28* (10), 1921–1938.

Harrison, D., Silva, M. D., Wu, C., Oliveira, M. D., Ravelombola, F., Florez-Palacios, L., Acunã, A., and Mozzoni, L. (2020). Effect of light wavelength on soybean growth and development in the context of speed breeding. *Crop Science*, *61* (2), 917–928.

Hatfield, J. L. and Prueger, J. H. (2015). Temperature extremes: Effect on plant growth and development. *Weather and Climate Extremes*, *10*, 4–10.

Hickey, L. T., Germán, S., Pereyra, S., Díaz, J. F., Ziems, L., Fowler, R., Platz, G. J., Franckowiak, J. D., and Dieters, M. J. (2017). Speed breeding for multiple disease resistance in barley. *Euphytica*, *213* (3), 1–14.

Hickey, L. T., N. Hafeez, A., Robinson, H., Jackson, S. A., Leal-Bertioli, S. C. M., Tester, M., Gao, C., Godwin, I. D., Hayes, B. J., and Wulff, B. B. H. (2019). Breeding crops to feed 10 billion. *Nature Biotechnology*, *37* (7), 744–754.

Hilker, M. and Schmülling, T. (2019). Stress priming, memory, and signaling in plants. *Plant, Cell & Environment*, *42*, 753–761.

Huang, C.-Y. and Jin, H. (2022). Coordinated epigenetic regulation in plants: A potent managerial tool to conquer biotic stress. *Frontiers in Plant Science*, *12*, 795274.

Hufford, M. B., Seetharam, A. S., Woodhouse, M. R., Chougule, K. M., Ou, S., Liu, J., Ricci, W. A., Guo, T., Olson, A., Qiu, Y., Della Coletta, R., Tittes, S., Hudson, A. I., Marand, A. P., Wei, S., Lu, Z., Wang, B., Tello-Ruiz, M. K., Piri, R. D., Wang, N., Kim, D. won, Zeng, Y., O'Connor, C. H., Li, X., Gilbert, A. M., Baggs, E., Krasileva, K. V., Portwood, J. L., Cannon, E. K. S., Andorf, C. M., Manchanda, N., Snodgrass, S. J., Hufnagel, D. E., Jiang, Q., Pedersen, S., Syring, M. L., Kudrna, D. A., Llaca, V., Fengler, K., Schmitz, R. J., Ross-Ibarra, J., Yu, J., Gent, J. I., Hirsch, C. N., Ware, D., and Dawe, R. K. (2021). De novo assembly, annotation, and comparative analysis of 26 diverse maize genomes. *Science*, *373* (6555), 655–662.

Impens, L., Jacobs, T. B., Nelissen, H., Inzé, D., and Pauwels, L. (2022). Mini-review: Transgenerational CRISPR/Cas9 gene editing in plants. *Frontiers in Genome Editing*, *4*, 825042.

Jagtap, U. B., Gurav, R., and Bapat, V. A. (2011). Role of RNA interference in plant improvement. *The Science of Nature*, *98*(6), 473–492.

Jähne, F., Hahn, V., Würschum, T., and Leiser, W. L. (2020). Speed breeding short-day crops by LED-controlled light schemes. *Theoretical and Applied Genetics*, *133*(8), 2335–2342.

Jamali, S. H., Cockram, J., and Hickey, L. T. (2020). Is plant variety registration keeping pace with speed breeding techniques? *Euphytica*, *216* (8), 131.

Jin, S. and Daniell, H. (2015). The engineered chloroplast genome just got smarter. *Trends in Plant Science*, *20* (10), 622–640.

Kafer, J. M., Molinari, M. D. C., Henning, F. A., Koltun, A., Marques, V. V., Marin, S. R. R., Nepomuceno, A. L., and Mertz-Henning, L. M. (2023). Transcriptional profile of soybean seeds with contrasting seed coat color. *Plants*, *1555* (7), 1555.

Khahani, B., Tavakol, E., Shariati, V., and Rossini, L. (2021). Meta-QTL and ortho-MQTL analyses identified genomic regions controlling rice yield, yield-related traits, and root architecture under water deficit conditions. *Scientific Reports*, *11* (1), 6942.

Khan, M. H. U., Wang, S., Wang, J., Ahmar, S., Saeed, S., Khan, S. U., Xu, X., Chen, H., Bhat, J. A., and Feng, X. (2022). Applications of artificial intelligence in climate-resilient smart-crop breeding. *International Journal of Molecular Sciences*, *11156* (19), 11156.

Komatsu, S., Mock, H.-P., Yang, P., and Svensson, B. (2013). Application of proteomics for improving crop protection/artificial regulation. *Frontiers in Plant Science*, *4*, 76250.

Kumar, S. and Mohapatra, T. (2021). Dynamics of DNA methylation and its functions in plant growth and development. *Frontiers in Plant Science*, *12*, 596236.

Lenaerts, B., de Mey, Y., & Demont, M. (2018). Global impact of accelerated plant breeding: Evidence from a meta-analysis on rice breeding. *PLoS One*, *13*, e0199016.

Lenaerts, B., Collard, B. C. Y., and Demont, M. (2019). Review: Improving global food security through accelerated plant breeding. *Plant Science*, *287*, 110207.

Li, C., Fu, Y., Liu, Q., Du, L., and Trotsenko, V. (2020). A review of genetic mechanisms of early maturity in cotton (Gossypium hirsutum L.). *Euphytica*, *216* (7), 1–15.

Li, H., Xu, Z., and Tang, C. (2010). Effect of light-emitting diodes on growth and morphogenesis of upland cotton (Gossypium hirsutum L.) plantlets in vitro. *Plant Cell, Tissue and Organ Culture (PCTOC)*, *216*(2), 155–163.

Liang, X., Liu, S., Wang, T., Li, F., Cheng, J., Lai, J., Qin, F., Li, Z., Wang, X., and Jiang, C. (2021). Metabolomics-driven gene mining and genetic improvement of tolerance to salt-induced osmotic stress in maize. *New Phytologist*, *230* (6), 2355–2370.

Liotino, M., Varangot, A., Beznec, A., Auteri, M., Bossio, A. E., Lewi, D., and Faccio, P. (2019). Shortening of generation cycles in inbred lines of maize (Zea mays L.) through embryo rescue technique. *Maydica*, *64* (2), 1–6.

Litvinov, D. Y., Karlov, G. I., and Divashuk, M. G. (2021). Metabolomics for crop breeding: General considerations. *Genes*, *12* (10), 1602.

Maqbool, M. A., Beshir, A., and Khokhar, E. S. (2020). Doubled haploids in maize: Development, deployment, and challenges. *Crop Science*, *60* (6), 2815–2840.

Marcolino-Gomes, J., Rodrigues, F. A., Fuganti-Pagliarini, R., Nakayama, T. J., Ribeiro Reis, R., Bouças Farias, J. R., Harmon, F. G., Correa Molinari, H. B., Correa Molinari, M. D., and Nepomuceno, A. (2015). Transcriptome-wide identification of reference genes for expression analysis of soybean responses to drought stress along the day. *PLoS One*, *10* (9), e0139051.

Marsh, J. I., Hu, H., Gill, M., Batley, J., and Edwards, D. (2021). Crop breeding for a changing climate: Integrating phenomics and genomics with bioinformatics. *Theoretical and Applied Genetics*, *134* (6), 1677–1690.

McCaw, M., Lee, K., Kang, M., Zobrist, J. D., Azanu, M. K., Birchler, J. A., and Wang, K. (2021). Development of a transformable fast-flowering mini-maize as a tool for maize gene editing. *Frontiers in Genome Editing*, *2*, 622227.

Michereff, M. F. F., Grynberg, P., Togawa, R. C., Costa, M. M. C., Laumann, R. A., Zhou, J., Schimmelpfeng, P. H. C., Borges, M., Pickett, J. A., Birkett, M. A., and Blassioli-Moraes, M. C. (2021). Priming of indirect defense responses in maize is shown to be genotype-specific. *Arthropod-Plant Interactions*, *15* (3), 313–328.

Molinari, H. B. C., Vieira, L. R., Volpi e Silva, N., Filho, J. H. L., and Prado, G. S. (2020). Tecnologia CRISPR na edição genômica de plantas. Brasília, DF, Brasil.

Molinari, M. D. C., Fuganti-Pagliarini, R., Angelotti, J., Barbosa, D. A., Marin, D. R., Mertz-Henning, L. M., and Nepomuceno, A. L. (2021a). Transcriptome analysis using RNA-seq from experiments with and without biological replicates: A review. *Revista Ciência Agrária*, *64*, 13.

Molinari, M. D. C., Fuganti-Pagliarini, R., Marcolino-Gomes, J., de Amorim Barbosa, D., Rockenbach Marin, S. R., Mertz-Henning, L. M., Nepomuceno, A. L., and Rech Filho, E. L. (2021b). Flower and pod genes involved in soybean sensitivity to drought. *Journal of Plant Interactions*, *16* (1), 187–200.

Molinari, M. D. C., et al. (2021c). Flowering process in soybean under water deficit conditions: A review on genetic aspects. *Genetics and Molecular Biology*, *45*, 10.1590/1678-4685-gmb-2021-0016.

Molinari, M. D. C., Fuganti-Pagliarini, R., de Amorim Barbosa, D., Barbosa, E. G. G., Kafer, J. M., Marin, D. R., Marin, S. R. R., Mertz-Henning, L. M., and Nepomuceno, A. L. (2022). Comparative ABA-responsive transcriptome in soybean cultivars submitted to different levels of drought. *Plant Molecular Biology Reporter*, *41* (2), 260–276.

Molinari, M. D. C., Fuganti-Pagliarini, R., Yoshida, Y., Florentino, L. H., Mertz-Henning, L. M., Lima, R. N., Bittencourt, D., Freire, M., and Rech, E. L. (2023). Exploring the proteomic profile of soybean bran: Unlocking the potential for improving protein quality and quantity. *Plants*, *2704* (14), 2704.

Molinari, M. D. C., et al. (2024). *Plant Genome Editing Technologies. Interdisciplinary Biotechnological Advances*. Singapore: Springer. https://doi.org/10.1007/978-981-99-9338-3_2

Mubarik, M. S., Ma, C., Majeed, S., Du, X., and Azhar, M. T. (2020). Revamping of cotton breeding programs for efficient use of genetic resources under changing climate. *Agronomy*, *1190*(8), 1190.

Muhammad, T., Zhang, F., Yan, Z., and Liang, Y. (2019). RNA interference: A natural immune system of plants to counteract biotic stressors. *Cells*, *38* (1), 38.

Mustafa, G. and Komatsu, S. (2021). Plant proteomic research for improvement of food crops under stresses: A review. *Molecular Omics*, *17* (6), 860–880.

Nagatoshi, Y. and Fujita, Y. (2019). Accelerating Soybean Breeding in a CO_2-Supplemented Growth Chamber. *Plant and Cell Physiology*, *60* (1), 77–84.

Owusu, A. G., Lv, Y.-P., Liu, M., Wu, Y., Li, C., Guo, N., Li, D., and Gao, J. (2023). Transcriptomic and metabolomic analyses reveal the potential mechanism of waterlogging resistance in cotton (Gossypium hirsutum L.) [online]. *Frontiers in Plant Science*. Available from: 10.3389/fpls.2023.1088537.

Pallaoro, D. S., Camili, E. C., Guimarães, S. C., and De Figueiredo E Albuquerque, M. C. (2016). Methods for priming maize seeds. *Journal of Seed Science*, *38* (2), 148–154.

Pazos-Navarro, M., Castello, M., Bennett, R. G., Nichols, P., and Croser, J. (2017). In vitro-assisted single-seed descent for breeding-cycle compression in subterranean clover (Trifolium subterraneum L.). *Crop and Pasture Science*, *68* (11), 958.

Rajcan, I., Boersma, J. G., and Shaw, E. J. (2011). Plant genetic techniques: Plant breeder's toolbox. In: *Comprehensive biotechnology*. Elsevier, pp. 133–147. https://www.researchgate.net/profile/Istvan-Rajcan/publication/257526830_Plant_SystemsPlant_Genetic_Techniques_Plant_Breeder's_Toolbox/links/0c9605255b95391b16000000/Plant-Systems-Plant-Genetic-Techniques-Plant-Breeders-Toolbox.pdf

Rajput, M., Choudhary, K., Kumar, M., Vivekanand, V., Chawade, A., Ortíz, R., and Pareek, N. (2021). RNA interference and CRISPR/Cas gene editing for crop improvement: Paradigm shift towards sustainable agriculture. *Plants*, *1914* (9), 1914.

Ravelombola, W. S., Qin, J., Shi, A., Nice, L., Bao, Y., Lorenz, A., Orf, J. H., Young, N. D., and Chen, S. (2019). Genome-wide association study and genomic selection for soybean chlorophyll content associated with soybean cyst nematode tolerance. *BMC Genomics*, *20* (1), 1–18.

Rolland, V., Farazi, M., Conaty, W. C., Cameron, D. R., Liu, S., Petersson, L., and Stiller, W. N. (2022). HairNet: A deep learning model to score leaf hairiness, a key phenotype for cotton fiber yield, value, and insect resistance. *Plant Methods*, *18* (1), 1–19.

Sakurai, N. (2022). Recent applications of metabolomics in plant breeding. *Breeding Science*, *72* (1), 56–65.

Saxena, K. B., Saxena, R. K., Hickey, L. T., and Varshney, R. K. (2019). Can a speed breeding approach accelerate genetic gain in pigeon peas? *Euphytica*, *215* (12), 1–7.

Singh, B., Kumar, S., Elangovan, A., Vasht, D., Arya, S., Duc, N. T., Swami, P., Pawar, G. S., Raju, D., Krishna, H., Sathee, L., Dalal, M., Sahoo, R. N., and Chinnusamy, V. (2023). Phenomics based prediction of plant biomass and leaf area in wheat using machine learning approaches. *Frontiers in Plant Science*, *14*, 1214801.

Singh, I., Sheoran, S., Kumar, B., Kumar, K., and Rakshit, S. (2021). Speed breeding in maize (Zea mays) vis-à -vis in other crops: Status and prospects. *The Indian Journal of Agricultural Sciences*, *91* (9), 1267–1273.

Sintaha, M., Man, C.-K., Yung, W., Duan, S., Li, M.-W., and Lam, H. (2022). Drought stress priming improved the drought tolerance of soybean. *Plants*, *2954* (21), 2954.

Statista [online] (2023). Available from: https://www.statista.com/statistics/732247/worldwide-research-and-development-gross-expenditure-top-countries/ [Accessed 13 Sep 2023].

Syaiful, S. A., Dungga, N. E., Riadi, M., and Ridwan, I. (2015). Seed priming with PEG 8000 for improving drought stress tolerance of soybean (Glycine max). *International Journal of Agriculture System*, *1* (1), 19–26.

Tao, Y., Jiang, L., Liu, Q., Zhang, Y., Zhang, R., Ingvardsen, C. R., Frei, U. K., Wang, B., Lai, J., Lübberstedt, T., and Xu, M. (2013). Combined linkage and association mapping reveal candidates for Scmv1, a major locus involved in resistance to sugarcane mosaic virus (SCMV) in maize. *BMC Plant Biology*, *13* (1), 1–13.

Tao, Y., Zhao, X., Mace, E., Henry, R., and Jordan, D. (2019). Exploring and Exploiting pan-genomics for crop improvement. *Molecular Plant*, *12* (2), 156–169.

Távora, F. T. P. K., De Assis Dos Santos Diniz, F., De Moraes Rêgo-Machado, C., Freitas, N. C., Arraes, F. B. M., De Andrade, E. C., Furtado, L. L., Osiro, K. O., De Sousa, N. L., Cardoso, T. B., Mertz-Henning, L. M., De Oliveira Molinari, P. A., Feingold, S. E., Hunter, W. B., De Sá, M. F. G., Kobayashi, A. K., Nepomuceno, A. L., Santiago, T. R., and Molinari, H. B. C. (2022). CRISPR/Cas- and topical RNAi-based technologies for crop management and improvement: Reviewing the risk assessment and challenges towards a more sustainable agriculture. *Frontiers in Bioengineering and Biotechnology*, *10*, 913728.

Tyagi, P., Singh, D., Mathur, S., Singh, A., and Ranjan, R. (2022). Upcoming progress of transcriptomics studies on plants: An overview. *Frontiers in Plant Science*, *13*, 1030890.

Verma, D. and Daniell, H. (2007). Chloroplast vector systems for biotechnology applications. *Plant Physiology*, *145* (4), 1129–1143.

Wan, Y., Duncan, D. R., Rayburn, A. L., Petolino, J. F., and Widholm, J. M. (1991). The use of antimicrotubule herbicides for the production of doubled haploid plants from anther-derived maize callus. *Theoretical and Applied Genetics*, *81* (2), 205–211.

Wang, X., Wang, Y., Zhang, G., and Ma, Z. (2011). An integrated breeding technology for accelerating generation advancement and trait introgression in cotton. *Plant Breeding*, *130* (5), 569–573.

Watson, A., Ghosh, S., Williams, M. J., Cuddy, W. S., Simmonds, J., Rey, M.-D., Asyraf Md Hatta, M., Hinchliffe, A., Steed, A., Reynolds, D., Adamski, N. M., Breakspear, A., Korolev, A., Rayner, T., Dixon, L. E., Riaz, A., Martin, W., Ryan, M., Edwards, D., Batley, J., Raman, H., Carter, J., Rogers, C., Domoney, C., Moore, G., Harwood, W., Nicholson, P., Dieters, M. J., DeLacy, I. H., Zhou, J., Uauy, C., Boden, S. A., Park, R. F., Wulff, B. B. H., and Hickey, L. T. (2018). Speed breeding is a powerful tool to accelerate crop research and breeding. *Nature Plants*, *4* (1), 23–29.

Xu, Y., Luo, H., Zhang, H., Yung, W.-S., Li, M.-W., Lam, H.-M., and Huang, C. (2023). Feeding the world using speed breeding technology. *Trends in Plant Science*, *28* (3), 372–373.

Yasir, M., Kanwal, H. H., Hussain, Q., Riaz, M. W., Sajjad, M., Rong, J., and Jiang, Y. (2022). Status and prospects of genome-wide association studies in cotton. *Frontiers in Plant Science*, *13*, 1019347.

Yung, W.-S., Wang, Q., Huang, M., Wong, F.-L., Liu, A., Ng, M.-S., Li, K.-P., Sze, C.-C., Li, M.-W., and Lam, H.-M. (2021). Priming-induced alterations in histone modifications modulate transcriptional responses in soybean under salt stress. *The Plant Journal*, *109* (6), 1575–1590.

Zhang, Y., Zhou, X., Dong, Y., Zhang, F., He, Q., Chen, J., Zhu, S., and Zhao, T. Z. (2021). Seed priming with melatonin improves salt tolerance in cotton by regulating photosynthesis, scavenging reactive oxygen species, and coordinating with phytohormone signal pathways. *Industrial Crops and Products*, *169*, 113671.

Zhao, T., Deng, X., Xiao, Q., Han, Y., Zhu, S., and Chen, J. (2020). IAA priming improves the germination and seedling growth in cotton (Gossypium hirsutum L.) by regulating the endogenous phytohormones and enhancing the sucrose metabolism. *Industrial Crops and Products*, *155*, 112788.

Zila, C. T., Samayoa, L. F., Santiago, R., Butrón, A., and Holland, J. B. (2013). A genome-wide association study reveals genes associated with fusarium ear rot resistance in a maize core diversity panel. *G3: Genes, Genomes, Genetics*, *3* (11), 2095–2104.

2 Speed Breeding in the Era of High-Throughput OMICS

Recent Developments and Future Outlook

Furqan Tayyab
Centre of Agricultural Biochemistry and Biotechnology (CABB), University of Agriculture, Faisalabad, Pakistan

Momna Mehmood
Department of Agricultural Biotechnology, Faculty of Agriculture, Ondokuz Mayis University, Samsun, Turkey

Aqsa Parvaiz
Department of Biochemistry and Biotechnology, The Women's University Multan, Multan, Pakistan

Ghulam Mustafa, Muhammad Sarwar Khan, and Faiz Ahmad Joyia
Centre of Agricultural Biochemistry and Biotechnology (CABB), University of Agriculture, Faisalabad, Pakistan

2.1 INTRODUCTION

Plant breeding has long been instrumental in ensuring food security and addressing global challenges related to food quantity and quality (Tester & Langridge, 2010; Shiferaw et al., 2011; Ahmar et al., 2020). Plant breeding plays a crucial role in developing crops with desired traits to meet the increasing demand for plant-based products (Cheema, 2018). Humans have been practicing artificial plant selection for thousands of years, resulting in the cultivation of plants with higher nutritional value (Moose & Mumm, 2008; Ahmar et al., 2020). The conventional approach to plant breeding has been pivotal in ensuring global food and nutritional security by harnessing genetic diversity and uncovering novel genes (Ahmar et al., 2020; Rai, 2022). Conventional breeding involves several stages, including the selection of parent plants, controlled pollination, evaluation of progeny, and repeated cycles of selection and backcrossing to stabilize desired traits. However, the time-consuming nature of this process, involving hybridization and intercrossing of cultivars, poses a significant bottleneck in developing desired crop varieties (Al-Khayri et al., 2015; Rai, 2022).

In addition, the global population is rapidly increasing, creating a greater demand for food production (Ahmar et al., 2020). Simultaneously, climate change-induced fluctuations, such as rising temperatures, frequent floods, and droughts, pose additional challenges by leading to novel diseases and pest outbreaks (Samantara et al., 2022; Mehmood, 2023). The projected population

DOI: 10.1201/b23372-2

growth to 10 billion by 2050 raises concerns about meeting the food requirements of this massive population (Ahmar et al., 2020).

Recent scientific advancements in plant breeding techniques offer a wide range of possibilities and innovations to address the challenges of food security (Ahmar et al., 2020). Early in the eighteenth century, NASA and Utah State University collaborated to investigate the potential of growing rapid-cycle wheat in space stations under continuous light. Their collaborative efforts resulted in the creation of the "USU-Apogee", a dwarf wheat line (Bula et al., 1991; Bugbee & Koerner, 1997; Ahmar et al., 2020). In the mid-nineteenth century, several techniques, most prominently, shuttle breeding and single-seed descent (SSD), were being utilized to accelerate plant lifecycle. However, the focus was shifted toward reducing the variety of development time incorporating controlled growth conditions(Samantara et al., 2022). This lead to the development of non-GMO modern technology, known as speed breeding (SB), which offers great opportunities for rapid generation advancement and shortened breeding cycles (Watson et al., 2018; Hickey et al., 2019; Voss-Fels et al., 2019; Ahmar et al., 2020; Wanga et al., 2021).

SB offers several key opportunities for pre-breeding and breeding programs (Wanga et al., 2021). This groundbreaking approach involves various techniques as mentioned by Ghosh et al. (2018) including accelerated single-seed descent (aSSD), rapid generation cycling (RGC) using DNA marker technology, fast generation cycling (FGC) using stressed conditions and in vitro culture of immature embryos, and rapid generation turnover (RGT) through immature seed harvest and photoperiod response (Ghosh et al., 2018; Samantara et al., 2022). All the above-mentioned techniques allow early flowering as well as seed maturity, thus enabling faster progress by cutting down the length of selection cycles (Hickey et al., 2019). Additionally, it addresses challenges associated with field selection processes, expert phenotyping, and resource management, facilitating more efficient breeding (Mobini et al., 2020). Plants of various important crops such as canola wheat and barley have been rapidly developed with desired phenotypes using SB (Ahmar et al., 2020). SB techniques have been successfully applied to important plant families like *Poaceae, Fabaceae*, and *Brassicaceae*, resulting in a significant improvement in annual generation turnover compared to traditional methods (Goulden, 1939; Brim, 1966; Borlaug, 1968; Ghosh et al., 2018; Samantara et al., 2022).

Most recently, SB combined with advanced OMICS techniques has gained prominence (Harfouche et al., 2019; Sartor et al., 2019). SB manipulates the growing environment to expedite the breeding process by enhancing flowering and seed set rates (Harfouche et al., 2019; Razzaq et al., 2021). Moreover, the integration of classical breeding with OMICS technologies has revolutionized data-driven decision-making and enabled unprecedented improvements in breeding programs (Sartor et al., 2019).

This chapter aims to provide an extensive overview of SB in the context of high-throughput OMICS technologies. By examining the evolution of breeding techniques, exploring the crops amenable to SB, and discussing the integration of high-throughput techniques, we aim to equip readers with a holistic understanding of this innovative approach. Moreover, by addressing the challenges and opportunities on the horizon, we seek to inspire further advancements and propel SB toward its full potential as a transformative tool for crop improvement in the future.

2.2 EVOLUTION OF BREEDING TECHNIQUES: FROM CONVENTIONAL BREEDING TO SPEED BREEDING

2.2.1 Conventional Breeding: Tapping into Natural Variation

Conventional breeding, also known as classical breeding, has been the foundation of plant breeding for centuries. Conventional plant breeding has long been recognized for its ability to increase genetic resources for crop improvement by incorporating desired traits (Rai, 2022). Since ancient times, breeders have been creating new varieties with improved traits such as yield, disease resistance, and quality by controlled cross-pollination and then careful selection. The conventional approach to crop breeding involves a series of steps, including parental selection, crosses, progeny development, trait selection, screening in various environments, and commercialization (Shimelis & Laing, 2012; Wanga et al., 2021). These procedures aim to develop crop varieties having improved agronomic traits mainly yield, nutrition, and ability to withstand biotic and abiotic stresses (Tester & Langridge, 2010; Breseghello & Coelho, 2013; Wanga et al., 2021).

Conventional plant breeding has been instrumental in improving crop quality by introducing beneficial traits from wild relatives. These traits include increased yield, tolerance against abiotic and biotic stresses, and improved and higher nutritional value (Dias et al., 2016; Niu et al., 2017). Breeders can have crop populations with desired traits simply by identifying, selecting, and combining them in familiar genotypes. However, it is important to note that this approach may occasionally raise concerns related to food and environmental safety due to unpredictable effects (Shi et al., 2016; Qaim, 2020). Selecting hybrid progeny with traits of interest following cross-breeding genetically different genotypes is another way to improve specific traits in crops (Han et al., 2017). Backcrossing, on the other hand, allows the introduction of desired traits into an elite line that has overall good characters but lacks one or more desired characters. This is achieved by repeatedly crossing the donor cultivar (having the desired character) with the recipient line over several generations to attain maximum homozygosity (Hou et al., 2019). However, conventional breeding methods are time-consuming, often taking over a decade to release a new variety (de Pater et al., 2013; Gerald et al., 2013; Ahmar et al., 2020). The long duration of breeding cycles, particularly in stages 3 and 4, significantly hinders the pace of cultivar development (Shimelis & Laing, 2012; Wanga et al., 2021).

Plant breeding has evolved significantly over time, starting with ancient farmers practicing selection and crossing to increase the domestication of plants in their surroundings (Ahmar et al., 2020). Early breeders selected wild cultivars with desirable agronomic traits and crossed them with local cultivars, followed by rigorous selection for several generations (Al-Khayri et al., 2015; Thudi et al., 2021). Ancient farmers used simple selection and crossing methods to domesticate regional crops. This selection of identifying ideal agronomic characters was based on the morphology of the crop's plant height, number of tillers, and yield per plant. However, as the knowledge of molecular science was incorporated into plant breeding, the selection process of desired characters was revolutionized. The integration of molecular markers with classical breeding facilitated a more efficient assessment of genetic variation in large plant populations (Gupta et al., 2010). Genetic markers have become invaluable tools for understanding genetic variation, expediting breeding programs, and aiding in efficient conservation efforts (Gupta et al., 2010; Al-Khayri et al., 2015). This integration can help researchers unlock the hidden genetic potential of wild relatives, leading to the identification of new genes that eventually can be incorporated into elite cultivars, therefore expanding the genetic variation (Gupta et al., 2010; Rana et al., 2019).

Various desired genetic variations in many important crops have been identified using genetic mapping with various molecular markers such as single nucleotide polymorphisms (Gupta et al., 2010). Whole-genome sequencing and RNA sequencing have provided insights into genome-wide interpretation, facilitating the identification of high-resolution quantitative trait loci maps (Gupta et al., 2010; Rana et al., 2019; Dormatey et al., 2020). The integration of molecular techniques with classical breeding has revolutionized the identification of novel genes and facilitated the construction of high-resolution genetic maps. While conventional breeding continues to be widely used, newer breeding techniques offer promising alternatives that can accelerate crop improvement efforts (Dormatey et al., 2020).

2.2.2 Mutation Breeding: Inducing Genetic Variation

Mutation breeding is a technique that harnesses natural or induced mutations to introduce genetic variation into crops. It involves exposing plants to radiation or chemical mutagens to induce random changes in their DNA. The resulting mutations can lead to alterations in traits, some of which may be desirable for crop improvement. After mutagenesis, large populations of mutated plants are generated, and individual plants with desired traits are selected through careful screening and evaluation. Mutation breeding has been successful in generating novel traits such as disease resistance, improved yield, and altered flowering time. It has provided breeders with a broader spectrum of genetic variation to work with, enhancing the diversity of breeding programs.

Although plant mutation has a very old history dating back to 300 BC, the term "mutation" was introduced just over a century ago in 1901 by Hugo De Vries. The mutation was described as the heritable changes in genetic material that do not occur through recombination and segregation during crossing-over (Van Harten, 1998). Mutation breeding offers a unique approach to plant improvement by selecting mutants having useful variation in subsequent generations (Kharkwal, 2012; Roychowdhury & Tah, 2013). It complements other breeding methods such as SB, rapid generation advance (RGA), and traditional breeding by introducing new genes and enhancing genetic variability (Hickey et al., 2017; Watson et al., 2018; Samineni et al., 2020).

Natural genetic variation emerging from spontaneous mutation is crucial for developing new plant varieties or improving existing ones. The spontaneous mutation rates have been estimated to be approximately 20 billion mutations per year in a one-hectare wheat field (Ossowski et al., 2010; Oladosu et al., 2016). Harnessing this natural genetic variation is essential in plant breeding programs.

Mutation breeding involves inducing heritable changes in the genetic material of organisms through exposure to various mutagens. These mutagens can be classified into physical (radiation-induced mutagenesis) and chemical mutagens, as well as insertional mutagenesis caused by DNA insertions (Forster & Shu, 2012; Roychowdhury & Tah, 2013). Radiation-induced mutagenesis can result from exposure of plant material to several radiations especially gamma, X-rays, or ion beams, while chemically induced mutagenesis uses chemical agents to give a similar effect. However, insertional mutagenesis can occur through the insertion of genetically transformed T-DNA or the activation of transposable elements that jump from one place to another causing frameshift or insertional mutations (Forster & Shu, 2012; Oladosu et al., 2016).

Seeds are commonly used plant materials for inducing mutation, but almost all plant materials can be utilized. Recently, other propagules like tubers, bulbs, and rhizomes have increasingly been employed due to their totipotency properties (Mba, 2013). M82 tomato mutant that was created with ethyl methanesulfonate (EMS) and fast neutrons had over 3,000 phenotypic changes (Menda et al., 2004). Micro-Tom, a dwarf tomato cultivar, created with EMS is another important resource for genetic studies of tomato (Watanabe et al., 2007).

Despite all the significant achievements, high-throughput phenotyping (HTP) remains an important bottleneck in the further advancement of mutation and traditional breeding; therefore, the yield of many crops has become stagnant (Acreche et al., 2008; Sadras & Lawson, 2011; Achigan-Dako et al., 2014; Blum, 2014; Tardieu et al., 2017). HTP can accelerate SSD, which is ideal for handling large segregating populations (White & Conley, 2013; Deery et al., 2014; Saxena et al., 2017; Shakoor et al., 2017). By optimizing resource distribution and reducing time and cost, SSD programs can progress through multiple generations annually (Sarutayophat & Nualsri, 2010).

Mutation breeding, with its ability to introduce heritable changes, complements traditional breeding methods in enhancing genetic variability for plant improvement. Harnessing natural genetic variation through spontaneous mutations is vital for developing new varieties and improving existing ones. The use of mutagens, including physical and chemical agents, offers an effective means to induce mutations. Furthermore, HTP platforms facilitate accelerated breeding programs, such as SSD, by optimizing resource distribution and enabling the rapid cycling of multiple lines. By addressing challenges in phenotyping efficiency, these advancements in mutation breeding and HTP contribute to the continued progress in plant breeding practices.

2.2.3 Molecular Breeding: Unleashing the Power of Genes

Molecular breeding represents a paradigm shift in plant breeding, as it enables the precise manipulation of genes associated with desirable traits. This approach relies on the identification and utilization of DNA markers linked to specific traits of interest. In this regard, techniques such as marker-assisted selection (MAS) and genetic engineering have propelled the field forward.

The field of molecular biology plays a crucial role in identifying and manipulating genes associated with human, animal, and plant disorders. To achieve this goal, researchers employ various genomic tools that aid in understanding gene function, developing biofortified crops, and enhancing plant traits. However, genome sequencing of the crops is a prerequisite in molecular breeding. Next-generation sequencing (NGS) is then used by researchers to develop genome, exome, or transcriptome libraries (Metzker, 2010; Wendler et al., 2014; Ahmar et al., 2020).

In molecular biology studies, a range of genomic tools are employed to investigate gene structure, function, and regulation. These tools involve restriction enzymes, molecular glue (ligases), molecular markers, as well as transcription and post-translational modification machinery (Huang, 2016). They enable researchers to manipulate genes involved in various disorders in humans, animals, and plants (Ahmar et al., 2020; Rai, 2022).

Genomic-assisted breeding combines genomics and traditional breeding to expedite the improvement of crop plants. DNA-based markers and insertion/deletion markers aid in selecting crossbreeds via marker-aided selection increasing the efficiency of selection. MAS involves the use of DNA markers that are genetically linked to target traits. By analyzing these markers, breeders can identify individuals carrying the desired traits at an early stage, accelerating the selection process. This technique allows for more precise and efficient trait selection, reducing the reliance on time-consuming phenotypic evaluation.

Molecular biology integrated with traditional plant breeding has revolutionized the development of improved crop cultivars. Now, phenotypic changes that are related to plant genotypes can be addressed by researchers (M Perez-de-Castro et al., 2012). NGS has changed the genome sequencing game as it allows a faster, more efficient, and more reliable way of decoding the genome of crops (M Perez-de-Castro et al., 2012; Wendler et al., 2014). It enables scientists and researchers to generate gene libraries to improve our understanding of the genome on the gene level (Metzker, 2010; Wendler et al., 2014). Identifying sources of cytoplasmic male sterility is an essential part of hybrid breeding. Now, molecular biology techniques are being used by researchers to identify these sources. Fertility restorer genes have been successfully cloned in maize, rice, and sorghum (Dwivedi et al., 2008).

The advent of genome editing (GE) techniques has revolutionized the field of plant breeding. Genetic engineering techniques have further expanded the possibilities of molecular breeding. By introducing specific genes into plants, breeders can confer novel traits or enhance existing ones. Genetic engineering has enabled the development of crops with traits like pest resistance, herbicide tolerance, enhanced nutritional content, and improved post-harvest qualities that otherwise were improbable. However, it has also raised concerns regarding biosafety and public acceptance.

Zinc finger nucleases (ZFNs) and transcription activator-like effector nucleases (TALENs) are two of the three primary systems used in molecular plant biotechnology for editing the genome of crop plants (Lloyd et al., 2005). However, each of these techniques had a limitation. ZFNs are limited by design complexities and non-specific nucleotide recognition, while TALENs are characterized by large repetitive constructs that require precision and time (Li et al., 2011; de Pater et al., 2013; Char et al., 2015; Schneider et al., 2016).

Then, the third system known as the clustered regularly interspaced short palindromic repeats/associated protein 9 (CRISPR/Cas9) system emerged as a groundbreaking tool for GE (Lloyd et al., 2005; Wood et al., 2011; Mao et al., 2013; Sprink et al., 2015). CRISPR/Cas9 was accepted widely and quickly became a preferred choice over other techniques as it offered increased efficiency and accuracy in inducing site-directed DNA breaks (Zhang et al., 2017). Recently, many new CRISPR-associated endonucleases have been reported which offer a wide range of uses. For example, Cpf1 from Prevotella and Francisella (Zaidi et al., 2017; Li et al., 2018), CRISPR/nCas9 (Cas9 nickase) or dCas9 (deactivated Cas9) fused with cytidine deaminase enables the creation of single base mutations in plant genomes (Qin et al., 2020).

2.2.4 Speed Breeding: Revolutionizing the Breeding Process

Plant breeding and research programs often face bottlenecks that impede their progress. Therefore, it becomes essential to develop technologies to accelerate plant growth and generation cycle (Slama-Ayed et al., 2019; Rai, 2022). SB is considered a groundbreaking technique that revolutionized the traditional breeding timeline. It combines optimized growth conditions to dramatically lower the generation time of plants. By manipulating factors like photoperiods and environmental conditions, breeders can cycle through multiple generations in just one year. The term "speed breeding" was coined in 2003 by a team of researchers at the University of Queensland (Hickey et al., 2019; Thudi et al., 2021). SB protocols are currently being developed for various commercial crops (Hickey et al., 2019; Thudi et al., 2021; Rai, 2022). Unlike other techniques, SB does not require specific equipment, making it suitable for a wide range of germplasm (Slama-Ayed et al., 2019).

Generally, SB involves the exposure of plants to artificial lights of varying wavelengths and controlled temperature for 22 hours (Ghosh et al., 2018; Ahmar et al., 2020). These artificial conditions mimicking natural day and night conditions result in increased photosynthesis and physiological, and biochemical processes leading to early fruit setting and seed development, thus shorter generation time (Chiurugwi et al., 2019; Ahmar et al., 2020; Thudi et al., 2021). By modifying these conditions, SB triggers rapid reproductive development in plants, effectively reducing breeding time and resources (Ghosh et al., 2018). Photoreceptors, which regulate the natural circadian rhythm, play a crucial role in perceiving light and responding to changing environmental conditions (Leal Filho et al., 2022). Optimizing the duration of light exposure for accelerated generation advancement depends on plant physiology. For example, short-day, long-day, and day-neutral plants will have different light exposure duration (Ghosh et al., 2018; Godwin et al., 2019).

The integration of OMICS techniques, including phenomics, genomics, transcriptomics, proteomics, and metabolomics, has transformed plant breeding by providing breeders with comprehensive data to make informed decisions (Sartor et al., 2019). These techniques generate vast amounts of complex data that present a challenge in data handling and analysis. By deciphering the genomic information of plants, breeders can predict and select the best lines or cultivars for hybridization, with a focus on increasing yield and resistance (Sartor et al., 2019; Rai, 2022).

SB has demonstrated its efficacy in accelerating crop breeding and achieving multiple generations within a year. It has been successfully applied to various crops, including peas, chickpeas, fava beans, lupins, peanuts, amaranth, rice, maize, and oilseed crops like canola, soybean, and sunflower (Sysoeva et al., 2010; Achigan-Dako et al., 2014; Stetter et al., 2016; Ghosh et al., 2018; Chiurugwi et al., 2019; Ahmar et al., 2020). For example, SB protocols have been developed for soybean, enabling four to five generations per year instead of one to two, and for canola, where pod-shattering traits were phenotyped, achieving four generations per year (Sysoeva et al., 2010; Achigan-Dako et al., 2014). The early breaking of seed dormancy through cold stratification or germination-promoting hormones has further expedited breeding cycles (Yao et al., 2017; Watson et al., 2018). In addition to improving agronomic traits, SB has facilitated phenotyping, transformation, and mutant studies (Achigan-Dako et al., 2014; Ghosh et al., 2018).

SB continues to evolve to further reduce generation times and accelerate plant growth and reproduction. Innovations in LED lighting systems, combined with extended photoperiods and early seed harvest, have broadened the application of SB (Chiurugwi et al., 2019; Jähne et al., 2020). Breaking seed dormancy immediately after harvest through techniques like cold stratification or embryo rescue has proven effective in shortening generation times (Watson et al., 2018).

Manipulating vernalization and adjusting the temperature at key growth stages have also shown potential for accelerating plant growth and reproductive processes (Ribalta et al., 2017; Watson et al., 2018; Cazzola et al., 2020; Rai, 2022). Optimizing day length, and light quality, and even exploring the use of laser light are areas of ongoing research to enhance the efficiency of SB (Jähne et al., 2020; Samineni et al., 2020; González-Barrios et al., 2021).

SB facilitates the rapid development of new varieties and the evaluation of genetic combinations, particularly for crops with long generation times. SB has opened up new avenues for the rapid introduction of genetic variation, accelerating the identification of desirable traits and the development of improved crop varieties (Figure 2.1). SB represents a revolution in plant breeding, offering significant potential to expedite plant growth, development, and commercialization.

2.2.4.1 Technologies and Infrastructure Required for Speed Breeding

SB is a method for accelerating crop breeding through the establishment of controlled conditions, such as light and temperature, that encourage speedy development and growth (Watson et al., 2018). This method takes advantage of organic processes to generate new plant generations quickly. Photoperiod, temperature, humidity, light quality and intensity, planting density, and other internal as well as external factors all have an impact on the development and growth of a plant. Developing and enhancing SB techniques for various plant species requires a thorough understanding of the fundamental principles behind plant growth and development (Ghosh et al., 2018; Bhatta et al., 2021). The following equipment and technologies are often needed to implement SB.

2.2.4.2 Manipulation of Photoperiod Regime

SB is a technique that modifies the growth circumstances to speed up the plant breeding period in a controlled environment. Using LED-supplemented glasshouse arrangements, which offer ideal lighting conditions for plant growth while enabling greater control over temperature and humidity, is one method of accelerating breeding (Watson et al., 2018). Using high-intensity light to give plants the best possible situations for growth is one of the essential prerequisites for SB. For plant SB, a variety of light-controlling strategies can be applied. Among the most popular methods are discussed below.

Artificial lighting: One of the most common methods is to create longer daylight hours using artificial lights. Fluorescent lights, high-intensity discharge (HID) lamps, and light-emitting diodes (LEDs) can all be used to control lighting (Ouzounis et al., 2015). In order to create the ideal

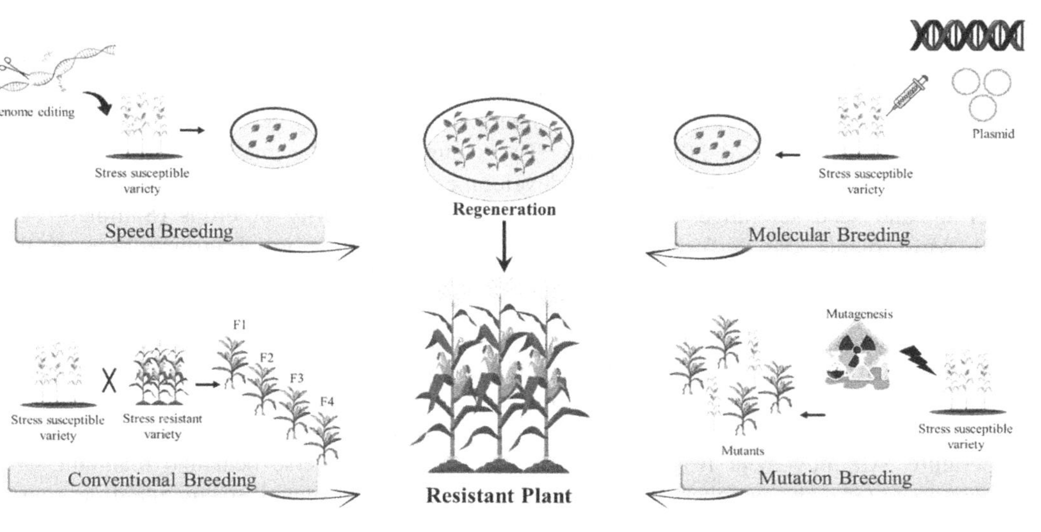

FIGURE 2.1 Emergence of the innovative approach of speed breeding from conventional breeding methods.

lighting conditions for plant growth, many light sources like LEDs, high-pressure sodium lamps, and metal halide lamps are employed in SB. Utilizing LEDs is especially beneficial because they generate particular wavelengths of light that can be customized to encourage the development and growth of plants (Hickey et al., 2017; Alahmad et al., 2018). Because it can be customized to provide particular wavelengths of light that are most beneficial to plant growth, fluorescent lighting is especially well suited for this purpose. Wheat and barley plants' development rates can be considerably accelerated by LED lighting, which enables breeders to create novel cultivars much more quickly (Watson et al., 2018).

Light quality: To accelerate plant growth, the quality of light can also be changed. To meet the unique requirements of their plants, growers might change the light's spectral makeup. Red and blue light are the two most crucial spectra for photosynthesis, and adjusting their intensity and ratio can have a big impact on the speed at which plants grow and develop (Massa et al., 2008). It is well known that red and blue fluorescent lights exhibit distinct effects on the growth and development of plants. Red light encourages plant growth, while blue light controls the morphology and physiology of plants, including the enlargement of leaves, the lengthening of stems, and photomorphogenesis. Red and blue fluorescent lights can improve plant development and growth when used in combination, resulting in larger yields as well as faster crop turnover (Li & Kubota, 2009).

Light duration: Additionally, prolonging the light period may accelerate plant growth. This can be accomplished by increasing the amount of time each day that plants are exposed to light. According to the length of the day and the amount of light, plants can either blossom or remain in a vegetative condition due to distinct photoperiodic reactions (Kang et al., 2013). For example, to start flowering, short-day plants need a time of darkness that is longer than a crucial length, whereas long-day plants need a time frame of darkness that is less than a critical length. However, day-neutral plants don't respond to the photoperiod and are capable of flowering at any time of day (Imaizumi & Kay, 2006). Understanding these responses is critical in manipulating photoperiods for SB.

Light intensity: To accelerate plant growth, growers might raise the light intensity. This can be achieved by using more lamps or by moving the lamps closer to the plants. For SB, it is possible to control the amount of light by utilizing high-intensity artificial light sources, extending the amount of time that plants are exposed to light, adding more light, employing light reflectors, and altering the distance between plants and light sources (Li & Kubota, 2009). These techniques are crucial tools for SB since they can encourage quicker plant growth and development.

2.2.4.3 Regulation of the Temperature

The effects of temperature on development, growth, flowering, seed set, and maturity in plants are significant. The rapid growth that shifts a plant from the vegetative to productive stages can be induced by temperature changes (Hatfield & Prueger, 2015; McClung et al., 2016). According to the plant species being cultivated, different temperature regimes may be best for SB, but usually a constant temperature of 20–30°C (68-86 °F) is suggested. The majority of crops needed temperatures between 12 and 30°C for development, whereas the ideal range for growth, flowering, and seed set ranges from 25 to 30°C for the majority of crops (Samineni et al., 2020). Most plants can perform photosynthesis and maintain normal growth rates at these temperatures. To grow and flourish at their best, some plants, however, could need various temperature ranges (Aftab & Hakeem, 2022).

Numerous techniques can be utilized to manage the temperature in a controlled environment for plant growth. Solar or battery-powered air conditioning systems may provide a dependable and affordable solution for maintaining a constant temperature in indoor SB programs (Wanga et al., 2021). HVAC systems and growth chambers are made to keep a steady atmosphere at a particular temperature. Air movement from cooling fans can help to disperse heat and maintain steady temperatures. Growth chambers and HVAC systems are designed to maintain a constant environment at a specific temperature. Circulation of air by cooling fans can aid in heat transfer and constant temperature maintenance (Bot et al., 1995).

2.2.4.4 Regulation of Humidity

SB is a method for accelerating plant breeding by creating the best possible growing conditions, including humidity regulation as a key component. To regulate humidity, the amount of moisture in the air must be kept at an ideal level to stop unnecessary evaporation and plant water loss (Atkinson & Urwin, 2012). Humidity control is a key component of SB since it is necessary to sustain such conditions. There are a few techniques for controlling humidity in SB.

Fogging: Fogging is a method for raising humidity in the air by misting it with a thin mist of water. This method has been modified for use in facilities for rapid breeding and is frequently employed in greenhouses. In conditions where rapid breeding occurs, fogging has been proven to be beneficial in preserving a high level of humidity (Devasirvatham et al., 2016).

Evaporative cooling: To cool and humidify the air, evaporative cooling requires blowing air over a wet surface, like a wet pad. This method has been demonstrated to be effective in keeping SB establishments' humidity levels at a high level.

Drip irrigation: Drip irrigation includes supplying the roots of the plants with small amounts of water. It has been demonstrated that using this method will keep the area around the plants very humid (Wang et al., 2019).

Covering: Maintaining humidity concentrations in the atmosphere for rapid breeding can be achieved by wrapping the plants with a plastic sheet or other material (Devasirvatham et al., 2016).

2.2.4.5 Modifying Carbon Dioxide Levels

In order to alter CO_2 levels for SB, the growth chamber's CO_2 concentration must be raised. The amount of carbon dioxide can be maintained using sensors and modified with air flow controls or automated systems to keep up the desired concentration (Pandey et al., 2022). Numerous sensors, including non-dispersive infrared (NDIR), electrochemical, and photoacoustic ones, can measure CO_2 levels. These sensors can detect the amount of carbon dioxide in the air while giving real-time information on the amounts of carbon dioxide in the surrounding area (Wang et al., 2016).

The amount of carbon dioxide can be managed by automated devices like HVAC systems, as well as by modifying airflow utilizing ventilation or circulation. The amount of CO_2 can be lowered by increasing ventilation or adding fresh air, while levels can be raised by increasing recirculation or reducing ventilation. In order to keep the amount of CO_2 within a particular range for the best quality of indoor air, automated devices can be set to alter ventilation rates based on the concentration of CO_2 (Nagatoshi & Fujita, 2019).

Other techniques, such as establishing carbon capture technology or utilizing sources of renewable energy, can be employed in addition to modifying the airflow to lower CO_2 levels in the atmosphere. To regulate indoor air quality and lessen the hazards related to high CO_2 concentrations, CO_2 levels can be monitored and controlled using air flow modifications and automated systems (Pandey et al., 2022).

2.2.4.6 Speed Breeding I—Controlled-Environment Chamber Speed Breeding Conditions

Accelerated and effective development of plants in SB systems requires a controlled-environment chamber. Thermoregulation, lighting, humidity, levels of CO_2, and nutrition are some of these factors. Depending on the particular plant species, temperature-controlled SB rooms are normally kept between 20°C and 28°C to maximize plant growth (Watson et al., 2018). SB, which accelerates plant growth, also needs intense and accurate lighting, which is frequently given by high-intensity lighting sources like LED or fluorescent lamps. Regulating the amount of moisture in the air can be used to adjust humidity levels, which are equally significant (Hickey et al., 2012). To promote the best possible plant growth, levels of carbon dioxide are measured and managed. Plants receive fertilizers using hydroponic or aeroponic systems, which supply exact dosages of nutrients immediately to the plants. SB methods can produce many generations of crops in a single year by regulating and optimizing certain environmental factors. This enables quick genetic advancement and the emergence of novel plant varieties (Reynolds et al., 2016).

2.2.4.7 Speed Breeding II—Glasshouse Speed Breeding Conditions

The environmental factors are optimized in glasshouse SB systems to produce quick and effective plant growth. These systems keep the temperature between 18°C and 25°C, which can be changed both during the day and at night in order to promote plant development. The major source of light in glasshouses is sunlight, but during the darker winter months, supplemental lighting, frequently utilizing high-pressure sodium or metal halide lamps, can be employed to add additional light. Humidity levels are regulated by controlling the amount of moisture in the air through misting systems or ventilation. Carbon dioxide is often supplied to the greenhouse to increase CO_2 concentrations and promote plant growth. Nutrients are provided through soil or hydroponic systems (Watson et al., 2018). Multiple crop generations can be produced in a single year because of greenhouse-SB, which encourages the production of new plant varieties and rapid genetic advancement. Due to the use of natural sunshine as a light source, it may be more affordable than controlled-environment chamber SB (Ghosh et al., 2018). However, environmental ambiguity and disease pressures might be present, which might have an impact on plant development and productivity.

2.2.4.8 Speed Breeding III—Homemade Growth Room Design for Low-Cost Speed Breeding

The SB III technology is an economical way to speed up the plant breeding cycle. Several essential parts must be installed in order to set up a homemade growth room for inexpensive SB. To keep a constant temperature and moisture level, the space should be well insulated and built utilizing insulated sandwich paneling. To maintain a comfortable temperature inside the space, a split system air conditioner is required. Additionally, fluorescent light panels should be installed to supply the ideal lighting for plant growth. Each pot should be given its watering through the use of an irrigation regulator and spike drippers. For the first four weeks of operation, the fluorescent bulbs should be set to an initial 12-hour photoperiod and then increased to an 18-hour photoperiod with 6 hours of darkness (Watson et al., 2018). For the species of plants being produced, the amount of moisture should be kept at a suitable level. To promote optimum growth and development, the environment will be carefully watched over and modified as necessary. Plants can be grown more rapidly and effectively by providing a controlled environment that offers ideal growing conditions, decreasing the time needed for breeding and selection.

2.2.4.9 Physical Factors Affecting Plant Responses to Speed Breeding

SB is a plant breeding technique that uses controlled environments to accelerate plant growth and development. The technique involves optimizing environmental factors to increase plant growth rate and shorten the time required for seed production. SB is a valuable tool in plant breeding programs for crop improvement as it allows for the rapid generation of new cultivars with desired traits. However, the success of SB largely depends on the optimization of physical factors such as light, temperature, humidity, and nutrient availability. In this chapter, we will discuss the physical factors that affect plant responses to SB.

2.2.4.10 Light

Photoperiod is one of the key factors used in SB to accelerate plant growth and development (Watson et al., 2018). By manipulating the duration and quality of light exposure, SB can promote early flowering and seed production, resulting in faster plant generations and accelerated breeding cycles (Ghosh et al., 2018). Plants use light for photosynthesis, and the amount of light they receive affects the rate of photosynthesis, which in turn affects plant growth rate. In SB, high-intensity lighting is used to accelerate plant growth and shorten the time required for seed production. High-intensity lighting also promotes plant growth by stimulating cell division and elongation (Taiz & Zeiger, 2010). In SB, red and blue LED lights are used to optimize plant growth and development. The intensity and ratio of these spectra can be manipulated to suit the specific needs of the plant species being grown. For example, increasing the blue-light intensity can promote the development of leaves and stems, while increasing the red-light intensity can promote flowering and fruiting (Folta & Carvalho, 2015). For example, red and blue LED lights significantly increased the yield of lettuce plants grown in a growth chamber. Blue light was effective in promoting the growth and development of potato plantlets during SB (Kim et al., 2019).

2.2.4.11 Temperature

Temperature is another critical environmental factor that affects plant growth and development. Optimal temperatures promote plant growth and shorten the time required for seed production (Clarke et al., 2005). In SB, temperature is controlled by maintaining a consistent temperature throughout the growth chamber. The temperature can be adjusted according to the specific requirements of the plant species being grown (Watson et al., 2018). High temperatures promote faster plant growth, but they can also result in heat stress and damage to the plant. Conversely, low temperatures can slow down plant growth and delay seed production. Extreme temperatures, whether low or high, can have significant effects on the rate of plant development, including triggering transitions from vegetative to reproductive stages in plants (Clarke et al., 2005). The use of temperature-controlled growth chambers is therefore essential in SB to ensure optimal growth conditions for the plants (Watson et al., 2018).

2.2.4.12 Humidity

Humidity is another critical environmental factor affecting plant growth and development. The optimal humidity level for plants varies depending on the plant species being grown. High humidity can promote plant growth and reduce water stress, but it can also lead to the development of fungal diseases. Conversely, low humidity can lead to water stress and reduced plant growth (Ghosh et al., 2018). In SB, humidity is controlled by maintaining a consistent relative humidity level throughout the growth chamber. The use of humidifiers and dehumidifiers is also employed to adjust humidity levels in the growth chamber as needed (Chiurugwi et al., 2019).

2.2.4.13 Nutrient Availability

Nutrient availability is another critical factor affecting plant growth and development. Plants require specific nutrients in specific quantities for optimal growth and development. In SB, hydroponics and aeroponics are employed to optimize nutrient availability for the plants (Watson et al., 2018). These techniques allow for precise control of the nutrients available to the plants, resulting in faster growth and increased seed production. The use of nutrient-rich growing media such as coconut coir and rockwool is also employed in SB to provide optimal nutrient availability to the plants (Ghosh et al., 2018).

2.2.4.14 Crops Acquiescent to Speed Breeding

It has been demonstrated that several crop species, including barley, wheat, oat, triticale, canola, chickpeas, pea, lentil, brassica, and tomato, are susceptible to SB. These crops may be cultivated more quickly and effectively in controlled situations with ideal light and temperature situations, allowing rapid breeding cycles and improved genetics.

2.2.4.15 Cereal Crops

To speed up the breeding of cereal varieties, novel techniques have been developed to obtain homozygous lines more quickly after hybridization. These include treating immature seeds with H_2O_2 at low temperatures (Richard et al., 2018). which has been successful in reducing the generation time of wheat by 12–23 days. In rice, the biotron breeding system has been developed to regulate temperature, photoperiod, and CO_2 levels to accelerate the breeding cycle. This system has been successful in reducing the generation intervals of 'Nipponbare' by up to three months. The SB technique has been used in wheat and barley to rapidly screen for disease resistance and evaluate plant traits (Robertson & Curtis, 1967). Other methods such as increasing the photoperiod and using a foliar mineral supplement have also been successful in reducing the time to anthesis for higher generation turnover in oats. Overall, these techniques have shown promise in expediting the breeding cycle and developing improved cereal varieties more quickly.

2.2.4.16 Oilseed Crops

Viable production of seed through precocious germination has been demonstrated in soybean (Burris, 1973), resulting in a significant reduction in crop duration from 130–140 days to 65–70 days using immature, pre-treated pods (Roumet & Morin, 1997). Similarly, a standardized rapid generation advancement protocol was developed for Japanese soybean cultivars by reducing crop duration from 102–132 days to 70 days, allowing for five generations per year instead of one to two generations (Nagatoshi & Fujita, 2019). Similar approaches have been optimized for canola to enhance generation turnover and facilitate phenotyping of specific traits, and for sunflower using the embryo rescue technique (Watson et al., 2018). These methods have been employed in controlled environment growth chambers, allowing for shortened generation times and accelerated breeding programs.

2.2.4.17 Legumes Crops

An approach known as in vitro-assisted short-stemmed seed (SSD) in clover (*Trifolium subterraneum* L.) has been developed to accelerate the generation cycle by cutting the time until flowering through controlled temperature and extended photoperiod during growth, shortened seed-filling period, and embryo rescue. To get around problems caused by seed dormancy, the immature seed is used. This approach permits 2.7–6.1 generations per year for various clover genotypes (Pazos-Navarro et al., 2017). Several legume species have had their protocols for speeding up generation cycles refined, especially temperate pulses, which show an adaptive long-day response, making them extremely responsive to photoperiod extent (Pazos-Navarro et al., 2017). Peanuts grew faster than expected because of a greenhouse facility's continuous lighting, ideal temperature, and suitable humidity levels (Pazos-Navarro et al., 2017). Combining in vitro methods with in vivo manipulations was found to be more efficient than greenhouse settings for shortening generation cycles in pea and bambara groundnut. However, a strategy that only uses in vivo manipulation in the case of peas and grass peas has produced promising results (Varshney et al., 2021). Legume breeding programs have effectively used in vitro techniques including tissue culture and hydroponics to shorten generation times and increase the number of generations each year (Varshney et al., 2021). These methods have been applied to the production of recombinant inbred lines (RILs) for gene-trait interactions as well as rapid generation progression in legume species such as lentil, fava bean, chickpea, pigeon pea, and soybean. To speed up the development of novel plant cultivars for peanuts, the single-seed chipping (SSC) procedure has been combined with high-throughput genotyping and seed banking (Varshney et al., 2021). These techniques have resulted in significant time-saving, space-saving, and resource-saving benefits, making them valuable tools in legume breeding programs.

2.2.4.18 Fruit Crops

Several fruit crops require over 20 years of their juvenile stage before they begin to flower (van Nocker & Gardiner, 2014). By using SB approaches, apple and chestnut trees have experienced rapid vegetative development and early flowering (thereby reducing the duration from five years to 10 months), respectively (Baier et al., 2012; van Nocker & Gardiner, 2014). Transgenic apple plants with early flowering characteristics have been merged with MAS technology using the SB technique to produce a new cultivar with desirable attributes (Flachowsky et al., 2011). To reduce blooming time frames, increase flowering rates, and improve flowering predictability, clonally propagated crops—aside from fruit crops—such as bananas, roots, and tubers, have increasingly adopted SB. This approach has tremendously helped the introduction of disease resistance traits such as those for bacterial wilt in bananas (Vira et al., 2015; Souza et al., 2018).

2.2.4.19 Vegetable Crops

Some vegetable crops, such as pepper, tomato, and amaranth, have a significant response to prolonged photoperiods, and it has been demonstrated that manipulating the length of daylight can shorten generation intervals in these plants (Demers et al., 1998; Stetter et al., 2016). In tomatoes, the germination of immature seeds at different stages of maturity has created new opportunities for potentially growing up to five generations of plants, which is more than the usual three generations attained by conventional cultivation techniques (Bhattarai et al., 2009). It has been discovered that altering the photoperiod in grain amaranth is useful for synchronizing flowering in various germplasm lines. This strategy has made it easier to create and recognize real hybrids when combined with DNA marker technology, which has sped up the breeding program (Stetter et al., 2016). Upregulating flowering-related genes, like the CaFT-LIKE gene in peppers, to encourage early flowering, is one alternative strategy to improve the flowering process and speed up the vegetable generation cycle (Borovsky et al., 2020). Tomato plants may be able to adapt to longer photoperiods by introducing the CAB-13 gene, which can offer tolerance to prolonged exposure to continuous light (Velez-Ramirez et al., 2014).

2.2.4.20 High-Throughput Technologies Augmenting Speed Breeding

High-throughput technologies are instrumental in augmenting the SB process, revolutionizing crop improvement and accelerating genetic gain in agricultural procedures. The integration of high-throughput techniques with SB offers several advantages in plant breeding programs. Here are some key points extracted from the given text that highlight the contributions of high-throughput technologies to SB.

Genome editing and SB: GE techniques, such as CRISPR/Cas9, combined with SB allow for rapid generation of modified plants and accelerated production of modified seeds. For example, researchers have successfully applied GE in combination with SB in crops like *Brassica napus*, *B. oleracea*, and soybean (Yang et al., 2017; Murovec et al., 2018). This integration enables the development of superior plant varieties with desired traits, such as improved disease resistance or enhanced nutritional value, in a shorter time frame (Bao et al., 2020).

Marker-assisted backcrossing (MABC) and SB: SB can be integrated into MABC to facilitate the introgression of beneficial alleles in crops. This approach minimizes linkage drag, which refers to the unintended incorporation of unfavorable alleles during backcrossing. By using SB, the progress of MABC can be accelerated, allowing for quick transfer of target traits (Varshney et al., 2021). For instance, researchers have successfully integrated SB into MABC programs to introduce genes like pi21, which confers quantitative resistance to rice blast, into selected rice cultivars. This enables the development of improved rice varieties with enhanced disease resistance (Angeles-Shim et al., 2020).

Genomic selection (GS) and SB: The combination of GS and SB leads to increased genetic gain in complex traits. SB can be used in the phenotyping of specific traits during the population development, selection candidate identification, and phenotyping steps. This approach allows for accurate prediction of traits comparable to direct field selection (Watson et al., 2019). For example, in spring wheat breeding, GS combined with SB has been shown to enhance genetic gain by accurately predicting traits such as plant height and flowering time (Voss-Fels et al., 2019). This accelerates the breeding process by enabling the selection of superior genotypes at early stages.

Multi-trait phenotyping and MAS: SB optimizes multi-trait phenotyping protocols, facilitating the evaluation of multiple traits simultaneously. For instance, in wheat breeding under SB conditions, researchers have optimized protocols to evaluate traits like crown root resistance and leaf rust tolerance (Croser et al., 2021). This approach allows for the efficient analysis of multiple traits in the same group of plants, leading to improved genetic gain. Through the integration of MAS, the selection of genotypes with desirable traits can be further enhanced, ensuring efficient utilization of resources and reducing the time required for trait incorporation (Ortiz et al., 2007).

High-throughput phenotyping: HTP under SB situations offers new possibilities for the identification and implementation of beneficial traits. Rapid selection for improved root structure of mature plants can be achieved by focusing on indirect factors like seminal root number and angle of wheat seedlings under SB conditions (greater planting density, controlled temperature, and prolonged photoperiod) (Richard et al., 2015; Al-Tamimi et al., 2016).

Single-seed descent (SSD) and SB: Under SB conditions, the SSD method proves to be a valuable approach in generating nearly homozygous lines within a single year, preserving genetic diversity while expediting the breeding process. This approach has proven beneficial in various crop breeding programs (Dadu et al., 2021; Khoo et al., 2021). For instance, the development of rapid RILs using SB technology has permitted the identification of such a gene-trait associated with breeding applications. This allows the development of crop varieties with desired traits (Zaman et al., 2019; Taylor et al., 2021).

2.2.4.21 Future Challenges

Modern SB techniques use controlled settings, like growth chambers or greenhouses, to accelerate the plant breeding process. SB has demonstrated a lot of promise in accelerating the pace and effectiveness of plant breeding. Future SB may face the following difficulties, to mention:

Cost: Many plant breeding programs may not be able to afford the pricey equipment and specialized facilities needed for SB. In order to make SB more widely available to researchers, cost-effective ways of performing it must be developed.

Standardization: To ensure the accuracy and comparability of findings, SB procedures and methods must be standardized. Collaboration between researchers and breeders is necessary to develop standardized techniques for various crops and environmental circumstances.

Genetic diversity: SB may speed up the process of breeding, but it can also lead to less genetic diversity, which can restrict the possibilities for subsequent breeding programs. The genetic variety of the eventual crop must be preserved, according to breeders.

Regulatory approval: The safety of GM crops and their possible effects on the environment are both issues that are being raised due to SB technology, which is still relatively recent. SB techniques may therefore have trouble receiving regulatory approval, particularly in areas with strong guidelines.

Ethical considerations: Crops can be developed by SB to become resistant to diseases, insects, and adverse environmental conditions. However, ethical factors must be taken into account, such as the potential impact on regional communities and their ecosystems. It's critical to weigh the advantages of rapid breeding against any potential drawbacks.

2.3 CONCLUSION

Recent innovations in plant breeding technologies have widened the options to address food security challenges around the globe. In this context, SB has emerged as a groundbreaking approach contributing to pre-breeding and breeding programs. It involves novel technologies like aSSD, RGC, FGC and in vitro culturing, and RGT. These technologies permit early flowering, thus enabling cutting down the length of selection cycles. Moreover, it addresses challenges associated with field selection processes, expert phenotyping, and resource management, facilitating more efficient breeding. Recently, SB has been effectively employed for important crops resulting in significant improvements in annual generation turnover. Most recently, SB combined with advanced OMICS techniques has gained prominence. As we know, NGS allows faster, more efficient, and more reliable way of decoding the crops' genomes. Likewise, the advent of GE has entirely transformed the field of plant breeding. GE techniques including ZFNs, transcription activator-like effector nucleases (TALENs), and clustered regularly interspaced short palindromic repeats/associated protein 9 (CRISPR/Cas9) system have further expanded the possibilities of molecular breeding. This method takes advantage of organic processes to generate new plant generations quickly. Photoperiod, temperature, humidity, light quality and intensity, planting density, and other internal as well as external factors all have an impact on the development and growth of a plant. Developing and improving SB techniques for various plant species will help to ensure food security for the ever-increasing population on this planet.

REFERENCES

Achigan-Dako, E., Sogbohossou, O., & Maundu, P. (2014). Current knowledge on Amaranthus spp.: Research avenues for improved nutritional value and yield in leafy amaranths in sub-Saharan Africa. *Euphytica*, *197*, 303–317.

Acreche, M. M., Briceño-Félix, G., Sánchez, J. M., & Slafer, G. A. (2008). Physiological bases of genetic gains in Mediterranean bread wheat yield in Spain. *European Journal of Agronomy*, *28*(3), 162–170.

Aftab, T. & Hakeem, K. R. (2022). Plant abiotic stress physiology: Volume 2: Molecular advancements. Apple Academic Press.

Ahmar, S., Gill, R. A., Jung, K.-H, Faheem, A., Qasim, M. U., Mubeen, M., & Zhou, W. (2020). Conventional and molecular techniques from simple breeding to speed breeding in crop plants: Recent advances and future outlook. *International Journal of Molecular Sciences*, *21*(7), 2590.

Al-Khayri, J. M., Jain, S. M., & Johnson, D. V. (2015). *Advances in plant breeding strategies: Breeding, biotechnology and molecular tools.*

Al-Tamimi, N., Brien, C., Oakey, H., Berger, B., Saade, S., Ho, Y. S., Schmöckel, S. M., Tester, M., & Negrão, S. (2016). Salinity tolerance loci revealed in rice using high-throughput non-invasive phenotyping. *Nature Communications*, *7*, 13342.

Alahmad, S., Dinglasan, E., Leung, K. M., Riaz, A., Derbal, N., Voss-Fels, K. P., Able, J. A., Bassi, F. M., Christopher, J., & Hickey, L. T. (2018). Speed breeding for multiple quantitative traits in durum wheat. *Plant Methods*, *14*, 1–15.

Angeles-Shim, R. B., Reyes, V. P., del VALLE, M. M., Lapis, R. S., Shim, J., Sunohara, H., Jena, K. K., Ashikari, M., & Doi, K. (2020). Marker-assisted introgression of quantitative resistance gene pi21 confers broad spectrum resistance to rice blast. *Rice Science*, *27*, 113–123.

Atkinson, N. J. & Urwin, P. E. (2012). The interaction of plant biotic and abiotic stresses: From genes to the field. *Journal of Experimental Botany*, *63*, 3523–3543.

Baier, K. M., Maynard, C., Powell, W., Barakat, A., Staton, M., & Cheng, C. H. (2012). Early flowering in chestnut species induced under high dose light in growth chambers. *Journal of the American Chestnut Foundation*, *26*, 8–10.

Bao, A., Zhang, C., Huang, Y., Chen, H., Zhou, X., & Cao, D. (2020). Genome editing technology and application in soybean improvement. *Oil Crop Science*, *5*, 31–40.

Bhatta, M., Sandro, P., Smith, M. R., Delaney, O., Voss-Fels, K. P., Gutierrez, L., & Hickey, L. T. (2021). Need for speed: Manipulating plant growth to accelerate breeding cycles. *Current Opinion in Plant Biology*, *60*, 101986.

Bhattarai, S. P., de la Pena, R. C., Midmore, D. J., & Palchamy, K. (2009). In vitro culture of immature seed for rapid generation advancement in tomato. *Euphytica*, *167*, 23–30.

Blum, A. (2014). Genomics for drought resistance – getting down to earth. *Functional Plant Biology*, *41*(11), 1191–1198.

Borlaug, N. E. (1968). *Wheat breeding and its impact on world food supply.* CIMMYT.

Borovsky, Y., Mohan, V., Shabtai, S., & Paran, I. (2020). CaFT-LIKE is a flowering promoter in pepper and functions as florigen in tomato. *Plant Science*, *301*, 110678.

Bot, G. P. A., Challa, H., & Van de Braak, N. J. (1995). *Greenhouse climate control: An integrated approach.* Wageningen Academic Publishers.

Breseghello, F. & Coelho, A. S. G. (2013). Traditional and modern plant breeding methods with examples in rice (Oryza sativa L.). *Journal of Agricultural and Food Chemistry*, *61*(35), 8277–8286.

Brim, C. A. (1966). A modified pedigree method of selection in soybeans 1. *Crop Science*, *6*(2), 220.

Bugbee, B. & Koerner, G. (1997). Yield comparisons and unique characteristics of the dwarf wheat cultivar 'USU-Apogee'. *Advances in Space Research*, *20*(10), 1891–1894.

Bula, R. J., Morrow, R. C., Tibbitts, T. W., Barta, D. J., Ignatius, R. W., & Martin, T. S. (1991). Light-emitting diodes as a radiation source for plants. *HortScience: A Publication of the American Society for Horticultural Science*, *26*(2), 203–205.

Burris, J. S. (1973). Effect of seed maturation and plant population on soybean seed quality 1. *Agronomy Journal 65*, 440–441.

Cazzola, F., Bermejo, C. J., Guindon, M. F., & Cointry, E. (2020). Speed breeding in pea (Pisum sativum L.), an efficient and simple system to accelerate breeding programs. *Euphytica*, *216*(11), 178.

Char, S. N., Unger-Wallace, E., Frame, B., Briggs, S. A., Main, M., Spalding, M. H., …, & Yang, B. (2015). Heritable site-specific mutagenesis using TALEN s in maize. *Plant Biotechnology Journal*, *13*(7), 1002–1010.

Cheema, K. S. (2018). Plant breeding its applications and future prospects. *International Journal of Engineering Technology Science and Research*, *5*(3), 88–94.

Chiurugwi, T., Kemp, S., Powell, W., & Hickey, L. T. (2019). Speed breeding orphan crops. *Theoretical and Applied Genetics*, *132*, 607–616.

Clarke, F. R., Knox, R. E., & DePauw, R. M. (2005). Expression of dormancy in a spring wheat cross grown in field and controlled environment conditions. *Euphytica,143*, 297–300.

Croser, J., Mao, D., Dron, N., Michelmore, S., McMurray, L., Preston, C., Bruce, D., Ogbonnaya, F. C., Ribalta, F. M., & Hayes, J. (2021). Evidence for the application of emerging technologies to accelerate crop improvement: A collaborative pipeline to introgress herbicide tolerance into chickpea. *Frontiers in Plant Science, 2839*.

Dadu, R. H. R., Bar, I., Ford, R., Sambasivam, P., Croser, J., Ribalta, F., Kaur, S., Sudheesh, S., & Gupta, D. (2021). Lens orientalis contributes quantitative trait loci and candidate genes associated with ascochyta blight resistance in lentil. *Frontiers in Plant Science*, *12*, 703283.

Deery, D., Jimenez-Berni, J., Jones, H., Sirault, X., & Furbank, R. (2014). Proximal remote sensing buggies and potential applications for field-based phenotyping. *Agronomy*, *4*(3), 349–379.

Demers, D.-A., Dorais, M., Wien, C. H., & Gosselin, A. (1998). Effects of supplemental light duration on greenhouse tomato (Lycopersicon esculentum Mill.) plants and fruit yields. *Scientia Horticulturae (Amsterdam)*, *74*, 295–306.

Devasirvatham, V., Tan, D. K. Y., & Trethowan, R. M. (2016). Breeding strategies for enhanced plant tolerance to heat stress BT. In: J. M. Al-Khayri, S. M. Jain, & D. V. Johnson (eds), (pp. 447–469). Advances in plant breeding strategies: Agronomic, abiotic and biotic stress traits. Cham: Springer International Publishing.

de Pater, S., Pinas, J. E., Hooykaas, P. J., & van der Zaal, B. J. (2013). ZFN-mediated gene targeting of the Arabidopsis protoporphyrinogen oxidase gene through Agrobacterium-mediated floral dip transformation. *Plant Biotechnology Journal*, *11*(4), 510–515.

Dias, M. I., Sousa, M. J., Alves, R. C., & Ferreira, I. C. (2016). Exploring plant tissue culture to improve the production of phenolic compounds: A review. *Industrial Crops and Products*, *82*, 9–22.

Dormatey, R., Sun, C., Ali, K., Coulter, J. A., Bi, Z., & Bai, J. (2020). Gene pyramiding for sustainable crop improvement against biotic and abiotic stresses. *Agronomy*, *10*(9), 1255.

Dwivedi, S., Perotti, E., & Ortiz, R. (2008). Towards molecular breeding of reproductive traits in cereal crops. *Plant Biotechnology Journal*, *6*(6), 529–559.

Flachowsky, H., Le Roux, P., Peil, A., Patocchi, A., Richter, K., & Hanke, M. (2011). Application of a high-speed breeding technology to apple (Malus× domestica) based on transgenic early flowering plants and marker-assisted selection. *New Phytologist*, *192*, 364–377.

Folta, K. M. & Carvalho, S. D. (2015). Photoreceptors and control of horticultural plant traits. *HortScience*, *50*, 1274–1280. 10.21273/hortsci.50.9.1274

Forster, B. & Shu, Q. (2012). Plant mutagenesis in crop improvement: Basic terms and applications. In Shu QY, Forster BP, Nakagawa H (eds), *Plant mutation breeding and biotechnology* (pp. 9–20). Wallingford UK: CABI.

Gerald, N., Frei, U. K., & Lübberstedt, T. (2013). Accelerating plant breeding. *Trends in Plant Science*, *18*(12), 667–672.

Ghosh, S., Watson, A., Gonzalez-Navarro, O. E., Ramirez-Gonzalez, R. H., Yanes, L., Mendoza-Suárez, M., …, & Green, P. (2018). Speed breeding in growth chambers and glasshouses for crop breeding and model plant research. *Nature Protocols*, *13*(12), 2944–2963.

Godwin, I. D., Rutkoski, J., Varshney, R. K., & Hickey, L. T. (2019). Technological perspectives for plant breeding. *Theoretical and Applied Genetics*, *132*(3), 555–557.

González-Barrios, P., Bhatta, M., Halley, M., Sandro, P., & Gutiérrez, L. (2021). Speed breeding and early panicle harvest accelerates oat (Avena sativa L.) breeding cycles. *Crop Science*, *61*(1), 320–330.

Goulden, C. (1939). Problems in plant selection. Burnett, R. C. (ed.), Proceedings of the 7th International Genetic Congress. Cambridge University Press, Cambridge.

Gupta, P., Kumar, J., Mir, R., & Kumar, A. (2010). 4 Marker-assisted selection as a component of conventional plant breeding. *Plant Breeding Reviews*, *33*, 145.

Han, Q., Wang, Z., He, Y., Xiong, Y., Lv, S., Li, S., ..., & Zeng, H. (2017). Transgenic cotton plants expressing the HaHR3 gene conferred enhanced resistance to Helicoverpa armigera and improved cotton yield. *International Journal of Molecular Sciences*, *18*(9), 1874.

Harfouche, A. L., Jacobson, D. A., Kainer, D., Romero, J. C., Harfouche, A. H., Mugnozza, G. S., ..., & Altman, A. (2019). Accelerating climate resilient plant breeding by applying next-generation artificial intelligence. *Trends in Biotechnology*, *37*(11), 1217–1235.

Hatfield, J. L. & Prueger, J. H. (2015). Temperature extremes: Effect on plant growth and development. *Weather and Climate Extremes*, *10*, 4–10. 10.1016/j.wace.2015.08.001.

Hickey, L. T., Wilkinson, P. M., Knight, C. R., Godwin, I. D., Kravchuk, O. Y., Aitken, E. A. B., Bansal, U. K., Bariana, H. S., De Lacy, I. H., & Dieters, M. J. (2012). Rapid phenotyping for adult-plant resistance to stripe rust in wheat. *Plant Breeding*, *131*, 54–61.

Hickey, L. T., Germán, S. E., Pereyra, S. A., Diaz, J. E., Ziems, L. A., Fowler, R. A., Platz, G. J., Franckowiak, J. D., & Dieters, M. J. (2017). Speed breeding for multiple disease resistance in barley. *Euphytica*, *213*, 1–14.

Hickey, L. T., Germán, S. E., Pereyra, S. A., Diaz, J. E., Ziems, L. A., Fowler, R. A., ..., & Dieters, M. J. (2017). Speed breeding for multiple disease resistance in barley. *Euphytica*, *213*, 1–14.

Hickey, L. T., N. Hafeez, A., Robinson, H., Jackson, S. A., Leal-Bertioli, S. C., Tester, M., ..., & Wulff, B. B. (2019). Breeding crops to feed 10 billion. *Nature Biotechnology*, *37*(7), 744–754.

Hou, Q., Xu, L., Liu, G., Pang, X., Wang, X., Zhang, Y., ..., & Liang, R. (2019). Plant-mediated gene silencing of an essential olfactory-related Gqα gene enhances resistance to grain aphid in common wheat in greenhouse and field. *Pest Management Science*, *75*(6), 1718–1725.

Huang, X. (2016). From genetic mapping to molecular breeding: Genomics have paved the highway. *Molecular Plant*, *9*(7), 959–960.

Imaizumi, T. & Kay, S. A. (2006). Photoperiodic control of flowering: Not only by coincidence. *Trends in Plant Science*, *11*, 550–558.

Jähne, F., Hahn, V., Würschum, T., & Leiser, W. L. (2020). Speed breeding short-day crops by LED-controlled light schemes. *Theoretical and Applied Genetics*, *133*(8), 2335–2342.

Kang, J. H., KrishnaKumar, S., Atulba, S. L. S., Jeong, B. R., & Hwang, S. J. (2013). Light intensity and photoperiod influence the growth and development of hydroponically grown leaf lettuce in a closed-type plant factory system. *Horticulture, Environment, and Biotechnology*, *54*, 501–509.

Kharkwal, M. (2012). A brief history of plant mutagenesis. In Shu QY, Forster BP, Nakagawa H (eds), *Plant mutation breeding and biotechnology* (pp. 21–30). Wallingford, UK: CABI.

Khoo, K. H. P., Sheedy, J. G., Taylor, J. D., Croser, J. S., Hayes, J. E., Sutton, T., Thompson, J. P., & Mather, D. E. (2021). A QTL on the Ca7 chromosome of chickpea affects resistance to the root-lesion nematode Pratylenchus thornei. *Molecular Breeding*, *41*, 78.

Kim, D. W., Chang, C., Chen, X., Doran, A. C, Gaudreault, F., Wager, T., DeMarco, G. J, & Kim, J. K. (2019). Systems approach reveals photosensitivity and PER 2 level as determinants of clock-modulator efficacy. *Molecular Systems Biology*, *15*. 10.15252/msb.20198838

Leal Filho, W., Wall, T., Mucova, S. A. R., Nagy, G. J., Balogun, A.-L, Luetz, J. M., ..., & Alves, F. (2022). Deploying artificial intelligence for climate change adaptation. *Technological Forecasting and Social Change*, *180*, 121662.

Li, Q. & Kubota, C. (2009). Effects of supplemental light quality on growth and phytochemicals of baby leaf lettuce. *Environmental and Experimental Botany*, *67*, 59–64.

Li, S., Zhang, X., Wang, W., Guo, X., Wu, Z., Du, W., ..., & Xia, L. (2018). Expanding the scope of CRISPR/Cpf1-mediated genome editing in rice. *Molecular Plant*, *11*(7), 995–998.

Li, T., Huang, S., Jiang, W. Z., Wright, D., Spalding, M. H., Weeks, D. P., & Yang, B. (2011). TAL nucleases (TALNs): hybrid proteins composed of TAL effectors and FokI DNA-cleavage domain. *Nucleic Acids Research*, *39*(1), 359–372.

Lloyd, A., Plaisier, C. L., Carroll, D., & Drews, G. N. (2005). Targeted mutagenesis using zinc-finger nucleases in Arabidopsis. *Proceedings of the National Academy of Sciences*, *102*(6), 2232–2237.

M Perez-de-Castro, A., Vilanova, S., Cañizares, J., Pascual, L., M Blanca, J., J Diez, M., ..., & Picó, B. (2012). Application of genomic tools in plant breeding. *Current Genomics*, *13*(3), 179–195.

Mao, Y., Zhang, H., Xu, N., Zhang, B., Gou, F., & Zhu, J.-K. (2013). Application of the CRISPR–Cas system for efficient genome engineering in plants. *Molecular Plant*, *6*(6), 2008–2011.

Massa, G. D., Kim, H-H., Wheeler, R. M., & Mitchell, C. A. (2008). Plant productivity in response to LED lighting. HortScience, *43*, 1951–1956. 10.21273/hortsci.43.7.1951.

Mba, C. (2013). Induced mutations unleash the potentials of plant genetic resources for food and agriculture. *Agronomy*, *3*(1), 200–231.

McClung, C. R., Lou, P., Hermand, V., & Kim, J. A. (2016). The importance of ambient temperature to growth and the induction of flowering. *Frontiers in Plant Science*, *7*. 10.3389/fpls.2016.01266.

Mehmood, U. (2023). Environmental sustainability through renewable energy and banking sector development: Policy implications for N-11 countries. *Environmental Science and Pollution Research*, *30*(9), 22296–22304.

Menda, N., Semel, Y., Peled, D., Eshed, Y., & Zamir, D. (2004). In silico screening of a saturated mutation library of tomato. *The Plant Journal*, *38*(5), 861–872.

Metzker, M. L. (2010). Sequencing technologies—the next generation. *Nature Reviews Genetics*, *11*(1), 31–46.

Mobini, S., Khazaei, H., Warkentin, T. D., & Vandenberg, A. (2020). Shortening the generation cycle in faba bean (Vicia faba) by application of cytokinin and cold stress to assist speed breeding. *Plant Breeding*, *139*(6), 1181–1189.

Moose, S. P., & Mumm, R. H. (2008). Molecular plant breeding as the foundation for 21st century crop improvement. *Plant Physiology*, *147*(3), 969–977.

Murovec, J., Guček, K., Bohanec, B., Avbelj, M., & Jerala, R. (2018). DNA-free genome editing of Brassica oleracea and B. rapa protoplasts using CRISPR-Cas9 ribonucleoprotein complexes. *Frontiers in Plant Science*, *9*, 1594.

Nagatoshi, Y. & Fujita, Y. (2019). Accelerating soybean breeding in a CO2-supplemented growth chamber. *Plant and Cell Physiology*, *60*, 77–84.

Niu, X., Kassa, A., Hu, X., Robeson, J., McMahon, M., Richtman, N. M., …, & Sandahl, G. (2017). Control of western corn rootworm (Diabrotica virgifera virgifera) reproduction through plant-mediated RNA interference. *Scientific Reports*, *7*(1), 12591.

Oladosu, Y., Rafii, M. Y., Abdullah, N., Hussin, G., Ramli, A., Rahim, H. A., …, & Usman, M. (2016). Principle and application of plant mutagenesis in crop improvement: a review. *Biotechnology and Biotechnological Equipment*, *30*(1), 1–16.

Ortiz, R., Trethowan, R., Ferrara, G. O., Iwanaga, M., Dodds, J. H., Crouch, J. H., Crossa, J., & Braun, H.-J. (2007). High yield potential, shuttle breeding, genetic diversity, and a new international wheat improvement strategy. *Euphytica*, *157*, 365–384.

Ossowski, S., Schneeberger, K., Lucas-Lledó, J. I., Warthmann, N., Clark, R. M., Shaw, R. G., …, & Lynch, M. (2010). The rate and molecular spectrum of spontaneous mutations in Arabidopsis thaliana. *Science*, *327*(5961), 92–94.

Ouzounis, T., Rosenqvist, E., & Ottosen, C.-O. (2015). Spectral effects of artificial light on plant physiology and secondary metabolism: A review. *HortScience*, *50*, 1128–1135.

Pandey, S., Singh, A., Parida, S. K., & Prasad, M. (2022). Combining speed breeding with traditional and genomics-assisted breeding for crop improvement. *Plant Breeding*, *141*, 301–313.

Pazos-Navarro, M., Castello, M., Bennett, R. G., Nichols, P., & Croser, J. (2017). In vitro-assisted single-seed descent for breeding-cycle compression in subterranean clover (Trifolium subterraneum L.). *Crop and Pasture Science*, *68*, 958–966.

Qaim, M. (2020). Role of new plant breeding technologies for food security and sustainable agricultural development. *Applied Economic Perspectives and Policy*, *42*(2), 129–150.

Qin, L., Li, J., Wang, Q., Xu, Z., Sun, L., Alariqi, M., …, & Ding, X. (2020). High-efficient and precise base editing of C• G to T• A in the allotetraploid cotton (Gossypium hirsutum) genome using a modified CRISPR/Cas9 system. *Plant Biotechnology Journal*, *18*(1), 45–56.

Rai, K. K. (2022). Integrating speed breeding with artificial intelligence for developing climate-smart crops. *Molecular Biology Reports*, *49*(12), 11385–11402.

Rana, M., Sood, A., Hussain, W., Kaldate, R., Sharma, T. R., Gill, R., …, & Singh, S. (2019). Gene pyramiding and multiple character breeding. In Mohar Singh (Editor) *Lentils* (pp. 83–124): Elsevier.

Razzaq, A., Kaur, P., Akhter, N., Wani, S. H., & Saleem, F. (2021). Next-generation breeding strategies for climate-ready crops. *Frontiers in Plant Science*, *12*, 620420.

Reynolds, M. P., Quilligan, E., Aggarwal, P. K., Bansal, K. C., Cavalieri, A. J., Chapman, S. C., Chapotin, S. M., Datta, S. K., Duveiller, E., & Gill, K. S. (2016). An integrated approach to maintaining cereal productivity under climate change. *Global Food Security*, *8*, 9–18.

Ribalta, F. M., Pazos-Navarro, M., Nelson, K., Edwards, K., Ross, J. J., Bennett, R. G., …, & Croser, J. S. (2017). Precocious floral initiation and identification of exact timing of embryo physiological maturity facilitate germination of immature seeds to truncate the lifecycle of pea. *Plant Growth Regulation*, *81*, 345–353.

Richard, C., Hickey, L., Fletcher, S., Chenu, K., Borrell, A., & Christopher, J. (2015). High-throughput phenotyping of wheat seminal root traits in a breeding context. *Procedia Environmental Sciences*, *29*, 102–103.

Richard, C., Christopher, J., Chenu, K., Borrell, A., Christopher, M., & Hickey, L. (2018). Selection in early generations to shift allele frequency for seminal root angle in wheat. *Plant Genome*, *11*, 170071.

Robertson, L. D. & Curtis, B. C. (1967). Germination of Immature Kernels of Winter Wheat 1. *Crop Science*, *7*, 269–270.

Roumet, P. & Morin, F. (1997). Germination of immature soybean seeds to shorten reproductive cycle duration. *Crop Science*, *37*, 521–525.

Roychowdhury, R. & Tah, J. (2013). Mutagenesis—a potential approach for crop improvement. In: Hakeem KR, Ozturk M, and Ahmad P. (Ediotrs) *Crop improvement: New approaches and modern techniques*, pp. 149–187.

Sadras, V. O. & Lawson, C. (2011). Genetic gain in yield and associated changes in phenotype, trait plasticity and competitive ability of South Australian wheat varieties released between 1958 and 2007. *Crop and Pasture Science*, *62*(7), 533–549.

Samantara, K., Bohra, A., Mohapatra, S. R., Prihatini, R., Asibe, F., Singh, L., …, & Croser, J. S. (2022). Breeding more crops in less time: A perspective on speed breeding. *Biology*, *11*(2), 275.

Samineni, S., Sen, M., Sajja, S. B., & Gaur, P. M. (2020). Rapid generation advance (RGA) in chickpea to produce up to seven generations per year and enable speed breeding. *The Crop Journal*, *8*(1), 164–169.

Sartor, R. C., Noshay, J., Springer, Nathan M., & Briggs, S. P. (2019). Identification of the expressome by machine learning on omics data. *Proceedings of the National Academy of Sciences*, *116*(36), 18119–18125.

Sarutayophat, T. & Nualsri, C. (2010). The efficiency of pedigree and single seed descent selections for yield improvement at generation 4 (F4) of two yardlong bean populations. *Agriculture and Natural Resources*, *44*(3), 343–352.

Saxena, K., Saxena, R. K., & Varshney, R. K. (2017). Use of immature seed germination and single seed descent for rapid genetic gains in pigeonpea. *Plant Breeding*, *136*(6), 954–957.

Schneider, K., Schiermeyer, A., Dolls, A., Koch, N., Herwartz, D., Kirchhoff, J., …, & Corbin, D. R. (2016). Targeted gene exchange in plant cells mediated by a zinc finger nuclease double cut. *Plant Biotechnology Journal*, *14*(4), 1151–1160.

Shakoor, N., Lee, S., & Mockler, T. C. (2017). High throughput phenotyping to accelerate crop breeding and monitoring of diseases in the field. *Current Opinion in Plant Biology*, *38*, 184–192.

Shi, L., Zhang, J., Shen, G., Xu, Z., Xu, Q., & He, L. (2016). Collaborative contribution of six cytochrome P450 monooxygenase genes to fenpropathrin resistance in Tetranychus cinnabarinus (Boisduval). *Insect Molecular Biology*, *25*(5), 653–665.

Shiferaw, B., Prasanna, B. M., Hellin, J., & Bänziger, M. (2011). Crops that feed the world 6. Past successes and future challenges to the role played by maize in global food security. *Food Security*, *3*, 307–327.

Shimelis, H. & Laing, M. (2012). Timelines in conventional crop improvement: pre-breeding and breeding procedures. *Australian Journal of Crop Science*, *6*(11), 1542–1549.

Slama-Ayed, O., Bouhaouel, I., Ayed, S., De Buyser, J., Picard, E., & Amara, H. S. (2019). Efficiency of three haplomethods in durum wheat (Triticum turgidum subsp. durum Desf.): Isolated microspore culture, gynogenesis and wheat× maize crosses. *Czech Journal of Genetics and Plant Breeding*, *55*(3), 101–109.

Souza, L. S., Diniz, R. P., Neves, R. de Jesus, Alves, A. A. C., & de Oliveira, E. J. (2018). Grafting as a strategy to increase flowering of cassava. *Scientia Horticulturae (Amsterdam)*, *240*, 544–551.

Sprink, T., Metje, J., & Hartung, F. (2015). Plant genome editing by novel tools: TALEN and other sequence specific nucleases. *Current Opinion in Biotechnology*, *32*, 47–53.

Stetter, M. G, Zeitler, L., Steinhaus, A., Kroener, K., Biljecki, M., & Schmid, K. J. (2016). Crossing methods and cultivation conditions for rapid production of segregating populations in three grain amaranth species. *Frontiers in Plant Science*, *7*, 816.

Sysoeva, M. I., Markovskaya, E. F., & Shibaeva, T. G. (2010). Plants under continuous light: A review. *Plant Stress*, *4*(1), 5–17.

Taiz, L. & Zeiger, E. (2010). *Plant physiology*. Sunderland, MA: Sinauer Associates Inc.

Tardieu, F., Cabrera-Bosquet, L., Pridmore, T., & Bennett, M. (2017). Plant phenomics, from sensors to knowledge. *Current Biology*, *27*(15), R770–R783.

Taylor, C. M., Garg, G., Berger, J. D., Ribalta, F. M., Croser, J. S., Singh, K. B., Cowling, W. A., Kamphuis, L. G., & Nelson, M. N. (2021). A Trimethylguanosine Synthase1-like (TGS1) homologue is implicated in vernalisation and flowering time control. *Theoretical and Applied Genetics*, *134*, 3411–3426.

Tester, M. & Langridge, P. (2010). Breeding technologies to increase crop production in a changing world. *Science*, *327*(5967), 818–822.

Thudi, M., Palakurthi, R., Schnable, J. C., Chitikineni, A., Dreisigacker, S., Mace, E., …, & Tiwari, V. K. (2021). Genomic resources in plant breeding for sustainable agriculture. *Journal of Plant Physiology*, *257*, 153351.

Van Harten, A. M. (1998). *Mutation breeding: Theory and practical applications* (Vol. 1). Cambridge University Press, Cambridge UK.

van Nocker, S. & Gardiner, S. E. (2014). Breeding better cultivars, faster: applications of new technologies for the rapid deployment of superior horticultural tree crops. *Horticulture Research, 1.*

Varshney, R. K., Bohra, A., Yu, J., Graner, A., Zhang, Q., & Sorrells, M. E. (2021). Designing future crops: Genomics-assisted breeding comes of age. *Trends in Plant Science, 26*, 631–649.

Velez-Ramirez, A. I., Van Ieperen, W., Vreugdenhil, D., Van Poppel, P. M. J. A., Heuvelink, E., & Millenaar, F. F. (2014). A single locus confers tolerance to continuous light and allows substantial yield increase in tomato. *Nature Communications, 5*, 4549.

Vira, B., Wildburger, C., & Mansourian, S. (2015). Forests, trees and landscapes for food security and nutrition: A global assessment report. IUFRO World Series, *33*, 14–24.

Voss-Fels, K. P., Herzog, E., Dreisigacker, S., Sukumaran, S., Watson, A., Frisch, M., Hayes, B., & Hickey, L. T. (2019). "SpeedGS" to accelerate genetic gain in spring wheat. *Applications of genetic and genomic research in cereals*. Elsevier. pp. 303–327.

Voss-Fels, K. P., Stahl, A., & Hickey, L. T. (2019). Q&A: Modern crop breeding for future food security. *BMC Biology, 17*(1), 1–7.

Wang, J., Niu, X., Zheng, L., Zheng, C., & Wang, Y. (2016). Wireless mid-infrared spectroscopy sensor network for automatic carbon dioxide fertilization in a greenhouse environment. *Sensors, 16*, 1941.

Wang, Z., Fan, B., & Guo, L. (2019). Soil salinization after long-term mulched drip irrigation poses a potential risk to agricultural sustainability. *European Journal of Soil Science, 70*, 20–24. 10.1111/ejss.12742

Wanga, M. A., Shimelis, H., Mashilo, J., & Laing, M. D. (2021). Opportunities and challenges of speed breeding: A review. *Plant Breeding, 140*(2), 185–194.

Watanabe, S., Mizoguchi, T., Aoki, K., Kubo, Y., Mori, H., Imanishi, S., ..., & Ezura, H. (2007). Ethylmethanesulfonate (EMS) mutagenesis of Solanum lycopersicum cv. Micro-Tom for large-scale mutant screens. Plant Biotechnology, *24*(1), 33–38.

Watson, A., Ghosh, S., Williams, M. J., Cuddy, W. S., Simmonds, J., Rey, M.-D, ..., & Reynolds, D. (2018). Speed breeding is a powerful tool to accelerate crop research and breeding. *Nature Plants, 4*(1), 23–29.

Watson, A., Hickey, L. T., Christopher, J., Rutkoski, J., Poland, J., & Hayes, B. J. (2019). Multivariate genomic selection and potential of rapid indirect selection with speed breeding in spring wheat. *Crop Science, 59*, 1945–1959.

Wendler, N., Mascher, M., Nöh, C., Himmelbach, A., Scholz, U., Ruge-Wehling, B., & Stein, N. (2014). Unlocking the secondary gene-pool of barley with next-generation sequencing. *Plant Biotechnology Journal, 12*(8), 1122–1131.

White, J. W. & Conley, M. M. (2013). A flexible, low-cost cart for proximal sensing. *Crop Science, 53*(4), 1646–1649.

Wood, A. J., Lo, T.-W, Zeitler, B., Pickle, C. S., Ralston, E. J., Lee, A. H., ..., & Meng, X. (2011). Targeted genome editing across species using ZFNs and TALENs. *Science, 333*(6040), 307.

Yang, H., Wu, J.-J., Tang, T., Liu, K.-D., & Dai, C. (2017). CRISPR/Cas9-mediated genome editing efficiently creates specific mutations at multiple loci using one sgRNA in Brassica napus. *Scientific Reports, 7*, 7489.

Yao, Y., Zhang, P., Liu, H., Lu, Z., & Yan, G. (2017). A fully in vitro protocol towards large scale production of recombinant inbred lines in wheat (Triticum aestivum L.). *Plant Cell, Tissue and Organ Culture (PCTOC), 128*, 655–661.

Zaidi, Syed Shan-e-Ali, Mahfouz, M. M., & Mansoor, S. (2017). CRISPR-Cpf1: A new tool for plant genome editing. *Trends in Plant Science, 22*(7), 550–553.

Zaman, M. S. U., Malik, A. I., Kaur, P., Ribalta, F. M., & Erskine, W. (2019). Waterlogging tolerance at germination in field pea: Variability, genetic control, and indirect selection. *Frontiers in Plant Science, 10*, 953.

Zhang, Y., Ma, X., Xie, X., & Liu, Y.-G. (2017). CRISPR/Cas9-based genome editing in plants. *Progress in Molecular Biology and Translational Science, 149*, 133–150.

3 Accelerate Crop Breeding Utilizing Modern Genomics and Phenomics Tools

Amooru Harika, Sivapragasam Ezhumalai, Biswabiplab Singh, Elangovan Allimuthu, Sukumar Taria, Dhandapani Raju, Sudhir Kumar, Mahesh Kumar, Madhurima Das, and Viswanathan Chinnusamy

Division of Plant Physiology and Nanaji Deshmukh Plant Phenomics Center (NDPPC), Division of Plant Physiology, ICAR-Indian Agricultural Research Institute (IARI), New Delhi, India

3.1 INTRODUCTION

Researchers were compelled by the growing human population to explore faster food production techniques and varieties with larger yields. Estimates indicate that to nourish a population of 10 billion people by the year 2050, we would have to increase food production by approximately 60–80% (Hickey et al., 2019). Most agricultural cultivar improvement efforts employ traditional breeding techniques (Rommens, 2007). Without the implementation of a comprehensive pre-breeding initiative, conventional breeding methods can demand more than a decade to generate and introduce an upgraded variety (Ahmar et al., 2020). In most variety development programs, the bulk of resources, time, and physical space are primarily allocated to the selection and genetic advancement of the most promising offspring, with a focus on desired traits. Additionally, the selection of top-performing and consistent candidate cultivars is carried out through screening in diverse target production environments as an integral part of the breeding process (Cobb et al., 2019). The length of time spent in these stages greatly slows down the rate of cultivar commercialization (Wanga et al., 2021). Field selection methods take a whole season, and the conventional breeding method's poor pace of generational development is caused by the crop cycles themselves (Samantara et al., 2022). Typically, for commonly grown grain and legume crops, the standard crop cycle involves a duration of 3–6 months for each crop, generation, and per year (Beebe et al., 2008). One breeding generation lasts 15–18 months in other crops like cassava (Okogbenin et al., 2013). In certain agroecology, unpredictable meteorological factors such as extreme temperatures, irregular and insufficient rainfall patterns, and varying day lengths can result in only a single crop cycle occurring each year (Shahzad et al., 2021). Two generations can be produced annually in some tropical agricultural environments. By employing modern methods like speed breeding (Hickey et al., 2019) and doubled haploid breeding (Dwivedi et al., 2015), it becomes possible to reduce the duration of each crop breeding cycle. In contrast to the 1–2 generations per year typically achieved through traditional selection methods, speed breeding allows for the production of 3–9 generations each year (Watson et al., 2018; Begna, 2022). Speed breeding involves the use of controlled environments like greenhouses or growth chambers to hasten plant growth and development (Ghosh et al., 2018). By exposing plants to ideal growth conditions characterized by consistent light, controlled temperature, and humidity, scientists can generate numerous crop generations within a single year, substantially diminishing the time frame

DOI: 10.1201/b23372-3

necessary for conventional breeding processes (Pandey et al., 2022). Speed breeding can be employed to swiftly create novel crop varieties and expedite the overall crop improvement process (Ghosh et al., 2018; Watson et al., 2018; Abdul-Fiyaz et al., 2020; Begna, 2022). This innovative approach to plant breeding holds the potential to have a significant impact on addressing the challenge of feeding an ever-expanding global population.

3.1.1 Need for Accelerating Crop Improvement

The need for accelerating crop improvement is driven by several factors, including population growth, climate change, and food security concerns (Misra, 2014). As the global population is projected to reach 9.7 billion by 2050, the demand for enhanced food production is steadily rising, underscoring the imperative for innovative approaches like speed breeding (Elferink & Schierhorn, 2016; Lal, 2016; Plesse, 2020). Simultaneously, climate change is contributing to more frequent occurrences of droughts, floods, and extreme weather events, all of which have the potential to harm crops and diminish agricultural yields (Ghosh et al., 2018). In addition, food security concerns are heightened by the fact that millions of people around the world still suffer from hunger and malnutrition. Conventional plant breeding methods often require several years to develop new crop varieties that exhibit sought-after traits like increased yield, disease resistance, or tolerance to drought. Speed breeding techniques provide a hopeful solution to these challenges by expediting the breeding process, thus aiding in the creation of crops that display greater resilience to the shifting climate and are better aligned with the requirements of a burgeoning population (Lenaerts et al., 2019).

Speed breeding represents an innovative approach to plant breeding with the potential to bring about a transformative revolution in crop improvement (Ghosh et al., 2018; Watson et al., 2018; Abdul-Fiyaz et al., 2020; Begna, 2022). By harnessing controlled environments to expedite plant growth and development, scientists can generate multiple crop generations within a single year, consequently substantially curtailing the timeline traditionally associated with breeding. This expedited process can facilitate the swift creation of novel crop varieties possessing coveted attributes like heightened yield, disease resistance, or drought tolerance (Watson et al., 2018). Beyond its role in expediting the breeding process, speed breeding can also serve as a valuable tool for researchers to gain deeper insights into the genetic underpinnings of desirable traits in plants. This knowledge can, in turn, pave the way for the formulation of more precise and targeted breeding strategies in the future (Bhat et al., 2023). Speed breeding has already demonstrated its effectiveness in a variety of crop species, including wheat, barley, chickpea, pea, and several others (Shivakumar et al., 2018). In summary, speed breeding stands as a promising tool for hastening crop improvement, with the potential to hold a crucial role in addressing the challenge of nourishing a continuously expanding global population (Begna et al., 2022).

The objective of this chapter is to explore the concept of speed breeding techniques as a new tool for accelerating crop improvement. The review aims to provide an overview of the various methods used to accelerate crop breeding, including growth chambers, greenhouses, hydroponics, aeroponics, tissue culture, and marker-assisted selection (MAS). It also discusses the principles of speed breeding, including the manipulation of environmental factors, selection of plant genotypes, and rapid generation advancement (RGA). The review highlights the advantages and disadvantages of each method and provides examples of their applications in crop improvement. The review also explores the potential of speed breeding techniques in addressing some of the critical challenges confronting agriculture, including climate change, food security, and the burgeoning global population.

3.2 SPEED BREEDING TECHNIQUES

Speed breeding techniques encompass a collection of methods employed to expedite the growth and maturation of plants, with the primary goal of diminishing the time needed to generate new plant varieties (Chirugwi et al., 2019). These techniques serve to enhance the efficiency of breeding programs, enabling the creation of new plant varieties within a more condensed time frame. One pivotal method employed in speed breeding involves manipulating environmental variables such as light, temperature, and humidity to stimulate accelerated growth and development (Supriya et al., 2023). For example, plants can be grown under conditions of continuous light, which can speed up their growth and development by up to three times compared to traditional growth conditions. Another technique used in speed breeding is the use of genetic modification to introduce specific traits into plants. This can be achieved through a variety of methods, including gene editing and transgenic approaches.

3.2.1 Growth Chamber and Greenhouse Environments

Speed breeding techniques, such as the utilization of growth chambers and controlled greenhouse environments, are methodologies employed to hasten the growth and development of plants, primarily for research, breeding, and the advancement of crop varieties (Watson et al., 2018; Wanga et al., 2021).

3.2.1.1 Growth Chambers

Growth chambers are enclosed spaces specifically designed to offer meticulously controlled environmental conditions conducive to the growth and development of plants. These chambers can simulate a wide range of environmental factors, including temperature, humidity, light intensity and quality, and CO_2 concentration. There are different types of growth chambers, such as reach-in or walk-in chambers, multi-tiered chambers, and phytotron facilities, which can accommodate different sizes and numbers of plants and provide varying levels of environmental control (Ghosh et al., 2018). The advantages of using growth chambers for speed breeding include precise control over environmental factors, reproducibility of experimental conditions, and the ability to manipulate environmental factors to accelerate plant growth and development. Growth chambers additionally enable year-round experimentation and serve as valuable tools for investigating the impacts of environmental stressors on plants. However, growth chambers can be expensive to purchase and maintain, and they may not always replicate natural growing conditions (Kapiel et al., 2018). Additionally, the size of the chamber may limit the number of plants that can be grown, and the need for frequent monitoring and maintenance can be time-consuming. Growth chambers are widely used in plant research and breeding, as they allow for the manipulation of environmental factors to speed up plant growth and development. An instance of the effective application of growth chambers for speed breeding is the "speed breeding" technique pioneered by researchers at the University of Queensland. This approach entails the manipulation of light intensity and duration to expedite the growth and maturation of plants (Edet & Ishii, 2022). This technique has been used to produce six generations of wheat, chickpea, and barley plants in a year, compared to the traditional two or three generations that can be produced in the field.

3.2.1.2 Greenhouses

Greenhouses are enclosed structures that provide a controlled environment for plant growth and development (Shamshiri et al., 2018). They can be made of different materials, such as glass or plastic, and can vary in size and shape. Greenhouses can also have different types of environmental control systems, such as ventilation, heating, and cooling, to maintain optimal growing conditions (Abdul Fiyaz et al., 2020). Greenhouses present several advantages for speed breeding, including their capacity to cultivate a substantial number of plants within a relatively compact area, the potential for more faithful replication of field conditions compared to growth chambers, and the utilization of natural sunlight for supporting plant growth. Greenhouses can also be less expensive to set up and maintain than growth chambers, and they can provide a more realistic testing environment for plant breeding (Hohmann et al., 2016). However, greenhouses can still be expensive to build and maintain, and they may not offer the same level of environmental control as growth chambers. They can also be subject to pests and diseases, which can impact plant growth and productivity. Greenhouses are commonly used for plant breeding and production, as they can provide a more realistic growing environment than growth chambers. An illustration of the effective application of greenhouses for speed breeding is the "fast-track breeding" method formulated by researchers at the International Maize and Wheat Improvement Center (CIMMYT). This approach employs greenhouses for the screening and selection of wheat and maize varieties that possess sought-after traits, such as drought tolerance or disease resistance (Rai et al., 2022). This approach has resulted in the creation of novel wheat and maize varieties that have been embraced by farmers in diverse regions around the world.

Some examples of applications of growth chambers and greenhouses for speed breeding include:

Developing new cultivars: Researchers can utilize growth chambers and greenhouses to expedite the screening and selection of plants with desired traits, such as disease resistance or enhanced yield, within a shorter time frame (Samantara et al., 2022).

Studying plant physiology: Both growth chambers and greenhouses serve as valuable tools for investigating the influence of environmental factors on plant growth and development. These controlled environments facilitate the study of variables like light quality or temperature and their effects on processes such as photosynthesis and flowering (Paradiso & Proietti, 2022).

Climate change research: Growth chambers are suitable for examining the consequences of shifting environmental conditions on plant growth and productivity. For instance, they can be employed to investigate the influence of elevated CO_2 concentrations on crop development (Poorter et al., 2012).

Year-round production: Greenhouses offer the advantage of year-round crop production, irrespective of weather conditions or seasonal variations, making them highly advantageous for commercial crop cultivation (Massel et al., 2021).

3.2.2 Other Techniques

3.2.2.1 Hydroponics

Hydroponics is a cultivation method that involves growing plants in a nutrient-rich water solution, without the use of soil. This technique is widely favored in speed breeding due to its capacity for precise regulation of nutrient levels and environmental conditions in the growth environment, thereby expediting plant growth and development (Cazzola et al., 2020; Samantara et al., 2022). In hydroponics-based speed breeding, plants are cultivated within a meticulously controlled environment where they receive a continuous supply of a nutrient solution to support their growth and development. This method can increase the rate of growth and reduce the time required for plant development, allowing for more generations per year. Hydroponics can also help to improve the quality of the plant material by providing optimal nutrient levels and reducing the risk of soil-borne diseases. Furthermore, hydroponics can be employed to assess the performance of various plant varieties under diverse environmental conditions, including conditions simulating drought or salt stress. Hydroponics is particularly useful in speed breeding because it allows breeders to bypass the traditional constraints of soil-based growth such as limited space and nutrient availability (Song et al., 2022). This method can also be applied to cultivate plants in regions characterized by subpar soil quality or restricted availability of arable land.

3.2.2.2 Aeroponics

Aeroponics is an alternative method for plant cultivation suitable for speed breeding. It entails growing plants in a misty environment where their roots are suspended in the air, and they receive a nutrient-rich water supply through a misting system (Loman, 2018). One of the advantages of aeroponics is that it provides plants with higher oxygen levels, which can promote faster growth and development. This technique provides improved control over nutrient delivery to the plants, consequently leading to enhanced plant performance and overall quality (Sharathkumar et al., 2020; Tiwari et al., 2021). Furthermore, aeroponics can mitigate the risk of disease transmission among plants and offer advantages in terms of water and nutrient conservation. In speed breeding, aeroponics can be used to accelerate plant growth and development and to produce more generations of plants in a shorter period. It is particularly useful for crops that require a large root zone or that grow slowly in soil-based environments. Aeroponics is also useful for genetic transformation studies and for the production of high-value crops such as medicinal herbs and specialty vegetables. The technique can help to increase the yield and quality of these crops while reducing production costs.

3.2.2.3 Tissue Culture

Tissue culture, also known as micropropagation, is a method of plant propagation that can be used in speed breeding (Bridgen et al., 2018). Tissue culture is a method that entails cultivating plant tissue in a controlled, sterile environment while supplying the necessary nutrients, hormones, and growth regulators to stimulate cell division and plant growth. This technique is highly valuable in speed breeding as it can yield substantial quantities of genetically identical plantlets within a relatively brief time frame (Ahmar et al., 2020; Samantara et al., 2022). This method allows for the rapid propagation of desirable traits, such as disease resistance, and the production of large numbers of plants for testing in field trials. Tissue culture can also be used to produce plants from somatic cells, which allows for the regeneration of plants from individual cells or plant parts. This technique can be harnessed to generate plants exhibiting desired traits such as disease resistance or high yield potential. Additionally, tissue culture can be instrumental in preserving rare or endangered plant species. In speed breeding, tissue culture can be used to rapidly multiply plantlets, which can then be transferred to hydroponic or aeroponic systems for further growth and development (Naik & Buckseth, 2018). This can accelerate the production of large numbers of plants for breeding programs and the selection of desirable traits.

3.2.3 Genetic Engineering Techniques

Indeed, genetic engineering techniques can be employed to hasten the development of new crop varieties endowed with desirable traits (Tester & Langridge, 2010). Genetic engineering involves the manipulation of plant DNA to introduce new genes or modify existing genes to achieve specific traits (Gepts, 2002). Genetic engineering techniques have the potential to expedite the development of novel crop varieties featuring desirable traits, and they can also contribute to addressing critical agricultural challenges like food security, climate change, and environmental sustainability (Ahmar et al., 2020). Nonetheless, the application of genetic engineering in agriculture is a subject of controversy, with concerns regarding potential risks to both human health and the environment. Consequently, it is imperative to take these risks into account and adopt responsible and transparent practices when utilizing genetic engineering techniques in speed breeding. Various genetic engineering techniques can be employed in this context, including:

Gene editing: This technique involves making precise changes to specific genes in the plant DNA using tools like CRISPR-Cas9 (Wolter et al., 2019). Gene editing can serve as a powerful tool to introduce precise traits into crops such as disease resistance, drought tolerance, or enhanced nutritional content.

Transgenic technology: This technique involves the insertion of foreign genes into the plant DNA to introduce new traits. Transgenic plants can be created with traits such as insect resistance, herbicide tolerance, or improved yield potential.

RNA interference (RNAi): This technique involves the suppression of specific genes in the plant DNA using small RNA molecules. RNA interference (RNAi) can be harnessed to silence genes that contribute to undesirable traits such as susceptibility to pests or diseases (Rai et al., 2022).

3.3 PRINCIPLES OF SPEED BREEDING

3.3.1 Manipulation of Environmental Factors

Environmental factors indeed exert a significant influence on plant growth and development, and the deliberate manipulation of these factors can be a pivotal strategy for expediting the breeding process. Some of the key environmental factors that can be manipulated for speed breeding are:

Light intensity and duration: Light is vital for photosynthesis, the process through which plants harness energy from sunlight. Modulating the intensity and duration of light exposure can be effectively employed to expedite plant growth and development. Elevating both the intensity and duration of light exposure can contribute to the acceleration of plant growth and development, while also stimulating flowering and enhancing seed production.

Temperature and humidity: Temperature and humidity are indeed significant factors influencing plant growth and development. The deliberate adjustment of temperature and humidity levels can be employed to facilitate more rapid growth and development of plants (Chiurugwi et al., 2019). Higher temperatures can promote faster growth, while lower temperatures can promote flowering and seed production. Maintaining optimal humidity levels can also help to reduce water stress and improve plant growth.

Carbon dioxide (CO_2) concentration: Carbon dioxide (CO_2) is indeed a crucial component for photosynthesis, and elevating its concentration in the growth environment can stimulate accelerated plant growth and development (Wanga et al., 2021). The current atmospheric CO_2 concentration stands at approximately 400 ppm, but augmenting it to around 1000 ppm can effectively encourage swifter plant growth.

Nutrient availability: Nutrient availability plays a pivotal role in plant growth and development, and the deliberate adjustment of nutrient levels can be a valuable strategy for expediting plant growth and development. Heightening the availability of essential nutrients like nitrogen, phosphorus, and potassium can facilitate quicker growth and development while also enhancing the quality of the end product (Virk et al., 2021). Additionally, providing plants with a complete nutrient solution that contains all the essential nutrients can help to ensure that plants have everything they need for optimal growth and development (Figure 3.1).

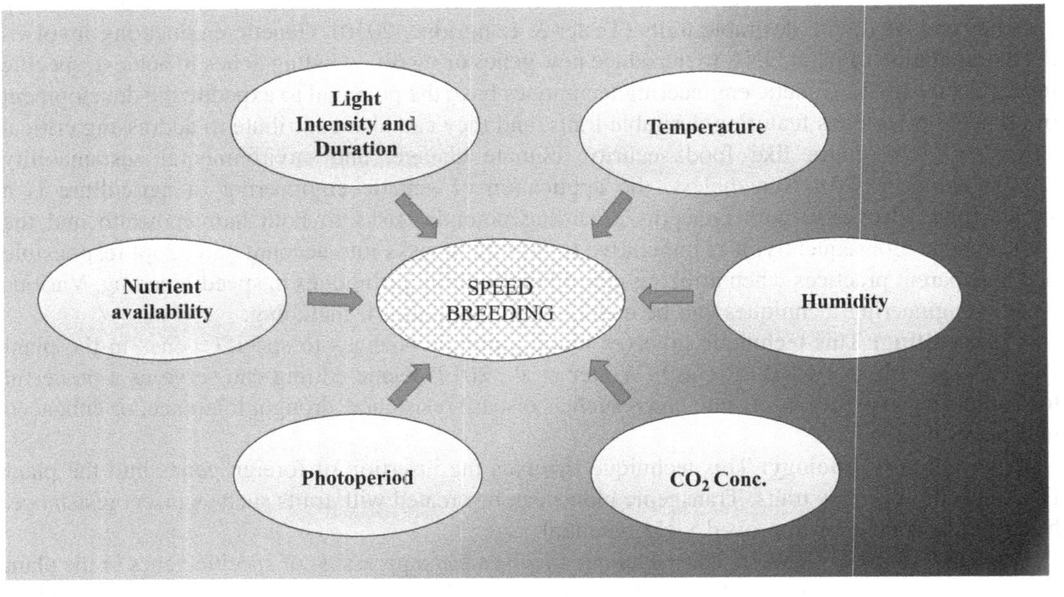

FIGURE 3.1 Key factors altering speed breeding for crop improvement.

3.3.2 Selection of Plant Genotypes

The selection of plant genotypes is a crucial aspect of speed breeding. By selecting plants with desirable traits, breeders can reduce the time required for plant development. Some key factors to consider when selecting plant genotypes for speed breeding include:

Desirable traits: Breeders should select plants with desirable traits that can be improved through breeding such as disease resistance, higher yield, improved quality, and reduced time to maturity. These traits can be selected through phenotypic or genotypic screening.

Genetic diversity: Maintaining genetic diversity is important for the long-term success of breeding programs. Breeders should select a diverse range of plant genotypes to ensure that genetic variation is preserved and to prevent the accumulation of deleterious mutations.

Hybridization: Hybridization refers to the practice of mating two distinct plant genotypes to generate progeny with sought-after traits. This approach is instrumental in amalgamating desirable characteristics from diverse genotypes, thereby expediting the creation of novel plant varieties.

Genome editing: Genome editing technologies like CRISPR/Cas9 offer the capability to make precise modifications to plant genomes. This innovative technology can be harnessed to introduce desired traits into plant genotypes more rapidly and with greater precision than conventional breeding methods.

3.3.3 Rapid Generation Advancement

RGA stands as a pivotal component of speed breeding, focused on expediting the turnover rate of plant generations. By diminishing the time necessary for each generation, breeders can hasten the breeding process and expedite the development of new plant varieties (Bhargava et al., 2023). Some key strategies for RGA include:

Shortening the vegetative phase: The vegetative phase denotes the stage of plant growth that occurs before flowering. By abbreviating the vegetative phase, the time needed for overall plant development can be reduced. This can be accomplished through the selection of plant genotypes with naturally shorter vegetative phases or by manipulating environmental factors, such as adjusting light intensity and duration, to expedite this phase.

Increasing the number of generations per year: Augmenting the number of generations per year is another approach to accelerate the breeding process. This can be realized through the application of techniques like hydroponics or tissue culture, which can curtail the duration needed for plant growth and development

Early generation selection: Early generation selection entails the process of identifying and selecting plants with desirable traits at an early phase of their development, often during the seedling stage. This practice contributes to the acceleration of the breeding process by decreasing the number of generations needed to produce a desired plant variety.

Marker-assisted selection: MAS is a technique that involves the utilization of genetic markers to identify and select plants with specific desirable traits. This method streamlines the breeding process by allowing breeders to more precisely and efficiently identify individuals carrying the desired genetic characteristics (Xu & Crouch, 2008). This can help to identify desirable traits more quickly and accurately than traditional breeding methods, reducing the time required for plant development.

3.4 APPLICATIONS OF SPEED BREEDING

3.4.1 Crops Successfully Improved Using Speed Breeding

Speed breeding has been successfully applied to a diverse array of crops, including:

Wheat: Certainly, speed breeding has been employed to generate multiple wheat generations within a single year, resulting in the development of novel wheat varieties that exhibit enhanced yield, heightened disease resistance, and increased tolerance to abiotic stress (Riaz et al., 2016; Alahmad et al., 2018; Ghosh et al., 2018).

Barley: Speed breeding techniques have been applied to expedite the breeding of barley as well. This approach has contributed to the creation of new barley varieties characterized by enhanced yield, improved quality, and heightened disease resistance (Hickey et al., 2017).

Chickpea: Speed breeding has been effectively utilized to develop novel chickpea varieties characterized by improved yield, enhanced drought tolerance, and heightened resistance to diseases (Samineni et al., 2020).

Pea: Speed breeding techniques have been successfully employed to produce new pea varieties that demonstrate enhanced yield, increased disease resistance, and improved seed quality (Cazzola et al., 2020).

Canola: Speed breeding techniques have played a role in the development of new canola varieties with improved yield and enhanced oil content (Watson et al., 2018).

Tomato: Speed breeding has been effectively employed to expedite the breeding of tomato varieties. This has led to the development of new tomato cultivars with improved yield, enhanced fruit quality, and heightened resistance to diseases (Ayenan et al., 2019).

Brassica: Speed breeding techniques have been successfully applied to accelerate the breeding of Brassica varieties, resulting in the development of new cultivars with improved yield and quality (Ferrie et al., 2022).

Soybean: Speed breeding has been implemented to augment the amount of light exposure for short-day plants by utilizing supplementary lighting, including LED lights. This practice helps create extended daylight hours, which can stimulate the growth and development of these plants and accelerate the breeding process (Jahne et al., 2020; Nagatoshi et al., 2019) (Table 3.1).

3.4.2 Desirable Traits Targeted in Speed Breeding

Speed breeding can be used to target a range of desirable traits in crops (Karthika et al., 2020), including:

Yield: Enhancing crop yield stands as a primary objective of speed breeding. By expediting the growth and development of plants, researchers can generate new crop varieties that exhibit increased yields within a reduced time frame. This is crucial in addressing the global demand for food production and ensuring food security for growing populations.

Disease resistance: Speed breeding can also be used to develop crops that are more resistant to diseases such as rust or powdery mildew. This can help reduce the use of pesticides and other chemicals, which can be harmful to the environment and human health.

Drought tolerance: Developing crops that are more tolerant to drought is another important application of speed breeding, particularly in regions that are prone to drought or water scarcity.

Nutrient content: Speed breeding techniques can be employed not only to enhance crop yield but also to develop crops that are enriched in specific nutrients, including essential vitamins and minerals. This targeted approach can help address nutritional deficiencies in specific populations, contributing to improved food quality and nutritional health.

Abiotic stress tolerance: Speed breeding techniques can be harnessed to cultivate crops that possess increased resilience to abiotic stresses like extreme heat or cold. This resilience can serve as a crucial strategy to alleviate the adverse impacts of climate change on crop productivity.

By targeting these and other desirable traits in crops, speed breeding has the potential to play an important role in addressing global food security challenges and promoting sustainable agriculture.

3.4.3 Potential for Addressing Global Food Security Challenges

Speed breeding holds immense potential in addressing the pressing global food security challenges that stem from the need to enhance crop productivity and resilience. With the world's population

TABLE 3.1
Application of Speed Breeding in Different Crops

Crop	Goal	Techniques	Number of Generations Per Year	Selection Method	References
Wheat	Resistance to yellow spot, crown rot, stem rust, and stripe rust	Temperature, immature seed germination, soil fertility, photoperiod, and embryo rescue	4-6	SSD	Ghosh et al., 2018; Riaz et al., 2016; Alahmad et al., 2018; Zheng et al., 2013
Barley	Resistance to leaf rust	Photoperiod, temperature, soil fertility, immature seed germination and embryo rescue	5-6	SSD	Hickey et al., 2017; Zheng et al., 2013
Pea	Rapid generation advance	Plant hormones, photoperiod, and immature seed germination	2–3	SSD	O'Connor et al., 2013
Chickpea	Rapid generation advance	Photoperiod and immature seed germination	4-6	SPD	Samineni et al., 2020
Soybean	Rapid generation advance	Photoperiod, temperature, and immature seed germination	4-5	SSD	Jähne et al., 2020
Canola	Pod shattering	Photoperiod, light intensity, temperature, immature seed germination, and soil moisture	4-6	SSD	Watson et al., 2018
Rice	Rapid generation advance	Photoperiod, temperature, and high-density planting	3-4	SSD	Collard et al., 2017
Amaranth	Rapid generation advance	Photoperiod and temperature	6	SSD	Stetter et al., 2016

Abbreviations: SPD: single pod descent, SSD: single seed descent.

projected to reach 9.7 billion by 2050, there is a daunting requirement for a 70% increase in food production (Hickey et al., 2019). Traditional breeding methods, however, can be sluggish and inefficient, demanding years or even decades to develop new crop varieties with desired traits. Speed breeding emerges as a solution by expediting the breeding process, enabling the generation of multiple crop generations within a single year, and culminating in the creation of novel crop varieties distinguished by improved yield, disease resistance, and abiotic stress tolerance (Watson et al., 2018). Beyond its contribution to crop productivity enhancement, speed breeding also carries the potential to diminish the environmental impact of agriculture. By developing crops with heightened resistance to diseases and pests, speed breeding can curtail the utilization of pesticides and other chemicals, which can have detrimental consequences on both the environment and human health. Moreover, by fostering the development of crops with improved drought tolerance and resistance to abiotic stresses, speed breeding can aid in conserving water and other valuable resources, thereby promoting long-term sustainability in agriculture.

3.5 ADVANTAGES AND LIMITATIONS OF SPEED BREEDING

3.5.1 Advantages of Speed Breeding over Traditional Breeding Methods

Speed breeding, a revolutionary approach to crop breeding, has emerged as a potent alternative to conventional breeding techniques, offering a multitude of advantages that accelerate and improve the process of creating new crop varieties (Gudi et al., 2022). One of the most notable benefits of speed breeding is the accelerated rate at which results can be achieved. Compared to conventional breeding, speed breeding significantly shortens the breeding cycle, allowing researchers to generate crops with desirable traits in a remarkably reduced time frame. This swift turnaround is invaluable in meeting the urgent global challenges of food security and climate change.

Efficiency is another hallmark of speed breeding. By employing controlled and optimized growth conditions, researchers can screen and identify plants with desired traits with unprecedented speed and accuracy. This efficiency streamlines the selection process, reducing the labor and resources required while increasing the likelihood of identifying the most promising candidates for further development. Moreover, speed breeding enables greater precision in manipulating and targeting specific traits. Researchers can fine-tune the breeding process to emphasize traits that are crucial for improved crop performance, be it resistance to diseases, pests, or abiotic stresses (Pandey et al., 2022). This level of precision ensures that the resulting crop varieties are tailored to meet specific needs, thereby enhancing overall crop yield and quality. Furthermore, speed breeding proves to be cost-effective in the long run. The reduction in breeding cycle duration and optimized resource utilization lowers operational costs. As a result, researchers can allocate their limited resources more efficiently, making crop breeding programs economically sustainable and potentially accessible to a broader range of researchers and farmers. One critical aspect of speed breeding is its contribution to maintaining genetic diversity. By incorporating wild relatives and other genetic resources, speed breeding can introduce novel traits and genetic variability into cultivated crop populations. Preserving genetic diversity is essential for building resilient crops that can adapt to changing environmental conditions, pests, and diseases, ultimately contributing to long-term food security.

In summary, speed breeding presents substantial advantages over traditional breeding approaches (Watson et al., 2018; Wanga et al., 2021), empowering researchers to efficiently and precisely cultivate new crop varieties with sought-after traits within a reduced time frame.

3.5.2 Limitations and Challenges of Speed Breeding

Although speed breeding presents considerable benefits in crop breeding, it does pose various challenges and considerations that require attention. A critical aspect pertains to the equipment and infrastructure prerequisites. Speed breeding relies on specialized growth chambers or greenhouses, which can be costly to set up and maintain. Additionally, operating such facilities requires technical expertise, making it a resource-intensive endeavor for researchers and institutions. Another concern is the potential impact on genetic diversity. Focusing on specific traits during speed breeding may lead to reduced genetic diversity within the crop population. A lack of genetic variation can make crops more vulnerable to diseases, pests, and other environmental stresses, potentially compromising long-term crop resilience. Regulatory and public acceptance pose further hurdles to the adoption of speed breeding techniques (Song et al., 2022). The use of genetic engineering and other advanced breeding methods may face strict regulations and resistance from certain regions or public perceptions, limiting their widespread application and acceptance.

Not all crops may be amenable to speed breeding due to their unique growth requirements or biological limitations. Some crops may not respond well to the accelerated growth conditions, making this technique unsuitable for their breeding. Moreover, there are risks associated with the use of genetic engineering and advanced breeding techniques. Unintended consequences, such as unintended gene editing or unintended changes to the genome, could lead to unpredictable outcomes that may impact crop performance or safety. Ethical concerns also come to the forefront concerning the potential repercussions on non-target organisms and the environment (Sharma et al., 2022). Manipulating genes and traits through advanced breeding techniques raises questions about the potential unintended consequences on ecosystems and biodiversity.

Addressing these challenges and concerns requires thorough research, responsible application, and adherence to strict safety protocols. Collaborative efforts among researchers, policymakers, and stakeholders are crucial to ensure that speed breeding is implemented responsibly and ethically, harnessing its potential while safeguarding the environment and promoting sustainable agriculture.

3.6 FUTURE DIRECTIONS

3.6.1 Emerging Technologies and Innovations in Speed Breeding

Emerging technologies and innovations in speed breeding include:

Automated phenotyping: The utilization of automated phenotyping systems, employing sensors and imaging technologies, can significantly enhance the speed and accuracy of phenotypic analysis within speed breeding.

Machine learning algorithms: Machine learning algorithms offer the capacity to analyze extensive datasets derived from speed-breeding experiments. This enables the identification of intricate traits and the development of predictive models for future breeding endeavors.

Genome editing: Advances in genome editing techniques, such as CRISPR-Cas9, empower precise and targeted genetic modifications to enhance crop traits (Haroon et al., 2019).

Optogenetics: Optogenetics, a technology employing light to regulate gene expression in plants, allows for the precise manipulation of plant growth and development.

Controlled environment agriculture: Progress in controlled environment agriculture, encompassing vertical farming and aeroponics, offers a meticulously controlled and efficient growth environment conducive to speed breeding.

Overall, these emerging technologies and innovations hold great promise for further improving the efficiency and effectiveness of speed breeding for crop improvement.

3.6.2 Challenges to Wider Adoption of Speed Breeding

Challenges to the wider adoption of speed breeding include:

Regulatory barriers: The use of genetic engineering and other advanced techniques in speed breeding may face regulatory barriers, which can limit the adoption of these methods by growers and breeders.

Technical expertise: The implementation of speed breeding often requires technical expertise and specialized equipment, which can be a barrier to adoption, particularly for smaller farms or breeding programs.

Infrastructure: The implementation of speed breeding may require specialized facilities and equipment, which can be costly to set up and maintain.

Public acceptance: The use of genetic engineering and other advanced techniques in crop improvement may be met with public resistance, which can limit the adoption of these methods.

Intellectual property: The ownership and control of intellectual property related to speed breeding techniques and resulting plant varieties can be a complex issue, which can impact the adoption of these methods by smaller growers and breeders.

Addressing these challenges will be critical to the wider adoption and implementation of speed breeding, and will require collaboration and engagement among stakeholders in government, industry, and academia.

3.6.3 Future Research Directions

Future directions for research on speed breeding include:

Advancements in gene editing and genetic engineering techniques: Future research should focus on the continual development of gene editing and genetic engineering techniques. This will expand the scope of traits that can be targeted and modified through speed breeding.

Exploration of innovative growth conditions and environments: Researchers should explore innovative growth conditions and environments such as hydroponics, aeroponics, and vertical farming. These explorations can optimize speed breeding techniques for different crops and varying environmental conditions.

Integration with traditional breeding methods: There is potential in integrating speed breeding with traditional breeding methods such as MAS. This integration can result in even more efficient and effective strategies for crop improvement.

Standardization of protocols and methodologies: The development of standardized protocols and methodologies for speed breeding is crucial. This standardization can enhance the reproducibility and comparability of speed breeding experiments across different research groups.

Long-term impact assessment: Future research should also investigate the long-term impact of speed breeding on crop performance. This includes assessing aspects like yield, quality, and the crop's resilience to environmental stresses.

3.7 CONCLUSION

3.7.1 Summary of Speed Breeding as a Promising Tool for Crop Improvement

In conclusion, speed breeding stands out as a promising instrument for expediting crop enhancement, presenting substantial advantages over conventional breeding techniques. Through the fine-tuning of growth conditions and the incorporation of tissue culture and genetic engineering methods, speed breeding facilitates accelerated generation turnover and the precise targeting of a broader spectrum of sought-after traits. The successful deployment of speed breeding has already been evident in crops like wheat, barley, and chickpea, and this innovative approach holds the potential to make a substantial impact on the challenges surrounding global food security. While there are challenges and limitations to the adoption and implementation of speed breeding, including technical expertise requirements and regulatory barriers, continued research and development in this area has the potential to further optimize this approach for crop improvement. Thus, speed breeding stands as an exciting and promising realm of research for tackling the global food security challenges we face. Continued investment in this field holds the potential to bring about a revolution in the field of crop improvement in the upcoming years.

3.7.2 Potential Impact on Global Food Security

Speed breeding has the potential to make a significant impact on global food security challenges by accelerating the development of improved crop varieties. The projected global population surpassing 9 billion by 2050 signifies a substantial challenge in meeting the escalating demand for food. It is estimated that food production will need to surge by approximately 50% to adequately address this demand (Fang et al., 2021). Speed breeding indeed presents a promising solution to this formidable challenge. It facilitates the rapid and efficient development of crop varieties that boast heightened productivity, resilience, and adaptability to the ever-changing environmental conditions, ultimately contributing to addressing the increased demand for food in the coming decades. Additionally, speed breeding can help address challenges such as climate change, pest and disease pressures, and limited arable land, by developing crops that are more resistant to these challenges and can be grown in non-traditional environments. By improving crop yields and quality, speed breeding can also help to reduce food waste and increase food availability, particularly in regions that are most vulnerable to food insecurity. Thus, speed breeding holds substantial potential for making a considerable impact on global food security challenges. Sustained investment and ongoing research in this domain will play a pivotal role in cultivating enhanced crop varieties capable of meeting the escalating food demand anticipated in the forthcoming decades.

3.7.3 Call for Continued Research and Development of Speed Breeding Techniques

In summary, the potential advantages of speed breeding position it as an immensely promising tool for enhancing crop quality and addressing global food security concerns. Nevertheless, it is vital to acknowledge and tackle the associated challenges and constraints. Continuous research and development efforts dedicated to refining speed breeding techniques are pivotal in harnessing its full potential and ensuring its effectiveness in the realm of agriculture and food security. Research efforts should focus on optimizing growth conditions, improving the efficiency of tissue culture and genetic engineering techniques, and developing new approaches for the targeting of specific desirable traits. Additionally, efforts should be made to address regulatory and public acceptance barriers to the adoption of speed breeding, while also building capacity and technical expertise among researchers, breeders, and farmers. Committing resources to research and development in the field of speed breeding has the potential to expedite the creation of enhanced crop varieties. This, in turn, can effectively tackle the pressing issues surrounding global food security while also contributing to the promotion of sustainable agriculture and economic development. Therefore, there is a pressing need for continued investment and research in this area to ensure that speed breeding can be fully harnessed as a powerful tool for crop improvement and global food security.

REFERENCES

Abdul Fiyaz, R., Ajay, B. C., Ramya, K. T., Aravind Kumar, J., Sundaram, R. M., & Subba Rao, L. V. (2020). Speed breeding: Methods and applications. *Accelerated Plant Breeding, Volume 1: Cereal Crops*, *1*, 31–49.

Ahmar, S., Gill, R. A., Jung, K. H., Faheem, A., Qasim, M. U., Mubeen, M., & Zhou, W. (2020). Conventional and molecular techniques from simple breeding to speed breeding in crop plants: Recent advances and future outlook. *International Journal of Molecular Sciences*, *21*(7), 2590.

Alahmad, S., Dinglasan, E., Leung, K. M., Riaz, A., Derbal, N., Voss-Fels, K. P., …, Hickey, L. T. (2018). Speed breeding for multiple quantitative traits in durum wheat. *Plant Methods*, *14*(1), 1–15.

Ayenan, M. A. T., Danquah, A., Hanson, P., Ampomah-Dwamena, C., Sodedji, F. A. K., Asante, I. K., & Danquah, E. Y. (2019). Accelerating breeding for heat tolerance in tomato (Solanum lycopersicum L.): An integrated approach. *Agronomy*, *9*(11), 720.

Beebe, S. E., Rao, I. M., Cajiao, C., & Grajales, M. (2008). Selection for drought resistance in common beans also improves yield in phosphorus-limited and favorable environments. *Crop Science*, *48*(2), 582–592.

Begna, T. (2022). Speed breeding to accelerate crop improvement. *International Journal of Agricultural Science and Food Technology*, *8*(2), 178–186.

Bhargava, K., Abhishek, E., Madhusudhan, B., Naveen, A., Akhil, V. S., Yadav, T. V., …, & Kumar, G. A. (2023). A review of rapid generation advancement (RGA) in crop improvement. *International Journal of Plant & Soil Science*, *35*(7), 138–145.

Bhat, J. A., Feng, X., Mir, Z. A., Raina, A., & Siddique, K. H. (2023). Recent advances in artificial intelligence, mechanistic models and speed breeding offer exciting opportunities for precise and accelerated genomics-assisted breeding. *Physiologia Plantarum*, *175*, e13969.

Bridgen, M. P., Van Houtven, W., & Eeckhaut, T. (2018). Plant tissue culture techniques for breeding. *Ornamental Crops*, *11*, 127–144.

Cazzola, F., Bermejo, C. J., Guindon, M. F., & Cointry, E. (2020). Speed breeding in pea (Pisum sativum L.), is an efficient and simple system to accelerate breeding programs. *Euphytica*, *216*(11), 178.

Chiurugwi, T., Kemp, S., Powell, W., & Hickey, L. T. (2019). Speed breeding orphan crops. *Theoretical and Applied Genetics*, *132*, 607–616.

Cobb, J. N., Juma, R. U., Biswas, P. S., Arbelaez, J. D., Rutkoski, J., Atlin, G., …, & Ng, E. H. (2019). Enhancing the rate of genetic gain in public-sector plant breeding programs: Lessons from the breeder's equation. *Theoretical and Applied Genetics*, *132*, 627–645.

Collard, B. C., Beredo, J. C., Lenaerts, B., Mendoza, R., Santelices, R., Lopena, V., …, & Islam, M. R. (2017). Revisiting rice breeding methods – evaluating the use of rapid generation advance (RGA) for routine rice breeding. *Plant Production Science*, *20*(4), 337–352.

Dwivedi, S. L., Britt, A. B., Tripathi, L., Sharma, S., Upadhyaya, H. D., & Ortiz, R. (2015). Haploids: Constraints and opportunities in plant breeding. *Biotechnology Advances*, *33*(6), 812–829.

Edet, O. U., & Ishii, T. (2022). Cowpea speed breeding using regulated growth chamber conditions and seeds of oven-dried immature pods potentially accommodates eight generations per year. *Plant Methods*, *18*(1), 106.

Elferink, M., & Schierhorn, F. (2016). Global demand for food is rising Can we meet it? *Harvard Business Review*, *7*(4), 2016.

Fang, Y., Wang, L., Sapey, E., Fu, S., Wu, T., Zeng, H., …, & Han, T. (2021). Speed-breeding system in soybean: Integrating off-site generation advancement, fresh seeding, and marker-assisted selection. *Frontiers in Plant Science*, *12*, 717077.

Ferrie, A. M., & Polowick, P. L. (2022). Accelerated breeding for brassica crops. *Accelerated plant breeding, Volume 4: Oil crops*, pp. 139–163. https://doi.org/10.1007/978-3-030-81107-5_5

Gepts, P. (2002). A comparison between crop domestication, classical plant breeding, and genetic engineering. *Crop Science*, *42*(6), 1780–1790.

Ghosh, S., Watson, A., Gonzalez-Navarro, O. E., Ramirez-Gonzalez, R. H., Yanes, L., Mendoza-Suárez, M., …, & Hickey, L. T. (2018). Speed breeding in growth chambers and glasshouses for crop breeding and model plant research. *Nature Protocols*, *13*(12), 2944–2963.

Gudi, S., Kumar, P., Singh, S., Tanin, M. J., & Sharma, A. (2022). Strategies for accelerating genetic gains in crop plants: Special focus on speed breeding. *Physiology and Molecular Biology of Plants*, *28*(10), 1921–1938.

Haroon, M., Afzal, R., Idrees, F., Sunny, A., & Khan, A. S. (2019). Genome editing and speed breeding; game changers to boost the crop production. *International Journal of Biological Research*, *2*, 295–300.

Hickey, L. T., Germán, S. E., Pereyra, S. A., Diaz, J. E., Ziems, L. A., Fowler, R. A., …, & Dieters, M. J. (2017). Speed breeding for multiple disease resistance in barley. *Euphytica*, *213*, 1–14.

Hickey, L. T., N. Hafeez, A., Robinson, H., Jackson, S. A., Leal-Bertioli, S. C., Tester, M., …, & Wulff, B. B. (2019). Breeding crops to feed 10 billion. *Nature Biotechnology*, *37*(7), 744–754.

Hohmann, M., Stahl, A., Rudloff, J., Wittkop, B., & Snowdon, R. J. (2016). Not a load of rubbish: Simulated field trials in large-scale containers (Vol. 39, No. 9, pp. 2064–2073). 10.1111/pce.12737

Jähne, F., Hahn, V., Würschum, T., & Leiser, W. L. (2020). Speed breeding short-day crops by LED-controlled light schemes. *Theoretical and Applied Genetics*, *133*(8), 2335–2342.

Kapiel, T. Y. S. (2018). Speed breeding: A powerful innovative tool in agriculture. *Innovative Techniques in Agriculture*, *2*(3), 413–415.

Karthika, V., Babitha, K. C., Kiranmai, K., Shankar, A. G., Vemanna, R. S., & Udayakumar, M. (2020). Involvement of DNA mismatch repair systems to create genetic diversity in plants for speed breeding programs. *Plant Physiology Reports*, *25*, 185–199.

Lal, R. (2016). Feeding 11 billion on 0.5 billion hectare of area under cereal crops. *Food and Energy Security*, *5*(4), 239–251.

Laux, P., Jäckel, G., Tingem, R. M., & Kunstmann, H. (2010). Impact of climate change on agricultural productivity under rainfed conditions in Cameroon—a method to improve attainable crop yields by planting date adaptations. *Agricultural and Forest Meteorology*, *150*(9), 1258–1271.

Lenaerts, B., Collard, B. C., & Demont, M. (2019). Improving global food security through accelerated plant breeding. *Plant Science*, *287*, 110207.

Loman, L. J. (2018). *Vertical Farming Can It Change the Global Food Production Landscape?* Nuffield International Scholar.

Massel, K., Lam, Y., Wong, A. C., Hickey, L. T., Borrell, A. K., & Godwin, I. D. (2021). Hotter, drier, CRISPR: The latest edit on climate change. *Theoretical and Applied Genetics*, *134*, 1691–1709.

Misra, A. K. (2014). Climate change and challenges of water and food security. *International Journal of Sustainable Built Environment*, *3*(1), 153–165.

Mobini, S. H., Lulsdorf, M., Warkentin, T. D., & Vandenberg, A. (2015). Plant growth regulators improve in vitro flowering and rapid generation advancement in lentil and faba bean. *In Vitro Cellular & Developmental Biology – Plant*, *51*, 71–79.

Nagatoshi, Y., & Fujita, Y. (2019). Accelerating soybean breeding in a CO2 supplemented growth chamber. *Plant and Cell Physiology*, *60*(1), 77–84.

Naik, P. S., & Buckseth, T. (2018). Recent advances in virus elimination and tissue culture for quality potato seed production. *Biotechnologies of crop improvement, Volume 1: Cellular approaches*, pp. 131–158. https://doi.org/10.1007/978-3-319-78283-6_4

O'Connor, D. J., Wright, G. C., Dieters, M. J., George, D. L., Hunter, M. N., Tatnell, J. R., & Fleischfresser, D. B. (2013). Development and application of speed breeding technologies in a commercial peanut breeding program. *Peanut Science*, *40*(2), 107–114.

Okogbenin, E., Setter, T. L., Ferguson, M., Mutegi, R., Ceballos, H., Olasanmi, B., & Fregene, M. (2013). Phenotypic approaches to drought in cassava. *Frontiers in Physiology*, *4*, 93.

Pandey, S., Singh, A., Parida, S. K., & Prasad, M. (2022). Combining speed breeding with traditional and genomics-assisted breeding for crop improvement. *Plant Breeding*, *141*(3), 301–313.

Paradiso, R., & Proietti, S. (2022). Light-quality manipulation to control plant growth and photo-morphogenesis in greenhouse horticulture: The state of the art and the opportunities of modern LED systems. *Journal of Plant Growth Regulation*, *41*(2), 742–780.

Plesse, M. (2020). *Global food and water security in 2050: Demographic change and increased demand.* Perth W.A: Future Directions International.

Poorter, H., Fiorani, F., Stitt, M., Schurr, U., Finck, A., Gibon, Y., …, & Pons, T. L. (2012). The art of growing plants for experimental purposes: A practical guide for the plant biologist. *Functional Plant Biology*, *39*(11), 821–838.

Rai, A. C., & Rai, K. K. (2022). Speed breeding for rapid cycling of crops for stress management and global food security. In *Microbial biocontrol: Food security and post harvest management: Volume 2* (pp. 23–37). Cham: Springer International Publishing. https://doi.org/10.1007/978-3-030-87289-2_2

Rai, K. K. (2022). Integrating speed breeding with artificial intelligence for developing climate-smart crops. *Molecular Biology Reports*, *49*(12), 11385–11402.

Rana, M. M., Takamatsu, T., Baslam, M., Kaneko, K., Itoh, K., Harada, N., …, & Mitsui, T. (2019). Salt tolerance improvement in rice through efficient SNP marker-assisted selection coupled with speed-breeding. *International Journal of Molecular Sciences*, *20*(10), 2585.

Riaz, A., Periyannan, S., Aitken, E., & Hickey, L. (2016). A rapid phenotyping method for adult plant resistance to leaf rust in wheat. *Plant Methods*, *12*(1), 1–10.

Rommens, C. M. (2007). Intragenic crop improvement: Combining the benefits of traditional breeding and genetic engineering. *Journal of Agricultural and Food Chemistry*, *55*(11), 4281–4288.

Samantara, K., Bohra, A., Mohapatra, S. R., Prihatini, R., Asibe, F., Singh, L., …, & Varshney, R. K. (2022). Breeding more crops in less time: A perspective on speed breeding. *Biology*, *11*(2), 275.

Samineni, S., Sen, M., Sajja, S. B., & Gaur, P. M. (2020). Rapid generation advance (RGA) in chickpea to produce up to seven generations per year and enable speed breeding. *The Crop Journal*, *8*(1), 164–169.

Shahzad, A., Ullah, S., Dar, A. A., Sardar, M. F., Mehmood, T., Tufail, M. A., …, & Haris, M. (2021). Nexus on climate change: Agriculture and possible solution to cope future climate change stresses. *Environmental Science and Pollution Research*, *28*, 14211–14232.

Shamshiri, R., Kalantari, F., Ting, K. C., Thorp, K. R., Hameed, I. A., Weltzien, C., …, & Shad, Z. M. (2018). Advances in greenhouse automation and controlled environment agriculture: A transition to plant factories and urban agriculture. *Int J Agric & Biol Eng*,*11*, 1–22.

SharathKumar, M., Heuvelink, E., & Marcelis, L. F. (2020). Vertical farming: Moving from genetic to environmental modification. *Trends in Plant Science*, *25*(8), 724–727.

Sharma, S., Kumar, A., Dhakte, P., Raturi, G., Vishwakarma, G., Barbadikar, K. M., …, & Deshmukh, R. (2022). Speed breeding opportunities and challenges for crop improvement. *Journal of Plant Growth Regulation*, *42*, 1–14.

Shivakumar, M., Nataraj, V., Kumawat, G., Rajesh, V., Chandra, S., Gupta, S., & Bhatia, V. S. (2018). Speed breeding for Indian Agriculture: A rapid method for development of new crop varieties. *Current Science 115*(7), 1241.

Song, Y., Duan, X., Wang, P., Li, X., Yuan, X., Wang, Z., …, & Hong, D. (2022). Comprehensive speed breeding: A high-throughput and rapid generation system for long-day crops. *Plant Biotechnology Journal*, *20*(1), 13.

Stetter, M. G., Zeitler, L., Steinhaus, A., Kroener, K., Biljecki, M., & Schmid, K. J. (2016). Crossing methods and cultivation conditions for rapid production of segregating populations in three grain amaranth species. *Frontiers in Plant Science*, *7*, 816.

Supriya, K., Praveen, P., & Basandrai, D. (2023). Speed breeding: To expedite research and breeding in rice. *Himachal Journal of Agricultural Research*, *49*(1), 120–123.

Tester, M., & Langridge, P. (2010). Breeding technologies to increase crop production in a changing world. *Science*, *327*(5967), 818–822.

Tiwari, J. K., Changan, S. S., Buckseth, T., Singh, R. K., Singh, B., Luthra, S. K., …, & Kumar, M. (2021). High-throughput phenotyping in potato breeding. *High-throughput crop phenotyping*, 165–182. https://doi.org/10.1007/978-3-030-73734-4_8

Vermeulen, S. J., Challinor, A. J., Thornton, P. K., Campbell, B. M., Eriyagama, N., Vervoort, J. M., …, & Laderch, P. (2022). Speed breeding: Accelerated plant breeding. *Journal of Agriculture Research and Technology*, *36*, 36–39.

Virk, P. S., Andersson, M. S., Arcos, J., Govindaraj, M., & Pfeiffer, W. H. (2021). Transition from targeted breeding to mainstreaming of biofortification traits in crop improvement programs. *Frontiers in Plant Science*, *12*, 703990.

Wanga, M. A., Shimelis, H., Mashilo, J., & Laing, M. D. (2021). Opportunities and challenges of speed breeding: A review. *Plant Breeding*, *140*(2), 185–194.

Watson, A., Ghosh, S., Williams, M. J., Cuddy, W. S., Simmonds, J., Rey, M. D., …, & Hickey, L. T. (2018). Speed breeding is a powerful tool to accelerate crop research and breeding. *Nature Plants*, *4*(1), 23–29.

Wolter, F., Schindele, P., & Puchta, H. (2019). Plant breeding at the speed of light: The power of CRISPR/Cas to generate directed genetic diversity at multiple sites. *BMC Plant Biology*, *19*(1), 1–8.

Xu, Y., & Crouch, J. H. (2008). Marker-assisted selection in plant breeding: From publications to practice. *Crop Science*, *48*(2), 391–407.

Zheng, Z., Wang, H. B., Chen, G. D., Yan, G. J., & Liu, C. J. (2013). A procedure allowing up to eight generations of wheat and nine generations of barley per annum. *Euphytica*, *191*, 311–316.

4 Implementation of Novel Genomic and Biotechnological Interventions for Accelerated Breeding of Crops

Muhammad Ramzan Khan, Nazia Rehman, Safeena Inam, Muhammad Kashif Naeem, Amna Muhammad, Muhammad Uzair, Aamir Riaz, Obaid Ur Rehman, Fara Muqaddas, Maryam Murtaza, Bilal Saleem, and Muhammad Aqeel

Functional Genomics and Bioinformatics Program, National Institute for Genomics and Advanced Biotechnology, National Agricultural Research Centre, Park Road, Islamabad, Pakistan

Muhammad Zahid

Wheat Program, Crop Sciences Institute, National Agricultural Research Centre, Park Road, Islamabad, Pakistan

4.1 INTRODUCTION

From the current trends of population growth, it is anticipated that a one-third increase in population will be eminent by 2050. The situation is further aggravated by the fact that developing countries will share a major portion of this population explosion. To feed this population – with an unprecedented growth trend of current income and consumption – the agricultural production growth of 60% is inevitable. The agricultural research needs a thorough revamping plan as major transformations are needed for economic growth based on sustainable agricultural production. Another obstacle in the way to sustainable agricultural production is the imminent challenge of climate change. Weather extremes and aberrant weather patterns have adversely impacted agriculture and food security. The outcome is a reduction in production leading to lower incomes of resource-poor farmers in vulnerable areas. Hence, climate change is an utter reality now. It seems to be the biggest danger to global food security, sustainable development, and poverty reduction. Plant breeders and scientists are leaving no stone unturned in developing improved crops in apropos of higher yield, better nutrient values, resistance to pests and pathogens, and climate resilience. Up till now, conventional breeding has played a pivotal role in developing improved crop varieties. But it involved time-consuming laborious processes including growing and examining large populations of crops over several generations often coupled with the co-integration of lethal genes, i.e., linkage drag. The limitation of linkage drags demands multiple times of backcrossing and selection to re-establish the elite background (Lidder & Sonnino, 2012).

DOI: 10.1201/b23372-4

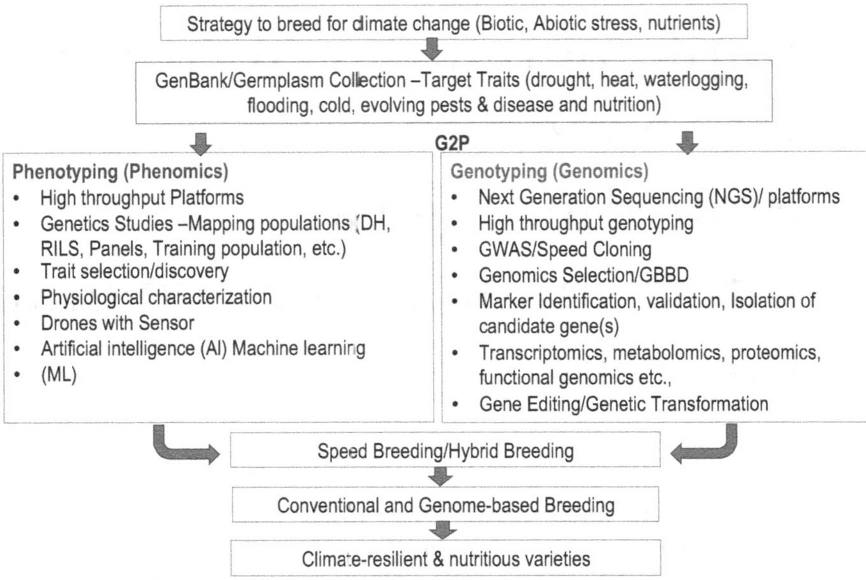

FIGURE 4.1 Strategy for implementation of novel breeding technologies (NBTs) in agriculture. All the technologies required from genomics to phenomics (G2P) are enlisted.

There is a dire need to employ new technologies of carbon farming, cover cress, regenerative agriculture, hybrid, biological alternative, and digital solutions in agriculture systems. Concerning genetic improvement coupled with climate resilience, genome-based Novel Breeding Techniques (NBTs) offer substantial advantages over conventional as well as other breeding techniques (Figure 4.1). The genome-based NBTs can lessen the stress of the farmers and plant breeders in offsetting the impact of climate change and adapting to its effects, helping to improve agriculture and the resilience of food production systems, thus sustainably leading to food security.

Recently, a plenitude of NBTs has culminated the modern agriculture. One of the phenomics-related technologies is high-throughput phenotyping using drones and thermal sensors with accurate data recording devices (Rahman et al., 2021). The availability of cheaper and cheaper ultra-high-throughput DNA sequencing with a number of breeding applications has facilitated the breeders to have a complete profile of DNA mutations in the traits of interest (https://www.idtechex.com/en/research-report/dna). Pre-breeding and GWAS have emerged out of the next-generation sequencing (NGS) platforms leading to the discovery of new traits controlled by new genes (Reddy et al., 2023). The K-mers-based association genetic has resulted in the speedy cloning of important genes from wild-type relatives (Aylward et al., 2023). Comparative genomics specifically pan-genomics has resulted in the discovery of larger mutations underlying the important traits for improvement in breeding climate and disease-resistant varieties. Through genomic selection (GS), breeders get emancipation from laborious time-consuming manually obtained phenotyping data. Estimation of breeding values through different algorithms provides accurate information about crops without the requirement of phenotyping. Genome-based breeding by design using GS and genome editing has helped to target the focused traits with precise manipulation for desirable phenotypes (Chen et al., 2023). Remarkably, the advent of speed breeding (SB) in the form of getting 5–6 generations per year by using grow lights has reduced the variety developing time to less than half (Swami et al., 2023).

This genomics to phenomics transition augmented with SB is the most ideal scenario to transform more productive agriculture systems, use inputs more efficiently, have less variability and greater stability in their outputs, and are more resilient to risks, shocks, and long-term climate alterations. This will lead to a path of sustainable agriculture for food and feed security of this planet in a very short period.

4.2 HIGH-THROUGHPUT PHENOTYPING FOR PHENOMICS DATA COLLECTION

The G2P and digitalization projects have instigated the requirement of high-throughput phenotyping. Off-late, both the ground and unmanned aerial vehicles (UAVs) are in regular use in phenomics data collection.

4.2.1 Unmanned Aerial Vehicles

UAVs are one component of a suite of smart-farming technologies being evaluated for use in agriculture (Table 4.1). Due to their versatility, agricultural UAVs are now widely used for everything from spraying pesticides and fertilizers to planting seeds and measuring and mapping crop development (Rahman et al., 2021). Autonomously flown UAVs with both multispectral/hyperspectral and RGB sensors capture aerial views of the crops. Images are handled either after capture or while in transit. Images after processing may be used to spot sick plants. Both UAVs and UGVs may benefit from aerial data for tasks like crop monitoring, harvest prediction, and more (Bhandari et al., 2017). The Association for Unmanned Vehicle Systems International (AUVSI) estimates that by 2020, more than 2,900 UAVs will be in operation across more than 900 businesses in over 100 countries. AV technology advancements have enabled manufacturers to create a diverse lineup of models in a variety of sizes, weights, and forms, all with the capacity to carry a broad variety of sensor payloads (del Cerro et al., 2021). Adapted to improve quality by employing different software to search the target region normally and accurately and properly detect unusual things. An agricultural UAV makes use of sophisticated image processing to keep track of crucial data like plant growth, soil quality, insect management, and more. UAVs can swiftly scan a region larger than a few kilometers and offer sufficient long-range connectivity, with or without payloads. There is a wide range of UAVs available, each with its own unique set of capabilities and aerodynamic designs. Capacity for speed and load varies widely amongst UAV types (Rahman et al., 2021).

4.2.2 Drones

Drones, also known as UAVs, unmanned aircraft systems (UAS), and remotely piloted aircraft, are becoming more important in the remote-sensing industry for a number of reasons (Zhang & Zhu, 2023). For instance, even on overcast days, drones can provide high-quality, high-resolution imagery. Drones are gaining traction in the farming sector as a result of the data they can collect, analyze, and provide to farmers (Abbas et al., 2023). Some examples of drones being put to use in the farming industry are as follows: Drones with cameras and other sensors may be used for remote sensing to learn more about fields, animals, and the environment. By collecting and analyzing this information, farmers can track crop development, pinpoint nutritional deficits, and spot pests and illnesses. Third, precision agriculture: by flying drones to collect data, farmers can make more informed decisions about watering, fertilizing, and otherwise managing their crops. Precision agriculture describes this method (Delavarpour et al., 2021).

Drones may also keep tabs on animals by following their movements and even counting them. Farmers may use this information to foresee issues and fine-tune their management strategies.

Saving time: By giving farmers a bird's-eye view of their crops and cattle, drones may help them save time by enabling them to immediately see problems and make educated choices. Larger farms may benefit greatly from this.

TABLE 4.1
Unmanned Aerial Vehicles (UAVs) Used in Agriculture Sectors

No.	Author	UAV Type	Research Work	Year Published	Possible Future Agrarian Opportunity
1	(Kim et al., 2019)	–	Investigated enhanced learning and applications for agricultural UAVs, control processes, facilities, and upgrades. The development of UAVs and their use in agriculture are also discussed.	2019	Crop monitoring and agricultural spraying
2	(Popescu et al., 2020)	Quad-rotor	Proposed a hierarchical agricultural crop monitoring system based on UAV-WSN cooperation.	2019	Crop monitoring
3	(Giordan et al., 2020)	–	Monitoring systems, standard methods, and practical engineering geology goals were described.	2020	Crop monitoring
4	(Spoorthi et al., 2017)	Quadcopter	Created a refurbished UAV to smoothly spread pesticides and reduce producer stress.	2017	Agricultural spraying
5	(Shilin et al., 2017)	3 Single rotor UAVs and a six-rotor UAV	Examined four Chinese pesticide-applying UAVs for enforcement.	2017	Agricultural spraying
6	(Tsouros et al., 2019)	–	UAV precision agricultural applications were examined.	2019	Crop monitoring and agricultural spraying
7	(Fernández-Guisuraga et al., 2018)	Octocopter	Assessed the pros and cons of using UAVs to produce ultra-high-resolution multispectral ortho-mosaics for monitoring large burned regions.	2018	Crop monitoring
8	(Hartanto et al., 2019)	–	Focused on the logical potential of UAVs for farming and agro-industry to boost economic production and reduce costs.	2019	Crop monitoring and livestock farming
9	(Chen et al., 2020b)	Quadcopter and fixed-wing	Used a UAV to photograph harvest fields in high detail. Areas were allocated using software-based image analysis.	2020	Crop monitoring
10	(Mammarella et al., 2018)	Fixed-wing	A UAV gyro-pilot direction and tracing control method is suitable for multi-scenario mapping.	2018	Crop monitoring
11	(Zhang et al., 2017)	Small UAV	Created a reliable UAV platform to detect rice sheath blight.	2017	Crop monitoring
12	(Yi et al., 2017)	Large Fixed-wing UAV	Used a multidimensional technique to precisely analyze the UAV system to demonstrate its versatility in overhead transmission line applications.	2017	Crop monitoring
13	(Olejnik et al., 2020)	Small Helicopter	Utilized remote sensing to demonstrate the possibility of using a small aircraft for certain purposes. Additionally, this technology is used to monitor gas transportation and network design.	2020	Crop monitoring and livestock farming

14	(Kim et al., 2020)	Hexacopter	Airborne mixed reality for first-person vision UAV soaring is carefully engineered to provide an enthusiastic and safe experience for UAV users by adding practical impediments to open field operations.	2020	Crop monitoring
15	(Budiharto et al., 2019)	Multi-rotor (quad-rotor)	Described UAV research and applications in agriculture, delivery, and GIS. Field mapping and pesticide spraying were routine.	2019	Crop monitoring and agricultural spraying
16	(Jeziorska, 2019)	Fixed-wing and multi-rotor	The operating system, enactment, scientific adaptability, and information processing techniques needed for wetland monitoring and hydrology were reviewed.	2019	Crop monitoring
17	(Hashimoto et al., 2019)	–	Highlighted realistic paddy field reflective and vegetation indicators simulation.	2019	Crop monitoring
18	(Poley & McDermid, 2020)	–	Explained that spectral and structural data imaging may accurately assess vegetation biomass from several angles.	2020	Crop monitoring
19	(Messina & Modica, 2020)	–	Revised agricultural thermal remote sensing and UAV use and prospects.	2020	Crop monitoring
20	(Ayhan et al., 2020)	–	Suggested using NDVI calculation and machine learning to identify vegetation.	2020	Crop monitoring
21	(Chen et al., 2020)	Quadrotor	Examined that a tiny UAV nozzle may improve planthopper control.	2020	Agricultural spraying
22	(Agapiou, 2020)	–	Explored visible vegetation indices in different situations using UAV sensors and cameras to verify execution.	2020	Crop monitoring
23	(Gao et al., 2020)	Quad-rotor	Outlined a system to explain the relationship between pests, illnesses, and weather.	2020	Crop monitoring
24	(Singh & Frazier, 2018)	–	Field-based plant phenomics remote sensing applications.	2022	Crop monitoring

4.2.3 Sensors and Tools

Drones used in the agricultural sector may be outfitted with a diverse array of sensors and equipment, enabling them to collect data and provide farmers with vital insights (Table 4.1). The following enumeration comprises a selection of sensors and technologies that are applicable in the field of drone-based agriculture.

- Visual (RGB) sensors: RGB sensors, also known as visual sensors, are capable of capturing pictures within the visible spectrum. These sensors have the potential to be used in monitoring crop development and identifying changes in vegetation (Jafarbiglu & Pourreza, 2022).
- Multispectral sensors: These sensors are capable of capturing pictures over many narrow bands of the electromagnetic spectrum. They have the potential to be used for the identification of nutrient shortages, the detection of pests and illnesses, and the monitoring of crop health (Lu et al., 2020).
- Hyperspectral sensors: The mentioned sensors can collect pictures over many narrow bands within the electromagnetic spectrum. This functionality enables their use in the identification of distinct plant species and the detection of minute alterations in vegetation.
- Thermal cameras: These cameras can measure crop temperature and water stress by capturing heat radiation.
- Lidar: Lasers make 3D maps of fields to assess crop height and detect topographical characteristics (Mertens et al., 2021).
- Optical sensors: These sensors can identify crop disease and other issues.
- Liquid level sensors: These sensors can be utilized to apply substances such as pesticides and fertilizers to crops (Fahey et al., 2020).

4.3 ACCELERATING THE CONVENTIONAL BREEDING THROUGH GWAS AND DNA MARKERS

Conventional breeding, known as classic or traditional breeding, is a method of crossing parents with desirable traits to develop new crop varieties. In the past, conventional breeding played a key role in improving agricultural productivity by modulating the canopy architecture and improving nutrient-use-efficiency, resulting in the "Green Revolution". Development of high-yielding crop varieties through a conventional breeding approach is a laborious, time (10–13 years), and resource-consuming process. Current breeding strategies for developing new crop varieties rely on selection and crossing within the native gene pool. These breeding methods negatively impact crop diversity, resulting in a narrow genetic base. Domestication and further deliberate human-mediated selection to develop high-yielding and adaptability to diverse climates results in dramatic loss of genetic variation (Figure 4.2). Genome-wide comparative analysis of wild, landraces, and cultivated populations revealed a dramatic genetic loss occurred in rice, cotton, and soybean domestication (Zhou et al., 2015; Wang et al., 2016; Fang et al., 2017).

To improve crop productivity, there is a need to adopt the pre-breeding strategy to broaden the gene pools by collecting diverse germplasms. Collection and exploitation of diverse germplasm is the key to any breeding program for enriching the crop germplasm. Keeping in view a large number of gene banks have been curated among them CIMMYT in Mexico, the National Small Grains Collection in the United States, and the National Genbank in China. In an effort, CIMMYT has a wheat collection of over 3,000 germplasm including diploid and tetraploid wild species and pre-breeding cultivars, and recently has a core collection of 600 durum and bread wheat accessions collected from over 25 countries representing major wheat dense cultivation regions (Velu et al., 2011). Enriching the gene pools with wild species and

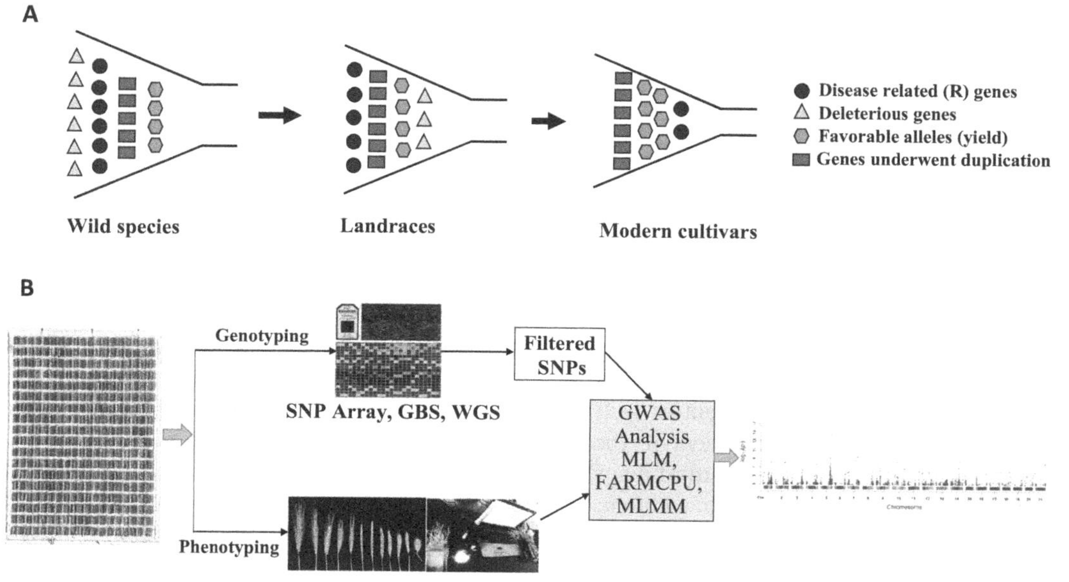

FIGURE 4.2 (A) Narrow genetic base of modern cultivars due to impact of domestication and selection. (B) GWAS pipeline for hunting novel candidate genes.

landraces from major and diversified growing areas, especially the Fertile Crescent area is a useful source. First plant domestication occurred in a few discrete regions known as the center of origin or Fertile Crescent area and later widespread via human migration and/or trade (Luo et al., 2007). These germplasm resources reveal the large genetic variations for valuable traits such as disease resistance, high grain quality (Fe and Zn enriched), drought, and heat tolerance (Figure 4.2A). Exploitation and evaluation of these valuable genetic resources will help the breeders to improve the crop yield, disease resistance, quality, and resilience to harsh environmental conditions of commercial crop varieties. Recent advancement in pre-breeding information rapidly improves the genetic stocks for breeders to develop and improve the trait-specific varieties.

SB is a recently emerging technique to improve the rapid advancement of breeding cycles enabling the wheat breeder to attain up to six generations in one year for spring wheat (Watson et al., 2018). Employing SB in backcrossing or synthetic development breeding programs for rapid generation will significantly speed up the development of new breeding lines. Further combining the SB and marker-assisted selection (MAS) approaches for crop improvement will revolutionize the conventional breeding program. Integration of these techniques will leverage the cost and time for generation turnover and will open new endeavors for directional breeding programs.

Integration of advanced genomic technologies, i.e., MAS, quantitative trait loci(QTL) mapping, genome-wide association study (GWAS), GS, and SB in current breeding pipelines speed up the breeding programs. MAS is a potent tool in contemporary breeding for the development of functional markers. MAS profoundly improves the selection efficiency and reduces the field cost. In developing countries, the integration of MAS in breeding significantly impacts genetic gains by improving economically important traits in major food crops. MAS is employed in different breeding strategies including germplasm screening, marker-assisted backcrossing, marker-assisted recurrent selection, and gene pyramid (Rossetto & Henry, 2014). In the past introgression of natural variations for quality traits through MAS has been evidenced successfully in wheat (Kumar et al., 2010; Kumar et al., 2011; Tyagi et al., 2014). The fundamental application of marker-assisted breeding is the rapid improvement of underpin traits in breeding lines.

4.3.1 Advancement in Marker Platforms

In the past low throughput markers such as AFLPs (amplified fragment length polymorphisms), SSRs (single sequence repeats), and ESTs (expressed sequence tags) markers were employed for QTL mapping. With the advancement and availability of crop reference genomes, high-throughput single-nucleotide polymorphism (SNP) markers were employed for genetic studies. SNP markers have become essential for both genetic studies and breeding programs over the past 10 years. GWAS and QTL mapping exponentially proliferated accompanied by the rapid development of breeding chips initiated, but still, many crops' genome coverage of these chips is quite low. There is an urgent need to develop and employ new breeding chips that have dense coverage of the genome, are effective in capturing the major and minor alleles, and more importantly cost-effective (as this issue addressed in detail by Rasheed et al. (2017)). In the case of wheat, a complete annotated reference genome recently changed the scenario for wheat research and breeding (Rasheed, 2024), acting as a tool for disruptive innovation in wheat improvement as well as a base for expediting wheat research and genomic-assisted breeding. The development of wheat high-density SNP array, i.e., 35 K, 55 K, 90 K, 384 K, 660 K, and 820 K, provided the most innovative tool for candidate gene searching by QTL mapping or GWAS (Wang et al., 2014). In addition, there are different NGS sequencing-based platforms like WGS, GBS, DArT-seq, REST-seq, and SBG (sequence-based genotyping) implemented in diversity studies, crop genomics, and population genomics to unravel the crop domestication, evolution, gene flow, and selective pressure (Table 4.2). Further, to reduce the cost, and burden of handling large SNPs targeted sequencing approaches were implemented in crop genomics studies such as GBTS, SLAF-seq, and exome capture.

The freshly released 18 common wheat and four wheat progenitor genomes will open new horizons in wheat genomics. These genomes may also help decipher complicated marker-trait correlations, and hidden genetic variations, making it easier to search for and identify the genes responsible for key agronomic features (Clavijo et al., 2017). Development of pangenome sequence based on these available genome resources will help to effectively capture rare alleles in wild and landraces diverse genetic stocks.

4.3.2 Linkage Mapping Success in Fast-Forwarding Genetic Gain in Cereals

At the beginning of the 21st century, association mapping started to be employed in crop plants. Recent advances in genotyping and sequencing technologies have resulted in the development of an

TABLE 4.2
SNP Platforms Utilized in Crops

Platforms	Technologies	Marker Density
Chip array (in Wheat)	BeadXpress	≤384 SNPs
	GoldenGate	≤3072 SNPs
	Illumina Infinium	55 K
	Illumina iSelect	90 K
	Affymetrix Axiom	35 K, 55 K, 660 K, 820 K
Sequence-based genotyping	WGS	Depends on coverage depth
	GBS	Same as above
	Exome capture	Same as above
Targeted sequencing	GBTS	≥ 3 K
	SLAF-seq (Illumina)	≥ 2 K
	Exome capture	60–150 K

obvious approach, GWAS, which overcomes the drawbacks of QTL mapping (Figure 4.2B). GWAS provides an effective strategy to fine mapping the QTL region by using a huge number of natural recombinant events, which cause rapid linkage disequilibrium (LD) decay and ease in identifying the candidate gene. LD is the foundation of QTL and GWAS mapping, biparental and natural populations are used as resources for linkage mapping, respectively (Yu et al., 2006; Zhang et al., 2010; Zhou & Stephens, 2012). LD is the feature of SNPs on a genomic region that explains the extent to of an allele of one SNP is correlated with an allele of another SNP within a population. LD extent explains the range of genomic stretch, carrying the candidate gene. LD decay mainly depends on the existence of natural recombinant events and population size. Population size increases as a result of rapid LD decay. QTL and GWAS mapping studies help to exploit important natural variations and haplotypes. Identification of major QTLs or natural variations with larger phenotypic effects will accelerate the development of new crop varieties. GWAS has emerged as a powerful approach to understanding the genetic architecture of traits and identifying genes underlying complex traits (Wang et al., 2016; Yano et al., 2016; Mwadzingeni et al., 2017).

Natural diverse populations are used to detect the association between SNP markers and traits. The power of GWAS depends on the phenotypic variance within a population (either common or rare variant), marker density, and genome coverage. Accurate and precision phenotyping with normal distribution and accompanied by high heritability is essential for detecting accurate marker-trait associations. Field experimentation must account for the environmental impact on the target trait. For this purpose, field experiments should be conducted with three replications, multiple locations, and at least two consecutive years. Association among SNP markers and traits was evaluated using different statistical approaches like GLM, MLM, CMLM, MLMM, EMMA, FarmCPU, etc., having different algorithms and power. Improvement in statistical approaches also resolves the issue of statistical power. With time, robust statistical methods and approaches have been employed to handle the large genotyping data and improve the statistical power (Table 4.3). The mixed model approach is the most routinely used GWAS tool in crop plants and it counts genetic relatedness and population structure. These mixed linear models are single variant methods, while recently Liu et al. (2016) developed the new multi-variant test model "FarmCPU" and considered the population structure. It has the power to detect two or more independent causal SNPs, which exist together in the LD block and inherit together. Multi-variant tests have the potential to detect rare variants by avoiding synthetic association (Lee et al., 2014). In the future, the development of statistical methods would help to overcome genetic confounding, and complex population structure and understand the genetic architecture of the underlying trait.

Previous studies of GWAS provide dense genome coverage and allow for accurate identification of loci in different plant species (Atwell et al., 2010; Xuehui et al., 2010; Tian et al., 2011). Several GWAS studies identified the significant SNPs and functionally evaluated the causal gene(s) in rice and maize. An example from rice GWAS, Yano et al. (2016) performed the GWAS for heading date

TABLE 4.3
GWAS Progress in the Development of New Methods and Statistical Approaches

	Method	Year	Computational Speed	Statistical Power
Single variant tests	MLM	2006	Low	High
	EMMA	2008	Intermediate	High
	FAST-LMM	2011	Fast	High/ intermediate
	MLMM	2012		
	SUPER	2014	Fast	High
Multi-variant tests	FarmCPU	2016	Fast	High

and identified two novel QTL regions and functional studies proved the involvement of these candidate genes in plant heading date. Further, the exploitation of wild progenitors and landraces of crops will identify the novel genes. These novel genes help to understand the selection sweeps, crop domestication patterns, and geographic adaptation loci.

4.3.3 Natural Variation-Based KASP Markers for Genes Pyramiding

GWAS paved the way to utilize the genomic-informed pre-breeding information for the traits of agricultural importance and utilized such natural variations to improve the crops. The association SNP of underlying trait also helps breeders to detect the potential parents for the future crossing of specific underlying traits. Parents carrying the same haplotype and belonging to different breeding pools should be crossed to enhance the expression of the trait. GWAS gained information on desirable haplotype prevalence in different breeding pools and provided reliable and direct information regarding the selection of appropriate parents for breeding. SNP-based marker systems, i.e., CAPS, TaqMan, STARP, and KASP were used for introgressing the desirable trait in respective elite varieties. Taking maize pro-vitamin, A biofortification is a successful example of the utilization of natural variation from GWAS. Genome-wide association analysis of vitamin-A identified the two candidate genes LcyE and crtRB1, favorable alleles controlled by rare variants (Harjes et al., 2008; Yan et al., 2010). Introgression of these rare alleles into cultivated high-yielding varieties with the help of marker-assisted breeding, results in elevated levels of pro-vitamin A in maize. These lines help to decline the vitamin A deficiency in African children.

Side by side such information about the prevalence of desirable natural variation (SNP)/haplotypes for multiple traits helps the breeder for gene pyramiding. A pile of desirable haplotypes in a single cultivar easily boosts the trait of interest significantly. In this regard, recently scientists have adopted the breeding by design strategy in rice to pyramid the multiple traits using genomic information and develop the "Green Super Rice" population. These lines showed tolerance against biotic and abiotic stresses along with stay-green traits and high yielding.

4.4 SPEED CLONING TO DEVELOP THE CLIMATE-RESILIENT CROPS

Improvement in agriculturally important traits is among the topmost breeding priorities of crops for agriculture scientists. Advancements in genomic technologies have upgraded our understanding of the genetics of complex traits, and genomic regions associated with traits were used in marker-assisted breeding and cloning of genes. Among multiple sequencing methods, the whole-exome sequencing approach is more promising for identifying the variations in the exonic regions and consequently increases the identification of target gene mutations responsible for particular traits (Jaganathan et al., 2020). Presently several targeted genome sequencing methods are available to accelerate gene identification and cloning.

4.1 Generation of Bait Library and Capture-Based Target Enrichment

NGS technology has also revolutionized the target capture sequencing technique. Several approaches related to the library preparation and capturing of target genomic regions with different bait categories are now accessible on a commercial level. The main focus of target capture sequencing is on functional regions of the genome and currently, whole-exome sequencing is routinely being used with more coverage, cost-effectiveness, and ease of managing the data (Zhou et al., 2021). Generally, for target enrichment, the two projecting methodologies are currently being employed: (1) PCR-based or amplicon-based; (2) hybridization capture-based. These target enrichment methods are based on the same basic principles, however, with time multiple improvements and variations have evolved and coupled with their advantages and limitations (Singh, 2022). At present, various alternative platforms are available for capturing the targets based on bait types, such as single and double-stranded DNA, single and double-stranded RNA baits, and mixed bait comprising of dsRNA and ssDNA (Zhou et al., 2021). Increasing genomic knowledge and procedures have unraveled the genetics of complex traits and identified the association of genomic regions with particular traits. This trait-based genomic information has been utilized for marker-assisted breeding and accelerated cloning of QTLs and genes. Till now, disease-resistant traits conferred by resistance (R) genes have been preferably studied for targeted whole-exome sequencing to detect the variabilities and hence identify targeted mutations in disease-causing genes.

Since the last decade, a number of targeted genome sequencing approaches are available for speedy gene identification and cloning such as (a) RenSeq, Resistance gene enrichment, and sequencing method that enables the identification of disease resistance NBS-LRR gene family members in plants by targeted re-sequencing. This approach was based on the preparation of a customized NBS-LRR RNA bait-library followed by enrichment and re-annotation of target sequences (Figure 4.3A). Jupe et al. (2013) utilized this technique in potato and tomato for rapid mapping of resistant loci (b) MutRenSeq, mutational R gene enrichment sequencing exploited by

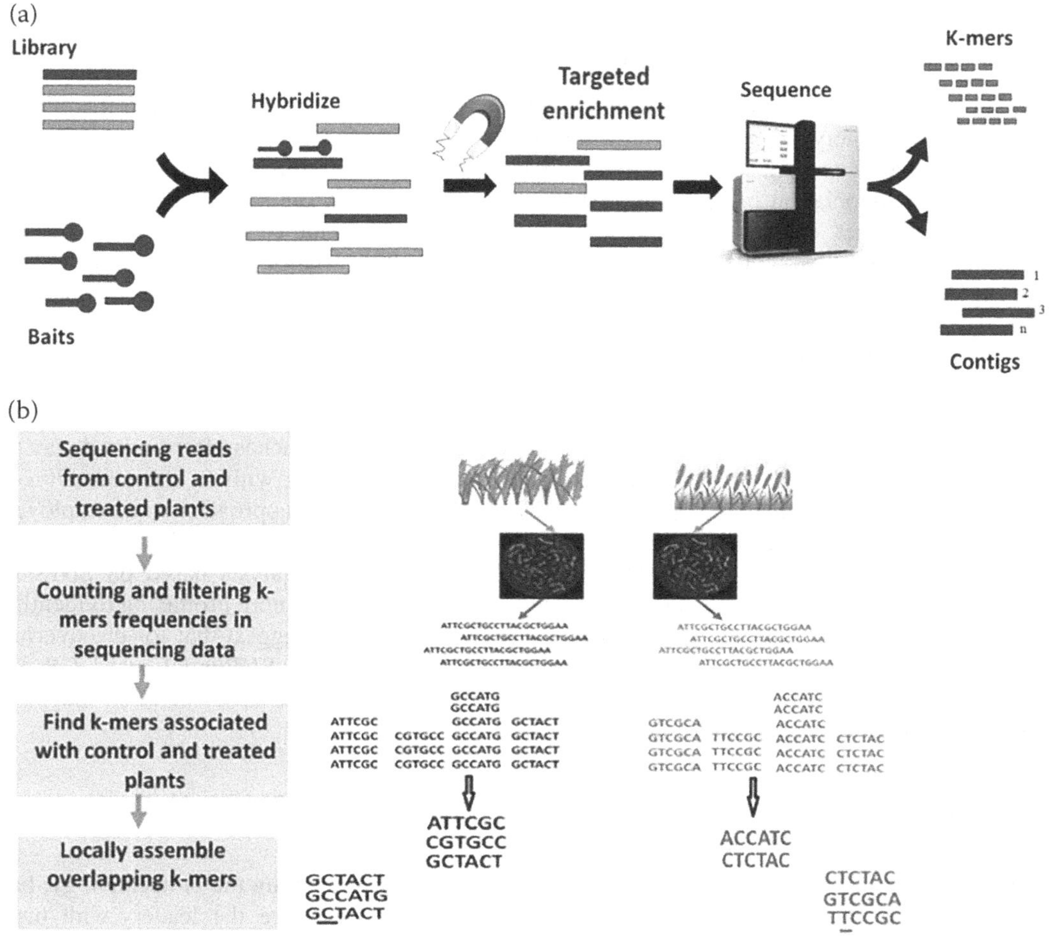

FIGURE 4.3 (A) Schematic representation of the generation of bait library and capture-based target enrichment for fine mapping and candidate genes cloning. (B) Workflow of k-mer-based association mapping for rapid gene cloning.

Steuernagel et al. (2016) which comprises chemical mutagenesis accompanied with targeted exome capture and sequencing for faster cloning of rust resistance genes in wheat. (c) SMART-RenSeq, single-molecule real-time RenSeq was employed by Witek et al. (2016) for cloning a resistant gene (Rpi-amr3i) for conferring resistance against late blight in potato (d) AgRenSeq, a technique for prompt cloning of R genes consisting of combined approach of association genetics and enriched R genes sequencing (Arora et al., 2019).

This strategy encompasses screening of wild relatives for disease resistance and sequencing to discover the resistance genes. AgRenSeq was applied in wheat for the rapid cloning of uncharacterized stem rust resistance R genes by using variability of wheat pan-genome. Apart from wheat, this strategy can be applied to other crops having germplasm diversity for speedy cloning of genes related to complex traits. All these studies have shown the effectiveness of RenSeq for enhancing disease resistance in plants. Applications of this technique for rapid cloning of abiotic stress-resistance genes will effectively improve crop tolerance against abiotic stress. These targeted sequencing methodologies reduce the utilization of resources and time for rapid gene cloning compared to whole-genome resequencing if prior knowledge about the gene or loci of interest is available. The projects that were accomplished in 10–15 years to fine-mapping, and cloning of QTL/gene now can be ended within 10–20 months.

4.2 K-mer-Based Association Mapping

Association mapping is generally carried out by GWAS through the detection of SNPs. Organisms are genotyped based on a group of SNPs using an SNP array which is ultimately verified for significant phenotypic association for particular traits through statistical analysis. However, this association mapping approach relies on a reference genome for designing SNP arrays and requires prior knowledge of SNPs location (Rahman et al., 2018). Moreover, structural variations like copy number variations and insertion-deletions (indels) are generally overlooked and there is a chance of rare variants for a phenotype that is not present on the SNP chip. These limitations are overcome by the k-mer-based association approach which has presently been developed. The bioinformatics term k-mers refers to the subset of sequences of a fixed length *k* acquired from sequencing dataset for instance, from the nucleotide string GACCTCA the following *k*-mers of length four (4-mers) can be obtained: GACC, ACCT, CCTC, and CTCA (Luo et al., 2007). *K*-mer-based association approaches offer a broad way to evaluate variations based on SNPs, indels, and structural variations without the necessity of specifically identifying and genotyping these variants. Moreover, this approach can be employed for those species whose genomes have not been sequenced. Lemay et al. (2023) conducted a study to compare k-mer-based GWAS and conventional GWAS analysis based on SNPs in soybean for 13 traits. The significantly associated K-mer results were more promising to identify underlying variants or candidate genes. Taken together, the results suggest that for discovering candidate genes for particular traits in crops, coupling k-mer- and SNP/indel-based GWAS approaches is a powerful method. A general workflow for K-mers-based association mapping analysis is presented in Figure 4.3B.

4.5 NEXT-GENERATION SEQUENCING, GENOMIC SELECTION, AND BREEDING BY DESIGN

With the passage of time, cheaper and cheaper NGS, besides culminating the biotechnology, has thrived in plant breeding with a greater pace. Illumina and MGI are the leaders with their NovaSeqX and DNBSEQ-T10x4RS, respectively as their ultra-high throughput sequencing platforms in the market. New advancements in sequencing technologies, chemistry, and read sizes have emerged at a rapid pace. Particularly, the single-cell sequencing is a new addition in sequencing technologies. Interestingly, different commercial companies are offering solutions to agricultural genomics, which are rapid, cost-effective, and have higher accuracy of sampling, library preparation, and eventual sequencing. They are sponsoring competitive grants in agricultural genomics applications, i.e., Whole genome re-sequencing, Exome sequencing, GBTS and Low pass, and various other applications. Some companies such as MGI are even offering customized ATOPlex kits for PCR-based NGS library for indigenous crops and livestock breeds. The principle behind developing these kits is that variations in crop and animal genomes contain rich potential as they control not only crop productivity but also resistance against diseases. A software program is developed for the screening of SNP-rich genomic regions. There are the following steps to capture the landscape of SNP-rich genomic regions in crops and animals:

- It scans all chromosomes one by one, covering the whole genome.
- During the scan, it cuts the genomic regions initially up to 100 bp nucleotides.
- Then it successfully locates SNPs in each 100 bp genomic segment and memorizes the SNP sum implicitly.
- It ranks the genomic segments while keeping the highest SNPs containing segments on top and similarly keeps them in decreasing order.
- SNP collection in each crop or animal accounted for the top 12,000 genomic segments, each of 100 bp.

Then these soft files are sent to sequencing companies such as MGI to synthesize the oligos and develop the PCR-based NGS indigenous library kit for genotyping of crop and animal samples. For larger genomes, GBTS is the best cost-effective application to be used in plant breeding through GS or genome-assisted breeding. There is a difference between the terms "genomic selection" and "genome-wide association studies". The latter aims to provide information on pre-breeding, genetic signatures, or the discovery of new genes associated with a trait.

GS over the past two decades has revolutionized modern plant breeding research. It encompasses modeling and statistical implements to appraise which combinations of genetic variants can lead to the highest yield (or any other relevant trait) under any set of environmental regimes. There are obvious differences between marker-assisted breeding and GS. Simple marker-assisted selection (sMAS), which is the study of a few genes or traits based on linked DNA markers, has been successfully applied in plant breeding. However, it is more suitable for traits with mutations of larger effect. GS, by contrast, uses genome-wide DNA markers such as SNPs to predict the genetic merit of breeding individuals for complex traits. This technology was established to discern complex traits including flowering, yield, and grain quality that are influenced by variants in a large number of genes and/or *cis*-regulatory elements. Nevertheless, these elements typically impart an incremental effect.

While implementing a GS scheme (Figure 4.4A–C, the following steps need to be undertaken.

- Training populations (TPs) are prerequisites of GS that have genotype and phenotype data of multiple quantitative traits.
- Dense genome-wide genotype and precise high throughput phenotype data of multiple quantitative traits result in higher prediction accuracy.
- These genotype and phenotype data are used to train the statistical models.
- Whole-genome regression models for single or multiple traits were used for GS to obtain higher prediction accuracy.
- The trained model facilitates the prediction of genomic estimated breeding values (GEBVs) of quantitative traits in a validating population (VP).
- GEBVs were cross-validated using the genotype and phenotype data of VP.
- After cross-validation, these GEBVs were implemented for the selection of superior parents from breeding populations (BPs) that have only genotype data.

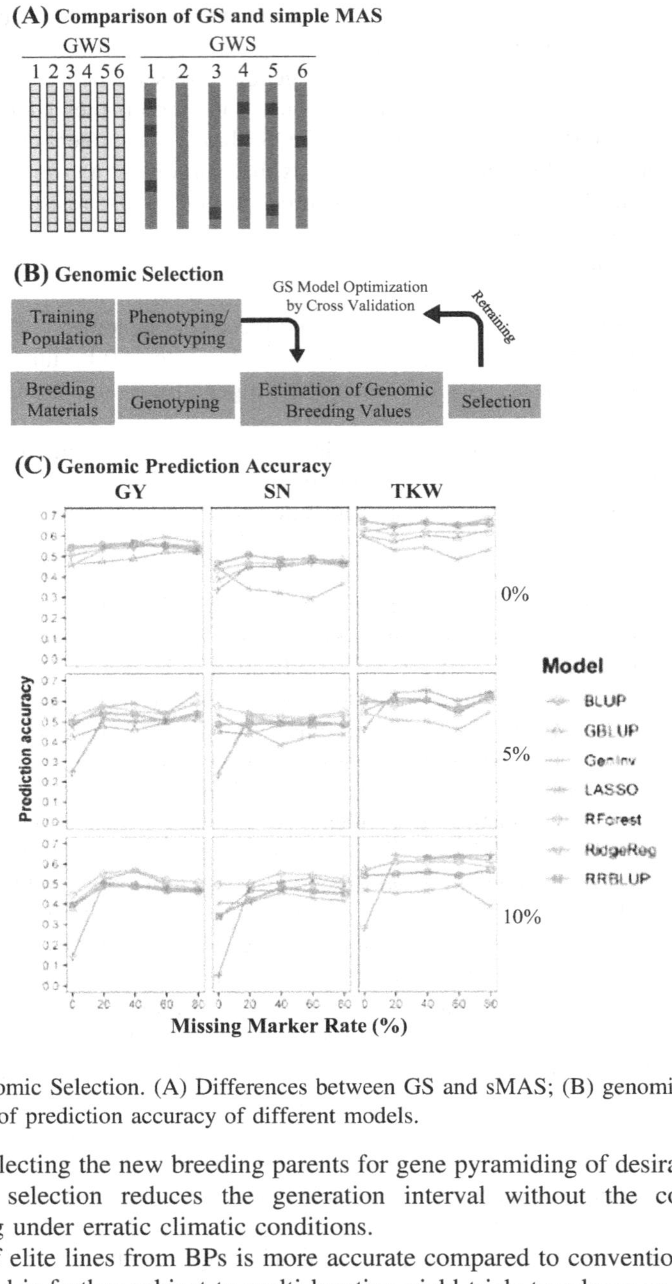

FIGURE 4.4 Genomic Selection. (A) Differences between GS and sMAS; (B) genomic selection scheme; and (C) evaluation of prediction accuracy of different models.

- Precisely selecting the new breeding parents for gene pyramiding of desirable alleles and subsequent selection reduces the generation interval without the cost of precise phenotyping under erratic climatic conditions.
- Selection of elite lines from BPs is more accurate compared to conventional phenotypic selection and is further subject to multi-location yield trials to release as a new variety.

There are a number of factors that affect the GS. These include the number and types of markers and genome coverage; single markers or in the form of haplotypes; structure and size of BP; relationship between TP and BP; engagement of already characterized genes; quality of phenomics data for development training model and finally the heritability attributes of traits under investigation.

A new concept of genome-based breeding by design has emerged out of the availability of NGS/GBTS data of crop lines. To implement the genome-based breeding by design strategy, the following information is a pre-requisite: To obtain accurate genetic/genomic information for target traits in the target genetic background (identification traits genes and their functional genomics) and target environments for developing traits specific varieties for Pakistan by BBD; It is necessary to define clearly the two new concepts of "Core Adaptive Genomic Constitutions (CAGCs)" and "Target Ideal Genomic Constitutions (TIGCs)" related to future varieties to be developed for the target environments (TEs) by BBD. For any specific TE, the CAGC(s) is (are) defined as the whole set of genes/alleles or genome constitution(s) required for the general adaptation (minimum required yield level) in the TE. The second concept, the TIGCs is defined as the genomic constitution(s) of the target varieties to be developed by BBD. Thus, designing TIGCs for a specific TE according to breeding objectives is a critical step for BBD. Given the fact that the CAGCs will be specific for each of the TEs and genes/alleles underlying the target traits in each of the TEs may not be the same (this is true even the target traits are the same for some of the TEs), the TIGCs would vary for different TEs. To design the BBD procedures implementable for developing HT wheat varieties for each of the TEs including identification of best ILs as parents to make crosses that have the highest probability to generate the expected TIGCs for each of the major wheat TEs of Pakistan, design of best selection scheme(s) for accurate identification and quick fixation of the TIGCs from segregating BPs, and validation of selected progenies for their phenotypic performances with the breeding objectives and for their genomic constitutions with the designed TIGCs. Finally, the BBD procedures are implemented for the generation of desired crop varieties.

One advantage of GS over traditional breeding methods is that lines can be selected and used as parents early in the variety development pipeline (Figure 4.4B), and multiple breeding cycles based on GEBV can be achieved in the same amount of time as a single cycle of traditional breeding was achieved previously. The potential for GS to save time and resources is greatest for traits that are typically measured late in the variety development pipeline (evaluation phase, Figure 4.4B and C) and are costly to phenotype, such as yield. GS is independent of the necessity of functional characterization of genes underlying traits. It is specifically suitable for polygenic traits including yield, as it is a simultaneous selection of a large number of favorable genomic variants which are being modeled. Globally, GS has had a significant and consistent effect on the yields of major food crops. The GS is not entirely an independent solution for fruitful plant breeding. It has to be associated with other innovations aiming at reducing time interval generation, linkage drag, and emancipation from LD by including the precise location of causative mutations underlying the target traits. Hence, there should be a close collaboration with molecular biologists involved in delineating the system biology including pathways, gene ontology, and KEGG that could deliver the biggest contributions to a crop's yield and its ability to adapt to aberrant environmental patterns. When combined with SB, this stacking approach could be used to rapidly develop new crop varieties with high performance across multiple traits.

4.6 GENOME EDITING EMERGING AS AN ADVANCED TOOL FOR CROP IMPROVEMENT

The process of genome editing encompasses the deliberate manipulation of genomic DNA at a designated target location inside various cell types and organisms. This manipulation entails the introduction, removal, or substitution of genetic material to deactivate certain genes, acquire distinctive genetic traits, and correct detrimental variations. Due to significant advancements in life sciences, genome editing technology has emerged as a highly effective approach for studying gene function, exploring the underlying mechanisms of genetic diseases, identifying potential targets for gene therapy, and enhancing crop breeding techniques (Kim et al., 2012; Knott & Doudna, 2018).

The RNA-guided CRISPR-Cas systems, zinc finger nucleases (ZFNs), and transcription activator-like effector nucleases (TALENs) are widely employed genome editing tools (Miller et al., 2007; Christian et al., 2010; Hsu et al., 2014). Meganucleases are another genome editing method, these are enzymes that occur naturally and can cleave DNA at specific loci (Silva et al., 2011). The CRISPR-Cas systems have gained significant popularity over others in molecular biology as a genome editing technique (Figure 4.5A). This is primarily attributed to its straightforward design, cost-effectiveness, high efficacy, reliable reproducibility, and rapid turnaround time (Wang et al., 2018).

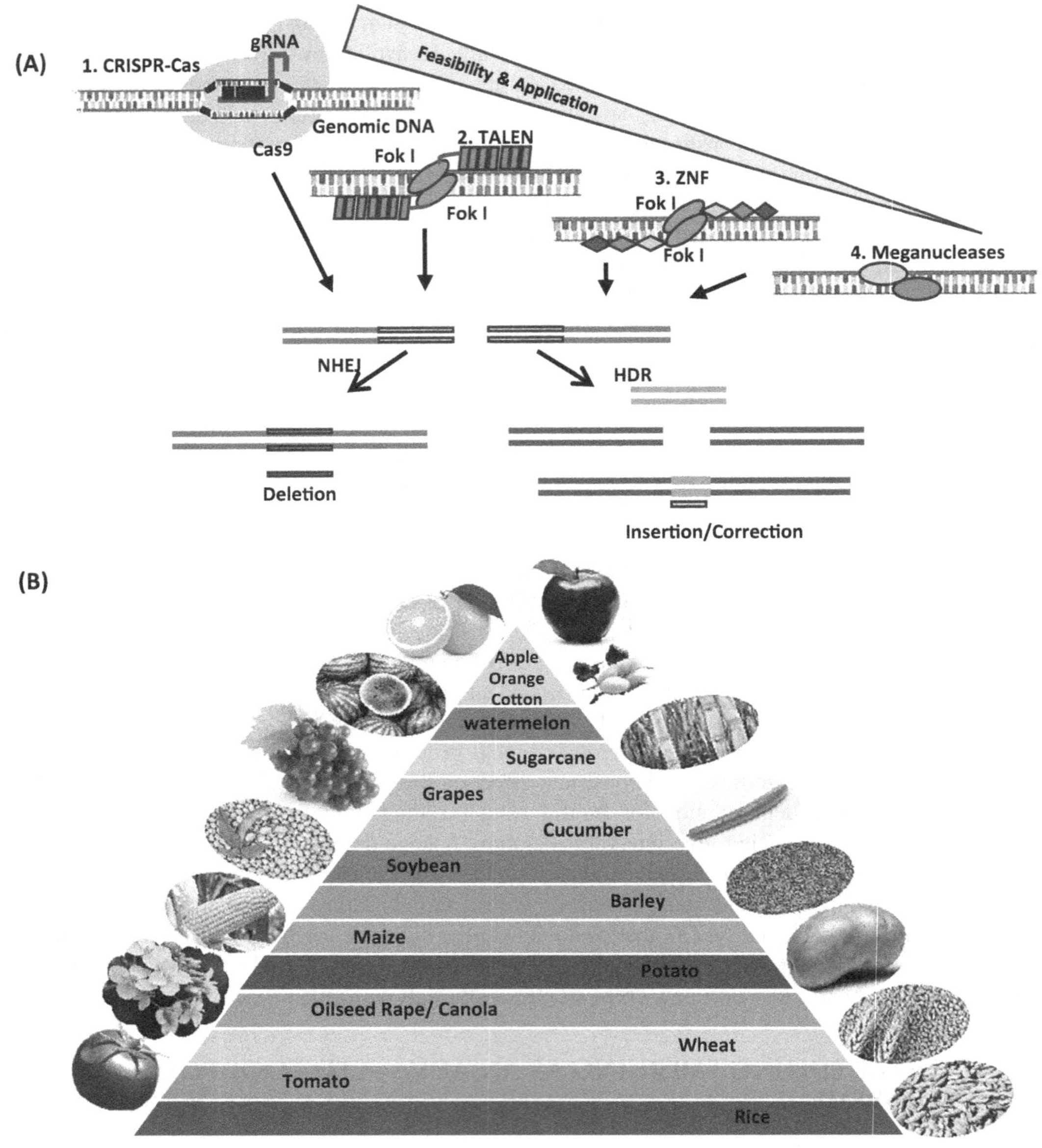

FIGURE 4.5 (A) The principal mechanism of the principal genome-editing technologies. In contrast to ZFNs and TALENs, CRISPR targeting specificity is determined by simple RNA-DNA base pairing and the PAM sequence. Each zinc finger nuclease recognizes a triple DNA code, whereas each TALE recognizes a single DNA base. In contrast to ZFNs and TALENs, CRISPR targeting specificity is determined by simple RNA-DNA base pairing and the PAM sequence. Meganucleases are restriction enzymes designed to recognize extensive DNA sequences. All of these instruments result in DNA double-strand breaks, which are repaired through error-prone non-homology end joining or homology-directed repair. HDR can incorporate a single- or double-stranded DNA template into the target region for precise gene editing, unlike NHEJ, which results in random indels and disruption of the gene. (B) Genome editing in various plant species, with rice being one of the most genome-edited crops globally. As we ascend the hierarchy, the amount of genome editing work performed decreases.

4.6.1 CRISPR/Cas

CRISPR-Cas serves as a protective mechanism against phages, viruses, and other exogenous genetic elements in the majority of bacteria and archaea (Deveau et al., 2010; Mali et al., 2013). CRISPR repeat-spacer arrays are responsible for the production of two types of RNA molecules, namely crRNA and tracrRNA. Additionally, CRISPR-associated (Cas) genes are responsible for the expression of Cas proteins that possess endonuclease activity. Upon infection of prokaryotes by foreign genetic elements, Cas proteins cleave the DNA into smaller fragments, which are then incorporated into the CRISPR array as novel spacers (Makarova et al., 2011). Upon the re-entry of the identical intruder, the crRNA promptly identifies and forms a connection with the exogenous DNA, thereby directing the Cas protein to cleave specific sections of interest and safeguard the host organism. The CRISPR/Cas9 system employs brief synthetic guide RNAs (gRNAs) to direct the Cas9 endonuclease to specific genomic regions, in contrast to the ZFN and TALEN methodologies. The introduction of gRNA induces a state of selectivity, hence necessitating only a minor alteration in the RNA sequence to modify the target (Deltcheva et al., 2011). CRISPR-associated (Cas) genes encode Cas proteins and have various orientations. The Cas core, which encompasses Cas1-Cas9, is made up of 11 of 35 sequence-similar families of Cas proteins. CRISPR repeat-spacer arrays often have small Cas gene clusters. The random distribution of CRISPR/Cas subtypes suggests horizontal gene transfer throughout the microbial evolution (Redman et al., 2016).

The innovative genetic engineering tool CRISPR-Cas is divided into Class I and Class II. Type I, III, and IV systems. These types of Cas detect and hydrolyze DNA and RNA molecules. The Type II system, which uses Cas9, is known for its precision in DNA selection and cleavage. Type V contains the DNA-cleaving Cas12a mechanism, also known as Cpf1. Type VI is unique for its RNA modification via the Cas13 mechanism. Type IV is less understood, with its roles in DNA or RNA generally unknown. Fundamental to this genetic modifying and biological research tool are CRISPR-Cas classes and subtypes, each with unique traits and applications (Jiang & Doudna, 2017). Overall, the CRISPR-Cas system has the unique ability to induce DSBs at specific target loci. Any target organism, including crop plants, can have its genome modified using a variety of approaches. After double-strand breaks (DSBs), homology-directed repair and non-homologous end joining (NHEJ) begin. Non-homologous end joining (NHEJ) is a common DNA error-prone repair mechanism in higher eukaryotes (Maruyama et al., 2015). During NHEJ-mediated DSB repair, the cellular apparatus improperly incorporates small insertions or deletions (indel) at the break site. Interrupting the open reading frame renders genes inactive. CRISPR-Cas facilitates gene deletion (Allen et al., 2019; Asmamaw & Zawdie, 2021).

Despite the significant advancements in CRISPR-Cas9 gene editing, a substantial number of biological science laboratories have not yet included the requisite experimental methodologies associated with this powerful technology. The achievement of successful gene editing by CRISPR technology is contingent upon several crucial elements. These factors include the effective designing of guide RNA (gRNA), the careful selection of suitable nucleases and base editors, and the efficient distribution of Cas9 and gRNA vectors. The careful selection and regeneration of modified organisms, and the accurate detection of gene editing events are also deemed important for the successful gene editing experiment. The frequency of gene editing is contingent upon the discussed optimal conditions.

In 2012, the applicability of the CRISPR-Cas system as a precise genome-editing tool was reported Since its application to plants in 2013, the utilization of CRISPR-Cas has significantly impacted the field of plant biology and biotechnology. The utilization of CRISPR-Cas systems for plant genome editing has proven to be a straightforward and highly efficient approach. The advancement in this field has facilitated the emergence of novel crop varieties characterized by enhanced attributes such as heightened resistance against pests and diseases, increased yield potential, and improved nutritional composition. In the last ten years, the genomes of several crops such as rice (Li et al., 2020), wheat (Hamada et al., 2018), maize (Li et al., 2020), soybeans (Li et al., 2020), potatoes (Hou et al., 2023), tomatoes (Tiwari et al., 2023), and fruits (Pompili et al., 2020) have been modified using CRISPR/Cas technology.

CRISPR-Cas is being employed for the investigation of plant biology and the advancement of novel plant-derived products, in addition to its agricultural uses e.g., the use of Cas9 (specifically dCas9) has been demonstrated to involve the binding of DNA without inducing any modifications to it, Cas9 is employed in conjunction with a fluorescing protein (FP). The effectiveness of this approach in the observation of DNA-protein interactions in live plant cells was demonstrated (Dreissig et al., 2017). The initial instance of genome-edited crops that were exempted from USDA restrictions involved a white button mushroom that had been modified using CRISPR/Cas9 technology to exhibit resistance against browning. A lot of other crops have been modified using CRISPR Cas and approved by USDA. Many countries are accepting the usability of these genome-edited crops and several have made concurrent regulations and policies for genome-edited crops. Although CRISPR/Cas technology is still in its infancy, it has already revolutionized biotechnology and plant biology. As technological advancements progress, the applications of CRISPR/Cas in plants are expected to demonstrate greater levels of innovation and efficiency.

4.6.2 Applications of CRISPR/Cas in Agriculture

CRISPR/Cas-based genome editing has been utilized to enhance agriculturally important crop traits like quality, yield, drought resistance, heat resistance, salinity resistance, disease, and herbicide sensitivity. CRISPR/Cas systems have been established as versatile genome-editing technologies in plants with a wide range of potential applications. The CRISPR/Cas9 system has been described as the third generation of targeted genome editing technology (Zhang et al., 2019). ZFNs and TALENs have answers to fundamental plant biology concerns. CRISPR/Cas9 has gained rapid and extensive adaptability in plant science due to its simplicity, efficacy, and specificity (Zhang et al., 2019). This approach has been extensively applied in gene knockout, knocking, and regulation for yield enhancement, quality, and resistance to biotic and abiotic stresses in essential crops (Figure 4.5B). Crop improvement initiatives will benefit greatly from the use of CRISPR-Cas technology in conjunction with modern breeding techniques (Gao et al., 2020).

4.6.3 Application of CRISPR in Different Crops

CRISPR/Cas genome editing has been extensively used for yield enhancement in cereals as well as fruit crops, including rice, maize, wheat, soybean, barley, sorghum, potato, tomato, flax, rapeseed, Camelina, cotton, cucumber, lettuce, grapes, grapefruit, apple, oranges, and watermelon (Table 4.4). Yield is the prime trait and is of high consideration in crop breeding. Knockout of LAZY1 gene in rice enhanced crop production. The disruption of Gn1a, DEP1, and GS3 genes using CRISPR/Cas9 produced rice mutants having high grain numbers, dense erect panicles, and bigger grain sizes. High-yielding wheat lines were produced by knocking out the *TaCKX2-1*, *TaGLW7*, and *TaGW2* genes through CRISPR. Likewise, genome editing of an Excess number of floral genes (ENO) gene in tomatoes produced a large number of locules resulting in larger fruits (Yuste-Lisbona et al., 2020).

Crop nutritional value can be enhanced using CRISPR/Cas9 technology. The FAD2 gene was targeted in the oil seed plant Camelina to increase oleic acid content while decreasing polyunsaturated fatty acids. Sun et al. (2016) induced specific mutations in the SBEIIb gene in rice for the enhancement of starch fine structure and nutritional quality. Grain with low amylose in wheat rice and maize is not only for eating and cooking purposes but it has wide applications in the textile and adhesive manufacturing industries. The disruption of the granule-bond starch synthase 1(*GBSS1*) gene in maize resulted in a low amylose content (Zhu et al., 2020).

TABLE 4.4
List of Targeted Gene and Trait Improvement in Plants through Different CRISPR/Cas-based Genome Editing Approaches

Plant Species	Gene Function	Target Genes	Gene Editor	sgRNA Promoter Used	References
Rice	Early flowering and seed setting	OsHd2, OsHd4, and OsHd5	CRISPR-Cas-9	OsU3, OsU6	(Li et al., 2020)
Rice	Higher yield and Blast resistance	OsRLCK185 and OsCERK1	Base Editor	OsU6 Pro	(Ahmar et al., 2023)
Rice	Increasing plant yield and photosynthetic activity.	OsHXK1	CRISPR-Cas-9	OsU3	(Zheng et al., 2021)
Rice	Herbicides resistance	HPTII, OsEPSPS	Prime Editor	OsU6	(Li et al., 2020)
Rice	Low amylose content	OsWaxy, OsEUI1	Base Editor	OsU3, OsU6	(Liu et al., 2022)
Rice	Herbicide Resistance	OsALS, OsCDC, OsPDS	Prime Editor	OsU3	(Jin et al., 2023)
Maize	Male-sterility	ZmMs13	CRISPR-Cas-9	TaU3	(Fang et al., 2023)
Maize	High yield	ZmCENH3	Cytosine Base Editor	OsUbi1	(Zong et al., 2017)
Maize	Opaque kernels	ZmO2	CRISPR-Cpf1	OsUbi1	(Gong et al., 2021)
Wheat	Increased tolerance to drought	TaDREB2 and TaERF3	CRISPR-Cas9	TaU6	(Kumar et al., 2019)
Wheat	Resistance to fusarium head blight	TaHRC	CRISPR-Cas9	TaU6	(Su et al., 2019)
Tomato	Fruit development and ripening	AP2a, NOR, FUL1, FUL2	CRISPR-Cas9	U6	(Wang et al., 2019)
Canola	Low erucic acid	BnaAFAE1	CRISPR-Cas9	AtU6-26, AtU6-29	(Liu et al., 2022)
Sweet Potato	High amylose content	IbGBSSI, IbSBEII	CRSIPR-Cas9	AtU6	(Wang et al., 2019)
Soyabean	Flower induction	GmFT2a, GmFT4	Base Editing	AtU6	(Cai et al., 2020)
Watermelon	Herbicide resistance	ALS	Base Editing	U6-26	(Tian et al., 2018)
Cotton	GhCLA functions in chloroplast development. GhPEBP participates in the multiplex-branch developmental process	GhCLA, GhPEBP	Base Editing	GhU6-7	(Qin et al., 2020)

Different approaches have been used to study the resistance of crops to disease through the CRISPR-Cas system inactivation of susceptibility factor encoding gene, knockout, disruption of promoters, and introduction of precise mutations in targeting regions (Schenke & Cai, 2020). The mildew locus (MLO) gene in crops is susceptible to fungal diseases. While CRISPR-Cas-based mutations in MLO genes stop the penetration of fungus thus providing resilient resistance in tomato, wheat (TaMLO-A1), rice, grapes, and maize against powdery mildew disease (Pessina et al., 2016; Schenke & Cai, 2020; Wan et al., 2020; Yan et al., 2021). Using the high-efficacy CRISPR/Cas9-FLP/FRT system fire blight resistance was introduced in the apple (Pompili et al., 2020). Genome-edited lines of rice were generated by targeting the promoter region of SWEET 13 and SWEET 14 genes in rice to confer resistance against bacterial blight (Oliva et al., 2019). Also, in citrus, the promoter region of the *LOB1* gene was mutated which resulted in decreased subspeciality for *Xanthomonas citri* subsp. citri in the citrus (Peng et al., 2017). The plant immune system against viruses can be enhanced through inactivation of plant susceptibility genes. Disrupting the eIF4E gene showed resistance to potyviruses in cucumber (Hu et al., 2021; Fidan et al., 2023).

Herbicide-resistant crops become an efficient weed control method due to reduced phytotoxicity. The emerging CRISPR/Cas-based genome editing is a reliable approach for the production of herbicide-resistant plants (Hu et al., 2021). In order to confer herbicide resistance in rice, certain base transitions were introduced into the *OsALS* gene in rice through base editing (Kuang et al., 2020; Hu et al., 2021).

Advancements in CRISPR/Cas-based genome editing include the production of male sterile lines. Male sterility was introduced in wheat cultivars by targeting male sterile MS1 and Ms45 genes. (Okada et al., 2019). This strategy has also been applied to produce a male sterile line in tomato and rice by targeting the strictosidine synthase gene and thermosensitive genic male-sterile respectively (Du et al., 2020).

4.6.4 New Thrusts or Advancements in CRISPR/Cas System

Numerous attempts have been made to increase the possibilities offered by the CRISPR/cas9 system. Its efficacy has been further increased through the engineering of Cas9, Cas12a, Cas12b, and Cas12f proteins (Matiello et al. 2022). Evolving technologies of CRISPR/Cas are Cas 14, dCas9, base editing, prime editing, and PASTE (programmable addition via site-specific targeting elements) (PASTE) (Yarnall et al., 2023). Base editing (BE)allows for selective and irreversible base conversion without the formation of double-strand breaks (DSBs). This technique was recently developed. Because of its precision, simplicity, and multiplexing, it received immediate adoption and adaptation. It can make precise changes in plant genomes. It plays a vital role in crop improvement as most of the significant agronomic features are governed by SNP (Azameti & Dauda, 2021). Prime editing (PE) is a precise system of CRISPR/Cas-mediated genome editing that is capable of performing small insertions and deletions. PE does not need double-stranded breaks and makes accurate editing of target sequence so it can be applied in plants for crop improvement. Recent advancements and variations in the CRISPR/Cas system can be used for advanced applications such as genomes of mitochondria and chloroplasts, tracing cell lineages to elucidate the patterns underlying plant development, building genetic circuits to integrate and transduce signals, developing plant biosensors to detect internal and external signals, and other plant synthetic biology applications. CRISPR/Cas has been a revolutionary tool for both fundamental and practical plant research. A number of CRISPR-derived editors have been developed that can precisely manipulate genomes in addition to the indel mutations caused by the CRISPR/Cas nuclease. These tools have revolutionized breeding techniques and enabled the development of hundreds of crop varieties with enhanced agronomic performance thanks to their unmatched capacity to modify genes. CRISPR/Cas increases the possibility of quickly domesticating wild orphan species or crops to ensure world-wide food security (Zhu et al., 2020).

4.7 SPEED BREEDING: A NEW AVENUE FOR CROP IMPROVEMENT

Various techniques, procedures, and approaches have been created since ancient times to increase crop output. Despite sophisticated and efficient molecular techniques, new approaches to fulfill future food demands are still required. Traditional breeding takes a long time to develop crops because of the long generation advancement times. The "speed breeding" method shortens generation time and accelerates breeding and research operations. NASA has noticed rapid plant growth and development under continuous photoperiod in order to cultivate plants in space. Following that, methods for this technique were brought by Australian and British experts from The University of Queensland and The John Innes Center, respectively. NASA's observation of accelerated growth led Amy Watson and her colleagues to conduct an experiment with Dr. Lee Hickey on SB procedures (Watson et al., 2018). They discovered that implementing this method can significantly cut the generation time of major crops such as wheat, chickpea, canola, and barley (Figure 4.6). Six generations of spring, durum wheat (Triticum aestivum, Triticum durum), barley (Hordeum vulgare), chickpea (Cicer arietinum), and pea (Pisum sativum) and four generations of canola crop (Brassica napus) can be achieved per year through SB, which can normally be grown up to 2–3 generations under glasshouse conditions. Plants were grown in completely contained chambers with regulated environmental conditions to speed up plant growth and development for SB reasons.

The SB technique has already been used successfully to speed up breeding goals for pea (*Pisum sativum*) (Mobini & Warkentin, 2016), amaranth (Amaranthus spp.) (Stetter et al., 2016), and peanut (*Arachis hypogaea*) (O'Connor et al., 2013). Other crop plants, including pepper, radish, and sunflower, respond better to longer photoperiods and SB is probably the best way to shorten

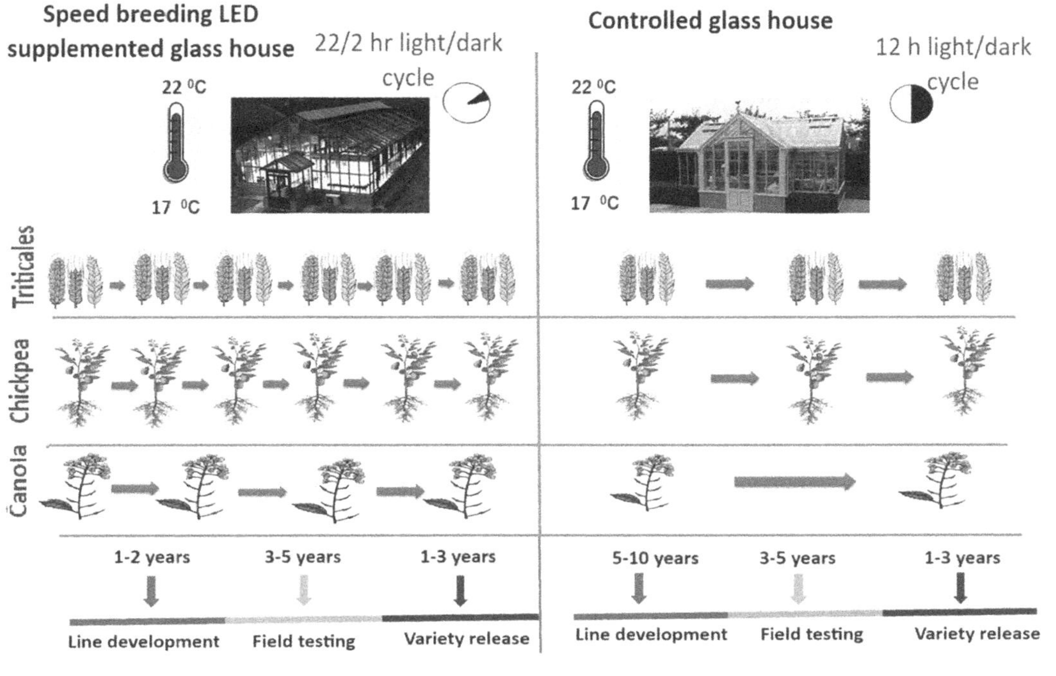

FIGURE 4.6 A schematic representation of speed breeding (Watson et al., 2018) speeds the production time of main agricultural plants for study and breeding. Under normal circumstances, a fast-breeding glasshouse is compared to a standard glasshouse. Annually, 2–3 generations of wheat, barley, chickpea, and canola can be cultivated (right), and by speed breeding annually, 4–6 generations of these crops can be attained (left). These values are representative of relatively rapid cycling cultivars of each crop.

their generation time (Sysoeva et al., 2010). Plant scientists are now able to focus on crops instead of model plants thanks to improvements in genomic resources and methodologies as well as lower sequencing costs. Even with these advancements, many crops' extended generation durations still impose a hurdle. For many crops, SB methods have been devised by various researchers to shorten the generation time (Watson et al., 2018). In contrast to the double haploid approach, which rescues haploid embryos for chromosomal doubling in order to produce homozygous lines (Laurie & Bennett, 1988). SB works with a variety of germplasm and doesn't require in-vitro culture in specific labs. Photoperiod and regulated temperature are utilized in SB technology to reduce the generation time quality and intensity of light, together with early seed production, by accelerating photosynthesis and flower growth rate. Specialized procedures are available for plants that require particular environmental circumstances, such as vernalization or short days, in order to flower. These methods allow cereals to be produced at high densities, such as 1,000 plants/m^2, in a small amount of space and at a low cost (Ghosh et al., 2018). The attempts to meet the growing population's food needs will be strengthened by combining cutting-edge technologies with rapid breeding (Hickey et al., 2017). To improve crops more quickly, there is a great deal of promise for combining SB with contemporary tools and technologies like CRISPR gene editing, high throughput genotyping, GS, speed cloning, MAS, and so on.

4.8 INTEGRATION OF NOVEL BREEDING STRATEGIES

It is therefore possible to reduce the generation advancement time from 4–8 years to 1–2 years by combining speed cloning, gene editing, and SB procedures. Three main approaches are included in this 5- to 6-year rapid crop improvement program. First, through speed cloning, desirable genes can be extracted from a diverse population in 6 months. Second, crop genomes can be manipulated using various genome editing techniques such as expressEdit. Third, through SB techniques, 4–6 generations can be produced in 1 year. Lastly, field trials require 2–4 years (Figure 4.7).

According to Watson et al. (2018), SB technology has the potential to expedite progress in various plant research domains, such as crossing, population mapping, and plant phenotyping for certain traits. SB can further hasten characteristics pyramiding and backcrossing (Hickey et al., 2017).

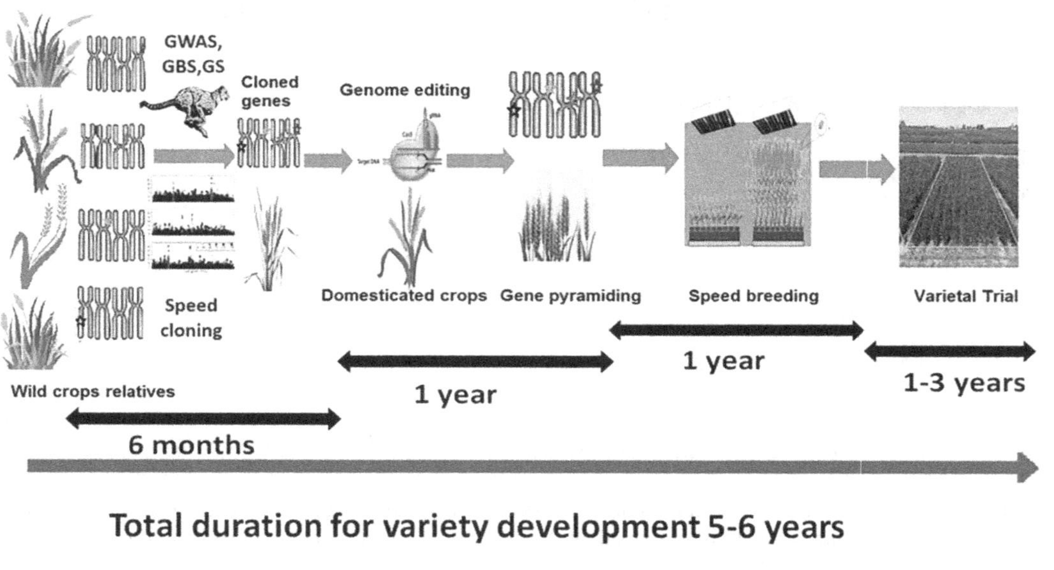

FIGURE 4.7 A schematic representation of advanced technical breakthroughs to enhance crop improvement.

4.9 OUTLOOK

The past few decades have seen significant advancements in agriculture thanks to conventional breeding, which depends on diverse plant populations. Still, the majority of population changes result from infrequent, random events like spontaneous mutations, chemically induced mutations, or physical irradiation. Furthermore, it may not be practical to vary complex traits in elite varieties through extensive breeding labor. As a result, cutting-edge technologies can quickly and precisely manipulate crop plants. The technological developments and their incorporation into breeding programs to shorten the generation period from 10–15 years to 5–6 years have been covered in this review. The most effective way to produce a generation rapidly is to combine genome editing and rapid cloning technologies to create SB. By changing the gRNA sequence, the system known as CRISPR/Cas is a precise, quick, and adaptable tool in the gene editing process that may be used to target any region of an organism's genome. It will be helpful in annotating a vast amount of sequencing data, which will further make it easier to find the genomes of the various crops by getting over the restrictions of looking through a variety of plant populations and traditional breeding techniques. Consequently, in order to take advantage of the vast amount of genome sequence data that is currently available and produce improved crop varieties to meet the needs of an ever-increasing population, crop improvement programs must incorporate speed cloning and the CRISPR/Cas system, which are easier, safer, and more precise techniques. However, because plants are subjected to a variety of stresses, the paradigm for research needs to shift in order to adopt a broader perspective of agriculture and food production and achieve sustainable food security.

REFERENCES

Abbas, A., Zhang, Z., Zheng, H., Alami, M. M., Alrefaei, A. F., Abbas, Q., …, & Abbas, Q. (2023). Drones in plant disease assessment, efficient monitoring, and detection: A way forward to smart agriculture. *Agronomy*, *13*(6), 1524.

Agapiou, A. (2020). Vegetation extraction using visible-bands from openly licensed unmanned aerial vehicle imagery. *Drones*, *4*(2), 27.

Ahmar, S., Hensel, G., & Gruszka, D. (2023). CRISPR/Cas9-mediated genome editing techniques and new breeding strategies in cereals – current status, improvements, and perspectives. *Biotechnology Advances*, *69*, 108248.

Allen, F., Crepaldi, L., Alsinet, C., Strong, A. J., Kleshchevnikov, V., De Angeli, P., …, & Kosicki, M. (2019). Predicting the mutations generated by repair of Cas9-induced double-strand breaks. *Nature Biotechnology*, *37*(1), 64–72.

Arora, S., Steuernagel, B., Gaurav, K., Chandramohan, S., Long, Y., Matny, O., …, & Singh, N. (2019). Resistance gene cloning from a wild crop relative by sequence capture and association genetics. *Nature Biotechnology*, *37*(2), 139–143.

Asmamaw, M., & Zawdie, B. (2021). Mechanism and applications of CRISPR/Cas-9-mediated genome editing. *Biologics: Targets and Therapy*, 353–361.

Atwell, S., Huang, Y. S., Vilhjálmsson, B. J., Willems, G., Horton, M., Li, Y., …, & Hu, T. T. (2010). Genome-wide association study of 107 phenotypes in Arabidopsis thaliana inbred lines. *Nature*, *465*(7298), 627–631.

Ayhan, B., Kwan, C., Budavari, B., Kwan, L., Lu, Y., Perez, D., …, & Vlachos, M. (2020). Vegetation detection using deep learning and conventional methods. *Remote Sensing*, *12*(15), 2502.

Aylward, A. J., Petrus, S., Mamerto, A., Hartwick, N. T., & Michael, T. P. (2023). PanKmer: k-mer based and reference-free pangenome analysis. *Bioinformatics*, *39*, btad621.

Azameti, M. K., & Dauda, W. P. (2021). Base editing in plants: Applications, challenges, and future prospects. *Frontiers in Plant Science*, *12*, 664997.

Bhandari, S., Raheja, A., Green, R. L., & Do, D. (2017). Towards collaboration between unmanned aerial and ground vehicles for precision agriculture. In *Autonomous Air and Ground Sensing Systems for Agricultural Optimization and Phenotyping II* (Vol. 10218, pp. 26–39). SPIE.

Budiharto, W., Chowanda, A., Gunawan, A. A. S., Irwansyah, E., & Suroso, J. S. (2019). A review and progress of research on autonomous drone in agriculture, delivering items and geographical information systems (GIS). In 2019 2nd world symposium on communication engineering (WSCE), Nagoya, Japan, 20–23 December 2019; pp. 205–209.

Cai, Y., Chen, L., Zhang, Y., Yuan, S., Su, Q., Sun, S., …, & Hou, W. (2020). Target base editing in soybean using a modified CRISPR/Cas9 system. *Plant Biotechnology Journal*, *18*(10), 1996.

Chen, P., Lan, Y., Huang, X., Qi, H., Wang, G., Wang, J., …, & Xiao, H. (2020a). Droplet deposition and control of planthoppers of different nozzles in two-stage rice with a quadrotor unmanned aerial vehicle. *Agronomy*, *10*(2), 303.

Chen, P.-C., Chiang, Y.-C., & Weng, P.-Y. (2020b). Imaging using unmanned aerial vehicles for agriculture land use classification. *Agriculture*, *10*(9), 416.

Chen, Z., Bu, Q., Liu, G., Wang, M., Wang, H., Liu, H., …, & Chu, C. (2023). Genomic decoding of breeding history to guide breeding-by-design in rice. *National Science Review*, *10*(5), nwad029.

Christian, M., Cermak, T., Doyle, E. L., Schmidt, C., Zhang, F., Hummel, A., …, & Voytas, D. F. (2010). Targeting DNA double-strand breaks with TAL effector nucleases. *Genetics*, *186*(2), 757–761.

Clavijo, B. J., Venturini, L., Schudoma, C., Accinelli, G. G., Kaithakottil, G., Wright, J., …, & Chapman, H. (2017). An improved assembly and annotation of the allohexaploid wheat genome identifies complete families of agronomic genes and provides genomic evidence for chromosomal translocations. *Genome Research*, *27*(5), 885–896.

del Cerro, J., Cruz Ulloa, C., Barrientos, A., & de León Rivas, J. (2021). Unmanned aerial vehicles in agriculture: A survey. *Agronomy*, *11*(2), 203.

Delavarpour, N., Koparan, C., Nowatzki, J., Bajwa, S., & Sun, X. (2021). A technical study on UAV characteristics for precision agriculture applications and associated practical challenges. *Remote Sensing*, *13*(6), 1204.

Deltcheva, E., et al. (2011). CRISPR RNA maturation by trans-encoded small RNA and host factor RNase III. *Nature*, *471*, 602–607. 10.1038/nature09886

Deveau, H., Garneau, J. E., & Moineau, S. (2010). CRISPR/Cas system and its role in phage-bacteria interactions. *Annual Review of Microbiology*, *64*, 475–493.

Dreissig, S., Schiml, S., Schindele, P., Weiss, O., Rutten, T., Schubert, V., … & Houben, A. (2017). Live-cell CRISPR imaging in plants reveals dynamic telomere movements. *The Plant Journal*, *91*(4), 565–573.

Du, M., Zhou, K., Liu, Y., Deng, L., Zhang, X., Lin, L., …, & Xing, J. (2020). A biotechnology-based male-sterility system for hybrid seed production in tomato. *The Plant Journal*, *102*(5), 1090–1100.

Fahey, T., Pham, H., Gardi, A., Sabatini, R., Stefanelli, D., Goodwin, I., & Lamb, D. W. (2020). Active and passive electro-optical sensors for health assessment in food crops. *Sensors*, *21*(1), 171.

Fang, C., Wu, S., Niu, C., Hou, Q., An, X., Wei, X., …, & Wan, X. (2023). Triphasic regulation of ZmMs13 encoding an ABCG transporter is sequentially required for callose dissolution, pollen exine and anther cuticle formation in maize. *Journal of Advanced Research*, *49*, 15–30.

Fang, L., Wang, Q., Hu, Y., Jia, Y., Chen, J., Liu, B., …, & Zhou, B. (2017). Genomic analyses in cotton identify signatures of selection and loci associated with fiber quality and yield traits. *Nature Genetics*, *49*(7), 1089–1098.

Fernández-Guisuraga, J. M., Sanz-Ablanedo, E., Suárez-Seoane, S., & Calvo, L. (2018). Using unmanned aerial vehicles in postfire vegetation survey campaigns through large and heterogeneous areas: Opportunities and challenges. *Sensors*, *18*(2), 586.

Fidan, H., Calis, O., Ari, E., Atasayar, A., Sarikaya, P., Tek, M. I., …, & Firat, G. (2023). Knockout of elF4E using CRISPR/Cas9 for large-scale production of resistant cucumber cultivar against WMV, ZYMV, and PRSV. *Frontiers in Plant Science*, *14*, 1143813.

Gao, D., Sun, Q., Hu, B., & Zhang, S. (2020a). A framework for agricultural pest and disease monitoring based on internet-of-things and unmanned aerial vehicles. *Sensors*, *20*(5), 1487.

Gao, H., Gadlage, M. J., Lafitte, H. R., Lenderts, B., Yang, M., Schroder, M., …, & Feigenbutz, L. (2020b). Superior field performance of waxy corn engineered using CRISPR–Cas9. *Nature Biotechnology*, *38*(5), 579–581.

Ghosh, S., Watson, A., Gonzalez-Navarro, O. E., Ramirez-Gonzalez, R. H., Yanes, L., Mendoza-Suárez, M., …, & Green, P. (2018). Speed breeding in growth chambers and glasshouses for crop breeding and model plant research. Nature Protocols, *13*(12), 2944.

Giordan, D., Adams, M. S., Aicardi, I., Alicandro, M., Allasia, P., Baldo, M., …, & Hobbs, P. (2020). The use of unmanned aerial vehicles (UAVs) for engineering geology applications. *Bulletin of Engineering Geology and the Environment*, *79*, 3437–3481.

Gong, C., Huang, S., Song, R., & Qi, W. (2021). Comparative study between the CRISPR/Cpf1 (Cas12a) and CRISPR/Cas9 systems for multiplex gene editing in maize. *Agriculture*, *11*(5), 429.

Hamada, H., Liu, Y., Nagira, Y., Miki, R., Taoka, N., & Imai, R. (2018). Biolistic-delivery-based transient CRISPR/Cas9 expression enables in planta genome editing in wheat. *Scientific Reports*, *8*(1), 14422.

Harjes, C. E., Rocheford, T. R., Bai, L., Brutnell, T. P., Kandianis, C. B., Sowinski, S. G., ..., & Wurtzel, E. T. (2008). Natural genetic variation in lycopene epsilon cyclase tapped for maize biofortification. *Science*, *319*(5861), 330–333.

Hartanto, R., Arkeman, Y., Hermadi, I., Sjaf, S., & Kleinke, M. (2019). Intelligent unmanned aerial vehicle for agriculture and agroindustry. In *IOP Conference Series: Earth and Environmental Science* (Vol. 335, No. 1, p. 012001). IOP Publishing.

Hashimoto, N., Saito, Y., Maki, M., & Homma, K. (2019). Simulation of reflectance and vegetation indices for unmanned aerial vehicle (UAV) monitoring of paddy fields. *Remote Sensing*, *11*(18), 2119.

Hickey, L. T., Germán, S. E., Pereyra, S. A., Diaz, J. E., Ziems, L. A., Fowler, R. A., ..., & Dieters, M. J. (2017). Speed breeding for multiple disease resistance in barley. *Euphytica*, *213*(3), 64.

Hou, X., Guo, X., Zhang, Y., & Zhang, Q. (2023). CRISPR/Cas genome editing system and its application in potato. *Frontiers in Genetics, 14*, 1017388.

Hsu, P. D., Lander, E. S., & Zhang, F. (2014). Development and applications of CRISPR-Cas9 for genome engineering. *Cell*, *157*(6), 1262–1278.

Hu, C., Sheng, O., Deng, G., He, W., Dong, T., Yang, Q., ..., & Liu, S. (2021). CRISPR/Cas9-mediated genome editing of MaACO1 (aminocyclopropane-1-carboxylate oxidase 1) promotes the shelf life of banana fruit. *Plant Biotechnology Journal*, *19*(4), 654.

Jafarbiglu, H., & Pourreza, A. (2022). A comprehensive review of remote sensing platforms, sensors, and applications in nut crops. *Computers and Electronics in Agriculture*, *197*, 106844.

Jaganathan, D., Bohra, A., Thudi, M., & Varshney, R. K. (2020). Fine mapping and gene cloning in the post-NGS era: Advances and prospects. *Theoretical and Applied Genetics*, *133*(5), 1791–1810.

Jeziorska, J. (2019). UAS for wetland mapping and hydrological modeling. *Remote Sensing*, *11*(17), 1997.

Jiang, F., & Doudna, J. A. (2017). CRISPR–Cas9 structures and mechanisms. *Annual Review of Biophysics*, *46*, 505–529.

Jin, S., Lin, Q., Gao, Q., & Gao, C. (2023). Optimized prime editing in monocot plants using PlantPegDesigner and engineered plant prime editors (ePPEs). *Nature Protocols*, *18*(3), 831–853.

Jupe, F., Witek, K., Verweij, W., Śliwka, J., Pritchard, L., Etherington, G. J., ..., & Bryan, G. J. (2013). Resistance gene enrichment sequencing (R en S eq) enables reannotation of the NB-LRR gene family from sequenced plant genomes and rapid mapping of resistance loci in segregating populations. *The Plant Journal*, *76*(3), 530–544.

Kim, D.-H., Go, Y.-G., & Choi, S.-M. (2020). An aerial mixed-reality environment for first-person-view drone flying. *Applied Sciences*, *10*(16), 5436.

Kim, E., Kim, S., Kim, D. H., Choi, B.-S., Choi, I.-Y., & Kim, J.-S. (2012). Precision genome engineering with programmable DNA-nicking enzymes. *Genome Research*, *22*(7), 1327–1333.

Kim, J., Kim, S., Ju, C., & Son, H. I. (2019). Unmanned aerial vehicles in agriculture: A review of perspective of platform, control, and applications. *IEEE Access*, *7*, 105100–105115.

Knott, G. J., & Doudna, J. A. (2018). CRISPR-Cas guides the future of genetic engineering. *Science*, *361*(6405), 866–869.

Kuang, Y., Li, S., Ren, B., Yan, F., Spetz, C., Li, X., ..., & Zhou, H. (2020). Base-editing-mediated artificial evolution of OsALS1 in planta to develop novel herbicide-tolerant rice germplasms. *Molecular Plant*, *13*(4), 565–572.

Kumar, J., Jaiswal, V., Kumar, A., Kumar, N., Mir, R., Kumar, S., ..., & Prabhu, K. (2011). Introgression of a major gene for high grain protein content in some Indian bread wheat cultivars. *Field Crops Research*, *123*(3), 226–233.

Kumar, J., Mir, R., Kumar, N., Kumar, A., Mohan, A., Prabhu, K., ..., & Gupta, P. (2010). Marker-assisted selection for pre-harvest sprouting tolerance and leaf rust resistance in bread wheat. *Plant Breeding*, *129*(6), 617–621.

Kumar, R., Kaur, A., Pandey, A., Mamrutha, H., & Singh, G. (2019). CRISPR-based genome editing in wheat: A comprehensive review and future prospects. *Molecular Biology Reports*, *46*(3), 3557–3569.

Laurie, D. A., & Bennett, M. (1988). The production of haploid wheat plants from wheat x maize crosses. *Theoretical and Applied Genetics*, *76*(3), 393–397.

Lee, S., Abecasis, G. R., Boehnke, M., & Lin, X. H. (2014). Rare-variant association analysis: Study designs and statistical tests. *American Journal of Human Genetics*, *95*(1), 5–23. 10.1016/j.ajhg.2014.06.009

Lemay, M.-A., de Ronne, M., Bélanger, R., & Belzile, F. (2023). k-mer-based GWAS enhances the discovery of causal variants and candidate genes in soybean. *The Plant Genome*, *16*, e20374.

Li, H., Li, J., Chen, J., Yan, L., & Xia, L. (2020). Precise modifications of both exogenous and endogenous genes in rice by prime editing. *Molecular Plant*, *13*(5), 671–674.

Lidder, P., & Sonnino, A. (2012). Biotechnologies for the management of genetic resources for food and agriculture. In *Advances in genetics* (Vol. 78, pp. 1–167). Elsevier.

Liu, X., Huang, M., Fan, B., Buckler, E. S., & Zhang, Z. (2016). Iterative usage of fixed and random effect models for powerful and efficient genome-wide association studies. *PLoS Genetics*, *12*(2), e1005767.

Liu, Y., Du, Z., Lin, S., Li, H., Lu, S., Guo, L., & Tang, S. (2022). CRISPR/Cas9-targeted mutagenesis of BnaFAE1 genes confers low-erucic acid in Brassica napus. *Frontiers in Plant Science*, *13*, 848723.

Lu, B., Dao, P. D., Liu, J., He, Y., & Shang, J. (2020). Recent advances of hyperspectral imaging technology and applications in agriculture. *Remote Sensing*, *12*(16), 2659.

Luo, M.-C., Yang, Z.-L., You, F., Kawahara, T., Waines, J., & Dvorak, J. (2007). The structure of wild and domesticated emmer wheat populations, gene flow between them, and the site of emmer domestication. *Theoretical and Applied Genetics*, *114*, 947–959.

Makarova, K. S., Haft, D. H., Barrangou, R., Brouns, S. J., Charpentier, E., Horvath, P., ..., & Yakunin, A. F. (2011). Evolution and classification of the CRISPR–Cas systems. *Nature Reviews Microbiology*, *9*(6), 467–477.

Mali, P., Yang, L., Esvelt, K. M., Aach, J., Guell, M., DiCarlo, J. E., ..., & Church, G. M. (2013). RNA-guided human genome engineering via Cas9. *Science*, *339*(6121), 823–826.

Mammarella, M., Capello, E., Dabbene, F., & Guglieri, G. (2018). Sample-based SMPC for tracking control of fixed-wing UAV. *IEEE Control Systems Letters*, *2*(4), 611–616.

Maruyama, T., Dougan, S. K., Truttmann, M., Bilate, A. M., Ingram, J. R., & Ploegh, H. L. (2015). Inhibition of non-homologous end joining increases the efficiency of CRISPR/Cas9-mediated precise [TM: inserted] genome editing. *Nature Biotechnology*, *33*(5), 538.

Mattiello, L., Rütgers, M., Sua-Rojas, M. F., Tavares, R., Soares, J. S., Begcy, K., & Menossi, M. (2022). Molecular and computational strategies to increase the efficiency of CRISPR-based techniques. *Frontiers in Plant Science*, *13*. 10.3389/fpls.2022.868027

Mertens, S., Verbraeken, L., Sprenger, H., Demuynck, K., Maleux, K., Cannoot, B., ..., & Bonaventure, G. (2021). Proximal hyperspectral imaging detects diurnal and drought-induced changes in maize physiology. *Frontiers in Plant Science*, *12*, 640914.

Messina, G., & Modica, G. (2020). Applications of UAV thermal imagery in precision agriculture: State of the art and future research outlook. *Remote Sensing*, *12*(9), 1491.

Miller, J. C., Holmes, M. C., Wang, J., Guschin, D. Y., Lee, Y.-L., Rupniewski, I., ..., & Kim, K. A. (2007). An improved zinc-finger nuclease architecture for highly specific genome editing. *Nature Biotechnology*, *25*(7), 778–785.

Mobini, S. H., & Warkentin, T. D. (2016). A simple and efficient method of in vivo rapid generation technology in pea (Pisum sativum L.). *In Vitro Cellular & Developmental Biology – Plant*, *52*(5), 530–536.

Mwadzingeni, L., Shimelis, H., Rees, D. J. G., & Tsilo, T. J. (2017). Genome-wide association analysis of agronomic traits in wheat under drought-stressed and non-stressed conditions. *PloS One*, *12*(2), e0171692.

O'Connor, D., Wright, G., Dieters, M., George, D., Hunter, M., Tatnell, J., & Fleischfresser, D. (2013). Development and application of speed breeding technologies in a commercial peanut breeding program. *Peanut Science*, *40*(2), 107–114.

Okada, A., Arndell, T., Borisjuk, N., Sharma, N., Watson-Haigh, N. S., Tucker, E. J., ..., & Whitford, R. (2019). CRISPR/Cas9-mediated knockout of Ms1 enables the rapid generation of male-sterile hexaploid wheat lines for use in hybrid seed production. *Plant Biotechnology Journal*, *17*(10), 1905–1913.

Olejnik, A., Kiszkowiak, Ł., Rogolski, R., Chmaj, G., Radomski, M., Majcher, M., & Omen, Ł. (2020). The use of unmanned aerial vehicles in remote sensing systems. *Sensors*, *20*(7), 2003.

Oliva, R., Ji, C., Atienza-Grande, G., Huguet-Tapia, J. C., Perez-Quintero, A., Li, T., ..., & Liu, B. (2019). Broad-spectrum resistance to bacterial blight in rice using genome editing. *Nature Biotechnology*, *37*(11), 1344–1350.

Peng, A., Chen, S., Lei, T., Xu, L., He, Y., Wu, L., ..., & Zou, X. (2017). Engineering canker-resistant plants through CRISPR/Cas9-targeted editing of the susceptibility gene Cs LOB 1 promoter in citrus. *Plant Biotechnology Journal*, *15*(12), 1509–1519.

Pessina, S., Angeli, D., Martens, S., Visser, R. G., Bai, Y., Salamini, F., ..., & Malnoy, M. (2016). The knock-down of the expression of MdMLO19 reduces susceptibility to powdery mildew (Podosphaera leucotricha) in apple (Malus domestica). *Plant Biotechnology Journal*, *14*(10), 2033–2044.

Poley, L. G., & J. McDermid, G. (2020). A systematic review of the factors influencing the estimation of vegetation aboveground biomass using unmanned aerial systems. *Remote Sensing*, *12*(7), 1052.

Pompili, V., Dalla Costa, L., Piazza, S., Pindo, M., & Malnoy, M. (2020). Reduced fire blight susceptibility in apple cultivars using a high-efficiency CRISPR/Cas9-FLP/FRT-based gene editing system. *Plant Biotechnology Journal*, *18*(3), 845–858.

Popescu, D., Stoican, F., Stamatescu, G., Ichim, L., & Dragana, C. (2020). Advanced UAV–WSN system for intelligent monitoring in precision agriculture. *Sensors*, *20*(3), 817.

Qin, L., Li, J., Wang, Q., Xu, Z., Sun, L., Alariqi, M., …, & Ding, X. (2020). High-efficient and precise base editing of C center dot G to T center dot A in the allotetraploid cotton (Gossypium hirsutum) genome using a modified CRISPR/Cas9 system. *Plant Biotechnology Journal*, *18*(1), 45–56.

Rahman, A., Hallgrímsdóttir, I., Eisen, M., & Pachter, L. (2018). Association mapping from sequencing reads using k-mers. *eLife*, *7*, e32920.

Rahman, M. F. F., Fan, S., Zhang, Y., & Chen, L. (2021). A comparative study on application of unmanned aerial vehicle systems in agriculture. *Agriculture*, *11*(1), 22.

Rasheed, A., Hao, Y., Xia, X., Khan, A., Xu, Y., Varshney, R. K., & He, Z. (2017). Crop breeding chips and genotyping platforms: Progress, challenges, and perspectives. *Molecular Plant*, *10*(8), 1047–1064.

Rasheed, A., Qayyum, H., & Appels, R. (2024). *The Wheat Genome. Compendium of Plant Genomes*. Springer, Cham. https://doi.org/10.1007/978-3-031-38294-9_9

Reddy, S. S., Saini, D. K., Singh, G. M., Sharma, S., Mishra, V. K., & Joshi, A. K. (2023). Genome-wide association mapping of genomic regions associated with drought stress tolerance at seedling and reproductive stages in bread wheat. *Frontiers in Plant Science*, *14*, 1166439.

Redman, M., King, A., Watson, C., & King, D. (2016). What is CRISPR/Cas9? *Archives of Disease in Childhood-Education and Practice*, *101*(4), 213–215.

Rossetto, M., & Henry, R. J. (2014). Escape from the laboratory: New horizons for plant genetics. *Trends in Plant Science*, *19*(9), 554–555.

Sánchez-Martín, J., Steuernagel, B., Ghosh, S., Herren, G., Hurni, S., Adamski, N., …, & Wicker, T. (2016). Rapid gene isolation in barley and wheat by mutant chromosome sequencing. *Genome Biology*, *17*(1), 221.

Schenke, D., & Cai, D. (2020). Applications of CRISPR/Cas to improve crop disease resistance: Beyond inactivation of susceptibility factors. *Iscience*, *23*(9), 101478.

Shilin, W., Jianli, S., Xiongkui, H., Le, S., Xiaonan, W., Changling, W., …, & Yun, L. (2017). Performances evaluation of four typical unmanned aerial vehicles used for pesticide application in China. *International Journal of Agricultural and Biological Engineering*, *10*(4), 22–31.

Silva, G., Poirot, L., Galetto, R., Smith, J., Montoya, G., Duchateau, P., & Paques, F. (2011). Meganucleases and other tools`2` for targeted genome engineering: Perspectives and challenges for gene therapy. *Current Gene Therapy*, *11*, 11–27. 10.2174/156652311794520111

Singh, K. K., & Frazier, A. E. (2018). A meta-analysis and review of unmanned aircraft system (UAS) imagery for terrestrial applications. *International Journal of Remote Sensing*, *39*(15–16), 5078–5098.

Singh, R. R. (2022). Target enrichment approaches for next-generation sequencing applications in oncology. *Diagnostics*, *12*(7), 1539.

Spoorthi, S., Shadaksharappa, B., Suraj, S., & Manasa, V. (2017). Freyr drone: Pesticide/fertilizers spraying drone-an agricultural approach. 2017 2nd International Conference on Computing and Communications Technologies (ICCCT) (pp. 252–255). IEEE.

Stetter, M. G., Zeitler, L., Steinhaus, A., Kroener, K., Biljecki, M., & Schmid, K. J. (2016). Crossing methods and cultivation conditions for rapid production of segregating populations in three grain amaranth species. *Frontiers in Plant Science*, *7*, 816.

Steuernagel, B., Periyannan, S. K., Hernández-Pinzón, I., Witek, K., Rouse, M. N., Yu, G., …, & Jones, J. D. (2016). Rapid cloning of disease-resistance genes in plants using mutagenesis and sequence capture. *Nature Biotechnology*, *34*(6), 652–655.

Su, Z., Bernardo, A., Tian, B., Chen, H., Wang, S., Ma, H., …, & Li, T. (2019). A deletion mutation in TaHRC confers Fhb1 resistance to Fusarium head blight in wheat. *Nature Genetics*, *51*(7), 1099–1105.

Sun, Y., et al. (2016). Engineering herbicide-resistant rice plants through CRISPR/Cas9-mediated homologous recombination of acetolactate synthase. *Molecular Plant*, *9*, 628–631. 10.1016/j.molp.2016.01.001

Swami, P., Deswal, K., Rana, V., Yadav, D., & Munjal, R. (2023). Speed breeding—A powerful tool to breed more crops in less time accelerating crop research. In *Abiotic stresses in wheat* (pp. 33–49). Academic Press.

Sysoeva, M. I., Markovskaya, E. F., & Shibaeva, T. G. (2010). Plants under continuous light: A review. *Plant Stress*, *4*(1), 5–17.

Thind, A. K., Wicker, T., Šimková, H., Fossati, D., Moullet, O., Brabant, C., …, & Krattinger, S. G. (2017). Rapid cloning of genes in hexaploid wheat using cultivar-specific long-range chromosome assembly. *Nature Biotechnology*, *35*(8), 793.

Tian, F., Bradbury, P., Brown, P., Sun, Q., Flint-Garcia, S., Rocheford, T., …, & Buckler, E. (2011). Genome-wide association study of maize identifies genes affecting leaf architecture. *Nature Genetics*, *43*, 159–162.

Tian, S., Jiang, L., Cui, X., Zhang, J., Guo, S., Li, M., …, & Zong, M. (2018). Engineering herbicide-resistant watermelon variety through CRISPR/Cas9-mediated base-editing. *Plant Cell Reports*, *37*(9), 1353–1356.

Tiwari, J. K., Singh, A. K., & Behera, T. K. (2023). CRISPR/Cas genome editing in tomato improvement: Advances and applications. *Frontiers in Plant Science*, *14*, 1121209.

Tsouros, D. C., Bibi, S., & Sarigiannidis, P. G. (2019). A review on UAV-based applications for precision agriculture. *Information*, *10*(11), 349.

Tyagi, S., Mir, R., Kaur, H., Chhuneja, P., Ramesh, B., Balyan, H., & Gupta, P. (2014). Marker-assisted pyramiding of eight QTLs/genes for seven different traits in common wheat (Triticum aestivum L.). *Molecular Breeding*, *34*, 167–175.

Velu, G., Singh, R., Huerta-Espino, J., Peña-Bautista, R., & Ortiz-Monasterio, I. (2011). Breeding for enhanced zinc and iron concentration in CIMMYT spring wheat germplasm.

Wan, D.-Y., Guo, Y., Cheng, Y., Hu, Y., Xiao, S., Wang, Y., & Wen, Y.-Q. (2020). CRISPR/Cas9-mediated mutagenesis of VvMLO3 results in enhanced resistance to powdery mildew in grapevine (Vitis vinifera). *Horticulture Research*, *7*. https://doi.org/10.1038/s41438-020-0339-8

Wang, H., Wu, Y., Zhang, Y., Yang, J., Fan, W., Zhang, H., …, & Zhang, P. (2019). CRISPR/Cas9-based mutagenesis of starch biosynthetic genes in sweet potato (Ipomoea Batatas) for the improvement of starch quality. *International Journal of Molecular Sciences*, *20*(19), 4702.

Wang, H., Xu, X., Vieira, F. G., Xiao, Y., Li, Z., Wang, J., …, & Chu, C. (2016). The power of inbreeding: NGS-based GWAS of rice reveals convergent evolution during rice domestication. *Molecular Plant*, *9*(7), 975–985.

Wang, M., Wang, S., Liang, Z., Shi, W., Gao, C., & Xia, G. (2018). From genetic stock to genome editing: gene exploitation in wheat. *Trends in Biotechnology*, *36*(2), 160–172.

Wang, S., Wong, D., Forrest, K., Allen, A., Chao, S., Huang, B., …, & Cattivelli, L. (2014). International Wheat Genome Sequencing C. Lillemo M, Mather D, Appels R, Dolferus R, Brown-Guedira G, Korol A, Akhunova AR, Feuillet C, Salse J, Morgante M, Pozniak C, Luo MC, Dvorak J, Morell M, Dubcovsky J, Ganal M, Tuberosa R, Lawley C, Mikoulitch I, Cavanagh C, Edwards KJ, Hayden M, Akhunov E, 787–796.

Watson, A., Ghosh, S., Williams, M. J., Cuddy, W. S., Simmonds, J., Rey, M.-D., …, & Reynolds, D. (2018). Speed breeding is a powerful tool to accelerate crop research and breeding. *Nature Plants*, *4*(1), 23–29.

Watson, A., Ghosh, S., Williams, M. J., Cuddy, W. S., Simmonds, J., Rey, M.-D., …, & Reynolds, D. (2018). Speed breeding is a powerful tool to accelerate crop research and breeding. *Nature Plants*, *4*(1), 23.

Witek, K., Jupe, F., Witek, A. I., Baker, D., Clark, M. D., & Jones, J. D. (2016). Accelerated cloning of a potato late blight–resistance gene using RenSeq and SMRT sequencing. *Nature Biotechnology*, *34*(6), 656–660.

Xuehui, H., Xinghua, W., Tao, S., Qiang, Z., Qi, F., Yan, Z., …, & Zhiwu, Z. (2010). Genome-wide association studies of 14 agronomic traits in rice landraces. *Nature Genetics*, *42*(11), 961–967.

Yan, J., Kandianis, C. B., Harjes, C. E., Bai, L., Kim, E.-H., Yang, X., …, & Li, Q. (2010). Rare genetic variation at Zea mays crtRB1 increases β-carotene in maize grain. *Nature Genetics*, *42*(4), 322–327.

Yan, Z., Appiano, M., van Tuinen, A., Meijer-Dekens, F., Schipper, D., Gao, D., …, & Wolters, A.-M. A. (2021). Discovery and characterization of a novel tomato mlo mutant from an ems mutagenized micro-tom population. *Genes*, *12*(5), 719.

Yano, K., Yamamoto, E., Aya, K., Takeuchi, H., Lo, P.-c., Hu, L., …, & Hirano, K. (2016). Genome-wide association study using whole-genome sequencing rapidly identifies new genes influencing agronomic traits in rice. *Nature Genetics*, *48*(8), 927–934.

Yarnall, M. T., Ioannidi, E. I., Schmitt-Ulms, C., Krajeski, R. N., Lim, J., Villiger, L., …, & Roberts, N. (2023). Drag-and-drop genome insertion of large sequences without double-strand DNA cleavage using CRISPR-directed integrases. *Nature Biotechnology*, *41*(4), 500–512.

Yi, W., Liming, C., Lingyu, K., Jie, Z., & Miao, W. (2017). Research on application mode of large fixed-wing UAV system on overhead transmission line. 2017 IEEE International Conference on Unmanned Systems (ICUS) (pp. 88-91). IEEE.

Yu, J., Pressoir, G., Briggs, W. H., Vroh Bi, I., Yamasaki, M., Doebley, J. F., …, & Holland, J. B. (2006). A unified mixed-model method for association mapping that accounts for multiple levels of relatedness. *Nature Genetics*, *38*(2), 203–208.

Yuste-Lisbona, F. J., Fernández-Lozano, A., Pineda, B., Bretones, S., Ortíz-Atienza, A., García-Sogo, B., …, & Moreno, V. (2020). ENO regulates tomato fruit size through the floral meristem development network. *Proceedings of the National Academy of Sciences*, *117*(14), 8187–8195.

Zhang, D., Zhou, X., Zhang, J., Huang, L., & Zhao, J. (2017). Developing a small UAV platform to detect sheath blight of rice. 2017 IEEE International Geoscience and Remote Sensing Symposium (IGARSS) (pp. 3190–3193). IEEE.

Zhang, Z., Ersoz, E., Lai, C.-Q., Todhunter, R. J., Tiwari, H. K., Gore, M. A., …, & Ordovas, J. M. (2010). Mixed linear model approach adapted for genome-wide association studies. *Nature Genetics*, *42*(4), 355–360.

Zhang, Z., Hua, L., Gupta, A., Tricoli, D., Edwards, K. J., Yang, B., & Li, W. (2019). Development of an Agrobacterium-delivered CRISPR/Cas9 system for wheat genome editing. *Plant Biotechnology Journal*, *17*(8), 1623–1635.

Zhang, Z., & Zhu, L. (2023). A review on unmanned aerial vehicle remote sensing: Platforms, sensors, data processing methods, and applications. *Drones*, *7*(6), 398.

Zheng, S., Ye, C., Lu, J., Liufu, J., Lin, L., Dong, Z., …, & Zhuang, C. (2021). Improving the rice photosynthetic efficiency and yield by editing OsHXK1 via CRISPR/Cas9 system. *International Journal of Molecular Sciences*, *22*(17), 9554.

Zhou, J., Zhang, M., Li, X., Wang, Z., Pan, D., & Shi, Y. (2021). Performance comparison of four types of target enrichment baits for exome DNA sequencing. *Hereditas*, *158*(1), 1–12.

Zhou, X., & Stephens, M. (2012). Genome-wide efficient mixed-model analysis for association studies. *Nature Genetics*, *44*(7), 821–824.

Zhou, Z., Jiang, Y., Wang, Z., Gou, Z., Lyu, J., Li, W., …, & Ma, Y. (2015). Resequencing 302 wild and cultivated accessions identifies genes related to domestication and improvement in soybean. *Nature Biotechnology*, *33*(4), 408–414.

Zhu, H., Li, C., & Gao, C. (2020). Applications of CRISPR–Cas in agriculture and plant biotechnology. *Nature Reviews Molecular Cell Biology*, *21*(11), 661–677. 10.1038/s41580-020-00288-9

Zong, Y., Wang, Y., Li, C., Zhang, R., Chen, K., Ran, Y., …, & Gao, C. (2017). Precise base editing in rice, wheat and maize with a Cas9-cytidine deaminase fusion. *Nature Biotechnology*, *35*(5), 438–440.

5 Recent Advances in Plant Phenomics and Speed Breeding for Climate-Smart Agriculture

Rubab Zahra Naqvi[a], *Muhammad Farooq*[a], *Rida Zahid, and Ifrah Imran*

Agricultural Biotechnology Division, National Institute for Biotechnology and Genetic Engineering (NIBGE), Constituent College of Pakistan Institute of Engineering and Applied Sciences, Jhang Road, Faisalabad, Pakistan

Muhammad Arslan Mahmood

Agricultural Biotechnology Division, National Institute for Biotechnology and Genetic Engineering (NIBGE), Constituent College of Pakistan Institute of Engineering and Applied Sciences, Jhang Road, Faisalabad, Pakistan

Plant Sciences Division, Research School of Biology, The Australian National University, Canberra, ACT, Australia

Muhammad Jawad Akbar Awan

Agricultural Biotechnology Division, National Institute for Biotechnology and Genetic Engineering (NIBGE), Constituent College of Pakistan Institute of Engineering and Applied Sciences, Jhang Road, Faisalabad, Pakistan

Shahid Mansoor

Agricultural Biotechnology Division, National Institute for Biotechnology and Genetic Engineering (NIBGE), Constituent College of Pakistan Institute of Engineering and Applied Sciences, Jhang Road, Faisalabad, Pakistan

International Center for Chemical and Biological Sciences, University of Karachi, Karachi, Pakistan

Imran Amin and Muhammad Asif

Agricultural Biotechnology Division, National Institute for Biotechnology and Genetic Engineering (NIBGE), Constituent College of Pakistan Institute of Engineering and Applied Sciences, Jhang Road, Faisalabad, Pakistan

[a] Both authors contributed equally

DOI: 10.1201/b23372-5

5.1 INTRODUCTION

One of the prime global concerns of the present day is mitigating the interconnected challenges of agriculture production, food security, and climate change. In this connection, population expansion is a complicating factor, aggravating worries about hunger, and malnutrition of millions of people worldwide. According to the United Nations (2022), it is projected that the global population will witness a growth of approximately 2 billion people within the next 30 years, resulting in a total population of 9.7 billion by the year 2050 (Pison et al., 2022). Due to the rising need for food, the agricultural industry is under a great deal of pressure (Ehrlich & Holdren, 1971). To end hunger and malnutrition, which still exist in many regions of the world, it is necessary to produce more food while maintaining equitable distribution. Furthermore, the rising middle class in many countries is shifting dietary patterns toward resource-intensive animal-based diets, further straining food resources. Climate change is undeniably affecting agriculture on a global scale (Hutchins et al., 2019). It has disrupted traditional weather patterns, leading to more frequent and severe weather events such as droughts, floods, and heatwaves (Brás et al., 2021). These events harm crop yields and livestock production, causing food shortages and increasing food prices. Rising temperatures can reduce crop yields and alter growing seasons, affecting the suitability of certain regions for agriculture (Hatfield et al., 2011). For instance, some crops may become unviable in areas where they were traditionally grown. These warmer temperatures also create more favorable conditions for pests and diseases (Bebber et al., 2013). Farmers are compelled to increase pesticide utilization and adopt new pest control methods (Mamta & Rajam, 2017). The phenomenon of soil erosion and degradation worsens by rising temperatures and unpredictable rainfall patterns that has a significant effect on the fertility of the soil, thereby posing increased difficulties in cultivating crops. Alterations in precipitation patterns lead to the occurrence of extended periods of drought and resulting scarcity of water in numerous geographical regions (Corwin, 2021).

The complicated relationship between climate change, food security, and agriculture underscores the need for a comprehensive approach to tackling these pressing issues. By allocating resources toward the cultivation of resilient crop varieties and implementing sustainable farming practices, it is possible to improve food security and reduce the negative impacts of climate change on agricultural systems (Lipper et al., 2014). This includes the development of precise and high-throughput cutting-edge technologies for germplasm development and innovative approaches for environmental monitoring and farm management such as precision farming and speed breeding systems. These smart farming solutions involve using various software and hardware to optimize and automate the routine breeding processes. The widespread adoption of the internet and mobile devices has facilitated the accessibility of these technologies, extending their availability beyond big businesses to encompass small-scale private farms (Akkem et al., 2023). This chapter discusses these technologies with descent detail.

5.2 EMERGENCE OF NEW PLANT BREEDING TECHNOLOGIES

The emergence of new plant breeding technologies (NPBTs) signifies a revolutionary shift in the domain of plant breeding (Brookes & Barfoot, 2013). These advancements encompass cutting-edge molecular tools and methodologies that are fundamentally transforming the accuracy and effectiveness of crop enhancement. Significantly, methodologies such as CRISPR-Cas9 genome editing facilitate the precise manipulation of plant genomes, allowing the introduction, deletion, alteration, or suppression of specific genes (Shan et al., 2013). Moreover, high-throughput sequencing and phenotyping, facilitated by automated imaging and sensing technologies, yield extensive data that accelerates the process of trait selection and characterization.

5.3 OMICS-DRIVEN BREEDING

The ability to generate vast amounts of data from whole-genome DNA sequencing and whole-phenome phenotyping in a high-throughput fashion has given rise to the use of terms -omics. For instance, the former is referred to as genomics and accordingly, the latter is referred as "phenomics", respectively. The same goes true for various molecular phenotypes, e.g. transcriptomics, proteomics, and metabolomics. Integration of the big data from these widespread omics technologies through novel bioinformatics methodologies has positively influenced the breeding strategies. Moreover, this promises to provide a thorough understanding of an organism's molecular components and their interactions. For instance, proteomics largely examines the profiles of proteins, whereas transcriptomics primarily studies the patterns of gene expression (Dong & Chen, 2013; Aslam et al., 2017). Omics data are essential for deciphering intricate biological processes and locating biomarkers associated with desirable features (Chao et al., 2023). The combination of omics data and genomes has been shown to enhance the precision of breeding methods (Henry, 2012). The field of genomics includes the comprehensive analysis of an organism's complete DNA sequence, facilitating the identification of genes associated with certain features. Researchers can uncover the genetic framework of several traits, including yield, disease resistance, and nutritional content, by employing techniques like whole-genome sequencing and high-throughput genotyping. The utilization of genomic data enables breeders to enhance their ability to choose plants possessing desirable traits, hence providing significant advantages for targeted breeding (Varshney et al., 2011). On the other hand, utilization of bioinformatics is crucial in the management, examination, and elucidation of biological data, particularly in the realm of genetics. The present modernization in the field involves the construction of databases and computer tools in order to assess and extract significant insights from vast quantities of genetic data. The utilization of bioinformatics tools is crucial for gene identification, functional annotation, and comparative genomics (Kashyap et al., 2015). By empowering breeders to make informed choices throughout their breeding operations, these tools offer breeders an invaluable support.

5.4 PLANT PHENOMICS

Similar to the genome, which refers to the recording of each DNA base in the correct sequence, phenome refers to measuring the precise response of plants due to each genetic and environmental influence, along with their interactions. To understand how a plant's genetic makeup and molecular components affect its observable characters, high-throughput and high-resolution characterization of the phenome is important. In turn, phenomics refers to a detailed analysis of a plant's physiological, morphological, and biochemical response, measured using high-throughput methods at a large spatio-temporal scale. Phenomics analyzes and focuses on the production, modification, and regulatory processes of phenotypic manifestations of biological systems (Lobos et al., 2007; Skelly et al., 2013; Orgogozo et al., 2015). The field initially arose at the junction and integration of biology, engineering, computer science, and physics. It utilizes a range of submissive and experimental observational methods to depict plant phenotype through panels of graphical, physiological, and biochemical data. The necessity for non-invasive evaluation of a diverse array of plant characteristics in extensive screening trials and the introduction of reasonably priced digital photography and videography both contributed to the significant development of phenomics (Gerlai, 2002; Deans et al., 2015; Kochinke et al., 2016; Awada et al., 2018). The effectiveness of breeding and molecular genetics efforts, mainly in field crops, is now significantly influenced by the application of phenomics (Afonnikov et al., 2016). According to some researchers, a more "digital" and practical version of plant physiology might be observed in plant phenotyping (Furbank & Tester, 2011). The quantity and size of plant organs are greatly influenced by drought, nitrogen shortage, and poor light, and a researcher without access to contemporary digital technologies would have challenges to describe the plant phenotypes under these factors. The phenomics approaches are particularly distinctive in this aspect as they offer a novel opportunity to comprehensively and objectively comprehend (including the identification of patterns) a broad spectrum of plant phenotypes when subjected to complex and constantly changing environmental conditions (Timple et al., 2014; Afonnikov et al., 2016).

Currently, the agricultural industry generates a considerable amount of data from a variety of sources, including sensors, satellites, drones, and field tests (Alreshidi, 2019). The administration and understanding of this vast amount of information are made easier by the use of big data analytics. It makes it easier to track agricultural development, anticipate disease outbreaks, and improve resource allocation effectiveness. Big data-driven breeding combines computational approaches with genomes, phenomics, and omics data. It has been demonstrated that using big data-driven approaches in precision agriculture can significantly increase crop yields while consuming fewer resources (da Silveira et al., 2021). With this system, breeders can select plants by taking into account both their genetic and phenotypic potential. Machine learning algorithms can assess big datasets and predict the performance of traits. This makes it possible to identify parent lines that display exceptional traits. Big data use in breeding procedures has been proven to hasten crop variety progress, leading to improvements in yield, nutritional composition, and stress tolerance (Huang et al., 2012). Together, these cutting-edge technology and approaches increase crop enhancement efforts. Technologies like phenomics and omics provide a wealth of plant data. The discovery of genes with a particular interest depends critically on the field of genomics. Through the use of insightful knowledge, the use of big data in breeding processes enables the creation of improved crop varieties. In conclusion, a significant shift to modern agriculture can be seen in the integration of genomics, big data, bioinformatics, phenomics, omics technologies, and data-driven breeding.

5.5 ADVANCING AGRICULTURE THROUGH PLANT PHENOMICS

With the addition of standardized, statistically analyzed, and highly detailed "quantitative" factual data, plant phenomics builds on the accomplishments of the genomic era and traditional bioinformatics. One of its exciting aspect is to take digital phenotypic information and link it to "homological" comparison standards, which is conceptually comparable to the practice of comparing nucleotide and protein sequences (Schnable & Freeling, 2011; Endara et al., 2017). The phenomics technologies' data arrays are automatically captured and processed. As a result, the aforementioned methodologies do not exhibit the subjectivity and statistical unreliability concerns encountered by previous methods (Flood et al., 2016). Phenomics technologies are more accurate than what a human operator can do and are very instructive. They include thermal imaging (Jones et al., 2009), X-ray tomography (Hughes et al., 2017), hyperspectral analysis (Henebry, 2011; Thorp et al., 2015), and many other minimally invasive techniques. Open agricultural land, woodlands, and other natural phytocenoses, alongside limited zones, offer the potential for highly effective phenotyping (Warner et al., 2015). The area of phenomics is advanced by the progression of programmable sovereign agricultural machinery, rail systems, and specialized unmanned aerial vehicles (UAVs), including drones, aerostats, multicopters, and similar technologies (Gago et al., 2015; Lee & Persson, 2016; Ludovisi et al., 2017).

The advancement of the digital image process, the growth of computers and system biology, rise and edifice of plant phenomics are all intimately connected. The capacity to acquire, examine, store, and process RGB digital images served as the foundation for the development of the initial phenotyping systems, which were software and hardware ensembles tailored for certain experimental requirements (Pereyra-Irujo et al., 2012). Phenotyping options have been further increased by moving outside of the RGB spectrum, or the visible portion of the electromagnetic spectrum (Yang et al., 2013; Thorp et al., 2015). The correlation between the progress in phenotyping and the emergence of fixed systems, characterized by the immobilization of plants and cameras or the exclusive movement of recording equipment, is evident. Conventional conveyor systems, characterized by the transportation of plants over a mobile belt or rollers, are currently being substituted by fixed systems. In order to ensure consistency in observations, it is necessary for stationary systems to limit the mechanical effects caused by the shaking of plants during their mobility. In conjunction with this, there has been notable advancement in software designed for the purpose of effectively managing phenomics complexes and facilitating data processing (Paproki et al., 2012; White et al., 2012; Yang et al., 2013). As a result of the progress of machine learning techniques and artificial neural networks (NN), there is a current technological revolution that enables us to advance picture analysis to a qualitatively new level. The utilization of NN has greatly enhanced the bioinformation capacities of phenomics. These networks provide novel insights into the underlying mechanisms governing organismal functioning, particularly the processes involved in organogenesis. Additionally, they enable the assessment of quantitative and qualitative indicators of phenotypes, specifically in relation to biomass accumulation and resistance to stress (Gandomi & Haider, 2015). Plant phenotyping is a scientific procedure used to assess the observable characteristics, or phenotype, of a plant. These characteristics include its size, shape, physiological functions, biochemical composition, and genetic activity. This evaluation is conducted within specific environmental conditions. Phenotyping is a capability exhibited by various organizational levels within an organism, encompassing the subcellular, cellular, tissue, and organism levels (Granier & Vile, 2014; Fahlgren et al., 2015).

Modern phenotyping or high-throughput phenotyping (HTP) techniques are very effective and enable the collection of real-time data and the analysis of details about the whole range of physiological indicators (Figure 5.1). These techniques allow for the detailed investigation of photosynthesis, respiration, water exchange, mineral nutrition, productivity regulatory mechanisms, and stress tolerance in plants as well as their growth, development, and reproductive activities (Furbank & Tester, 2011; Li et al., 2014; Dutta et al., 2015). Phenomics investigations now prioritize the examination of physiological and morphometric characteristics of the aerial components of plants, including stems, leaves, fruits, and, to a lesser extent, flowers (Gregory et al., 2009; Plazas et al., 2014; Vanhaeren et al., 2015). The existing frameworks for studying the structure of root systems lack a solid foundation, despite the considerable development that has been shown in this area over the previous 8–10 years (Galkovskyi et al., 2012; Guan et al., 2013; Pound et al., 2013). Phenotyping mostly has been focused on small models and agriculturally important higher plant species, along with specific strains of algae. *Arabidopsis thaliana*, the most prominent model organism, has been associated with the development of the largest number of platforms and is akin to "Plant *Drosophila*". More than 20 different varieties of grains, legumes, and vegetables, as well as roughly ten different kinds of woody plants, have their platforms.

Limited sites are also being developed for the cultivation of ornamental herbaceous plants. The main objective of shoot phenotyping is to acquire knowledge pertaining to the morphological attributes, spatial arrangement, and movement patterns of a plant's foliage and other aboveground structures, as well as to gain insights into fundamental physiological processes such as photosynthesis, respiration, and transpiration (Furbank & Tester, 2011). In recent years, the utilization of advanced analysis techniques has facilitated the acquisition of chlorophyll fluorescence images. These images have played a crucial role in the development of vital models elucidating the chloroplasts' behavior in response to photochemical activity. Additionally, they have provided insights into the molecular mechanisms underlying the impact of environmental stressors and pathogens on photosynthetic processes (Rolfe & Scholes, 2010; Plazas et al., 2014; Dutta et al., 2015). A model for the development and expansion of the Arabidopsis flower was successfully developed by quantitatively phenotyping changes in leaf

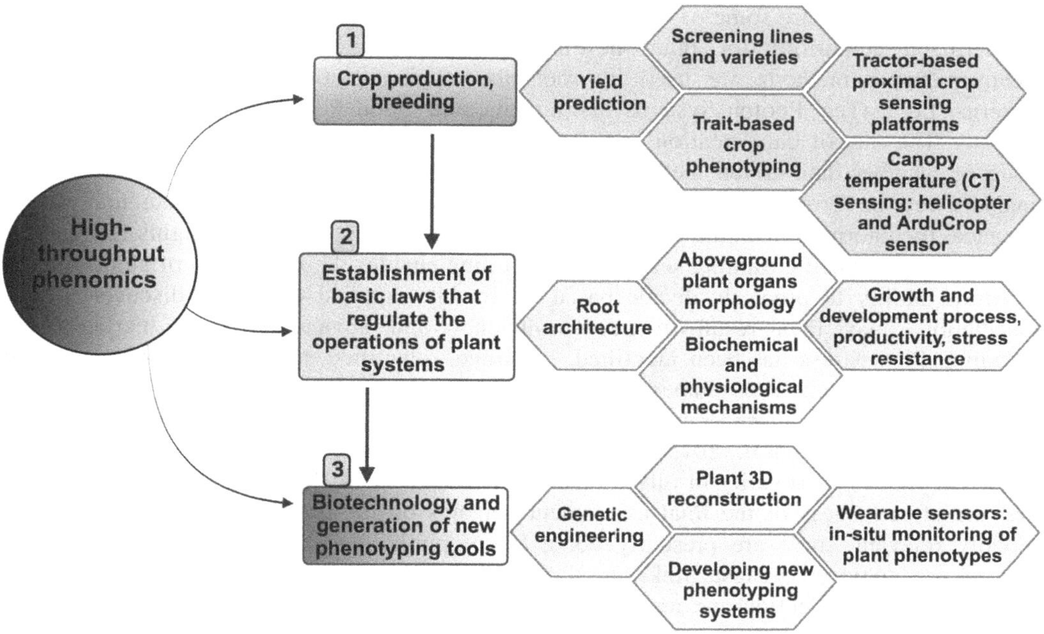

FIGURE 5.1 Utilization of high-throughput phenomics in plant biology. (1) High-throughput phenotyping (HTP) utilizes phenomics technology to assess plant traits such as yield prediction and canopy temperature sensing through remote sensing. (2) HTP in plant operation systems revolutionizes agriculture by improving the efficiency and productivity of crop cultivation such as precision agriculture (information about crop health, nutrient status, and growth patterns), deeper understanding of aboveground and below-ground plant architecture by employing HTP technologies (for instance, hyperspectral imaging, thermal cameras, etc.). (3) HTP in biotechnology and new phenotyping tools is transformative. It expedites research, enables precise breeding, facilitates the development of innovative phenotyping technologies (such as 3D reconstruction, and plant wearable sensors), and contributes significantly to enhancing crop productivity and resilience in the face of global agricultural challenges.

area. This approach allowed for the identification of key genetic factors associated with this process and facilitated the establishment of a relationship between increasing leaf area, parameters related to photosynthetic activity, and biomass accumulation (Dutta et al., 2015). The root system's phenotyping enables the observation of macroscopic (architectonics, branching, and dimensions) and microscopic (trichoblast structure, tissues, stele) traits. Modern phenotyping has shown the association between these factors, which were considerably more complex than first thought (Afonnikov et al., 2016). In recent years, it has been effective to combine the study of receptors encoding genes and the other systems that offer quick root development responses with 3D modeling of root architectonics (in order to discover correlations) (Clark et al., 2011). It is important to admit the remarkable intricacies involved in learning the spread and formation of root networks in the fields.

5.6 DATA-DRIVEN APPROACHES FOR PHENOTYPING

To streamline data collection and processing for phenotyping, automated "phenotyping platforms" include both hardware and software technologies. Growth boxes, sensors, cameras, and control systems are some of these platforms that make it easier to gather data and provide standardized conditions for plant development. Systems like vertical hydroponics and electromagnetic methods are used by root phenotyping platforms to map different root patterns. LemnaTec, Photon Systems Instruments, and Qubit Phenomics are a few well-liked systems. The rate of data creation in plant phenotyping has increased to the point where the issue of whether data-driven methods may take the place of conventional hypothesis-driven analysis can be brought up. By using machine learning techniques in data analysis, for instance, the enormous volume of data may in fact provide us with new insights (Singh et al., 2016). Using a specific training set, machine learning enables the instigation of algorithms that can have ability to precisely decide based on freshly supplied data. The discovery of data associations may then result in new proficiency and advancements in marker-assisted breeding. Plant stress has been identified, classified, quantified, and predicted using machine learning, where each level shapes the previous one. For example, three *Alternaria* species' disease indicators in oilseed rape have been categorized based on hyperspectral and thermal imaging (Baranowski et al., 2015). These imaging techniques have also been used to quantify the *Verticillium wilt* severity in olives (Calderón et al., 2015). However, one should be aware that the effectiveness of the final assessment depends on the training dataset. For this to be broadly relevant, there are presently not a lot of publicly accessible datasets. Progressing beyond associations to causal linkages is needed for well-planned trials, including thorough time series data collection and analysis.

5.7 PLANT PHENOTYPING WITH SENSORS

Complex platforms aim to increase plant/genotype phenotyping output, but expenses often make them inaccessible to researchers on a tight budget. Due to the capacity limitations of static facilities, inexpensive, portable sensors are becoming more popular. Throughput, cost-effectiveness, precision, resolution, accuracy, simplicity of use, automation, and data complexity are all aspects of user needs. Commercial sensor choices are inadequate and upscaling, downscaling, cost-cutting, and technique streamlining are impending issues. Instrumentation capabilities are constantly expanding in size and throughput to better represent more complex processes. The primary tendency at first has been to adapt commercially available technology for phenotyping purposes. More sophisticated acquisition solutions have started to emerge lately. Nowadays, sensor manufacturers are being compelled to modify their products for measuring plant features due to the rising awareness of the HTP area and significant technical investment from research teams. For instance, the arrangement of green tissues within a canopy can be assessed by employing commercially available light detection and ranging (LiDAR) instruments that operate within the red band (Rebetzke et al., 2016). The LiDAR technology provides a comprehensive three-dimensional representation of the canopy. However, it does not capture data regarding the whole canopy's density, which is essential for accurately calculating the actual biomass (Jimenez-Berni et al., 2018). Several techniques have been proposed to enhance the accuracy of biomass estimation. These include the integration of LiDAR-based 3D plant reconstruction with aerial imagery, as well as the utilization of spectral or microwave sensing (Eitel et al., 2014; Geipel et al., 2014; Kaasalainen et al., 2015; Tilly et al., 2015). Additionally, the application of RGB photography coupled with a void fill technique, or the manipulation of gradients among dark and bright colors, has shown potential in improving biomass estimation. There have been overture at various levels of conceptualization (Douglas et al., 2012; Wei et al., 2012; Wallace et al., 2013; Danson et al., 2014; Dai et al., 2018) to integrate LiDAR and multispectral imaging into a single sensor, allowing for the concurrent extraction of both structural and biological characteristics without the limitations associated with passive remote sensing. The categorization of land cover and airborne vegetation mapping is now both possible (van Rees, 2015; Morsy et al., 2017)

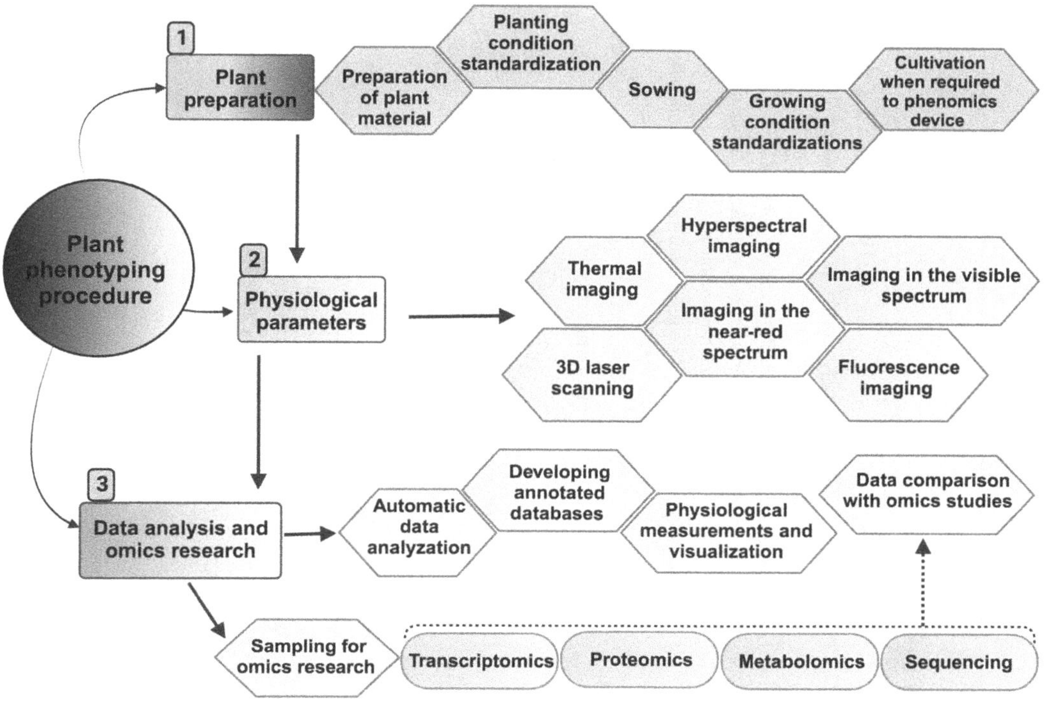

FIGURE 5.2 Overview of plant phenotyping procedure. The procedure involves several steps, from experimental design to data analysis. (1) To set up an experimental design, the determination of objectives, and selecting plant material is necessary. Other steps involve plant preparation (species, variety, or genotype), and experimental layouts (environmental conditions, replication for statistical analysis). (2) Collection of data by employing HTP phenotyping tools such as imaging, remote sensing, and sensors without harming the plant population. Additionally, for detailed analysis, plants are harvested and traits like root structure, and biomass are measured in the laboratory setting. (3) Then, data processing and analysis can be carried out in the context of research questions and objectives. Visualization can be drawn by utilizing bioinformatics and statistical tools followed by the integration of data with genomic techniques such as transcriptomics, proteomics, and sequencing technologies for better understanding. Furthermore, plant phenomics procedure is continually evolving with advancements in technology enabling researchers to gain deeper insights into plant biology, genetics, and environmental interactions.

(Figure 5.2). Such ground-based imaging methods, however, are still scarce but some of them are mentioned in (Table 5.1) (Demidchik et al., 2020).

When LiDAR is impractical or inaccurate, video game laser scanners may analyze crop architecture (Paulus et al., 2014) and ultrasonic distance sensors can be used for crop height (Barmeier et al., 2016). Cheap or DIY sensors' origin and component quality vary, making data exchange and comparability difficult without worldwide calibration. As DIY sensors are used in research, calibration documentation is essential before publishing as most commercial manufacturers conform. A larger protocol sharing, sensor development, calibration, and data processing community would solve most low-cost techniques. Open-source software, hardware, and innovative initiatives have transformed sensor production by decreasing cost and skill barriers (Tables 5.2 and 5.3) (Demidchik et al., 2020). Scientific and academic organizations may quickly develop sensor technologies using platforms like Arduino, Raspberry Pi, and other strong connected communities (Dobrescu et al., 2017; Valle et al., 2017). Public Lab follows the idea of DIY thermal cameras, spectrophotometers, and multispectral sensors.

TABLE 5.1
Crop Characteristics Using Technological Advancements and Sensor-Based Methods

Imaging Techniques	Sensor	Target Trait	Scale	Limitations
Visible light imaging/ RGB imaging	Cameras sensitive in the visible spectral range	Heading and maturity, winter hardiness	Plant	Only provides plant physiological information
Fluorescence imaging	Fluorescence cameras and setups	Photosynthesis transpiration	Leaf, plant, canopy	Difficult to analyze complicated whole-shoot of non-rosette species; pre-acclimation conditions required
Thermal imaging	Near-infrared cameras	Water use efficiency	Plant, canopy	Imaging sensor calibration and atmospheric correction are often required; sound physics-based results interpretation needed
Near-infrared imaging	Near-infrared cameras, multispectral line scanning cameras, active thermography (SWIR)	Biomass	Plant, canopy	Low sensitivity
Hyperspectral imaging	Near-infrared instruments, spectrometers, hyperspectral cameras, thermal cameras	Nitrogen uptake efficiency, shoot nitrogen content, stem carbohydrate, grain protein content, pathogen infection, weed detection, senescence	Plant, stem, grain, organ	Cost and complexity
Laser imaging	Laser scanning instruments with widely different ranges (LiDAR)	Biomass, senescence	Plant, canopy	Specific illumination is required for some laser scanning instruments
MRI	Magnetic resonance imagers	Lodging	Plant	Cost and complexity
PET	Positron emission detectors for short-lived isotopes (e.g., 11CO2)	Nitrogen uptake efficiency, water use efficiency	Plant, canopy	Can give false results if chemical balances within the body are not normal
CT	X-ray computed tomography and X-ray digital radiography	Growth stage determination, tuber development, root development	Plant	Cost and scanning time

TABLE 5.2
Plant Phenotyping Software Platforms

Software Platform	Company	Type/Sensor
Small plants		
PhenoCenter	LemnaTec (Aachen, Germany)	Laboratory conveyor system comprised of robotic arm; LS, PM/FL, VIS, NIR,
PlantScreen™	Photon Systems Instruments (Drasov, Czech Republic)	Medium-sized laboratory system; HS, MBS/FL, VLS, TH
Phenopsis	Optimalog (Tour, France)	Medium-sized laboratory system; NIR, VIS
MobileDevice	SMO ND VIB (Eklo, Ghent, Belgium)	Small-size laboratory system; NIR, MBS/VIS, LS
DroughtSpotter		Gravimetric system
HIS-PP	Environment for Visualizing Images, Inc., Boulder, CO, USA	Hyperspectral imaging
Cereals		
Phenospex FieldScan	Phenospex (Maastricht, Netherlands)	Large-scale field system; MBS/VIS, NIR, LS, FL, TH
TraitMill	CropDesign (Ghent, Belgium)	Smart greenhouse/VIS, NIR, LS

TABLE 5.3
Plant Phenotyping Hardware Platforms

Hardware Platform	Company	Type/Sensors	Objects
Lab Scanalyzer	LemnaTec (Aachen, Germany)	Compact modular laboratory system; HS, MBS/FL, NIR, VIS, TH	Small plants, seeds, fruits, algae
Scanalyzer HTS		Medium-sized modular laboratory system; LS, HS, MBS/FL, NIR, VIS, TH	Leaf plants, mushrooms, nematodes, larvae
Germination scanalyzer		Small- or medium-sized modular laboratory system; VIS, MBS/LS	Plants' seeds
Scanalyzer FIELD		Large-scale field system; HS, MBS/FL, NIR, VIS, TH	Cereals and forage crops, plant populations
PlantScreen™ Modular Systems	Photon Systems Instruments (Drasov, Czech Republic)	Medium-sized laboratory conveyor system; PM/FL, HS, LS, NIR, TH, VIS	Herbaceous plants
PlantScreen™ Compact Systems		Compact laboratory conveyor system; PM/FL, HS, LS, NIR, TH, VIS	Small plants
PlantScreen™ Self Contained Systems		Compact robotic system for greenhouses; PM/FL, HS, TH, VIS	Plants up to 50 cm long
PlantScreen™ Field		Field system; MBS/FL, HS, TH, VIS	Cereals and forage crops
PlantEye F500	Phenospex (Maastricht, Netherlands)	Multispectral 3D scanner; VIS, NIR, LS	Small plants
WIWAM XY	SMO ND VIB (Eklo, Ghent, Belgium)	Small-size laboratory system; VIS, HS, NIR	Small plants, Arabidopsis

(Continued)

TABLE 5.3 ***(Continued)***
Plant Phenotyping Hardware Platforms

Hardware Platform	Company	Type/Sensors	Objects
WIWAM Line		Medium-sized laboratory system; VIS, HS, NIR	Small plants
WIWAM Conveyor		Greenhouse conveyor system; VIS, HS, NIR	Large plants, cereals
PlantEye F500	Phenoscope (Heerlen, Netherlands)	Multispectral 3D scanner; LS, VIS, NIR	Small plants
FieldScan		Multispectral 3D-scanner; field system; MBS/VIS, NIR, LS	Large plants
FieldScale		Field gravimetric system	Small plants

5.8 DATA PROCESSING, HANDLING, AND HARMONIZING

Integration of factors is needed to phenotype complex phenotypes (Großkinsky et al., 2015; Reynolds & Schurr, 2019). Increased demand for high-resolution data is driving more sensors and huge databases. HTP makes data management and processing difficult. Data might be imaging or non-imaging, where non-imaging data offer single values per variable but need quick data transport and synchronization (Fukatsu & Hirafuji, 2005; Salehi et al., 2015; Utsushi et al., 2015). Imaging data provide geographical and temporal details but are data intensive. Real-time data transmission works in greenhouses but becomes limited in the field, particularly with aerial platforms. Images need calibration, geo-referencing, and additional processing. Automation relies on picture quality and computational power. Controlled settings have improved, but field applications require more automation (Zhou et al., 2015; Dorh et al., 2016; Valle et al., 2017; Ngo et al., 2018).

Manual tasks like ground control point identification and plot border design hinder the process. Growing datasets restrict computational power, hence, cloud processing with several suppliers handles huge data efficiently. Advanced phenotyping methods allow the collection of large datasets from RGB and hyperspectral cameras at various sizes and levels of plant organization. If made accessible to scientists, these records on plant attributes may be utilized for agricultural adaptation research and expedited breeding. Phenotyping facilities create massive amounts of data, which presents issues (Adam-Blondon et al., 2016; Tardieu et al., 2017). The major problem is managing and analyzing heterogeneous, huge data owing to its amount, diversity, velocity, value, and authenticity. Existing solutions have prioritized image analysis tools over sensor-derived data information systems. This task requires rapid, sophisticated data visualization and analysis tools (Hartmann et al., 2011; Klukas et al., 2014). Dataset and infrastructure interoperability, including Open Data debates, is the second obstacle. The International Agro Informatics Alliance and Open Ag Data Alliance promote data interoperability and security. EPPN^{2020} (http://eppn2020.plantphenotyping.eu/), Trans-PLANT (http://transplantdb.eu/), ELIXIR-EXCELERATE (https://www.elixir-europe.org/excelerate/plants), and EMPHASIS (http://emphasis.plant-phenotyping.eu/) have created FAIR data-based protocols (Findable, Accessible, Interoperable and Reusable) and software frameworks for European plant phenotyping (Junker et al., 2015; Krajewski et al., 2015; Ćwiek-Kupczyńska et al., 2016). Recently, the PHIS (www.phis.inra.fr) Information System and Planteome project formalized connections and integrated data using ontologies. Data management initiatives allow for the collection, combination, and analysis of large datasets from numerous sources to develop new knowledge (Neveu et al., 2019).

5.9 COMBINING SPEED BREEDING AND NPBTS FOR PRECISE AND FAST CULTIVAR CREATION

Speed breeding aims to shorten the time needed for the development of new crop cultivars (Watson et al., 2018). The method is referred to as controlled environment agriculture, in which different elements including light intensity, temperature, and daylight hours are meticulously regulated within controlled growth chambers to allow rapid plant growth and reproductive activities. It enables the production of several plant generations in a single year by providing an environment with extended daylight and ideal temperature conditions (Wanga et al., 2021). The combination of NPBTs with speed breeding is situated at the cutting edge of endeavors aimed at cultivating crop varieties that exhibit enhanced traits essential for mitigating global food security concerns amidst a rapidly evolving climate (Figure 5.3). The utilization of NPBTs is effectively enhancing the application of speed breeding techniques, thereby collectively advancing the field of crop improvement toward a new era characterized by increased and precise development (Whelan & Lema, 2015). By integrating NPBTs with speed breeding methodologies, breeders have the capacity to accelerate the production of numerous plant generations within the span of one year (Comeau et al., 2001). Additionally, this approach enables breeders to carefully manipulate and selectively favor specific traits with unparalleled efficiency (Table 5.4).

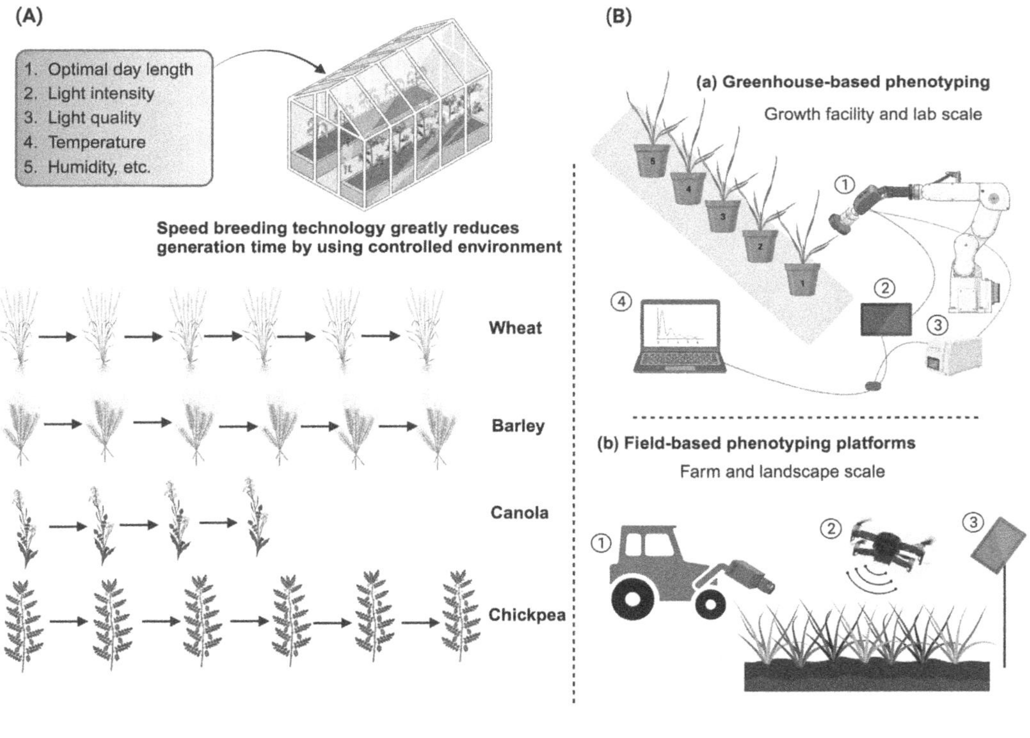

FIGURE 5.3 HTP combined with speed breeding: a revolutionary approach. The combination of controlled environment speed breeding with HTP facilitates the development of new crop varieties with improved traits such as disease resistance and high yield. (A) A Controlled environment involves extended photoperiods, controlled humidity, and temperature as well as optimized nutrient supply, leading to a significant reduction in the generation time of various crop species such as wheat, barley, canola, and chickpea. (B)The advancement in high-throughput plant phenotyping involves greenhouse-based phenotyping and field-based phenotyping. (a) In greenhouse phenotyping, a full robotics platform comprises of standard conveyor belt system for circulating plants, (1) a programmable image-driven robot arm used to take reading, (2 and 3) devices for rapid analysis, and (4) a computer system for robust data collection and interpretation. (b) In field-based phenotyping platforms, (1) tractor-based proximal crop sensing platform, (2) drones, satellite, and other remote sensing technologies and (3) ArduCrop wireless infrared sensors for continuous CT measurement are employed.

TABLE 5.4
Applications of Speed Breeding

Crops	Approach	Days to Flowering	No of Generations per Year	Source
Method of Selection: Single-Seed Descent (SSD)				
Amaranth	Photoperiod and temperature	28	6	Stetter et al. (2016)
Canola	Photoperiod, temperature, light intensity, soil moisture, immature seed germination	73	4	Watson et al. (2018)
Soybean	Photoperiod, immature seed, germination temperature	23	5	Jähne et al. (2020)
Sorghum		40–50	6	Forster et al. (2014)
Radish	Extending light exposure through supplemental light	–	NA	Ghosh et al. (2018)
Alfalfa		–	NA	Ghosh et al. (2018)
Flax		–	NA	Watson et al. (2018)
Arabidopsis		–	NA	Watson et al. (2018)
Rose		–	NA	Ghosh et al. (2018)
Spring wheat		–	4–6	Ghosh et al. (2018), Riaz et al. (2016)
Rice	Photoperiod, temperature, and high-density planting	75–85	4	Collard et al. (2017)
Method of Selection: Single-Pod Descent (SPD)				
Barley	Extending light exposure through supplemental light	–	4–6	Zheng et al. (2013), Hickey et al. (2017)
Chickpea	Photoperiod and immature seed germination	33	7	Samineni et al. (2020)
Faba bean	Plant hormones, photoperiod, light intensity, and immature seed	29–32	7	Mobini et al. (2015)
Groundnut	Photoperiod and temperature	25–27	3	O'Connor et al. (2013)
Lentil	Plant hormones, photoperiod, light intensity, and immature seed	31–33	8	Mobini et al. (2015)
Pigeon pea	Photoperiod, temperature, immature seed germination	50–56	4	Saxena et al. (2019)
Pea	Plant hormones, photoperiod, and immature seed germination	33	5	Mobini and Warkentin (2016)
Hemp	Photoperiod, short-day conditions, and water stress	44–70	1	Schilling et al. (2023)
Lupin	Early flowering and in vitro germination of immature seeds	–	5	Croser et al. (2016)
Durum wheat	Constant light and controlled temperature	–	6	Alahmad et al. (2018)

5.10 REDEFINING BREEDING WITH MULTI-OMICS, AI, AND NEXT-GENERATION SEQUENCING

The combination of breeding methods with OMICS techniques has had a significant impact on the amount and quality of data available to breeders. This has enabled breeders to achieve unprecedented success in their breeding programs (Sartor et al., 2019). Nevertheless, the management of multi-omics data poses an important challenge. The vast and complex network of data presents a significant challenge in accurately predicting and identifying the most suitable lines or cultivars for breeding programs (Rajasundaram & Selbig, 2016; Sartor et al., 2019). Hence, the integration of artificial intelligence (AI) in the field of agriculture signifies a cutting-edge approach that enables efficient analysis of large-scale multi-omics data and their correlation with fundamental biological mechanisms in diverse environmental conditions (Rajasundaram & Selbig, 2016).

The field of plant "OMICS" research has undergone significant technological advancements, resulting in the generation of large and intricate datasets. The precise interpretation of these intricate datasets holds great significance in the process of characterizing crop plants based on a particular trait (Rajasundaram & Selbig, 2016). Additionally, NGS technology has greatly accelerated the accessibility of the entire genome sequence of targeted plant species, resulting in the generation of huge datasets (Godwin et al., 2019). Moreover, when combined with transcriptomics and proteomics analysis, the marker-assisted selection (MAS) and genome-wide association studies (GWAS) approaches provide an in-depth analysis of plant genotypes and phenotypes. AI, specifically NN and deep learning (DL), is currently being utilized to enhance the effectiveness and precision of multi-omics data analysis (Parmley et al., 2019). The opacity of the mechanisms underlying these two AI functions often involves the utilization of multiple nonlinear hierarchical methods to construct nodes that facilitate the classification of datasets, resembling the behavior of neurons in the brain (Parmley et al., 2019; Niazian & Niedbała, 2020). It will undergo iterative learning and improvement processes in order to enhance the accuracy and efficiency of data mining. This will enable better prediction of the factors that contribute to disease resistance and agronomic traits, eventually accelerating breeding programs.

Currently, researchers are incorporating genotypic and environmental data, along with the observed phenotype, in order to enhance the effectiveness of the agronomic abiotic stress breeding program. The goal is to locate the genotype that possesses the most important agronomic features. It is crucial to have a comprehensive monitoring system capable of catching even the smallest modifications or variations that take place in plants since these qualities are multifactorial. Researchers have created an AI-based physiologically based gravimetric system to address this problem. This system, known as the soil-plant-atmosphere continuum (SPAC) (Godwin et al., 2019; Harfouche et al., 2019), can detect even the most minute changes in plants with regard to their surrounding soil and environment. This technique offers plant scientists an easy way to phenotype even the most minute differences in complex features throughout a range of plant growth and development phases (Harfouche et al., 2019). Furthermore, the consistent and thorough surveillance of these phenotypic data, coupled with the utilization of the next-generation AI methodology, can aid in the detection of stress-responsive quantitative trait loci (QTLs) associated with significant agronomic characteristics (Parmley et al., 2019; Niazian & Niedbała, 2020). The data produced is subsequently examined by AI or specialist software, facilitating breeders in identifying superior genotypes that exhibit ideal agronomic characteristics and resistance to diseases. The integration of this sophisticated phenotypic tool with the methodologies of MAS and GWAS enables the comprehensive analysis of plants at the molecular level, facilitating the discovery of previously unidentified genes and QTLs (Esposito et al., 2020; van Dijk et al., 2021).

5.11 CONCLUSION

Developing high-yielding cultivars that can tolerate abiotic and biotic stressors has been difficult for plant breeders in recent decades. The field of plant phenomics is having a transformative impact on the agricultural industry. The precision, scalability, and real-time data provided by phenomics enable the observation and analysis of plant features and behaviors that were previously unobservable. The incorporation of this technology streamlines the process of crop monitoring and assessment across various climatic conditions, hence enhancing our comprehension of intricate biological mechanisms. Similarly, the technique of speed breeding offers advantages in terms of time efficiency, spatial requirements, and cost-effectiveness when it comes to the development, distribution, and commercialization of improved cultivars. This approach enables the development of new and superior plant varieties with enhanced precision and predictability. The process of growth and development has a pivotal role in determining the performance of plants in response to fluctuations in their surrounding environmental conditions. Molecular breeding techniques such as MAS and GWAS can be effectively employed in conjunction with the speed breeding process to identify genes or QTLs associated with the enhancement of biotic and abiotic stress tolerance, nutritional attributes, and crop production. The advent of Next-Gen AI has revolutionized the field of breeding and agriculture, providing a platform for efficient decision-making and effective administration of extensive OMICS data. This technological advancement has facilitated a deeper understanding of plant functionalities in the face of harsh climatic conditions. In the current era of climate-smart agriculture, the utilization of phenomics and speed breeding techniques has become crucial. These methods facilitate informed decision-making in breeding programs and enable precise selection of traits. This, in turn, ensures the development of crop varieties that are resistant to various challenges, thus contributing to global food security.

REFERENCES

Adam-Blondon, A.-F., Alaux, M., Pommier, C., Cantu, D., Cheng, Z. M., Cramer, G. R., Davies, C., Delrot, S., Deluc, L., Di Gaspero, G., & Grimplet, J. (2016). Towards an open grapevine information system. *Horticulture Research*, *3*, 16056.

Afonnikov, D. A., Genaev, M. A., Doroshkov, A. V., Komyshev, E. G., & Pshenichnikova, T. A. (2016). Methods of high-throughput plant phenotyping for large-scale breeding and genetic experiments. *Russian Journal of Genetics*, *52*, 688–701.

Akkem, Y., Biswas, S. K., & Varanasi, A. (2023). Smart farming using artificial intelligence: A review. *Engineering Applications of Artificial Intelligence*, *120*, 105899, doi: 10.1016/j.engappai.2023.105899.

Alahmad, S., Dinglasan, E., Kung, K. M., Riaz, A., Derbal, N., Voss-Fels, K. P., Able, J. A., Bassi, F. M., Christopher, J., & Hickey, L. T. (2018). Speed breeding for multiple quantitative traits in durum wheat. *Plant Methods*, *14*(1), 36, doi: 10.1186/s13007-018-0302-y.

Alreshidi, E. (2019). Smart sustainable agriculture (SSA) solution underpinned by internet of things (IoT) and artificial intelligence (AI). *International Journal of Advanced Computer Science and Applications*, *10*(5), doi: 10.14569/IJACSA.2019.0100513.

Aslam, B., Basit, M., Nisar, M. A., Khurshid, M., & Rasool, M. H. (2017). Proteomics: Technologies and their applications. *Journal of Chromatographic Science*, *55*(2), 182–196, doi: 10.1093/chromsci/bmw167.

Awada, L., Phillips, P. W., & Smyth, S. J. E. (2018). The adoption of automated phenotyping by plant breeders. *Euphytica*, *214*, 1–15.

Baranowski, P., Jedryczka, M., Mazurek, W., Babula-Skowronska, D., Siedliska, A., & Kaczmarek, J. J. P. o. (2015). Hyperspectral and thermal imaging of oilseed rape (Brassica napus) response to fungal species of the genus Alternaria. *PloS One*, *10*(3), e0122913.

Barmeier, G., Mistele, B., Schmidhalter, U. J. C., & Science, P. (2016). Referencing laser and ultrasonic height measurements of barleycultivars by using a herbometre as standard. *Crop and Pasture Science*, *67*(12), 1215–1222.

Bebber, D. P., Ramotowski, M. A. T., & Gurr, S. J. (2013). Crop pests and pathogens move polewards in a warming world. *Nature Climate Change*, *3*(11), 985–988, doi: 10.1038/nclimate1990.

Brás, T. A., Seixas, J., Carvalhais, N., & Jägermeyr, J. (2021). Severity of drought and heatwave crop losses tripled over the last five decades in Europe. *Environmental Research Letters*, *16*(6), 065012, doi: 10.1088/1748-9326/abf004.

Brookes, G. & Barfoot, P. (Apr–Jun 2013). Key environmental impacts of global genetically modified (GM) crop use 1996–2011. (in Eng). *GM Crops Food*, *4*(2), 109–119, doi: 10.4161/gmcr.24459.

Calderón, R., Navas-Cortés, J. A., & Zarco-Tejada, P. J. J. R. S. (2015). Early detection and quantification of *Verticillium* wilt in olive using hyperspectral and thermal imagery over large areas. *Remote Sensing*, *7*(5), 5584–5610.

Chao, H., Zhang, S., Hu, Y., Ni, Q., Xin, S., Zhao, L., Ivanisenko, V. A., Orlov, Y. L., & Chen, M. (2023). Integrating omics databases for enhanced crop breeding. *Journal of Integrative Bioinformatics*, *20*(4), 20230012, doi: 10.1515/jib-2023-0012.

Clark, R. T., MacCurdy, R. B., Jung, J. K., Shaff, J. E., McCouch, S. R., Aneshansley, D. J., & Kochian, L. V. (2011). Three-dimensional root phenotyping with a novel imaging and software platform. *Plant Physiology*, *156*(2), 455–465.

Collard, B. C., Beredo, J. C., Lenaerts, B., Mendoza, R., Santelices, R., Lopena, V., Verdeprado, H., Raghavan, C., Gregorio, G. B., Vial, L., & Demont, M. (2017). Revisiting rice breeding methods – evaluating the use of rapid generation advance (RGA) for routine rice breeding. *Plant Production Science*, *20*(4), 337–352, doi: 10.1080/1343943X.2017.1391705.

Comeau, A., Caetano, V. R., St-Pierre, C. A., & Haber, S. (2001). Accelerated breeding: Dream or reality?. in Z. Bedö and L. Láng (Eds.), Wheat in a global environment, Proceedings of the 6th International Wheat Conference, 5–9 June 2000, Budapest, Hungary (pp. 671–679). Dordrecht: Springer Netherlands.

Corwin, D. L. (2021). Climate change impacts on soil salinity in agricultural areas. *European Journal of Soil Science*, *72*(2), 842–862, doi: 10.1111/ejss.13010.

Croser, J. S., Pazos-Navarro, M., Bennett, R. G., Tschirren, S., Edwards, K., Erskine, W., Creasy, R., & Ribalta, F. M. (2016). Time to flowering of temperate pulses in vivo and generation turnover in vivo–in vitro of narrow-leaf lupin accelerated by low red to far-red ratio and high intensity in the far-red region. Plant Cell, Tissue and Organ Culture (PCTOC), *127*(3), 591–599, doi: 10.1007/s11240-016-1092-4.

Ćwiek-Kupczyńska, H., Altmann, T., Arend, D., Arnaud, E., Chen, D., Cornut, G., Fiorani, F., Frohmberg, W., Junker, A., Klukas, C., & Lange, M. (2016). Measures for interoperability of phenotypic data: minimum information requirements and formatting. *Plant Methods*, *12*, 1–18.

da Silveira, F., Lermen, F. H., & Amaral, F. G. (2021). An overview of agriculture 4.0 development: Systematic review of descriptions, technologies, barriers, advantages, and disadvantages. *Computers and Electronics in Agriculture*, *189*, 106405, doi: 10.1016/j.compag.2021.106405.

Dai, W., Yang, B., Dong, Z., & Shaker, A. (2018). A new method for 3D individual tree extraction using multispectral airborne LiDAR point clouds. *ISPRS Journal of Photogrammetry and Remote Sensing*, *144*, 400–411, doi: 10.1016/j.isprsjprs.2018.08.010.

Danson, F. M., Gaulton, R., Armitage, R. P., Disney, M., Gunawan, O., Lewis, P., Pearson, G., & Ramirez, A. F. (2014). Developing a dual-wavelength full-waveform terrestrial laser scanner to characterize forest canopy structure. *Agricultural and Forest Meteorology*, *198*, 7–14, doi: 10.1016/j.agrformet.2014.07.007.

Deans, A. R., Lewis, S. E., Huala, E., Anzaldo, S. S., Ashburner, M., Balhoff, J. P., Blackburn, D. C., Blake, J. A., Burleigh, J. G., Chanet, B., & Cooper, L. D. (2015). Finding our way through phenotypes. *PLoS Biology*, *13*(1), e1002033.

Demidchik, V. V., Shashko, A. Y., Bandarenka, U.Y., Smolikova G. N., Przhevalskaya, D. A., Charnysh, M. A., Pozhvanov, G. A., Barkosvkyi, A. V., Smolich, I. I., Sokolik, A. I., & Yu, M. (2020). Plant phenomics: Fundamental bases, software and hardware platforms, and machine learning. *Russian Journal of Plant Physiology*, *67*, 397–412.

Dobrescu, A., Scorza, L. C., Tsaftaris, S. A., & McCormick, A. J. (2017). A "Do-It-Yourself" phenotyping system: Measuring growth and morphology throughout the diel cycle in rosette shaped plants. *Plant Methods*, *13*, 1–12.

Dong, Z. & Chen, Y. (2013). Transcriptomics: Advances and approaches. *Science China Life Sciences*, *56*(10), 960–967, doi: 10.1007/s11427-013-4557-2.

Dorh, N., Sarua, A., Stokes, J., Hueting, N. A., & Cryan, M. J. (2016). Fluorescent emission enhancement by aluminium nanoantenna arrays in the near UV. *Journal of Optics*, *18*(7), 075008.

Douglas, E. S., Strahler, A., Martel, J., Cook, T., Mendillo, C., Marshall, R., Chakrabarti, S., Schaaf, C., Woodcock, C., Li, Z., Yang, X., Culvenor, D., Jupp, D., Newnham, G., & Lovell, J. (2012). DWEL: A dual-wavelength echidna LiDAR for ground-based forest scanning. in *2012 IEEE International Geoscience and Remote Sensing Symposium*, IEEE, 4998–5001.

Dutta, S., Cruz, J. A., Jiao, Y., Chen, J., Kramer, D. M., & Osteryoung, K. W. (2015). Non-invasive, whole-plant imaging of chloroplast movement and chlorophyll fluorescence reveals photosynthetic phenotypes independent of chloroplast photorelocation defects in chloroplast division mutants. *The Plant Journal*, *84*(2), 428–442.

Ehrlich, P. R. & Holdren, J. P. (1971). Impact of population growth. *Science*, *171*(3977), 1212–1217, doi: 10.1126/science.171.3977.1212.

Eitel, J. U., Magney, T. S., Vierling, L. A., Brown, T. T., & Huggins, D. R. (2014). LiDAR based biomass and crop nitrogen estimates for rapid, non-destructive assessment of wheat nitrogen status. *Field Crops Research*, *159*, 21–32.

Endara, L., Cole, H. A., Burleigh, J. G., Nagalingum, N. S., Macklin, J. A., Liu, J., Ranade, S., & Cui, H. (2017). Building the "Plant glossary"—a controlled botanical vocabulary using terms extracted from the floras of North America and China. *Taxon*, *66*(4), 953–966.

Esposito, S., Carputo, D., Cardi, T., & Tripodi, P. (2020). Applications and trends of machine learning in genomics and phenomics for next-generation breeding. *Plants*, *9*(1), 34. [Online]. Available: https://www.mdpi.com/2223-7747/9/1/34.

Fahlgren, N., Gehan, M. A., & Baxter, I. (2015). Lights, camera, action: high-throughput plant phenotyping is ready for a close-up. *Current Opinion in Plant Biology*, *24*, 93–99.

Flood, P. J., Kruijer, W., Schnabel, S. K., van der Schoor, R., Jalink, H., Snel, J. F., Harbinson, J., & Aarts, M. G. (2016). Phenomics for photosynthesis, growth and reflectance in Arabidopsis thaliana reveals circadian and long-term fluctuations in heritability. *Plant Methods*, *12*(1), 1–14.

Forster, B., Till, B., Ghanim, A., Huynh, H., Burstmayr, H., & Caligari, P. (2014). Accelerated plant breeding. *CABI Reviews*, *9*(043), 1–16.

Fukatsu, T. & Hirafuji, M. (2005). Field monitoring using sensor-nodes with a web server. *Journal of Robotics Mechatronics*, *17*(2), 164–172.

Furbank, R. T. & Tester, M. (2011). Phenomics – technologies to relieve the phenotyping bottleneck. *Trends in Plant Science*, *16*(12), 635–644.

Gago, J., Douthe, C., Coopman, R. E., Gallego, P. P., Ribas-Carbo, M., Flexas, J., Escalona, J., & Medrano, H. (2015). UAVs challenge to assess water stress for sustainable agriculture. *Agricultural Water Management*, *153*, 9–19.

Galkovskyi, T., Mileyko, Y., Bucksch, A., Moore, B., Symonova, O., Price, C. A., Topp, C. N., Iyer-Pascuzzi, A. S., Zurek, P. R., Fang, S., & Harer J. (2012). GiA roots: Software for the high throughput analysis of plant root system architecture. *BMC Plant Biology*, *12*(1), 1–12.

Gandomi, A. & Haider, M. (2015). Beyond the hype: Big data concepts, methods, and analytics. *International Journal of Information Management*, *35*(2), 137–144.

Geipel, J., Link, J., & Claupein, W. (2014). Combined spectral and spatial modeling of corn yield based on aerial images and crop surface models acquired with an unmanned aircraft system. *Remote Sensing*, *6*(11), 10335–10355.

Gerlai, R. (2002). Phenomics: Fiction or the future? *TRENDS in Neurosciences*, *25*(10), 506–509.

Ghosh, S., Watson, A., Gonzalez-Navarro, O. E., Ramirez-Gonzalez, R. H., Yanes, L., Mendoza-Suárez, M., . . . Hickey, L. T. (2018). Speed breeding in growth chambers and glasshouses for crop breeding and model plant research. *Nature Protocols*, *13*(12), 2944–2963, doi: 10.1038/s41596-018-0072-z.

Godwin, I. D., Rutkoski, J., Varshney, R. K., & Hickey, L. T. (Mar 2019). Technological perspectives for plant breeding. (in Eng). *Theoretical and Applied Genetics*, *132*(3), 555–557, doi: 10.1007/s00122-019-03321-4.

Granier, C. & Vile, D. (2014). Phenotyping and beyond: Modelling the relationships between traits. *Current Opinion in Plant Biology*, *18*, 96–102.

Gregory, P. J., Bengough, A. G., Grinev, D., Schmidt, S., Thomas, W. B. T., Wojciechowski, T., & Young, I. M. (2009). Root phenomics of crops: Opportunities and challenges. *Functional Plant Biology*, *36*(11), 922–929.

Großkinsky, D. K., Svensgaard, J., Christensen, S., & Roitsch, T. (2015). Plant phenomics and the need for physiological phenotyping across scales to narrow the genotype-to-phenotype knowledge gap. *Journal of Experimental Botany*, *66*(18), 5429–5440.

Guan, Q., Wu, J., Yue, X., Zhang, Y., & Zhu, J. (2013). A nuclear calcium-sensing pathway is critical for gene regulation and salt stress tolerance in Arabidopsis. *PLoS Genetics*, *9*(8), e1003755.

Harfouche, A. L., Jacobson, D. A., Kainer, D., Romero, J. C., Harfouche, A. H., Mugnozza, G. S., Moshelion, M., Tuskan, G. A., Keurentjes, J. J., & Altman, A. (Nov 2019). Accelerating climate resilient plant breeding by applying next-generation artificial intelligence. (in Eng). *Trends in Biotechnology. Trends in Biotechnology*, *37*(11), 1217–1235, doi: 10.1016/j.tibtech.2019.05.007.

Hartmann, A., Czauderna, T., Hoffmann, R., Stein, N., & Schreiber, F. (2011). HTPheno: An image analysis pipeline for high-throughput plant phenotyping. *BMC Bioinformatics*, *12*(1), 1–9.

Hatfield, J. L., Boote, K. J., Kimball, B. A., Ziska, L. H., Izaurralde, R. C., Ort, D., Thomson, A. M., & Wolfe, D. (2011). Climate impacts on agriculture: Implications for crop production. *Agronomy Journal*, *103*(2), 351–370, doi: 10.2134/agronj2010.0303.

Henebry, G. M. (2011). Making sense of remotely sensing vegetation. *BioScience*, *61*(7), 568–569.

Henry, R. J. (Jan 2012). Next-generation sequencing for understanding and accelerating crop domestication. (in Eng). *Briefings in Functional Genomics*, *11*(1), 51–56, doi: 10.1093/bfgp/elr032.

Hickey, L. T., Germán, S. E., Pereyra, S. A., Diaz, J. E., Ziems, L. A., Fowler, R. A., . . . Dieters, M. J. (2017). Speed breeding for multiple disease resistance in barley. *Euphytica*, *213*(3), 64, doi: 10.1007/s10681-016-1803-2.

Huang, B. E., George, A. W., Forrest, K. L., Kilian, A., Hayden, M. J., Morell, M. K., & Cavanagh, C. R. (2012). A multiparent advanced generation inter-cross population for genetic analysis in wheat. *Plant Biotechnology Journal*, *10*(7), 826–839.

Hughes, A., Askew, K., Scotson, C. P., Williams, K., Sauze, C., Corke, F., Doonan, J. H., & Nibau, C. (2017). Non-destructive, high-content analysis of wheat grain traits using X-ray micro computed tomography. *Plant Methods*, *13*(1), 1–16.

Hutchins, D. A., Jansson, J. K., Remais, J. V., Rich, V. I., Singh, B. K., & Trivedi, P. (2019). Climate change microbiology—problems and perspectives. *Nature Reviews Microbiology*, *17*(6), 391–396.

Jähne, F., Hahn, V., Würschum, T., & Leiser, W. L. (2020). Speed breeding short-day crops by LED-controlled light schemes. *Theoretical and Applied Genetics*, *133*(8), 2335–2342, doi: 10.1007/s00122-020-03601-4.

Jimenez-Berni, J. A., Deery, D. M., Rozas-Larraondo, P., Condon, A. T. G., Rebetzke, G. J., James, R. A., Bovill, W. D., Furbank, R. T., & Sirault, X. R. (2018). High throughput determination of plant height, ground cover, and above-ground biomass in wheat with LiDAR. *Frontiers in Plant Science*, *9*, 237.

Jones, H. G., Serraj, R., Loveys, B. R., Xiong, L., Wheaton, A., & Price, A. H. (2009). Thermal infrared imaging of crop canopies for the remote diagnosis and quantification of plant responses to water stress in the field. *Functional Plant Biology*, *36*(11), 978–989.

Junker, A., Muraya, M. M., Weigelt-Fischer, K., Arana-Ceballos, F., Klukas, C., Melchinger, A. E., Meyer, R. C., Riewe, D., & Altmann, T. (2015). Optimizing experimental procedures for quantitative evaluation of crop plant performance in high throughput phenotyping systems. *Frontiers in Plant Science*, *5*, 770.

Kaasalainen, S., Holopainen, M., Karjalainen, M., Vastaranta, M., Kankare, V., Karila, K., & Osmanoglu, B. (2015). Combining LiDAR and synthetic aperture radar data to estimate forest biomass: Status and prospects. *Forests*, *6*(1), 252–270.

Kashyap, H., Ahmed, H. A., Hoque, N., Roy, S., & Bhattacharyya, D. K. (2015). Big data analytics in bioinformatics: A machine learning perspective. arXiv preprint arXiv:1506.05101.

Klukas, C., Chen, D., & Pape, J. M. (2014). Integrated analysis platform: An open-source information system for high-throughput plant phenotyping. *Plant Physiology*, *165*(2), 506–518.

Kochinke, K., Zweier, C., Nijhof, B., Fenckova, M., Cizek, P., Honti, F., Keerthikumar, S., Oortveld, M. A., Kleefstra, T., Kramer, J. M., & Webber, C. (2016). Systematic phenomics analysis deconvoluted genes mutated in intellectual disability into biologically coherent modules. *The American Journal of Human Genetics*, *98*(1), 149–164.

Krajewski, P., Chen, D., Ćwiek, H., van Dijk, A. D., Fiorani, F., Kersey, P., Klukas, C., Lange, M., Markiewicz, A., Nap, J. P., & van Oeveren, J. (2015). Towards recommendations for metadata and data handling in plant phenotyping. *Journal of Experimental Botany*, *66*(18), 5417–5427.

Lee, S. & Persson, P. (2016). Financing from family and friends. *The Review of Financial Studies*, *29*(9), 2341–2386.

Li, L., Zhang, Q., & Huang, D. J. S. (2014). A review of imaging techniques for plant phenotyping. *Sensors*, *14*(11), 20078–20111.

Lipper, L., Thornton, P., Campbell, B. M., Baedeker, T., Braimoh, A., Bwalya, M., Caron, P., Cattaneo, A., Garrity, D., Henry, K., & Hottle, R. (2014). Climate-smart agriculture for food security. *Nature Climate Change*, *4*(12), 1068–1072, doi: 10.1038/nclimate2437.

Lobos, G. A., Camargo, A. V., del Pozo, A., Araus, J. L., Ortiz, R., & Doonan, J. H. (2007). Plant phenotyping and phenomics for plant breeding. *Frontiers in Plant Science*, *8*, 313980.

Ludovisi, R., Tauro, F., Salvati, R., Khoury, S., Mugnozza Scarascia, G., & Harfouche, A. (2017). UAV-based thermal imaging for high-throughput field phenotyping of black poplar response to drought. *Frontiers in Plant Science*, *8*, 252873.

Mamta, B. & Rajam, M. V. (2017). RNAi technology: A new platform for crop pest control. *Physiology and Molecular Biology of Plants*, *23*(3), 487–501, doi: 10.1007/s12298-017-0443-x.

Mobini, S. H., Lulsdorf, M., Warkentin, T. D., & Vandenberg, A. (2015). Plant growth regulators improve in vitro flowering and rapid generation advancement in lentil and faba bean. *In Vitro Cellular & Developmental Biology – Plant*, *51*(1), 71–79, doi: 10.1007/s11627-014-9647-8.

Mobini, S. H. & Warkentin, T. D. (2016). A simple and efficient method of in vivo rapid generation technology in pea (*Pisum sativum* L.). *In Vitro Cellular & Developmental Biology – Plant*, *52*(5), 530–536, doi: 10.1007/s11627-016-9772-7.

Morsy, S., Shaker, A., & El-Rabbany, A. J. S. (2017). Multispectral LiDAR data for land cover classification of urban areas. *Sensors*, *17*(5), 958.

Neveu, P., Tireau, A., Hilgert, N., Nègre, V., Mineau-Cesari, J., Brichet, N., Chapuis, R., Sanchez, I., Pommier, C., Charnomordic, B. , & Tardieu, F. (2019). Dealing with multi-source and multi-scale information in plant phenomics: The ontology-driven phenotyping hybrid information system. *New Phytologist*, *221*(1), 588–601.

Ngo, Q. M., Ho, Y.-L. D., Pugh, J. R., Sarua, A., & Cryan, M. J. (2018). Enhanced UV/blue fluorescent sensing using metal-dielectric-metal aperture nanoantenna arrays. *Current Applied Physics*, *18*(7), 793–798, doi: 10.1016/j.cap.2018.04.007.

Niazian, M. & Niedbała, G. (2020). Machine learning for plant breeding and biotechnology. *Agriculture*, *10*(10), 436. [Online]. Available: https://www.mdpi.com/2077-0472/10/10/436.

O'Connor, D. J., Wright, G. C., Dieters, M. J., George, D. L., Hunter, M. N., Tatnell, J. R., & Fleischfresser, D. B. (2013). Development and application of speed breeding technologies in a commercial peanut breeding program. *Peanut Science*, *40*(2), 107–114, doi: 10.3146/ps12-12.1.

Orgogozo, V., Morizot, B., & Martin, A. (2015). The differential view of genotype–phenotype relationships. *Frontiers in Genetics*, *6*, 144986.

Paproki, A., Sirault, X., Berry, S., Furbank, R., & Fripp, J. (2012). A novel mesh processing based technique for 3D plant analysis. *BMC Plant Biology*, *12*, 1–13.

Parmley, K. A., Higgins, R. H., Ganapathysubramanian, B., Sarkar, S., & Singh, A. K. (2019). Machine learning approach for prescriptive plant breeding. *Scientific Reports*, *9*(1), 17132, doi: 10.1038/s41598-019-53451-4.

Paulus, S., Behmann, J., Mahlein, A.-K., Plümer, L., & Kuhlmann, H. (2014). Low-cost 3D systems: Suitable tools for plant phenotyping. *Sensors*, *14*(2), 3001–3018. [Online]. Available: https://www.mdpi.com/1424-8220/14/2/3001.

Pereyra-Irujo, G. A., Gasco, E. D., Peirone, L. S., & Aguirrezábal, L. A. (2012). GlyPh: A low-cost platform for phenotyping plant growth and water use. *Functional Plant Biology*, *39*(11), 905–913.

Pison, G., Couppié, E., & Caporali, A. (2022). The population of the world, 2022. (in Eng). *Population & Societies*, *603*(8), 1–8, doi: 10.3917/popsoc.603.0001.

Plazas, M., Andújar, I., Vilanova, S., Gramazio, P., Herraiz, F. J., & Prohens, J. (2014). Conventional and phenomics characterization provides insight into the diversity and relationships of hypervariable scarlet (*Solanum aethiopicum* L.) and gboma (*S. macrocarpon* L.) eggplant complexes. *Frontiers in Plant Science*, *5*, 318.

Pound, M. P., French, A. P., Atkinson, J. A., Wells, D. M., Bennett, M. J., & Pridmore, T. (2013). RootNav: Navigating images of complex root architectures. *Plant Physiology*, *162*(4), 1802–1814.

Rajasundaram, D. & Selbig, J. (Apr 2016). More effort – more results: Recent advances in integrative 'omics' data analysis. (in Eng). *Current Opinion in Plant Biology*, *30*, 57–61, doi: 10.1016/j.pbi.2015.12.010.

Rebetzke, G. J., Jimenez-Berni, J. A., Bovill, W. D., Deery, D. M., & James, R. A. (2016). High-throughput phenotyping technologies allow accurate selection of stay-green. *Journal of Experimental Botany*, *67*(17), 4919–4924, doi: 10.1093/jxb/erw301.

Reynolds, M. & Schurr, U. (2019). The 4th International Plant Phenotyping Symposium. *Plant Science: An International Journal of Experimental Plant Biology*, *282*, 1–1.

Riaz, A., Periyannan, S., Aitken, E., & Hickey, L. (2016). A rapid phenotyping method for adult plant resistance to leaf rust in wheat. *Plant Methods*, *12*(1), 17, doi: 10.1186/s13007-016-0117-7.

Rolfe, S. A. & Scholes, J. D. (2010). Chlorophyll fluorescence imaging of plant–pathogen interactions. *Protoplasma*, *247*, 163–175.

Salehi, A., Jimenez-Berni, J., Deery D. M., Palmer, D., Holland, E., Rozas-Larraondo, P., Chapman, S. C., Georgakopoulos, D., & Furbank R. T. (2015). SensorDB: A virtual laboratory for the integration, visualization and analysis of varied biological sensor data. *Plant Methods*, *11*(1), 1–14.

Samineni, S., Sen, M., Sajja, S. B., & Gaur, P. M. (2020). Rapid generation advance (RGA) in chickpea to produce up to seven generations per year and enable speed breeding. *The Crop Journal*, *8*(1), 164–169, doi: 10.1016/j.cj.2019.08.003.

Sartor, R. C., Noshay, J., Springer, N. M., & Briggs, S. P. (Sep 3 2019). Identification of the expressome by machine learning on omics data. (in Eng). *Proceedings of the National Academy of Sciences of the United States of America*, *116*(36), 18119–18125, doi: 10.1073/pnas.1813645116.

Saxena, K. B., Saxena, R. K., Hickey, L. T., & Varshney, R. K. (2019). Can a speed breeding approach accelerate genetic gain in pigeonpea? *Euphytica*, *215*(12), 202, doi: 10.1007/s10681-019-2520-4.

Schilling, S., Melzer, R., Dowling, C. A., Shi, J., Muldoon, S., & McCabe, P. F. (2023). A protocol for rapid generation cycling (speed breeding) of hemp (Cannabis sativa) for research and agriculture. *The Plant Journal*, *113*(3), 437–445, doi: 10.1111/tpj.16051.

Schnable, J. C. & Freeling, M. (2011). Genes identified by visible mutant phenotypes show increased bias toward one of two subgenomes of maize. *PloS One*, *6*(3), e17855.

Shan, Q., Wang, Y., Li, J., Zhang, Y., Chen, K., Liang, Z., Zhang, K., Liu, J., Xi J. J., Qiu, J. L., & Gao, C. (Aug 2013). Targeted genome modification of crop plants using a CRISPR-Cas system. (in Eng). *Nature Biotechnology*, *31*(8), 686–688, doi: 10.1038/nbt.2650.

Singh, A., Ganapathysubramanian, B., Singh, A. K., & Sarkar, S. (2016). Machine learning for high-throughput stress phenotyping in plants. *Trends in Plant Science*, *21*(2), 110–124.

Skelly, D. A., Merrihew, G. E., Riffle, M., Connelly, C. F., Kerr, E. O., Johansson, M., Jaschob, D., Graczyk, B., Shulman, N. J., Wakefield, J. , & Cooper, S. J. (2013). Integrative phenomics reveals insight into the structure of phenotypic diversity in budding yeast. *Genome Research*, *23*(9), 1496–1504.

Stetter, M. G., Zeitler, L., Steinhaus, A., Kroener, K., Biljecki, M., & Schmid, K. J. (2016). Crossing methods and cultivation conditions for rapid production of segregating populations in three grain amaranth species. (in Eng). *Frontiers in Plant Science, Original Research,7*, 816, doi: 10.3389/fpls.2016.00816.

Tardieu, F., Cabrera-Bosquet, L., Pridmore, T., & Bennett, M. (2017). Plant phenomics, from sensors to knowledge. *Current Biology*, *27*(15), R770–R783.

Thorp, K. R., Gore, M. A., Andrade-Sanchez, P., Carmo-Silva, A. E., Welch, S. M., White, J. W., & French, A. N. (2015). Proximal hyperspectral sensing and data analysis approaches for field-based plant phenomics. *118*, 225–236.

Tilly, N., Hoffmeister, D., Cao, Q., Lenz-Wiedemann, V., Miao, Y., & Bareth, G. (2015). Transferability of models for estimating paddy rice biomass from spatial plant height data. *Agriculture*, *5*(3), 538–560.

Timple, S., Hay, F. R., & Van Duijn, B. (2014). Can chlorophyll flourescence be used to determine the optimal time to harvest rice seeds for long-term genebank storage. *Philippine Journal of Crop Science (Philippines)*, *19*(1), 403.

United Nations. (2022). GLOBAL ISSUES: Population Available: https://www.un.org/en/global-issues/population.

Utsushi, H., Abe, A., Tamiru, M., Ogasawara, Y., Obara, T., Sato, E., Ochiai, Y., Terauchi, R., & Takagi, H. (2015). WIPPER: An accurate and efficient field phenotyping platform for large-scale applications. *Breeding Science*, *65*(3), 285–289.

Valle, B., Simonneau, T., Boulord, R., Sourd, F., Frisson, T., Ryckewaert, M., Hamard, P., Brichet, N., Dauzat, M., & Christophe, A. (2017). PYM: A new, affordable, image-based method using a Raspberry Pi to phenotype plant leaf area in a wide diversity of environments. *Plant Methods*, *13*, 1–17.

van Dijk, A. D. J., Kootstra, G., Kruijer, W., & de Ridder, D. (Jan 22 2021). Machine learning in plant science and plant breeding. (in Eng). *iScience*, *24*(1), 101890, doi: 10.1016/j.isci.2020.101890.

van Rees, E. J. G. (2015). The first multispectral airborne LiDAR sensor. *GeoInformatics*, *18*(1), 10.

Vanhaeren, H., Gonzalez, N., & Inzé, D. (2015). A journey through a leaf: Phenomics analysis of leaf growth in Arabidopsis thaliana. *The Arabidopsis Book/American Society of Plant Biologists*, *13*, e0181.

Varshney, R. K., Chen, W., Li, Y., Bharti, A. K., Saxena, R. K., Schlueter, J. A., Donoghue, M. T., Azam, S., Fan, G., Whaley, A. M., & Farmer, A. D. (Nov 6 2011). Draft genome sequence of pigeonpea (Cajanus cajan), an orphan legume crop of resource-poor farmers. (in Eng). *Nature Biotechnology*, *30*(1), 80, doi: 10.1038/nbt.2022.

Wallace, A. M., McCarthy, A., Nichol, C. J., Ren, X., Morak, S., Martinez-Ramirez, D., Woodhouse, I. H., & Buller, G. S. (2013). Design and evaluation of multispectral LiDAR for the recovery of arboreal parameters. *IEEE Transactions on Geoscience and Remote Sensing*, *52*(8), 4942–4954.

Wanga, M. A., Shimelis, H., Mashilo, J., & Laing, M. D. (2021). Opportunities and challenges of speed breeding: A review. *Plant Breeding*, *140*(2), 185–194.

Warner, J. L., Denny, J. C., Kreda, D. A., & Alterovitz, G. (2015). Seeing the forest through the trees: Uncovering phenomic complexity through interactive network visualization. *Journal of the American Medical Informatics Association: JAMIA*, *22*(2), 324.

Watson, A., Ghosh, S., Williams, M. J., Cuddy, W. S., Simmonds, J., Rey, M. D., Asyraf Md Hatta, M., Hinchliffe, A., Steed, A., Reynolds, D. , & Adamski, N. M. (2018). Speed breeding is a powerful tool to accelerate crop research and breeding. *Nature Plants*, *4*(1), 23–29.

Wei, G., Shalei, S., Bo, Z., Shuo, S., Faquan, L. , & Xuewu, C. (2012). Multi-wavelength canopy LiDAR for remote sensing of vegetation: Design and system performance. *ISPRS Journal of Photogrammetry and Remote Sensing*, *69*, 1–9.

Whelan, A. I. & Lema, M. A. (2015). Regulatory framework for gene editing and other new breeding techniques (NBTs) in Argentina. (in Eng). *GM Crops & Food*, *6*(4), 253–265, doi: 10.1080/21645698.2015.1114698.

White, J. W., Andrade-Sanchez, P., Gore, M. A., Bronson, K. F., Coffelt, T. A., Conley, M. M., Feldmann, K. A., French, A. N., Heun, J. T., Hunsaker, D. J. , & Jenks, M. A. (2012). Field-based phenomics for plant genetics research. *Field Crops Research*, *133*, 101–112.

Yang, W., Duan, L., Chen, G., Xiong, L., & Liu, Q. (2013). Plant phenomics and high-throughput phenotyping: Accelerating rice functional genomics using multidisciplinary technologies. *Current Opinion in Plant Biology*, *16*(2), 180–187.

Zheng, Z., Wang, H. B., Chen, G. D., Yan, G. J., & Liu, C. J. (2013). A procedure allowing up to eight generations of wheat and nine generations of barley per annum. *Euphytica*, *191*(2), 311–316, doi: 10.1007/s10681-013-0909-z.

Zhou, B., Elazab, A., Bort, J., Vergara, O., Serret, M. D. , & Araus, J. L. (2015). Low-cost assessment of wheat resistance to yellow rust through conventional RGB images. *Computers and Electronics in Agriculture*, *116*, 20–29.

GLOSSARY

Artificial intelligence (AI) The development of computer systems that can do tasks that typically require human intellect.

Deep learning (DL) A subfield of machine learning that focuses on the development and use of artificial neural networks with several layers. These neural networks are designed to

Food security The ongoing and reliable access to an adequate quantity of food that is safe for consumption and provides necessary nutrients for individuals.

Genome editing A technique used to introduce precise modifications to the DNA of a cell or organism.

Genome-Wide Association Studies (GWAS) A kind of research methodology that aims to identify genetic variants associated with various traits or diseases throughout the whole genome.

Genotype The specific variation existing at a particular site, known as a locus, within the genome.

High-throughput phenotyping (HTP) Use of many instruments to characterize plant phenotypes. These tools may range from imaging techniques conducted on the ground to aerial phenotyping methods and even remote sensing approaches.

High-throughput sequencing The simultaneous reading of a large number of DNA molecules.

Hyperspectral imaging Process of capturing and analyzing the spatial and spectral properties of an object via the acquisition of images at several wavelengths.

Light Detection and Ranging (LiDAR) Optical remote sensing technology that is used for measuring the distance to a certain object.

Marker-Assisted Selection (MAS) Use of DNA markers as a means of indirect selection criteria in the process of choosing agriculturally significant features during crop breeding.

Morphometric features The features measured every week such as body weight, beak length, drumstick length, shank length, drumstick circumference, shank circumference, body length, and wing spread.

Multi-spectral imaging A set of techniques used to capture pictures that correspond to a minimum of two spectral channels.

Neural networks (NN) Computational models that aim to mimic the information-processing mechanisms of the human brain.

New Plant Breeding Technologies (NPBTs) Novel methods and techniques used in the field of plant breeding.

Next-generation sequencing (NGS) Sequencing method that enables the simultaneous sequencing of many DNA fragments in a massively parallel manner.

Phenome Characteristics that are shown by a cell, tissue, organ, organism, or species.

Phenomics Involves a comprehensive examination of the physical and biological characteristics of plants.

Remote sensing The seeing and analyzing of the physical attributes of a certain geographical region using measuring the radiation that is reflected or emitted from that location, often achieved via the use of satellites or airplanes.

Smart farming The use of emerging technology in the field of agriculture coincides with the advent of the Fourth Industrial Revolution.

Speed breeding Practice used by plant breeders to enhance crop yield by the manipulation of temperature, light duration, and intensity, hence facilitating accelerated plant growth.

The soil-plant-atmosphere continuum (SPAC) The interconnected process via which water is transported from the soil, passes through plants and eventually reaches the atmosphere.

Thermal imaging The conversion of infrared (IR) radiation into visual pictures.

Vertical hydroponics Cultivation method that enables the growth of plants in a vertical orientation.

X-ray digital radiography An upgraded iteration of x-ray examination that generates a digital radiographic picture, which can be promptly seen on a computer screen.

X-ray tomography Used for the three-dimensional imaging of plant tissues and organs.

6 Genome Editing Technologies for Accelerating Crop Improvement

An Updated Overview

Anil Namdev Kale, Raj Kumar, Rushikesh Sanjay Mane, Sagar Krushnaji Rangari, Praful Mukund Damdar, Zeba Quaiyum, and Karma L. Bhutia
Department of Agricultural Biotechnology and Molecular Biology, Dr. Rajendra Prasad Central Agricultural University, Pusa, Bihar, India

Akbar Hossain
Division of Soil Science, Bangladesh Wheat and Maize Research Institute, Dinajpur, Bangladesh

6.1 INTRODUCTION

In the era of a rapidly growing population, global hunger and food insecurity have been increasing at an alarming rate since 2015. Over 2 billion people do not have regular access to safe, nutritious, and sufficient food in the world. Given this, the United Nations has established a goal of zero hunger as one of its sustainable development goals by 2030 (Blesh et al., 2019). Advanced technologies are required to mitigate food and nutritional insecurity (Pachapur et al., 2020). Crop production systems must use modern effective strategies to increase their outputs so that they can sustainably feed a population that is expanding quickly. Conventional crop breeding techniques often raise yields too slowly to keep up with the rising demand for food. The lengthy generation times of crop plants are one reason for this delayed rate of progress. Also, global warming and climate change have had a significant impact on plant growth and food production. Therefore, reducing such limitations and consequences is essential for sustainable agricultural output. Speed breeding is a method that significantly shortens the generation period and accelerates plant breeding programs worldwide. With the advancement in crop improvement technologies, the best way to create a pipeline of new crops is through speed breeding in conjunction with other cutting-edge technologies like genome editing (GE).

GE is one of the important strategies to maintain global food security and achieve sustainable agricultural development. It gives promising output with high speed and precision through targeted mutagenesis to obtain new varieties with improved traits (Ahmad, 2023). Moreover, it has revolutionized the agricultural field by fetching more profit in the form of increased yield, tolerance against pesticides as well as climate change-driven different types of biotic and abiotic stresses. Also, gene editing became the base of precision plant breeding programs. A precision breeding program can accelerate the development of new genetic variability and make cultivar development more efficient without sexual crossings by accumulating favorable alleles (Farinati et al., 2023). Therefore, GE technology attracted researchers as a safer and easier tool for GE in several living organisms

DOI: 10.1201/b23372-6

mostly in plants. This technology, developed fast lately to avoid the obstacles that face genetically modified (GM) crops. The development of transgenic-free edited plants without the introduction of foreign DNA has met with regulatory approval in various countries (Abdallah et al., 2021). Taking into consideration, GE is treated as a system for speed breeding in the modern agricultural field. A plethora of approaches have been incorporated to speed up the breeding program with the association of GE tools. Some of these involve *de novo* domestication of wild cultivars, gene stacking of homeologs in polyploid crops, and the development of haploid inducer lines and male sterile (MS) lines. There are various GE techniques, all of which are modifications of the DNA break and repair mechanisms that occur naturally. In a DNA-protein interaction, Xenopus first found zinc finger proteins that cling to the DNA with the help of repetitive zinc-binding domains. Later on, the combination of zinc finger with DNA cutting function consisting of specific DNA (or RNA) sequences originates zinc finger nuclease (ZFN) (Bhutia & Tyagi, 2017). Another analogous strategy is used by transcription activator-like effector nucleases (TALEN), first discovered in Xanthomonas secreted effector protein.

A DNA-recognition domain on the effector protein can be used to modify its association with endonuclease (such as Fok1) for gene/GE (Bhutia & Tyagi, 2017). Both ZFN and TLEN are efficient tools for recognizing specified DNA sequences and carrying out double-stranded DNA cleavage operations. The main drawback of these approaches, however, is the laborious process of creating a fresh pair of ZFNs and TALEN each time one wants to edit several sites and the high likelihood of off-target, which could have a negative impact on the cell (Zhang et al., 2023a). Thus, the more contemporary CRISPR-mediated improvements in crops are promising among all approaches. The doctrine behind the CRISPR system is the recognition of targets by RNA and DNA complementarity, which has become an integral part of agricultural research in life science (Chen et al., 2021a). It can target any form of genetic information either in the form of DNA or RNA. Additionally, it can manipulate them in several ways, like knock-in, knock-out, activation or repression, base editing, prime editing, epigenome engineering, and nuclear rearrangement. (Ahmad et al., 2021). Among all gene editing tools, CRISPR-Cas is now widely used because of its precision and high throughput. The workings of CRISPR-Cas9-mediated DNA cleavage rely on the 20-bp sequence called guide RNA (gRNA) and the protospacer adjacent motif (PAM) which is one of the preferred variants among all the CRISPR Cas systems (Allemailem et al., 2023). The *in vitro* and *in vivo* synthesis of ribonucleoprotein complex (RNP-complex) is easier than the synthesis of protein homing modules in TALENs and ZFNs tools (Voytas, 2014). The applications of the CRISPR-Cas technique have been reported in major crops (Gan et al., 2022). All these approaches speed up the breeding approaches and thus expedite crop production and productivity to secure foods for all.

The natural genetic diversity is low in plants, which limits crop improvement. This bottleneck can be overcome by directed evolution and random mutation in the target gene of host plants. The directed evolution aims to produce diverse sequence variants with desirable traits under selective pressure. Accordingly, the CRISPR system can be programmed to target any genomic locus and thus perform a targeted CRISPR-mediated directed evolution (Butt et al., 2020). Alternatively, plant domestication is the process of genetically altering a wild species to produce a new kind of plant tailored to human needs (Doebley et al., 2006). It is mostly seen in mustard and cauliflower. It occurs due to gain or loss of function of an allele (Meyer & Purugganan, 2013). However, it causes a loss of genetic diversity which is important for crop improvement. On the contrary, *de novo* domestication or "Genome Editing accelerated Re-Domestication" (GEaReD) of wild relatives has come into existence (Yu & Li, 2022). This approach helps in retaining genetic diversity and domestication-related traits in wild relatives. Moreover, the development of MS lines using GE-based CRISPR technology aids in the production of hybrid seeds against the conventional approach in a short time with improved yield. It modified the genes of some crops of economic interest like rice, wheat, maize, sorghum, tomato, and cotton. Doubled haploid (DH) development from the haploid speeds up the production of homozygous lines in the minimum number of generations. *In vivo* haploid induction, based on DH technology, is a better alternative than other techniques to develop completely homozygous individuals only in two generations (Forster & Thomas, 2005; Gordillo & Geiger, 2008; Chang & Coe, 2009a). This technology has been applied in many cereals like rice, wheat, and maize; some legumes; and even millets. Therefore, DH production is very straightforward and time-saving, once haploid inducer lines develop in any crop.

6.2 GENOME-EDITING-MEDIATED EARLY VARIETAL DEVELOPMENTS WITH YIELD OPTIMIZATION

Genetic engineering is an effective approach to combating insect pests and weeds for which natural variation is insufficient (Rafalski, 2017). However, the farmers of developed countries have realized the benefits of GM crops produced through genetic engineering (Klümper & Qaim, 2014). The prices of the seed and consumer opposition to GM crops have prevented them from entering developing nations, especially Africa (Rafalski, 2017). Prior to the advent of targeted GE, the majority of genetic engineering approaches used to introduce novel features into organisms depended on stable yet random genetic change (such as traditional mutagenesis) or foreign genetic material (transgenesis). This randomness makes it possible for undesirable changes to occur in the genome, such as (unwanted) disruption of genes and/or regulatory elements or the emergence of new undesirable open reading frames (e.g., with a sequence similar to known toxins or allergens), which explains why the screening and selection procedures for GMOs with desirable traits are frequently laborious and time-consuming. Site-directed nucleases (SDN) are enzymes that can be used to target a specific locus of DNA, which makes it possible to lessen the likelihood of these undesirable deleterious consequences. CRISPR-Cas system is the most promising among all SDN approaches. On the contrary, backcrossing required introducing the newly produced feature from

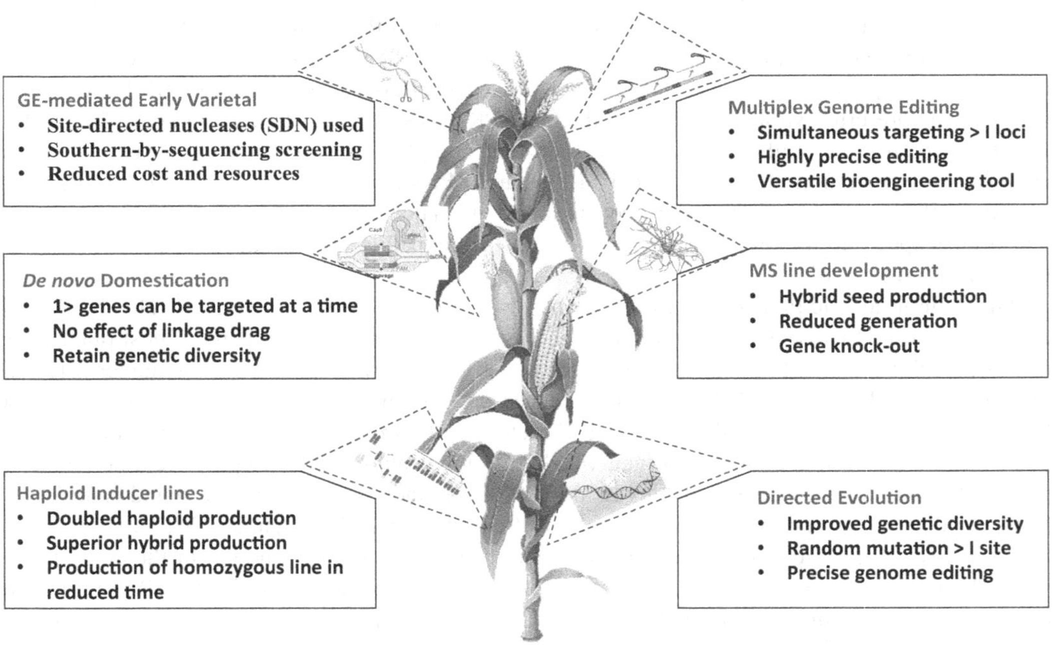

FIGURE 6.1 Different approaches of genome editing system for plant speed breeding. Each approach's peculiar features are given an inbox with the connected image of technology being used or improved crop improvement. The maize plant is taken here just for representing an edited improved plant showing robust yield and tolerant to abiotic stress and resistance to biotic stress.

an experimental line into elite varieties, negating the benefit of gene editing in terms of speeding up product development and removing yield drag. Therefore, editing a gene directly into elite varieties eliminates the need for backcrossing (Dhugga, 2022). No components of the vector backbone should remain in the edited plants after self-pollinating or out-crossing the edited plants to non-edited plants with the same genetic make-up, along with concurrent screening for any unintended changes in the genome with highly sensitive molecular tools (Zastrow-Hayes et al., 2015). Southern-by-sequencing is a robust screening approach for molecular characterization of GM crops. It replaces Southern blot analysis for event selection in a high-throughput molecular characterization setting with sequence capture combined with next-generation sequencing (NGS) technology. As a result, compared to forward breeding, the number of generations needed for the gene-altered plants to reach commercialization is only 2–3 as opposed to 5–6 (Zastrow-Hayes et al., 2015). Ultimately, it reduced cost with lesser utilization of resources (the most expensive components of varietal development) as well as time (Figure 6.1).

6.3 CONQUERING THE BOTTLENECK OF GENOME EDITING FOR SPEED BREEDING OF CROP PLANTS

The GE potential in improving agronomic performances of crops is mostly restricted by two important factors, namely, regeneration efficiency and transformable genotypes (Matres et al., 2021). In that context, cell morphogenesis genes are associated with transformation vectors to ease the transformation in elite varieties of major crops. It involves both monocot and dicot plants like rice, maize, wheat, and citrus. The principle behind this transformation is the use of chimeric protein, i.e., protein factor with its cofactor to improve efficiency and regeneration potential. It involves growth-regulating factor 4 (GRF4) and GRF-interacting factor 1 (GIF1) present in wheat greatly improves the regeneration efficiency of triticale as well as rice (Luo & Palmgren, 2021; Lee & Wang, 2023). Moreover, GRF4-GIF1 associated with CRISPR-Cas9 also generated wheat plants by interfering with the function of the *AP2L-A5* gene (Campos et al., 2021). Similarly, it improved the regeneration efficiency of dicot plants like citrus and lettuce (Bull et al., 2023). The crop improvement can be expedited for many of the traits but mostly for biofortification and biotic resistance such as rancidity in pearl millet can be reduced by suppressing the fatty acid desaturase (*FAD2*) gene (Damude & Kinney, 2008); single gene mutation in sorghum produces strigolactones which provide resistance against striga(Gobena et al., 2017); adult plant resistance (APR) conferred by *APR* genes in wheat against broad spectrum fungal pathogens (Moore et al., 2015).

6.4 DOMESTICATION OF WILD CULTIVARS THROUGH GENOME EDITING

Domestication of crops was started thousands of years ago (Tubb & Seba, 2021). During the process of domestication, divergent selection pressure was employed on cereal crops for crop improvement (Rangan et al., 2023), and developed elite cultivars on the basis of selecting natural variation and recombining important genes from wild progenitors (Lidder & Sonnino, 2012). Genes in the QTL region that had a significant impact were selected and fixed in the crop genome during the process of domestication. This led to crop productivity and yield being successfully increased, but crucial wild crop characteristics like disease resistance and stress tolerance were lost (van de Wouw et al., 2010). Selection during domestication is responsible for reducing the genetic diversity of the crop (Wolter et al., 2019) and making them susceptible to various biotic and abiotic stress (Bohra et al., 2022; Kumar et al., 2022). Conventional breeding is possible with monogenic traits and Mendelian inheritance (Zsögön et al., 2018). For the introduction of polygenic traits such as disease resistance and stress tolerance, linkage drag is the main barrier (Collard & Mackill, 2008). The linkage drag of wild alleles is fixed and difficult to break (Figure 6.1) while focusing on the introgression of important wild alleles for a particular trait into elite cultivars (Huang et al., 2022). Various types of linkage drag as a barrier for crop improvement have already been reported in different crops (Lin et al., 2014; Wei et al., 2021), and a cross between wild and exotic breeding material is sterile because of the post-zygotic incompatibility barrier and to preserve embryos derived from such a cross, need of the embryo rescue (Kumar et al., 2022). The use of wild and exotic germplasm is avoided in breeding programs due to several domestication syndromes and unwanted traits (Kumar et al., 2022).

Several genes involved in domestication were already known, map-based cloning and GWAS provide knowledge of key genes involved in the domestication of crops (Huang et al., 2022). Domestication occurs due to the gain or loss in function of an allele (Meyer & Purugganan, 2013). Such types of mutations in the genes responsible for domesticating wild crops have been made in wild progenitors using a novel GE tool, i.e., CRISPR-Cas9. The improved CRISPR-Cas9 technique known as multiplex CRISPR-Cas9 is employed to insert or edit a set of domesticated genes in wild relatives (Shen et al., 2017; Li et al., 2018; Yu et al., 2022). The rapid development of climate-resilient varieties will be enhanced by the domestication of wild crop relatives with the help of improved GE techniques. Groundcherry (Lemmon et al., 2018) and tomato (Li et al., 2018; Zsögön et al., 2018) are two predominant examples of re or *de novo* domestication. By using CRISPR-Cas9 to introduce this gene into wild relatives, the breeding process can be completed more rapidly.

Modification into crucial genes that lead to the domestication of wild cultivars will satisfy the need for polygenic traits and agronomically important traits. So, a new approach called Re or *de novo* domestication or GEaReD of wild relatives has come into existence (Yu & Li, 2022). This approach helps retain domestication-related traits in wild relatives. The basic difference between domestication and *de novo* domestication is represented in Table 6.1.

6.4.1 Genome Editing Accelerated Re-Domestication (GEaReD)

In conventional breeding, frequent crossing over many generations is essential to achieve domestication with interesting characteristics (Haroon et al., 2019). While GEaReD functions in a highly automated environment, with the aid of automated screening facilities, omics facilities, and evaluating the obtained data with the help of algorithms it examines wild progenitors with desirable traits to discover promising candidates. The GE process is applied to identify promising candidates, and after altering major domestication genes, a new cultivar is created with formerly non-existent properties (Hanak et al., 2022). Advanced GE technologies, like CRISPR-Cas9, are employed to introduce crucial agronomic traits into the promising candidate of a wild relative (Lemmon et al., 2018; Li et al., 2018; Zsögön et al., 2018; Yu et al., 2022). Some of the *de novo* domesticated crops are briefly described.

TABLE 6.1
Basic Differences between Domestication and *de novo* Domestication

Domestication	*De novo* Domestication
Long-term artificial selection leads to the accumulation of beneficial mutation	Selection of promising candidates with the help of omics, AI and mutated by genome editing systems
Loss in genetic diversity of crops	Retain genetic diversity and domesticated genes
No use of genome editing tools	Use of advanced genome editing tools to develop a new cultivar
Need more generations to fix traits of interest	Need a few generations to fix traits of interest
Linkage drag is a barrier to crop improvement	No effect of linkage drags
Few numbers of genes can be targeted at a time	Many numbers of genes can be targeted at a time

6.4.1.1 Tomato

Solanum pimpinellifolium, the wild progenitor of the tomato, was chosen as a promising candidate for *de novo* domestication. Six loci that are crucial for the yield and productivity of tomato were targeted in wild progenitor and edited with the help of CRISPR-Cas9, the developed T_1 generation plants displayed large fruit-size tomatoes (Table 6.2) and plants were able to withstand stress conditions as compared to parental wild type (Zsögön et al., 2018).

TABLE 6.2
List of Traits Engineered in Wild Crops for de novo Domestication

Sl. No.	Crop	Targeted Genes or Loci	Targeted Trait	Type of Mutation	Altered Phenotype	Reference
1.	Tomato (*Solanum pimpinellifolium*)	self-pruning	Plant growth habit	Loss in function	Reduced plant height and number of sympodial units	Zsögön et al., 2018
		ovate	Fruit shape		Oval fruit shape	
		multiflora	Number of flowers		Increase in the number of flowers	
		fasciated (fas)	Size of fruit		Increase in fruit size	
		lycopene beta cyclase	Nutritional quality		Flowers with orange antheridial cones (deep red fruits)	
2.	Tomato (*Solanum galapagense*)	*fw2.2*	Fruit size	Loss in function	30% increase in fruit size	Li et al., 2018
		locule number	Locule number		Increase the number of locules	
		fasciated (fas)	Fruit of size		Increase in fruit size	
		Cyc-B	Fruit color		Red color fruits	
		SlWUS	Flower and fruit production		Increased fruit size	
4.	Groundcherry (*Physalis pruinosa*)	*Ppr-SP*	Fruit production	Loss in function	Sympodial growth termination	Lemmon et al., 2018
		Ppr-SP5G	Number of flowers		More floral plant	
		Ppr-CLV1	Fruit size		Produced flowers with more organs	
5.	Tetraploid wild rice (*Oryza alta*)	*qSH1*	Seed shattering	Loss in function	Prevent seed shattering	Yu et al., 2021
		An-1 & *An-2*	Awn length		Short awn length	
		Sd1	Height of plant		Reduced height of the plant	
		GS3	Seed morphology in diploid rice		Increase in grain length	
		OalPA1	Stem diameter		Increase in stem diameter	
		Ghd7	Grain number, plant height and heading date		Reduced heading date	

Table 6.2 provides a detailed explanation of the targeted genes and altered phenotype. Two separate *S. pimpinellifolium* accession, LA1589 and LA1357 which are resistant to bacterial spot disease (T3 strain) and salt tolerant, respectively, were successfully domesticated by targeting cis-regulatory regions of the gene with the use of CRISPR-Cas9. All modified lines of T_1 progenies exhibit strong salt tolerance and hypersensitivity for bacterial spots (Li et al., 2018).

6.4.1.2 Groundcherry

Groundcherry (*Physalis pruinosa)* is a type of sweet berry native to Central and South America that is more distinctly relative to tomato (Särkinen et al., 2013). The orthologous genes for tomato domestication were identified and altered in groundcherry. That leads to improvement in the plant architecture, flower production, and fruit size traits observed in transgenic plants (Table 6.2).

6.4.1.3 Rice

A technique for the first time *de novo* domestication of polyploidy rice (*Oryza alta*) was developed (Yu et al., 2021). They established a *de novo* domestication roadmap for domestication and transformation systems for polyploidy rice and also generated a high-quality genome assembly for the allotetraploid wild rice. By targeting six agronomically important traits and altering homologs of the genes controlling domestication trait in diploid rice, *Oryza alta* (CCDD), an allotetraploid rice, was *de novo* domesticated (Yu et al., 2021).

Though the *de novo* domestication technique has the potential to quickly domesticate novel crops from their wild relatives, there is a need to focus on many non-agronomically important traits of wild relatives. Many hurdles in this process need to be taken into account, non-agronomically important traits in wild relatives of crops such as non-compact growth habitat and a high tendency of seed shattering are not appropriate for agronomic applications (Koinange et al., 1996). Additionally, a multiplex GE system targeting a few domestication genes at a time is probably necessary for the successful domestication of crop wild relatives within one generation and a brief period (Yu et al., 2022). So, creating new crop species with varied gene pools with the help of *de novo* domestication is an innovative approach to crop improvement.

6.5 CRISPR SCREEN, GENE STACKING IN CROPS, AND GENE DRIVE

CRISPR screens are an effective tool for biological discovery because they allow for the impartial investigation of gene function across a wide range of species and applications (Bock et al., 2022). It is useful in characterizing genes at the genome level and identifying the relationship between genes and phenotypes (forward genetics) in many biological systems. CRISPR gRNA libraries help in functional genomics studies of plants by enabling genome-scale screening of mutations in the population (Gaillochet et al., 2021). A pooled CRISPR screen is an extension of a CRISPR screen in which a different gRNA consisting library of cells is created (Bock et al., 2022). Selection pressure is employed on the gene-edited cells for screening of the fit cells and with the help of high throughput sequencing their encoded gRNA is read out (Bock et al., 2022). In soybean, to target 102 potential genes in 407 transgenic lines and induce multiplex mutations inside a single transgenic plant, a pooled CRISPR screen was used. This screen used a small-scale CRISPR library (5-10gRNA) with several alleles per gRNA (Bai et al., 2020).

Wild relatives of crops serve as natural repositories for diverse alleles, many of which are advantageous from an agronomical perspective. Such natural alleles however are only found in a few species or in species that are closely related to those species, and are not able to be used in other cross-genus crops. Linkage drag is a major barrier to introgressing such alleles even within closely related species (Hao et al., 2020). Synthetically generating alleles through GE is a technique to get around these introgression issues and allow for more regulated allele transfer to different crop systems. One of the successful examples of this approach is the cloning of the FT-D1 candidate gene in wheat linked to spikelet number and heading date. The generated allele is conserved in polyploidy wheat and to increase wheat yield it could be possibly pyramided with other yield-controlling genes (Chen et al., 2022). Multiplex CRISPR editing tool can target multiple genes at a time which helps in the accelerating crop breeding process.

One of the best examples of this is the elite winter wheat variety Zhengami 7698(ZM). This elite wheat was developed by using different six gRNA targeting three homologous to each gene, leading to the development of the desired phenotype within one year (Luo et al., 2021). Drought-tolerant hexaploid wheat was developed by designing three gRNA targeting to five active homologous genes of 3' (2'), 5' bisphosphate nucleotide genes (Abdallah et al., 2022). In cabbage, various genes that are responsible for male sterility were targeted and modified by a multiplex CRISPR editing tool to develop a pure F_1 hybrid seed (Ma et al., 2019).

These achievements have improved our ability to analyze and understand the regulation of paralogous genes in polyploidy crops, which raises the likelihood that efforts to stack genes in these extremely complicated genomes will be successful. Multiplex GE relies heavily on effective methods for gRNA delivery into the host plants and transgenic production Another approach to change genes allele into the desired version is gene drive. It can be accomplished using the CRISPR-Cas system, and the process afterward (Gentzel et al., 2022) converts heterozygote cells into homozygotes, causing all of the offspring to carry the desired allele at 100% frequency (Siddiqui et al., 2021).

The first successful attempt at gene drive was on *Anopheles gambia* the main vector of malaria to suppress the malaria mosquito population, by targeting genes that are responsible for the female sterility (Hammond et al., 2016). Other potential uses for CRISPR gene drive in plants include the management of weeds (Neve et al., 2018) and enhancement of yield attributes (Rodríguez-Leal et al., 2017).

6.6 DEVELOPMENT OF HAPLOID INDUCER LINES USING GENOME EDITING IN DIFFERENT CROPS

Single-cross hybrids (SCHs) are becoming popular in most crops, and the acquisition of elite inbred lines from segregating populations is a crucial component of crop breeding as it is time-consuming and requires 7–8 generations to achieve near homozygosity. In the current time, due to adverse climatic conditions and man-made calamities, it is important to speed up the process of superior hybrid production. DH development (Figure 6.1) from the haploids speeds up the production of homozygous lines in the minimum number of generations. *In vivo* haploid induction, based on DH technology, is a better alternative than other techniques to develop completely homozygous individuals only in two generations (Forster & Thomas, 2005; Gordillo & Geiger, 2008; Chang & Coe, 2009b). Cytoplasmic male sterility (CMS) is controlled by the mitochondrial DNA mutation; to streamline the CMS-based hybrid system, DH could be used (Jacquier et al., 2021). DH tools will secure the breeding of the crops for developing high-yielding superior varieties resistant to biotic and abiotic stresses.

In 1952, the utility of monoploids (Haploids) was proposed for the first time to develop homozygous lines with a short duration (Chase et al., 1952). Since then, plant breeders and molecular biologists have discovered various methods for stimulating haploids in several crop species. Haploids could be generated by various approaches such as wide hybridization, *in vitro* anther and microspore culture, pollen irradiation, and haploid inducer stocks (Rangari et al., 2023). In general, most of the genotypes are recalcitrant to tissue culture media, again the varietal development cost is very high (Ishida et al., 2020). Haploid induction through wide hybridization requires a well-controlled environment and proper availability of embryo rescue (Liu et al., 2020). Hence, *in vivo,* approaches have been developed for DH production in cereals and will be available soon in different groups of crops like legumes and millet. In maize, haploid production is easy due to the availability of the haploid inducer stock called Stock 6, (Coe, 1959). However, the haploid induction rate is very low (6.7%) through these haploid inducer lines. When haploid inducer lines are crossed with cultivated lines, haploid seeds develop, due to the failure of pollen nuclei to fuse in double fertilization with central nuclei or egg nuclei (Chen et al., 2020). According to a new study, chromosome elimination and haploid induction occur due to the fragmentation of chromosomes (Sun et al., 2022). Haploid induction in maize occurs because of 4 bp frame-shift mutation in *MATRILINEAL* (*ZmPLA1*), and new edits in *ZmPLA1* lead to a 6.7% haploid induction rate (Kelliher et al., 2017; Liu et al., 2017).

GE systems are now becoming popular to revolutionize molecular breeding with their broad applications in the life sciences (Shan et al., 2015). This powerful technique can be used to develop haploid inducer stocks in various crops by targeting the candidate genes, such as the *ZmPLA1*, *CENH3* genes in maize, wheat, and other crops as discussed below.

6.6.1 Wheat

The haploid inducer lines in wheat (*Triticum aestivum*) are developed by targeting genes like *CENH3* (Zargar et al., 2022). Introducing mutation in candidate genes through GE will develop a stable and promising haploid inducer line in wheat, as multiple copies of this gene are present in the wheat genome. 18.9% haploid induction rate was obtained by mutating the *TaMTL* gene in wheat through the CRISPR-Cas9 system (Liu et al., 2020). Two sgRNA were used to target the c-terminal region of the wheat *MTL* gene to prove that chromosome fragmentation leads to haploid induction, which is one of the reasons for haploid induction in the case of post-fertilization (Sun et al., 2022). In addition, a fast and convenient technique for the easy screening of wheat haploids through an anthocyanin marker was proposed (Tang et al., 2023). By utilizing all the available resources, tools, and techniques in wheat, a higher haploid induction rate line can be developed by using the CRISPR-Cas system.

6.6.2 Rice

By doubling the chromosomes of the rice hybrids, fully homozygous and true breeding lines are developed in a short period (Kyum et al., 2022). The orthologue gene *OsMLA1* shows 78–79% similarity to the *ZmPLA1* in maize (Kelliher et al., 2017). The recalcitrant nature of rice, the large number of albinos, and genotypic bias are the major limitations in the rice haploid inducer line development (Lantos et al., 2022). On the other hand, CRISPR-Cas technology has the potential to mutate any gene of interest; hence, the candidate genes for haploid induction in the rice genome have been targeted in a few studies. The knockdown of the *REC8*, *PAIR*, *OSD1*, and *MTL* genes has been done for the production of the haploid inducer with the fixation of the heterozygosity (Wang et al., 2019b). The *MTL* in the rice (*Oryza sativa*) was edited by using CRISPR-Cas technology to develop the haploid inducer line and achieved an HIR of more than 10% (Wang et al., 2022). The *OsMATL* gene was targeted to develop a haploid inducer line, which has a 2–5% HIR (Liu et al., 2021a). In the future, high HIR inducer lines will be developed with this technology.

6.6.3 Maize

Maize (*Zea mays*) has the most superior haploid induction systems along with the availability of phenotypic markers to identify the haploids (Sun et al., 2022). The orthologue of matrilineal (*MTL*) genes was obtained in wheat and rice. Maize has the potential haploid inducer stock for the temperate background with HIR improved up to 6.7% (Liu et al., 2017). To improve the rate of haploid induction, several efforts have been made. CRISPR-Cas approach has been used for haploid inducer development in different backgrounds (Wang et al., 2019a; Li et al., 2021). The standardization of the RNP complex-based DNA-free GE for haploid induction was done in the Indian background by targeting the *ZmPLA1* gene (Rangari et al., 2023). Around 2–3-fold HIR has been enhanced by one base change in the *ZmDMP* gene (Zhong et al., 2019). Therefore, in the maize crop, there is lots of scope to achieve the maximum HIR using gene editing tools like CRISPR Cas.

6.6.4 Legumes

Global warming and continuously rising temperatures hamper the production of important legumes like chickpea (Kumar et al., 2022). Hence, for rapid development of the heat-tolerant and other biotic and abiotic stress-tolerant high-yielding varieties, DH technology could be used in legumes (Wang et al., 2022). Due to the recalcitrant nature of most legume crops, DH production is very difficult; hence, after many attempts, scientists failed to recover a sufficient number of haploids (Bhowmik et al., 2021).

The haploid inducer line-based DH technology ensures the fixation of the recombinant haplotypes within a very short duration (Wang et al., 2022). This technology has not been explored much in the legume crops, CRISPR Cas technology could help to open new doors for DH applications by targeting candidate genes for haploid induction. A model legume, *Medicago truncatula,* has the haploid inducer gene *DMP,* which was successfully edited to get the haploids with 0.29% to 0.82% HIR (Wang et al., 2022). In other major legume crops like chickpea, pigeon pea, groundnut, and lentils, CRISPR-Cas technology could play an important role in capturing the haploid inducer lines by targeting the potential candidate genes.

6.6.5 Millets

Millets are an important group of food grains and have the potential to sustain in harsh and changing climates. It requires minimum water and nutrients to grow, yet it still produces a high yield. Due to this potential, the United Nations Assembly announced '2023' as a millet year and set a target to develop maximum high-yielding varieties. Given this, haploid inducer line-based DH technology can achieve this target within a short period. In foxtail millet (*Seteria italica*), haploid induction has been achieved by targeting the *SiMTL* gene with an HIR of 2.8% (Cheng et al., 2021). In other major and minor millets, this technology can reduce the breeding timing of hybrids by fixing it within two generations only with this haploid inducer line-based DH technology.

6.7 DEVELOPMENT OF MALE STERILE LINES USING GENOME EDITING

Hybrid seed production is one of the major achievements in the era of the modern agricultural system. It fetches more benefits to the farmers as well as seed companies by increasing yields, disease resistance, and tolerance against biotic and abiotic stress as compared to parental inbred lines. The generation of MS lines and the application of haploids are among the major approaches to hybrid seed production (Figure 6.1). The male sterility in plants comprehends the inability to produce functional male reproductive parts such as anthers, pollen, or male gametes. According to Horner and Palmer (1995), it was classified into CMS and genic male sterility (GMS), depending on the fertility gene source. Male sterility is one of the important traits that have been primarily used to produce hybrid seeds for the crop of economic interest. Conventionally, MS lines were produced either by manual emasculation (Figure 6.1) or by a mutagenesis approach (Hawkes et al., 2011). However, the traditional approach was restricted by many limitations: (i) labor and time-intensive (ii) availability of MS resources in agricultural systems (iii) Detection of MS only in the flowering stage (iv) influenced by environmental conditions like temperature, photo-period and, (v) breeders' dependency for choosing materials based on phenotypic analyses. By analyzing these limitations, the development of MS lines becomes one of the important challenges in the plant breeding community.

TABLE 6.3
Genome Editing Mediated Development of MS Lines in Important Agricultural Crops

Crops	Approach	Targeted Gene	Functional Modification	References
Rice	CRISPR-Cas9	*TMS5*	–	Barman et al., 2019
Maize	CRISPR-Cas9	*MS8*	The meiotic stage of anther development	Chen et al., 2018
Wheat	CRISPR-Cas9	*MS1, TaNP1, TaDCL4, TaDCL5, and TaRDR6*	–	Okada et al., 2019; Li et al., 2020; Zhang et al., 2023b
Soybean	CRISPR-Cas9	*GmAMS1*	Pollen wall and tapetum	Chen et al., 2021b
Mustard	CRISPR-Cas9	*BnaRFL11*	Fertility restoration	Farooq et al., 2022
Tomato	CRISPR-Cas9	*GSTAA and Ms10*[35]	–	Liu et al., 2021b
Melon	CRISPR-Cas9	*eif4e*	Microsporocytes and tapetum	Pechar et al., 2022
Watermelon	CRISPR-Cas9	*ClATM1*	Anther development	Zhang et al., 2021
Cotton	CRISPR-Cas9	*GhEMS1*	Middle layer and tapetum cells	Zhang et al., 2023c

On the contrary, GE-based development of MS lines diminished these limitations. GE method is particularly based on gene knockout mediated by CRISPR/Cas system, targeting nuclear genic loci. It is flexible and easy to induce MS by altering the function of key genes, thus easing the detection of the manipulated gene (Farinati et al., 2023). In view of this, the CRISPR/Cas system, mostly based on Cas9 protein, has been applied in many crops of economic interest for the generation of MS lines across the globe. Such functionally modified crops are rice, wheat, maize, tomato, soybean, cotton, and mustard (Table 6.3). Ultimately, as a result of the transformation/transfection of a ribonucleoprotein (RNP) complex or DNA vector, GE can be induced by a transient or stable expression of the Cas nuclease and guide RNA (gRNA).

Mostly, the MS lines in plants are produced by targeting the specific genes involved in microsporogenesis. Accordingly, it targets tapetum-specific, meiosis-related, and transcription-regulatory genes like tpd1, ms1, eme1/exs, and ams (Zhao et al., 2002; Yang et al., 2003). Moreover, many transcription factor genes responsible for pollen development are associated with classes of bHLH genes (bHLH10, bHLH89, and bHLH91), DYTI, MYB103, TDF1, and AMS in a model plant, Arabidopsis (Sorensen et al., 2003; Zhu et al., 2008; Zhang et al., 2007; Zhu et al., 2015).

6.7.1 Mechanism of Gene Regulation in MS Line Development

Generally, genic male sterility is produced using genome editing technology in most plants. In that context, several mechanisms are involved in different species. However, in many of them, genes encoding transcription factors are responsible for modifying the expression of genes involved in pollen development. Moreover, it involves altering the gamete formation, Failure of homologous chromosome separation during meiotic anaphase I, and even delay in programmed cell death of tapetum. Many genes responsible for producing transcription factor proteins are reported in *Arabidopsis.*

In particular, many nuclear genes have been reported to cause arresting microspore development and mapped using different types of molecular markers like SSR (simple sequence repeats), SNP (single nucleotide polymorphism), RFLP (restriction fragment length polymorphism), and SCAR (sequence characterized amplified region). However, in more than 610 species of flowering plants, the male sterility trait is under investigation, to date. Additionally, at least 40 genes associated with GMS have been identified by marker-assisted selection and characterized in Arabidopsis and rice (Chen & Liu, 2014; Wan et al., 2019).

6.8 SPEED BREEDING THROUGH CRISPR-MEDIATED DIRECTED EVOLUTION OF CROPS

The natural genetic diversity is low in plants, which limits crop improvement. This bottleneck can be overcome by directed evolution, random mutation in the target gene of host plants. The directed evolution aims to produce diverse sequence variants with desirable traits under selective pressure. Accordingly, CRISPR system can be programmed to target any genomic locus and thus perform a targeted CRISPR-mediated directed evolution (Butt et al., 2020). It mimics Darwinian evolution principles by applying selective pressure to a large plant population with random genetic diversity in DNA, resulting in the generation and selection of sequences with specific desirable features.

Directed evolution could help to enhance the affinity/resistance of ascorbate peroxidase (APX) to hydrogen peroxide which will increase plant resilience to stress, and Rubisco with improved carboxylation properties is predicted to increase agricultural yield (Gionfriddo et al., 2019). Depending on the identification of 120 suitable target genes and the capacity to create informative screens and selections, CRISPR-Cas mediated directed evolution (CDE) may be utilized to expedite this study. CDE can target APX and isoprene synthase for abiotic stress tolerance; nucleotide-binding leucine-rich repeat (NB-NLR)-containing proteins; FLAGELLIN-SENSING2 for increased pathogen resistance; and EPSPS (5-enolpyruvylshikimate 3-phosphate synthase) for improved resistance to the herbicide glyphosate. For CDE, other noncoding components that alter transcriptional activity, such as cis-elements and promoter sequences, can also be used. The plant's capacity to respond to nutritional stress can be enhanced by using CDE to develop protein variations that regulate responses to it (Butt et al., 2020).

6.9 MULTIPLEX GENOME EDITING TECHNOLOGIES FOR CROP IMPROVEMENT

Multiplex genome editing (MGE) technologies are versatile bioengineering tools that have recently been created to allow for the highly precise modification of two or more particular DNA loci in a genome. The possibility of incorporating desired alterations at various nucleotide levels into a target genome has significantly enhanced as a result of these genome editing techniques. Particularly, MGE methods based on the CRISPR/Cas system enable the simultaneous production of direct mutations exactly at several loci in a gene or genes (Abdelrahman et al., 2021). The ability to induce such kind of genomic alteration will facilitate the understanding and investigation of the interaction among redundantly functioning members of a gene family and may offer a chance to examine epistatic relationship between closely related genetic pathways (Xing et al., 2014) furthermore the most practical platform for metabolic pathway engineering in plants may be provided by these multiple genomic editing tools.

The first MGE tool used for the development of broad-spectrum and durable *Blumeria graminis f.sp. tritici (Bgt)* resistant wheat lines, the objective was to edit three homoealleles encoding mildew-resistant locus protein (MLOs) in hexaploidy wheat (*tritium aestivum* L.). Moreover, biotic stress-resistant lines against the fungus *Botryis cinerea* have been developed. In that effort to confer fungal resistance four guide RNA were used to target the grapevine *VvWRKY52* gene. *Indels* were introduced at four targets, and larger deletions affecting multiple targets were observed, suggesting that simultaneous targeting was successful (Wang et al., 2018).

6.10 CONCLUSION

GE technology is the most promising one in the era of the post-green revolution. It became the base of precision plant breeding programs. In a nutshell, all the speed breeding approaches mediated by GE play a pivotal role in accomplishing zero hunger. It has become familiar among researchers as it is increasing production and productivity. The goal of GE-based speed breeding is to provide disease resistance, and tolerance against herbicides and insecticides. Moreover, it is producing hybrid seeds and developing climate-resilient cultivars in a short period of time. With the use of CRISPR-Cas9, it is possible to incorporate a large superior gene pool from domesticated crops into their wild relatives, accelerating the breeding process such as speed breeding.

REFERENCES

Abdallah, N. A., Elsharawy, H., Abulela, H. A., Thilmony, R., Abdelhadi, A. A., & Elarabi, N. I. (2022). Multiplex CRISPR/Cas9-mediated genome editing to address drought tolerance in wheat. *GM Crops & Food*, 1–17. 10.1080/21645698.2022.2120313

Abdallah, N. A., Hamwieh, A., Radwan, K., Fouad, N., & Prakash, C. (2021). Genome editing techniques in plants: A comprehensive review and future prospects toward zero hunger. *GM Crops & Food*, *12*(2), 601–615. 10.1080/21645698.2021.2021724

Abdelrahman, M., Wei, Z., Rohila, J. S., & Zhao, K. (2021). Multiplex genome-editing technologies for revolutionizing plant biology and crop improvement. *Frontiers in Plant Science*, *12*, 721203. 10.3389/fpls.2021.721203

Ahmad, M. (2023). Plant breeding advancements with "CRISPR-Cas" genome editing technologies will assist future food security. *Frontiers in Plant Science*, *14*, 1133036. 10.3389/fpls.2023.1133036

Ahmad, S., Tang, L., Shahzad, R., Mawia, A. M., Rao, G. S., Jamil, S., & Tang, S. (2021). CRISPR-based crop improvements: A way forward to achieve zero hunger. *Journal of Agricultural and Food Chemistry, 69*(30), 8307–8323. 10.1021/acs.jafc.1c02653

Allemailem, K. S., Almatroodi, S. A., Almatroudi, A., Alrumaihi, F., Al Abdulmonem, W., Al-Megrin, W. A. I., & Khan, A. A. (2023). Recent advances in genome-editing technology with CRISPR/Cas9 variants and stimuli-responsive targeting approaches within tumor cells: A future perspective of cancer management. *International Journal of Molecular Sciences*, *24*(8), 7052. 10.3390/ijms24087052

Bai, M., Yuan, J., Kuang, H., Gong, P., Li, S., Zhang, Z., & Guan, Y. (2020). Generation of a multiplex mutagenesis population via pooled CRISPR-Cas9 in soya bean. *Plant Biotechnology Journal*, *18*(3), 721–731. 10.1111/pbi.13239

Barman, H. N., Sheng, Z., Fiaz, S., Zhong, M., Wu, Y., Cai, Y., et al. (2019). Generation of a new thermo-sensitive genic male sterile rice line by targeted mutagenesis of TMS5 gene through CRISPR/Cas9 system. *BMC Plant Biology*, *19*, 109. 10.1186/s12870-019-1715-0

Bhowmik, P., Konkin, D., Polowick, P., Hodgins, C. L., Subedi, M., Xiang, D., & Kagale, S. (2021). CRISPR/Cas9 gene editing in legume crops: Opportunities and challenges. *Legume Science*, *3*(3), e96. 10.1002/leg3.96

Bhutia, K. L., & Tyagi, W. (2017). Use of sequence specific nucleases for site specific modification of plant genome for crop improvement. *International Journal of Agricultural Science Research (IJASR)*, *7*, 491–502.

Blesh, J., Hoey, L., Jones, A. D., Friedmann, H., & Perfecto, I. (2019). Development pathways toward "zero hunger". *World Development*, *118*, 1–14. 10.1016/j.worlddev.2019.02.004

Bock, C., Datlinger, P., Chardon, F., Coelho, M. A., Dong, M. B., Lawson, K. A., & Zhuang, X. (2022). High-content CRISPR screening. *Nature Reviews Methods Primers*, *2*(1), 8. 10.1038/s43586-021-00093-4

Bohra, A., Kilian, B., Sivasankar, S., Caccamo, M., Mba, C., McCouch, S. R., & Varshney, R. K. (2022). Reap the crop wild relatives for breeding future crops. *Trends in Biotechnology*, *40*(4), 412–431. 10.1016/j.tibtech.2021.08.009

Bull, T., Debernardi, J., Reeves, M., Hill, T., Bertier, L., Van Deynze, A., & Michelmore, R. (2023). GRF–GIF chimeric proteins enhance in vitro regeneration and Agrobacterium-mediated transformation efficiencies of lettuce (*Lactuca spp.*). *Plant Cell Reports*, *42*(3), 629–643. 10.1007/s00299-023-02980-4

Butt, H., Zaidi, S. S. E. A., Hassan, N., & Mahfouz, M. (2020). CRISPR-based directed evolution for crop improvement. *Trends in Biotechnology*, *38*(3), 236–240. 10.1016/j.tibtech.2019.08.001

Campos, G., Chialva, C., Miras, S., & Lijavetzky, D. (2021). New technologies and strategies for grapevine breeding through genetic transformation. *Frontiers in Plant Science*, *12*, 767522. 10.3389/fpls.2021.767522

Chang, M and Coe, E (2009a). Molecular genetics approaches to maize improvement. In A. L. Kriz & B. A. Larkins (eds) *Biotechnology in agriculture and forestry* (pp 127–142). Berlin, Heidelberg, Germany: Springer.

Chang, M. T., & Coe Jr, E. H. (2009b). Doubled haploids. In A. L. Kriz & B. A. Larkins (eds) *Molecular genetic approaches to maize improvement* (pp. 127–142). Berlin, Heidelberg: Springer Berlin Heidelberg.

Chase, S. S. (1952). Monoploids in maize. In J. W. Gowen (eds), *Heterosis* (pp. 389–399). Ames: Iowa State College Press.

Chen, C., Xiao, Z., Zhang, J., Li, W., Li, J., Liu, C., & Chen, S. (2020). Development of in vivo haploid inducer lines for screening haploid immature embryos in maize. *Plants*, *9*(6), 739. 10.3390/plants9060739

Chen, G., Zhou, Y., Kishchenko, O., Stepanenko, A., Jatayev, S., Zhang, D., & Borisjuk, N. (2021a). Gene editing to facilitate hybrid crop production. *Biotechnology Advances*, *46*, 107676.10.1016/j.biotechadv.2020.107676

Chen, L., and Liu, Y. G. (2014). Male sterility and fertility restoration in crops. *Annual Review of Plant Biology*, *65*, 579–606. 10.1146/annurev-arplant-050213-040119

Chen, R., Xu, Q., Liu, Y., Zhang, J., Ren, D., Wang, G., & Liu, Y. (2018). Generation of transgene-free maize male sterile lines using the CRISPR/Cas9 system. *Frontiers in Plant Science*, *9*, 1180.10.3389/fpls.2018.01180

Chen, X., Yang, S., Zhang, Y., Zhu, X., Yang, X., Zhang, C., & Feng, X. (2021b). Generation of male-sterile soybean lines with the CRISPR/Cas9 system. *The Crop Journal*, *9*(6), 1270–1277.10.1016/j.cj.2021.05.003

Chen, Z., Ke, W., He, F., Chai, L., Cheng, X., Xu, H., & Ni, Z. (2022). A single nucleotide deletion in the third exon of FT-D1 increases the spikelet number and delays heading date in wheat (*Triticum aestivum* L.). *Plant Biotechnology Journal*, *20*(5), 920–933.10.1111/pbi.13773

Cheng, Z., Sun, Y., Yang, S., Zhi, H., Yin, T., Ma, X., & Sui, Y. (2021). Establishing in planta haploid inducer line by edited SiMTL in foxtail millet (*Setaria italica*). *Plant Biotechnology Journal*, *19*(6), 1089.10.1111/pbi.13584

Coe Jr, E. H. (1959). A line of maize with high haploid frequency. *The American Naturalist*, *93*(873), 381–382.

Collard, B. C., & Mackill, D. J. (2008). Marker-assisted selection: An approach for precision plant breeding in the twenty-first century. *Philosophical Transactions of the Royal Society B: Biological Sciences*, *363*(1491),557–572. 10.1098/rstb.2007.2170

Damude, H. G., & Kinney, A. J. (2008). Enhancing plant seed oils for human nutrition. *Plant Physiology*, *147*, 962–968. 10.1104/pp.108.121681

Dhugga, K. S. (2022). Gene editing to accelerate crop breeding. *Frontiers in Plant Science*, *13*, 889995.10.3389/fpls.2022.889995

Doebley, J. F., Gaut, B. S., & Smith, B. D. (2006). The molecular genetics of crop domestication. *Cell*, *127*(7), 1309–1321.10.1016/j.cell.2006.12.006

Farinati, S., Draga, S., Betto, A., Palumbo, F., Vannozzi, A., Lucchin, M., & Barcaccia, G. (2023). Current insights and advances into plant male sterility: New precision breeding technology based on genome editing applications. *Frontiers in Plant Science*, *14*. 10.3389/fpls.2023.1223861

Farooq, Z., Nouman Riaz, M., Farooq, M. S., Li, Y., Wang, H., Ahmad, M., & Yi, B. (2022). Induction of male sterility by targeted mutation of a restorer-of-fertility gene with CRISPR/Cas9-mediated genome editing in *Brassica napus* L. *Plants*, *11*(24), 3501.10.3390/plants11243501

Forster, B. P., & Thomas, W. T. (2005). Doubled haploids in genetics and plant breeding. *Plant Breeding Reviews*, *25*, 57–88.

Gaillochet, C., Develtere, W., & Jacobs, T. B. (2021). CRISPR screens in plants: Approaches, guidelines, and future prospects. *The Plant Cell*, *33*(4), 794–813. 10.1093/plcell/koab099

Gan, W. C., & Ling, A. P. (2022). CRISPR/Cas9 in plant biotechnology: Applications and challenges. *BioTechnologia*, *103*(1), 81–93.10.5114/bta.2022.113919

Gentzel, I. N., Ohlson, E. W., Redinbaugh, M. G., & Wang, G. L. (2022). VIGE: Virus-induced genome editing for improving abiotic and biotic stress traits in plants. *Stress Biology*, *2*, 1–10. 10.1007%2Fs44154-021-00026-x

Gionfriddo, M., De Gara, L., & Loreto, F. (2019). Directed evolution of plant processes: Towards a green (r) evolution? *Trends in Plant Science*, *24*(11), 999–1007.

Gobena, D., Shimels, M., Rich, P. J., Ruyter-Spira, C., Bouwmeester, H., & Kanuganti, S., (2017). Mutation in sorghum LOW GERMINATION STIMULANT 1 alters strigolactones and causes Striga resistance. *The Proceedings of the National Academy of Sciences*, *114*, 4471–4476. 10.1073/pnas.1618965114

Gordillo, G. A., & Geiger, H. H. (2008). Alternative recurrent selection strategies using doubled haploid lines in hybrid maize breeding. *Crop Science*, *48*(3), 911–922. 10.2135/cropsci2007.04.0223

Hammond, A., Galizi, R., Kyrou, K., Simoni, A., Siniscalchi, C., Katsanos, D., & Nolan, T. (2016). A CRISPR-Cas9 gene drive system targeting female reproduction in the malaria mosquito vector *Anopheles gambiae*. *Nature Biotechnology*, *34*(1), 78–83.10.1038/nbt.3439

Hanak, T., Madsen, C. K., & Brinch-Pedersen, H. (2022). Genome Editing-accelerated Re-Domestication (GEaReD): A new major direction in plant breeding. *Biotechnology Journal*, *17*(7), 2100545. 10.1002/biot.202100545

Hao, M., Zhang, L., Ning, S., Huang, L., Yuan, Z., Wu, B., & Liu, D. (2020). The resurgence of introgression breeding, as exemplified in wheat improvement. *Frontiers in Plant Science*, *11*, 252. 10.3389/fpls.2020.00252

Haroon, M., Afzal, R., Idrees, F., Sunny, A., & Khan, A. S. (2019). Genome editing and speed breeding; game changers to boost the crop production. *International Journal of Biological Research*, *2*, 295–300. 10.1111/pbi.13773

Hawkes, T., Pline-Srnic, W., Dale, R., Friend, E., Hollinshead, T., Howe, P., & Greenland, A. (2011). D-glufosinate as a male sterility agent for hybrid seed production. *Plant Biotechnology Journal*, *9*(3), 301–314. 10.1111/j.1467-7652.2010.00549.x

Horner, H.T., & Palmer, R.G. (1995). Mechanisms of genic male sterility.*Crop Science*, *35*, 1527–1535. 10.2135/cropsci1995.0011183X003500060002x

Huang, X., Huang, S., Han, B., & Li, J. (2022). The integrated genomics of crop domestication and breeding. *Cell*, *185*, 2828–2839. 10.1016/j.cell.2022.04.036.

Ishida, Y., Hiei, Y., & Komari, T. (2020). Tissue culture protocols for gene transfer and editing in maize (*Zea mays* L.). *Plant Biotechnology*, *37*(2), 121–128.10.5511/plantbiotechnology.20.0113a

Jacquier, N. M., Gilles, L. M., Martinant, J. P., Rogowsky, P. M., & Widiez, T. (2021). Maize in planta haploid inducer lines: A cornerstone for doubled haploid technology. In J. M. Segui-Simarro (eds) Doubled Haploid Technology. Methods in Molecular Biology, vol 2288. Humana, New York, NY., pp. 25–48.10.1007/978-1-0716-1335-1_2

Kelliher, T., Starr, D., Richbourg, L., Chintamanani, S., Delzer, B., Nuccio, M. L., & Martin, B. (2017). MATRILINEAL, a sperm-specific phospholipase, triggers maize haploid induction. *Nature*, *542*(7639), 105–109.10.1038/nature20827

Klümper, W., & Qaim, M. (2014). A meta-analysis of the impacts of genetically modified crops. *PloS One*, *9*(11), e111629.10.1371/journal.pone.0111629

Koinange, E. M., Singh, S. P., & Gepts, P. (1996). Genetic control of the domestication syndrome in common bean. *Crop Science*, *36*(4), 1037–1045. 10.2135/cropsci1996.0011183X003600040037x

Kumar, K., Mandal, S. N., Pradhan, B., Kaur, P., Kaur, K., & Neelam, K. (2022). From evolution to revolution: Accelerating crop domestication through genome editing. *Plant and Cell Physiology*, *63*(11), 1607–1623. 10.1093/pcp/pcac124

Kyum, M., Kaur, H., Kamboj, A., Goyal, L., & Bhatia, D. (2022). Strategies and prospects of haploid induction in rice (Oryza sativa). *Plant Breeding*, *141*(1), 1–11.10.1111/pbr.12971

Lantos, C., Jancsó, M., Székely, Á., Nagy, É., Szalóki, T., & Pauk, J. (2022). Improvement of anther culture to integrate doubled haploid technology in temperate rice (*Oryza sativa* L.) breeding. *Plants*, *11*(24), 3446.10.3390/plants11243446

Lee, K., & Wang, K. (2023). Strategies for genotype-flexible plant transformation. *Current Opinion in Biotechnology*, *79*, 102848.10.1016/j.copbio.2022.102848

Lemmon, Z. H., Reem, N. T., Dalrymple, J., Soyk, S., Swartwood, K. E., Rodriguez-Leal, D., & Lippman, Z. B. (2018). Rapid improvement of domestication traits in an orphan crop by genome editing. *Nature Plants*, *4*(10), 766–770.10.1016/j.jgg.2020.05.004

Li, J., Wang, Z., He, G., Ma, L., & Deng, X. W. (2020). CRISPR/Cas9-mediated disruption of TaNP1 genes results in complete male sterility in bread wheat. *Journal of Genetics and Genomics*, *47*(5), 263–272. 10.1016/j.jgg.2020.05.004

Li, T., Yang, X., Yu, Y., Si, X., Zhai, X., Zhang, H., & Xu, C. (2018). Domestication of wild tomato is accelerated by genome editing. *Nature Biotechnology*, *36*(12), 1160–1163.10.1038/nbt.4273

Li, Y., Lin, Z., Yue, Y., Zhao, H., Fei, X., E, L., & Song, W. (2021). Loss-of-function alleles of ZmPLD3 cause haploid induction in maize. *Nature Plants*, 7(12), 1579–1588.10.1038/s41477-021-01037-2

Lidder, P., & Sonnino, A. (2012). Biotechnologies for the management of genetic resources for food and agriculture. *Advances in Genetics*, *78*, 1–167.10.1016/B978-0-12-394394-1.00001-8

Lin, T., Zhu, G., Zhang, J., Xu, X., Yu, Q., Zheng, Z., & Huang, S. (2014). Genomic analyses provide insights into the history of tomato breeding. *Nature Genetics*, *46*(11), 1220–1226.

Liu, H., Ding, Y., Zhou, Y., Jin, W., Xie, K., & Chen, L. L. (2017). CRISPR-P 2.0: an improved CRISPR-Cas9 tool for genome editing in plants. *Molecular Plant*, *10*(3), 530–532. DOI: 10.1016/j.molp.2017.01.003

Liu, H., Wang, K., Jia, Z., Gong, Q., Lin, Z., Du, L., & Ye, X. (2020). Efficient induction of haploid plants in wheat by editing of *TaMTL* using an optimized Agrobacterium-mediated CRISPR system. *Journal of Experimental Botany*, *71*(4), 1337–1349. 10.1093/jxb/erz529

Liu, J., Liang, D., Yao, L., Zhang, Y., Liu, C., Liu, Y., & Bandyopadhyay, A. (2021a). Rice haploid inducer development by genome editing. *Rice Genome Engineering and Gene Editing: Methods and Protocols* (pp. 221–230). New York, NY: Springer US. 10.1007/978-1-0716-1068-8_14

Liu, J., Wang, S., Wang, H. et al. (2021b). Rapid generation of tomato male-sterile lines with a marker used for hybrid seed production by CRISPR/Cas9 system. *Molecular Breeding*, *41*, 25. 10.1007/s11032-021-01215-2

Luo, G., & Palmgren, M. (2021). GRF-GIF chimeras boost plant regeneration. *Trends in Plant Science*, *26*(3), 201–204. 10.1016/j.tplants.2020.12.001

Luo, J., Li, S., Xu, J., Yan, L., Ma, Y., & Xia, L. (2021). Pyramiding favourable alleles in an elite wheat variety in one generation by CRISPR-Cas9-mediated multiplex gene editing. *Molecular Plant*, *14*(6), 847–850. 10.1016/j.molp.2021.03.024

Ma, C., Zhu, C., Zheng, M., Liu, M., Zhang, D., Liu, B., & Song, H. (2019). CRISPR/Cas9-mediated multiple gene editing in *Brassica oleracea* var. capitata using the endogenous tRNA-processing system. *Horticulture Research*, *6*. 10.1038/s41438-018-0107-1

Matres, J. M., Hilscher, J., Datta, A., Armario-Nájera, V., Baysal, C., He, W., & Slamet-Loedin, I. H. (2021). Genome editing in cereal crops: An overview. *Transgenic Research*, *30*, 461–498. 10.1007/s11248-021-00259-6

Meyer, R. S., & Purugganan, M. D. (2013). Evolution of crop species: Genetics of domestication and diversification. *Nature Reviews Genetics*, *14*(12), 840–852. 10.1038/nrg3605

Moore, J. W., Herrera-Foessel, S., Lan, C., Schnippenkoetter, W., Ayliffe, M., Huerta-Espino, J., & Lagudah, E. (2015). A recently evolved hexose transporter variant confers resistance to multiple pathogens in wheat. *Nature Genetics*, *47*(12), 1494–1498. 10.1038/ng.3439.

Neve, P. (2018). Gene drive systems: Do they have a place in agricultural weed management? *Pest Management Science*, *74*(12), 2671–2679. 10.1002/ps.5137

Okada, A., Arndell, T., Borisjuk, N., Sharma, N., Watson-Haigh, N. S., Tucker, E. J., & Whitford, R. (2019). CRISPR/Cas9-mediated knockout of Ms1 enables the rapid generation of male-sterile hexaploid wheat lines for use in hybrid seed production. *Plant Biotechnology Journal*, *17*(10), 1905–1913. 10.1111/pbi.13106

Pachapur, P. K., Pachapur, V. L., Brar, S. K., Galvez, R., Le Bihan, Y., & Surampalli, R. Y. (2020). Food security and sustainability. *Sustainability: Fundamentals and Applications*, 357–374. 10.1002/9781119434016.ch17

Pechar, G. S., Donaire, L., Gosalvez, B., García-Almodovar, C., Sánchez-Pina, M. A., Truniger, V., & Aranda, M. A. (2022). Editing melon eIF4E associates with virus resistance and male sterility. *Plant Biotechnology Journal*, *20*(10), 2006–2022. 10.1111/pbi.13885

Rafalski, J. A. (2017). Biotechnology and bioeconomy of complex traits in crop plants. *Biotechnologia*, *98*, 67–71. 10.1111/gcbb.12610

Rangan, P., Henry, R., Wambugu, P., & Periyannan, S. (2023). Plant genetic and genomic resources for sustained crop improvement. *Frontiers in Plant Science*, *14*, 1266698. 10.3389/fpls.2023.1266698

Rangari, S. K., Kaur Sudha, M., Kaur, H., Uppal, N., Singh, G., Vikal, Y., & Sharma, P. (2023). DNA-free genome editing for ZmPLA1 gene via targeting immature embryos in tropical maize. *GM Crops & Food*, *14*, 1–7. 10.1080/21645698.2023.2197303

Rodríguez-Leal, D., Lemmon, Z. H., Man, J., Bartlett, M. E., & Lippman, Z. B. (2017). Engineering quantitative trait variation for crop improvement by genome editing. *Cell*, *171*(2), 470–480. 10.1016/j.cell.2017.08.030

Särkinen, T., Bohs, L., Olmstead, R. G., & Knapp, S. (2013). A phylogenetic framework for evolutionary study of the nightshades (*Solanaceae*): a dated 1000-tip tree. *BMC Evolutionary Biology*, *13*, 1–15. 10.1186/1471-2148-13-214

Shan, Q., Zhang, Y., Chen, K., Zhang, K., & Gao, C. (2015). Creation of fragrant rice by targeted knockout of the *OsBADH2* gene using TALEN technology. *Plant Biotechnology Journal*, *13*(6), 791–800. 10.1111/pbi.12312

Shen, L., Hua, Y., Fu, Y., Li, J., Liu, Q., Jiao, X., & Wang, K. (2017). Rapid generation of genetic diversity by multiplex CRISPR/Cas9 genome editing in rice. *Science China Life Sciences*, *60*, 506–515. 10.1007/s11427-017-9008-8

Siddiqui, H. A., Harvey-Samuel, T., & Mansoor, S. (2021). Gene drive: A faster route to plant improvement. *Trends in Plant Science*, *26*(12),1204–1206. 10.1016/j.tplants.2021.09.005

Sorensen, A. M., Krober, S., Unte, U. S., Huijser, P., Dekker, K., and Saedler, H. (2003). The arabidopsis ABORTED MICROSPORES (AMS) gene encodes a MYC class transcription factor. *The Plant Journal*, *33*, 413–423. 10.1046/j.1365-313X.2003.01644.x

Sun, G., Geng, S., Zhang, H., Jia, M., Wang, Z., Deng, Z., & Mao, L. (2022). Matrilineal empowers wheat pollen with haploid induction potency by triggering postmitosis reactive oxygen species activity. *New Phytologist*, *233*(6), 2405–2414. 10.1111/nph.17963

Tang, H., Wang, K., Zhang, S., Han, Z., Chang, Y., Qiu, Y., & Ye, X. (2023). A fast technique for visual screening of wheat haploids generated from TaMTL-edited mutants carrying anthocyanin markers. *Plant Communications*, *4*(3), 1–3. 10.1016/j.xplc.2023.100569

Tubb, C., & Seba, T. (2021). Rethinking food and agriculture 2020–2030: The second domestication of plants and animals, the disruption of the cow, and the collapse of industrial livestock farming. *Industrial Biotechnology*, *17*(2), 57–72. 10.1089/ind.2021.29240.ctu

Van de Wouw, M., Kik, C., van Hintum, T., van Treuren, R., & Visser, B. (2010). Genetic erosion in crops: Concept, research results and challenges. *Plant Genetic Resources*, *8*(1), 1–15. 10.1017/S14792621 09990062

Voytas, D. F., & Gao, C. (2014). Precision genome engineering and agriculture: Opportunities and regulatory challenges. *PLoS Biology*, *12*(6), e1001877. 10.1371/journal.pbio.1001877

Wan, X., Wu, S., Li, Z., Dong, Z., An, X., Ma, B., & Li, J. (2019). Maize genic male-sterility genes and their applications in hybrid breeding: Progress and perspectives. *Molecular Plant*, *12*(3), 321–342. 10.1016/ j.molp.2019.01.014

Wang, B., Zhu, L., Zhao, B., Zhao, Y., Xie, Y., Zheng, Z.,& Wang, H. (2019a). Development of a haploid-inducer mediated genome editing system for accelerating maize breeding. *Molecular Plant*, *12*(4), 597–602. 10.1016/j.molp.2019.03.006

Wang, C., Liu, Q., Shen, Y., Hua, Y., Wang, J., Lin, J., & Wang, K. (2019b). Clonal seeds from hybrid rice by simultaneous genome engineering of meiosis and fertilization genes. *Nature Biotechnology*, *37*(3), 283–286. 10.1038/s41587-018-0003-0

Wang, J., Cao, Y., Wang, K., & Liu, C. (2022). Development of multiple-heading-date mtl haploid inducer lines in rice. *Agriculture*, *12*(6), 806. 10.3390/agriculture12060806

Wang, N., Xia, X., Jiang, T., Li, L., Zhang, P., Niu, L., & Lin, H. (2022). In planta haploid induction by genome editing of DMP in the model legume *Medicago truncatula*. *Plant Biotechnology Journal*, *20*(1), 22. 10.1111/pbi.13740

Wang, Z., Cheng, J., Fan, A., Zhao, J., Yu, Z., Li, Y., … ... & Wang, X. (2018). LecRK-V, an L-type lectin receptor kinase in Haynaldia villosa, plays positive role in resistance to wheat powdery mildew. *Plant Biotechnology Journal*, *16*(1), 50–62. 10.1111/pbi.12748

Wei, X., Qiu, J., Yong, K., Fan, J., Zhang, Q., Hua, H., & Huang, X. (2021). A quantitative genomics map of rice provides genetic insights and guides breeding. *Nature Genetics*, *53*(2), 243–253. 10.1038/s41588-020-00769-9

Wolter, F., Schindele, P., & Puchta, H. (2019). Plant breeding at the speed of light: the power of CRISPR/Cas to generate directed genetic diversity at multiple sites. *BMC Plant Biology*, *19*(1), 1–8. 10.1186/s12870-019-1775-1

Xing, HL., Dong, L., Wang, ZP. *et al.* A CRISPR/Cas9 toolkit for multiplex genome editing in plants. *BMC Plant Biology*, *14*, 327 (2014). 10.1186/s12870-014-0327

Yang, S. L., Xie, L. F., Mao, H. Z., Puah, C. S., Yang, W. C., Jiang, L., & Ye, D. (2003). Tapetum determinant1 is required for cell specialization in the Arabidopsis anther. *The Plant Cell*, *15*(12), 2792–2804. 10.1105/tpc.016618

Yu, H., & Li, J. (2022). Breeding future crops to feed the world through *de novo* domestication. *Nature Communications*, *13*(1), 1171. 10.1038/s41467-022-28732-8

Yu, H., Lin, T., Meng, X., Du, H., Zhang, J., Liu, G., & Li, J. (2021). A route to *de novo* domestication of wild allotetraploid rice. *Cell*, *184*(5), 1156–1170. 10.1016/j.cell.2021.01.013

Zargar, M., Zavarykina, T., Voronov, S., Pronina, I., & Bayat, M. (2022). The recent development in technologies for attaining doubled haploid plants in vivo. *Agriculture*, *12*(10), 1595. 10.3390/ agriculture12101595

Zastrow-Hayes, G. M., Lin, H., Sigmund, A. L., Hoffman, J. L., Alarcon, C. M., Hayes, K. R., Richmond, T. A., Jeddeloh, J. A., May, G. D., Beatty, M. K. (2015). Southern-by-sequencing: A robust screening approach for molecular characterization of genetically modified crops. *Plant Genome*, *8*(1), 1–15. 10.3 835/plantgenome2014.08.0037

Zhang, F., Neik, T. X., Thomas, W. J., & Batley, J. (2023a). CRISPR-Based Genome Editing Tools: An Accelerator in Crop Breeding for a Changing Future. *International Journal of Molecular Sciences*, *24*(10), 8623. 10.3390/ijms24108623

Zhang, J., Wu, P., Li, N., Xu, X., Wang, S., Chang, S., & Guo, X. (2023c). A male-sterile mutant with necrosis-like dark spots on anthers was generated in cotton. *Frontiers in Plant Science*, *13*, 1102196. 10.3389/fpls.2022.1102196

Zhang, R., Chang, J., Li, J., Lan, G., Xuan, C., Li, H. (2021). Disruption of the bHLH transcription factor abnormal tapetum 1 causes male sterility in watermelon. *Horticulture Research*, *8*, 258. 10.1038/s41438-021-00695-9

Zhang, R., Zhang, S., Li, J., Gao, J., Song, G., Li, W., & Li, G. (2023b). CRISPR/Cas9-targeted mutagenesis of *TaDCL4, TaDCL5* and *TaRDR6* induces male sterility in common wheat. *Plant Biotechnology Journal*, *21*(4), 839–853.10.1111/pbi.14000

Zhang, Z. B., Zhu, J., Gao, J. F., Wang, C., Li, H., &Li, H., (2007). Transcription factor AtMYB103 is required for anther development by regulating tapetum development, callose dissolution and exine formation in Arabidopsis. *The Plant Journal*, *52*, 528–538. 10.1111/j.1365-313X.2007.03254.x

Zhao, D. Z., Wang, G. F., Speal, B., & Ma, H. (2002). The EXCESS MICROSPOROCYTES1 gene encodes a putative leucine-rich repeat receptor protein kinase that controls somatic and reproductive cell fates in the Arabidopsis anther. *Genes & Development*,*16*, 2021–2031. 10.1101/gad.997902

Zhong, Y, Liu, C, Qi X, Jiao Y, Wang D, Wang Y, Liu Z, Chen C, Chen B, Tian X, Li J, Chen M, Dong X, Xu X, Li L, Li W, Liu W, Jin W, Lai J and Chen S (2019). Mutation of ZmDMP enhances haploid induction in maize. *Nature Plants*, *5*(6): 575–580. 10.1038/s41477-019-0443-7

Zhu, E., You, C., Wang, S., Cui, J., Niu, B., Wang, Y., & Chang, F. (2015). The DYT 1-interacting proteins b HLH 010, b HLH 089 and b HLH 091 are redundantly required for Arabidopsis anther development and transcriptome. *The Plant Journal*, *83*(6), 976–990.10.1111/tpj.12942

Zhu, J., Chen, H., Li, H., Gao, J. F., Jiang, H., Wang, C., & Yang, Z. N. (2008). Defective in tapetal development and function 1 is essential for anther development and tapetal function for microspore maturation in Arabidopsis. *The Plant Journal*, *55*(2), 266–277.10.1111/j.1365-313x.2008.03500.x

Zsögön, A., Čermák, T., Naves, E. R., Notini, M. M., Edel, K. H., Weinl, S., & Peres, L. E. P. (2018). *De novo* domestication of wild tomato using genome editing. *Nature Biotechnology*, *36*(12), 1211–1216. 10.1038/nbt.4272

7 The CRISPR/Cas System for Plant Speed Breeding

Advances and Prospects

Muhammad Jawad Akbar Awan, Muhammad Ismail Buzdar, Asma Zia, Aiman Raza, Rubab Zahra Naqvi, and Imran Amin
Agricultural Biotechnology Division, National Institute for Biotechnology and Genetic Engineering (NIBGE), Constituent College of Pakistan Institute of Engineering and Applied Sciences, Jhang Road, Faisalabad, Pakistan

Shahid Mansoor
Agricultural Biotechnology Division, National Institute for Biotechnology and Genetic Engineering (NIBGE), Constituent College of Pakistan Institute of Engineering and Applied Sciences, Jhang Road, Faisalabad, Pakistan

International Center for Chemical and Biological Sciences, University of Karachi, Karachi, Pakistan

7.1 INTRODUCTION

A large population of the world relies on agricultural products for food, nourishment, and other consumable products. On account of the increasing population around the globe, an inevitable threat of global food insecurity lurks. Traditional breeding practices have been utilized to revamp crops through the stacking of desired genes into the breeding pipeline and generation of high-yielding cultivars; however, such approaches lack the pace to meet future demands (Hickey et al., 2019). Furthermore, other persistent challenges such as rapid urbanization, shrinkage of arable land, and climate change demand urgent solutions to improve agricultural productivity (Chen et al., 2019; Awan et al., 2022d). Therefore, there is a dire need to shift toward sustainable agriculture to reproduce climate-smart crops having escalated yield potential, harnessing resistance against abiotic and biotic stresses in a fast-forward manner.

Mutation breeding mostly involves the employment of random mutagenesis through mutagens of a chemical nature such as di-ethyl-sulfate (DES) and ethyl methane-sulfate (EMS), and physical irradiations, for instance, gamma rays, X-rays, and fast neutrons. These randomly induced genetic variations confer novel genetic diversity that can result in elite agronomic characteristics. However, these practices require labor-intensive selection and large screening processes which are challenging and time-consuming. Conversely, transgenesis allows the introduction of desirable alleles into the elite cultivars through biolistic or *Agrobacterium*-mediated transformation. Although this approach generates desired plants in less time and bypasses the bottleneck of hybrid incompatibility or reproduction isolation, the regulatory procedures and public concerns about transgenic plants pose hindrances in their commercialization.

DOI: 10.1201/b23372-7

Advancements in sequencing technologies and open access to sequence databases for various plants provide genomic information that can be exploited for the employment of targeted mutagenesis to develop improved crops (Zhu et al., 2020; Awan et al., 2022e). For this purpose, scientists have utilized the target-specific binding ability of transcription activator-like effectors and zinc fingers to DNA that enabled the formation of transcription activator-like effector nucleases (TALENs) and zinc finger nucleases (ZFNs) after their fusion with the *FokI*-nuclease domain, respectively (Chen et al., 2019). These molecular scissors show structural similarities and can be customized for the target-specific double-stranded breaks (DSBs) induction in the genome. Following DSBs, the cell's repair machinery induces random nucleotide insertions and/or deletions (Indels) in the targeted region of DNA (Zafar et al., 2021). ZFNs have been deployed in allohexaploid wheat to generate heritable resistance against imidazolinone herbicide by targeting the *acetohydroxy acid synthase* (*AHAS*) gene (Ran et al., 2018). Similarly, TALENs are known to be utilized for the desired mutagenesis in all three homeologs of the *TaMLO* (*Mildew-Resistance Locus)* gene in hexaploid bread wheat that revoked heritable resistance against powdery mildew disease (Wang et al., 2014). Despite their customizability, several limiting factors such as off-site cleavage, chromosomal aberrations, and cell death hampered their utilization in crops (Awan et al., 2022d).

The revolutionary discovery of novel nucleases CRISPR (cluster regularly interspaced palindromic repeats) and associated Cas proteins in archaea and bacteria as means of an adaptive immunity system against foreign nucleic acids or invading bacteriophages (Garneau et al., 2010), opened new avenues for the programmable target-specific RNA-guided DSBs in the genome (Jinek et al., 2012; Charpentier & Doudna, 2013). To date, different CRISPR-Cas systems have been derived from bacteria, archaea, and viruses. Owing to their structural and functional versatility, different nucleic acid molecules like dsDNA, single-stranded DNA (ssDNA), and single-stranded RNA (ssRNA) are targets for various Cas endonucleases such as Cas9, Cas12, Cas13, and Cas14 (Jinek et al., 2012; Zetsche et al., 2015; Abudayyeh et al., 2016; Harrington et al., 2018). The CRISPR systems have been exploited in different crops to produce intended mutations in the target of interest due to their ease of programmability.

The improvement of agronomically significant traits through genome editing (GE) by utilizing CRISPR-Cas systems increases the process of attaining genetic diversity and provides novel germplasm that can be rapidly merged into hybrid breeding strategies (Awan et al., 2022d). For this purpose, the speed breeding technology enables the fast-forward production of up to six generations in major cereal crops like wheat (Watson et al., 2018). Subsequently, mutated alleles can be integrated into elite varieties which can be deployed into the farmers' field. Moreover, the eradication of transgene harboring CRISPR-Cas genes can be possible through speed breeding technology. This chapter will highlight the implications of CRISPR-Cas9 followed by its variants for the improvement of crops that can be a key player in the enhancement of crop yield and fend off the inevitable threat of global food insecurity.

7.2 CRISPR-CAS9 SYSTEMS FOR PLANT GENOME EDITING

Although it is evident that TALENs and ZFNs have been deployed for genomic modifications in various plants, GE by CRISPR-Cas9 posits simple and easy RNA-guided programmability even for targeting multiple genomic sites, simultaneously. Multiple sgRNAs can be used with a single Cas9 protein for creating DSBs at multiple sites. To date, Since its isolation from Streptococcus pyogenes (SpCas9), the CRISPR-Cas9 system has been extensively employed in plants for knock-in or knock-out mutagenesis (Gasiunas et al., 2012). The *Sp*Cas9 endonuclease recruits a sgRNA (single guide-RNA) composed of 20-nucleotides (nts), crRNA (long CRISPR RNA), and ~76-nts long (*trans*-activating) crRNA, that induce conformational alterations enabling Cas9 effector to carry out DSBs in the genomic region bearing complementary to crRNA sequence (Jinek et al., 2014). Upon binding with the target gene sequence in the genome, the *Sp*Cas9 effector induces DSBs at three-nts upstream to the protospacer adjacent motif (PAM; 5`-NGG) through two nuclease domains, termed HNH and RuvC domains. After DSBs, the cellular repairing machinery is triggered that initiates genome modifications through either a more accurate template-dependent homology-directed repair (HDR) mechanism or the error-prone non-homologous end-joining pathway (NHEJ) (San Filippo et al., 2008; Lieber, 2010).

7.3 CRISPR-CAS9-MEDIATED BASE TRANSITIONS AND TRANSVERSIONS

The CRISPR and associated proteins Cas9-mediated DSBs generate random indels or insertion/replacement of small chunks of DNA via NHEJ or HDR pathways, respectively (Figure 7.1A). However, these approaches lack the capability to introduce more precise base transitions and transversion at the targeted sites. The fusion of distinct catalytically independent domains such as deaminases and reverse transcriptase (RT) with Cas9 effector enhanced the functional tendency of fusion complexes that led to the formation of adenine base editor (ABE), cytosine base editor (CBE) and prime editor (PE) which have been employed for transition mutations and insertion of large DNA fragments in various crops (Zong et al., 2017; Zong et al., 2022; Ahmad et al., 2023). Interestingly, these fusion complexes evade dependency on DSBs and a template donor for the integration of targeted mutations with precision and accuracy in plant genomes. For instance, CBE, comprised of a fusion complex of Cas9 nickase (n-Cas9) with deactivated RuvC domain owing to D10A mutation, a cytidine deaminase, and at last uracil DNA glycosylase inhibitor (UGI) (Komor et al., 2016). By binding to the target site via sgRNA, CBE converts C:G to T:A by converting cytosine into uracil via a deamination reaction in the complementary to the target strand followed by replacement with thymine during the DNA replication (Figure 7.1B). The UGI averts uridine excision during the repair pathway and boosts base editing efficiency.APOBEC1 cytidine deaminase originated from rats, showed C to T conversion in the span of 3 to 9-nts upstream to PAM sequence (Zong et al., 2017), while human-originated APOBEC3A cytidine deaminase mediates C to T conversions in a large editing window ranging (1 to 17-nts) (Zong et al., 2018). CBEs have been widely utilized in plants and demonstrated high specificity and editing efficiency. Similarly, ABE, composed of nCas9 (D10A) and *Escherichia coli*-originated TadA (transfer RNA adenosine deaminase), conducts A:T to G:C conversions at the targeted place (Figure 7.1C). ABE has been deployed in wheat and rice to carry out A to G conversion with up to 59% editing

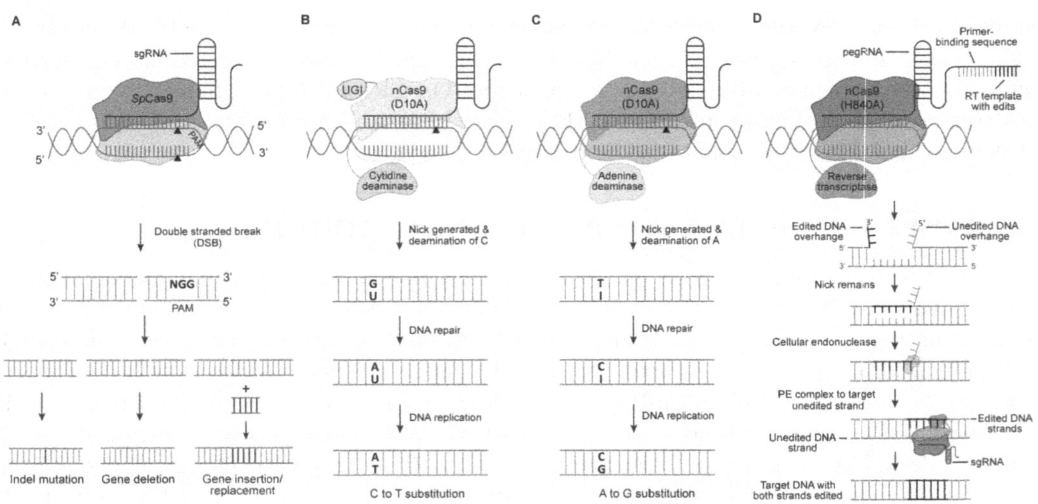

FIGURE 7.1 Schematic demonstration of the working mechanism of CRISPR-Cas9 and its variants for target-specific mutagenesis.

efficiency (Li et al., 2018a). Furthermore, modified ABE (ABE-7.10) demonstrated up to 42% editing efficiency in the rice genome with minimal to no off-target effect (Hua et al., 2018).

CBE and ABE are proven to be efficient for on-site base conversion and have been broadly utilized in plants for trait improvement, however, their efficacy is limited to base transition. PE offers versatile options for target-specific GE and expands the CRISPR toolbox from base transition to transversion. PE is composed of RT and Cas9 nickase having H840A mutation and operates via programmable PE guide RNA (pegRNA) (Figure 7.1D) (Anzalone et al., 2019). The prime editing gRNA carries a crRNA sequence which is complementary to the targeted region and primer binding sequence along with the RT template carrying desired edits. Contrary to base editors, PE holds the tendency to insert or delete DNA sequences at the target site with higher precision in the genome. The nCas9 induces a nick in the target DNA strand (strand with PAM) followed by hybridization of primer binding sequence with nicked DNA strand. RT incorporates the RT template sequence into the nicked strand. The unedited DNA overhang is removed by the cellular repair machinery while the edited DNA overhang ligates back with the unnicked DNA strand. Second sgRNA-mediated incision is introduced within the DNA strand that is unedited to initiate the process of chromosomal repairing, subsequently, the edited sequences are incorporated into both DNA strands. Plant prime editors (PPE) have been deployed in wheat and rice for insertions, deletions, and point mutations with an efficacy of up to 22% in the later (Lin et al., 2020). Recently, a pair of PEs, termed twin PEs, has been employed for the targeted indels of large DNA sequences (Anzalone et al., 2022; Awan et al., 2022a). Zong et al. (2022) reported an engineered PPE (ePPE) by the augmentation of RT and nucleocapsid protein derived from the modified *Moloney-murine leukemia virus* to develop imidazolinone and sulfonylurea herbicides tolerant rice plants with up to 11.3% editing frequency. This approach enables the selective switch-on or switch-off of the resistance or susceptibility genes, ultimately in the production of disease-resistant plants (Ahmad et al., 2023). Likewise, PASTE (the programmable addition via site-specific targeting element), consists of two units namely, PE and serine integrase for site-specific integration, demonstrated targeted deletion and insertion of up to 36 kb in the genome more efficiently and precisely (Yarnall et al., 2023). PASTE not only holds the potential to insert desired one gene or genes, simultaneously but can also remove the transgene from the genome by bypassing the induction of unwanted DSBs (Awan et al., 2023d).

7.4 DEPLOYMENT OF GENOME EDITING TOOLS IN PLANTS

The major objective for agriculture researchers is to exploit crop genetic diversity to sustain production with better agronomic traits and better nutritional values. Similarly, stress tolerance in the current climate change scenario is another key area to be tackled effectively. Despite the significant contribution of conventional breeding in maintaining crop yields and mitigating stresses, the growing need for food challenged the production rate of crops and global food security (Ma et al., 2018). It is imperative to promote efficient approaches in plant biology to enhance yield by simultaneously conserving natural resources. Over the past decade, various genetic loci have been identified that control different agronomically significant traits in plants. However, the exploitation of these genetic resources remained challenging (Saurabh, 2021).

Genetic transformation has gained significant attention in plant biology as a powerful tool to impart desirable characteristics in various crops (Lanigan et al., 2020; Yan et al., 2022). Deployment of specific genes of interest in plants through improved procedures of genetic transformation has resulted in better yields, enhanced nutritional values, and improved stress tolerance in rice, maize, barley, wheat, sorghum, soybean, and several other crops (Abdallah et al., 2015; Zhang et al., 2018b). Besides their role in the improvement of crops, genetically modified crops can be exploited for non-food uses, particularly in biofuel production, bioremediation, and pharmaceuticals. Crop improvement through targeted genetic modifications largely depends on efficient genetic transformation and regeneration-enhancing breakthroughs that can fast-forward the development of desired plants (Aglawe et al., 2018).

Different transformation methods including Agrobacterium-mediated transformation, polyethylene glycol, and particle bombardment are mainly utilized approaches for the stable plant transformation and CRISPR-Cas9 system delivery. Transformation methods such as particle bombardment and Agrobacterium-mediated transformation allow the integration of transgene in the host genome which confers continuous expression of Cas9 effector and sgRNAs to produce desirable traits. Among these transformation techniques, the latter one is widely utilized to transform the CRISPR-Cas9 system in various crops (Awan et al., 2023b). The simple procedure, capacity to harbor larger fragments, and reproducibility are the distinguishing features of the *Agrobacterium*-mediated transformation strategy favored over other approaches such as plant transformation via liposomes, electroporation method, microinjection-mediated plant transformation and through silicon-carbide-whiskers (TOURAEV et al., 2011; Su et al., 2023).

The physical method of plant transformation such as particle bombardment relies on the transfer of exogenous DNA directly into the host genome (TOURAEV et al., 2011; Ramkumar et al., 2020). This method utilizes the DNA-coated micro-carriers that are synthesized by employing gold particles or tungsten powder. The microcarrier is accelerated with high-pressure helium pulses using a pressurized helium gas stream or an electric discharge. The accelerated particles gained, pierce the recipient cells, which enables the target DNA to enter the cell (Ozyigit et al., 2020). The transgene eventually becomes part of the host genome by integrating into it and consequently, production of gene-edited plants. This technique is being applied to improve several crops but is considered less efficient compared to *Agrobacterium*-mediated transformation because of higher costs, lower transformation rate, and unprotected exogenous DNA. Additionally, this technique of transformation is only able to cargo smaller DNA sequences (up to 10 kb) (Ozyigit et al., 2020).

Moreover, there are certain limitations of conventional transformation methods like low transformation and regeneration efficiency, cell damage, and improper DNA integration. To overcome these issues, nanoparticles-mediated transformation is gaining the spotlight for efficient transportation of CRISPR Cas machinery in the plant tissues (Lv et al., 2020). Nanoparticle-mediated transfer of genetic material can be carried out through peptide nanoparticles, layered double hydroxide (LDH) nanosheets, magnetic nanoparticles, carbon nanotubes, and carbon DNA nanostructures (Cunningham et al., 2018). In contrast to conventional gene delivery systems, nanoparticles-mediated transformation shows low cytotoxicity and can be applied in various plant species. Though nanoparticle-mediated gene delivery has proved effective in various crop species certain limitations such as the synthesis and design of nanoparticles and their entry into plant cells challenge its efficiency (Xu et al., 2022). Apart from this, low regeneration efficiency limits their applicability for crop improvement.

In order to overcome the off-target effects (DSBs in unwanted or non-targeted regions of the genome) and develop transgene-free plants, the transient expression systems can be exploited (Tyurin et al., 2020). The plant tissues that can be utilized for the delivery of this system often include protoplasts (Lin et al., 2018), calluses, suspension cultures of plant cells (Zagorskaya & Deineko, 2017), intact plants (Jelly et al., 2014), as well as isolated plant organs (Guidarelli & Baraldi, 2015) or specialized plant tissues (Jia et al., 2018). Transient expression is carried out through expression vectors with binary Ti plasmid and viral vectors (Vyacheslavova et al., 2012; Awan et al., 2023a). The gene delivery for transient expression is carried out through transfection using PEG, electroporation, biolistic, and agroinfiltration (Wang & Jiang, 2011; Tsuda et al., 2012; Tan et al., 2013). In addition, as a valuable research tool transient expression can be effectively utilized to study the proteins localization in plant cells, physiological role of genes of interest in growth and stable development of host, and adaptability to biotic and abiotic stresses. It is also useful in studying regulatory metabolic pathways and protein-protein interactions (Tyurin et al., 2020). Despite the significance of this system, transient expression affects the host cell transcription and translation efficiency.

Plant transformation is highly affected by genotype dependence (Ye et al., 2022) and low regeneration efficiency during tissue culture. Moreover, regeneration post-transformation is a crucial

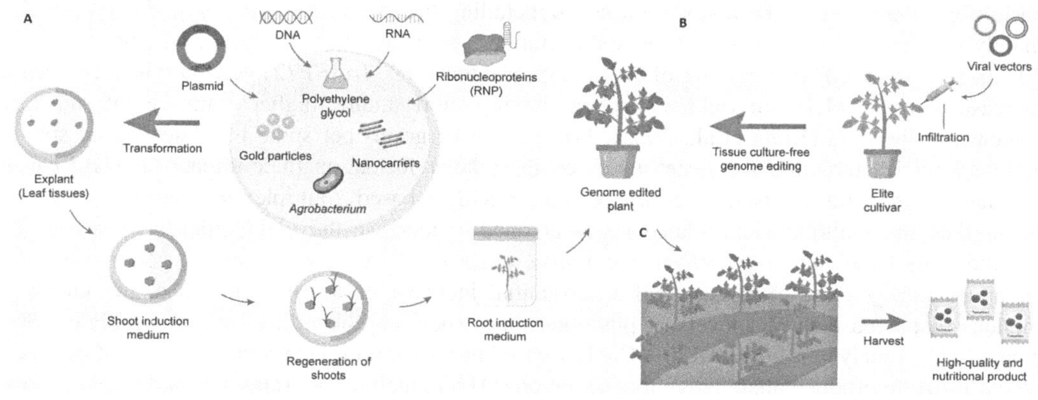

FIGURE 7.2 Graphical illustration of different approaches for the deployment of CRISPR-Cas system in plants. (A) Employment of stable and transient techniques involving plasmid and linear double-stranded DNA carrying T-DNA cassette, RNA transcripts encoding Cas9 endonuclease and single guide RNA (sgRNA), and ribonucleoproteins (RNPs) into the explants aided with *Agrobacterium*-mediated transformation, polyethylene glycol, nanoparticles, and gold particles to generate genome-edited plants via tissue culture. (B) Deployment of viral vectors harboring CRISPR-Cas system to generate tissue culture-independent heritable genome editing in plants. (C) The climate-smart and tailor-made plants, generated through genome editing, can be deployed into the field to harvest high-quality and nutritional products.

pace-limiting step in gene delivery systems. Therefore, it is imperative to overcome these limitations to upgrade genetic transformation efficiency and rate of regeneration of more transgenic events for the generation of a large number of edited plants (Saurabh, 2021). In this regard, various techniques such as the employment of morphogenic regulators (MRs) have been deployed to enhance regeneration efficiency. The ectopic expression of MRs such as *Bbm (Baby boom)*, *Wus2* (*Wuschel2*), *GRF* (*Growth-regulating factor)*, *LEC1* (*LEAFY COTYLEDON1*),and *LEC2* (*LEAFY COTYLEDON2)* genes hold the potential to improve the regeneration tendency in eudicots and monocots (Lotan et al., 1998; Stone et al., 2001; Boutilier et al., 2002; Zuo et al., 2002; Debernardi et al., 2020). The impact of various growth regulatory and morphogenic genes on plant transformation systems, their expression, and synergistic and additive effects can be critical for normal transformed fertile plants in different species. In contrast, other tissue culture-free methods are also in place that can potentially be leveraged for the cargo of editing machinery, e.g., CRISPR-Cas-based components via *in-plant* transformation to generate heritable mutagenesis (Lee & Wang, 2023). For this purpose, GE components can be augmented with mobile RNA elements to increase their movement to reproductive tissues in various plants to enhance heritable editing efficiency (Chen, 2020; Awan et al., 2023c) (Figure 7.2).

7.5 EXPLOITATION OF CRISPR-CAS MACHINERY FOR CROP IMPROVEMENT

7.5.1 Yield Improvement

The rise in the incidence of natural catastrophes and the effects of climate change has lately had a negative influence on agricultural systems. Looking ahead to the immediate future, the agricultural sector is prone to face significant hurdles in its effort to feed a population that is anticipated to increase to almost 10 billion by 2050 (Hickey et al., 2019). Improving the precision and efficiency of GE techniques for building novel cereal cultivars having specific desired genes and traits to remit the challenges of catastrophic climate change and increasing global population is imperative. The editing system based on CRISPR-Cas9 has developed into a powerful and versatile tool for precise GE, enabling the creation of desirable characteristics like increased yield, stress tolerance, and disease resistance in key cereal crops, even without the integration of transgenes. Yield enhancement is a basic objective in all of the major food crops including bread wheat (*Triticum aestivum*) which is mainly challenged by factors such as spike number, grain number, and grain weight (Liu et al., 2022b). In wheat, GE by targeting of *auxin response factor12* (*TaARF12*) gene in wheat showed an increase of up to 11.1% in yield with plants having short stature, higher grain weight, and large spikelets in the field (Kong et al., 2023). The number of grains per spike is influenced by spikelet architecture. By targeting the genes involved in spikelet formation their impact on yield can be evaluated. As yield is itself a complex trait, CRISPR-based multiplexing can be utilized to manipulate the multiple yield-related genes at once to increase the yield efficacy of wheat. The simultaneous knockout of *E3 RING-type ligase* and *Rht-B1* (*Reduced height 1)* genes resulted in enhanced nitrogen use efficiency and a substantial increase in wheat yield. Also, the knock-out mutants displayed balanced levels of phytohormones such as gibberellins and brassinolides (Song et al., 2023). Lately, a robust CRISPR-Cas12a-based promotor editing system has been introduced in *Oryza sativa* to produce quantitative trait variation (QTV) pipeline for grains size and starch contents by targeting genes such as *OsGS3* and *OsGBSS1* (Zhou et al., 2023). The *OsD1* promotor edited lines showed enhanced yield in field trials. CRISPR-based GE approaches hold the potential for precise integration of QTLs in elite crop cultivars as yield is a complex polygenic trait governed by both genetic and non-genetic factors. Recently, Wu and fellows (Wu et al., 2023) investigated knockout mutation in a yield-related gene, termed *LOGL5*, on GY3 QTL in *O. indica* demonstrated higher yield (7.4% to 16.3%) with more grains per panicle in the mutant lines. The *LOGL5* gene involved in the cytokinin pathway can prove to be a potential target for GE by the CRISPR/Cas system in other rice cultivars as well as its homologs in major crops. Another study in rice reported the *grain filling duration mutant-1* (*gfdm-1*) which exhibited large grain size and grain filling duration by controlling carbohydrate uptake in stems, grains, and leaves. This mutant along with sugar transporter *OsSWEET4* and *OsSUT2* attunes the number of grains per panicle (Sun et al., 2023).

In maize (*Zea mays*), determining yield is highly dependent on endosperm cell number. The transcription factor for grain size *ZmMYB74* was targeted by employing the CRISPR-Cas9 system, and mutant plants exhibited large grain size validating its regulation role in the proliferation of endosperm cells in comparison to control plants with suppressed expression (Wang et al., 2023). Introgression of THP-teosinte in modern maize increases the quantity of free amino acids in seeds thus increasing the yield and nitrogen use efficiency under low nitrogen scenarios (Huang et al., 2022). GE of these seed protein-related genes can significantly enhance the yield potential in maize in less time. The *CYP78A* gene family is involved in determining the yield in terms of grain/fruit weight and size in various major food crops. Similarly, in tomatoes (*Solanum lycopersicum*), CRISPR-Cas9-mediated introduction of single nucleotide polymorphism (SNP) in the *KLUH* promotor yielded an enormous increase of up to 16.3% in weight and reduction in the ratio of small fruits production (Li et al., 2022a). Likewise, Tang et al. (2022) validated that the CRISPR-Cas9-edited *StIT1* gene in potatoes (*Solanum tuberosum*) led toward stolon branching and increased yield in the field. Moreover, to overcome global food insecurity, it is mandatory to bend toward more integrated approaches such as CRISPR-based precise GE and transgene-free methods coupled with speed breeding.

7.5.2 Quality Enhancement

Recent research has indicated that CRISPR-based GE has the potential to upgrade nutritional value, quality, and biofortification of major cereal crops. The quality of food crops has a direct impact on human health as they are the main source of many nutrients, minerals, and bioactive compounds. Bread wheat is high in gluten proteins that are known to cause coeliac disease in approximately 2% of the world population. A recent research demonstrated that simultaneous knockout of two genes coding for γ-gliadin proteins resulted in decreased gluten contents in knockout mutants and improved the end-user quality (Kong et al., 2023). Similarly, multiple quality-related genes such as *puroindoline b* (*PINB*), *waxy gene encoding GBSS1 enzyme* (*WAXY*), *polyphenyl oxidase* (*PPO*), and *phytoene synthase* (*PSY*) were knocked out using CRISPR-based multiplexed GE significantly improved starch quality, dough color, and wheat grain hardness (Zhang et al., 2021). The CRISPR system has been utilized to manipulate the metabolic profile of rice grains to enhance the contents of essential minerals and vitamins such as vitamin A, zinc, and iron. Amylose contents and gel consistency are the two important parameters that influence the grain quality and cooking time of rice. Recently, *Os*ISA1 mutant plants were developed by CRISPR-Cas9-driven frameshift mutation in *Isoamylase 1* (*ISA1*) gene, which were rich in starch gel consistency and sugar contents with reduced contents of amylose as well as amylopectin (Shufen et al., 2019). ASD16 an aromatic rice cultivar has been developed using CRISPR by integrating the gene responsible for aroma *betaine aldehyde dehydrogenase 2* (*OsBADH2*). Many factors that have the potential to upgrade rice quality have been targeted using GE technologies such as oleic acid contents, anthocyanins, rice grain shape, and improving contents of various minerals and vitamins. This technology is also implicated to improve the quality and nutritional content of maize. Editing of *Waxy* genes in maize achieved improved storage capacity and water absorption imparting a positive impact on maize yield (Gao et al., 2020). Likewise, aromatic was produced by *ZmBADH2a* and *ZmBADH2b* modifications (Wang et al., 2021). The post-harvest shelf life, appearance, taste, and fruit quality are some of the most crucial traits concerning the improvement of tomatoes and fulfilling consumer demands. In a recent study, targeted mutations were induced upstream of the uORF (open reading frame) regions in the *SlBZIP* gene responsible for sucrose-induced repression of translation (SIRT) mechanism, the mutant genes influenced the amino acids and sugar metabolic pathways resulting in the production of sour tasting amino acids from 77 to 144% and sweet-tasting amino acids from 14 to 104% with no effect on plant yield. This strategy was able to produce fine-quality tomatoes with improved taste (Nguyen et al., 2023). In tomatoes, fruit color is an important trait determined by the presence of various carotenoids and anthocyanin pigments. The CRISPR-based multiplexing approach has been employed to target multiple color-controlling genes in tomatoes *phytoene synthase 1* (*PSY1*), *myeloblastosis 12* (*MYB12*), and *stay green 1* (*SGR1*)genes which resulted in the production of transgene-free variety of colored tomatoes genotypes ranging from light red, pink to yellow, yellow-green, and brown in less than one year without compromising agronomic traits (Huang et al., 2022). Potato tuber quality is determined by the presence of amylopectin and down-regulation of some anti-nutrients. Steroidal glycoalkaloids (SGAs) cause bitter taste in mature potatoes. The *St16DOX* gene encodes a steroid α-1,6-hydroxylase for SGA metabolism. Hence, this gene is a potential GE target for the production of SGA-free potato varieties. Knocking out *St16DOX* by CRISPR-Cas9, resulted in the elimination of the SGA contents accretion in potato hairy roots (Nakayasu et al., 2018). Cytidine base editor (CBE)-mediated amino acid substitution in the *GBSSI* (*granule bound starch synthase I*) gene-modified starch composition in potatoes (Veillet et al., 2019).

7.5.3 Disease Resistance

A diversity of microorganisms inhabits plants developing plant-microbe interaction which can be beneficial as well as pathogenic. The latter causes plant diseases (phytopathogens), responsible for major quality and yield losses, causing a threat to global food security such as pathogenic species including *Fusarium, Magnaporthe, Mycosphaerella, Phytophthora, Puccinia,* and *Xanthomonas*. CRISPR-Cas9 is utilized to develop resistance against biotic factors in major crops. In a recent report, wheat harnessing resistance to powdery mildew pathogen was developed by knocking out the *MLO* locus that activated *tonoplast monosaccharide transporter 3* without suppressing growth, development and inducing yield penalty (Li et al., 2022b). In another study, resistance against *yellow mosaic virus* in wheat is conferred by generating knock-out mutations in *PDIL5–1* and *eIF4E* genes (Kan et al., 2023). The employment of CRISPR in developing resistant rice cultivars contributed to the generation of high-yielding and fine-quality rice owing to the reduction in the losses caused by viral, fungal, and bacterial infections. By employing this system, sugar transporter genes such as SWEETs 11, 13, and 14 (*sugars will eventually be transported*) have been knocked out which resulted in resistant plants against *Xanthomonas oryzae*-mediated bacterial blight disease (Oliva et al., 2019). Similarly, *Pi21*, *ERF922,* and *Bsr-d1* genes were knocked out to produce bacterial blast-free rice cultivars (Zhou et al., 2022b).

Recent research conducted the successful development of broad-spectrum multi-pathogen resistance in *O. sativa* by editing the *resistance to blast 1*(*RBL1*) gene which transcribes a CDP-DAG (cytidine diphosphate diacylglycerol synthase) and has a vital function in plant immunity by controlling various biosynthetic pathways and cell death regulation. Mutants developed in recent research provide a broad-spectrum resistance window against fungal and bacterial diseases without compromising rice yield (You et al., 2023). Kong et al. (2023) discovered an allele for proteasome maturation factor in rice which is also known to cause multi-pathogen resistance by enhancing the activity of proteasome in releasing reactive oxygen species (ROS) and respective enzymes when encountered by a pathogen. Moreover, these broad-spectrum resistant rice varieties are highly desirable to counter the unalarmed biotic stresses and can provide essential gene targets to be translated into other major crops using GE. In maize, the CRISPR-mediated knocking out of the enhanced resistance against gibberella stalk rot (*COI1a*), increased virus resistance (*GDIα*) and resilience against herbicides chlorsulfuron (*ALS1*/*ALS2*) genes have contributed toward the production of resistant and high yielding plants (Liu et al., 2022a; Pramanik et al., 2021). The most devastating disease in tomatoes caused by *tomato yellow leaf curl virus* has been successfully controlled in various cultivars by deploying CRISPR-based GE to target and *MLO1* and *PELO* genes (Pramanik et al., 2021). Editing of *MYBS2* gene (*MYB transcription factor S2*)-induced resistance against *Phytophthora infestans* responsible for late blight in tomatoes (Liu et al., 2021a). However, modifying a single resistance (*R*) gene is not enough as new strains of this pathogen quickly overcome the resistance in tomatoes and potatoes. Knocking out susceptibility (*S*) genes and manipulation genes taking part in the metabolic pathways of these pathogens using GE is a smart approach for the development of more resilient crops. For instance, *S* genes such as *StDND1*, *StCHL1*, and *StDMR6* are potential knock-out targets to develop resistance against *P. infestans* (Kieu et al., 2021).

7.5.4 Tolerance against Abiotic Factors

In addition to biotic challenges, agriculture has perennially grappled with a range of abiotic stressors, capable of causing a potential 50% decline in average productivity (Younis et al., 2020). The primary sustainable approach to confront elevated temperatures, cold, salinity, drought, and heavy metals is the technique of breeding abiotic stress-tolerant varieties. Recently, the utilization of CRISPR-Cas has played a crucial role in mitigating these abiotic factors; however, the meticulous choice of candidate genes controls the desired characteristics.

In *Zea mays*, under drought stress conditions, the *ARGOS8* gene negatively regulates the ethylene response. Manipulation of the *ARGOS8* gene's promoter region through the introduction of the GOS2 promoter using CRISPR-Cas9-driven HDR, resulted in increased grain yield and decreased ethylene sensitivity in transgenic plants (Shi et al., 2017). The CRISPR machinery was employed to investigate potential ABRE and MYB recognition sites within the *ZmWRKY40* gene, a transcription factor belonging to the abiotic stress-related *WRKY* family. Its over-expression under drought conditions promotes plant survival rates by elevating the POD (peroxide dismutase) and CAT (catalase) activities (Wang et al., 2018a; Khoso et al., 2022). Tomato seedlings require the HSA1 (*heat-sensitive albino1*) gene for proper chloroplast development. Deletion mutants of *HSA1* produced by utilizing CRISPR-Cas technology, achieved higher heat sensitivity, leading to a substantial delay in chloroplast biogenesis in contrast to wild-type tomatoes (Qiu et al., 2018). Additionally, heat-tolerant tomatoes designed by CRISPR-Cas-mediated modification of the *SlAGL6* (*agamous-like 6)* gene, promotes the development of parthenocarpic fruit under heat stress (Klap et al., 2017). A BZR1-like protein is a key player in both brassinosteroid (BR) response and heat endurance by regulating the expression profiles of FERONIA (FER) homologs. In the *bzr1*-mutant, a reduced activation of RBOH1 (*respiratory burst oxidase homolog 1*) is observed resulting in a noticeable decrease in apoplastic hydrogen peroxide (H_2O_2) concentration, and also a decreased level of heat stress tolerance (Yin et al., 2018). CRISPR-

Cas-generated *cbf1* mutants unveiled the significance of CBF1 (*C-repeat binding factor 1*) in safeguarding crops against cold/chilling injury and minimizing electrolyte leakage levels. These tomato mutants exhibited elevated levels of IAA and H_2O_2 contributing to improved cold stress resilience (Ran et al., 2018). The TEY motifs within the *SlMAPK3* (*Mitogen-Activated Protein Kinase 3*) gene actively respond to abiotic stress, suggesting its potential role regarding drought stress response in tomatoes by modulating transcription of stress-related genes and protecting cell membranes from oxidative damage (Wang et al., 2017; Yu et al., 2019; Zhu et al., 2022). CRISPR-Cas edited tomato lines with knocked-out *SlSPL-CNR* (*Squamosa Promoter-Binding Protein-Like*) transcription factor, exhibited enhanced responses to iron deficiency, indicating its negative regulatory role in tomato's iron-deficiency response(Zhu et al., 2022).

In *O. sativa*, CRISPR-Cas9 induced gene knockout within sub-families of abscisic acid (ABA) receptor genes *PYL1/4/6* and *OsNAC006* has demonstrated significant heat stress tolerance (Miao et al., 2018; Nandy et al., 2019; Wang et al., 2020). The *OsANN3* gene, responsible for encoding annexin (a calcium-dependent phospholipid-binding protein) has been acknowledged as a critical factor in augmenting cold stress tolerance in rice. Mutant lines with altered *OsANN3* expression demonstrated enhanced relative electrical conductivity and bolstered resistance under cold conditions (Shen et al., 2017; Que et al., 2020; Le et al., 2022). Additionally, the Os*SAPK2* and *OsMYB30* edited lines yielded a significant increase in cold stress tolerance, achieving an overall efficiency of around 63% (Lou et al., 2017; Zeng et al., 2020). Salt and drought tolerance *OsDST* gene was targeted through GE in mega MTU1010 cultivar of the *O. indica* which resulted in the *dst* mutant, characterized by decreased stomatal density and broader leaves, facilitating improved leaf moisture retention (Santosh Kumar et al., 2020; Ganie et al., 2021). Furthermore, the soil's contamination with toxic heavy metals gradually builds up within plants. Certain rice genes such as *OsARM1, OsHAK1, OsLCT1, OsNramp5, OsPRX2,* and *OsATX1* have been targeted for frameshift mutations to produce varieties that exhibit resistance against the accumulation of harmful metals for instance cadmium, cesium, and arsenic (Zhang et al., 2018a; Songmei et al., 2019; Chang et al., 2020; Mirza et al., 2022; Saha et al., 2022; Sathee et al., 2022). Arabidopsis *Paraquat Tolerance 3* gene *AtPQT3* encodes a ligase enzyme (E3 ubiquitin ligase) that negatively regulates abiotic stress tolerance. Three CRISPR-Cas knockout mutants were designed by targeting conserved regions of the rice homolog *OsPQT3* gene, leading to improved resistance against salt and oxidative stress (Alfatih et al., 2020; Tang et al., 2022).

Similarly, in bread wheat, a transcription factor known as *DREB (dehydration-responsive element binding protein)*, within the *ERF (ethylene-responsive element-binding factor)* family, was subjected to CRISPR Cas9-based editing. By targeting the conserved regions of all homeologs of *TaDREB2* and *TaEFR3,* as a result, improved tolerance to cold and drought conditions was observed (Elsharawy & Refat, 2023). *Histone acetyltransferase* (*TaHAG1*) induces cytosolic enzymatic activity by elevating the expression levels of antioxidants like ascorbate peroxidase, catalase and superoxide dismutase. This enabled plants to maintain ROS homeostasis and develop salt stress resilience. The CRISPR system has been used to develop knock-out mutants of the *TaHAG1* which showed enhanced salinity stress tolerance (Zheng et al., 2021). The *Sal1* gene which encodes the enzyme 3'(2'),5'-bisphosphate nucleotidase, plays a crucial role in detoxifying cellular 3′-phosphoadenosine 5′-phosphate (PAP) during periods of drought and exposure to salt and light stresses. Also in wheat, the *TaSAL1* serves as a key negative regulator in pathways that enhance drought tolerance. Multiplex CRISPR-Cas9 system targeting conserved regions of five active *TaSal1* homologous genes served as an efficient tool to develop drought-resistant wheat (Abdallah et al., 2022).

Knocking out *NCED4*, a pivotal enzyme in ABA synthesis facilitates lettuce seed germination at higher temperatures. Consequently, *LsNCED4* mutants offer commercial potential in areas exceeding elevated temperature limits (Bertier et al., 2018). Arabidopsis displayed improved stomatal response to drought stress through CRISPR-Cas9, which utilized truncated gRNA targeting the *AtEF1* promoter and *OST2/AHA1* gene for the expression profiling of Cas9 endonuclease (Osakabe et al., 2016). Likewise, CRISPR-Cas utilization in activation of *ABRE1/ABF2* promoter upgraded drought endurance in Arabidopsis, by fusing dCas9 (catalytically inactive Cas9) with HAT (histone acetyltransferase) enzyme (Roca Paixão et al., 2019; de Melo et al., 2020). The CRISPR-Cas9 toolbox including MdU6 promoter-driven sgRNA and pHDE-35S-Cas9-mCherry-UBQ was utilized for mutagenesis of *MdCNGC2* (*cyclic nucleotide-gated ion channels)* gene in apple (*Malus domestica*) which resulted in continuous accumulation of SA (salicylic acid) with enhanced expression levels of defense and immunity-related genes, eventually conferring resistance against cankers caused by *Botryosphaeria dothidea* (Zhou et al., 2020).

7.5.5 Herbicide Resistance

Weeds pose a significant threat to agricultural productivity by depleting essential nutrients from arable land and reducing soil fertility. This results in delayed plant growth, and loss in grain yield as well as quality, which can be controlled by global use of herbicides. In a world with a growing population and escalating food demands, the ability to cultivate crops with resilience to herbicides has become indispensable for sustainable and high-yield agriculture. CRISPR-Cas platforms offer a means to genetically modify crops for herbicide resistance without the need for transgene incorporation. Previously, HDR-based modification of acetolactate synthase (*ALS*) gene resulted from the induction of a double-stranded break (Sun et al., 2016). Later, the CRISPR system was utilized to design a range of herbicide-tolerant crop varieties including watermelon, potato, soybean, and maize (Butler et al., 2015; Li et al., 2015; Svitashev et al., 2016; Tian et al., 2018). Two maize patents focused on herbicide-resistant germplasms produced via CRISPR-Cas-driven editing of *ALS1*, *ALS2*, and *EPSPS* (*5-enolpyruvylshikimate-3-phosphate synthase*) genes demonstrated promising commercial potential (Wang et al., 2022b).

Herbicides such as astriazolopyrimidines, imidazolinone, and sulfonylurea, inhibit the activity of ALS enzyme, causing aromatic amino acid deprivation in plants. Natural *ALS* gene variants characterized by point mutations, grant herbicide tolerance. Through CRISPR-nCas9 editing in *O. sativa*, substitutions of four amino acids are introduced to *OsALS1* proteins, leading to diverse herbicide tolerance levels against bispyribac sodium (Kuang et al., 2020). Moreover, the building of glyphosate-resistant rice, cassava, and flax plants involved the introduction of dual amino acid substitutions and alterations to the promoter of the *EPSPS* gene (Li et al., 2016; Sauer et al., 2016; Hummel et al., 2018). By employing the nCas9 or dCas9 variants fused with *PmCDA1* (*Petromyzon marinus* cytidine deaminase) point mutations were introduced, leading to the development of herbicide-resistant rice mutants (Shimatani et al., 2017). Rice plants resilient to herbicides were introduced against herboxidiene (GEX1A) by the utilization of CRISPR-Cas-based directed evolution strategy, which employs DSB-inducing multiplex Cas9, targeting coding sequences within the *SF3B1* gene (Butt et al., 2019).

In wheat, the *TaALS* gene proved to be an ideal candidate for base editing, allowing the targeted incorporation of point mutations that conferred herbicide tolerance with minimal impact on plant productivity (Zhang et al., 2019a). Moreover, the *ACCase* (*Acetyl-CoA carboxylase)* gene is an important player in the biosynthesis of fatty acids, which makes it a potential target against herbicides for instance, cyclohexanedione (CHD), aryloxyphenoxypropionate (APP), and phenylpyrazoline (PPZ). The CRISPR-based GE system has been employed to modify this gene in rice and wheat which resulted in herbicide resistance (Liu et al., 2020b; Dong et al., 2021; Rajak et al., 2023).

A single amino acid alteration fromglycine 336 to tryptophan (G336W) of *4-hydroxyphenyl pyruvate dioxygenase (HPPD)* decreased susceptibility to diketonitrile, the component of isoxaflutole herbicide with a mediocre impact on the enzymatic activity of HPPD. Similarly, the HPPD isozyme AvHPPD-03, derived from oat (*Avena sativa*), exhibited reduced binding affinity to mesotrione due to the deletion of a single amino acid at position 111 in its N-terminal domain. These genes *HPPD W336* and *AvHPPD-03*, have been instrumental in genetically modifying crops like soybean and cotton (Dreesen et al., 2018). Altogether, the CRISPR-Cas9-based GE and its variants demonstrate impeccable potential for improvement of crops by inducing targeted modifications in the genome and such modified crops can be incorporated into the speed breeding programs for their fast-forward deployment into the farmer's field to harvest their true potential on large scale (Figure 7.3) (Table 7.1).

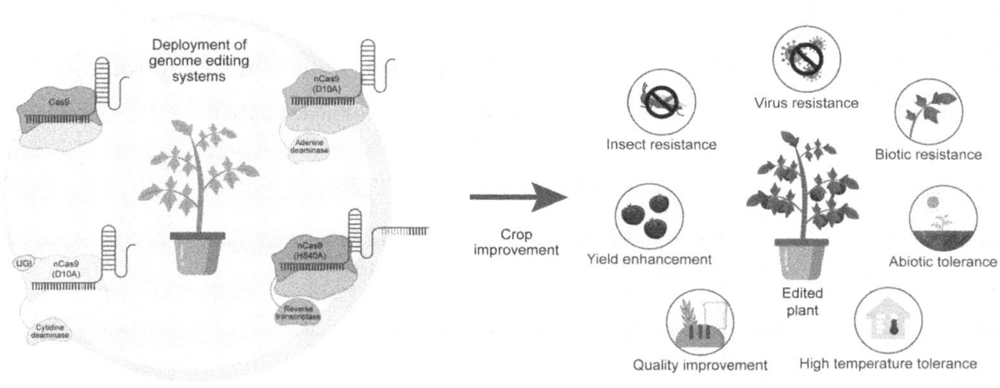

FIGURE 7.3 Schematic illustration of employment of various CRISPR-Cas9 variants for the improvement of crops.

TABLE 7.1
List of Genes That Have Been Targeted by CRISPR-Cas9 Systems for Crop Improvement

Gene name	Gene function	CRISPR-Cas9-based tool	Post-edited phenotype	Reference
Tomato (*Solanum lycopersicum*)				
SlAGL6	Promotes fruit development	CRISPR-Cas9	Heat tolerance	(Klap et al., 2017)
SlBZR1	Involve in BZR1-dependent brassinosteroid signaling by regulating the expression of FERONIA (FER2/3) receptors	CRISPR-Cas9	Heat tolerance	(Yin et al., 2018)
SlCBF1	Encode C-repeat binding factors, controlling postharvest chilling injury	CRISPR-Cas9	Cold Tolerance	(Li et al., 2018b)
SlMAPK3	Regulates the expression of ethylene signaling genes	CRISPR-Cas9	Salt, drought, and heat stress tolerance	(Wang et al., 2017; Yu et al., 2019; Shu et al., 2022)
SlMYBS2	Regulates expression levels of pathogenesis-related genes and ROS levels	CRISPR-Cas9	Resistance against *Phytophthora infestans*	(Liu et al., 2021a)
SlPELO	Involves in ribosome cycling during protein synthesis	CRISPR-Cas9	*Tomato yellow leaf curl virus* (TYLCV) resistance	(Pramanik et al., 2021; Liu et al., 2022a)
SlMLO1	Modulates Pathogen defense and leaf cell death	CRISPR-Cas9	Powdery mildew fungal resistance	(Pramanik et al., 2021; Liu et al., 2022a)
SlHyPRP1	Involves in proline-rich protein synthesis	CRISPR-Cas9	Salt stress tolerance	(Tran et al., 2021)
SlSPL-CNR	Interacts with the *SlbHLH101* promoter negatively regulates Fe-deficiency response	CRISPR-Cas9	Iron deficiency	(Zhu et al., 2022)

(*Continued*)

TABLE 7.1 ***(Continued)***
List of Genes That Have Been Targeted by CRISPR-Cas9 Systems for Crop Improvement

Gene name	Gene function	CRISPR-Cas9-based tool	Post-edited phenotype	Reference
SlKLUH	M9 SNP in *KLUH* promoter regulates fruit weight	CRISPR-Cas9	Improves fruit weight	(Li et al., 2022a)
SlBZIP	Involves in sucrose-induced repression of translation (SIRT) mechanism	CRISPR-Cas9	Improves quality and taste	(Nguyen et al., 2023)
SlPSY1 SlMYB12 SlSGR1	Involves in accumulation of pigments, including carotenoids and flavonoids	CRISPR-Cas-mediated multiplex gene editing	Affects tomato fruit color	(Yang et al., 2023)
Rice (*Oryza sativa*)				
OsSAPK1 OsSAPK2	Encodes non-sucrose fermenting protein kinases and plays a role in osmotic stress balance	CRISPR-Cas9	Salt stress tolerance	(Lou et al., 2018)
OsFAD2-1	Encodes fatty acid desaturase-2 enzyme	CRISPR-Cas9	Increased oleic acid and reduced linolic acid contents	(Abe et al., 2018)
OsACC-T1 OsALS	Encodes key enzymes in lipid biosynthesis	Adenine base editors	Herbicide resistance	(Li et al., 2018a)
OsCCD7	Carotenoid cleavage dioxygenase-7 enzyme activity and lowers seed strigolactone	CRISPR-Cas9	Increased tiller number per plant and short height	(Butt et al., 2018)
OsRLCK185 OsCERK1	Recognizes chitin and peptidoglycan signaling pathways	Cytosine base editors	Higher yield and blast resistance	(Ren et al., 2018)
OsISA1	Plays a role in amylose and soluble sugar contents in endosperm via isoamylase	CRISPR-Cas9	Reduced amylose contents	(Shufen et al., 2019)
Os8N3	The sucrose transporter factor and affects the grain-filling process	CRISPR-Cas9	Resistance to bacterial leaf bight and pollen development	(Kim et al., 2019)
OsGL2 OsGRF4 OsGRF3	Grain length and growth size factors	CRISPR-Cas9	Increased grain size and yield	(Shufen et al., 2019)
OsIPA1 OsBZIP5 OsSLR1	Controls spikelet development and role in environment induces stress	CGBEs	Increased nitrogen use efficiency	(Tian et al., 2022)
OsACCOsEPSPS	Role in glyphosate resistance	Prime editors	Herbicide resistance	(Jiang et al., 2022)
OsWRKY63	Encodes negative regulator WRKY transcription factor	CRISPR-Cas9	Increased chilling tolerance	(Zhang et al., 2022b)
OsCDC, OsPDS	Encodes carotenoid-induced chlorosis and phytoene desaturase	Prime editors	Herbicide resistance	(Liu et al., 2023)
OsPUB7	Negative regulator of salt stress and controls protein turnover and degradation	CRISPR-Cas9	Increased resistance to salinity and drought stress	(Kim et al., 2023)
OsFTL1	GFP protein	CRISPR-Cas9	Knock in precise GE	(Shen et al., 2023)

TABLE 7.1 ***(Continued)***
List of Genes That Have Been Targeted by CRISPR-Cas9 Systems for Crop Improvement

Gene name	Gene function	CRISPR-Cas9-based tool	Post-edited phenotype	Reference
Wheat (*Triticum aestivum*)				
TaGliadin-α	The Gliadin gene controls gluten accumulation in seed	CRISPR-Cas9	Low gluten wheat	(Sánchez-León et al., 2018)
TaGW2	Involved in grain weight determination	CRISPR-Cas9	Increased grain size and weight	(Wang et al., 2018b)
TaDREB2 TaERF3	Encodes dehydration-responsive element-binding protein	CRISPR-Cas9	Enhanced tolerance to drought stress	(Kim et al., 2018)
TaCKX2	Encodes cytokinin oxidase/dehydrogenase that degrades cytokinin hormone	CRISPR-Cas9	Increased number of grains	(Zhang et al., 2019b)
TaALS	Encodes acetolactate synthase (ALS) enzyme which is involved in the biosynthesis of branched-chain amino acids	Cytosine base editor	Herbicide tolerance	(Zhang et al., 2019a)
TaHRC	Induces resistance	CRISPR-Cas9	Resistance to fusarium head blight	(Su et al., 2019)
TaCM3 TaCM1	Amylase and trypsin inhibitors	CRISPR-Cas9	Transgene-free homozygous plants	(Camerlengo et al., 2020)
TaWaxy TaMTL	Matrilineal and waxy gene	CRISP-Cpf1	Better flour quality and haploid induction	(Liu et al., 2020a)
TaFt-D1	Encodes an enzyme that is involved in flowering time	CRISPR-Cas9	Enhanced number of spikelets	(Chen et al., 2022)
TaSal1	Encodes bisphosphate nucleosidase	CRISPR-Cas9	Increased drought tolerance	(Abdallah et al., 2022)
TaIPK1	Phytic acid biosynthesis	CRISPR-Cas9	Enhanced iron and zinc contents in grains	(Ibrahim et al., 2022)
TaMLO1	Susceptibility gene	CRISPR-Cas9	Powdery mildew resistance	(Li et al., 2022b)
TaTubulin-α	Encodes adenosine deaminase	Adenosine base editors	Herbicide resistance	(Han et al., 2022)
TaUAM3	Encodes UDP-Ara mutases and positively regulates resistance	CRISPR-Cas9	Increased susceptibility to strip rust fungus	(Huang et al., 2023)
TaTN1	Negative regulator of yield	CRISPR-Cas9	Reduced tiller number per plant	(Dong et al., 2023)
TaPpd-1	Encodes photoperiod-1 enzymes	CRISPR-Cas9	Increase in grain number, weight, and spikelets	(Errum et al., 2023)
TaSal1	Encodes bisphosphate nucleosidase	CRISPR-Cas9	Increased drought tolerance	(Abdallah et al., 2022)

(Continued)

TABLE 7.1 *(Continued)*
List of Genes That Have Been Targeted by CRISPR-Cas9 Systems for Crop Improvement

Gene name	Gene function	CRISPR-Cas9-based tool	Post-edited phenotype	Reference
Maize (*Zea mays*)				
ZmARGOS8	A negative regulator of ethylene responses	CRISPR-Cas9	Decreased ethylene sensitivity and increased grain yield	(Shi et al., 2017)
ZmWRKY40	Promotes peroxide dismutase and catalase activities, as well as osmolyte accumulation, including proline, soluble sugar, and soluble protein	CRISPR-Cas9	Drought and salt stress tolerance	(Wang et al., 2018a)
ZmTMS5	Involves in the processing of Ub_{L40} mRNA (*ubiquitin fusion* ribosomal protein *L40*) genes into fragments	CRISPR-Cas9	Thermotolerance	(Lou et al., 2017)
ZmRAVL1	Regulates *brassinosteroid C-6 oxidase 1 (brd1)* expression, controlling brassinosteroid content and leaf angle	CRISPR-Cas9	Increase yield	(Tian et al., 2019)
ZmALS1 *ZmALS2*	Encodes *ALS* enzyme (acetolactate synthase), responsible for aromatic amino acid production	Prime editing	Herbicide Tolerance	(Jiang et al., 2020)
ZmMYB74	Controls endosperm cell number	CRISPR-Cas9	Large grain size	(Wang et al., 2023)
ZmTHP9	Encodes asparagine synthetase 4 enzyme which regulates seed protein content	CRISPR-Cas9	Improves yield and nitrogen use efficiency	(Huang et al., 2022)
ZmBADH2a *ZmBADH2b*	Convert γ-amino butyraldehyde into γ-amino butyric acid (GABA), thus involved in the 2-acetyl-1-pyrroline accumulation	CRISPR-Cas9	Increase aromaticity	(Wang et al., 2021)
ZmCLE	Controls CLE peptide signals involved in the CLAVATA-WUSCHEL pathway	CRISPR-Cas9	Increase grain yield	(Liu et al., 2021b)
ZmCOI1a	Involves in jasmonates signaling mechanism	CRISPR-Cas9	Resistance againstGibberella stalk rot	(Ma et al., 2021)
*Zm*GDIα	Controls small Rab GTPases, essential for vesicle membrane trafficking	CRISPR-Cas9	Maize rough dwarf disease (MRDD) viral resistance	(Liu et al., 2022a)
ZmAGO18	Encodes an AGO18 protein, a key regulator in plant defense	CRISPR-Cas9	Southern leaf blight fungal resistance	(Dai et al., 2023)

TABLE 7.1 ***(Continued)***
List of Genes That Have Been Targeted by CRISPR-Cas9 Systems for Crop Improvement

Gene name	Gene function	CRISPR-Cas9-based tool	Post-edited phenotype	Reference
Potato (*Solanum tuberosum*)				
StSP3 *StSP4*	Self-incompatibility gene	CRISPR-Cas9	Improved breeding	(Ye et al., 2018)
StCCoAOMT	Susceptibility gene	CRISPR-Cas9	Resistance to Late Blight	(Hegde et al., 2021)
StMYB44	Transcription factor	CRISPR-Cas9	Suppression of *StPH02* gene	(Zhou et al., 2017)
StGBBS	Starch quality	CRISPR-Cas9	Amylopectin-rich waxy potato	(Andersson et al., 2017)
StSS6	Starch biosynthesis	CRISPR-Cas9	Enhanced starch properties	(Tuncel et al., 2019)
St16DOX	Glycoalkloides	CRISPR-Cas9	Decreased steroidal glycoalkaloids	(Nakayasu et al., 2018)
StS-RNase	Self-incompatibility controlling gene	CRISPR-Cas9	Induction of self-incompatibility	(Ye et al., 2018)
St1T1	*Anthocyanin synthesis*	CRISPR-Cas9	Improved yield	(Tang et al., 2022)
StSli	S-locus inhibitors	CRISPR-Cas9	Improved breeding	(Eggers et al., 2021)
StSBE *StSBE2*	Starch branching in tubers	CRISPR-Cas9	Improved quality	(Tuncel et al., 2019)
Soybean (*Glycine max*)				
GmWRI1a	Transcription factor that regulates oil accumulation and is related to fatty acid biosynthesis	CRISPR-Cas9	Increased oil content	(Wang et al., 2022c)
GmHSP17.9	Small heat shock protein	CRISPR-Cas9	Increased nodulation and seed yield	(Yang et al., 2022)
GmBICs	Blue-light inhibitors of cryptochromes	CRISPR-Cas9	Stem elongation	(Mu et al., 2022)
GmNMHC5	MADS-box gene	CRISPR-Cas9	Increase nodulation	(Wang et al., 2022a)
GmGAMYB	Positive regulator of gibberellic acid	CRISPR-Cas9	Increase flowering and plant height	(Yang et al., 2021)
GmKR3	TIR-NBS-LRR type resistance *gene*	CRISPR-Cas9	Resistance to multiple viruses in soybean	(Xun et al., 2019)
GmTNL16	NBS-LRR gene	CRISPR-Cas9	Resistance against *Phytophthorasojae*	(Zhou et al., 2022a)
GmEF8	Encodes elongation factor Tu (EF-Tu) protein	CRISPR-Cas9	Enhanced heat and drought tolerance	(Zhang et al., 2022a)
GmDREB2	DRE-transcription binding factor	CRISPR-Cas9	Enhanced drought tolerance	(Marinho et al., 2022)
GmP450-N-Z1	Involves in the biosynthesis of structural polymers and hormonal signaling	CRISPR-Cas9	Tolerance to multiple herbicides	(Zheng et al., 2022)

7.6 LIMITATION OF CRISPR IN PLANTS

The main challenge linked with CRISPR-Cas systems is the potential for induction of off-target mutations in the genome, which can lead to detrimental consequences. However, the advent of advanced technologies like next-generation sequencing (NGS) has expedited whole genome sequencing (WGS), making it affordable as well as accessible for diverse organisms. Moreover, it empowers scientists to validate and address expected off-target effects of CRISPR-Cas systems using comprehensive genome data (Tang et al., 2018). In addition, third-generation base editor (BE3)-treated rice plant genomes exhibit substantial off-target mutations, primarily characterized as C→T single-nucleotide variants (SNVs), with a noticeable enrichment in genic regions (Jin et al., 2019). This emphasizes the increased off-target potential of CBEs juxtaposed to ABEs (Doman et al., 2020). Cas endonucleases, with their considerable dimensions, impede the development and successful delivery of vectors into the plant (Lino et al., 2018). Furthermore, the inability to find an appropriate PAM sequence hinders target site selection, rendering the endonuclease ineffective (Ren et al., 2021; Tan et al., 2022). Recently, several Cas-9 variants, including SpG and SpRY, have been designed for PAMs recognition such as 5'-NGN/NRN and 5'-NYN (with G/R representing A/G nucleotide, Y representing C/T, and N representing T, A, C) in their targeting mechanisms (Walton et al., 2020). Moreover, miniature Cas endonucleases with varying dependency on PAM sequences have also been reported that can be employed for target-specific genome modifications in plants (Wu et al., 2021; Awan et al., 2022b; Awan et al., 2022c; Li et al., 2023).

Lastly, concerns within regulatory frameworks regarding commercial adoption of genome-edited cultivars in different countries serve as an impediment to the deployment of such crops within the farming sector. However, specific regions have authorized the large-scale farming of CRISPR-edited crops like maize, rice, and soybean (Waltz, 2018). Furthermore, various countries have shown an inclination toward the adoption of a lenient or flexible policy regarding the commercialization of genome-edited crops generated via site-directed nucleases 1 (SDN-1) (Jones et al., 2022). However, the regulatory framework varies from country to country, thus, there exists a dire need to come up with a uniform policy regarding the commercialization of edited crops around the globe and address the unwanted ambiguity.

REFERENCES

Abdallah, N. A., Elsharawy, H., Abulela, H. A., Thilmony, R., Abdelhadi, A. A., & Elarabi, N. I. (2022). Multiplex CRISPR/Cas9-mediated genome editing to address drought tolerance in wheat. *GM Crops & Food*, *13*, 1–17.

Abdallah, N. A., Prakash, C. S., & McHughen, A. G. (2015). Genome editing for crop improvement: Challenges and opportunities. *GM Crops & Food*, *6*, 183–205.

Abe, K., Araki, E., Suzuki, Y., Toki, S., & Saika, H. (2018). Production of high oleic/low linoleic rice by genome editing. *Plant Physiology and Biochemistry*, *131*, 58–62.

Abudayyeh, O. O., Gootenberg, J. S., Konermann, S., Joung, J., Slaymaker, I. M., Cox, D. B. T., Shmakov, S., Makarova, K. S., Semenova, E., Minakhin, L., et al. (2016). C2c2 is a single-component programmable RNA-guided RNA-targeting CRISPR effector. *Science*, *353*, aaf5573.

Aglawe, S. B., Barbadikar, K. M., Mangrauthia, S. K., & Madhav, M. S. (2018). New breeding technique "genome editing" for crop improvement: Applications, potentials and challenges. *3 Biotech*, *8*, 1–20.

Ahmad, N., Awan, M. J. A., & Mansoor, S. (2023). Improving editing efficiency of prime editor in plants. *Trends in Plant Science*, *28*, 1–3.

Alfatih, A., Wu, J., Jan, S. U., Zhang, Z. S., Xia, J. Q., & Xiang, C. B. (2020). Loss of rice PARAQUAT TOLERANCE 3 confers enhanced resistance to abiotic stresses and increases grain yield in field. *Plant, Cell & Environment*, *43*, 2743–2754.

Andersson, M., Turesson, H., Nicolia, A., Fält, A.-S., Samuelsson, M., & Hofvander, P. (2017). Efficient targeted multiallelic mutagenesis in tetraploid potato (Solanum tuberosum) by transient CRISPR-Cas9 expression in protoplasts. *Plant Cell Reports*, *36*, 117–128.

Anzalone, A. V., Gao, X. D., Podracky, C. J., Nelson, A. T., Koblan, L. W., Raguram, A., Levy, J. M., Mercer, J. A. M., & Liu, D. R. (2022). Programmable deletion, replacement, integration and inversion of large DNA sequences with twin prime editing. *Nature Biotechnology*, *40*, 731–740.

Anzalone, A. V., Randolph, P. B., Davis, J. R., Sousa, A. A., Koblan, L. W., Levy, J. M., Chen, P. J., Wilson, C., Newby, G. A., Raguram, A., et al. (2019). Search-and-replace genome editing without double-strand breaks or donor DNA. *Nature*, *576*, 149–157.

Awan, M. J. A., Akram, A., Amin, I., & Mansoor, S. (2023a). Viral vectors as carriers of genome-editing reagents. *Trends in Plant Science*, *28*, 981–983.

Awan, M. J. A., Ali, Z., Amin, I., & Mansoor, S. (2022a). Twin prime editor: Seamless repair without damage. *Trends in Biotechnology*, *40*, 374–376.

Awan, M. J. A., Amin, I., & Mansoor, S. (2022b). CRISPR-Cas12c: A noncleaving DNA binder with minimal PAM requirement. *Trends in Biotechnology*, *40*, 1141–1143.

Awan, M. J. A., Amin, I., & Mansoor, S. (2022c). Mini CRISPR-Cas12f1: A new genome editing tool. *Trends in Plant Science*, *27*, 110–112.

Awan, M. J. A., Anjum, N., Pervaiz, K., Ijaz, M. U., Khan, M. Z., Amin, I., & Mansoor, S. (2023b). Genome editing: Mechanism and utilization in plant breeding. In A. Raina, M. R. Wani, R. A. Laskar, N. Tomlekova, & S. Khan (eds) *Advanced crop improvement, Volume 1: Theory and practice*. Switzerland: Springer. pp. 457–488.

Awan, M. J. A., Aslam, M. Q., Naqvi, R. Z., Amin, I., & Mansoor, S. (2023c). A graft that crafts nontransgenic and genome-edited plants. *Trends in Plant Science*, *28*, 614–616.

Awan, M. J. A., Mahmood, M. A., Naqvi, R. Z., & Mansoor, S. (2023d). PASTE: A high-throughput method for large DNA insertions. *Trends in Plant Science*, *28*, 509–511.

Awan, M. J. A., Pervaiz, K., Rasheed, A., Amin, I., Saeed, N. A., Dhugga, K. S., & Mansoor, S. (2022d). Genome edited wheat – current advances for the second green revolution. *Biotechnology Advances*, *60*, 108006.

Awan, M. J. A., Rasheed, A., Saeed, N. A., & Mansoor, S. (2022e). Aegilops tauschii presents a genetic roadmap for hexaploid wheat improvement. *Trends in Genetics*, *38*, 307–309.

Bertier, L. D., Ron, M., Huo, H., Bradford, K. J., Britt, A. B., & Michelmore, R. W. (2018). High-resolution analysis of the efficiency, heritability, and editing outcomes of CRISPR/Cas9-induced modifications of NCED4 in lettuce (Lactuca sativa). *G3: Genes, Genomes, Genetics*, *8*, 1513–1521.

Boutilier, K., Offringa, R., Sharma, V. K., Kieft, H., Ouellet, T.r.s., Zhang, L., Hattori, J., Liu, C.-M., van Lammeren, A. A. M., Miki, B. L. A., et al. (2002). Ectopic expression of BABY BOOM triggers a conversion from vegetative to embryonic growth. *The Plant Cell*, *14*, 1737–1749.

Butler, N. M., Atkins, P. A., Voytas, D. F., & Douches, D. S. (2015). Generation and inheritance of targeted mutations in potato (Solanum tuberosum L.) using the CRISPR/Cas system. *PloS One*, *10*, e0144591.

Butt, H., Eid, A., Momin, A. A., Bazin, J., Crespi, M., Arold, S. T., & Mahfouz, M. M. (2019). CRISPR directed evolution of the spliceosome for resistance to splicing inhibitors. *Genome Biology*, *20*, 1–9.

Butt, H., Jamil, M., Wang, J. Y., Al-Babili, S., & Mahfouz, M. (2018). Engineering plant architecture via CRISPR/Cas9-mediated alteration of strigolactone biosynthesis. *BMC Plant Biology*, *18*, 1–9.

Camerlengo, F., Frittelli, A., Sparks, C., Doherty, A., Martignago, D., Larré, C., Lupi, R., Sestili, F., & Masci, S. (2020). CRISPR-Cas9 multiplex editing of the α-amylase/trypsin inhibitor genes to reduce allergen proteins in durum wheat. *Frontiers in Sustainable Food Systems*, *4*, 104.

Chang, J.-D., Huang, S., Konishi, N., Wang, P., Chen, J., Huang, X.-Y., Ma, J. F., & Zhao, F.-J. (2020). Overexpression of the manganese/cadmium transporter OsNRAMP5 reduces cadmium accumulation in rice grain. *Journal of Experimental Botany*, *71*, 5705–5715.

Charpentier, E., & Doudna, J. A. (2013). Rewriting a genome. *Nature*, *495*, 50–51.

Chen, K., Wang, Y., Zhang, R., Zhang, H., & Gao, C. (2019). CRISPR/Cas Genome Editing and Precision Plant Breeding in Agriculture. *Annual Review of Plant Biology*, *70*, 667–697.

Chen, L.-L. (2020). The expanding regulatory mechanisms and cellular functions of circular RNAs. *Nature Reviews Molecular Cell Biology*, *21*, 475–490.

Chen, Z., Ke, W., He, F., Chai, L., Cheng, X., Xu, H., Wang, X., Du, D., Zhao, Y., Chen, X., et al. (2022). A single nucleotide deletion in the third exon of FT-D1 increases the spikelet number and delays heading date in wheat (Triticum aestivum L.). *Plant Biotechnology Journal*, *20*, 920–933.

Cunningham, F. J., Goh, N. S., Demirer, G. S., Matos, J. L., & Landry, M. P. (2018). Nanoparticle-mediated delivery towards advancing plant genetic engineering. *Trends in Biotechnology*, *36*, 882–897.

Dai, Z., Yang, Q., Chen, D., Li, B., Que, J., Hu, L., Zhang, B., Zhang, Z., Chen, K., Zhang, S., et al. (2023). ZmAGO18b negatively regulates maize resistance against southern leaf blight. *Theoretical and Applied Genetics*, *136*, 158.

de Melo, B. P., Lourenço-Tessutti, I. T., Paixão, J. F. R., Noriega, D. D., Silva, M. C. M., de Almeida-Engler, J., Fontes, E. P. B., & Grossi-de-Sa, M. F. (2020). Transcriptional modulation of AREB-1 by CRISPRa improves plant physiological performance under severe water deficit. *Scientific Reports*, *10*, 16231.

Debernardi, J. M., Tricoli, D. M., Ercoli, M. F., Hayta, S., Ronald, P., Palatnik, J. F., & Dubcovsky, J. (2020). A GRF–GIF chimeric protein improves the regeneration efficiency of transgenic plants. *Nature Biotechnology*, *38*, 1274–1279.

Doman, J. L., Raguram, A., Newby, G. A., & Liu, D. R. (2020). Evaluation and minimization of Cas9-independent off-target DNA editing by cytosine base editors. *Nature Biotechnology*, *38*, 620–628.

Dong, C., Zhang, L., Zhang, Q., Yang, Y., Li, D., Xie, Z., Cui, G., Chen, Y., Wu, L., & Li, Z. (2023). Tiller Number1 encodes an ankyrin repeat protein that controls tillering in bread wheat. *Nature Communications*, *14*, 836.

Dong, H., Huang, Y., & Wang, K. (2021). The development of herbicide resistance crop plants using CRISPR/Cas9-mediated gene editing. *Genes*, *12*, 912.

Dreesen, R., Capt, A., Oberdoerfer, R., Coats, I., & Kenneth, E. P. (2018). Characterization and safety evaluation of HPPD W336, a modified 4-hydroxyphenylpyruvate dioxygenase protein, and the impact of its expression on plant metabolism in herbicide-tolerant MST-FGØ72-2 soybean. *Regulatory Toxicology and Pharmacology*, *97*, 170–185.

Eggers, E.-J., van der Burgt, A., van Heusden, S. A., de Vries, M. E., Visser, R. G., Bachem, C. W., & Lindhout, P. (2021). Neofunctionalisation of the Sli gene leads to self-compatibility and facilitates precision breeding in potato. *Nature Communications*, *12*, 4141.

Elsharawy, H., & Refat, M. (2023). CRISPR/Cas9 genome editing in wheat: Enhancing quality and productivity for global food security—a review. *Functional & Integrative Genomics*, *23*, 265.

Errum, A., Rehman, N., Uzair, M., Inam, S., Ali, G. M., & Khan, M. R. (2023). CRISPR/Cas9 editing of wheat Ppd-1 gene homoeologs alters spike architecture and grain morphometric traits. *Functional & Integrative Genomics*, *23*, 66.

Ganie, S. A., Wani, S. H., Henry, R., & Hensel, G. (2021). Improving rice salt tolerance by precision breeding in a new era. *Current Opinion in Plant Biology, 60*, 101996.

Gao, Q., Li, G., Sun, H., Xu, M., Wang, H., Ji, J., Wang, D., Yuan, C., & Zhao, X. (2020). Targeted mutagenesis of the rice FW 2.2-like gene family using the CRISPR/Cas9 system reveals OsFWL4 as a regulator of tiller number and plant yield in rice. *International Journal of Molecular Sciences*, *21*, 809.

Garneau, J. E., Dupuis, M.-È., Villion, M., Romero, D. A., Barrangou, R., Boyaval, P., Fremaux, C., Horvath, P., Magadán, A. H., & Moineau, S. (2010). The CRISPR/Cas bacterial immune system cleaves bacteriophage and plasmid DNA. *Nature*, *468*, 67–71.

Gasiunas, G., Barrangou, R., Horvath, P., & Siksnys, V. (2012). Cas9–crRNA ribonucleoprotein complex mediates specific DNA cleavage for adaptive immunity in bacteria. *Proceedings of the National Academy of Sciences*, *109*, E2579–E2586.

Guidarelli, M., & Baraldi, E. (2015). Transient transformation meets gene function discovery: The strawberry fruit case. *Frontiers in Plant Science*, *6*, 444.

Han, H., Wu, Z., Zheng, L., Han, J., Zhang, Y., Li, J., Zhang, S., Li, G., Ma, C., & Wang, P. (2022). Generation of a high-efficiency adenine base editor with TadA8e for developing wheat dinitroaniline-resistant germplasm. *The Crop Journal*, *10*, 368–374.

Harrington, L. B., Burstein, D., Chen, J. S., Paez-Espino, D., Ma, E., Witte, I. P., Cofsky, J. C., Kyrpides, N. C., Banfield, J. F., & Doudna, J. A. (2018). Programmed DNA destruction by miniature CRISPR-Cas14 enzymes. *Science*, *362*, 839–842.

Hegde, N., Joshi, S., Soni, N., & Kushalappa, A. C. (2021). The caffeoyl-CoA O-methyltransferase gene SNP replacement in Russet Burbank potato variety enhances late blight resistance through cell wall reinforcement. *Plant Cell Reports*, *40*, 237–254.

Hickey, L. T., N. Hafeez, A., Robinson, H., Jackson, S. A., Leal-Bertioli, S. C. M., Tester, M., Gao, C., Godwin, I. D., Hayes, B. J., & Wulff, B. B. H. (2019). Breeding crops to feed 10 billion. *Nature Biotechnology, 37*, 744–754.

Hua, K., Tao, X., Yuan, F., Wang, D., & Zhu, J.-K. (2018). Precise A·T to G·C Base Editing in the Rice Genome. *Molecular Plant*, *11*, 627–630.

Huang, X., Bai, X., Qian, C., Liu, S., Goher, F., He, F., Zhao, G., Pei, G., Zhao, H., & Wang, J. (2023). TaUAM3, a UDP-Ara mutases protein, positively regulates wheat resistance to the stripe rust fungus. *Food and Energy Security*, *12*, e456.

Huang, Y., Wang, H., Zhu, Y., Huang, X., Li, S., Wu, X., Zhao, Y., Bao, Z., Qin, L., Jin, Y., et al. (2022). THP9 enhances seed protein content and nitrogen-use efficiency in maize. *Nature*, *612*, 292–300.

Hummel, A. W., Chauhan, R. D., Cermak, T., Mutka, A. M., Vijayaraghavan, A., Boyher, A., Starker, C. G., Bart, R., Voytas, D. F., & Taylor, N. J. (2018). Allele exchange at the EPSPS locus confers glyphosate tolerance in cassava. *Plant Biotechnology Journal*, *16*, 1275–1282.

Ibrahim, S., Saleem, B., Rehman, N., Zafar, S. A., Naeem, M. K., & Khan, M. R. (2022). CRISPR/Cas9 mediated disruption of Inositol Pentakisphosphate 2-Kinase 1 (TaIPK1) reduces phytic acid and improves iron and zinc accumulation in wheat grains. *Journal of Advanced Research*, *37*, 33–41.

Jelly, N. S., Valat, L., Walter, B., & Maillot, P. (2014). Transient expression assays in grapevine: A step towards genetic improvement. *Plant Biotechnology Journal*, *12*, 1231–1245.

Jia, N., Zhu, Y., & Xie, F. (2018). An efficient protocol for model legume root protoplast isolation and transformation. *Frontiers in Plant Science*, *9*, 670.

Jiang, Y.-Y., Chai, Y.-P., Lu, M.-H., Han, X.-L., Lin, Q., Zhang, Y., Zhang, Q., Zhou, Y., Wang, X.-C., Gao, C., et al. (2020). Prime editing efficiently generates W542L and S621I double mutations in two ALS genes in maize. *Genome Biology*, *21*, 257.

Jiang, Y., Chai, Y., Qiao, D., Wang, J., Xin, C., Sun, W., Cao, Z., Zhang, Y., Zhou, Y., & Wang, X.-C. (2022). Optimized prime editing efficiently generates glyphosate-resistant rice plants carrying homozygous TAP-IVS mutation in EPSPS. *Molecular Plant*, *15*, 1646–1649.

Jin, S., Zong, Y., Gao, Q., Zhu, Z., Wang, Y., Qin, P., Liang, C., Wang, D., Qiu, J.-L., Zhang, F., et al. (2019). Cytosine, but not adenine, base editors induce genome-wide off-target mutations in rice. *Science*, *364*, 292–295.

Jinek, M., Chylinski, K., Fonfara, I., Hauer, M., Doudna, J. A., & Charpentier, E. (2012). A programmable dual-RNA–guided DNA endonuclease in adaptive bacterial immunity. *Science*, *337*, 816–821.

Jinek, M., Jiang, F., Taylor, D. W., Sternberg, S. H., Kaya, E., Ma, E., Anders, C., Hauer, M., Zhou, K., Lin, S., et al. (2014). Structures of Cas9 endonucleases reveal RNA-mediated conformational activation. *Science*, *343*, 1247997.

Jones, M. G. K., Fosu-Nyarko, J., Iqbal, S., Adeel, M., Romero-Aldemita, R., Arujanan, M., Kasai, M., Wei, X., Prasetya, B., Nugroho, S., et al. (2022). Enabling trade in gene-edited produce in Asia and Australasia: The developing regulatory landscape and future perspectives. *Plants*, *11*, 2538, doi: 10.3390/plants11192538.

Kan, J., Cai, Y., Cheng, C., Chen, S., Jiang, C., He, Z., & Yang, P. (2023). CRISPR/Cas9-guided knockout of eIF4E improves Wheat yellow mosaic virus resistance without yield penalty. *Plant Biotechnology Journal*, *21*, 893.

Khoso, M. A., Hussain, A., Ritonga, F. N., Ali, Q., Channa, M. M., Alshegaihi, R. M., Meng, Q., Ali, M., Zaman, W., & Brohi, R. D. (2022). WRKY transcription factors (TFs): Molecular switches to regulate drought, temperature, and salinity stresses in plants. *Frontiers in Plant Science*, *13*, 1039329.

Kieu, N. P., Lenman, M., Wang, E. S., Petersen, B. L., & Andreasson, E. (2021). Mutations introduced in susceptibility genes through CRISPR/Cas9 genome editing confer increased late blight resistance in potatoes. *Scientific Reports*, *11*, 4487.

Kim, D., Alptekin, B., & Budak, H. (2018). CRISPR/Cas9 genome editing in wheat. *Functional & Integrative Genomics*, *18*, 31–41.

Kim, M.-S., Ko, S.-R., Jung, Y. J., Kang, K.-K., Lee, Y.-J., & Cho, Y.-G. (2023). Knockout mutants of OsPUB7 generated using CRISPR/Cas9 revealed abiotic stress tolerance in rice. *International Journal of Molecular Sciences*, *24*, 5338.

Kim, Y.-A., Moon, H., & Park, C.-J. (2019). CRISPR/Cas9-targeted mutagenesis of Os8N3 in rice to confer resistance to Xanthomonas oryzae pv. oryzae. *Rice*, *12*, 1–13.

Klap, C., Yeshayahou, E., Bolger, A. M., Arazi, T., Gupta, S. K., Shabtai, S., Usadel, B., Salts, Y., & Barg, R. (2017). Tomato facultative parthenocarpy results from Sl AGAMOUS-LIKE 6 loss of function. *Plant Biotechnology Journal*, *15*, 634–647.

Komor, A. C., Kim, Y. B., Packer, M. S., Zuris, J. A., & Liu, D. R. (2016). Programmable editing of a target base in genomic DNA without double-stranded DNA cleavage. *Nature*, *533*, 420–424.

Kong, X., Wang, F., Wang, Z., Gao, X., Geng, S., Deng, Z., Zhang, S., Fu, M., Cui, D., & Liu, S. (2023). Grain yield improvement by genome editing of TaARF12 that decoupled peduncle and rachis development trajectories via differential regulation of gibberellin signalling in wheat. *Plant Biotechnology Journal*, 10, 1990–2001.

Kuang, Y., Li, S., Ren, B., Yan, F., Spetz, C., Li, X., Zhou, X., & Zhou, H. (2020). Base-editing-mediated artificial evolution of OsALS1 in planta to develop novel herbicide-tolerant rice germplasms. *Molecular Plant*, *13*, 565–572.

Lanigan, T. M., Kopera, H. C., & Saunders, T. L. (2020). Principles of genetic engineering. *Genes*, *11*, 291.

Le, V. T., Kim, M.-S., Jung, Y.-J., Kang, K.-K., & Cho, Y.-G. (2022). Research trends and challenges of using CRISPR/Cas9 for improving rice productivity. *Agronomy*, *12*, 164.

Lee, K., & Wang, K. (2023). Strategies for genotype-flexible plant transformation. *Current Opinion in Biotechnology*, *79*, 102848.

Li, C., Zong, Y., Wang, Y., Jin, S., Zhang, D., Song, Q., Zhang, R., & Gao, C. (2018a). Expanded base editing in rice and wheat using a Cas9-adenosine deaminase fusion. *Genome Biology*, *19*, 59.

Li, J., Meng, X., Zong, Y., Chen, K., Zhang, H., Liu, J., Li, J., & Gao, C. (2016). Gene replacements and insertions in rice by intron targeting using CRISPR–Cas9. *Nature Plants*, *2*, 1–6.

Li, Q., Feng, Q., Snouffer, A., Zhang, B., Rodríguez, G. R., & van der Knaap, E. (2022a). Increasing fruit weight by editing a cis-regulatory element in tomato KLUH promoter using CRISPR/Cas9. *Frontiers in Plant Science*, *13*, 879642.

Li, R., Zhang, L., Wang, L., Chen, L., Zhao, R., Sheng, J., & Shen, L. (2018b). Reduction of tomato-plant chilling tolerance by CRISPR–Cas9-mediated SlCBF1 mutagenesis. *Journal of Agricultural and Food Chemistry*, *66*, 9042–9051.

Li, S., Lin, D., Zhang, Y., Deng, M., Chen, Y., Lv, B., Li, B., Lei, Y., Wang, Y., & Zhao, L. (2022b). Genome-edited powdery mildew resistance in wheat without growth penalties. *Nature*, *602*, 455–460.

Li, Z., Liu, Z.-B., Xing, A., Moon, B. P., Koellhoffer, J. P., Huang, L., Ward, R. T., Clifton, E., Falco, S. C., & Cigan, A. M. (2015). Cas9-guide RNA directed genome editing in soybean. *Plant Physiology*, *169*, 960–970.

Li, Z., Zhong, Z., Wu, Z., Pausch, P., Al-Shayeb, B., Amerasekera, J., Doudna, J. A., & Jacobsen, S. E. (2023). Genome editing in plants using the compact editor CasΦ. *Proceedings of the National Academy of Sciences*, *120*, e2216822120.

Lieber, M. R. (2010). The mechanism of double-strand DNA break repair by the nonhomologous DNA end-joining pathway. *Annual Review of Biochemistry*, *79*, 181–211.

Lin, H.-Y., Chen, J.-C., & Fang, S.-C. (2018). A protoplast transient expression system to enable molecular, cellular, and functional studies in Phalaenopsis orchids. *Frontiers in Plant Science*, *9*, 843.

Lin, Q., Zong, Y., Xue, C., Wang, S., Jin, S., Zhu, Z., Wang, Y., Anzalone, A. V., Raguram, A., Doman, J. L., et al. (2020). Prime genome editing in rice and wheat. *Nature Biotechnology*, *38*, 582–585.

Lino, C. A., Harper, J. C., Carney, J. P., & Timlin, J. A. (2018). Delivering CRISPR: A review of the challenges and approaches. *Drug Delivery*, *25*, 1234–1257.

Liu, C., Kong, M., Yang, F., Zhu, J., Qi, X., Weng, J., Di, D., & Xie, C. (2022a). Targeted generation of null mutants in ZmGDIα confers resistance against maize rough dwarf disease without agronomic penalty. *Plant Biotechnology Journal*, *20*, 803.

Liu, C., Zhang, Y., Tan, Y., Zhao, T., Xu, X., Yang, H., & Li, J. (2021a). CRISPR/Cas9-mediated SlMYBS2 mutagenesis reduces tomato resistance to phytophthora infestans. *International Journal of Molecular Sciences*, *22*, 11423.

Liu, D., Yang, H., Zhang, Z., Chen, Q., Guo, W., Rossi, V., Xin, M., Du, J., Hu, Z., & Liu, J. (2023). An elite γ-gliadin allele improves end-use quality in wheat. *New Phytologist*, *239*, 87–101.

Liu, H., Wang, K., Jia, Z., Gong, Q., Lin, Z., Du, L., Pei, X., & Ye, X. (2020a). Efficient induction of haploid plants in wheat by editing of TaMTL using an optimized Agrobacterium-mediated CRISPR system. *Journal of Experimental Botany*, *71*, 1337–1349.

Liu, J., Yao, Y., Xin, M., Peng, H., Ni, Z., & Sun, Q. (2022b). Shaping polyploid wheat for success: Origins, domestication, and the genetic improvement of agronomic traits. *Journal of Integrative Plant Biology*, *64*, 536–563.

Liu, L., Gallagher, J., Arevalo, E. D., Chen, R., Skopelitis, T., Wu, Q., Bartlett, M., & Jackson, D. (2021b). Enhancing grain-yield-related traits by CRISPR–Cas9 promoter editing of maize CLE genes. *Nature Plants*, *7*, 287–294.

Liu, X., Qin, R., Li, J., Liao, S., Shan, T., Xu, R., Wu, D., & Wei, P. (2020b). A CRISPR-Cas9-mediated domain-specific base-editing screen enables functional assessment of ACCase variants in rice. *Plant Biotechnology Journal*, *18*, 1845.

Lotan, T., Ohto, M.-a., Yee, K. M., West, M. A. L., Lo, R., Kwong, R. W., Yamagishi, K., Fischer, R. L., Goldberg, R. B., & Harada, J. J. (1998). Arabidopsis LEAFY COTYLEDON1 is sufficient to induce embryo development in vegetative cells. *Cell*, *93*, 1195–1205.

Lou, D., Wang, H., Liang, G., & Yu, D. (2017). OsSAPK2 confers abscisic acid sensitivity and tolerance to drought stress in rice. *Frontiers in Plant Science*, *8*, 993.

Lou, D., Wang, H., & Yu, D. (2018). The sucrose non-fermenting-1-related protein kinases SAPK1 and SAPK2 function collaboratively as positive regulators of salt stress tolerance in rice. *BMC Plant Biology*, *18*, 1–17.

Lv, Z., Jiang, R., Chen, J., & Chen, W. (2020). Nanoparticle-mediated gene transformation strategies for plant genetic engineering. *The Plant Journal*, *104*, 880–891.

Ma, L., Sun, Y., Ruan, X., Huang, P.-C., Wang, S., Li, S., Zhou, Y., Wang, F., Cao, Y., Wang, Q., et al. (2021). Genome-wide characterization of jasmonates signaling components reveals the essential role of ZmCOI1a-ZmJAZ15 action module in regulating maize immunity to Gibberella stalk rot. *International Journal of Molecular Sciences*, *22*, 870, doi: 10.3390/ijms22020870.

Ma, X., Mau, M., & Sharbel, T. F. (2018). Genome editing for global food security. *Trends in Biotechnology*, *36*, 123–127.

Marinho, J. P., Pagliarini, R. F., Molinari, M. D. C., Marcolino-Gomes, J., Caranhoto, A. L. H., Marin, S. R. R., Oliveira, M. C. N., Foloni, J. S. S., Melo, C. L. P., & Kidokoro, S. (2022). Overexpression of full-length and partial DREB2A enhances soybean drought tolerance. *Agronomy Science and Biotechnology*, *8*, 1–21.

Miao, C., Xiao, L., Hua, K., Zou, C., Zhao, Y., Bressan, R. A., & Zhu, J.-K. (2018). Mutations in a subfamily of abscisic acid receptor genes promote rice growth and productivity. *Proceedings of the National Academy of Sciences*, *115*, 6058–6063.

Mirza, Z., Haque, M. M., & Gupta, M. (2022). WRKY transcription factors: A promising way to deal with arsenic stress in rice. *Molecular Biology Reports*, *49*, 10895–10904.

Mu, R., Lyu, X., Ji, R., Liu, J., Zhao, T., Li, H., & Liu, B. (2022). GmBICs modulate low blue light-induced stem elongation in soybean. *Frontiers in Plant Science*, *13*, 803122.

Nakayasu, M., Akiyama, R., Lee, H. J., Osakabe, K., Osakabe, Y., Watanabe, B., Sugimoto, Y., Umemoto, N., Saito, K., & Muranaka, T. (2018). Generation of α-solanine-free hairy roots of potato by CRISPR/Cas9 mediated genome editing of the St16DOX gene. *Plant Physiology and Biochemistry*, *131*, 70–77.

Nandy, S., Pathak, B., Zhao, S., & Srivastava, V. (2019). Heat-shock-inducible CRISPR/Cas9 system generates heritable mutations in rice. *Plant Direct*, *3*, e00145.

Nguyen, N. H., Bui, T. P., Le, N. T., Nguyen, C. X., Le, M. T. T., Dao, N. T., Phan, Q., Van Le, T., To, H. M. T., Pham, N. B., et al. (2023). Disrupting Sc-uORFs of a transcription factor bZIP1 using CRISPR/Cas9 enhances sugar and amino acid contents in tomato (Solanum lycopersicum). *Planta*, *257*, 57.

Oliva, R., Ji, C., Atienza-Grande, G., Huguet-Tapia, J. C., Perez-Quintero, A., Li, T., Eom, J.-S., Li, C., Nguyen, H., & Liu, B. (2019). Broad-spectrum resistance to bacterial blight in rice using genome editing. *Nature Biotechnology*, *37*, 1344–1350.

Osakabe, Y., Watanabe, T., Sugano, S. S., Ueta, R., Ishihara, R., Shinozaki, K., & Osakabe, K. (2016). Optimization of CRISPR/Cas9 genome editing to modify abiotic stress responses in plants. *Scientific Reports*, *6*, 26685.

Ozyigit, I. I., & Yucebilgili Kurtoglu, K. (2020). Particle bombardment technology and its applications in plants. *Molecular Biology Reports*, *47*, 9831–9847.

Pramanik, D., Shelake, R. M., Park, J., Kim, M. J., Hwang, I., Park, Y., & Kim, J.-Y. (2021). CRISPR/Cas9-mediated generation of pathogen-resistant tomato against tomato yellow leaf curl virus and powdery mildew. *International Journal of Molecular Sciences*, *22*, 1878.

Qiu, Z., Kang, S., He, L., Zhao, J., Zhang, S., Hu, J., Zeng, D., Zhang, G., Dong, G., & Gao, Z. (2018). The newly identified heat-stress sensitive albino 1 gene affects chloroplast development in rice. *Plant Science*, *267*, 168–179.

Que, Z., Lu, Q., Liu, T., Li, S., Zou, J., & Chen, G. (2020). The rice annexin gene *OsAnn5* is involved in cold stress tolerance at the seedling stage. *Plant Direct*, 7, e539.

Rajak, B. K., Rani, P., Singh, N., & Singh, D. V. (2023). Sequence and structural similarities of ACCase protein of Phalaris minor and wheat: An insight to explain herbicide selectivity. *Frontiers in Plant Science*, *13*, 1056474.

Ramkumar, T., Lenka, S., Arya, S., & Bansal, K. (2020). A short history and perspectives on plant genetic transformation. In S. Rustgi & H. Luo (eds) *Biolistic DNA delivery in plants: methods and protocols.* New York, USA: Springer, pp. 39–68.

Ran, Y., Patron, N., Kay, P., Wong, D., Buchanan, M., Cao, Y.-Y., Sawbridge, T., Davies, J. P., Mason, J., Webb, S. R., et al. (2018). Zinc finger nuclease-mediated precision genome editing of an endogenous gene in hexaploid bread wheat (Triticum aestivum) using a DNA repair template. *Plant Biotechnology Journal*, *16*, 2088–2101.

Ren, B., Yan, F., Kuang, Y., Li, N., Zhang, D., Zhou, X., Lin, H., & Zhou, H. (2018). Improved base editor for efficiently inducing genetic variations in rice with CRISPR/Cas9-guided hyperactive hAID mutant. *Molecular Plant*, *11*, 623–626.

Ren, Q., Sretenovic, S., Liu, S., Tang, X., Huang, L., He, Y., Liu, L., Guo, Y., Zhong, Z., & Liu, G. (2021). PAM-less plant genome editing using a CRISPR–SpRY toolbox. *Nature Plants*, *7*, 25–33.

Roca Paixão, J. F., Gillet, F.-X., Ribeiro, T. P., Bournaud, C., Lourenço-Tessutti, I. T., Noriega, D. D., Melo, B.P.d., de Almeida-Engler, J., & Grossi-de-Sa, M. F. (2019). Improved drought stress tolerance in Arabidopsis by CRISPR/dCas9 fusion with a Histone AcetylTransferase. *Scientific Reports*, *9*, 8080.

Saha, J., Chaudhuri, D., Kundu, A., Bhattacharya, S., Roy, S., & Giri, K. (2022). Phylogenetic, structural, functional characterisation and effect of exogenous spermidine on rice (Oryza sativa) HAK transporters under salt stress. *Functional Plant Biology*, *50*, 160–182

San Filippo, J., Sung, P., & Klein, H. (2008). Mechanism of Eukaryotic Homologous Recombination. *Annual Review of Biochemistry*, *77*, 229–257.

Sánchez-León, S., Gil-Humanes, J., Ozuna, C. V., Giménez, M. J., Sousa, C., Voytas, D. F., & Barro, F. (2018). Low-gluten, nontransgenic wheat engineered with CRISPR/Cas9. *Plant Biotechnology Journal*, *16*, 902–910.

Santosh Kumar, V., Verma, R. K., Yadav, S. K., Yadav, P., Watts, A., Rao, M., & Chinnusamy, V. (2020). CRISPR-Cas9 mediated genome editing of drought and salt tolerance (OsDST) gene in indica mega rice cultivar MTU1010. *Physiology and Molecular Biology of Plants*, *26*, 1099–1110.

Sathee, L., Jagadhesan, B., Pandesha, P. H., Barman, D., Adavi B, S., Nagar, S., Krishna, G., Tripathi, S., Jha, S. K., & Chinnusamy, V. (2022). Genome editing targets for improving nutrient use efficiency and nutrient stress adaptation. *Frontiers in Genetics, 13*, 900897.

Sauer, N. J., Mozoruk, J., Miller, R. B., Warburg, Z. J., Walker, K. A., Beetham, P. R., Schöpke, C. R., & Gocal, G. F. (2016). Oligonucleotide-directed mutagenesis for precision gene editing. *Plant Biotechnology Journal*, *14*, 496–502.

Saurabh, S. (2021). Genome editing: Revolutionizing the crop improvement. *Plant Molecular Biology Reporter*, *39*, 752–772.

Shen, C., Que, Z., Xia, Y., Tang, N., Li, D., He, R., & Cao, M. (2017). Knock out of the annexin gene OsAnn3 via CRISPR/Cas9-mediated genome editing decreased cold tolerance in rice. *Journal of Plant Biology*, *60*, 539–547.

Shen, R., Yao, Q., Zhong, D., Zhang, X., Li, X., Cao, X., Dong, C., Tian, Y., Zhu, J.-K., & Lu, Y. (2023). Targeted insertion of regulatory elements enables translational enhancement in rice. *Frontiers in Plant Science*, *14*, 1134209.

Shi, J., Gao, H., Wang, H., Lafitte, H. R., Archibald, R. L., Yang, M., Hakimi, S. M., Mo, H., & Habben, J. E. (2017). ARGOS 8 variants generated by CRISPR-Cas9 improve maize grain yield under field drought stress conditions.*Plant Biotechnology Journal*, *15*, 207–216.

Shimatani, Z., Kashojiya, S., Takayama, M., Terada, R., Arazoe, T., Ishii, H., Teramura, H., Yamamoto, T., Komatsu, H., & Miura, K. (2017). Targeted base editing in rice and tomato using a CRISPR-Cas9 cytidine deaminase fusion. *Nature Biotechnology*, *35*, 441–443.

Shu, P., Li, Y., Li, Z., Sheng, J., & Shen, L. (2022). SlMAPK3 enhances tolerance to salt stress in tomato plants by scavenging ROS accumulation and up-regulating the expression of ethylene signaling related genes. *Environmental and Experimental Botany*, *193*, 104698.

Shufen, C., Yicong, C., Baobing, F., Guiai, J., Zhonghua, S., Ju, L., Shaoqing, T., Jianlong, W., Peisong, H., & Xiangjin, W. (2019). Editing of rice isoamylase gene ISA1 provides insights into its function in starch formation. *Rice Science*, *26*, 77–87.

Song, L., Liu, J., Cao, B., Liu, B., Zhang, X., Chen, Z., Dong, C., Liu, X., Zhang, Z., & Wang, W. (2023). Reducing brassinosteroid signalling enhances grain yield in semi-dwarf wheat. *Nature*, 1–7.

Songmei, L., Jie, J., Yang, L., Jun, M., Shouling, X., Yuanyuan, T., Youfa, L., Qingyao, S., & Jianzhong, H. (2019). Characterization and evaluation of OsLCT1 and OsNramp5 mutants generated through CRISPR/Cas9-mediated mutagenesis for breeding low Cd rice. Rice *Science*, *26*, 88–97.

Stone, S. L., Kwong, L. W., Yee, K. M., Pelletier, J., Lepiniec, L., Fischer, R. L., Goldberg, R. B., & Harada, J. J. (2001). LEAFY COTYLEDON2 encodes a B3 domain transcription factor that induces embryo development. *Proceedings of the National Academy of Sciences*, *98*, 11806–11811.

Su, W., Xu, M., Radani, Y., & Yang, L. (2023). Technological Development and Application of Plant Genetic Transformation. *International Journal of Molecular Sciences*, *24*, 10646.

Su, Z., Bernardo, A., Tian, B., Chen, H., Wang, S., Ma, H., Cai, S., Liu, D., Zhang, D., & Li, T. (2019). A deletion mutation in TaHRC confers Fhb1 resistance to Fusarium head blight in wheat. *Nature Genetics*, *51*, 1099–1105.

Sun, C., Wang, Y., Yang, X., Tang, L., Wan, C., Liu, J., Chen, C., Zhang, H., He, C., & Liu, C. (2023). MATE transporter GFD1 cooperates with sugar transporters, mediates carbohydrate partitioning and controls grain-filling duration, grain size and number in rice. *Plant Biotechnology Journal*, *21*, 621–634.

Sun, Y., Zhang, X., Wu, C., He, Y., Ma, Y., Hou, H., Guo, X., Du, W., Zhao, Y., & Xia, L. (2016). Engineering herbicide-resistant rice plants through CRISPR/Cas9-mediated homologous recombination of acetolactate synthase. *Molecular Plant*, *9*, 628–631.

Svitashev, S., Schwartz, C., Lenderts, B., Young, J. K., & Mark Cigan, A. (2016). Genome editing in maize directed by CRISPR–Cas9 ribonucleoprotein complexes. *Nature Communications*, *7*, 13274.

Tan, B., Xu, M., Chen, Y., & Huang, M. (2013). Transient expression for functional gene analysis using Populus protoplasts. *Plant Cell, Tissue and Organ Culture (PCTOC)*, *114*, 11–18.

Tan, J., Zeng, D., Zhao, Y., Wang, Y., Liu, T., Li, S., Xue, Y., Luo, Y., Xie, X., & Chen, L. (2022). PhieABEs: A PAM-less/free high-efficiency adenine base editor toolbox with wide target scope in plants. *Plant Biotechnology Journal*, *20*, 934–943.

Tang, D., Jia, Y., Zhang, J., Li, H., Cheng, L., Wang, P., Bao, Z., Liu, Z., Feng, S., & Zhu, X. (2022). Genome evolution and diversity of wild and cultivated potatoes. *Nature*, *606*, 535–541.

Tang, X., Liu, G., Zhou, J., Ren, Q., You, Q., Tian, L., Xin, X., Zhong, Z., Liu, B., & Zheng, X. (2018). A large-scale whole-genome sequencing analysis reveals highly specific genome editing by both Cas9 and Cpf1 (Cas12a) nucleases in rice. *Genome Biology*, *19*, 1–13.

Tian, J., Wang, C., Xia, J., Wu, L., Xu, G., Wu, W., Li, D., Qin, W., Han, X., Chen, Q., et al. (2019). Teosinte ligule allele narrows plant architecture and enhances high-density maize yields. *Science*, *365*, 658–664.

Tian, S., Jiang, L., Cui, X., Zhang, J., Guo, S., Li, M., Zhang, H., Ren, Y., Gong, G., & Zong, M. (2018). Engineering herbicide-resistant watermelon variety through CRISPR/Cas9-mediated base-editing. *Plant Cell Reports*, *37*, 1353–1356.

Tian, Y., Shen, R., Li, Z., Yao, Q., Zhang, X., Zhong, D., Tan, X., Song, M., Han, H., & Zhu, J. K. (2022). Efficient C-to-G editing in rice using an optimized base editor. *Plant Biotechnology Journal*, *20*, 1238.

Touraev, A., Citovsky, V., & Tzfira, T. (2011). *Plant transformation technologies*. Danvers, USA: Wiley Online Library.

Tran, M. T., Doan, D. T. H., Kim, J., Song, Y. J., Sung, Y. W., Das, S., Kim, E. J., Son, G. H., Kim, S. H., Van Vu, T., et al. (2021). CRISPR/Cas9-based precise excision of SlHyPRP1 domain(s) to obtain salt stress-tolerant tomato. *Plant Cell Reports*, *40*, 999–1011.

Tsuda, K., Qi, Y., Nguyen, L. V., Bethke, G., Tsuda, Y., Glazebrook, J., & Katagiri, F. (2012). An efficient Agrobacterium-mediated transient transformation of Arabidopsis. *The Plant Journal*, *69*, 713–719.

Tuncel, A., Corbin, K. R., Ahn-Jarvis, J., Harris, S., Hawkins, E., Smedley, M. A., Harwood, W., Warren, F. J., Patron, N. J., & Smith, A. M. (2019). Cas9-mediated mutagenesis of potato starch-branching enzymes generates a range of tuber starch phenotypes. *Plant Biotechnology Journal*, *17*, 2259–2271.

Tyurin, A. A., Suhorukova, A. V., Kabardaeva, K. V., & Goldenkova-Pavlova, I. V. (2020). Transient gene expression is an effective experimental tool for the research into the fine mechanisms of plant gene function: Advantages, limitations, and solutions. *Plants*, *9*, 1187.

Veillet, F., Chauvin, L., Kermarrec, M.-P., Sevestre, F., Merrer, M., Terret, Z., Szydlowski, N., Devaux, P., Gallois, J.-L., & Chauvin, J.-E. (2019). The Solanum tuberosum GBSSI gene: A target for assessing gene and base editing in tetraploid potato. *Plant Cell Reports*, *38*, 1065–1080.

Vyacheslavova, A., Berdichevets, I., Tyurin, A., Shimshilashvili, K. R., Mustafaev, O., & Goldenkova-Pavlova, I. (2012). Expression of heterologous genes in plant systems: New possibilities. *Russian Journal of Genetics*, *48*, 1067–1079.

Walton, R. T., Christie, K. A., Whittaker, M. N., & Kleinstiver, B. P. (2020). Unconstrained genome targeting with near-PAMless engineered CRISPR-Cas9 variants. *Science*, *368*, 290–296.

Waltz, E. (2018). With a free pass, CRISPR-edited plants reach market in record time. *Nature Biotechnology*, *36*, 6–8.

Wang, B., Zhong, Z., Wang, X., Han, X., Yu, D., Wang, C., Song, W., Zheng, X., Chen, C., & Zhang, Y. (2020). Knockout of the OsNAC006 transcription factor causes drought and heat sensitivity in rice. *International Journal of Molecular Sciences*, *21*, 2288.

Wang, C.-T., Ru, J.-N., Liu, Y.-W., Yang, J.-F., Li, M., Xu, Z.-S., & Fu, J.-D. (2018a). The maize WRKY transcription factor ZmWRKY40 confers drought resistance in transgenic Arabidopsis. *International Journal of Molecular Sciences*, *19*, 2580.

Wang, H., & Jiang, L. (2011). Transient expression and analysis of fluorescent reporter proteins in plant pollen tubes. *Nature Protocols*, *6*, 419–426.

Wang, L., Chen, L., Li, R., Zhao, R., Yang, M., Sheng, J., & Shen, L. (2017). Reduced drought tolerance by CRISPR/Cas9-mediated SlMAPK3 mutagenesis in tomato plants. *Journal of Agricultural and Food Chemistry*, *65*, 8674–8682.

Wang, Q., Wan, J., Dang, K., Meng, S., Hu, D., Lin, Y., Qiu, X., Guo, Z., Fu, Z., Ding, D., et al. (2023). zma-miR159 targets ZmMYB74 and ZmMYB138 transcription factors to regulate grain size and weight in maize. *Plant Physiology*, kiad455.

Wang, W., Simmonds, J., Pan, Q., Davidson, D., He, F., Battal, A., Akhunova, A., Trick, H. N., Uauy, C., & Akhunov, E. (2018b). Gene editing and mutagenesis reveal inter-cultivar differences and additivity in the contribution of TaGW2 homoeologues to grain size and weight in wheat. *Theoretical and Applied Genetics*, *131*, 2463–2475.

Wang, W., Wang, Z., Hou, W., Chen, L., Jiang, B., Ma, W., Bai, L., Song, W., Xu, C., & Han, T. (2022a). GmNMHC5 may promote nodulation via interaction with GmGAI in soybean. *The Crop Journal*, *10*, 273–279.

Wang, Y., Cheng, X., Shan, Q., Zhang, Y., Liu, J., Gao, C., & Qiu, J.-L. (2014). Simultaneous editing of three homoeoalleles in hexaploid bread wheat confers heritable resistance to powdery mildew. *Nature Biotechnology, 32*, 947–951.

Wang, Y., Liu, X., Zheng, X., Wang, W., Yin, X., Liu, H., Ma, C., Niu, X., Zhu, J.-K., & Wang, F. (2021). Creation of aromatic maize by CRISPR/Cas. *Journal of Integrative Plant Biology*, *63*, 1664–1670.

Wang, Y., Tang, Q., Kang, Y., Wang, X., Zhang, H., & Li, X. (2022b). Analysis of the utilization and prospects of CRISPR-Cas technology in the annotation of gene function and creation new germplasm in maize based on patent data. *Cells*, *11*, 3471.

Wang, Z., Wang, Y., Shang, P., Yang, C., Yang, M., Huang, J., Ren, B., Zuo, Z., Zhang, Q., Li, W., et al. (2022c). Overexpression of soybean GmWRI1a stably increases the seed oil content in soybean. *International Journal of Molecular Sciences*, *23*, 5084.

Watson, A., Ghosh, S., Williams, M. J., Cuddy, W. S., Simmonds, J., Rey, M.-D., Asyraf Md Hatta, M., Hinchliffe, A., Steed, A., Reynolds, D., et al. (2018). Speed breeding is a powerful tool to accelerate crop research and breeding. *Nature Plants*, *4*, 23–29.

Wu, B., Meng, J., Liu, H., Mao, D., Yin, H., Zhang, Z., Zhou, X., Zhang, B., Sherif, A., & Liu, H. (2023). Suppressing a phosphohydrolase of cytokinin nucleotide enhances grain yield in rice. *Nature Genetics*, *55*, 1–9.

Wu, Z., Zhang, Y., Yu, H., Pan, D., Wang, Y., Wang, Y., Li, F., Liu, C., Nan, H., Chen, W., et al. (2021). Programmed genome editing by a miniature CRISPR-Cas12f nuclease. *Nature Chemical Biology*, *17*, 1132–1138.

Xu, N., Kang, M., Zobrist, J. D., Wang, K., & Fei, S.-z. (2022). Genetic transformation of recalcitrant upland switchgrass using morphogenic genes. *Frontiers in Plant Science*, *12*, 781565.

Xun, H., Yang, X., He, H., Wang, M., Guo, P., Wang, Y., Pang, J., Dong, Y., Feng, X., & Wang, S. (2019). Over-expression of GmKR3, a TIR–NBS–LRR type R gene, confers resistance to multiple viruses in soybean. *Plant Molecular Biology*, *99*, 95–111.

Yan, Y., Zhu, X., Yu, Y., Li, C., Zhang, Z., & Wang, F. (2022). Nanotechnology strategies for plant genetic engineering. *Advanced Materials*, *34*, 2106945.

Yang, T., Ali, M., Lin, L., Li, P., He, H., Zhu, Q., Sun, C., Wu, N., Zhang, X., Huang, T., et al. (2023). Recoloring tomato fruit by CRISPR/Cas9-mediated multiplex gene editing. *Horticulture Research*, *10*, uhac214.

Yang, X., Li, X., Shan, J., Li, Y., Zhang, Y., Wang, Y., Li, W., & Zhao, L. (2021). Overexpression of GmGAMYB accelerates the transition to flowering and increases plant height in soybean. *Frontiers in Plant Science*, *12*, 667242.

Yang, Z., Du, H., Xing, X., Li, W., Kong, Y., Li, X., & Zhang, C. (2022). A small heat shock protein, GmHSP17.9, from nodule confers symbiotic nitrogen fixation and seed yield in soybean. *Plant Biotechnology Journal*, *20*, 103–115.

Yarnall, M. T. N., Ioannidi, E. I., Schmitt-Ulms, C., Krajeski, R. N., Lim, J., Villiger, L., Zhou, W., Jiang, K., Garushyants, S. K., Roberts, N., et al. (2023). Drag-and-drop genome insertion of large sequences without double-strand DNA cleavage using CRISPR-directed integrases. *Nature Biotechnology*, *41*, 500–512.

Ye, M., Peng, Z., Tang, D., Yang, Z., Li, D., Xu, Y., Zhang, C., & Huang, S. (2018). Generation of self-compatible diploid potato by knockout of S-RNase. *Nature Plants, 4*, 651–654.

Ye, R., Yang, X., & Rao, Y. (2022). Genetic engineering technologies for improving crop yield and quality. *Agronomy*, *12*, 759.

Yin, Y., Qin, K., Song, X., Zhang, Q., Zhou, Y., Xia, X., & Yu, J. (2018). BZR1 transcription factor regulates heat stress tolerance through FERONIA receptor-like kinase-mediated reactive oxygen species signaling in tomato. *Plant and Cell Physiology*, *59*, 2239–2254.

You, X., Ning, Y., & Wang, G.-L. (2023). Editing a rice CDP-DAG synthase confers broad-spectrum resistance. *Trends in Plant Science*, *28*, 1344–1346.

Younis, A., Ramzan, F., Ramzan, Y., Zulfiqar, F., Ahsan, M., & Lim, K. B. (2020). Molecular markers improve abiotic stress tolerance in crops: A review. *Plants*, *9*, 1374.

Yu, W., Wang, L., Zhao, R., Sheng, J., Zhang, S., Li, R., & Shen, L. (2019). Knockout of SlMAPK3 enhances tolerance to heat stress involving ROS homeostasis in tomato plants. *BMC Plant Biology*, *19*, 1–13.

Zafar, K., Noureen, A., Awan, M. J. A., Anjum, N., Aslam, M. Q., Khan, M. Z., Amin, I., & Mansoor, S. (2021). Genome editing to develop disease resistance in crops. In S. K. Upadhyay (eds) *Genome engineering for crop improvement*. New Jersey, USA: John Wiley & Sons Ltd., pp. 224–252.

Zagorskaya, A., & Deineko, E. V. (2017). Suspension-cultured plant cells as a platform for obtaining recombinant proteins. *Russian Journal of Plant Physiology*, *64*, 795–807.

Zeng, Y., Wen, J., Zhao, W., Wang, Q., & Huang, W. (2020). Rational improvement of rice yield and cold tolerance by editing the three genes OsPIN5b, GS3, and OsMYB30 with the CRISPR–Cas9 system. *Frontiers in Plant Science*, *10*, 1663.

Zetsche, B., Gootenberg, J. S., Abudayyeh, Omar O., Slaymaker, I. M., Makarova, K. S., Essletzbichler, P., Volz, S. E., Joung, J., van der Oost, J., Regev, A., et al. (2015). Cpf1 Is a single RNA-guided endonuclease of a class 2 CRISPR-Cas system. *Cell*, *163*, 759–771.

Zhang, H.-Y., Hou, Z.-H., Zhang, Y., Li, Z.-Y., Chen, J., Zhou, Y.-B., Chen, M., Fu, J.-D., Ma, Y.-Z., & Zhang, H. (2022a). A soybean EF-Tu family protein GmEF8, an interactor of GmCBL1, enhances drought and heat tolerance in transgenic Arabidopsis and soybean. *International Journal of Biological Macromolecules*, *205*, 462–472.

Zhang, J., Zhang, H., Li, S., Li, J., Yan, L., & Xia, L. (2021). Increasing yield potential through manipulating of an ARE1 ortholog related to nitrogen use efficiency in wheat by CRISPR/Cas9. *Journal of Integrative Plant Biology*, *63*, 1649–1663.

Zhang, M., Zhao, R., Huang, K., Huang, S., Wang, H., Wei, Z., Li, Z., Bian, M., Jiang, W., & Wu, T. (2022b). The OsWRKY63–OsWRKY76–OsDREB1B module regulates chilling tolerance in rice. *The Plant Journal*, *112*, 383–398.

Zhang, R., Liu, J., Chai, Z., Chen, S., Bai, Y., Zong, Y., Chen, K., Li, J., Jiang, L., & Gao, C. (2019a). Generation of herbicide tolerance traits and a new selectable marker in wheat using base editing. *Nature Plants*, *5*, 480–485.

Zhang, Y., Chen, K., Zhao, F.-J., Sun, C., Jin, C., Shi, Y., Sun, Y., Li, Y., Yang, M., & Jing, X. (2018a). OsATX1 interacts with heavy metal P1B-type ATPases and affects copper transport and distribution. *Plant Physiology*, *178*, 329–344.

Zhang, Y., Massel, K., Godwin, I. D., & Gao, C. (2018b). Applications and potential of genome editing in crop improvement. *Genome Biology*, *19*, 1–11.

Zhang, Z., Hua, L., Gupta, A., Tricoli, D., Edwards, K. J., Yang, B., & Li, W. (2019b). Development of an Agrobacterium-delivered CRISPR/Cas9 system for wheat genome editing. *Plant Biotechnology Journal*, *17*, 1623–1635.

Zheng, M., Lin, J., Liu, X., Chu, W., Li, J., Gao, Y., An, K., Song, W., Xin, M., Yao, Y., et al. (2021). Histone acetyltransferase TaHAG1 acts as a crucial regulator to strengthen salt tolerance of hexaploid wheat. *Plant Physiology*, *186*, 1951–1969.

Zheng, T., Yu, X., Sun, Y., Zhang, Q., Zhang, X., Tang, M., Lin, C., & Shen, Z. (2022). Expression of a cytochrome P450 gene from bermuda grass Cynodon dactylon in soybean confers tolerance to multiple herbicides. *Plants*, *11*, 949.

Zhou, H., Bai, S., Wang, N., Sun, X., Zhang, Y., Zhu, J., & Dong, C. (2020). CRISPR/Cas9-mediated mutagenesis of MdCNGC2 in apple callus and VIGS-mediated silencing of MdCNGC2 in fruits improve resistance to Botryosphaeria dothidea. *Frontiers in Plant Science*, *11*, 575477.

Zhou, J., Liu, G., Zhao, Y., Zhang, R., Tang, X., Li, L., Jia, X., Guo, Y., Wu, Y., Han, Y., et al. (2023). An efficient CRISPR–Cas12a promoter editing system for crop improvement. *Nature Plants*, *9*, 588–604.

Zhou, L., Deng, S., Xuan, H., Fan, X., Sun, R., Zhao, J., Wang, H., Guo, N., & Xing, H. (2022a). A novel TIR-NBS-LRR gene regulates immune response to Phytophthora root rot in soybean. *The Crop Journal*, *10*, 1644–1653.

Zhou, X., Zha, M., Huang, J., Li, L., Imran, M., & Zhang, C. (2017). StMYB44 negatively regulates phosphate transport by suppressing expression of PHOSPHATE1 in potato. *Journal of Experimental Botany*, *68*, 1265–1281.

Zhou, Y., Xu, S., Jiang, N., Zhao, X., Bai, Z., Liu, J., Yao, W., Tang, Q., Xiao, G., & Lv, C. (2022b). Engineering of rice varieties with enhanced resistances to both blast and bacterial blight diseases via CRISPR/Cas9. *Plant Biotechnology Journal*, *20*, 876–885.

Zhu, H., Li, C., & Gao, C. (2020). Applications of CRISPR–Cas in agriculture and plant biotechnology.*Nature Reviews Molecular Cell Biology*, *21*, 661–677.

Zhu, H., Wang, J., Jiang, D., Hong, Y., Xu, J., Zheng, S., Yang, J., & Chen, W. (2022). The miR157-SPL-CNR module acts upstream of bHLH101 to negatively regulate iron deficiency responses in tomato. *Journal of Integrative Plant Biology*, *64*, 1059–1075.

Zong, Y., Liu, Y., Xue, C., Li, B., Li, X., Wang, Y., Li, J., Liu, G., Huang, X., Cao, X., et al. (2022). An engineered prime editor with enhanced editing efficiency in plants. *Nature Biotechnology*, *40*, 1394–1402.

Zong, Y., Song, Q., Li, C., Jin, S., Zhang, D., Wang, Y., Qiu, J.-L., & Gao, C. (2018). Efficient C-to-T base editing in plants using a fusion of nCas9 and human APOBEC3A. *Nature Biotechnology*, *36*, 950–953.

Zong, Y., Wang, Y., Li, C., Zhang, R., Chen, K., Ran, Y., Qiu, J.-L., Wang, D., & Gao, C. (2017). Precise base editing in rice, wheat and maize with a Cas9-cytidine deaminase fusion. *Nature Biotechnology*, *35*, 438–440.

Zuo, J., Niu, Q.-W., Frugis, G., & Chua, N.-H. (2002). The WUSCHEL gene promotes vegetative-to-embryonic transition in Arabidopsis. *The Plant Journal*, *30*, 349–359.

8 Genomic Selection Tools for Plant Speed Breeding

Sumi Sarkar, Marium Khatun, F. M. Era, and A. K. M. Aminul Islam

Department of Genetics and Plant Breeding, Bangabandhu Sheikh Mujibur Rahman Agricultural University, Gazipur, Bangladesh

8.1 INTRODUCTION

The world population is increasing day by day and the ultimate pressure is coming over agricultural production. However, climatic change has been providing a great challenge to the growers for several years. So conventional breeding strategies cannot provide adequate support to the breeders for overcoming the continuous pressure of crop improvement and development of novel varieties with higher yield, nutritional quality, and resistance to biotic and abiotic stresses (Begna, 2022). Crop improvement requires cycling and assessment of several generations in order to incorporate and evaluate many traits of interest. So, procedures of rapid generation advancement are the main focus of breeders nowadays. Modern plant breeding tools like genomic selection, speed breeding, mutagenic breeding, marker-assisted selection (MAS), doubled haploid (DH), and genome editing techniques like CRISPR and genome sequencing are providing great outcomes to the breeder, thus increasing the efficiency of crop breeding (Table 8.1) (Crossa et al., 2017; Shivakumar et al., 2018; Shimelis et al., 2019; Watson et al., 2019; Krishnappa et al., 2021; Venezia & Creasey Krainer, 2021; Wanga et al., 2021; Liu et al., 2022; Jones et al., 2022; Samantara et al., 2022).

Speed breeding is one of these effective techniques that is used by plant breeders mainly for rapid generation advancement. Due to the fast-rising population in the world, most of the conventional breeding approaches are found unsatisfactory to meet the augmented demand for food (Samantara et al., 2022). To solve the rising scarcity of food as well as ensure food security, the development of better crop varieties through rapid generation advancement is one of the best strategies of plant breeding. Speed breeding abridges the breeding cycle through rapid generation advancement where a wide range of germplasms can be improved and evaluated without any special in vitro tools, which provides accelerated crop research to the plant breeders (Ghosh et al., 2018; Wolf & Blankenship, 2019). This speed breeding technique supplements another molecular approach to improve a crop variety within a short period (Krishnappa et al., 2021).

Genomic selection however is a cost-effective and high-throughput genotyping technique that can efficiently be used in speed breeding programs to make the breeders' mission easier with time to develop crop variety, testing their performance, and commercial release of that variety (Ahmar et al., 2020). So scaling up research for crop improvement is possible by applying genomic selection tools in speed breeding programs. This genomic selection method is another form of MAS where the effects of several genome-wide markers are estimated as well as the genetic values are computed for an untested population (Annicchiarico et al., 2019; Cui et al., 2020). Therefore, the incorporation of genomic selection in speed breeding programs is a

DOI: 10.1201/b23372-8

TABLE 8.1
Modern Plant Breeding Tools for Crop Improvement

SL	Plant Breeding Tools and Major Crops to Improve	Advantages/Application	Limitations
1	Speed breeding • Wheat • Barley • Peanut • Chickpea • Canola • Pea • Sunflower • Pepper • Radish	• Speeding up the breeding cycle • Facilitates researchers to apply molecular approaches • Developing variety within a short period • Two to six generations per year based on crop • Easy to apply advanced breeding methods in a controlled environment • Crop improvement for the desired trait in a shorter period	• Requires expertise, effective, and complementary plant phenomics facilities • Appropriate infrastructure and continuous financial support for research and development • Requires skills and expertise in plant breeding and biotechnology, • Long-term funding and government policy support. For example, in sub-Saharan
2	Genomic selection • Maize • Rice • Wheat • Sorghum • Barley • Rapeseed • Cassava	• Potentially accelerate the rate of genetic gain in crops • Increase grain production in less time • Reduce the breeding interval cycle to at least half the conventional time • Produces lines that, in hybrid combinations, significantly increase grain yield performance over that of commercial checks • Develop in less time germplasm that is adapted to sudden climate change. • Rapid selection of superior genotypes accelerates the breeding cycle • Germplasm enhancement (i.e., prebreeding) programs could accelerate the flow of genes from gene bank accessions to elite lines • Predicting additive effects in early generations of a breeding program ($F_{2:3}$) to achieve a rapid selection cycle with a short interval	• The reference population needs to be of sufficient size to be able to estimate accurate associations between genotype and phenotype • Costly • Difficult to determine if sequence codes for a gene
3	Mutagenic breeding • Most of the cereals	• A cheap and rapid method of developing new varieties of crop • Induced mutagens are used for developing CMS • More effective in improving oligogenic character • A simple and quick way to induce a new character • This can apply to single cell • Screen very high population	• Random and unpredictable process • Useful mutants are prominently recessive • Mutants may have negative pleiotropic effects on other traits • Health risks by chemical mutagens
4	Marker-assisted selection • Cereals and vegetable crops	• A simpler method compared to phenotypic screening • Save time and resources • Increased reliability • No environmental effect • Can separate homozygote and heterozygote and select single plants	• Costly method • Required well-equipped laboratory • Required well-trained manpower to handle sophisticated equipment • Detection of various linked DNA markers is difficult

TABLE 8.1 *(Continued)*
Modern Plant Breeding Tools for Crop Improvement

SL	Plant Breeding Tools and Major Crops to Improve	Advantages/Application	Limitations
5	Double haploid • Rice • Wheat • Maize • Mustard • Potato • Onion	• Developing complete homozygous line in 2–3 generation • Perfect DUS • Require less time and labor • High frequency of homozygous plant	• Frequency of haploid is low • Selection cannot be imposed on population • Success is genotype dependent • Success is unpredictable and can consume valuable resources • May reduce genetic variation
6	Genome editing techniques • Wheat • Maize • Tomato • Watermelon • Soybean	• Potential to cure genetic diseases • Create crop resistance against environmental stresses • Improve a beneficial trait or deactivate an undesirable trait Improve resilience to increasingly extreme weather conditions • Reduce the need for inputs such as pesticides or fungicides	• Can create a range of changes to the genome in plants that pose risks to biodiversity, water and soil, human health, and organic food production • Low precision, high probability of off-target genome damage and tissue damage in biolistic bombardment • Low rate of stable transformation events occurs with both methodologies

promising method to attain food security as well as sustainable agriculture by reducing the breeding cycles of crops.

8.2 SPEED BREEDING

Speed breeding is one of the fastest plant breeding techniques that is used by scientists for pyramiding genes of different traits to achieve a novel crop variety. This breeding technique was first explained by NASA (National Aeronautics and Space Administration) of the United States to reduce the breeding cycle as fast as possible to grow 5–6 generations per year (Watson et al., 2018). To feed the fast-growing population of the world, speed breeding is one of the effective breeding methods that can be utilized to develop a crop variety within the lowest time duration based on required traits. This is hence the most powerful method of accelerating the breeding cycle of crops and effective technological evolution has made this method stronger to set up a breeding program of versatile crops (Figure 8.1) (Meena et al., 2022). Although this tool was first used for wheat, later this breeding technique was found to be perfect for advancing the generation of many other valuable crops (Ghosh et al., 2018). A breeding program can be set up with diverse germplasm in a speed breeding program suitably. Breeding of small grain crops can be facilitated with reduced space and cost in speed breeding programs for developing a large number of inbreed lines (Slama-Ayed et al., 2019). Different phenotypic gene insertion is possible by speed breeding and MAS in speed breeding helps to select elite hybrid lines (Hickey et al., 2017). Genomic selection in speed breeding contributes to improving quantitative as well as qualitative traits (Wolter et al., 2019).

Genetic advancement of crop varieties in association with reduced time and resources is possible through speed breeding and leads to high-performing cultivar development through marker-assisted traits (Samineni et al., 2020). Breeders can provide a new generation of varieties with disease-resistant and reduced salt sensitivity to augment the agricultural yield by modifying the light intensity and temperature.

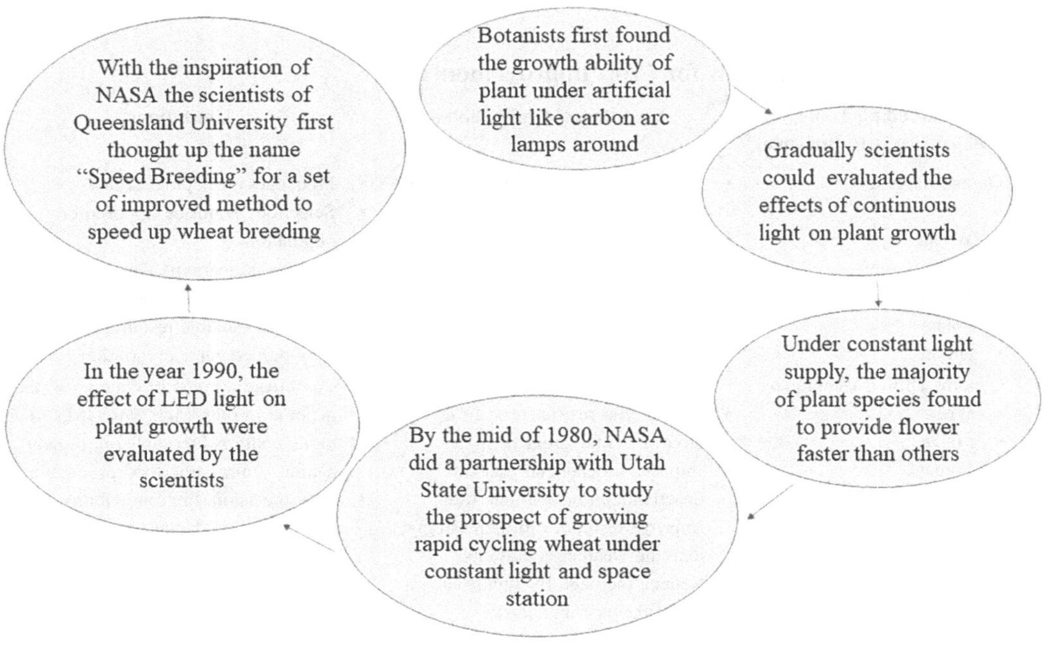

FIGURE 8.1 Evolution of speed breeding with time.

8.3 APPLICATION OF SPEED BREEDING AND GENOMIC SELECTION

Conventional breeding is used in developing most crop cultivars and it takes more than 10 years of effort to release an improved variety, whereas speed breeding is effectively applied to curtail the length of the breeding cycle to produce and release a crop variety within a short period (Hickey et al., 2019; Ahmar et al., 2020). The basic application of speed breeding is to speed up the breeding cycle of crops by enhancing the photosynthesis rate, thus facilitating early flowering and reducing the duration of generation. Controlled day length, light intensity, temperature, and plant hormone provide an optimum growth condition that regulates the flowering of plants and stimulates the maturity period thus reducing the time of the growth cycle (Table 8.2) (Chiurugwi et al., 2019), (Ghosh et al., 2018; Watson et al., 2018; Hickey et al., 2019; Ahmar et al., 2020; Pandey et al., 2022; Samantara et al., 2022).

The photosynthetic process is affected by light and controls Photomorphogenesis in plants through fast regulating the growth and development of plants. Plants can utilize the differential wavelength of the sun and track the daytime. Speed breeding stimulates this process by providing particular light intensity (Voss-Fels et al., 2019). The growth and structural development of plants is triggered by the specific spectrum of visible red/far-red and violet-blue light. Utilizing these types of spectrum early and late flowering germplasms were developed for legume crops like chickpea, pea, faba beans, and lupins (Croser et al., 2016). Small grain crops were found to reduce their generation length by extending the photoperiod (Heuschele et al., 2019). Some crops showed a positive response when their day length was increased such as sunflower, radish, amaranth, pepper, and some other leafy vegetables (Chiurugwi et al., 2019). Speed breeding has also been applied to short-day plants such as peanuts, pigeon pea, sorghum, and millet for early flowering and fast crop improvement. In major crops like rice and maize, speed breeding is done to shorten the photoperiod to start flowering and increase the number of generations (Collard et al., 2017). In the case of photo-insensitive crops around six generations are possible by speed breeding and two to three generations are possible in photosensitive crops (Wanga et al., 2021). Genomic selection of crops can be boosted along with the transgenic approaches and valuable physiological traits of crops can be studied through speed breeding (Watson et al., 2018).

TABLE 8.2
Increased Crop Generation through Speed Breeding

Image	Crop with Scientific Name	Number of Generations Possible per Year	Days to Flowering	Controlled Environment
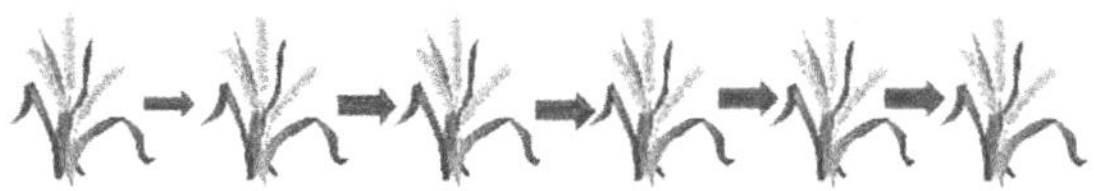	Wheat *Tritichum aestivum*	6-7	28-41	Photoperiod Temperature
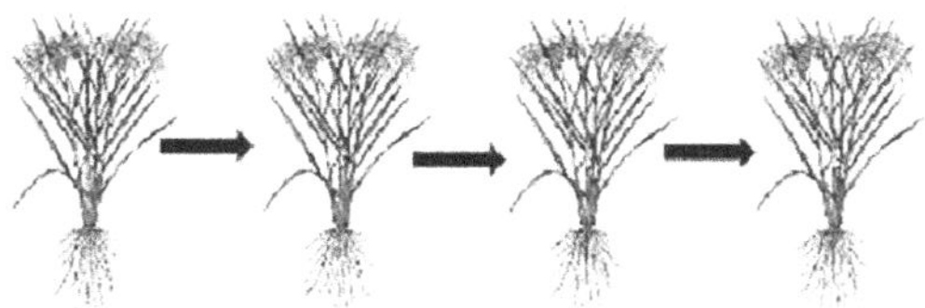	Rice *Oryza sativa*	4	75-85	Photoperiod Temperature
	Barley *Hordeum vulgare*	9	24-36	Photoperiod Temperature Soil fertility
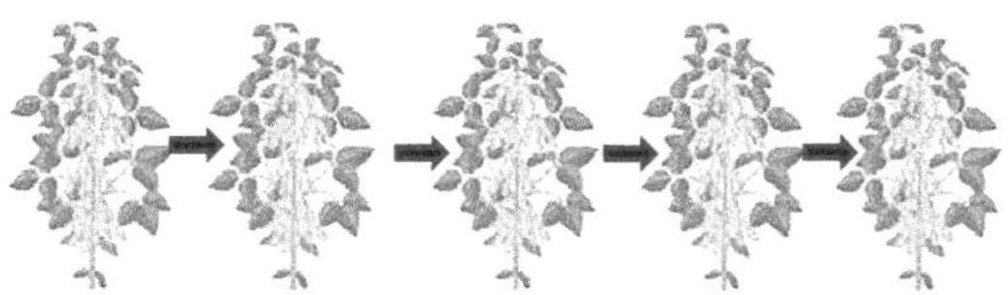	Soybean *Glycine max*	5	23	Photoperiod Temperature Immature seed germination
	Groundnut *Arachis hypogaea*	3	25-27	Photoperiod Temperature

(*Continued*)

TABLE 8.2 ***(Continued)***
Increased Crop Generation through Speed Breeding

Image	Crop with Scientific Name	Number of Generations Possible per Year	Days to Flowering	Controlled Environment
	Pea *Pisum sativum*	5	33	Plant hormones, photoperiod, Photoperiod Temperature
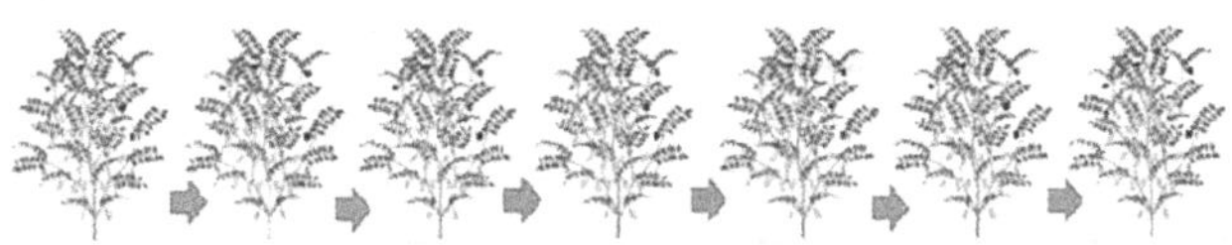	Chickpea *Cicer arietinum*	7	33	Photoperiod and immature seed germination
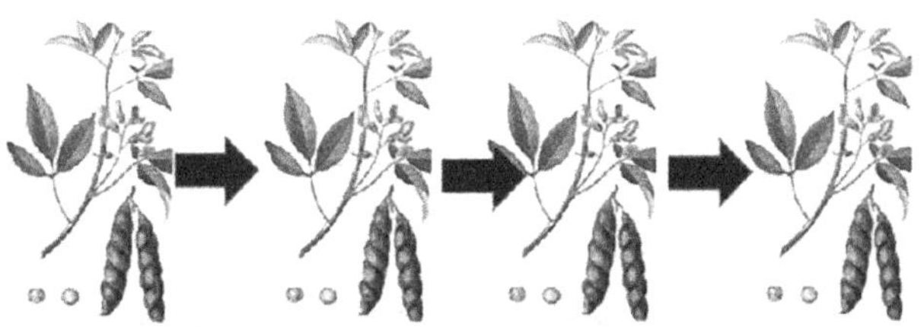	Pigeon pea *Cajanus cajan*	4	50-56	Photoperiod, temperature, immature seed germination
	Lentil *Lens culinaris*	8	31-33	Plant hormones, photoperiod, light intensity and immature seed
	Faba bean *Vicia faba*	7	29-32	Plant hormones, photoperiod, light intensity and immature seed

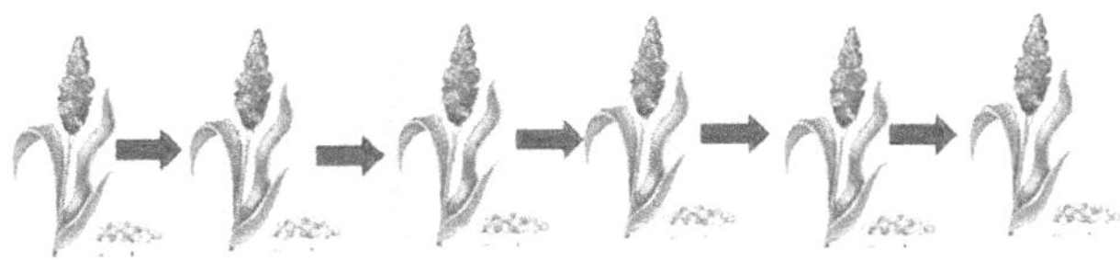	Sorghum *Sorghum bicolor*	6	40-50	Photoperiod, temperature and immature seed germination
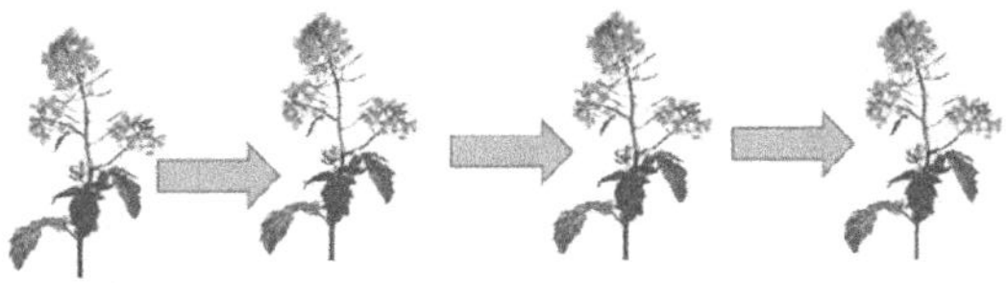	Canola *Brassica napus*	4	73	Photoperiod, light intensity, temperature, immature seed germination and soil moisture
	Amaranth *Amaranthus viridis*	6	28	Photoperiod and temperature

Novel traits of crops that are evolved with people's demand such as high yield, stress tolerance, and better nutritional quality can be achieved faster by the combination of modern genomic tools along with speed breeding compared to a sole breeding approach (Wanga et al., 2021). Particularly, genomic selection is one of the most effective approaches to breeding for quantitative traits. The breeding cycle of crops can be reduced to produce recombinant inbred lines faster by combining genomic selection with speed breeding and this combination also facilitates improvement of the key traits through indirect phenotypic selection. In multivariate genomic selection, improved yield prediction is possible through the assessment of speed breeding traits that are correlated with field-based yield (Watson et al., 2019). In speed breeding, a greater amount of genetic gain can be attained rather than in direct field-level phenotyping by greater generation turnover rate as well as higher section intensity. Various advantages can be achieved through speed breeding over conventional breeding including accelerated backcrossing, pyramiding traits, and transgenic approaches. This breeding technique requires a small region to apply on crops, which makes the integration of this method with other breeding methods easier (Begna, 2022).

8.4 SELECTION TOOLS FOR PLANT SPEED BREEDING

Selection of best genotypes in speed breeding can be made more effective if this breeding method can be combined with other modern breeding techniques such as genomic selection, gene editing, and MAS (Araus et al., 2018). These methods help to enhance the strength and accuracy of selection through the expansion of high-throughput phenotyping (HTP). On the other side, genome sequencing at the second and third generation of crops helps to discover new genes for both the quality and quantity of crop production through the utilization of more cost-effective DNA makers (Bassi et al., 2016; Ghosh et al., 2018). In the case of genomic selection, estimated breeding values (EBVs) are calculated by exploiting genome-wide markers that provide an evaluation of minor to large effects of genes throughout the whole genome of a plant (Figure 8.2) (Heffner et al., 2009; Begna, 2022). Selection of a number of traits together is possible through the genomic selection

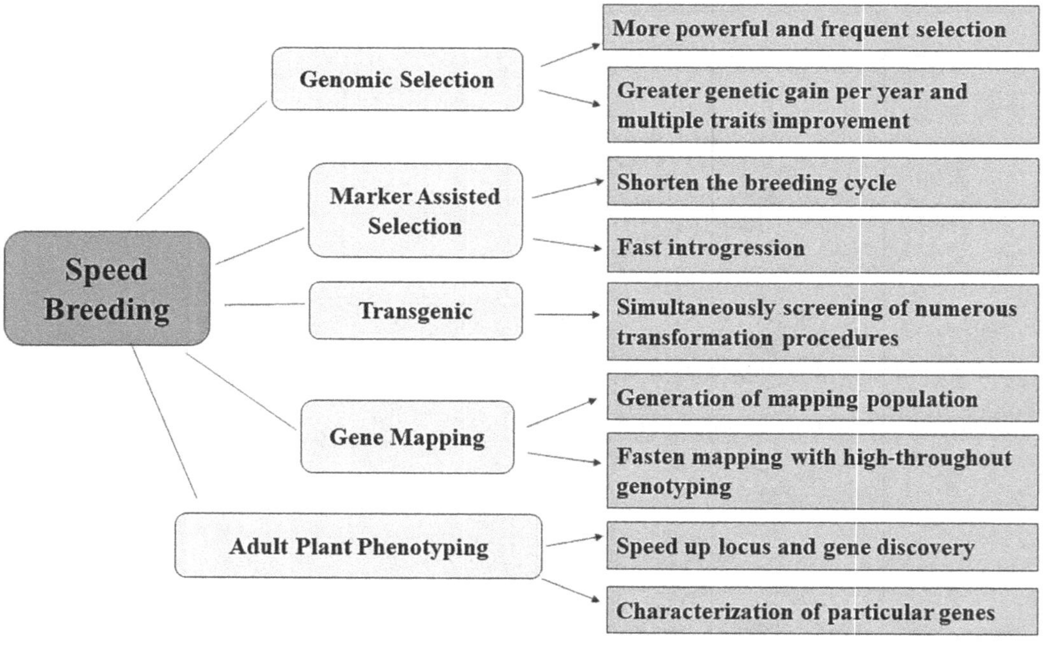

FIGURE 8.2 Selection methods used in plant speed breeding for the improvement of crops.

method. Basically in speed breeding, phenotyping techniques are applied to select important traits. However, genomic selection in the early generation of speed breeding helps breeders to screen a large number of valuable traits altogether thus allowing selection of desired plants (Krishnappa et al., 2021). Various small effect genes regulate different agronomically significant quantitative traits that are difficult to map even by MAS or genome editing techniques because in these techniques the first target is to find out large gene effects (Laidig et al., 2017). However, genomic selection has provided the solution to improve such traits associated with small gene effects as in the genomic section the genetic potential of an individual rather than identifying a particular quantitative trait loci (QTL) (Robertsen et al., 2019).

8.5 GENOMIC SELECTION

A plant breeding program is normally intended with fewer replications in initial segregating generations as well as additional replications in later generations with bigger-sized plots and testing at locations (Merrick et al., 2022). Researchers should consider this system when planning for the combination of genomic selection in their breeding program because it affects the genetic gain as well as implementation cost. In initial generations, more decline in the plant breeding cycle led by GS through the elimination of one or two selfing cycles (Krishnappa et al., 2021). Genotypes that are being selected for high GEBVs, are used as parents and cycled back to reduce the duration of the cycle. Selection accuracy may increase by incorporation of genomic selection at advanced generations to analyze the performance of lines, but in that case, the breeding cycle would not be decreased. Although the advantages over phenotypic selection are minimal in the later generations, plant breeders are interested more in using genomic selection at advanced generations due to less genotyping cost at a later stage (Fernandes et al., 2018). To identify the relation between the accuracy of selection and the duration of the cycle, different schemes of genomic selection should be assessed in active plant breeding programs in order to enhance genetic gain (Rutkoski et al., 2017). A scientist found 0.47 genetic gain in an early segregating generation which was the highest per year as well as six times higher than conventional breeding, whereas in early generations, the cost of genotyping was the highest due to a huge number of plants to be genotyped (Bassi et al., 2016). Intermediate and later-generation schemes were also assessed by the researchers and it was realized that practicing genomic selection in early generations like F2 to F4 was better than practicing it in late generations like F7 in terms of cost and genetic gain. To accomplish the stability between genomic selection and phenotypic selection in a breeding program it is suggested to perform one to two cycles of genomic selection and one cycle of phenotypic selection (Garewal et al., 2022). Several studies with various crops have indicated that genomic selection is progressively incorporated into advanced programs of crop breeding to attain fast expansion of genetic gain (Dang et al., 2023).

8.5.1 Models for Genomic Selection

In the case of estimating the association between independent variables and dependent variables, researchers depend basically on the statistical models as a fundamental tool (He & Wang, 2020). The major target of a statistical design is to confirm the accuracy, consistency, and genuineness of research outputs as well as decrease biases and errors. This is possible by decreasing the number of variables in a study. The principal perceptions of validity and consistency are considered as the conceptual foundations for the theoretical keystones of a statistical design (Reynolds, 2022). Establishing consistent data collection methods, utilizing authenticated and reliable gadget devices, reducing the effect of bias and misperceiving variables, and practicing appropriate statistical methods are highly recommended to the researchers to maintain the reliability and stability of the statistical design they use (K. Choi & Lee, 2018). The development of statistical design is a very crucial part of any research. It contains the recognition of the sample or population of the study, the design of research hypotheses, the formation of data collection approaches, and finally the assortment of suitable statistical analysis procedures (Khan et al., 2022). Selection techniques for variables like principal component analysis (PCA) or factor analysis may also be used by the researchers to conclude the most substantial variables. It is an imperative practice because having extraneous variables may result in overfitting in the statistical model (Sinha et al., 2023). The primary phase comprises a fundamental linear model in the case of genomic selection which is stated as OLS (ordinary least-squares regression). The formula $Y = 1\ n\mu + X\beta + \varepsilon$, represents this model where Y indicates the response variable, μ represents as overall mean, X denotes marker genotype, β effect vector of the marker and ε signifies the residual error. But, when the marker number (p) exceeds the observations number (n), particularly when thousands of genome-wide markers are exploited it creates a problem called over parameterization or the problem for p greater than n ($p \gg n$) (Budhlakoti et al., 2022). Ridge regression (RR) is a valuable method to solve the overparameterization problem in linear models. RR is a kind of penalized regression that presents a penalty term to the formula of least-squares regression. Situations with a large number of markers relative to the number of observations can efficiently be handled by this penalty and the estimates are stabilized by this (Saini et al., 2022). Every marker contributes uniformly to the variance in this model. However, this hypothesis is not applicable to all traits because different markers may have diverse effects on the trait variance. So prediction of variation among markers of the desirable traits is necessary based on the precise genetic construction (Sandhu et al., 2022). An additional method for marker-based models that suggests distinctive variations is the BLUP (best linear unbiased prediction) such as GBLUP (genomic BLUP), ssGBLUP (single-step GBLUP), RRBLUP (ridge regression BLUP) and rrGBLUP (GBLUP with linear ridge kernel regression) (Saini et al., 2022).

8.5.1.1 Multi-Environment Models

The precision of genomic selection is found in single environments, whereas most of the calculation models do not have the analytical power to perform selections throughout multiple environments or account for GE (genotype-by-environment) interaction (Costa-Neto et al., 2021). GE plays a key part in the variation of specific traits, such as grain yield. Phenotypic variation can be distributed into genetic effects and non-genetic effects. Heritability decreases when the phenotypic variation increases by GE without increasing genetic variation. While forming a prediction model, considering GE is important to optimize a breeding program such as using GBLUP (genomic best linear unbiased prediction) to have trials combination over several years and locations that at last intensify the accuracy of selection as well as genetic variance (Crossa et al., 2017). Before assimilating phenotypic data into the GS models, corrections are made to field and environmental data. Developing GS models for multi-environment was accounting for the consequences of ME (marker-environment) interaction. By modeling interactions between markers and environments GE can be demonstrated using ME in GBLUP (Lopez-Cruz et al., 2015). Additionally, using GBLUP (ME GBLUP) or variable selection methods ME can be modeled. However, there are limitations of the ME method on covariance patterns, making the model positive and homogenous, and for combined analysis of positively related environments this method is the best match.

8.5.1.2 Multi-Trait Model

For advancing lines, breeders select multiple traits in addition to GE. Genomic selection is used to predict single traits but the capacity to select multiple traits would be beneficial when attempting to assess and select genotypes based on combinations of yield components, quality of end-use, or disease traits. The complication of improving multiple traits arises due to the positive or negative correlation of the multiple traits (Montesinos-López et al., 2016). Genetic correlation between the lowly heritable traits would be advantageous which can augment the prediction accuracy through combined analysis of multiple traits. On the other side, the multiple traits may be genetically correlated with highly heritable traits which finally increases the genetic variance as well as the accuracy of selection similar to accounting for GE (Jiang et al., 2015). Unobserved traits and untested lines can also be predicted by multi-trait (MT) analysis. The multivariate trait models were found to perform the same as the single univariate models for polygenic traits (Merrick et al., 2022). When a highly heritable correlated trait is accessible, multivariate models can also considerably increase the accuracy of genomic selection for low heritable traits like grain yield. Indirect selection can be improved by multivariate trait models due to improved genetic correlation estimations (Montesinos-López et al., 2019). Higher accuracy of prediction is possible by the higher genetic correlation between traits that provides benefit to multivariate traits over single trait models (Michel et al., 2018). To select multiple traits, the prediction of selection indices using GS is another way. The creation of greater genetic gain may happen through the integration and weighs of multiple traits by a selection index compared to a selection of independent traits.

8.5.2 Genotyping in Genomic Selection

Implementation of genomic selection in a plant breeding program is an economical solution against the costly genotyping of the training population and breeding population in relation to the cost of pedigree selection. SNPs are the most plentiful type of marker among various molecular markers in a genome which is suitable for an extensive range of genomic studies (Krishnappa et al., 2021). For large numbers of SNPs developing HTG (high-throughput genotyping) would be time-consuming as well as expensive (He et al., 2014). The high-throughput next-generation sequencing (NGS) technologies developed by the necessity of generating cost-effective sequence data can provide huge sequence data that are economically viable. An effective genotyping method developed by the reduced cost, high accuracy, and higher throughput evaluation of NGS technologies is stated as the GBS (genotyping by sequencing) method (Figure 8.3) (Lipka et al., 2015). As a cost-effective

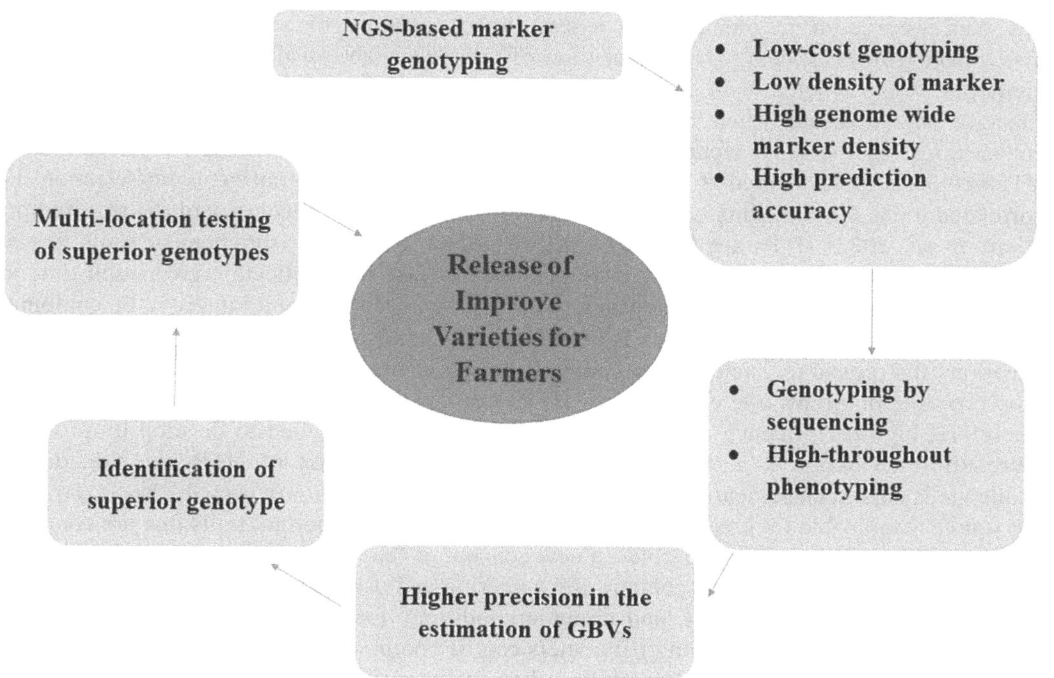

FIGURE 8.3 Genotyping in genomic selection for achieving higher genetic gain.

genotyping technique GBS has expanded acceptance in genomics-assisted breeding in recent years (He et al., 2014). To genotype large populations, it is one of the rapid as well as cheap tools that enable researchers to follow GS in larger pant breeding programs (Poland & Rife, 2012). Genotyping by sequencing can be practiced in crops where the genome sequencing has not been done because this technique doesn't need previous information on genome sequences for genotyping and discovery of SNP (Narum et al., 2013).

In segregating generation as well as non-segregating generation GBS can be applied as one of the genotyping tools for an expanded breeding population by enabling high-throughput NGS technologies. GS was found with comparatively better potential of intensifying the genetic gain with time as paralleled to phenotypic selection while the cost accompanying genotypic and phenotypic selection was the same (Bassi et al., 2016). The cost of genotyping per plant was decreased by a factor corresponding to the individual number combined in GBS proposing genotype attributing with genotyping by sequencing can be a cost-effective method for a big population in plant breeding (Technow & Gerke, 2017). Here the major concern for many scientists is a great quantity of absent data in GBS sets of data. For GS studies in crops, SNP chips are another implying genotyping tool because the problems of bias connected with SNP chips are solved at a great level with the application of wild relatives in a breeding program.

8.5.3 Phenotyping in Genomic Selection

Genomic selection modeling requires reliable phenotyping Apart from efficient genotyping, to attain high accuracy of prediction. There is a major constraining issue of precision phenotyping in the genetic assessment of quantitative traits. To obtain precision phenotypic data designing HTP approaches has recently turned into an effective part of research (Munjal et al., 2023). HTPs permit fast collection of phenotypic data with comparatively superior accuracy that is useful for genomic selection modeling. For plant height measurement, scientists used HTP and this data were used in modeling to assess the opportunity of substituting the traditional assessment with HTP data for the execution of GS (Watanabe et al., 2017). The accuracy of prediction was depleted with HTP data compared to traditional measurements. Significant capacity was detected when the errors of measurement were decreased from HTP with enhancement in the HTP technologies and experimental designs. Through correlated traits, HTPs could be used to predict complex traits, in exhibiting multi-trait genomic selection. For correlated traits, GS modeling can improve the genomic prediction accuracy of the required trait (Gaire et al., 2022) HTPs are more advantageous than manual measurement of traits that are sensitive environmentally like canopy temperature because data collection errors and time are reduced that were caused by the environment. Prediction accuracy could improve by combining data of HTP with genotyping which is highly advantageous in circumstances of harsh weather situations that constrain yield phenotyping (Rutkoski et al., 2016). For regular application in crop breeding programs, the expansion of HTP is dropped behind the genotyping technologies. Therefore, high-performance and cost-effective platforms are needed to develop by providing more effort (Shakoor et al., 2017). For the active combination of HTP into GS, another challenge is data management, data handling, and data processing (Araus et al., 2018). To detect the source responsible for creating variation in agronomically important traits that are connected to measurable environmental variables, a new concept of 'envirotyping' can be applied with the advances in HTP. Envirotyping contains the gathering and use of data on environmental factors like geographic conditions, soil, and climatic conditions for phenotypic prediction by multi-location assessments. Envirotyping parameters can be comprised in the linear mixed models for genomic prediction to augment heritability, thus the precisions of prediction for diverse traits (Van Eeuwijk et al., 2019).

8.6 IMPLEMENTATION OF GENOMIC SELECTION IN SPEED BREEDING

Genomic selection has been proven as one of the most effective methods to combine with speed breeding to accelerate the entire breeding program of a crop because of the cost-effectiveness of genotyping through high-density SNP (Single Nucleotide polymorphism) markers as well as EBVs that can precisely calculate the effect of marker on individual plant (Song et al., 2017). Both the breeding program of hybrid and inbred genomic selection can be practiced through accurate assumption of linkage equilibrium of the markers with the QTLs thus estimating the consequence of each marker in the entire set of markers to describe the total genetic variance of a crop (Robertsen et al., 2019). Both the genotyping and phenotyping data of a population called training population (TP) are used to estimate the effect of markers on plants and the maker data of the plants are combined to calculate GEBV (genomic estimated breeding value). On the other side, the formula of prediction is applied to the SC (selection candidate) population but phenotyping is not required here before the selection of next-generation potential genotypes or parents thus compensating for the time (Watson et al., 2019). The degree of achieving genetic gain depends on the accuracy of the genomic selection method which means higher genetic gain indicates the higher accuracy of the method. Phenotyping of the selection candidate population for the secondary trait like yield is possible and the result can be included in multivariate genomic selection models with grain yield of training population measured from the field to increase the predictability of the selection technique (Watson et al., 2019). This provides a way of combining selection candidate population development with phenotyping for the incorporation of genomic selection in a speed breeding program. Genomic selection is practiced in crops like rice, wheat, maize, and other important cereals to augment the genetic gain rate. When genomic selection is combined with a rapid generation advancement process like speed breeding, it influences greatly a crop improvement program (Jighly et al., 2019; Wolter et al., 2019).

8.6.1 Genomic Selection throughout the Speed Breeding Cycle

In basic breeding programs, the phenotypic selection method takes around 11 years, whereas genomic selection with speed breeding takes 4–9 years to complete a breeding cycle to achieve the desired traits as well as greater genetic gain for the traits most specifically traits that show low heritability (Schmidt et al., 2016). Genetic gain was found to become higher with each addition of round sped breeding whereas the accuracy of genomic selection declined with this addition. Utilization of combined modern breeding techniques exposes the capability of enhancing genetic gain for diverse crop species (Michel et al., 2017). Adoption of genomic selection and speed breeding can provide an opportunity to practice the selection of such traits where direct measurement of phenotype cannot take place (Figure 8.4).

One of the most important requirements for having higher accuracy in genomic selection throughout a speed breeding program is to get a high relationship between the training population and selection candidate population for the subsequent generations, which means either a number of parents are being used in crossing for the subsequent generations must be same or the progenies those are reused as parents must be the basis of the crossing plan (Michel et al., 2018).

8.6.2 Genomic Selection within the Breeding Cycle

Genomic selection can also be applied in the generations of a speed breeding cycle. In this process, lines are chosen as training populations from a similar breeding cycle for genomic selection such as estimating sister-lines with absent phenotypes. Such prediction can be enhanced by knowingly moderating phenotyping, overlooking environments, or assessing valuable traits in a portion of the offspring (Bassi et al., 2016). Additionally, the precision of line selection will improve by the combination of GEBV with phenotypes to persist in the breeding program, and with succeeding

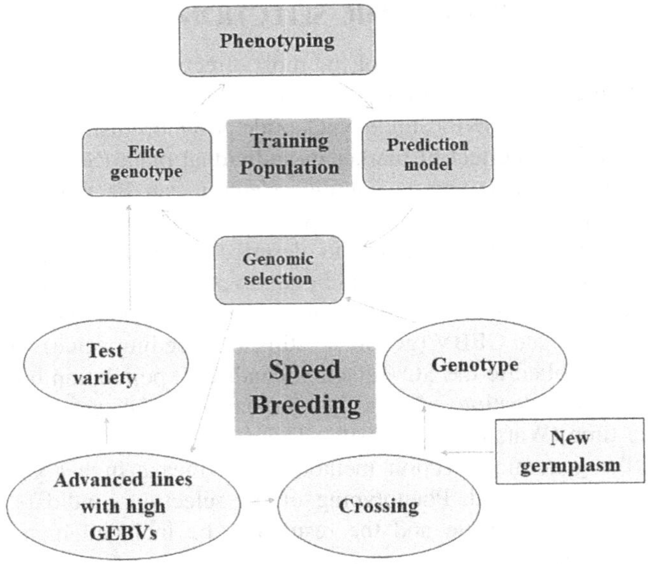

FIGURE 8.4 Genomic selection through the speed breeding cycle to reduce the time of variety release.

generations different phenotypic data will become accessible for the plants making GEBVs more precise, thereby extending supplementary in the selection procedure (Gaynor et al., 2017). A high relationship may be found between lines during practicing predictions within a generation because here multiple lines are verified from each family. An enhanced accuracy can be found when genomic selection is done within the generation of a speed breeding cycle most specifically in the early generations where the phenotypic copies are reduced for each line and data can be derived from both the full-sib and half-sib (Cericola et al., 2017).

8.6.3 Genomic Selection with Unknown Parents for Speed Breeding

Genomic selection can be done in another way like completely skipping the phenotyping for some portion of the breeding program and untested parent selection based on the GEBV value only. Using untested parents can considerably curtail the breeding cycle because of wide-ranging phenotypic analysis thus providing faster genetic gain per generation (Longin et al., 2015). Genomic selection with untested parents can revolutionize speed breeding programs. A boosted genetic gain was found when parents were selected based fully on GEBV for extremely heritable traits. A combination of such genomic selection with special generative techniques like speed breeding can decrease the generation time thus reducing the breeding cycle to less than two years (Watson et al., 2018). Speed breeding utilizing untested parents depends entirely on genomic selection with data from previous years.

8.7 CHALLENGES SPEED BREEDING AND GENOMIC SELECTION

Despite being an effective tool for speeding up the process of genetic gain, the main constraint of speed breeding is access to a controlled environment that is required for the rapid cycling of a plant species (Cazzola et al., 2020). The setup of speed breeding in having controlled environment is costly and additional facilities are needed for combining speed breeding with other in vitro techniques like embryo rescue or DH (Wanga et al., 2021). Regular use of speed breeding techniques is a challenge to the researchers of a country with resource deficiency. Limited seed yield may happen due to rigorous growth conditions in speed breeding where the phenotyping of seed traits may be affected by harvesting poor seed (González-Barrios et al., 2021). Execution of genomic selection in speed breeding is challenging when targeting traits present with genotype-by-environment interaction. The substitution effect of QTL alleles alters due to such interaction among the target environments of a breeding program. When there are modifications in positions of allelic effects that are concomitant with the alterations in substitution effects of environment, it creates the most challenging situation (Bustos-Korts et al., 2016). The presence of such interaction for different traits can influence genetic gain negatively in a plant breeding program (Araus et al., 2018).

8.8 CONCLUSION AND FUTURE PROSPECTIVE

To feed the continuously growing world population with nutritional food, crop researchers have been facing challenges to improve crop yield and nutrition against the changing climate for decades. However, modern plant breeding techniques are providing support to the breeders in their crop improvement. In a breeding program, a combination of genomic approach with rapid generation advancement process as well as rapid gene fixation can potentially provide a satisfactory way of increasing the rate of genetic gain. Speed breeding is such a method that is applied for accelerating the breeding of many valuable crops through attaining rapid homozygosity thus shortening the breeding cycle, whereas genomic selection can facilitate this breeding method with superior genotype selection as well as increased genetic gain (Varshney et al., 2021). Development of resistance and adaptation of crops through the identification and transfer of critical genes is possible by the combination of speed breeding with genomic selection. Although GS allows plant breeders genotypic selection earlier in the breeding cycle, selfing or DH techniques still be followed for line development of inbred crops to generate selection candidates until or still tested in multi-environment trials. However, the major problem accompanying DH technology is it is a costly process and basic traits cannot be selected by the breeders at the time of line development. So here speed breeding is achieving acceptance because the turnover of many plant generations is possible under glasshouse environments thus supporting the scientists to choose traits that perform high heritability. Therefore, a combination of speed breeding with other high throughput genotyping as well as phenotyping tools like GS can effectively be used for enhancing the adeptness of plant breeding programs for its potential applications in crop varietal development programs (Watson et al., 2018; Sharma et al., 2019; Ahmar et al., 2020).

REFERENCES

Ahmar, S., Gill, R., Jung, K., Faheem, A., Qasim, M., Mubeen, M. & Zhou, W. (2020). Conventional and molecular techniques from simple breeding to speed breeding in crop plants: Recent advances and future outlook. *International journal of Molecular Sciences*, *21*(7), 2590.

Annicchiarico, P., Nazzicari, N., Pecetti, L., Romani, M. & Russi, L. (2019). Pea genomic selection for Italian environments. *BMC Genomics*, *20*(1), 1–18.

Araus, J., Kefauver, S., Zaman-Allah, M., Olsen, M. & Cairns, J. (2018). Translating high-throughput phenotyping into genetic gain. *Trends in Plant Science*, *23*(5), 451–466.

Bassi, F., Bentley, A., Charmet, G., Ortiz, R. & Crossa, J. (2016). Breeding schemes for the implementation of genomic selection in wheat (Triticum spp.). *Plant Science*, *242*, 23–36.

Begna, T. (2022). Speed breeding to accelerate crop improvement. *International Journal of Agricultural Science and Food Technology*, *8*(2), 178–186.

Budhlakoti, N., Kushwaha, A., Rai, A., Chaturvedi, K., Kumar, A., Pradhan, A., Kumar, U., Kumar, R., Juliana, P., Mishra, D. & Kumar, S. (2022). Genomic selection: A tool for accelerating the efficiency of molecular breeding for development of climate-resilient crops. *Frontiers in Genetics*, *13*, 832153,

Bustos-Korts, D., Malosetti, M., Chapman, S. & van Eeuwijk, F. (2016). Modelling of genotype by environment interaction and prediction of complex traits across multiple environments as a synthesis of crop growth modelling, genetics and statistics. In Yin & Struik (eds) *Crop systems biology: Narrowing the gaps between crop modelling and genetics*, pp. 55–82.

Cazzola, F., Bermejo, C., Guindon, M. & Cointry, E. (2020). Speed breeding in pea (Pisum sativum L.), an efficient and simple system to accelerate breeding programs. *Euphytica*, *216*(11), 178.

Cericola, F., Jahoor, A., Orabi, J., Andersen, J., Janss, L. & Jensen, J. (2017). Optimizing training population size and genotyping strategy for genomic prediction using association study results and pedigree information. A case of study in advanced wheat breeding lines. *PloS One*, *12*(1), e0169606.

Chiurugwi, T. K. S., Powell, W. & Hickey, L. (2019). Speed breeding orphan crops. *Theoretical and Applied Genetics*, *132*, 607–616.

Collard, B., Beredo, J., Lenaerts, B., Mendoza, R., Santelices, R., Lopena, V., Verdeprado, H., Raghavan, C., Gregorio, G., Vial, L. & Demont, M. (2017). Revisiting rice breeding methods – evaluating the use of rapid generation advance (RGA) for routine rice breeding. *Plant Production Science*, *20*(4), 337–352.

Costa-Neto, G., Galli, G., Carvalho, H., Crossa, J. & Fritsche-Neto, R. (2021). EnvRtype: A software to interplay enviromics and quantitative genomics in agriculture. *G3: Genes, Genomes, and Genetics*, *11*(4), jkab040.

Croser, J., Pazos-Navarro, M., Bennett, R., Tschirren, S., Edwards, K., Erskine, W., Creasy, R. & Ribalta, F. (2016). Time to flowering of temperate pulses in vivo and generation turnover in vivo–in vitro of narrow-leaf lupin accelerated by low red to far-red ratio and high intensity in the far-red region. *Plant Cell, Tissue and Organ Culture (PCTOC)*, *127*, 591–599.

Crossa, J., Pérez-Rodríguez, P., Cuevas, J., Montesinos-López, O., Jarquín, D., De Los Campos, G., Burgueño, J., González-Camacho, J., Pérez-Elizalde, S., Beyene, Y. & Dreisigacker, S. (2017). Genomic selection in plant breeding: Methods, models, and perspectives. *Trends in Plant Science*, *22*(11), 961–975.

Cui, Y., Li, R., Li, G., Zhang, F., Zhu, T., Zhang, Q., Ali, J., Li, Z. & Xu, S. (2020). Hybrid breeding of rice via genomic selection. *Plant Biotechnology Journal*, *18*(1), 57–67.

Dang, D., Yuan, G., Hongjian, Z., Xuecai, Z., Ao, Z., Hui, W., Yanye, R. & Li, Q. (2023). Genome-wide association study and genomic prediction on plant architecture traits in sweet corn and waxy corn. *Plants*, *12*(2), 303.

Fernandes, S., Dias, K., Ferreira, D. & Brown, P. (2018). Efficiency of multi-trait, indirect, and trait-assisted genomic selection for improvement of biomass sorghum. *Theoretical and Applied Genetics*, *131*, 747–755.

Gaire, R., de Arruda, M., Mohammadi, M., Brown-Guedira, G., Kolb, F. & Rutkoski, J. (2022). Multi-trait genomic selection can increase selection accuracy for deoxynivalenol accumulation resulting from fusarium head blight in wheat. *The Plant Genome*, *15*(1), e20188.

Garewal, N., Joon, R., Kaur, R. & Singh, K. (2022). Genomic selection: Advances, applicability, and challenges. in *Genotyping by Sequencing for Crop Improvement*, 141–160.

Gaynor, R., Gorjanc, G., Bentley, A., Ober, E., Howell, P., Jackson, R., Mackay, I. & Hickey, J. (2017). A two-part strategy for using genomic selection to develop inbred lines. *Crop Science*, *57*(5), 2372–2386.

Ghosh, S., Watson, A., Gonzalez-Navarro, O., Ramirez-Gonzalez, R., Yanes, L., Mendoza-Suárez, M., Simmonds, J., Wells, R., Rayner, T., Green, P. & Hafeez, A. (2018). Speed breeding in growth chambers and glasshouses for crop breeding and model plant research. *Nature Protocols*, *13*(2), 2944–2963.

González-Barrios, P., Bhatta, M., Halley, M., Sandro, P. & Gutiérrez, L. (2021). Speed breeding and early panicle harvest accelerates oat (Avena sativa L.) breeding cycles. *Crop Science*, *61*(1), 320–330.

He, J., Zhao, X., Laroche, A., Lu, Z., Liu, H. & Li, Z. (2014). Genotyping-by-sequencing (GBS), an ultimate marker-assisted selection (MAS) tool to accelerate plant breeding. *Frontiers in Plant Science*, *5*, 484.

He, X. & Wang, J. (2020). Discovering model structure for partially linear models. *Annals of the Institute of Statistical Mathematics*, *72*, 45–63.

Heffner, E., Sorrells, M. & Jannink, J. (2009). Genomic selection for crop improvement. *Crop Science*, *49*(1), 1–12.

Heuschele, D., Case, A. & Smith, K. (2019). Evaluation of fast generation cycling in oat (Avena sativa). *Cereal Research Communications*, *47*(4), 626–635.

Hickey, L., Silvia, E., Silvia, A., Juan, E., Laura, A., Ryan, A., Greg, J., Jerome, D. & Mark, J. (2017). Speed breeding for multiple disease resistance in barley. *Euphytica*, *213*, 1–14.

Hickey, L., Hafeez, A., Robinson, H., Jackson, S., Leal-Bertioli, S., Tester, M., Gao, C., Godwin, I., Hayes, B. & Wulff, B. (2019). Breeding crops to feed 10 billion. *Nature Biotechnology*, *37*(7), 744–754.

Jiang, Y., Zhao, Y., Rodemann, B., Plieske, J., Kollers, S., Korzun, V., Ebmeyer, E., Argillier, O., Hinze, M., Ling, J. & Röder, M. (2015). Potential and limits to unravel the genetic architecture and predict the variation of Fusarium head blight resistance in European winter wheat (Triticum aestivum L.). *Heredity*, *114*(3), 318–326.

Jighly, A., Lin, Z., Pembleton, L., Cogan, N., Spangenberg, G., Hayes, B. & Daetwyler, H. (2019). Boosting genetic gain in allogamous crops via speed breeding and genomic selection. *Frontiers in Plant Science*, *10*, 1364.

Jones, T., Monaco, T., Larson, S., Hamerlynck, E. & Crain, J. (2022). Using genomic selection to develop performance-based restoration plant materials. *International Journal of Molecular Sciences*, *23*(5), 8275.

K. Choi, R. J. & Lee,S. (2018). Revisiting statistical design and analysis in scientific research. *Small*, *14*(40), 1802604.

Khan, A., Li, Q., Saqib, Z., Khan, N., Habib, T., Khalid, N., Majeed, M. & Tariq, A. (2022). MaxEnt modelling and impact of climate change on habitat suitability variations of economically important Chilgoza Pine (Pinus gerardiana Wall.) in South Asia. *Forests*, *13*(5), 715.

Krishnappa, G., Savadi, S., Tyagi, B., Singh, S., Mamrutha, H., Kumar, S., Mishra, C., Khan, H., Gangadhara, K., Uday, G. & Singh, G. (2021). Integrated genomic selection for rapid improvement of crops. *Genomics*, *113*(3), 1070–1086.

Laidig, F., Piepho, H., Rentel, D., Drobek, T., Meyer, U. & Huesken, A. (2017). Breeding progress, environmental variation and correlation of winter wheat yield and quality traits in German official variety trials and on-farm during 1983–2014. *Theoretical and Applied Genetics*, *130*, 223–245.

Lipka, A., Kandianis, C., Hudson, M., Yu, J., Drnevich, J., Bradbury, P. & Gore, M. (2015). From association to prediction: Statistical methods for the dissection and selection of complex traits in plants. *Current Opinion in Plant Biology*, *24*, 110–118.

Liu, G., Lin, Q., Jin, S. & Gao, C. (2022). The CRISPR-Cas toolbox and gene editing technologies. *Molecular Cell*, *82*(2), 333–347.

Longin, C., Mi, X. & Würschum, T. (2015). Genomic selection in wheat: Optimum allocation of test resources and comparison of breeding strategies for line and hybrid breeding. *Theoretical and Applied Genetics*, *128*, 1297–1306.

Lopez-Cruz, M., Crossa, J., Bonnett, D., Dreisigacker, S., Poland, J., Jannink, J.-L., Singh, R., Autrique, E. & de los Campos, G. (2015). Increased prediction accuracy in wheat breeding trials using a marker× environment interaction genomic selection model. *G3:Genes,Genomes,Genetics*, *5*(4), 569–582.

Meena, R. K., Deepak, K. K., Ganesh, K. K., Ram, K. M. & Annu (2022). Speed breeding: A powerful tool to accelerate crop research. *Biotica Research Today*, *4*, 382–384.

Merrick, L., Herr, A., Sandhu, K., Lozada, D. & Carter, A. (2022). Optimizing plant breeding programs for genomic selection. *Agronomy*, *12*(3), 714.

Michel, S., Ametz, C., Gungor, H., Akgöl, B., Epure, D., Grausgruber, H., Löschenberger, F. & Buerstmayr, H. (2017). Genomic assisted selection for enhancing line breeding: Merging genomic and phenotypic selection in winter wheat breeding programs with preliminary yield trials. *Theoretical and Applied Genetics*, *130*, 363–376.

Michel, S., Kummer, C., Gallee, M., Hellinger, J., Ametz, C., Akgöl, B., Epure, D., Löschenberger, F. & Buerstmayr, H. (2018). Improving the baking quality of bread wheat by genomic selection in early generations. *Theoretical and Applied Genetics*, *131*, 477–493.

Montesinos-López, O., Montesinos-López, A., Crossa, J., Toledo, F., Pérez-Hernández, O., Eskridge, K. & Rutkoski, J. (2016). A genomic Bayesian multi-trait and multi-environment model. *G3: Genes, Genomes, Genetics*, *6*(9), 2725–2744.

Montesinos-López, O., Montesinos-López, A., Crossa, J., Kismiantini, R.-A. J., Singh, R., Mondal, S. & Juliana, P. (2019). A singular value decomposition Bayesian multiple-trait and multiple-environment genomic model. *Heredity*, *122*(4), 381–401.

Munjal, R., Beniwal, J., Dhundwal, A., Goyal, A. & Kumari, A. (2023). Accelerating crop breeding in the 21st century: A comprehensive review of next generation phenotyping techniques and strategies. *Ekin Journal of Crop Breeding and Genetics*, *9*(2), 160–171.

Narum, S., Buerkle, C., Davey, J., Miller, M. & Hohenlohe, P. (2013). Genotyping-by-sequencing in ecological and conservation genomics. *Molecular Ecology*, *22*(1), 2841.

Pandey, S., Singh, A., Parida, S. & Prasad, M. (2022). Combining speed breeding with traditional and genomics-assisted breeding for crop improvement. *Plant Breeding*, *141*(3), 301–313.

Poland, J. & Rife, T. (2012). Genotyping-by-sequencing for plant breeding and genetics. *The Plant Genome*, *5*(3), 92–102.

Reynolds, P. (2022). Between two stools: Preclinical research, reproducibility, and statistical design of experiments. *BMC Research Notes*, *15*(1), 73.

Robertsen, C., Hjortshøj, R. & Janss, L. (2019). Genomic selection in cereal breeding. *Agronomy*, *9*(2), 95.

Rutkoski, J., Crain, J., Poland, J. & Sorrells, M. (2017). Genomic selection for small grain improvement. In Varshney et al. (eds) *Genomic selection for crop improvement: New molecular breeding strategies for crop improvement*. Springer International Publishing, pp. 99–130.

Rutkoski, J., Poland, J., Mondal, S., Autrique, E., Pérez, L. G., Crossa, J., Reynolds, M. & Singh R. (2016). Canopy temperature and vegetation indices from high-throughput phenotyping improve accuracy of pedigree and genomic selection for grain yield in wheat. *G3: Genes, Genomes, Genetics*, *6*(9), 2799–2808.

Saini, D., Chopra, Y., Singh, J., Sandhu, K., Kumar, A., Bazzer, S. & Srivastava, P. (2022). Comprehensive evaluation of mapping complex traits in wheat using genome-wide association studies. *Molecular Breeding*, *42*, 1–52.

Samantara, K., Bohra, A., Mohapatra, S., Prihatini, R., Asibe, F., Singh, L., Reyes, V., Tiwari, A., Maurya, A. & Croser, J. (2022). Breeding More Crops in Less Time: A Perspective on Speed Breeding. *Biology*, *11*, 275.

Samineni, S., Sen, M., Sajja, S. & Gaur, P. (2020). Rapid generation advance (RGA) in chickpea to produce up to seven generations per year and enable speed breeding. *The Crop Journal*, *8*(1), 164–169.

Sandhu, K., Shiv, A., Kaur, G., Meena, M., Raja, A., Vengavasi, K., Mall, A., Kumar, S., Singh, P. & Singh, J. H. G. (2022). Integrated approach in genomic selection to accelerate genetic gain in sugarcane. *Plants*, *11*(16), 2139.

Schmidt, M., Kollers, S., Maasberg-Prelle, A., Großer, J., Schinkel, B., Tomerius, A., Graner, A. & Korzun, V. (2016). Prediction of malting quality traits in barley based on genome-wide marker data to assess the potential of genomic selection. *Theoretical and Applied Genetics*, *129*, 203–213.

Shakoor, N., Lee, S. & Mockler, T. (2017). High throughput phenotyping to accelerate crop breeding and monitoring of diseases in the field. *Current Opinion in Plant Biology*, *38*, 184–192.

Sharma, A., Jones, J. & White, F. (2019). Recent advances in developing disease resistance in plants. *F1000Research*, *8*, 1934.

Shimelis, H., Gwata, E. & Laing, M. (2019). Crop improvement for agricultural transformation in Southern Africa. In Rechard et al. (eds) *Transforming agriculture in Southern Africa Agriculture*. Routledge: Tailor & Francis, pp. 97–103.

Shivakumar, M., Nataraj, V., Kumawat, G., Rajesh, V., Chandra, S., Gupta, S. & Bhatia, V. (2018). Speed breeding for Indian agriculture: A rapid method for development of new crop varieties. *Current Science*, *115*(7), 1241.

Sinha, D., Maurya, A., Abdi, G., Majeed, M., Agarwal, R., Mukherjee, R., Ganguly, S., Aziz, R., Bhatia, M., Majgaonkar, A. & Seal, S. (2023). Integrated genomic selection for accelerating breeding programs of climate-smart cereals. *Genes*, *14*(7), 1484.

Slama-Ayed, O., Bouhaouel, I., Ayed, S., De Buyser, J., Picard, E. & Amara, H. (2019). Efficiency of three haplomethods in durum wheat (Triticum turgidum subsp. durum Desf.): Isolated microspore culture, gynogenesis and wheatx maize crosses. *Czech Journal of Genetics and Plant Breeding*, *55*(3), 101–109.

Song, J., Carver, B., Powers, C., Yan, L., Klápště, J., El-Kassaby, Y. & Chen, C. (2017). Practical application of genomic selection in a doubled-haploid winter wheat breeding program. *Molecular Breeding*, *37*, 1–15.

Technow, F. & Gerke, J. (2017). Parent-progeny imputation from pooled samples for cost-efficient genotyping in plant breeding. *PLoS One*, *12*(12), e0190271.

Van Eeuwijk, F., Bustos-Korts, D., Millet, E., Boer, M., Kruijer, W., Thompson, A., Malosetti, M., Iwata, H., Quiroz, R., Kuppe, C. & Muller, O. (2019). Modelling strategies for assessing and increasing the effectiveness of new phenotyping techniques in plant breeding. *Plant Science*, *282*, 23–39.

Varshney, R., Bohra, A., Roorkiwal, M., Barmukh, R., Cowling, W., Chitikineni, A., Lam, H., Hickey, L., Croser, J., Bayer, P. & Edwards, D. (2021). Fast-forward breeding for a food-secure world. *Trends in Genetics*, *37*(12), 1124–1136.

Venezia, M. & Creasey Krainer, K. (2021). Current advancements and limitations of gene editing in orphan crops. *Frontiers in Plant Science*, *12*, 742932.

Voss-Fels, K., Herzog, E., Dreisigacker, S., Sukumaran, S., Watson, A., Frisch, M., Hayes, B. & Hickey, L. (2019). "SpeedGS" to accelerate genetic gain in spring wheat. In Miedaner et al. (eds) *Applications of genetic and genomic research in cereals*. Cambridge: Woodhead Publishing, pp. 303–327.

Wanga, M., Shimelis, H., Mashilo, J. & Laing, M. (2021). Opportunities and challenges of speed breeding: A review. *Plant Breeding*, *140*(2), 185–194.

Watanabe, K., Guo, W., Arai, K., Takanashi, H., Kajiya-Kanegae, H., Kobayashi, M., Yano, K., Tokunaga, T., Fujiwara, T., Tsutsumi, N. & Iwata, H. (2017). High-throughput phenotyping of sorghum plant height using an unmanned aerial vehicle and its application to genomic prediction modeling. *Frontiers in Plant Science*, *8*, 421.

Watson, A., Ghosh, S., Williams, M., Cuddy, W., Simmonds, J., Rey, M., Asyraf Md Hatta, M., Hinchliffe, A., Steed, A., Reynolds, D. & Adamski, N. (2018). Speed breeding is a powerful tool to accelerate crop research and breeding. *Nature Plants*, *4*(1), 23–29.

Watson, A., Hickey, L., Christopher, J., Rutkoski, J., Poland, J. & Hayes, B. (2019). Multivariate genomic selection and potential of rapid indirect selection with speed breeding in spring wheat. *Crop Science*, *59*(5), 1945–1959.

Wolf, B. & Blankenship, R. (2019). Far-red light acclimation in diverse oxygenic photosynthetic organisms. *Photosynthesis Research*, *142*(3), 349–359.

Wolter, F., Schindele, P. & Puchta, H. (2019). Plant breeding at the speed of light: The power of CRISPR/Cas to generate directed genetic diversity at multiple sites. *BMC Plant Biology*, *19*(1), 1–8.

9 Network of Low-Cost RGB-Depth Cameras and Mini-Computers for High-Throughput Monitoring of Plant Stress Response

Cordier Mathis
Laboratoire Angevin de Recherche en Ingénierie des Systèmes (LARIS), UMR INRAe-IRHS, Université d'Angers, 62 Avenue Notre Dame du Lac Angers, France

R&D Artificial Vision and automation, Vilmorin-Mikado, Rue du Manoir La Ménitré, France

Rasti Pejman
Laboratoire Angevin de Recherche en Ingénierie des Systèmes (LARIS), UMR INRAe-IRHS, Université d'Angers, 62 Avenue Notre Dame du Lac Angers, France

Centre d'Études et de Recherche pour l'Aide à la Décision (CERADE), ESAIP, 18 Rue du 8 Mai 1945 Saint-Barthélemy-d'Anjou, France

Torres Cindy
R&D Artificial Vision and automation, Vilmorin-Mikado, Rue du Manoir La Ménitré, France

Rousseau David
Laboratoire Angevin de Recherche en Ingénierie des Systèmes (LARIS), UMR INRAe-IRHS, Université d'Angers, 62 Avenue Notre Dame du Lac Angers, France

9.1 INTRODUCTION

The study of plant stress response is of particular interest for plant breeding. Tests are carried out to characterize the evolution of each variety under biotic or abiotic stresses. Current test protocols are based on visual diagnosis by an expert after a few days of growth. Monitoring their evolution over time offers a better understanding of plant responses. This is not possible for a human expert: automation of this task becomes necessary. To this end, a low-cost imaging system has been developed. It can monitor the evolution of a large number of plants at high throughput. Several modalities can also be acquired simultaneously such as color, infrared, and scene depth.

DOI: 10.1201/b23372-9

RGB-Depth cameras provide registered color and distance images (Shaikh & Chai, 2021; Li et al., 2022). The distance images correspond to the distance between the camera and the 3D scene in front of the camera. These sensors are widely used in robotics where they constitute basic devices accessible at very low cost (Shaikh & Chai, 2021). Plants are specifically well-suited to benefit from the information provided by RGB-Depth (Chéné et al., 2012). Indeed, while the contrast between leaves is often very poor in RGB, it appears stronger in Depth as leaves are mostly located at different heights along the stems of a plant (Mohammed Amean et al., 2021). Complementarily, color and infrared contrasts will appear on individual leaves when biotic or abiotic stress occurs. Since their introduction in 2012 (Chéné et al., 2012), plant phenotyping RGB-Depth sensors have therefore been established as a standard in plant imaging (Li et al., 2014, 2020). In the same trend of affordable phenotyping, mini-computers were introduced in plant phenotyping as a solution to perform time-lapse of growing plants with compact devices that can be deployed directly in the standard growing environment for plant phenotyping (Minervini et al., 2017). These mini-computers can even be replicated and connected to constitute a network of low-cost RGB-Depth cameras to monitor in parallel, and thus at a high-throughput, large population of plants. Such systems are efficient for various plant phenotyping questions (Samiei et al., 2020; Garbouge et al., 2021; Cordier et al., 2023a). These studies focus in particular on the biological questions raised and describe image processing methods to address these problems. As a companion material to (Samiei et al., 2020; Garbouge et al., 2021; Cordier et al., 2023a), we detail in this chapter how to settle a network of RGB-Depth cameras for high-throughput plant phenotyping, represented in Figure 9.1.

Both hardware architecture and software interactions between the various components are described. In the first section, the imaging system is introduced. The individual management of each sensor is explained, and then their networking is described. Adaptations of the imaging system to the growing environment are also presented. Second, data management is detailed, from acquisition to secure image storage. Then, error management is presented, with automatic and autonomous processing, and human intervention in the event of more severe problems. The last section details the plant analysis aspect and shows the interest in temporal and multimodal information on a wide variety of experiments.

9.2 SYSTEM ARCHITECTURE

The imaging system is detailed in this section, first individually at the level of each sensor/mini-computer pair, then globally at the level of camera networking. Finally, hardware adaptations to plant growth environments are presented.

9.2.1 Mini-Computer and Sensor Interactions

In this section, we focus on the interaction between one sensor and its master mini-computer. To acquire both color and depth information, Intel RealSense D435 cameras are used. They are low-cost sensors with output resolution up to 1280 × 720 on depth maps. The ideal depth range is 30 cm to 3 m which is particularly suitable for monitoring young plants from the top view. Color and depth frames can be natively aligned by the camera during the acquisition. The technology used by RealSense D435 is active infrared (IR) stereo vision which requires one IR projector and two IR sensors. Depth maps are then obtained by computing the time of flight of IR reflections.

As shown in Figure 9.2, each sensor is controlled by one low-cost mini-computer. As the mini-computers used are Raspberry Pi, the sensors must be able to run on Linux distributions, in particular on Raspberry Pi OS here. Intel RealSense cameras are all provided with the Intel RealSense Software Development Kit (SDK) which is cross-platform. For Linux distributions, *librealsense* provides all the requirements. The *pyrealsense2* official package enables control of sensors with Python3. A program was developed to acquire aligned color, depth, and infrared images.

FIGURE 9.1 Overview of the imaging system.

The acquisition is automated to monitor plant development under the cameras. In this case, the high frame rate of videos is not really suitable related to the natural plant growth speed. Here, the approach is to acquire images at a defined frame rate which can be configured according to the needs of the experiment. In practice, a period of 10 or 15 minutes between frames is used. This frame rate remains very highly related to plant growth speed and provides access to high-frequency

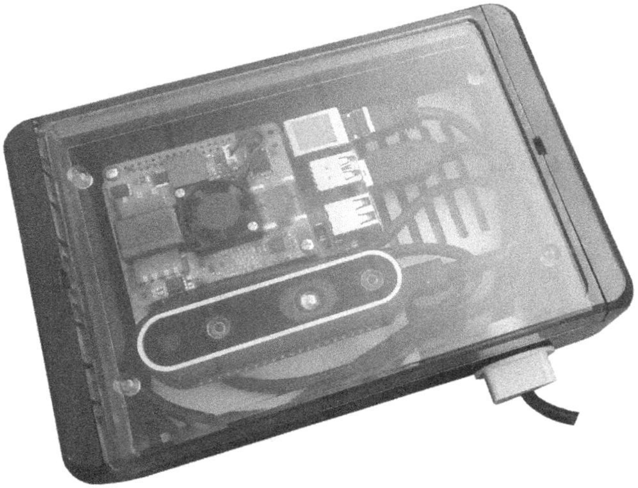

FIGURE 9.2 Camera case in top view.

growth patterns such as circadian cycles (McClung, 2006). It was then chosen as a compromise between temporal precision and storage volume.

To automate acquisitions, the mini-computer uses scheduled tasks. This functionality is offered by all operating systems and enables tasks to be triggered according to the value of the system clock. The functionality is called *crontab* under Linux. To add a scheduled task, use the *crontab -e* command to edit files in the */var/spool/cron/crontabs* directory. Each line corresponds to a scheduled task. For a task executed at regular intervals in minutes, the syntax should follow the following pattern:

**/[period (minutes)]   *   *   *   * [command]*

So, for an acquisition triggered every quarter of an hour with the bash program *take_image_rs.sh* by user *cam_driver*:

**/15   *   *   *   *   sh /home/cam_driver/Desktop/take_image_rs.sh*

The command to start acquisition is given by the bash script *take_image_rs.sh*. This defines the working directory and launches the main Python script if the configuration indicates that an experiment is currently running. This information is provided by the content of the *experiment_running_status* file. All subsequent actions are programmed in Python. Several scripts are required to separate the action of each program and its role. The main code then uses these programs as local Python modules. The file tree is then as follows:

```
* /home/cam_driver/Desktop/
    + take_image_rs.sh
    + Image_acquisition/
        > main.py
        > acquisition_management.py
        > alert_management.py
        > reset_usb_port_cam.py
        > last_acquisition_status
        > experiment_running_status
```

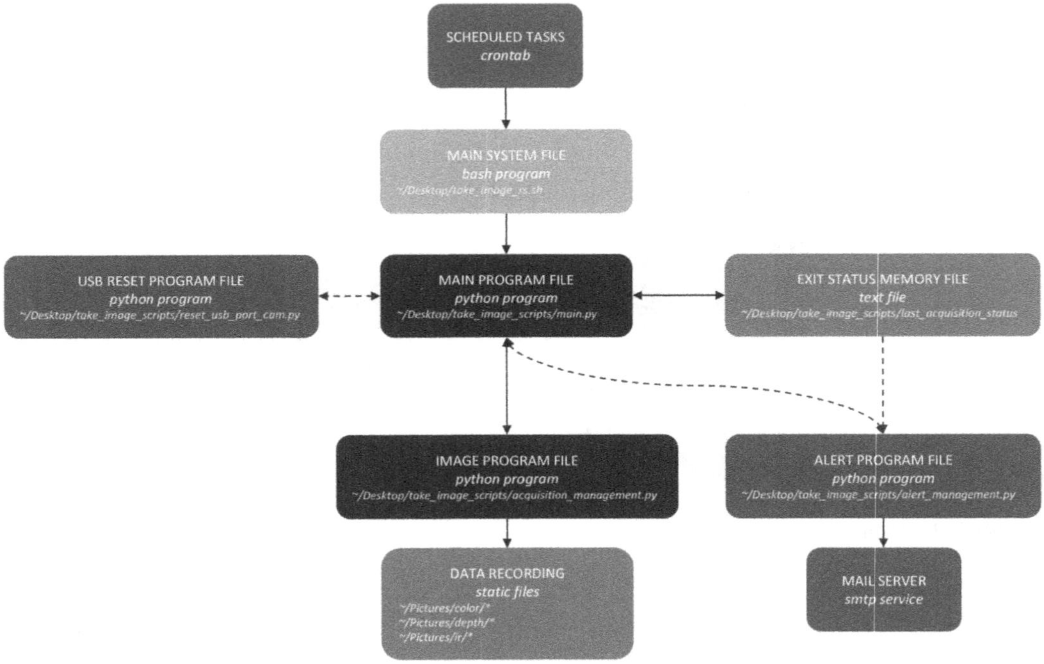

FIGURE 9.3 Software architecture diagram for robust RGB-Depth-Infrared acquisition.

The logical links between these different programs are shown in Figure 9.3. All the programs shown in the file tree above are supplied and available at https://github.com/Sithamfr/AcquisitionRS435.

The role of the *main.py* program is to manage interactions between all the different modules, order commands, and handle errors. The purpose of the *acquisition_management.py* program is to acquire a simultaneous image on each color, depth, and infrared imaging modality and save it. If an error occurs, the *reset_usb_port_cam.py* script is launched to restart the USB port to which the camera is connected. The acquisition is then restarted, and if an error persists, the alert program is run and the mini-computer is restarted. This alert program, which sends error notifications to users, is explained in detail in the Troubleshooting>Automatic control processes section.

Due to the physics of the sensors used, depth images can contain missing data. This problem arises mainly in the vicinity of tall objects, due to abrupt changes in height. The high object blocks the return of the IR waves to the sensor, so the time of flight cannot be calculated. The consequence is that the value on the pixel concerned is unknown, which corresponds to a lack of insight into the height for that particular location. The amount of missing values is also very dependent on the lighting conditions, e.g., natural or artificial lighting and lighting technology. Noise is also noticeable on flat surfaces. To counter these problems, we record two types of data from the depth modality: depth and smooth depth. Depth corresponds to the raw values obtained over one frame. Smooth depth is obtained by smoothing the values obtained over four consecutive frames at 30 frames per second. This smoothing reduces both noise and the number of missing data. However, it requires that there is no movement of the plants under the cameras during the short period of time required for the acquisition.

To prevent any storage issues, images are first saved locally into the mini-computer SD card. They must be saved so that each acquisition can be identified. To achieve this, we use the time/modality pair as an identifier, since it is unique. Each modality is separated into different folders and the time value is indicated in the file name, as shown below:

```
* /home/cam_driver/Pictures/
    + color/
        > color_[YEAR]-[MONTH]-[DAY]_[HOUR]_[MINUTE].png
    + depth/
        > depth_[YEAR]-[MONTH]-[DAY]_[HOUR]_[MINUTE].png
    + sdepth/
        > depth_[YEAR]-[MONTH]-[DAY]_[HOUR]_[MINUTE].png
    + ir/
        > ir_[YEAR]-[MONTH]-[DAY]_[HOUR]_[MINUTE].png
    + last_depth_colorized.png
    + last_sdepth_colorized.png
```

The *sdepth* folder corresponds to the smooth depth images. As depth images are saved as 16-bit images, their display is not native. We then save a temporary file displaying a clear colorized preview of the depth image at each acquisition in the *last_depth_colorized.png* and *last_sdepth_colorized.png* files for depth and smooth depth, respectively.

9.2.2 Global Network Structure

The imaging system is based on networking the mini-computers that control the sensors. Each camera in the system is connected to a single mini-computer. For an N-camera imaging system, N+1 mini-computers are required, since a local server is used. This server enables data to be stored locally on a hard disk, before being backed up remotely on a NAS. Usernames are set to *cam_driver* for the camera mini-computers and *system_driver* for the server.

Two different network architectures can be used. The choice of architecture depends on how the user wishes to integrate the local area network (LAN) into the overall corporate or institutional network, known as the intranet. The first solution, shown in Figure 9.2a, involves integrating all mini-computers into the global network via the switch. Each mini-computer is then directly connected to the intranet and has its IP address inside this network. To protect the system from possible intrusions, methods exist for isolating all its components within a virtual LAN (VLAN). In this way, although the mini-computers are physically connected to the global network, they are logically separated from it. The advantage of this setup is direct access to all LAN components from a workstation connected to the intranet. What's more, this ensures that the mini-computers remain synchronized via the network time protocol (NTP) of the global network (Figures 9.4a and 9.4b).

Another functional architecture consists of using the LAN server as a bridge between the local network and the intranet. In this case, access to a system component requires prior access to the local server. This is a way of isolating the LAN, so as not to monopolize an IP range on the intranet for all system components. However, this configuration implies additional constraints: Internet access must be shared by the LAN server. NTP must also be configured to ensure synchronization of the mini-computers.

Each mini-computer connected to a camera has an ID from 1 to N. Based on this ID, hostnames are set to [prefix]-[facility]-[room]-cam-[camera_ID]. To achieve the networking, each mini-computer is then configured to have a unique IP address. This avoids hostname addressing, which can be a source of connection errors. For a B+LAN architecture, local IP addresses are set to 192.168.0.[200+ID] for mini-computers connected to a camera and 192.168.0.101 for the local server. For a VLAN architecture, addresses are custom-set according to the network address range and the ID of each internal component. This configuration can be carried out either at the router level, by specifying a static IP address corresponding to the MAC address of the mini-computer ethernet port, or directly in the mini-computer system file */etc/dhcpcd.conf*. In this case, the syntax should follow the following template:

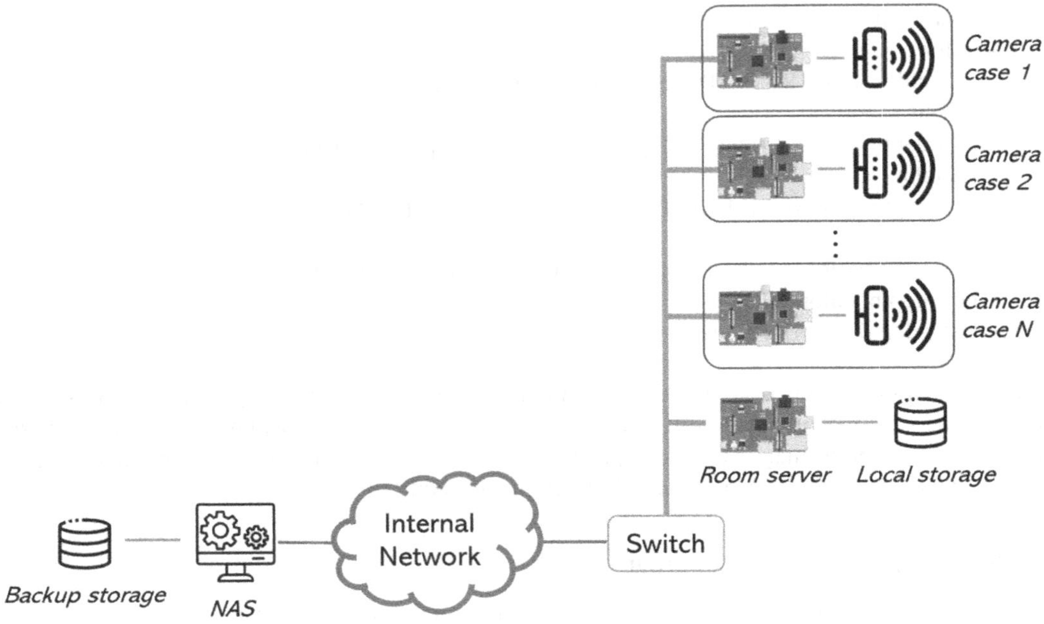

FIGURE 9.4a Hardware architecture for VLAN version.

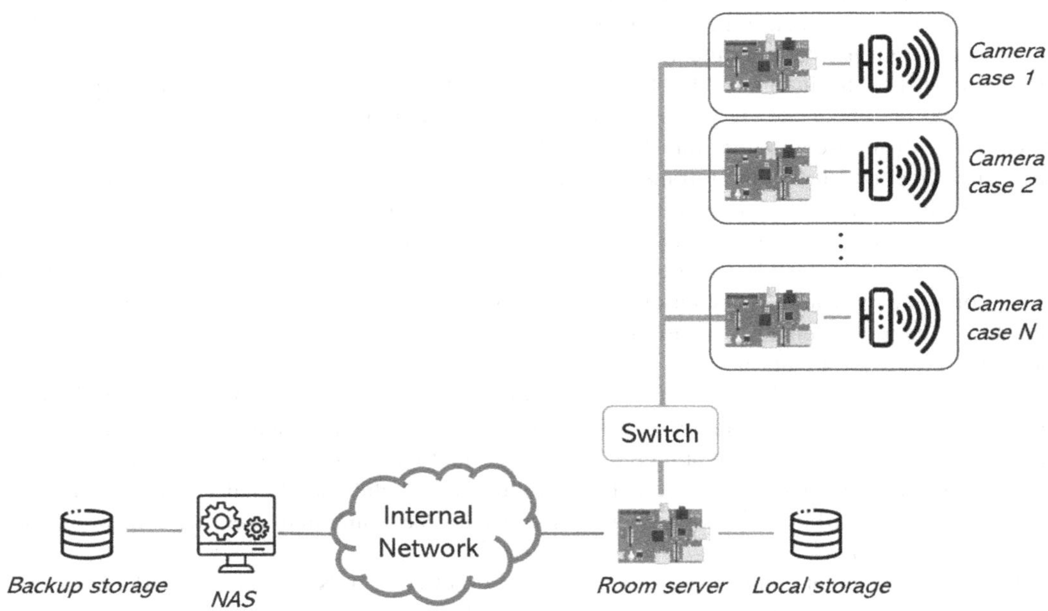

FIGURE 9.4b Hardware architecture for Bridge+LAN version.

interface [interface_name]
static ip_address=[ip_address]/24
static_routers=[router_ip_address]
static_domain_name_servers=[router_ip_address for B+LAN configuration, custom addresses for VLAN and institutional configuration]

9.2.3 Environmental Adaptations

These systems are designed to image plants in an environment with specific constraints. Temperature is a major constraint in greenhouses. Humidity is also a key factor since the system must be able to be deployed whatever the irrigation system such as a sprinkler system or watering ramp. Every piece of equipment must therefore be able to cope with high temperatures and humidity levels. To achieve this, each camera-computer pair is housed in a waterproof case, presented in Figure 9.5, which protects them from dust and humidity.

In phytopathology, this case also blocks the entry of pathogens and enables simple disinfection. Also, transparent plastic sheetings are sometimes placed above the plants to maintain constant humidity. In this case, the camera ramp of the system has to be positioned below it, as shown in Figure 9.6.

Cable routing between the inside and outside of the growth chamber is ensured by a cable gland. This is facilitated by the fact that a single cable provides both the power supply and the network connection. This functionality is called power over ethernet (PoE). The cable used is an RJ45 ethernet cable connecting the mini-computer to the room switch. To take advantage of this feature, you need to make sure that the switch is compatible and that a PoE hat is connected to the mini-computer. In the case of mini-computers without PoE hats, it is possible to add simple active ventilation to the processor via other modules.

As there is no contact between the inside and outside of the case, cooling inside the case becomes a challenge. The mini-computers used Raspberry Pi 4, are also known to run very hot. A minimum internal air volume must be maintained to avoid overheating. In addition, PoE hat fans are configured to circulate air continuously. This is done by declaring custom CPU warm thresholds for the 4 operating modes from *poe_fan_temp0* to *poe_fan_temp3* in the */boot/config.txt* system file. The default PoE hat fan configuration is set as below:

FIGURE 9.5 Overview of the imaging system in a humid environment.

FIGURE 9.6 Overview of the imaging system used under a transparent plastic sheeting.

```
dtoverlay=rpi-poe
dtparam=poe_fan_temp0=40000,poe_fan_temp0_hyst=2000
dtparam=poe_fan_temp1=45000,poe_fan_temp1_hyst=2000
dtparam=poe_fan_temp2=50000,poe_fan_temp2_hyst=2000
dtparam=poe_fan_temp3=55000,poe_fan_temp3_hyst=5000.
```

poe_fan_temp indicates the threshold temperature for mode change and *poe_fan_temp_hyst* indicates the temperature delta to downgrade mode. Temperatures are indicated in millicelsius. An example of a configuration that increases fan speed compared to the default configuration is provided below:

```
dtoverlay=rpi-poe
dtparam=poe_fan_temp0=10000,poe_fan_temp0_hyst=1000
dtparam=poe_fan_temp1=30000,poe_fan_temp1_hyst=5000
dtparam=poe_fan_temp2=40000,poe_fan_temp2_hyst=5000
dtparam=poe_fan_temp3=50000,poe_fan_temp3_hyst=5000.
```

Tests carried out under extreme heat and humidity conditions have proven the resistance of this hardware, with CPU temperatures of the mini-computers not exceeding 65°C. A significant margin is thus maintained to avoid reaching 80°C, which would send the mini-computer into safety mode, implying a reduction in performance.

9.3 DATA MANAGEMENT

In this section, the data flow is detailed, from data acquisition to secure storage. This can be split into two stages: the first consists of the flow from acquisition to local storage, and the second of the flow from local to remote, secure storage.

9.3.1 From Acquisition to Local Storage

As explained above, during acquisition, images are saved locally on the mini-computer. The first backup step consists of transferring these images to the local server since the local data backup is made on the server hard disk. This step also offloads the limited storage capacity of the mini-computer. The transfer is carried out every hour via scheduled tasks on the server using the *get_images_from_cam.sh* script.

```
[minute]    *    *    *    *    sh
/home/system_driver/Desktop/get_images_from_cam.sh [camera_ID]
```

For a 15-minute acquisition period, each transfer therefore corresponds to 12 images, 4 per imaging modality. To spread out the server task of importing images, the hourly transfer from each camera mini-computer takes place at a different time.

As we explained earlier, each image is identified within a camera mini-computer by the time/modality pair. During transfer to the local server, the camera identification information must be added so that the identifier remains unique. To guarantee the uniqueness of file names, the file tree on the hard disk is therefore:

```
* [hard_drive_path]/data/
    + [camera_ID]/
        + color/
            > color_[YEAR]-[MONTH]-[DAY]_[HOUR]_[MINUTE].png
        + depth/
            > depth_[YEAR]-[MONTH]-[DAY]_[HOUR]_[MINUTE].png
        + sdepth/
            > sdepth_[YEAR]-[MONTH]-[DAY]_[HOUR]_[MINUTE].png
        + ir/
            > ir_[YEAR]-[MONTH]-[DAY]_[HOUR]_[MINUTE].png
        + last_depth_colorized.png
        + last_sdepth_colorized.png
```

9.3.2 Remote Backup

Local data storage is very important, as it minimizes the problems of data transfer and loss. As the local storage environment is harsh, the risk of failure is increased. More durable storage and redundancy are needed to secure data storage. To this end, a copy of the data is made daily on a company or institutional NAS via the following scheduled tasks:

[minute] [hour] * * * [backup_command]

So, for an acquisition triggered every day at 1:50 with the bash program *make_backup_DD.sh*:

*50 150 * 50 * 50 * 50 sudo sh /home/system_driver/Desktop/make_backup_DD.sh*

Another advantage is improved data access. Several systems can be deployed at the same time, and all data produced is centralized. This central point, the NAS, must be chosen for easy access so that data can be quickly retrieved for processing.

The backup program is a two-stage bash process:

Mounting a remote folder via the *mount* command and CIFS protocol specifying encrypted credentials recognized by the institution.

sudo mount -t cifs -o username=[INSTITUTIONAL_USERNAME],password=[INSTITUTIONAL_PASSWORD] "[NAS_address]/[backup_path]" "/home/system_driver/BACKUP_folder"

Synchronizing the data with *rsync* and safe options: *rsync -av /[hard_drive_path]/data /home/system_driver/BACKUP_folder.* If the first step produces an error, the second is not launched. Obviously, the backup must not be run if the remote storage point is not mounted. This can lead to a very large amount of data being sent to the server's SD card, which can then be blocked.

9.4 TROUBLESHOOTING

Error handling is a very important part. Whatever the system structure, errors can occur during data acquisition or transfer. The system itself needs to be able to identify and handle errors, in order to guarantee its autonomy. This limits the need for user intervention.

9.4.1 Automatic Control Processes

A first way of acting on system autonomy is to implement solutions for generating electricity in the event of a power cut. Deploying systems in environments such as greenhouses can sometimes be made difficult due to electromagnetic noise. Micropower cuts can occur frequently. In addition to damaging the electronic components of systems, these outages can cause data loss or prevent data acquisition. An uninterruptible power supply (UPS) solution is essential in this context. In the event of an outage, the system is then powered by the UPS battery. In addition, this tool guarantees the quality of the alternating current sinusoid, providing maximum protection for electronic equipment.

At the software architecture level, many methods are used to detect anomalies, process them, and inform users. Many processes have been developed to maximize the system's self-sufficiency. Most error management processes take place at the image acquisition stage. Sometimes, the camera may produce an error during acquisition, preventing images from being captured. In this case, a simple means of restarting the sensor is used. This involves first detecting the USB 3 port to which the

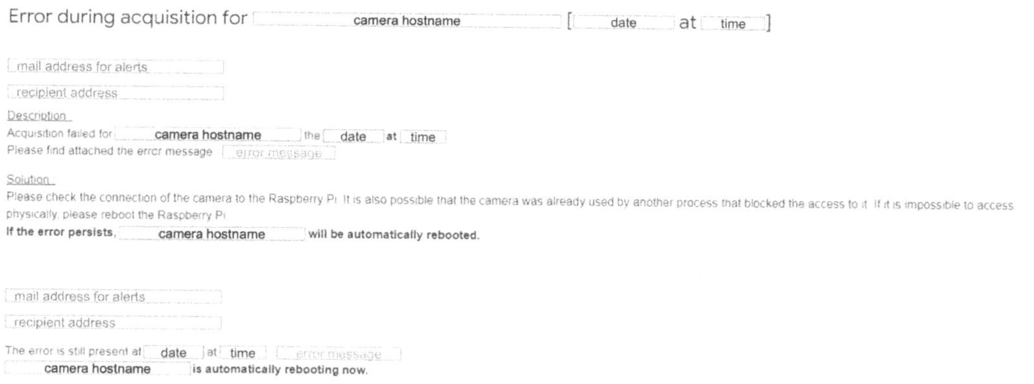

FIGURE 9.7 Mail template for camera alert triggered when acquisition failed.

camera is connected and then rebooting it. This task is performed by the *reset_usb_port_cam.py* script. Once this operation has been completed, the *main.py* file restarts image acquisition for each modality. If the error persists, the program takes the following three steps:

Send an alert to system users. This is done by the *alert_management.py* script. The protocol used is to send an alert email via simple mail transfer protocol (SMTP), specifying as much information as possible about the failure: date and time, name of the mini-computer, and error message from the acquisition. A mail template is shown in Figure 9.7. All users are thus informed in real time of an acquisition failure.

The *last_acquisition_status* file is overwritten with the output status indicating failure in the first line and the date and time afterward. This acts as a temporary log file and enables better monitoring of exceptions, particularly for sending alerts. In fact, this file memorizes whether the error was already present during the previous acquisition. In this case, an alert is still sent, but it is put in response to the first one to create a mail thread specifically linked to this error, as shown in Figure 9.7. As soon as an acquisition is successful, this thread will no longer be used, as the acquisition success status will be indicated in the *last_acquisition_status* file.

Finally, the mini-computer is restarted to fix the error if it originates from the mini-computer itself.

Errors can also occur in the connectivity of mini-computers within the local network. The need for connectivity arises when the server tries to retrieve images from a mini-computer in order to store them on the local hard disk. In the event of a connection failure, an alert is sent to users, also via SMTP.

Finally, all mini-computers are configured to undergo a daily reboot. This is done using a scheduled task with the following template:

*[minute] [hour] * * * sudo reboot*

The schedule must be carefully chosen so as not to conflict with an acquisition or a file transfer. A delay must be respected to allow the mini-computer time to become available again following the restart. For example, with acquisition every quarter of an hour, the following pattern

*50 0 * * * sudo reboot*

leaves 10 minutes between restart and the next image capture.

9.4.2 User Intervention

In the vast majority of cases, the error correction mechanisms work and no external intervention is required. In the case of more serious faults, however, the number of malfunction alerts sent to the user will multiply. It is therefore up to the user to intervene on the mini-computer indicated. Depending on the problem, this can be done remotely via several different protocols:

Secure Socket Shell (SSH) on the command line.
Virtual Networking Computing (VNC) to remotely take control of the graphical interface.
Remote Desktop Protocol (RDP) on Microsoft Windows with an xRDP configuration on the mini-computers to remotely take control of the graphical interface with native Windows software.

These protocols can be used at any time, for instance, to check that a camera is working properly. In general, mini-computers are available and accessible remotely at all times. A system error can break this functionality. Finally, it is sometimes necessary to have a direct view of the system, and this requires the physical presence of the user.

9.5 MODALITIES AND PLANT STRESS RESPONSE

The imaging system described in this chapter offers high-throughput temporal monitoring of color, depth, and infrared imaging modalities. These three imaging modalities were chosen for their complementarity in analyzing a wide range of plant stresses. Plant stresses can be divided into two categories: biotic and abiotic. Biotic stresses are the result of the harmful action of living organisms on plants. The main sources of biotic stress are parasites and pathogens such as bacteria, viruses, and fungi. Abiotic stresses are due to unfavorable growing conditions such as drought or excess water, extreme temperatures, salt stress, or mineral deficiencies. Drought, heat, and salt stress are the main abiotic stresses suffered by crops as a result of climate change. They are therefore the most studied abiotic stresses. A synthetic view of the stresses accessible by each of these imaging modalities is provided in Table 9.1.

TABLE 9.1
Reviews of Modalities Associated with Stress Detection in the Literature

		Imaging modality		
		Color	**Depth**	**Infrared**
Stress category	Biotic	(Humplík et al., 2015; Lowe et al., 2017; Gao et al., 2020; Zubler & Yoon, 2020; Galieni et al., 2021; Singh et al., 2021; Gill et al., 2022; Zahir et al., 2022; Ye et al., 2023; Al-Tamimi et al.)	(Galieni et al., 2021; Gill et al., 2022; Ye et al., 2023; Al-Tamimi et al.)	(Humplík et al., 2015; Lowe et al., 2017; Gao et al., 2020; Zubler & Yoon, 2020; Galieni et al., 2021; Gill et al., 2022; Zahir et al., 2022 Ye et al., 2023; Al-Tamimi et al.)
	Abiotic	(Humplík et al., 2015; Lowe et al., 2017; Gao et al., 2020; Galieni et al., 2021; Singh et al., 2021; Gill et al., 2022; Zahir et al., 2022; Zubler & Yoon, 2020; Ye et al., 2023; Al-Tamimi et al.)	(Galieni et al., 2021; Gill et al., 2022; Al-Tamimi et al.)	(Humplík et al., 2015; Lowe et al., 2017; Gao et al., 2020; Zubler & Yoon, 2020; Galieni et al., 2021; Gill et al., 2022; Zahir et al., 2022; Ye et al., 2023; Al-Tamimi et al.)

Concerning the color imaging modality, changes in reflectance over time can be particularly significant in detecting and characterizing plant stress. Color modality is captured on three 8-bit channels: Red, Green, and Blue. These channels correspond to a range of wavelengths and can provide significant contrasts for plant phenotyping. Some mathematical operations combining those color layers, known as vegetation indices, can also be beneficial for stress detection (Bannari et al., 1995). Under abiotic stress, photosynthesis is affected by unfavorable environmental conditions. This often leads to a loss of chlorophyll concentration. As chlorophyll molecules are green pigments, an overall dulling of the leaves appears in the visible range, such as yellowing or browning (Sakuraba et al., 2014; Liu et al., 2021). Under biotic stress, dulling can also appear. More locally, necrotic spots emerge and grow on leaf surfaces (Abdu et al., 2019; Gonçalves et al., 2021). Their color departs from the green of the leaf on which they grow as illustrated in the subsampled image time series in Figure 9.8. Plant responses to pathogen infection are highly pathosystem dependent (Ghatak et al., 2018). This results in a large number of different reactions, which vary further when stresses are combined. In all cases, the color modality remains a must for detecting plant stress as indicated by the large number of reviews (Humplík et al., 2015; Lowe et al., 2017; Gao et al., 2020; Zubler & Yoon, 2020; Galieni et al., 2021; Singh et al., 2021; Gill et al., 2022; Zahir et al., 2022; Ye et al., 2023; Al-Tamimi et al.).

Concerning the depth modality, the spatial evolution of the leaf canopy provides additional information on color imaging modality. Under both abiotic and biotic stresses, plant leaves are weakened and may point downwards. This morphological change can be captured via spatial features extracted from depth modality. Height maps of plants can be obtained from the depth maps by computing the difference in depth to the tray level. Canopy surface can also simply be computed with height thresholding since everything that appears above a certain height can only correspond to plants. Obtaining canopy surface from depth maps is therefore a trivial, low-energy, and physically meaningful operation. Without the use of the depth modality, it would require the use of a complex plant segmentation model. Similarly, a volume under the canopy can be studied. Such information cannot be obtained from any imaging modality other than depth and captures leaf decay. Spatial

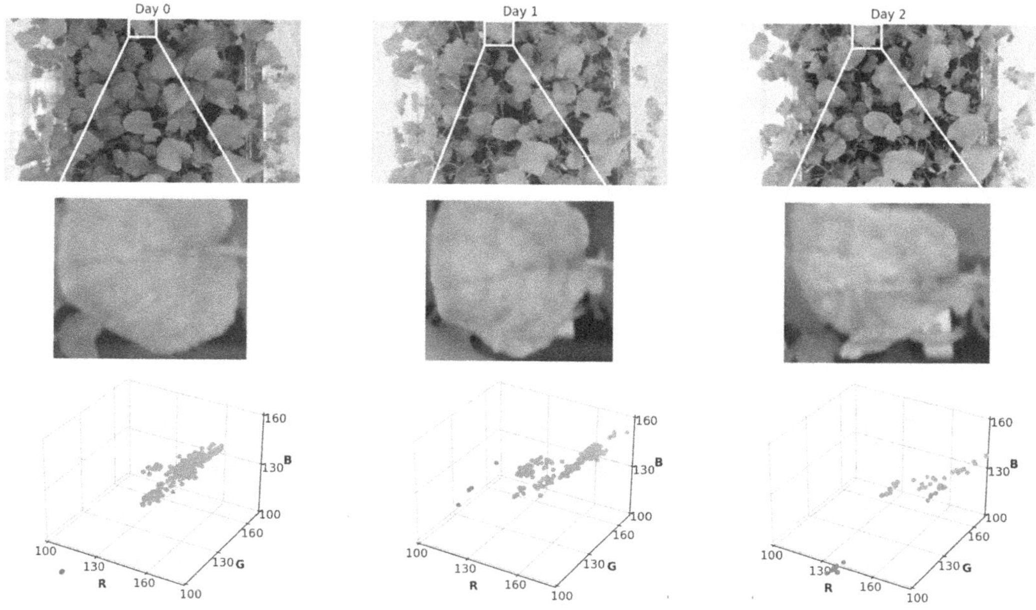

FIGURE 9.8 Overview of color time series of Brassica inoculated with Alternaria. An example of the targeted leaf is provided with its 3D color histogram revealing changes in color distribution over time.

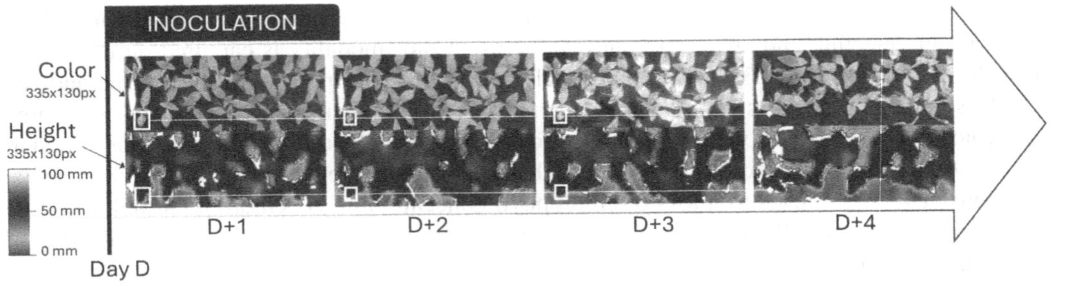

FIGURE 9.9 Overview of depth time series of Pepper inoculated with TSWV. Cotyledon senescence is revealed by the evolution of height maps.

information reveals stress when studied over time. Plant growth temporal signals can be analyzed into several features or components. First, the trend shows the global growth of plants. Growth trend is a great stress indicator over the medium and long term, as shown by Liu et al. (2021) for abiotic stress and by Raza et al. (2015) for biotic stress. For specific plant pathology studies, physiological mechanisms can be targeted to detect pathogen susceptibility (Cordier et al., 2023b), as illustrated in Figure 9.9. Second, some oscillations of leaves are visible at the exact day period. These oscillations allow plants, like many other living organisms, to synchronize their physiology with the daily period of light, maximizing their capability to benefit from lighting and minimizing energy loss when light is not available (McClung, 2006). They are called circadian rhythms or circadian cycles due to the periodic recurrence of day and night cycles. Many publications demonstrate that these cycles are affected by the exposure of plants to stress. In particular, it causes a phase shift in circadian cycles (Verma et al., 2022; Xu et al., 2022). The depth modality captured over time has access to circadian oscillation (Samiei et al., 2019). This oscillation creates periodic patterns that can be sensed via surface or height variations, as illustrated in Figure 9.10. A third component is visible at a higher frequency pattern that corresponds to plant mechanical movements and the appearance of new leaves.

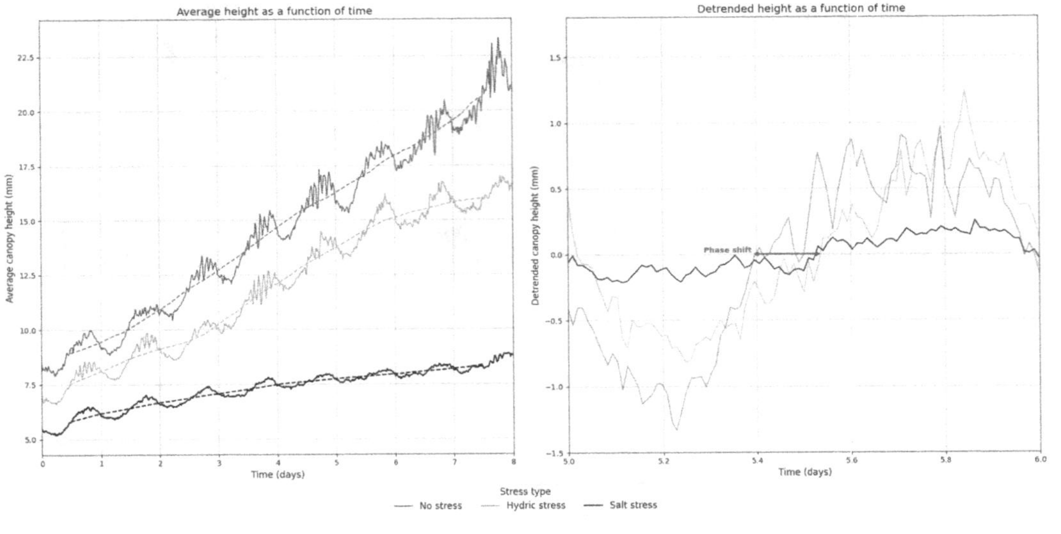

FIGURE 9.10 Comparison of height growth under control, water stress, and salt stress with data from Samiei et al. (2019). Both the growth trend (left) and circadian cycle (right) phase shift are relevant.

FIGURE 9.11 Example of the infrared map for a canopy of scab-infected young apple plants from http://eidolon.univ-lyon2.fr/cdouarre_ScabIrDataset/dataset.zip. Necrosis spots and diseased venules well contrast with the leaf surface intensities. Examples of infected leaves are framed in red.

All these growth mechanisms are continuous and appear both during the day and night. The depth modality, which operated day and night therefore appears to be a good alternative to color imaging which cannot be acquired at night.

Concerning the infrared modality, this type of imaging is recognized for its ability to detect stress at an early stage. The sensor used provides information in the near-infrared (NIR) range at 850 nm. The benefits of this range have been demonstrated for the observation of the health of leaf cells (Humplík et al., 2015; Lowe et al., 2017; Gao et al., 2020; Zubler & Yoon, 2020; Galieni et al., 2021; Gill et al., 2022; Zahir et al., 2022; Ye et al., 2023; Al-Tamimi et al.). A stress response results in a strong variation in the absorption spectrum over the corresponding wavelengths (Zahir et al., 2022). This imaging modality is all the more relevant as it enables early detection and diagnosis of stresses (Bleasdale et al., 2022). In particular, in the case of necrosis, a drop in intensity in the infrared will occur before the appearance of necrosis on the color modality (Chaerle & Straeten, 2000; Lowe et al., 2017), as illustrated in Figure 9.11 with apple scab. Many vegetation indices are based on this appearance of contrast in the infrared range. These indices are often obtained by mathematical operations between intensities obtained at wavelengths in both visible and infrared ranges to further enhance the contrast. Some vegetation indices are particularly suitable for the study of plant stress response (Lowe et al., 2017; Galieni et al., 2021).

9.6 DISCUSSION

The previous section highlighted the added value of depth imaging for studying plant stress responses. However, there are a couple of weaknesses in depth sensor technologies that need to be addressed. A major constraint is the height of the sensors in relation to the plants. A minimum height must be respected which corresponds to 28 cm for the Intel RealSense D435 at 1280 × 720 resolution. This minimum height causes troubles when the plants become quite tall. The use of the system in a growth chamber can be prevented by insufficient heights between shelves. Generally speaking, the height must also be sufficient to ensure full coverage of the scene under study and minimize cast shadows. Conversely, a position that is too high leads to a loss of accuracy on all captured modalities. The use of low-cost sensors can also lead to disturbances in in-depth modality acquisitions, regardless of sensor height. In particular, sensors are sometimes unable to recombine depth information locally. As a result, the intensity of certain pixels remains unknown, and a significant amount of data can sometimes be missed. This is usually due either to the blocking of the return of waves to the sensor where the depth contrast is too sharp, or to lighting conditions that interfere with the waves emitted by the sensor. Other technologies exist for the acquisition of depth modality (Chen et al., 2018). LiDAR (Lim et al., 2003) is one of them that also uses time of flight but does not require two receptors for stereo vision. This technology leads to better accuracy and a decrease in noise. The current cost of such sensors is incompatible with the low-cost constraint, but they will surely be more affordable in the near future.

Parallel acquisition of depth and infrared images is a complex task as it is done by the same sensor. Capturing depth modality requires the emission of an infrared pattern which is detrimental to infrared acquisition. During the day, the emitted pattern is drowned out by ambient light waves. At night, the pattern emitted for depth acquisition renders infrared acquisitions inoperable. If depth modality is not required, infrared emission can be switched off to obtain a much more accurate infrared modality. If both imaging modalities are required, a low-cost infrared sensor can be connected in parallel to the minicomputer, such as the Raspberry Pi IR module connected via the camera serial interface (CSI) port. In this case, the Intel RealSense D435 sensor must be configured for depth modality acquisition and an alignment process must be done to match the frames from each sensor.

The image recording format must be chosen carefully. Compressed image formats reduce the volume of data storage and flow. On the other hand, you need to be sure to use a format that does not imply any loss of data. We have chosen the *png* format here, as it enables lossless image compression. If storage and bandwidth are not an issue, consider a format without compression e.g., *raw* or *tif*. This can be modified in the acquisition program *acquisition_management.py*.

Other well-suited modalities can be exploited such as chlorophyll fluorescence (Krause & Weis, 1984; Rousseau et al., 2013; Kalaji et al., 2016; Bhagooli et al., 2021; Sapoukhina et al., 2022). This modality has proven its effectiveness in detecting plant stress and is widely used by biologists via tools that assess this imaging modality locally on leaves. These tools isolate the leaf in a black box to block ambient light and fully control the light spectrum used. However, they are currently incompatible with the constraints of high-throughput and parallel monitoring of a large number of plants. Another well-suited modality is thermal imaging. It is obtained from the long infrared range which is not acquired by our imaging system. It has proven its relevance for both biotic and abiotic stress detection (Chaerle & Straeten, 2000; Chaerle et al., 2009).

This chapter deals with the automation of plant breeding through the study of plant stress response. Other selection criteria exist such as plant robustness and germination speed. In this context, tests are carried out in greenhouses under conditions similar to those of the stress response study. The use of the imaging system remains highly relevant for this task and its usefulness has been demonstrated by Garbouge et al. (2021) for the detection of growth stages of seedlings. Other characteristic extractions can be automated such as the germination rate.

9.7 CONCLUSIONS

We have detailed the step-by-step construction of a network of low-cost sensors connected to minicomputers for high-throughput plant phenotyping. The imaging modalities captured enable monitoring of plant responses to stress. Finally, this chapter promotes a do-it-yourself reproducible approach which is specifically adapted for phenotyping groups who expect to have full control of their phenotyping systems with the capability of maintenance, upgrade, and adaptation of the acquisition parameters.

REFERENCES

Abdu, A. M., Mokji, M. M., & Sheikh, U. U. (August 2019). An automatic plant disease symptom segmentation concept based on pathological analogy. In Proceedings of the 2019 IEEE 10th Control and System Graduate Research Colloquium (ICSGRC), pp. 94–99.

Al-Tamimi, N., Langan, P., Bernád, V., Walsh, J., Mangina, E., & Negrão, S. Capturing crop adaptation to abiotic stress using image-based technologies. *Open Biology. 12*, 210353, doi:10.1098/rsob.210353.

Bannari, A., Morin, D., Bonn, F., & Huete, A. R. (1995). A review of vegetation indices. *Remote Sensing Reviews, 13*, 95–120, doi:10.1080/02757259509532298.

Bhagooli, R., Mattan-Moorgawa, S., Kaullysing, D., Louis, Y. D., Gopeechund, A., Ramah, S., Soondur, M., Pilly, S. S., Beesoo, R., Wijayanti, D. P., *et al.* (2021). Chlorophyll fluorescence – a tool to assess photosynthetic performance and stress photophysiology in symbiotic marine invertebrates and seaplants. *Marine Pollution Bulletin, 165*, 112059, doi:10.1016/j.marpolbul.2021.112059.

Bleasdale, A. J., Blackburn, G. A., & Whyatt, J. D. (2022). Feasibility of detecting apple scab infections using low-cost sensors and interpreting radiation interactions with scab lesions. *International Journal of Remote Sensing, 43*, 4984–5005, doi:10.1080/01431161.2022.2122895.

Chaerle, L., & Straeten, D. V. D. (2000). Imaging techniques and the early detection of plant stress. *Trends in Plant Science, 5*, 495–501, doi:10.1016/S1360-1385(00)01781-7.

Chaerle, L., Lenk, S., Leinonen, I., Jones, H. G., Van Der Straeten, D., & Buschmann, C. (2009). Multi-sensor plant imaging: Towards the development of a stress-catalogue. *Biotechnology Journal, 4*, 1152–1167, doi:10.1002/biot.200800242.

Chen, N., Zuo, C., Lam, E. Y., & Lee, B. (2018). 3D imaging based on depth measurement technologies. *Sensors, 18*, 3711, doi:10.3390/s18113711.

Chéné, Y., Rousseau, D., Lucidarme, P., Bertheloot, J., Caffier, V., Morel, P., Belin, É., & Chapeau-Blondeau, F. (2012). On the use of depth camera for 3D phenotyping of entire plants. *Computers and Electronics in Agriculture, 82*, 122–127, doi:10.1016/j.compag.2011.12.007.

Cordier, M., Torres, C., Rasti, P., & Rousseau, D. (2023a) On the use of circadian cycles to monitor individual young plants. *Remote Sensing, 15*, 2704, doi:10.3390/rs15112704.

Cordier, M., Rasti, P., Torres, C., & Rousseau, D. (2023b) High-throughput monitoring of plant-pathogen interactions based on low-cost imaging. 12th International Congress of Plant Pathology, Lyon, France.

Galieni, A., D'Ascenzo, N., Stagnari, F., Pagnani, G., Xie, Q., & Pisante, M. (2021). Past and future of plant stress detection: An overview from remote sensing to positron emission tomography. *Frontiers in Plant Science, 11*, 609155.

Gao, Z., Luo, Z., Zhang, W., Lv, Z., & Xu, Y. (2020). Deep learning application in plant stress imaging: A review. *AgriEngineering, 2*, 430–446, doi:10.3390/agriengineering2030029.

Garbouge, H., Rasti, P., & Rousseau, D. (2021). Enhancing the tracking of seedling growth using RGB-Depth fusion and deep learning. *Sensors, 21*, 8425, doi:10.3390/s21248425.

Ghatak, A., Chaturvedi, P., & Weckwerth, W. (2018). Metabolomics in Plant Stress Physiology. In R. K. Varshney, M. K. Pandey, A. Chitikineni (Eds.), *Plant genetics and molecular biology*; Advances in biochemical engineering/biotechnology; Cham: Springer International Publishing, pp. 187–236 ISBN 978-3-319-91313-1.

Gill, T., Gill, S. K., Saini, D. K., Chopra, Y., de Koff, J. P., & Sandhu, K. S. (2022). A Comprehensive review of high throughput phenotyping and machine learning for plant stress phenotyping. *Phenomics, 2*, 156–183, doi:10.1007/s43657-022-00048-z.

Gonçalves, J. P., Pinto, F. A. C., Queiroz, D. M., Villar, F. M. M., Barbedo, J. G. A., & Del Ponte, E. M. (2021). Deep learning architectures for semantic segmentation and automatic estimation of severity of foliar symptoms caused by diseases or pests. *Biosystems Engineering, 210*, 129–142, doi:10.1016/j.biosystemseng.2021.08.011.

Humplík, J. F., Lazár, D., Husičková, A., & Spíchal, L. (2015). Automated phenotyping of plant shoots using imaging methods for analysis of plant stress responses – a review. *Plant Methods, 11*, 29, doi:10.1186/s13007-015-0072-8.

Kalaji, H. M., Jajoo, A., Oukarroum, A., Brestic, M., Zivcak, M., Samborska, I. A., Cetner, M. D., Łukasik, I., Goltsev, V., & Ladle, R. J. (2016). Chlorophyll a fluorescence as a tool to monitor physiological status of plants under abiotic stress conditions. *Acta Physiologiae Plantarum, 38*, 102, doi:10.1007/s11738-016-2113-y.

Krause, G. H., & Weis, E. (1984). Chlorophyll fluorescence as a tool in plant physiology: II. Interpretation of fluorescence signals. *Photosynthesis Research, 5*, 139–157, doi:10.1007/BF00028527.

Li, J., Gao, W., Wu, Y., Liu, Y., & Shen, Y. (2022). High-quality indoor scene 3D reconstruction with RGB-D cameras: A brief review. *Computational Visual Media*, *8*, 369–393, doi:10.1007/s41095-021-0250-8.

Li, L., Zhang, Q., & Huang, D. (2014). A review of imaging techniques for plant phenotyping. *Sensors*, *14*, 20078–20111, doi:10.3390/s141120078.

Li, Z., Guo, R., Li, M., Chen, Y., & Li, G. (2020). A review of computer vision technologies for plant phenotyping. *Computers and Electronics in Agriculture*, *176*, 105672, doi:10.1016/j.compag.2020.105672.

Lim, K., Treitz, P., Wulder, M., St-Onge, B., & Flood, M. (2003). LiDAR remote sensing of forest structure. *Progress in Physical Geography: Earth and Environment*, *27*, 88–106, doi:10.1191/0309133303pp360ra.

Liu, W., Li, Y., Liu, J., & Jiang, J. (2021). Estimation of plant height and aboveground biomass of Toona sinensis under drought stress using RGB-D imaging. *Forests*, *12*, 1747, doi:10.3390/f12121747.

Lowe, A., Harrison, N., & French, A. P. (2017). Hyperspectral image analysis techniques for the detection and classification of the early onset of plant disease and stress. *Plant Methods*, *13*, 80, doi:10.1186/s13007-017-0233-z.

McClung, C. R. (2006). Plant circadian rhythms. *Plant Cell*, *18*, 792–803, doi:10.1105/tpc.106.040980.

Minervini, M., Giuffrida, M. V., Perata, P., & Tsaftaris, S. A. (2017). Phenotiki: An open software and hardware platform for affordable and easy image-based phenotyping of rosette-shaped plants. *Plant Journal: For Cell and Molecular Biology*, *90*, 204–216, doi:10.1111/tpj.13472.

Mohammed Amean, Z., Low, T., & Hancock, N. (2021). Automatic leaf segmentation and overlapping leaf separation using stereo vision. *Array*, *12*, 100099, doi:10.1016/j.array.2021.100099.

Raza, S.-A., Prince, G., Clarkson, J. P., & Rajpoot, N. M. (2015). Automatic detection of diseased tomato plants using thermal and stereo visible light images. *PLOS One*, *10*, e0123262, doi:10.1371/journal.pone.0123262.

Rousseau, C., Belin, E., Bove, E., Rousseau, D., Fabre, F., Berruyer, R., Guillaumès, J., Manceau, C., Jacques, M.-A., & Boureau, T. (2013). High throughput quantitative phenotyping of plant resistance using chlorophyll fluorescence image analysis. *Plant Methods*, *9*, 17, doi:10.1186/1746-4811-9-17.

Sakuraba, Y., Kim, D., Kim, Y.-S., Hörtensteiner, S., & Paek, N.-C. (2014). Arabidopsis STAYGREEN-LIKE (SGRL) promotes abiotic stress-induced leaf yellowing during vegetative growth. *FEBS Letters*, *588*, 3830–3837, doi:10.1016/j.febslet.2014.09.018.

Samiei, S., Rasti, P., Chapeau-Blondeau, F., & Rousseau, D. (2019) Cultivons Notre Jardin Avec Fourier. *27ème Colloque GRETSI sur Le Traitement du Signal et des Images*, Lille, France.

Samiei, S., Rasti, P., Ly Vu, J., Buitink, J., & Rousseau, D. (2020). Deep learning-based detection of seedling development. *Plant Methods*, *16*, 103, doi:10.1186/s13007-020-00647-9.

Sapoukhina, N., Boureau, T., & Rousseau, D. (2022). Plant disease symptom segmentation in chlorophyll fluorescence imaging with a synthetic dataset. *Frontiers in Plant Science*, *13*, 969205.

Shaikh, M. B., & Chai, D. (2021). RGB-D data-based action recognition: A review. *Sensors*, *21*, 4246, doi:10.3390/s21124246.

Singh, A., Jones, S., Ganapathysubramanian, B., Sarkar, S., Mueller, D., Sandhu, K., & Nagasubramanian, K. (2021). Challenges and opportunities in machine-augmented plant stress phenotyping. *Trends in Plant Science*, *26*, 53–69, doi:10.1016/j.tplants.2020.07.010.

Verma, S., Negi, N. P., Narwal, P., Kumari, P., Kisku, A. V., Gahlot, P., Mittal, N., & Kumar, D. (2022). Calcium signaling in coordinating plant development, circadian oscillations and environmental stress responses in plants. *Environmental and Experimental Botany 201*, 104935, doi:10.1016/j.envexpbot.2022.104935.

Xu, X., Yuan, L., & Xie, Q. (2022). The circadian clock ticks in plant stress responses. *Stress Biology, 2*, 15, doi:10.1007/s44154-022-00040-7.

Ye, D., Wu, L., Li, X., Atoba, T. O., Wu, W., & Weng, H. (2023). A synthetic review of various dimensions of non-destructive plant stress phenotyping. *Plants*, *12*, 1698, doi:10.3390/plants12081698.

Zahir, S. A. D. M., Omar, A. F., Jamlos, M. F., Azmi, M. A. M., & Muncan, J. (2022). A review of visible and near-infrared (vis-NIR) spectroscopy application in plant stress detection. *Sensors and Actuators A: Physical*, *338*, 113468, doi:10.1016/j.sna.2022.113468.

Zubler, A. V., & Yoon, J.-Y. (2020). Proximal methods for plant stress detection using optical sensors and machine learning. *Biosensors*, *10*, 193, doi:10.3390/bios10120193.

10 Speed Breeding and High-Throughout Phenotypic Platforms to Accelerate Improvements in Cereal Crops

Indra Rautela
Department of Biotechnology, School of Applied and Life Sciences, Uttaranchal University, Dehradun, India

Sonika Kalia
Department of Biotechnology, School of Applied and Life Sciences, Uttaranchal University, Dehradun, India

Department of Agricultural Biotechnology, CSK HPKV, Palampur, H.P., India

Megha Katoch
Department of Agricultural Biotechnology, CSK HPKV, Palampur, H.P., India

Shanvi Nautiyal and Natasha Charaya
Department of Biotechnology, School of Applied and Life Sciences, Uttaranchal University, Dehradun, India

Monika Kalia
Department of Entomology, CSK HPKV, Palampur, H.P., India

Himanshu Sharma
Agri-Biotechnology Division, National Agri-Food Biotechnology Institute, Knowledge City Sector-81, Mohali, Punjab, India

Vinay Sharma
International Crops Research Institute for the Semi-Arid Tropics (ICRISAT), Hyderabad, India

10.1 INTRODUCTION

With the continuous increase in population and to combat hunger, there is an increase in the demand for food. To feed everyone, the scientific community is working hard to develop new ideas

DOI: 10.1201/b23372-10

and methodologies for quality food production, which has put a lot of pressure on the agricultural system (Haroon et al., 2019). The lives of 7.5 billion people on the earth are mainly dependent on the growth and development of plants and their life processes involving photosynthesis, flowering, fruiting, and transpiration. Cereals are one such plant that is widely consumed by humans and contributes about 50% of the world's energy supply, making global food security greatly driven by cereal production. Unfortunately, increasing climate change, resource depletion, and increasingly growing population pose serious obstacles to the supply of this food staple. Growing populations lead to an extensive increase in demand for food, fuel, and clothing, a high rate of inflation, and food security challenges like malnutrition, especially in Asian countries like India and China. With the development of DNA technologies, high throughput genotyping, and next-generation sequencing, conventional breeding has been updated to molecular breeding. Adopting a successful and efficient speed breeding system that combines conventional and modern methods to identify desired traits will help overcome these difficulties while also increasing yield and providing additional benefits of tolerance to various stresses (Haroon et al., 2022). Determining the plant phenotype, with beneficial morphological, biochemical, and physiological characteristics, is the first important task in plant breeding for better crop selection for human consumption.

The selection and phenotyping of desired traits in a large population set grown in different environments was a critical and tedious task until the development of the High-Throughout Phenotypic Platforms (HT3P) technique (Araus & Cairns, 2014). Genome sequencing and the study of genomic information have significantly become cheaper and more accessible in the public domain, but phenotypic information and studies are still limited as phenotypes are a cumulative effect of interaction between the genes and the environment and differ from place to place. This technique has greatly revolutionized phenotyping and accelerated crop improvements (Williams, 2005; Ahmar et al., 2023).

10.2 TRADITIONAL BREEDING METHODS

Plant breeding techniques started and became prevalent in the 18th and 19th centuries with the onset of the green revolution and the discovery of fundamental principles of genetics by Gregor Mendel in the mid-1800s. Soon after the early cultivation and domestication of cereal crops, people were able to differentiate certain excellent crop varieties exhibiting a high degree of desired traits and select their seeds for future crop production (Haroon et al., 2022). The new varieties of crops produced by using these traditional plant breeding approaches were observed to be strikingly different from their wild parent type and made it difficult to recognize their ancestral lines. Traditionally, breaded crop varieties lacked desired traits but possessed better appearance, shape, size, flavors, and other observable characteristics (Laskowski et al., 2019). Plant breeding along with genetic engineering aids in the production of cereal crops with desired traits such as high yield, crop resistance to antibiotics and pests, enhanced flavor, and nutritional profile.

10.2.1 Cereal Crop Breeding Technique Based on Observable Variations

It is the most prehistoric method of plant breeding based on the selection of naturally occurring variants due to continuous selection pressure that resulted in significant changes in the phenotype of the plant, as seen in the ancestry of the maize plant from Teosinte (Haroon et al., 2022). The use of these methods lasted from the beginning of agricultural practices until the development of hybridization techniques in the 19th and 20th centuries (Figure 10.1).

10.2.1.1 Origin of Crops and Plant Domestication

The basis of this method was the theory of natural selection as stated by Darwin in the later years. Mutations occur at a greater rate among large numbers of plants grown in a field making the conditions unfavorable for the survival of the wild type and ultimately resulting in its extinction.

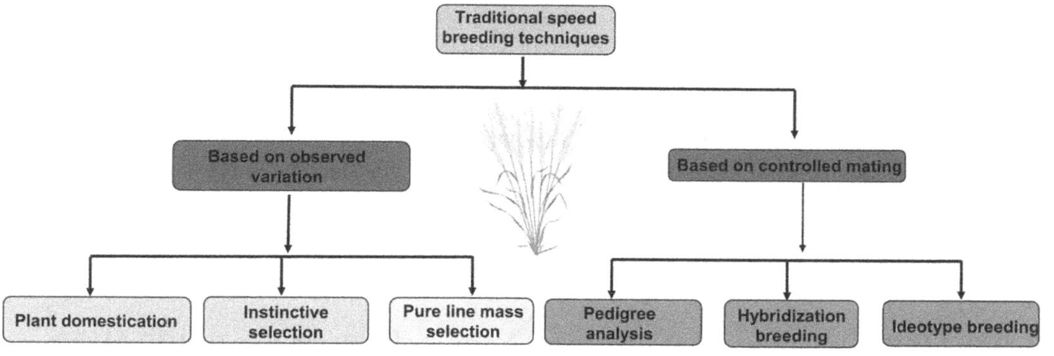

FIGURE 10.1 Traditional speed breeding methods.

However, it was observed that certain mutations in the genes of the crop variety could lead to the development of desirable traits increasing either the crop yield, flavor, phenotype, or food grain quality. The seeds of such highly beneficial crop varieties were extracted and conserved by the farmers for future grain production (Murphy et al., 2005; Breseghello et al., 2013).

The evolution of such novel varieties of the plant was greatly influenced by a change in their habitat from being grown in wild fields to cultivated fields under proper monitoring and care given by the farmer and let of development of the irreversibly interdependent relationship between humans and cultivated crop species for their survival, known as domestication syndrome (Murphy et al., 2005). The long-term consequence of this process of plant breeding was the loss of molecular variability of the wild type by completely dropping the way of variation in the form of valuable genes such as those carrying pest resistance and eliminating a larger part of the gene pool (Zeven et al., 2000).

10.2.1.2 Instinctive Selection by Farmers for Building Landraces

Landraces refer to a certain population of cereal plants cultivated for various generations in a particular habitat under the influence of various environmental stresses, seed manipulation, and seedling management (Murphy et al., 2005). The seed grains are either randomly subjected to selection, pollen exchange, variation, cross-pollination, or are intentionally induced, leading to the development of novel phenotypes and characters as observed dominantly when the habitat of a landrace is shifted or a new landrace is cultivated in close proximity to the original landrace (Breseghello et al., 2013). Landraces are cultivated due to the high range of genetic diversity, demand for edible parts by local people, and stable crop yield, ensuring food security in the area where they are sown (Haroon et al., 2022). This is carried out by an instinctive farmer who is well aware of the local habitat and environmental conditions and capable of cultivating crops based on local consumption and demand.

The major drawback of this process is that the selection of a desired trait may induce unwanted changes or mutations in other characters or traits. The best example illustrating this effect was the loss of harvest index of the cereal landraces due to the selection of their edible parts, such as ears and panicles, by humans. The seed samples collected from landraces are stored in seed banks across the world and are also under ex-situ conservation. With developing technologies and interconnectivity, various social media platforms are also being created for germplasm conservation in traditional communities (Ellen et al., 2011).

10.2.1.3 Mass Selection and Pure-Line Selection

Traditional farmers applied the basic knowledge and information available locally and assumed that the selection of seeds of plants exhibiting superior traits could be a way of developing plant species with novel functions and qualities. Mass selection, or phenotypic selection, is a process in which seeds isolated from desirable traits of plant varieties are mixed, and the next generation of crops is cultivated using seeds from this mixture. The primary basis of selection is the external appearance of the crop, and this method works on the ideology that this selection would result in the production of a progeny with superior characteristics as compared to the original plant line. Rice and wheat are self-pollinating species and therefore their landraces are almost pure lines mixtures. Applying the principle of mass selection, it was easy to separate pure lines having new enhanced features. This novelty comes at a price as the pure lines obtained are comparatively less stable and lack features of disease resistance and pest resistance as they are deficient in diversity.

10.2.2 Other Traditional Plant Breeding Methods Based on Controlled Mating

The true potential of cereal plant breeding resides in the shuffling of alleles to produce different combinations and progeny. An infinite number of combinations can be produced by utilizing highly adaptive parents grown under targeted environmental conditions and crossing them over. This can be achieved by placing emphasis on seed planting, harvest, and handling. These methods are mentioned below.

10.2.2.1 Pedigree Analysis

It involves crossing over the parent population and producing selected populations by conducting a series of self-pollinations and selections until the desirable characteristics of both parents are incorporated into the offspring generated (Valentine, 1984).

The pedigree breeding method is widely accepted by the cultivators as the qualitative traits of resistance to various stresses are accessed due to complementary traits, which allows producing varieties to have a cumulative effect of all the desired traits in a single offspring. Pedigree breeding is not successful for breeding self-pollinating species with smaller populations and limited genetic variation (Breseghello et al., 2013). The major drawback associated with this method is that progeny is only available for evaluation at the end of the process.

10.2.2.2 Hybridization Breeding

Heterosis, in simple words, means that progeny obtained after conducting several crosses among species of different varieties exhibit greater potential in terms of traits, biomass, and speed of re-development as compared to the original progeny, mainly due to the presence of heterogenous loci. Heterosis is directly linked to the diversity of the parents and increases with an increase in parent diversity. Currently, grown cereal varieties like maize and rice are therefore greatly determined to develop breeding programs capable of producing such hybrids (Hill et al., 2019).

The primary concern related to this method was developing at least two such adaptive varieties to boost agronomic adaptation while maintaining their genetic variability and developing enough hybrids for seed production. In maize plants, improving the parental lines eliminated the risk of coinciding gene pools in the two varieties and additionally resulted in greater yield (Chaudhary et al., 2014).

10.2.2.3 Idiotype Plant Breeding

This method enhances the pedigree breeding method by promoting the superiority of qualitative traits and yield. The main idea behind ideotype breeding is that we can improve the desired traits by manipulating and enhancing other traits that are indirectly associated with their functioning. An ideotype refers to a biological system that is designed and predicted to perform and adapt to a particular environment with enhanced capacities such as yield, greater taste texture of the grains, oil, greater resistance to various kinds of biotic and abiotic stresses, or other desired traits of the plant. The word ideotype was first coined by Donald while he was conducting experiments on wheat plants. Crop ideotype is a wider term that includes its morphological, biological, and

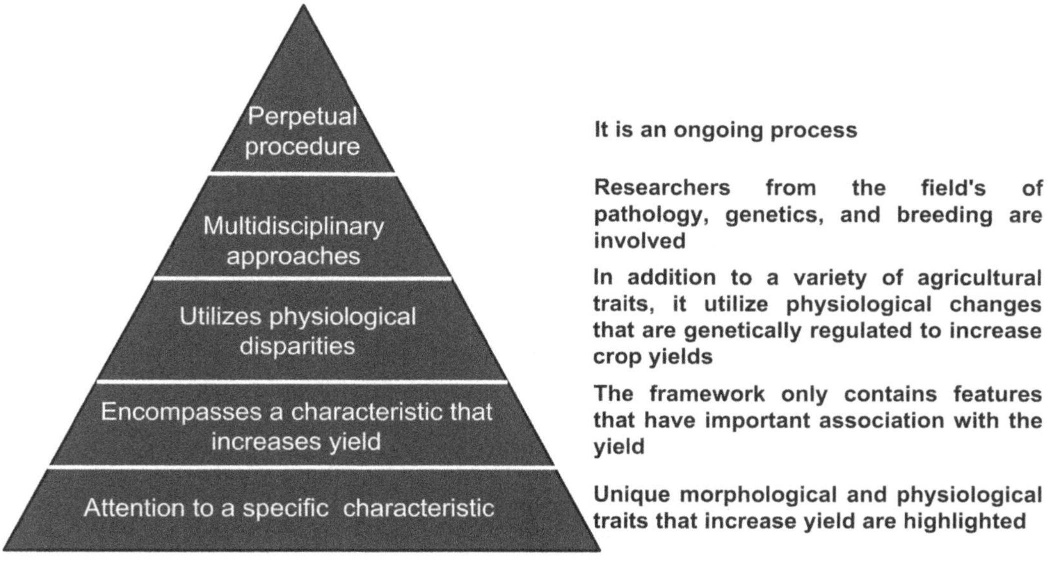

FIGURE 10.2 Different features of Ideotype breeding.

biochemical traits. To develop an ideotype the breeder is expected to have detailed knowledge about the traits associated with the desired characters, their linkages with other characters, their mode of inheritance, and how they can be manipulated to have greater yield (Figure 10.2) (Breseghello et al., 2013).

To summarize, the field of plant breeding demonstrates a wide range of traditional approaches, each customized to unique breeding aims (Table 10.1).

While pure line selection is praised for creating cultivars with consistently good yield and quality, pedigree breeding excels at capturing a wide variety of quantitative attributes (Begna, 2021). Contrarily, hybrid breeding makes use of genetic variation between different lines to produce hybrids with improved vigor, resilience, and productivity (Chen, 2010). While this is happening, population breeding strategically modifies allele frequencies within populations of average size, enlarging the genetic pool (Breseghello & Sorrells, 2006). A careful decision must be made based on the unique features and breeding program objectives because each method has its advantages and limits. Breeders can better utilize these traditional methods by comprehending their nuances.

By understanding the complexities of these traditional practices, breeders can make well-informed choices to advance agricultural growth and satisfy the growing demands of global food security (Rutkoski et al., 2022). The incorporation of molecular approaches and genomic data with these established methods promises to boost plant breeding even further as technology develops (Varshney et al., 2021), providing a more promising future for resilient and sustainable agriculture.

Multiple generations of sequencing techniques have emerged because of the development of sequencing technologies. Plasmid constructions and PCR products can be verified using the first-generation sequencing technique known as Sanger sequencing (Tahir ul Qamar et al., 2020). Short-read sequencing with high throughput and low cost are both advantageous for resequencing and SNP calling in second-generation sequencing, which is represented by Illumina technology (Kerstens et al., 2009). While third-generation sequencing, as demonstrated by PacBio and Oxford Nanopore, generates lengthy reads that are essential for completing genome assembly and identifying significant structural differences (Wenger et al., 2019). According to Xu and Dixon (2020), hi-C sequencing, which is based on chromosome conformation capture, assists in resolving difficult genomic areas and improves genome assembly. To improve genome assembly, identify

TABLE 10.1
Overview of Conventional Methods of Plant Breeding

S. No	Conventional Methods of Plant Breeding	Description	Advantage	Limitations	Reference
1	Pedigree breeding	Crossing parents with desirable characteristics to obtain separate inbred lines with combined favorable traits	Effective for quantitative traits controlled by multiple genes. Favorable characteristics from both parents	Inability to estimate total yield during the process. Low success potential in terms of yield, approximately 1% per year. Evaluation only possible at the end of the breeding process	Rutkoski, et al., 2022.
2	Pure line selection	Crossbreeding closely related pure lines with desirable traits, resulting in property screening for desired characters	Development of valuable cultivars with high yield or improved quality. Increased homozygosity	Does not produce new genotypes. Restricted to self-pollinated crops. Risk of inbreeding depression due to closely related parents	Begna, 2021.
3	Hybrid Breeding	Crossing individuals from two different lines to achieve heterosis or hybrid Vigor, resulting in more vigorous and tolerant hybrids	Improve traits, faster growth, and higher reproductive output. Tolerance to biotic and abiotic stresses. Potential for new species	This may lead to the extinction of some varieties or the loss of specific genes. Limited control over genetic mutations inbred progenies less stable in adverse condition	Chen, 2010.
4	Population Breeding	Enhancing favorable allele frequency through selection and recombination cycle on an intermediate-sized population	Increased probability of producing new favorable alleles. Combines desirable alleles from different sources	Requires large and effective population size. Depletion of genetic variance in small populations	Breseghello & Sorrells, 2006.

structural variation, and phase out haplotypes, optical mapping, a light microscope-based approach, produces vast DNA sequence fingerprints (Yuan et al., 2020). Genome-wide linkage mapping is also useful for identifying genes linked to important traits, including grain yield, stress tolerance, and disease resistance, through controlled crosses (Gao et al., 2015). Utilizing various germplasm, genome-wide association studies (GWAS) offer a way to find marker-trait relationships with improved mapping precision and shorter research times (Muñoz-Amatriaín et al., 2014).On the other hand, pangenome analysis investigates genic presence-absence variations (PAVs) within a species while employing a variety of genotypes to find relationships with critical agronomic traits (Liu & Tian, 2022). Together, these sequencing methods enable researchers to dive deeply into genetic architecture, opening the door for developments in areas ranging from genomics to agriculture (Table 10.2).

TABLE 10.2
Different Genome Editing Techniques

S. No.	Types	Zinc Finger Nucleases (ZFNs)	Transcription Activator-Like Effector (TALENs)	Base Editing	Clustered Regularly Interspaced Short Palindromic Repeats (CRISPR)/Cas9	CRISPR/Cpf1
1	**Definition**	Artificially designed zinc finger DNA binding nuclease that cleaves a specific gene of the target genome	Restriction enzymes with two domains to cleave DNA at specific target sites	Insertion, deletion, or substitution of bases at a particular location	Comprises Cas9 restriction enzyme to cut dsDNA at a specific location generally near the PAM region	It acts as a double nuclease, functions as an endoribonuclease, and allows multiplexed genome editing
2	**Mode of action**	Creates dsDNA breaks in a targeted region of the genome	Creates dsDNA breaks in a targeted region of the genome	ssDNA bases are changed	Induces both dsDNA and ssDNA breaks in target DNA	Creates dsDNA breaks
3	**Rate of mutation**	High	Average mutations	Very low	Low	Significantly high
4	**Length of target DNA sequence**	25–36 base pairs	26–35 base pairs	Point mutation	21–22 base pairs	20–25 base pairs
5	**Target recognizing capacity**	Fair	Good	Very good and specific	Good	Very enhanced
6	**Cloning**	Must be done	Necessary	Can not be done	Not required	Not required

10.3 MODERN SPEED BREEDING METHODS

Traditional speed breeding methods consume a large amount of time, energy, and workspace in order to achieve the desired results. With the invention of modern means of speed breeding by utilizing genetic engineering tools and techniques, these drawbacks of conventional speed breeding methods are eliminated. The introduction and utilization of gene editing tools and techniques have greatly hampered the process of gene editing. These techniques include primer editing, base editing, CRISPR technology, and dcas9 (Chen et al., 2019). The basic principle of these techniques suggests that any significant or minute change in the DNA, such as single or multiple nucleotide insertion, deletion, or substitution of the target organism, may lead to the development of novel characters (Ahmar et al., 2023).

A combined approach utilizing both traditional and modern plant breeding techniques is being used for the production of improved cereals with high yield capacity and resistance to stresses within a time frame. A brief history of the evolution of molecular genetics and plant breeding techniques has been presented below in the form of a flow diagram (Finckh, 2008; Lee et al., 2009) (Figure 10.3).

New breeding techniques (NBTs) are potent crop improvement techniques that make site-specific changes in the DNA of the target species in order to recover the desired product. Some of these techniques utilized for this purpose have been listed in Table 10.3 (Lee et al., 2009). These transgenic processes are regulated and approved by the Food and Drug Administration (FDA) in association with various environmental protection agencies to counter the ethical and legal issues related to these recombinant products before they are released into the market for global consumption (Finckh, 2008).

10.3.1 Rapid Generation Advancement Speed Breeding Techniques

Rapid generation advancement (RGA) is a bunch of breeding tools for faster development of fixed lines from the segregating generations (Figure 10.4).

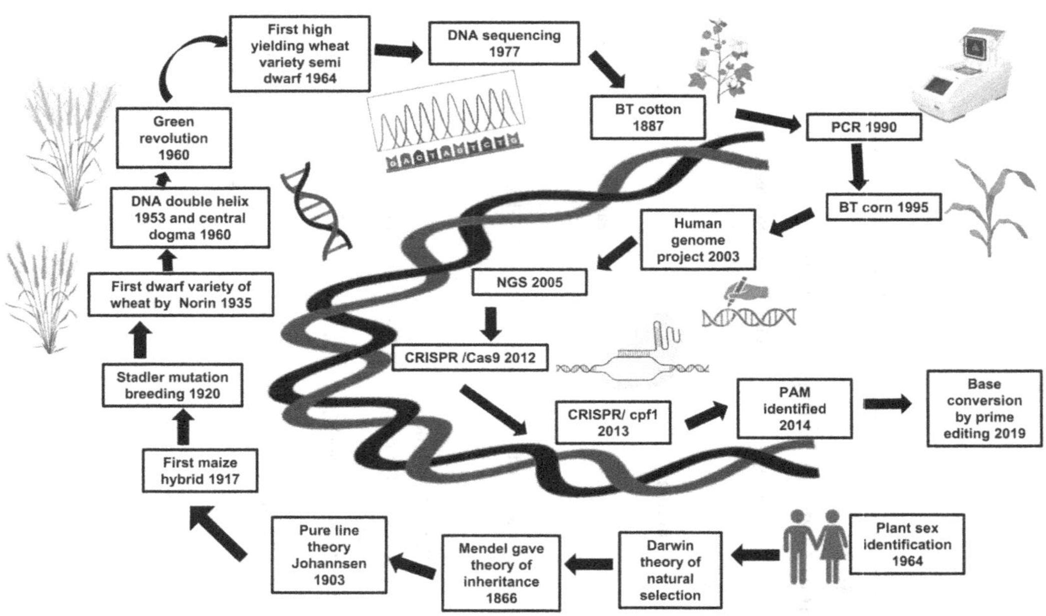

FIGURE 10.3 Evolution of plant breeding techniques.

TABLE 10.3
Various Omics-Based Strategies Used for Crop Improvement

S. No	Category of Method	Name of Technique	Description	Application	References
1	Genome sequencing	First generation (Sanger)	Traditional sequencing method, suitable for verifying plasmid constructs and for PCR products	Limited	Tahir ul Qamar et al., 2020.
		Second generation (illumine)	Short read sequencing, high throughputs, cost-effective method	Resequencing, SNP calling	Kerstens et al., 2009.
		Third generation (PacBio, Oxford Nanopore)	Long read sequencing, generates long reads for complete genomic assembly, spans large structural variation	Denovo genome assembly, SV detection	Wenger et al., 2019.
		Hi-C sequencing	Chromosome conformation capture, Aids in resolving complex genomic regions and improving genome assembly	Genome assembly, structure variation detection	Xu & Dixon 2020.
		Optical mapping	Light microscope-based technology, generates large DNA sequence fingerprints, spans challenging genomic regions	Genome assembly improvement, SV identification, haplotype phasing	Yuan et al., 2020.
		Genome-wide linkage mapping	Mapping QTL in segregating populations created through controlled crosses, controlled populations suitable for detecting gene underlying quantitative traits	Mapping genes related to grain yield, stress tolerance, disease resistance, etc.	Gao et al., 2015.
		Genome-wide association studies (GWAS)	Utilizing natural population to identify marker-trait associations, broad genetic diversity, higher mapping resolution, reduced research time	Study both qualitative and quantitative traits in diverse germplasm	Muñoz-Amatriaín et al., 2014.
		Pangenome analysis	Studying genic presence-absence Variations (PAVs) within a species using diverse genotypes, identifying core genomes and variable genomes suitable for breeding programs	Detecting genic PAVs and their associations with important agronomic traits	Liu et al., 2022.
2	Global proteome profiling	Two-dimensional electrophoresis (2-DE)	Separates proteins based on charge (isoelectric point) and size. Spots on gels correspond to individual proteins	For the visualization of protein through	Beckett, 2012.
		Gel-based approaches	Two-dimensional polyacrylamide gel electrophoresis (2D-PAGE)-Difference in Gel electrophoresis (DIGE) and three-dimensional gel electrophoresis (3-DGE)	Characterizing protein expression changes in response to various stresses. Identifying low-abundance proteins	Kaur et al., 2021.

(*Continued*)

TABLE 10.3 *(Continued)*
Various Omics-Based Strategies Used for Crop Improvement

		Gel free approaches	Isotope-coded affinity tagging (ICAT), isobaric tagging for relative and absolute quantification (iTRAQ), stable isotope labeling by amino acid in cell culture (SILAC), and multidimensional protein identification technology (MudPIT)	Quantitative analysis of protein expression changes in stress responses. Studying dynamic proteome changes in stress conditions	Tan et al., 2017
3	Targeted proteome profiling		Specific types of protein are isolated, enriched, and purified through immunoprecipitation (IP), strong cation exchange (SCX), Q-trap, LTQ-orbitrap, etc.	Identifying specific proteins involved in stress response. Studying protein interactions and post-translational modifications (PTMs) in response to stress and enhance the efficiency of protein	Rabilloud, 2013.
		Mass spectrometry (MS)	Identifies proteins based on mass-to-charge ratio (m/z) and fragmentation patterns	Quantifying and identifying proteins in stress-related pathways. Identifying PTMs and protein isoforms	Yates, 2011; Mastrobuoni et al., 2012.
		Peptidomics	Identification and analysis of peptides smaller than 10 KDa	Characterizing novel peptides derived from non-functional and functional precursors. Studying physiological and pathological peptides	Farrokhi et al., 2008.
		Phosphoproteomics	Identification of protein phosphorylation events and substrates generally through PepBase, Pep2Pro, and PhosPhat databases	Identifying uncharacterized protein kinases (PKs) and their substrates. Studying phosphorylation events in response to stress	Nakagami et al., 2012
		Redox proteomics	Identification of protein oxidation and reduction events	Studying oxidative stress-induced protein modification, investigating redox PTMs involved in plant development and stress responses	Navrot et al., 2011

4	Proteomics	Bioinformatic tools	Omic technologies, i. e. Proteomics IDEntification database (PRIDE), peptide Atlas, MassIVE, PPDB, 1001 Proteomes, Pep2 Pro Database, DIPOS, PASSEL, and MRMaid have been developed for data analysis	To achieve high-quality protein data expression	Kusebauch et al., 2014, Sun, et al., 2009, Joshi et al., 2012, Sapkota et al.,. 2011, Hirsch-Hoffmann, et al., 2012, Senkler and Braun, 2012, Fan et al., 2012, Farrah, et al., 2012.
5	Data processing techniques	Liquid chromatography-mass spectrometry (LC-MS), capillary electrophoresis mass spectrometry (CE-MS), nuclear magnetic resonance (NMR)	Separates complex protein mixture into individual components for protein identification	Useful in crop metabolic studies	Cao et al., 2017, Zhang et al., 2020, Jones et al., 2011. Khakimov et al.,. 2014
	Advanced analytical techniques	Gas chromatography time of flight mass spectrometry (GC-TOF-MS), Ultra performance liquid chromatography-mass spectrometry (UPLC-MS), and Liquid chromatography high-resolution mass spectrometry (LC-HRMS)	Series of isolation to the identification of analytical analysis of bioactive compounds	For metabolomic studies	Witt et al., 2012; Zhao et al., 2014; Suharti, et al., 2016.
6	Pre-processing and data assessment	Principal component analysis (PCA), multiple curve resolution (MCR), hierarchical cluster analysis (HCA), and batch learning self-organizing map (BL-SOM)	Bioinformatic tools used to develop metabolomics data mining, assessment, and data processing	To make meaningful profiling of a large metabolomic database	Balmer et al., 2013; Chun and Keleş 2010; Xu and Goodacre 2012
		Annotation software	KEGG and KNApSAcK pathway	Used for the reconstruction of the gene map	Afendi et al., 2012; Kanehisa, 2016.

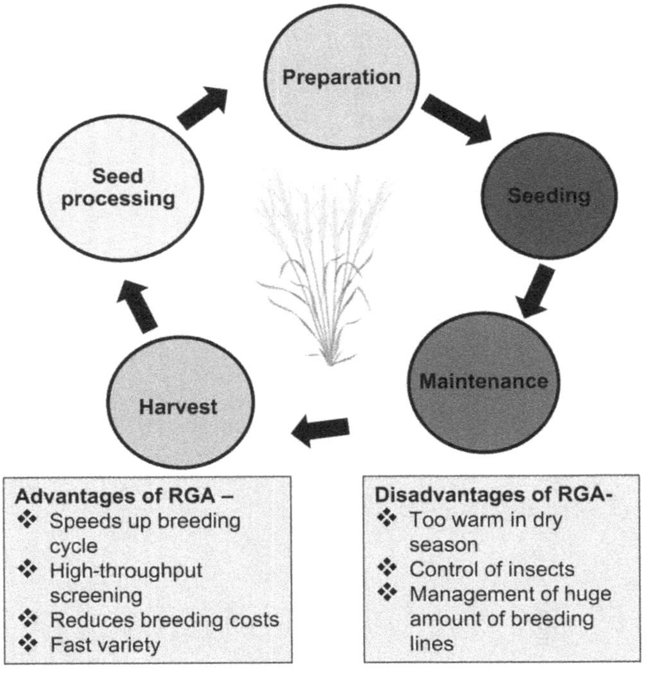

FIGURE 10.4 RGA cycle and its various advantages and disadvantages.

The RGA allows several generations per year by manipulating growth conditions of plants such as temperature, relative humidity, photoperiod, and day length for rapid induction of flowering and seed set. Often, this is achieved by shortening crop growth. It includes speed breeding, shuttle breeding, marker-assisted breeding, and single seed breeding for the production of agronomically superior high-yielding varieties within short time intervals. RGA was first used at the International Rice Research Institute (IRRI) in 1977 by Dr Ikehashi and Hille Rislambers for rice breeding (Tanaka et al., 2016; Ghosh et al., 2022). This method is used for its high efficiency, less time-consuming and simple technical procedure and requirements, and enhanced genetic gains. The primary aim of RGA is to reduce generation time and enhance yield (Valentine, 1984).

A comprehensive discussion on various speed breeding techniques under RGA has been done below.

10.3.1.1 Single-Seed Descent Method

It is a revised version of preexisting bulking methods for managing various segregating populations. Single-seed descent (SSD) refers to the process of extracting single seeds from each generation and combining them to produce the next generation (Tanaka et al., 2016). The technique is greatly focused on advancing crosses to homozygosity as rapidly as possible while maintaining the gene pool and biodiversity for future selections and is carried out either in a screen house or glass house space (Mobini et al., 2015). A large amount of genetic variation is observed among each homozygous line after the sixth generation. Plants are grown under mild stress conditions by raising the temperature, providing less space for flowering, and hastening flowering as compared to the original field. Ultimately, breeding lines develop approximately 4–6 generations each year with less time and cost as compared to fields where only 1 or 2 lines could be produced (Figure 10.5).

SSD continues to be used today for breeding rice varieties, photoperiod sensitivity, and other tolerances (Lupton et al., 1957).

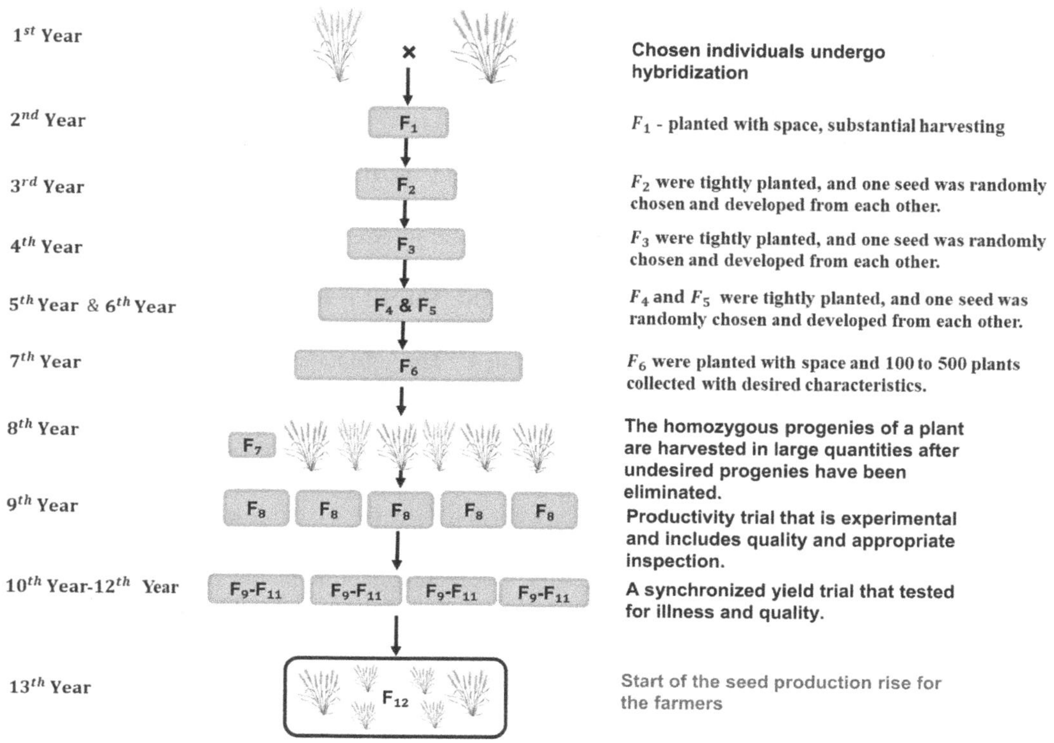

FIGURE 10.5 Diagrammatic representation of the single-seed descent technique in self-pollinating crops.

10.3.1.2 Speed Breeding

Generally focused on long day and neutral day crops, this technique involves increasing the photoperiod and regulating environmental conditions like temperature, high-density planting, soil, nutrients, and sustainable spacing in glasshouses to shorten generation time and promote RGA (Figure 10.6) (Prashanth et al., 2022).

Increasing levels of plant growth regulators enhance the production of carbon dioxide, promoting the growth of vegetative cells to reproductive cells. Light of wavelength 400–700 nm accompanied with LED light at temperatures 17–22°C and moderate humidity enhances plant growth. Conventional speed breeding techniques accompanied by single seed descent, single pod

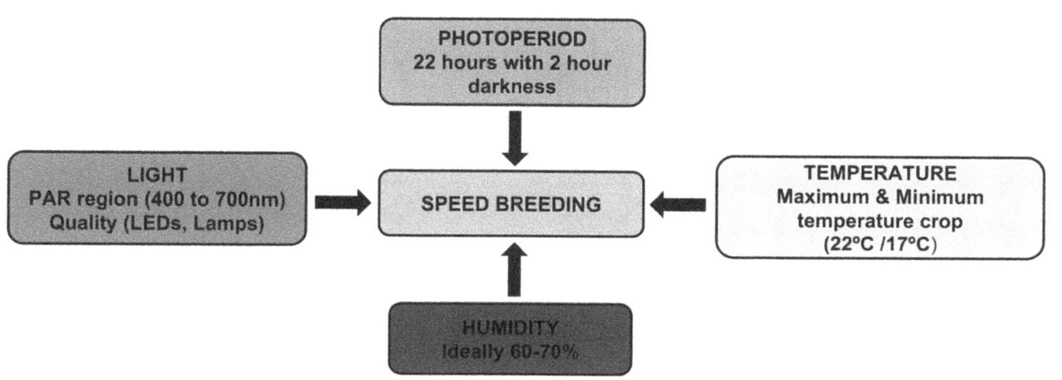

FIGURE 10.6 Factors affecting speed breeding.

descent, or single plant selection used in combination with this technique can greatly promote RGA (Wanga et al., 2021).

10.3.1.3 Shuttle Breeding

The development of a successful plant breeding exercise involves exposure to stress conditions like drought, saline conditions, and nutrient deficiency, and for this purpose, shuttle breeding was evolved to facilitate breeding under different environmental conditions. This technique was greatly utilized by IRRI in collaboration with other national and international organizations.

The procedure involves crossing the improved plant variety with a locally adapted variety to generate the first generation (Srivastava & Bains, 2018). This F1 generation is grown under desired environmental conditions in an experimental laboratory. This process is carried out multiple times until a progeny with desired traits is segregated. The ultimate product of this breeding methodology is greater yield, enhanced gain quality, and tolerance to stresses, diseases, and pests (Wanga et al., 2021) (Figure 10.7).

10.3.1.4 Double Haploidy

The plant breeding method involves doubling the haploid set of chromosomes, making them diploid, and achieving completely homozygous lines in a shorter time interval (Basu et al., 2011). Although there are plenty of techniques available for creating haploid plants, like anther and pollen tissue culture, many crops are susceptible to these techniques, and actual production is not achieved. Artificial chromosome duplication can be assisted by treatment with certain chemicals, like colchicine. This method greatly promotes the elimination of unwanted traits and induces the production of desired combinations of traits (Snape, 1976; Bordes et al., 2007).

10.3.1.5 Marker-Assisted Breeding and Genomic Editing

The above-discussed plant breeding techniques were primarily focused on the phenotypic evaluation of desired traits, making the process time-consuming and ineffective. The use of

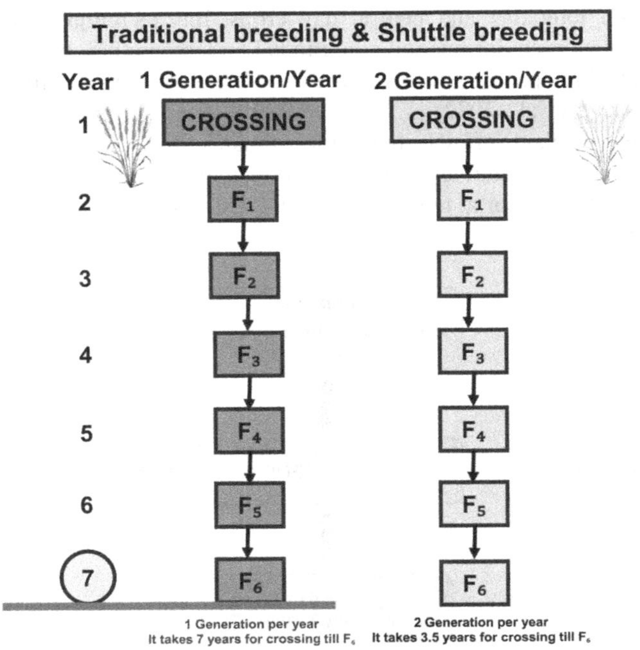

FIGURE 10.7 Difference in traditional breeding and shuttle breeding.

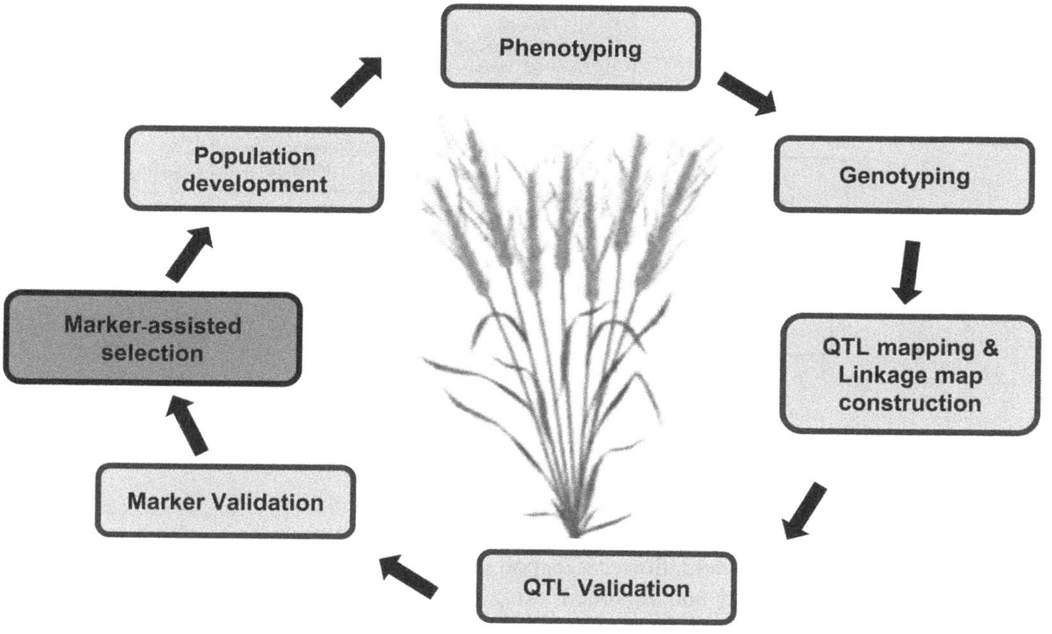

FIGURE 10.8 Different stages of marker-assisted selection.

selectable markers present on desirable genes and their alleles tightly linked to them using marker-assisted breeding revolutionized this conventional breeding method. This process involves incorporating characters into a plant variety using marker-assisted backcrossing (Knott & Kumar, 1975; Shivakumar et al., 2018) (Figure 10.8).

Genome editing techniques such as single base altering have become a valued tool for precisely editing DNA sequences. CRISPR-associated nucleases are the most recently used to achieve this goal in crops like cereals, horticulture, and cash crops (Raina, 1997). Nucleases help in cleaving the DNA at specific sites followed by a non-homologous end joining mechanism. These reformed lines are used for additional study of particular gene families and species (Knott & Kumar, 1975; Williams, 2005).

10.3.1.6 High-Throughput Screening or Phenotyping of Field Crop Lines

Precise use of technology for accurate and high throughput screening of various grown crop varieties in the field is necessary for crop improvement, greater yield, and tolerance to various stresses. The high cost of these techniques has greatly hampered the process of evaluation and critical monitoring of crops under various environments. Recent development in technology has produced combinations such as high resolution and high throughput phenotyping using plant science, computation, engineering, etc. (Prasanna et al., 2013; Neilson et al., 2015).

Data collection and analysis have become greatly efficient and accurate with the use of these technologies. The spectrum of phenotyping ranges from proximal to remote and aerial sensors for regular monitoring of plant organs, leaves, growth, and resistance to stresses (Araus & Cairns, 2014).

Certain satellite platforms and drones are introduced such as NASA's Ecosystem Spaceborne Thermal Radiometer Experiment, and hyperspectral infrared imager to collect high-resolution data and images of the environment where drought-like conditions prevail. The primary aim of using high-throughput phenotyping (HTP) in crop breeding is for the detection of architectural traits and desired phenotypes resulting from a genotype (Prasanna et al., 2013; Araus & Cairns, 2014; Neilson et al., 2015).

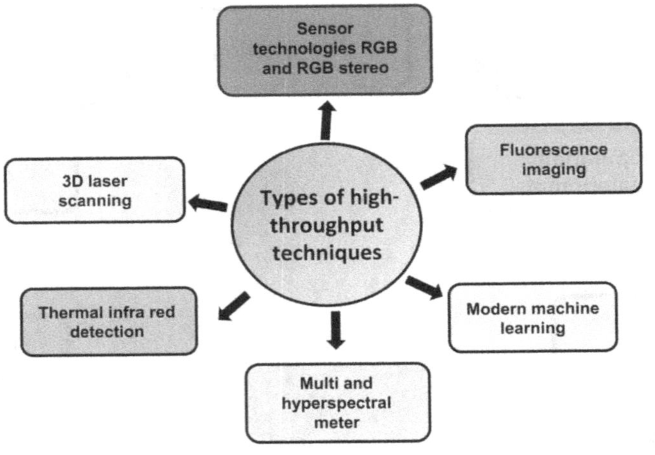

FIGURE 10.9 Various types of high-throughput techniques.

Desired traits such as crop yield, flower size, pod size, biomass, canopy formation, and length of stem are accurately measured and recorded using HTP. Recognition of plant diseases which traditionally was observed through visible traits and signs, can now be done at an early with the use of HTP methods such as 3D imaging, infrared sensing, semi-automatic and automatic systems, and hyperspectral meters (Figure 10.9).

Low-cost and low-maintenance technologies like RGB cameras are used for continuous screening of plant phenotypes and collecting pictures daily. Sensor-based technologies like stereo RGB provide high-resolution noninvasive screening of the crop field. Electromagnetic radiation reflected by growing crops can be detected and measured as a parameter for differentiating segregating varieties.

Handling of raw data collected from the fields, its processing, standardization, decision-making, and conclusions based on observations is the integral application of HTP (Araus & Cairns, 2014).

10.4 PROSPECTS

Traditional speed breeding techniques eliminated the key elements of genetic diversity due to manual selection resulting in the development of phenotypic changes. Another major limitation of conventional practices was long working hours and comparatively less yield and desired traits. The advent of modern plant breeding techniques greatly accelerated the plant breeding process by the use of molecular tools leading to genomic recombinations and gene-to-gene manipulation. These methods are expected to resolve the growing global food security issue with a growing population. Latest techniques such as CRISPR/Cas9 have enabled site-specific modifications in the DNA and joining by non-homologous end joining method (Mobini et al., 2015). Despite its applications, it possesses great limitations like off-target binding effects, unwanted mutations, and the presence of single targeted genomic sites. This limitation was overcome by the advent of CRISPR/ Cpf1 with multiple targeted genomic sites for crop improvement. With the increase in the use of RGA techniques comes the need for effective analysis and screening of crop genomes which was limited until the development of HTP tools. With the multidimensional phenotyping approach provided by HTP, data storage, management, and analysis remain a big challenge. Incorporating these phenotyping techniques at a large scale will require the development of less costly, complex, and easily accessible infrastructure (Lupton et al., 1957).

REFERENCES

Ahmar, S., Hensel, G., & Gruszka, D. (2023). CRISPR/Cas9-mediated genome editing techniques and new breeding strategies in cereals – current status, improvements, and perspectives. *Biotechnology Advances*, *69*, 108248.

Afendi, F. M., Okada, T., Yamazaki, M., Hirai-Morita, A., Nakamura, Y., Nakamura, K., Ikeda, S., Takahashi, H., Altaf-Ul-Amin, M., Darusman, L. K., & Saito, K. (2012). KNApSAcK family databases: Integrated metabolite–plant species databases for multifaceted plant research. *Plant and Cell Physiology*, *53*(2), e1–e1.

Araus, J. L., & Cairns, J. E. (2014). Field high-throughput phenotyping: The new crop breeding frontier. *Trends in Plant Science*, *19*(1), 52–61.

Basu, S. K., Datta, M., Sharma, M., & Kumar, A. (2011). Haploid production technology in wheat and some selected higher plants. *Australian Journal of Crop Science*, 5(9), 1087–1093.

Balmer, D., Flors, V., Glauser, G., & Mauch-Mani, B. (2013). Metabolomics of cereals under biotic stress: Current knowledge and techniques. Frontiers in Plant Science, *4*, 47367.

Beckett, P. (2012). The basics of 2D DIGE, *Methods in molecular biology,difference gel electrophoresis (DIGE)*, pp. 9–19.

Begna, T. (2021). Combining ability and heterosis in plant improvement. *Open Journal of Political Science*, *6*(1), 108–117.

Bordes, J., Charmet, G., De Vaulx, R. D., Lapierre, A., Pollacsek, M., Beckert, M., & Gallais, A. (2007). Doubled-haploid versus single-seed descent and S1-family variation for testcross performance in a maize population. *Euphytica*, *154*(1–2), 41–51.

Breseghello, F., & Coelho, A. S. G. (2013). Traditional and modern plant breeding methods with examples in rice (Oryza sativa L.). *Journal of Agricultural and Food Chemistry*, *61*(35), 8277–8286.

Breseghello, F., & Sorrells, M. E. (2006). Association mapping of kernel size and milling quality in wheat (*Triticum aestivum* L.) cultivars. *Genetics*, *172*(2), 1165–1177.

Cao, D., Lutz, A., Hill, C. B., Callahan, D. L., & Roessner, U. (2017). A quantitative profiling method of phytohormones and other metabolites applied to barley roots subjected to salinity stress. *Frontiers in Plant Science*, *7*, 233111.

Chaudhary, H. K., Vineeta, K., Shoukat A. R., & Tisu, T. (2014). Distant hybridisation and doubled-haploidy breeding. In A. Pratap & J. Kumar (eds) *Alien gene transfer in crop plants, Volume 1: Innovations, methods and risk assessment*. New York, NY: Springer, pp. 143–164.

Chen, K., Wang, Y., Zhang, R., Zhang, H., & Gao, C. (2019). CRISPR/Cas genome editing and precision plant breeding in agriculture. *Annual Review of Plant Biology*, *70*, 667–697.

Chen, Z. J. (2010). Molecular mechanisms of polyploidy and hybrid vigor. *Trends in Plant Science*, *15*(2), 57–71.

Chun, H., & Keleş, S. (2010). Sparse partial least squares regression for simultaneous dimension reduction and variable selection. *Journal of the Royal Statistical Society Series B: Statistical Methodology*, *72*(1), 3–25.

Ellen, R., & Platten, S. (2011). The social life of seeds: The role of networks of relationships in the dispersal and cultural selection of plant germplasm. *Journal of the Royal Anthropological Institute*, *17*(3), 563–584.

Fan, J., Mohareb, F., Bond, N. J., Lilley, K. S., & Bessant, C. (2012). MRMaid 2.0: Mining PRIDE for evidence-based SRM transitions. *OMICS: A Journal of Integrative Biology*, *16*(9), 483–488.

Farrah, T., Deutsch, E. W., Kreisberg, R., Sun, Z., Campbell, D. S., Mendoza, L., Kusebauch, U., Brusniak, M. Y., Hüttenhain, Ruth, Schiess, R., & Selevsek, N. (2012). PASSEL: The PeptideAtlas SRMexperiment library. *Proteomics*, *12*(8), 1170–1175.

Farrokhi, N., Whitelegge, J. P., & Brusslan, J. A. (2008). Plant peptides and peptidomics. *Plant Biotechnology Journal*, *6*(2), 105–134.

Finckh, M. R. (2008). Integration of breeding and technology into diversification strategies for disease control in modern agriculture. In D. B. Collinge, L. Munk, & B. M. Cooke (eds) *Sustainable disease management in a European context*. Dordrecht: Springer, pp. 399–409.

Gao, F., Wen, W., Liu, J., Rasheed, A., Yin, G., Xia, X., Wu, X., & He, Z. (2015). Genome-wide linkage mapping of QTL for yield components, plant height and yield-related physiological traits in the Chinese wheat cross Zhou 8425B/Chinese Spring. *Frontiers in Plant Science*, *6*, 1099.

Ghosh, S., Roy, A., & Dutta, S. (2022). Rapid generation advance methods to fast-track crop breeding: A review. *Agricultural Reviews*, *10*, 18805.

Haroon, M., Wang, X., Afzal, R., Zafar, M. M., Idrees, F., Batool, M., Khan, A. S., & Imran, M. (2022). Novel plant breeding techniques shake hands with cereals to increase production. *Plants*, *11*, 1052. 10.3390/plants11081052

Haroon, M., Idrees, F., Naushahi, H. A., Afzal, R., Usman, M., Qadir, T., & Rauf, H. (2019). Nitrogen use efficiency: Farming practices and sustainability. *Journal of Experimental Agriculture International*, *36*(3), 1–11.

Hill, C. B., Angessa, T. T., McFawn, L. A., Wong, D., Tibbits, J., Zhang, X. Q., Forrest, K., Moody, D., Telfer, P., Westcott, S., & Diepeveen, D. (2019). Hybridisation-based target enrichment of phenology genes to dissect the genetic basis of yield and adaptation in barley. *Plant Biotechnology Journal*, *17*(5), 932–944.

Hirsch-Hoffmann, M., Gruissem, W., & Baerenfaller, K. (2012). pep2pro: The high-throughput proteomics data processing, analysis, and visualization tool. *Frontiers in Plant Science*, *3*, p. 26730.

Jones, O. A., Maguire, M. L., Griffin, J. L., Jung, Y. H., Shibato, J., Rakwal, R., Agrawal, G. K., & Jwa, N. S. (2011). Using metabolic profiling to assess plant-pathogen interactions: An example using rice (Oryza sativa) and the blast pathogen Magnaporthe grisea. *European Journal of Plant Pathology*, *129*, 539–554.

Joshi, H. J., Christiansen, K. M., Fitz, J., Cao, J., Lipzen, A., Martin, J., Smith-Moritz, A. M., Pennacchio, L. A., Schackwitz, W. S., Weigel, D., & Heazlewood, J. L. (2012). 1001 Proteomes: A functional proteomics portal for the analysis of *Arabidopsis thaliana* accessions. *Bioinformatics*, 28(10), 1303–1306.

Kanehisa, M. (2016). KEGG bioinformatics resource for plant genomics and metabolomics. *Plant Bioinformatics, Methods in Molecular Biology*, pp. 55–70.

Kaur, B., Sandhu, K. S., Kamal, R., Kaur, K., Singh, J., Röder, M. S., & Muqaddasi, Q. H. (2021). Omics for the improvement of abiotic, biotic, and agronomic traits in major cereal crops: Applications, challenges, and prospects. *Plants*, *10*(10), 1989.

Kerstens, H. H., Crooijmans, R. P., Veenendaal, A., Dibbits, B. W., den Dunnen, J. T., & Groenen, M. A. (2009). Large scale single nucleotide polymorphism discovery in unsequenced genomes using second generation high throughput sequencing technology: Applied to turkey. *BMC Genomics*, *10*(1), 1–11.

Khakimov, B., Bak, S., & Engelsen, S. B. (2014). High-throughput cereal metabolomics: Current analytical technologies, challenges and perspectives. *Journal of Cereal Science*, *59*(3), pp. 393–418.

Knott, D. R., & Kumar, J. (1975). Comparison of early generation yield testing and a single seed descent procedure in wheat breeding. *Crop Science*, *15*(3), 295–299.

Kusebauch, U., Deutsch, E. W., Campbell, D. S., Sun, Z., Farrah, T., & Moritz, R. L. (2014). Using PeptideAtlas, SRMAtlas, and PASSEL: Comprehensive resources for discovery and targeted proteomics. *Current Protocols in Bioinformatics*, *46*(1), 1–25.

Laskowski, W., Górska-Warsewicz, H., Rejman, K., Czeczotko, M., & Zwolińska, J. (2019). How important are cereals and cereal products in the average Polish diet? *Nutrients*, *11*(3), 679. doi: 10.3390/nu11030679

Lee, E. A., & William, F. T. (2009). Modern maize breeding. In J. L. Bennetzen & S. Hake (eds) *Handbook of maize: Genetics and genomics*. New York, NY: Springer, pp. 141–160.

Liu, Y., & Tian, Z. (2022). Super graph-based pan-genome: Bringing rice functional genomic study into a new dawn. *Molecular Plant*, *15*(9), 1409–1411.

Lupton, F. G. H., & Whitehouse R. N. H. (1957). Studies on the breeding of self-pollinating cereals: I. Selection methods in breeding for yield. *Euphytica*, *6*(2), 169–184.

Mastrobuoni, G., Irgang, S., Pietzke, M., Aßmus, H. E., Wenzel, M., Schulze, W. X., & Kempa, S. (2012). Proteome dynamics and early salt stress response of the photosynthetic organism Chlamydomonas reinhardtii. *BMC Genomics*, *13*, 1–13.

Mobini, S. H., Lulsdorf, M., Warkentin, T. D., & Vandenberg, A. (2015). Plant growth regulators improve in vitro flowering and rapid generation advancement in lentil and faba bean. *In Vitro Cellular & Developmental Biology – Plant*, *51*, 71–79.

Muñoz-Amatriaín, M., Cuesta-Marcos, A., Endelman, J. B., Comadran, J., Bonman, J. M., Bockelman, H. E., Chao, S., Russell, J., Waugh, R., Hayes, P. M., & Muehlbauer, G. J. (2014). The USDA barley core collection: Genetic diversity, population structure, and potential for genome-wide association studies. *PloS One*, *9*(4), 1–13.

Murphy, K., Lammer, D., Lyon, S., Carter, B., & Jones, S. S. (2005). Breeding for organic and low-input farming systems: An evolutionary–participatory breeding method for inbred cereal grains. *Renewable Agriculture and Food Systems*, *20*(1), 48–55.

Nakagami, H., Sugiyama, N., Ishihama, Y., & Shirasu, K. (2012). Shotguns in the front line: Phosphoproteomics in plants. *Plant and Cell Physiology*, *53*(1), 118–124.

Navrot, N., Finnie, C., Svensson, B., & Hägglund, P. (2011). Plant redox proteomics. *Journal of Proteomics*, *74*(8), 1450–1462.

Neilson, E. H., Edwards, A. M., Blomstedt, C. K., Berger, B., Møller, B. L., & Gleadow, R. M. (2015). Utilization of a high-throughput shoot imaging system to examine the dynamic phenotypic responses of a C4 cereal crop plant to nitrogen and water deficiency over time. *Journal of Experimental Botany*, *66*(7), 1817–1832.

Prasanna, B. M., Araus, J. L., Crossa, J., Cairns, J. E., Palacios, N., Das, B., & Magorokosho, C. (2013). High-throughput and precision phenotyping for cereal breeding programs. In P. K. Gupta & R. K. Varshney (eds) *Cereal genomics II*. Dordrecht: Springer, pp. 341–374.

Prashanth, B. H., Kumar, M., Gaikwad, K. B., Kumar, R., Kumar, N., Palaparthi, D., Bharti, H., Kamre, K., & Yadav, R. (2022). Rapid generation advancement and fast-track breeding approaches in wheat improvement. In M. G. Mallikarjuna, S. C. Nayaka, & T. Kaul (eds) *Next-generation plant breeding approaches for stress resilience in cereal crops*. Singapore: Springer, pp. 241–262.

Rabilloud, T. (2013). When 2D is not enough, go for an extra dimension. *Proteomics*, *13*(14), 2065–2068.

Raina, S. K. (1997). Doubled haploid breeding in cereals. *Plant Breeding Reviews*, *15*, 141–186.

Rutkoski, J. E., Krause, M. R., & Sorrells, M. E. (2022). Breeding methods: Population improvement and selection methods. In M. P. Reynolds & H.-J. Braun (eds) *Wheat improvement: Food security in a changing climate*. Cham: Springer International Publishing, pp. 83–96.

Sapkota, A., Liu, X., Zhao, X. M., Cao, Y., Liu, J., Liu, Z. P., & Chen, L. (2011). DIPOS: Database of interacting proteins in Oryza sativa. *Molecular BioSystems*, *7*(9), 2615–2621.

Senkler, M., & Braun, H. P. (2012). Functional annotation of 2D protein maps: The GelMap portal. *Frontiers in Plant Science*, *3*, 87.

Shivakumar, M., Nataraj, V., Kumawat, G., Rajesh, V., Chandra, S., Gupta, S., & Bhatia, V. S. (2018). Speed breeding for Indian agriculture: A rapid method for development of new crop varieties. *Current Science*, *115*(7), 1241.

Snape, J. W. (1976). A theoretical comparison of diploidised haploid and single seed descent populations. *Heredity*, *36*(2), 275–277.

Srivastava, P., & Bains, N. S. (2018). Accelerated wheat breeding: Doubled haploids and rapid generation advance. In S. S. Gosal & S. H. Wani (eds) *Biotechnologies of crop improvement, Volume 1: Cellular approaches*. Cham: Springer, pp. 437–461.

Suharti, W. S., Nose, A., & Zheng, S. H. (2016). Metabolomic study of two rice lines infected by Rhizoctonia solani in negative ion mode by CE/TOF-MS. *Journal of Plant Physiology*, *206*, 13–24.

Sun, Y., Cai, Y., Liu, L., Yu, F., Farrell, M. L., McKendree, W., & Farmerie, W. (2009). ESPRIT: Estimating species richness using large collections of 16S rRNA pyrosequences. *Nucleic Acids Research*, *37*(10), e76–e76.

Tahir ul Qamar, M., Faryad, A., Bari, A., Zahid, B., Zhu, X., & Chen, L. L. (2020). Effectiveness of conventional crop improvement strategies vs. omics. In S. Fahad, M. Hasanuzzaman, M. Alam, H. Ullah, M. Saeed, I. A. Khan, & M. Adnan (eds) *Environment, climate, plant and vegetation growth*. Cham: Springer, pp. 253–284.

Tanaka, J., Hayashi, T., & Iwata, H. (2016). A practical, rapid generation-advancement system for rice breeding using simplified biotron breeding system. *Breeding Science*, *66*(4), 542–551.

Tan, B. C., Lim, Y. S., & Lau, S. E. (2017). Proteomics in commercial crops: An overview. *Journal of Proteomics*, *169*, 176–188.

Valentine, J. (1984). Accelerated pedigree selection: An alternative to individual plant selection in the normal pedigree breeding method in the self-pollinated cereals. *Euphytica*, *33*(3), 943–951.

Varshney, R. K., Bohra, A., Yu, J., Graner, A., Zhang, Q., & Sorrells, M. E. (2021). Designing future crops: Genomics-assisted breeding comes of age. *Trends in Plant Science*, *26*(6), 631–649.

Wanga, M. A., Shimelis, H., Mashilo, J., & Laing, M. D. (2021). Opportunities and challenges of speed breeding: A review. *Plant Breeding*, *140*(2), 185–194.

Wenger, A. M., Peluso, P., Rowell, W. J., Chang, P. C., Hall, R. J., Concepcion, G. T., Ebler, J., Fungtammasan, A., Kolesnikov, A., Olson, N. D., & Töpfer, A. (2019). Accurate circular consensus long-read sequencing improves variant detection and assembly of a human genome. *Nature Biotechnology*, *37*(10), 1155–1162.

Williams, J. L. (2005). The use of marker-assisted selection in animal breeding and biotechnology. *Revue Scientifique et Technique-Office International des Epizooties*, *24*(1), 379.

Witt, S., Galicia, L., Lisec, J., Cairns, J., Tiessen, A., Araus, J. L., Palacios-Rojas, N., & Fernie, A. R. (2012). Metabolic and phenotypic responses of greenhouse-grown maize hybrids to experimentally controlled drought stress. *Molecular Plant*, *5*(2), 401–417.

Xu, Z., & Dixon, J. R. 2020. Genome reconstruction and haplotype phasing using chromosome conformation capture methodologies. *Briefings in Functional Genomics*, *19*(2), 139–150.

Xu, Y., & Goodacre, R. (2012). Multiblock principal component analysis: An efficient tool for analyzing metabolomics data which contain two influential factors. *Metabolomics*, *8*, 37–51.

Yates III, J. R (2011). A century of mass spectrometry: From atoms to proteomes. *Nature Methods*, *8*(8), 633–637.

Yuan, Y., Chung, C. Y. L., & Chan, T. F. 2020. Advances in optical mapping for genomic research. *Computational and Structural Biotechnology Journal*, *18*, 2051–2062.

Zeven, A. C. (2000). Traditional maintenance breeding of landraces: 1. Data by crop. *Euphytica*, *116*, 65–85.

Zhang, Y., Liu, L., & Ren, L. (2020). Liquid chromatography-tandem mass spectrometry (LC-MS/MS) determination of cantharidin in biological specimens and application to postmortem interval estimation in cantharidin poisoning. *Scientific Reports*, *10*(1), 10438.

Zhao, J., Xiuwen, W., Hu, J., Liu, Q., Shen, D., & Xiao, R. (2014). Thermal degradation of softwood lignin and hardwood lignin by TG-FTIR and Py-GC/MS. *Polymer Degradation and Stability*, *108*, 133–138.

11 Proteomics

Aims at Stress Mitigation in Soybean under Abiotic Stress

Farhat Yasmeen and Rizwana Parveen
Department of Biosciences, University of Wah, Wah Cantt., Pakistan

Setsuko Komatsu
Faculty of Life and Environmental Sciences, Fukui University of Technology, Fukui, Japan

11.1 INTRODUCTION

Soybeans are of great importance in the world due to their tremendous use as a cheap protein, unsaturated fat, and carbohydrate source for the human diet, animal feed, aquaculture, and biofuels (Yuhong et al., 2021; Rasheed et al., 2022). Soybean typically contains 30–40% protein, 20% fats, 9% dietetic fiber, and 8.5% moisture based on the dry mass of mature unprocessed seeds (Tidke et al., 2015). It is not only an excellent source of proteins, minerals, vitamins, and fibers but also low in saturated fats (Jayachandran & Xu, 2019). The polyunsaturated fatty acids and high-quality fat content in soybeans are crucial from the nutraceutical standpoint (Kamshybayeva et al., 2017). Soybean-processing waste has potential as a sustainable source of a variety of phytochemicals, nutrients, and minerals (Nile et al., 2022). With such prevailing values, soybean has a long history of cultivation. However, soybean has modified itself with time to cope with the environmental variations.

The phenotypic makeup of soybeans has naturally changed during the process of domestication and the role of genetic alteration is critical for the breeding of soybeans and the advancement of crops (Zhou et al., 2015). The ability for quantitative genetics to unravel complicated attributes has greatly increased by a variety of high throughput techniques, considering omics, system biology, and genome-editing methods (Xu et al., 2020) which revolutionized the traditional breeding processes and enabled marked improvement in crop breeding (Chen et al., 2019). Adaptive modification in the soybean planting area could contribute to the improvement of crop production (Yue et al., 2017). Other approaches such as heterosis (Liu et al., 2019), mutation breeding techniques (Bhuiyan et al., 2022), and biotechnological advances in molecular breeding techniques (Ainsworth et al., 2012) brought genome and transcriptome alteration with a significant increase in yield. Using advanced techniques to improve soybean yield is an important direction for enhancing soybean production.

Edaphoclimatic factors could affect the protein content of grains and meals (Assefa et al., 2019). Abiotic stresses, in particular, that produce reactive oxygen species, harm cell membranes, interfere with protein structure, and cause moisture stresses at the cellular level are examples of secondary stresses (Staniak et al., 2023). Conduction through stomata, activities of photosynthesis and enzymes, and exact proportion of hormones, excessive development of reactive oxygen species, and inadequate immunity via antioxidant system become crucial under stressful conditions, with decreased consumption of water and nutrients (Sathi et al., 2022). However,

DOI: 10.1201/b23372-11

through intrinsic adaptation mechanisms facilitated by various physiological and metabolic processes at the cellular level, crops can withstand specific environmental challenges (Mickelbart et al., 2015). An extensive network of defenses at the cellular, physiological, and morphological levels is involved in how plants respond to harmful abiotic circumstances (de Melo et al., 2022). To understand the defense response, there is a need to interpret the molecular mechanism to mitigate the stress response.

Abiotic stresses such as drought, salinity, flooding, heat, cold, and heavy metals that impede plant growth and development are linked to climate variation and affect crop yield and quality (Khalid et al., 2022). Poor seed germination under drought and salinity growing conditions adversely affects crop establishment, growth, and ultimately yield. Within the plant life cycle, seed germination is the first event to be adversely influenced by drought, flooding, and salinity (Lamichhane et al., 2018). Due to future climate abnormalities, flooding is a big concern that will likely result in a significant loss in crop production in various areas of the world (Komatsu et al., 2021). Flooding causes various issues such as sluggish gas exchange and dim underwater lightening, which ultimately leads to plant death because of carbon and energy depletion (Sasidharan et al., 2018). Analysis of tolerance mechanisms to withstand unexpected flooding is critical for the development of new varieties (Jia et al., 2021). Focusing on the development of integrated crop management strategies could mitigate the negative effects of abiotic stresses on plant growth and yield.

Antioxidants are the essential elements that defend plants from the damage of abiotic stresses (Moustafa-Farag et al., 2020). The genetic control of multiple genes and the significant influence of the environment make stress-tolerant breed development a cumbersome process (Ratnaparkhe et al., 2022). By adopting precision genome engineering for specific crop attributes, new options for crop enhancement are made possible by advances in genome editing technologies (Zafar et al., 2020). To improve the ability to withstand biotic and abiotic stress conditions as well as to increase yield and seed composition, breeders apply various genomic-based techniques such as genome assistance, selection, and sequencing (Ratnaparkhe et al., 2022). However, developing tolerance against abiotic stresses using a breeding strategy is a challenging task because of the complex genetic makeup and environmental factors.

11.2 SOYBEAN AND ITS IMPORTANCE

Many food items made from soybeans, such as tofu, tempeh, soymilk, soy sauce, and natto, can be produced (Song et al., 2023). With a contribution of 61% of the world's oilseed production and 28% of the overall consumption of vegetable oil, soybean is acknowledged as the largest oil seed crop (Miao et al., 2020). Soybeans, which are rich in isoflavonoids and folic acid, are used around the world as part of a healthy diet. Soybean and its derivatives are considered an important source of vegetable protein because they contain sufficient essential amino acids. (Tidke et al., 2015). Soybean protein, which contains histidine, is a complete protein source since it is rich in the eight necessary amino acids that the human body needs (Kim et al., 2021). The top vegetable protein is soybean protein which is comparable to the proteins found in milk and eggs. Soybean milk has less lactose and cholesterol than cow milk (He & Chen, 2013). The consumption of soybean and its products is a healthy dietary choice due to the high content of essential amino acids as well as a good alternative to animal-based protein.

Soybean seed is a perfect feedstock for the food and feed industries since it accumulates around 20% fat, 40% protein, and about 30% carbohydrate (Song et al., 2023). Oils with a high monounsaturated fat content are healthier for human use, have better oxidative stability, and have a longer shelf life. (Clemente & Cahoon, 2009). Vegetable soybean acreage has increased in Asia, the Americas, and Europe (Dong et al., 2014). This growth is attributed to increased consumer needs from interior and global markets, as well as increased consumer awareness of the personal well-being and nutritional goodness of soybeans (Djanta et al., 2020). There is a need to find ways to increase soybean production to reduce the import cost.

11.3 SOYBEAN OIL AND HUMAN HEALTH

11.3.1 Types of Oils and Oil Quality in Soybean

The continuous process to produce soybean products used in the food, industrial, and livestock feed sectors is referred to as the processing of soybeans (Haun et al., 2014). Pure soybean oil, glycerol, free fatty acids, lecithins, and protein-rich soy meal all composed of different soy protein concentrates and isolates are all forms of soy proteins that can be produced through advanced processing (Johnson & White, 2008). The process recovers 95% of the soybean protein and 95–98% of the soybean oil using organic solvents like hexane (Ashlock et al., 2000). Normally, one bushel (60 pounds) of soybean grains can produce 48 pounds (80%) of protein-rich meal and 11 pounds (18.3%) of soybean oil. Hulls make up the remaining 1.7% of the total (Ashlock et al., 2000). To get refined soybean oil, crude soybean oil must undergo additional processing after extraction. Several volatile oils were compared using a solid phase micro-extraction mass-spectrometer, and the results showed that olive oil had the highest concentration of esters and the other edible plant oils had the highest concentrations of aldehydes (Hu et al., 2018). However, there is a dire need to maximize the efficiency of oil extraction to reduce the level of impurities (Figure 11.1).

With only a small amount of monounsaturated fatty acids and no trans fats, soybean oil has a high content of polyunsaturated fatty acids (Abdelghany et al., 2020). Around 40% of the necessary fatty acids, including linoleic acid, linolenic acid, and omega-3 fatty acids are found in soybean oil (Deol et al., 2017). Soybeans are a rich source of micronutrients, including vitamins and minerals, and have a very low glycaemic index compared to most other crops and legumes (Zhu et al., 2022). Traditional soybean oil has five major fatty acids, making it one of the most common edible oils. About 10% of them are palmitic, 4% are stearic, 20% are oleic, 55% are linoleic, and 10% are linolenic acids (Song et al., 2023). A byproduct of soybean oil is lecithin, which is composed mainly of phospholipids, and plays a crucial role as a natural emulsifier in the food and pharmaceutical industries (Johnson & White, 2008). Overall, soybean and its byproducts offer a wide range of important nutritional benefits making them a valuable source for both human health and commercial use.

In many nations throughout the world, the consumption of dietary polyunsaturated fatty acids is greatly influenced by soybean oil. It accounts for over 30% (Parcell et al., 2018) and 57% (Wünsche, 2021) of all edible oil consumption, respectively, making it the most extensively consumed oil both globally and in the United States. Around 7% of the country's total calorie intake and more than 40% of the consumption of both essential fatty acids come from soybean oil (Blasbalg et al., 2011). Soybean oil is primarily composed of linoleic acid, which is a significant polyunsaturated fatty acid belonging to the n-6 group, and constitutes approximately 51% of the total fat content (Messina et al., 2021). The n-3 fatty acids that include α-linolenic acid, docosahexaenoic acid, and eicosapentaenoic acid prevent the production of triglycerides and have a significant impact on reducing blood triglyceride levels (Han, 2012). However, investigation of using unsaturated fatty acids from a medical point of view is recommended.

Soybean protein is highly digestible, and its protein quality is comparable to that of beef, fish, milk, and eggs, and as good as that of animal protein (Ashlock et al., 2000). During processing, protein-rich soybean meal can be created, and it may be the only source of proteins for both people and animals (Johnson & White, 2008). The enhanced β-sheet content, better zeta potential, and reduced particle size of proteins recovered by ultrasonication make them viable components for specific food applications (Eze et al., 2022). One of the most readily available sources of bioactive nitrogenous chemicals and byproducts high in protein is soybean meal (Rayaprolu et al., 2013). Bioactive peptides with varied advantageous health qualities have been created by the utilization of soybean proteins (Kudełka et al., 2021). There is increasing interest in alternative and more sustainable ways of protein consumption which paved the way for soybean magnified utilization.

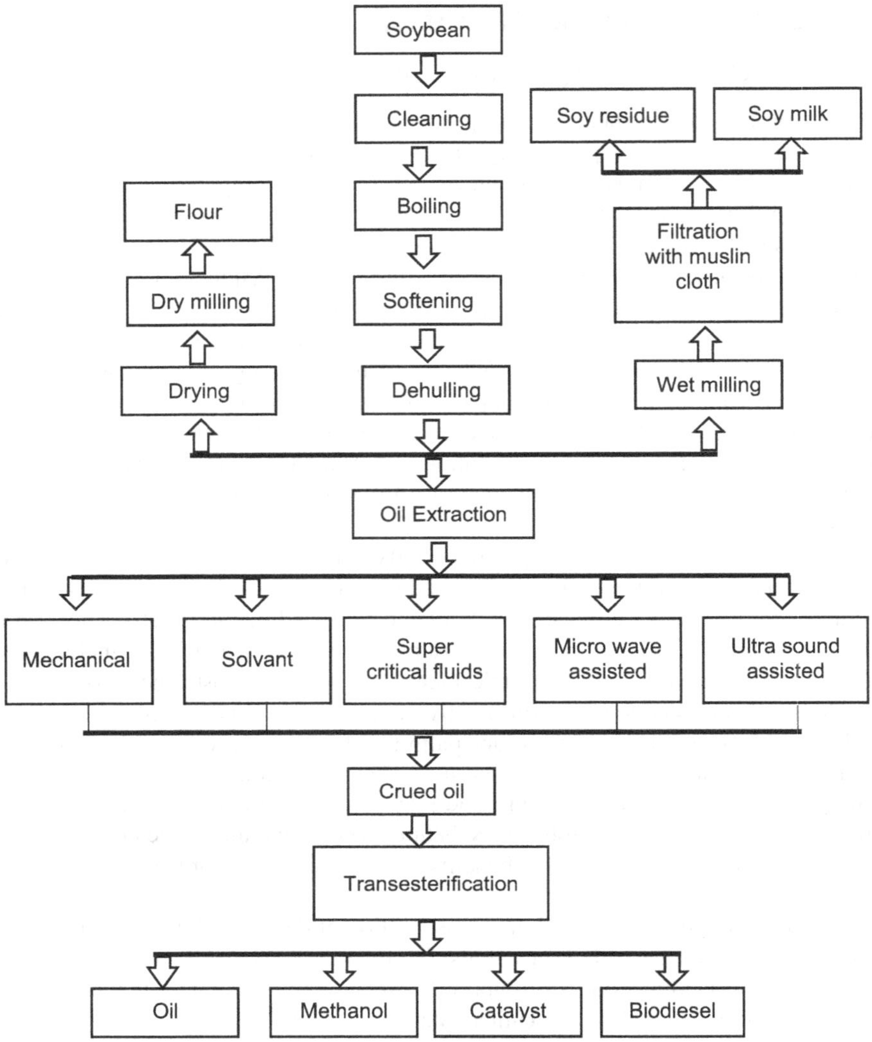

FIGURE 11.1 Processing of soybean. This figure provides a comprehensive overview of different methods employed in soybean processing. There are three basic methods which include dry milling, wet milling, and oil extraction. The crude oil could be esterified to be used as biodiesel, catalyst, as well as oil.

Soybean isoflavones can replace antibiotics in the production of chicken (Liu et al., 2020). By lowering the lipid content and cholesterol level in the food of poultry treated with isoflavones from vegetable soybean, broiler chicken carcass quality has improved (Utami & Hertamawati, 2020). To manufacture chips from vegetable soybean to be used as a source of chicken food, the pods and beans taken from the processing industry are fermented. In black-boned hens, fermented vegetable soybean meal promotes muscle weight and immunity (Buwjoom et al., 2016; Sukit et al., 2017). Probiotic bacteria were said to be delivered by fermented soymilk made from vegetable soybeans, as well as by soybean products with reduced concentrations of non-digestible oligosaccharides (Battistini et al., 2018). Further, investigation of fermented vegetable soybean meal on human and animal consumption should be the prime focus as they could contribute positively to enhancing chicken production.

11.3.2 Benefits and Hazards Associated with Soybean

Soybeans have an outstanding nutrient and dietary profile (Zhu et al., 2022). Soybean contains very large amounts of proteins, carbohydrate complexes, fatty acids, amino acids, and soybean oil (Rafiee-Yarandi et al., 2016). With the possible anti-hypertensive, anti-cancer, and anti-inflammatory capabilities, the bioactive peptides present in soybeans have been regarded as advantageous for human health (Chatterjee et al., 2018). Human health is affected by every byproduct of soybeans (Figure 11.2), and they can all be removed using various methods (Chen et al., 2021).

Isoflavonoids, the most known group of chemicals present in all portions of the soybean, have been investigated for their potential preventive properties in relation to chronic illnesses, cancer, osteoporoses, and menopausal conditions (Thangavel et al., 2019). Other phytochemicals found in soybean seeds, like saponin, lunasin, and sterols, include significant nutrients and neutraceuticals that have an impact on human health (Rizzo & Baroni, 2018). There is a need to investigate the nutritional profile of every component of soybean so that it can better be exploited for human health.

Menopausal symptoms, osteoporosis, cardiovascular illnesses, and hormone-dependent malignancies can all be reduced by consuming phytoestrogen-rich soybeans and herbal derivatives (De Franciscis et al., 2019). Soybean oil contains significant amounts of phosphate, magnesium,

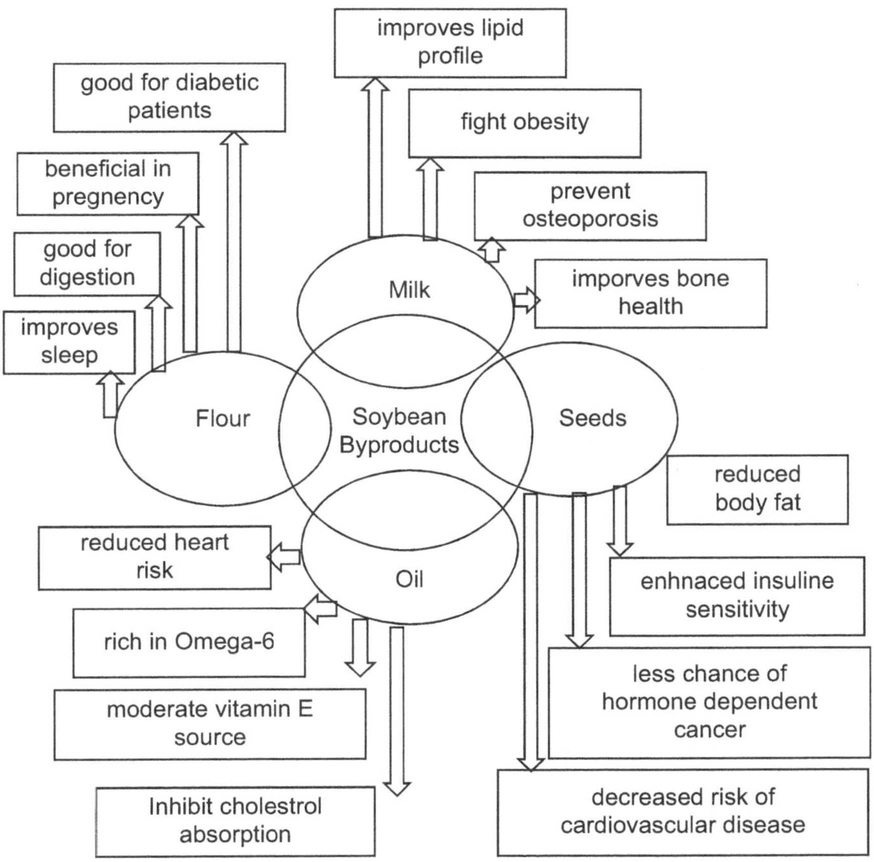

FIGURE 11.2 Beneficial and healthy impacts of different byproducts of soybean intake on the human body. This figure represents the important effects of using soybean byproducts on human health.

and potassium (Taketani et al., 2015). More consumption of soybeans is strongly correlated with a lower risk of cancer (Coward et al., 1998). A study indicated that soy proteins helped reduce blood cholesterol levels (Coward et al., 1998). Especially, the existence of isoflavones in soybeans may be connected with the ability to bring a biological change to the target with consequent advantages of soy production (Johnson & White, 2008). As phosphorus and fat interact in the blood, phospholipids are created, and this substance functions as a structural and metabolic component of human cell membranes as well as a component of bodily tissue structure (Taketani et al., 2015).

The American Heart Association and the National Lipid Association advise substituting sources of saturated fatty acids with soybean oil to increase circulating lipid and lipoprotein levels, which are the key targets for the prevention and management of cardiovascular illness (Arnett et al., 2019). Soybeans contain phytochemicals that possess anticancer, anti-aging, anti-renal failure, anti-obesity, and anti-cholesterolemic activities (Kim et al., 2021). These phytochemicals limit the growth of HIV, stop the development of gallstones, senile dementia, and hyperlipidemia, as well as stimulate diuretic activity, decrease arteriosclerosis, relieve constipation, and guard against cardiovascular disease (Hao et al., 2020). Additionally, soybeans attenuate fibrinolytic activity, improve lipid metabolism, control blood pressure, and exert anti-mutagenic, anti-carcinogenic, and antibacterial effects (Peighambardoust et al., 2021). Soybean is a beneficial dietary choice for the prevention and management of cardiovascular disease due to its ability to increase circulating lipids and lipoprotein levels.

11.4 ABIOTIC STRESSES AND SOYBEAN

Under harsh climatic conditions, abiotic stresses continually challenge our ability to boost crop productivity (Kairam & Sran, 2022) with ultimate reduction in the production area worldwide (Yijun et al., 2022). Environmental elements that negatively impact plant growth and productivity include drought, salinity, and extremely high temperatures (Shahid et al., 2020). These environmental factors affect the soybean plant at morphological, biochemical, and physiological level and result in reduced yield (Table 11.1).

11.4.1 Flooding

Flooding stress slows down the growth, productivity, and quality of soybeans, which endangers the world food supply chain (Yijun et al., 2022). Flooding stress resulted in chlorosis, necrosis, decreased photosynthesis, growth, and nitrogen fixation during the vegetative and reproductive stages of the plant, which ultimately led to plant mortality (Hasanuzzaman et al., 2016) with increased mortality rate and electrolyte leakage as well as delayed flowering ornamentation and matureness in comparison to control environment (Sathi et al., 2022). Flooding stress reduced the productivity of overall soybeans by 17–43% and 50–56% at the vegetative as well as the reproductive stage, respectively (Adegoye et al., 2023). The stress of flooding in soybeans primarily inhibits the growth of the plant through reduced hypocotyl pigmentation (Ma et al., 2020), which results in reduced oxygen levels within cells and the production of proteins associated with oxygen-free metabolic pathway (Pan et al., 2021). The biochemical components within soybean, particularly linolic/linolenic acid, decreased by flooding stress (Yijun et al., 2022). There is a need to develop flooding-tolerant soybeans as soon as possible for sustainable agriculture.

Flooding tolerance is a multigenetic characteristic (De Oliveira, 2021), and its mechanism in soybean is not completely understood due to the limitations of soybean germplasm (Wang & Komatsu, 2020). Soybean growth under flooding conditions could be increased using conventional techniques such as breeding methods and the utilization of various fertilizers and hormones (Kaur et al., 2017). Morpho-anatomical variations such as the development of adventitious roots in their stem (Sathi et al., 2022) and the presence of aerenchyma tissues in root, nodules, and hypocotyl may reduce oxygen deficiency by facilitating oxygen delivery (Ploschuk et al., 2022) and contributed to the tolerance mechanism. The application of plant growth regulators is a long-familiar agronomic

TABLE 11.1
Effects of Abiotic Stresses on Morpho-Physiology of Soybean

Stresses	Morpho-Physiology	References
Flooding	increased antioxidant enzyme activity, decreased the contents of non-enzymatic antioxidants	Wang et al., 2022
	suppressed the root growth	Wang et al., 2017
	reduced shoot growth/leaf size, caused a nutritional imbalance	Zandalinas et al., 2020
	induced chlorosis, necrosis, reduced photosynthesis rate, reduced growth, decreased nitrogen fixation and plant death during vegetative/reproductive phases	Hamayun et al., 2021
	promoted anaerobic conditions, ethanol accumulation, decreased sugar contents, and increased cell conductivity	Wang et al., 2018b
	reduced root growth	Sathi et al., 2022
	reduced root dry weight, total root length/ surface area	Sakazono et al., 2014
	enhanced root elongation, suppressed cell death in the root tip	Oh & Komatsu, 2015
	increased cytosolic calcium content	Oh et al., 2014
	regulated aquaporins	Wang et al., 2017
	suppressed hypocotyl	Maurel et al., 2015
	increased fresh weight, shortened adventitious root	Yin et al., 2014
	increased adventitious root	Xin et al., 2015
	increased fresh weight of germinating seed, decreased growth	Yin et al., 2014
	suppressed cell-wall lignification	Wang et al., 2016
Salinity	decreased percentage of phospholipids and lecithin in oil	Parveen et al., 2016
	reduced chlorophyll a,b, total chlorophyll, and carotenoid contents	Hashem et al., 2019
	reduced chlorophyll a, b, a + b, and carotenoids	Osman et al., 2021
	reduced photosynthetic rate and stomatal conductance	Liu et al., 2019
	decreased growth, yield, as well as quality	Afrin et al., 2021
	increased seed vulnerability to pathogen attacks, reduced economic value of seed	Wang et al., 2018c
	reduced chlorophyll and carotenoid photosynthetic rate, intercellular CO2 concentration, and stomatal conductance	Soliman et al., 2020
	reduced chlorophyll a, b, and carotenoid contents	Sadak et al., 2020
	reduced leaf chlorophyll, carotenoids, and anthocyanin's contents	Ghassemi-Golezani et al., 2018
	reduced chlorophyll a and b content	Abulfaraj et al., 2021
	reduced total chlorophyll and carotenoid content	Tamang et al., 2021
	reduced chlorophyll a, b, a/b, and total chlorophyll contents, reduced RuBisCO activity	Farhangi-Abriz et al., 2018
	decreased germination, seedling fresh weight, and radicle length	Shi et al., 2017
	reduced antioxidants activities	Begum et al., 2022
	increased ionic imbalance and electrolyte leakage	Nigam et al., 2022
	reduced growth and biomass	Sheteiwy et al., 2021
	enhanced chlorophyll contents, soluble sugar, catalase activity, plant height, and pods per plant	Shi et al., 2017
	enhanced salt tolerance	Li et al.,2017
	reduced seed germination percentage and seedling growth	Çirka et al., 2021

(Continued)

TABLE 11.1 ***(Continued)***
Effects of Abiotic Stresses on Morpho-Physiology of Soybean

Stresses	Morpho-Physiology	References
	reduced the nitrogen, phosphorus, and potassium contents	Otie et al., 2021
	reduced total phenolic contents	Osman et al., 2021
	decreased contents of protein and phospholipids	Liu et al., 2021
	decreased yield by affecting number of seeds/plant	Abdelghany et al., 2020
	decreased level of abscisic acid and hydrogen peroxide	Sadak et al., 2020
	reduced oil contents	Kataria et al., 2019
	reduced stomatal conductance	Parveen et al., 2016
	suppressed gibberellic acid and abscisic acid levels in cell	He et al., 2016
	decreased leaf reflectance and chlorophyll contents	Whitham et al., 2016
	reduced leaf length, fresh weight, and plant height	Komatsu et al., 2015
	enhanced germination and vigor	Kargar & Kareh, 2016
	decreased seed germination	Hashem et al., 2019
	reduced number of leaves	Farhoudi et al., 2015
	inhibited seed germination	Kumar, 2017
	decreased number of flowers, pod length, seed yield, and weight	Mangena et al., 2020
	decreased pod number, seed number, and weight	Sadak et al., 2020
	reduced number of pods plant^{-1}and harvest index	Kataria et al., 2019
	decreased number of pods per plant, seeds, and seed weight	Baghel et al., 2016
	reduced pods and seeds per plant	Ghassemi-Golezani et al., 2016
	reduced seed yield, increased soluble sugar and protein contents	Farhangi-Abriz et al., 2018
	reduced number of pods per plant, protein and oil contents, and seed yield	Akram et al., 2017
	increased total protein content	Tamang et al., 2014
	reduced soluble protein and sugar content	Zaki et al., 2019
	decreased number of pods per plant	Khan et al., 2019
	reduced protein and oil content	El-Esawi et al., 2018
	reduced number of pods plant	Parveen et al., 2016
	decreased number of pods plant	Khan et al., 2016
	reduced physiological response of plants	Feng et al., 2020
	reduced seedling length and fresh weight, and shorter roots	Putri et al., 2017
	reduced hydrolysis and translocation of nutrient reserves, decreased seedling vigor	Parveen et al., 2016
	reduced root/shoot length, dry weight, and seedling height,	Agarwal et al., 2015
	reduced water absorption	Kumar, 2017
	decreased yield traits and quality	El Sabagh et al., 2015
	enhanced germination	Zuffo et al., 2020
Chilling	reduced growth and development	Kuczyaski et al., 2020
	inhibited growth of seedlings, biomass accumulation, reduced photosynthesis	Feng et al., 2020
Drought	enhanced germination	Zuffo et al., 2020
	enhanced hormone synthesis in the root	Zhang et al., 2020
	increased abscisic acid accumulation	Roychoudhury & Basu, 2021

TABLE 11.1 *(Continued)*
Effects of Abiotic Stresses on Morpho-Physiology of Soybean

Stresses	Morpho-Physiology	References
	reduced growth of meristematic regions	Tiwari et al., 2022
	decreased transpiration rate	Mak et al., 2014
	decreased root lengths and dry biomass accumulation	Thu et al., 2014
	reduced root depth, root branching density, and root angle together with the partitioning of root to shoot biomass	Fenta et al., 2014
	changed shoot biomass and root structure	Dubey et al., 2021
	inhibited root growth	Michaletti et al., 2018
	enhanced nitrogen metabolism	Cao et al., 2018
	increased shelf life, improved oxidative stability	William et al., 2014
	increased storage proteins, namely water-soluble albumins, saline-solution soluble globulins, alcohol-soluble prolamins, and alkali-soluble glutelins	Arrutia et al. 2020
	reduced yield	Chen et al., 2021
	reduced rate of net photosynthesis	Farooq et al., 2017
	reduced growth, yield, and protein/oil content of grain	Miransari et al., 2016
	reduced yield components such as the number of flowers, total number of pods, percentage aborted flowers, harvest index, biomass, and grain yields	Mustafa et al., 2014
	injured effect of cell death in root	Hashimoto et al., 2020
Temperature	increased rate of aging	Chen et al., 2018
	suppressed seed germination	Hamayun et al., 2021
	decreased yield	Li et al., 2022
Pb, Cd	inhibited enzymes and growth	Tripathi et al., 2016
Heat waves combined with drought	decreased yield	Zandalinas et al., 2020
Heat stress	reduced seed development, reduced yield and quality	Cohen et al., 2021
	reduced growth and development	Bellaloui et al., 2017
	reduced seeds germination	Chebrolu et al., 2016
	altered vacuole structure and membrane integrity, reduced the seed protein contents	Krishnan et al., 2020
	decreased the contents of salicylic acid	Khan et al., 2022
	reduced soluble sugar and proline contents	Wang et al., 2018a
	reduced root growth and structure	Valdés-López et al., 2016
	reduced accessibility of assimilates to grains reduced sink-grain metabolization	Sharifi et al., 2022
	reduced yield	Thomey et al., 2019
	reduced seed oil and protein concentration	Ortiz et al., 2022
	reduced antioxidant enzyme activity and epicuticle wax content	Jumrani & Bhatia, 2019
	reduced concentration of lipids, metabolites, carbohydrates, and amino acids	Das et al., 2017
	reduced the seed coat wrinkling, coat boron, lignin, hard seed	Bellaloui et al., 2017
Climatic change	reduced growth and yield	Silva et al., 2021

practice to mitigate the toxicant aspects of flooding stress in soybean (Kang et al., 2021; Pan et al., 2021; Wang & Komatsu, 2022) such as abscisic acid, ethylene, and auxin serve as a mean of signaling component for controlling more precisely submergence stress response (Tamang et al., 2021). Morpho-anatomical variations and hormonal applications could serve as signaling factors that differentiate the stress-controlling reaction in soybeans, which is peculiar for submergence.

11.4.2 Drought

The scarcity of water is one of the most widespread environmental stresses that negatively impact the growth and productivity of crops (Du et al., 2020). Drought alters the allorhizic root architecture of soybean, including changes to root depth, branching, density, and angle as well as the segmentation of root/shoot biomass with an addition in root mass (Fenta et al., 2014). Plant physiological processes associated with the growth, development, and production of crops are impacted by drought (Wijewardana et al., 2017; Toscano et al., 2019) by hampering meristematic regions (Tiwari et al., 2022). Osmotic stress and reactive-oxygen species (Sarker & Oba, 2020) cause oxidative damage and alter the various functions of physiology and biochemistry, which include variations in the membrane, DNA, protein, and nutrient (Sarker & Oba, 2018) and ultimately reduce the photosynthetic efficiency and pigmentation (Hossain et al., 2022). Drought has a negative impact on growth, nodulation, blooming, seed quality, and seed number, which can result in a 40% reduction in soybean output (Li et al., 2021). Under drought conditions, grain yield is the appropriate characteristic for the selection of a variety for its ability to tolerate water stress.

In plants, tolerance to drought is a highly complex phenomenon that involves a variety of genetic, morphological, physiological, biochemical, and molecular pathways (Mishra et al., 2021). Among an array of morpho-physiological characteristics including root-shoot length/biomass, vigor index, relative water content, turgidity, and stomatal conductance, all have a significant role in determining the degree of drought adaptation (Kachare et al., 2019; Sahu et al., 2020). Latest technology such as the development of transgenic soybean enhanced stress tolerance such as the determination and depiction of soybean GmLOR Gene Family via genomics indicated that *GmLOR4/10/14/19* were significantly up-regulated after polyethylene glycol and salinity applications and contributed crucial roles in the betterment of plant tolerance against abiotic stress (Fang et al., 2021). The upstream regions of GmVOZs and GsVOZs include stress-related cis-elements, which emphasize the importance of these regions in abiotic stress tolerance (Rehman et al., 2021). The use of transgenic soybeans has been found to enhance stress tolerance, which may be a promising approach to improving drought tolerance in soybeans.

11.4.3 Salinity

Salinity stress, a troubling environmental condition, has antagonistic outcomes on the growth, yield, and quality of soybeans (Battisti et al., 2021). Since soybean is categorized as a salt-sensitive crop, all developmental processes are impacted during salt stress (Egamberdieva et al., 2017). Salinity stress causes a 40% reduction in soybean output and can result in complete plant loss (Pathanet et al., 2007). This is because of the absorption and accumulation of salt ions sodium ions (Al-Zahrani et al., 2022), and chloride ions in large concentrations (Li et al., 2022). Under salinity stress, soybeans were found to have shorter leaves, plant height, and reduced fresh weight (Amirjani et al., 2010) and caused oxidative stress by producing reactive oxygen species (Weisany et al., 2012). Different cultivars of soybean suffer a reduction in growth and total biomass under saline conditions (Sheteiwy et al., 2021). Further extensive studies are very crucial in understanding the soybean's responses to salt stress.

Saline tolerance is a multigenic trait, and classical breeding techniques take a long time and money to enhance its tolerance (Shahbaz & Ashraf, 2013). Transcription factors such as GmNAC06, a NAC domain enhanced tolerance to salinity in soybean by the assemblage of proline and glycine betaine which in return relieved or obviated the bad impacts of reactive oxygen species and controlled the ratio of Na+/K+ ions in hairy roots to keep ions in balance condition (Li et al., 2021). Similarly, through overexpression and RNAi analyses, GmWRKY27 enhanced transgenic soybean hairy root resistance to salt and drought (Wang et al., 2015). By reducing malondialdehyde buildup and increasing peroxidase and superoxide dismutase activity in soybean's hairy roots GmVOZIG positively controls their ability to withstand drought and salt stress (Li et al., 2020). The GsPRX9 gene was overexpressed, which improved the soybean's ability to withstand salt stress and respond to antioxidants. Root fresh weight, primary root length, peroxidase/ superoxide dismutase activity, and glutathione levels increased; however, H_2O_2 content decreased (Jin et al., 2019). The up-regulation of the above-mentioned genes could ameliorate tolerance against abiotic stress in transgenic plants through regulating ion homeostasis, accumulating compatible solutes, reducing oxidative damage, and enhancing antioxidant response.

11.4.3.1 Temperature

Soybean is a leguminous crop, which provides several essential dietary compounds. Extreme temperature at the germination stage has a negative impact on soybean production and quality (Jianing et al., 2022). Heat stress significantly affected the seed germination of soybean (Veas et al., 2021; Hamayun et al., 2021) with inflated seed danger to pathogen attack, and decreased economic value (Wang et al., 2018). Heat stress has detrimental effects on the growth, biochemical, and physiology parameters of soybeans (Sharifi et al., 2022). Heat stress promotes the closing of stomata and membrane damage, inhibits the actions of numerous enzymes, and decreases the metabolic activity of sink grains from immature leaves with reduced access to assimilates to soybean grains (Ergo et al., 2018). Heat stress caused oxidative damage, decreased stomatal conductance, and diminished the shoot/root dry weight (Sharifi et al., 2022). Metabolic pathways were impaired because of heat stress (Kosová et al., 2011), with some growth alteration (Liu et al., 2014). To cope with the stress alterations in soybeans, there is a need to find ways to develop any tolerance mechanism within the plant.

Global warming and climate change increased the episodes of heat waves (Balfagón et al., 2019; Ault, 2020) and modern agriculture requires enhanced tolerance to heat stress via breeding the crops (Bailey-Serres et al., 2019; Zandalinas et al., 2020). Hormones show a captious function in regulating the plant's response to heat stress since even little changes in hormone concentration can have a significant impact on cellular dynamics (Singh & Meena, 2020). Auxin, gibberellins, and cytokinin are generally involved in the development of heat tolerance during heat waves (Ozga et al., 2017). Abscisic acid and salicylic acid concentrations in soybeans increased as a consequence of heat stress, exhibiting that these phytohormones are participating in promoting tolerance from heat to support growth and productivity (Khan et al., 2020). Soybean seedlings were treated with auxin, which enhanced the germination rate by accelerating the production of abscisic acid (Shuai et al., 2017). Further research is required to clarify the changes in hormone content and signaling for seed germination in soybeans under heat waves (Sabagh et al., 2020). High heat tension changes the density of phytohormones and signaling response which could contribute to increasing tolerance from heat in soybean.

11.4.3.2 Others

Throughout their course of life, plants are regularly subjected to such environmental circumstances, which have a negative impact on their growth and development. One of the main factors contributing to global agriculture loss is abiotic stress (Wang et al., 2018). Rapid industrialization created the issue of heavy metals accumulation within the plants. When soybean is cultivated in soils with high levels of heavy metals, it ultimately carries those toxins into the human food chain, causing danger to human health (Schwalbert et al., 2021). Lead levels were taken down than cadmium in seeds, and the aggregation was localized in the vegetative organs, with 93% of the lead integrated according to an assessment of high carbon dioxide concentrations and occurrence of heat stress in soybean cultivars thriving in heavy metal-polluted soils. Additionally, pods accumulated 11 times more lead than seeds, suggesting that they operate as a barrier to the transmission of lead to their offspring (Blanco et al., 2022). Soil disturbance with the addition of heavy metals could change the composition and quantity of soil microorganisms with ultimate impact on the availability of other nutrients in the soil and decrease the crop yield.

Photoperiodic flowering is one of the key elements influencing regional modifications and soybean production (Lin et al., 2021). A longer photoperiod can result in a longer time to flower, with delayed maturity, and the crop becomes vulnerable to fall frost, which can diminish seed output and quality (Ort et al., 2022). Soybean was initially a short-day plant and as a result, requires short day lengths to induce flowering, photoperiod is a crucial consideration in soybean agriculture (Hamner, 1940). Mutation in the J locus, which offers a longer flowering phase and consequently increases yield in short-day plants, provides adaptation to low-latitude situations (Bu et al., 2021). However, because of breeding initiatives aimed at this reaction, cultivars are now accessible and are minutely delicate to long-day conditions, allowing soybean cultivation between the north and south poles (Zhang et al., 2020). Overall, balancing photoperiod requirements with regional adaptation as well as incorporation of the latest technologies is crucial for maximizing soybean yield and quality.

11.5 PROTEIN VARIATIONS IN SOYBEAN

Climate change is posing a terror to the world's food supply and food security (Jianing et al., 2022). Drought, salinity, and cold are just a few of the stress factors that seriously hamper the growth of the crops (Wang et al., 2016). Low yields are typically caused by environmental conditions including extreme cold or heat, prolonged drought or flooding, too much salt in the soil, and other abiotic variables (Taheri et al., 2022). To combat stresses, plants have internal defense systems that are coordinated by suitable solutes, reactive-oxygen species scavengers, and phytohormones (Bhat et al., 2021). Signaling molecules known as phytohormones govern essential physiological processes and influence plant's reactions to biotic and abiotic challenges (Waśkiewicz et al., 2016). The majority of the known expressed proteins in soybeans were shown to participate in photosynthesis, glucose metabolism, signal transduction, protein metabolism, defense, and stress tolerance under abiotic stress conditions, according to the proteomic study (Jan et al., 2022) (Table 11.2).

11.5.1 Flooding

Soybean is a significant crop that contains a lot of protein, vegetable oil, and phytohormones (Wang & Komatsu, 2018). Soybean growth is adversely impacted by flooding, and various stress reactions are unique to flood such as glycosylation (Wang et al., 2016), and phosphorylation was identified (Yin et al., 2016). The rRNA processing protein Rrp5 and zinc finger/BTB domain-containing protein 47, which both activate the abscisic acid, were phosphorylated as a consequence of the excessive water stress (Yin & Komatsu, 2015). Flooding predominantly impacted the various physiological processes in all parts of soybean (Wang et al., 2017). Nuclear proteomics revealed that 17 chromatin structure-related proteins decreased in response to flooding stress (Yin & Komatsu, 2016). Understanding the flooding-specific response at the organelle level would help to mitigate the stress consequences.

Proteins involved in protein synthesis and RNA regulation are crucial for soybeans to become tolerant to the initial flooding stress (Yin et al., 2016). During recuperation, the peroxidase superfamily protein was increased at the mRNA level and its abundance rose consistently (Khan, 2014). The ratios of RNA binding/processing-related proteins and flooding stress indicator proteins were strongly connected with the flooding tolerance index, according to the analyses of the flooding tolerance of soybean varieties (Nanjo et al., 2014). Tolerance from flooding stress was enhanced by the secondary metabolites such as isoflavonoids in soybeans. Further extensive studies on secondary metabolites and indicator proteins could lead to the development of flooding-tolerant soybean varieties (Li et al., 2019). There are several key factors involved in the initiation of tolerance to scarce water stress in soybeans, including protein synthesis and secondary metabolites.

TABLE 11.2
Proteomic Analysis of Soybean Exposed to Various Treatments

Stress	Organ	Subcellular	Main Findings	References
Flooding	Root	mitochondria	reduced the RNA binding/ processing related proteins	Nanjo et al., 2014
	Seedling	mitochondria	increased phospho-peptides enriched protein	Yin et al., 2014
	Seedlings	mitochondria	reduced the protein concentration and deactivated APX	Damanik et al., 2016
	Root and hypocotyl	endoplasmic reticulum	enhanced biochemical changes to counter toxic effects	Khan et al., 2022
	Radicle/root/hypocotyl	endoplasmic reticulum	increased glycoproteins	Komatsu et al., 2021
	radicle/root/hypocotyl	RNA	increased growth of pods, upregulated the gene expression of beta-glucosidase	Wang et al., 2021
	Seeds	–	decreased abundance of proteins mainly involved in the primary metabolism	Yin et al., 2014
	Radicle/root/hypocotyl	RNA	increased proteins	Wang et al., 2017
	Roots	RNA, DNA,	decreased energy resources, loss of carbon, deposited noxious compounds, mainly lactate	Tamang et al., 2014
	Radicle/root/hypocotyl	mitochondria	increased mortality rate and electrolyte leakage	Sathi et al., 2022
	Radicle/root/hypocotyl	reactive oxygen species-induced oxidative stress	inhibited antioxidant enzyme activity	Anjum et al., 2015
	Radicle/root/hypocotyl	mRNA	reduced root number/root surface/leaf area, inhibited plant growth	Yin et al., 2014
	Root	RNA in mitochondria	activated auxin-triggered transcriptomic modifications and suppressed cell death	Shimamura et al., 2014
	Root	RNA in mitochondria	increased expression of *TOC1*coincided with elevated abscisic acid content	Kim et al., 2018
	Leaves	chloroplast	enhanced energy conservation, cell wall integrity, inhibited cytochrome	Wang et al., 2018b
	Root tip	RNA	increased phosphorylation of the BTB domain-containing protein 47, increased glycine protein, and rRNA processing protein Rrp5	Yin et al., 2015
	Leaves	chloroplast	enhanced glutathione activity, chlorophyll content, no signaling	Khan et al., 2018

(Continued)

TABLE 11.2 ***(Continued)***
Proteomic Analysis of Soybean Exposed to Various Treatments

Stress	Organ	Subcellular	Main Findings	References
	Root	mitochondria	improved root anatomy, photosynthetic pigment, reactive oxygen species scavenging, increased biomass	Pereira et al., 2020
	Leaf and root of the seedling	RNA, DNA,	suppressed gamma-aminobutyric acid shunt, reactive oxygen species scavenging, mitochondrial impairment, ubiquitin/proteasome-mediated proteolysis, and cell-wall loosening	Wang et al., 2018b
	Root	RNA in mitochondria	inhibited abscisic acid elongation of cells derived from phellogen	Mutava et al., 2015
	Leaves	RNA in mitochondria	reduced abscisic acid content	Kim et al., 2018
	Root	RNA in mitochondria	downregulated abscisic acid	Syed et al., 2015
	Root tips	mitochondria	increased alcohol dehydrogenase mRNA and its protein	Komatsu et al., 2015
	Seed, Root, and hypocotyl	mitochondria	increased protein metabolism, cell, and amino acid metabolism, localized protein in cytosol, mitochondria, and plastids	Mustafa et al., 2016
	Root/hypocotyl	mitochondria	increased 327 proteins	Komatsu et al., 2015
	Soyabean	cell wall and plasma membrane	reduced protein synthesis-related proteins in the nucleus that respond to abscisic acid, increased lipoxygenases, four germin-like proteins, three stem glycoproteins, and a Cu-Zn superoxide dismutase	Yin et al., 2016
	Whole plant	nucleus	reduced nuclear proteins	Yi et al., 2017
	Whole plant	nucleus	reduced chromatin structure-related nuclear proteins	Yin et al., 2014
	Root, hypocotyl	mitochondria	reduced mitochondrial membrane protein spots and matrix protein spots, effected electron transport chains, through NADII production	Komatsu et al., 2014
	Root	endoplasmic reticulum	decreased calnexin, heat shock protein, and luminal binding protein	Mustafa et al., 2014
	Root and hypocotyl	mitochondria	decreased voltage-dependent anion channel 1, serine hydroxymethyltransferase 2, NAD(P)-binding Rossmann fold superfamily protein, and isocitrate dehydrogenase	Mustafa et al., 2016

Drought	Root	mitochondria	increased proteins related to energy metabolism, amino acid, and carbohydrate metabolism	Yu et al., 2017
	Seedling	cell membranes	decreased Ferro proteins	Gharechahi et al., 2020
	Seed filling	chloroplast	increased chlorophyll a-b binding proteins	Xie et al., 2016
	Leaf	chloroplast	decreased plant morphological, physiological, and biochemical levels	Michaletti et al., 2018
	Soybean meal	mitochondria	decreased hormone metabolism, secondary metabolism, cell wall, redox, and glycolysis	Khan et al., 2016
	Soybean meal	mitochondria	enhanced proteins involved in photosynthesis, metabolism, stress response, and defensive reactions, and that redox-induced post-translational modifications	Xie et al., 2016
	Soybean meal	mitochondria	decreased protein folding/degradation, and signal conduction	Khan et al., 2016
	Root including hypocotyl	mitochondria	increased length of root including hypocotyl	Oh & Komatsu, 2015
	Leaf Hypocotyl Root	chloroplast, cytoplasm, nucleus, mitochondria	produced methionine, enolase, protein synthesis	Hossain et al., 2014
	Root including hypocotyl	chloroplast	effected stromal 70 kDa heat shock-related protein, ATP-dependent zinc metalloprotease, RuBisCO large subunit-binding protein	Das et al., 2017
combined stresses (flooding and drought)	Leaves, hypocotyls, and roots	mitochondria	suppressed protein synthesis	Wang et al., 2016b
	Leaves, hypocotyls, and roots	cytoplasm, plasma membrane, cytosol, endoplasmic reticulum	reduced signal transduction and metabolic processes	Wang et al., 2017
Salinity	Leaves, hypocotyls, and roots	mitochondria	downregulated glyceraldehyde-3-phosphate dehydrogenase and fructokinase 2	Widyasari et al., 2022
	Leaves	cytoplasm, plasma membrane, cytosol, endoplasmic reticulum	activated absorption of K+, regulated ions homeostasis, and several stress-responsive genes	Shu et al., 2017
	Hypocotyls and roots	cytoplasm, plasma membrane, cytosol, endoplasmic reticulum	enhanced proline and glycine betaine contents	Ke et al., 2022

(*Continued*)

TABLE 11.2 *(Continued)*
Proteomic Analysis of Soybean Exposed to Various Treatments

Stress	Organ	Subcellular	Main Findings	References
	Hypocotyls and roots	cytoplasm, plasma membrane, cytosol, endoplasmic reticulum	mediated salt stress by the complex regulatory network	Zhang et al., 2022
	Seeds, root	cytoplasm, plasma membrane, cytosol, endoplasmic reticulum	overexpressed GmMYB84, increased germination, and root elongation	Li et al., 2021
	Leaves	Chloroplast	enhanced content of chlorophyll and proline	Zhang et al., 2017
	Leaves, hypocotyls, and roots	mitochondria	increased flavonoid biosynthesis	Li et al., 2019
	Leaves, hypocotyls, and roots	mitochondria	increased proline contents	Pi et al., 2019
	Leaves, hypocotyls, and roots	mitochondria	enhanced activation of GmDREB6 genes for improved proline content	Nguyen et al., 2019
	Whole plant	mitochondria	regulated the 14-3-3, MMK2, PP1, TRX-h, proteins	Whitham et al., 2016
Al2O3-, ZnO-, Ag-NPs treatment	Root, leaf	ribosomes	increased aerobic pathway, and ascorbate glutathione pathway	Mustafa et al., 2016
Al_2O_3-NP	Root including hypocotyl	–	activated aerobic pathway, ascorbate glutathione pathway	Yasmeen et al., 2016
Ag-NPs treatment	Root, cotyledon	endoplasmic reticulum	regulated protein quality control of misfolded proteins	Hashimoto et al., 2020
Al_2O_3-, ZnO-, Ag-NPs treatment	Root, leaf	–	suppressed ubiquitin proteasome	Zhong et al., 2020
Water logging	Leaf petiole	–	mutated genes	Gao et al., 2017
Overexpression of the *GMRNF1* gene	Embryo frame	phagocytes and nonphagocytic cells	increased MADS-box domain-containing protein	Yang et al., 2022
overexpression of transcription factor	Branch and node	DNA and RNA	overexpressed squamosal promoter binding protein	Sun et al., 2019
Overexpression of transcription factor	Seeds	RNA	increased aquaporin	Lu et al., 2017
Overexpression of AP2	Whole plant	mitochondria	generated EREB-box binding protein	Guo et al., 2020

Overexpression of cytochrome P450	Seeds	mitochondria	increase seed size	Zhao et al., 2016
near-isogenic lines	Seeds	mitochondria	increased cytochrome protein	Wang et al., 2015
Artificial aging	Seeds	mitochondria	depleted protamine sulfate precipitation	Min et al., 2016
Controlled deterioration	Seeds	mitochondria	regulated Urease, Ni-binding urease accessory protein	Min et al., 2017
Millimeter-wave treatment	Oil	mitochondria	increased proteins related to glycolysis and ascorbate/glutathione metabolism	Zhong et al., 2020
Smoke treatment	Seed	mitochondria	regulated carbohydrate metabolism, glycolysis, and cell-wall components	Lin et al., 2018
Heat stress	Seeds	RNA, DNA	upregulated *GmCLC1* expression increased accumulation of Cl– and NO3–	Liu et al., 2022
	Seeds	RNA, DNA	enhanced yield, greater epicotyl diameter, and xylem area	Ribichich et al., 2020
	Seeds	RNA, DNA	unregulated and expressed in prolonged heat stress conditions	Song et al., 2017
	Seeds	RNA, DNA	activated GmPYL21 gene	Kidokoro et al., 2015

11.5.2 Drought

On drought-treated soybean plants, the use of label-free and stable isotope labeling-based quantitative proteomics and subcellular proteomics has revealed that the drought response involves osmotic adjustment, redox signaling, calcium signaling, biotinylation-regulated energy management, and programmed cell death (Wang et al., 2017; Wang & Komatsu, 2018), where S-adenosylmethionine synthetase play a crucial function in redox signaling and calcium signaling (Oh & Komatsu, 2015; Khan & Komatsu, 2016). Many stress responses in soybeans have been linked to several common proteins that are concerned with the redox system, and metabolism of carbon and amino acids as well as in photosynthesis and signaling (Qin et al., 2013). These findings provide valuable insight into the complex molecular mechanism underlying the drought response in soybean plants and highlight potential targets for future research and agricultural practices aimed at improving plant resilience to environmental stress within a short period.

Under drought conditions, the agrobacterium-mediated transformation of FvC5SD in soybean developed the transgenes of soybean, which concentrated less amount of reactive oxygen species with increased activity of enzymes and cells than wild-type soybean (Zhang et al., 2019). Transgenic plants of soybean with ectopic expression of AtABF3 were able to survive even the 20-day drought treatment (Kim et al., 2018). As part of signaling through phosphorylation, which controls the functions of ion channel activities, calcium-dependent protein kinase is thought to function in the establishment of soybean drought tolerance throughout the process of water deprivation (Moradi et al., 2021). The over-expression of certain genes via genetic engineering and manipulation of specific proteins or pathways could be effective approaches for improving soybean drought tolerance.

11.5.3 Salinity

Crop productivity is reduced by soil salinity, which eventually deteriorates arable land and causes desertification (Zorb et al., 2019). In plants, the net photosynthetic rate decreases, and photosynthetic assimilates are used for survival and maintenance rather than increasing the biomass (Munns & Gilliham, 2015). Three key functional categories were identified from a shotgun proteome investigation of soybean embryonic axis during salt stress germination, including response to disease defense, protein destination/ storage, and primary metabolism (Fercha et al., 2016) while, under salinity stress, the endoplasmic-reticulum protein involved in protein processing and energy metabolism were amplified (Yin et al., 2022). An increase in methionine synthase or SAM synthase was found in soybeans on exposure to salinity (Oh & Komatsu, 2015). Exogeneous calcium enhanced signal transduction, energy pathways, and transportation, promoted protein biosynthesis, inhibited proteolysis, redistributed storage proteins, controlled endoplasmic reticulum protein processing, enriched antioxidant enzymes, accumulated secondary metabolites and osmolytes, and other adaptive responses (Yin et al., 2015). Further research is needed to elucidate the complex interactions between the different functional categories of protein and their roles in the soybean response to salt stress.

To compete the salinity stress, soybean plants developed intricate metabolic and signaling mechanisms at the level of cell, organ, and whole plant (Ji et al., 2016). Soybean lines that could tolerate salinity stress, showed an improved ability to maintain the high quantities of nutrients needed to carry out metabolic processes (Ning et al., 2018). The high concentration of nitrogen, phosphorus, and potassium under saline stress could be antagonized by high nutrient content (Ondrasek et al., 2022). Salinity tolerance is improved by the osmotic accumulation of compatible solutes and scavenging reactive oxygen species (Flowers & Colmer, 2015). By promoting the buildup of dihydroxy B-ring flavonoids, salt-induced phosphorylation of GmMYB173, an increase in its affinity for the GmCHS5 promoters, and increased transcription of GmCHS5 all likely to contribute to soybean salt tolerance (Pi et al., 2018). The involvement of various osmolytes paved the way for understanding the tolerance mechanism in soybeans (Figure 11.3).

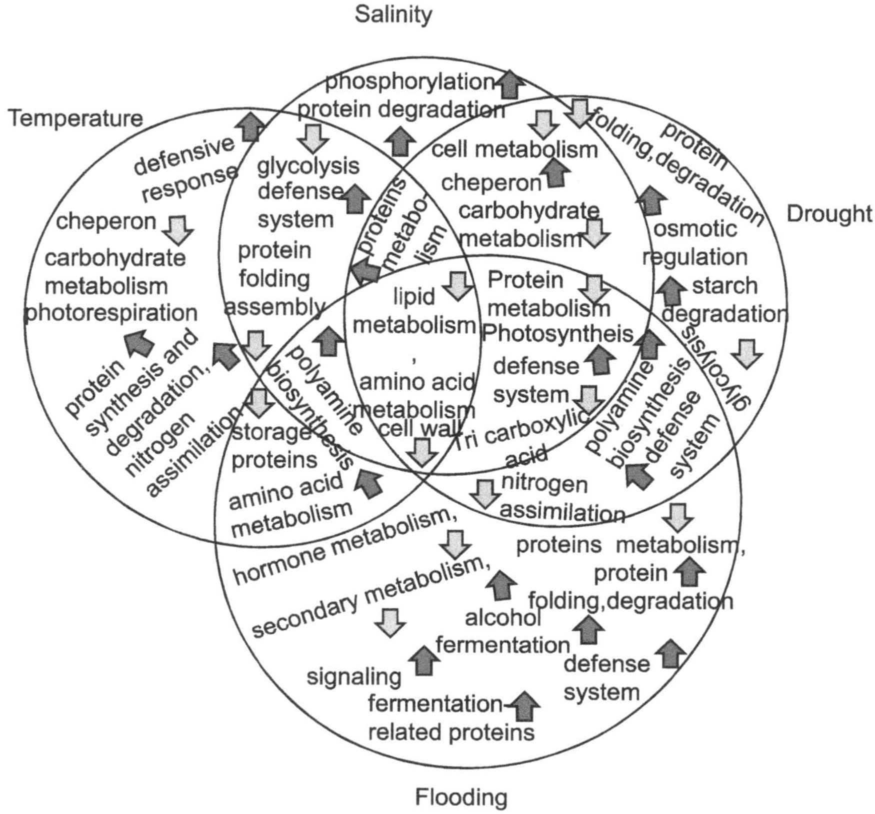

FIGURE 11.3 Proteomic variation in soybean under exposure to various external agents. The imposition of stressful conditions leads to alteration in the proteome of soybeans. Drought, salinity, temperature, and flooding greatly impact the metabolic processes of soybeans by altering the proteome under abiotic stress conditions. Red arrows indicate an increase while blue arrows indicate a decrease in the protein during stress conditions.

11.5.4 Temperature

Climate change associated with high temperatures frequently causes anthesis to occur, which affects the function and growth of soybean male reproductive organs and negatively affects soybean productivity (Li et al., 2020). The major mechanisms in pre-harvest soybean seed deterioration under high-temperature stress include the production of secondary metabolites, and the processing of protein in the endoplasmic reticulum (Shu et al., 2020). The high-temperature stress affects the development of vigor in a soybean cultivar to develop resistance to pre-harvest seed deterioration (Ma et al., 2016). The attributes such as a strong, stable, and viable embryo with potent photosynthetic ability and nutrition delivery in the leaf, and stronger protection in the pod during the high-temperature conditions all contribute to the production of high seed vigor (Wei et al., 2020). These results are important for improving crop resilience and productivity in the face of climate change-induced stresses. Overall, the insights gained from these studies render precious information for the evolution of soybean cultivars that are more resilient to environmental stress.

DS25-1 is a publically released line known for its ability to tolerate temperature stress (Smith et al., 2017) and was examined for the identification of various tolerance mechanisms which indicated that the high temperature did not affect the germination of DS25-1 seed; however, they had a 50% negative impact on the germination of seed from a representative of cultivars cultivated in the mid-south (Chebrolu et al., 2016; Gillman et al., 2019). Heat stress mainly impacts the process of photosynthesis by altering the photosynthetic apparatus (Zahra et al., 2023), and signaling about heat stress is pivotal for the endorsement of photosynthesis apparatus for the development of a heat tolerant line (Ohama et al., 2017). It was suggested that the transduction of heat signals involves a number of signaling molecules, including guanosine triphosphate-binding protein, nucleoside diphosphate kinase, annexin, and brassinosteroid-insensitive I-kinase domain interacting protein (Wang et al., 2017). The identification of a tolerance mechanism could lead to the development of heat-tolerant soybean varieties that maintain high germination rates and photosynthetic efficiency under elevated temperature conditions.

11.5.5 Others

Abiotic stresses brought unexpected levels of changes in the soybean proteins and altered the proteins involved in photosynthesis, redox systems, carbon metabolism, signaling pathways, and amino acid metabolism (Qin et al., 2013). A deficiency of phosphorus impacts photosynthesis mainly by changing the leaf structure, ATP synthesis, absorption/transport of carbon dioxide, photosynthesis, electron transport, production of assimilatory power, and levels of the Calvin cycle attached to enzymes (Chu et al., 2018). The expression patterns of several photosynthesis-related genes, such as RuBisCO activase, were characterized to elucidate the possible mechanism of low phosphorus in soybeans (Zhang et al., 2017). Furthermore, to reduce metal toxicity, soybean roots greatly increased the activities of stress-related proteins, such as glutathione S-transferase L3, isoflavone reductase-like, chalcone isomerase A, NAD(P)H dehydrogenase (quinone), FQR1-like 1 isoform X2, and Peroxidase 3 (Bilal et al., 2019). Further research is needed to better understand the molecular mechanisms underlying these responses and develop practical solutions for enhancing soybean growth and productivity under stress conditions.

The role of intrinsically disordered proteins in numerous crucial physiological processes including the efficient protection of lactate dehydrogenase from the inactivation induced by the freeze-thaw process; however, this activity was not associated with the number of abnormalities in protein during desiccation stress (Liu et al., 2017). Nanoparticles such as biosynthesized silver nanoparticles increased the growth of soybeans by controlling proteins involved in protein breakdown and ATP levels (Mustafa et al., 2020). These results underscore the need for interdisciplinary and holistic approaches to understanding and addressing plant stress responses, and for developing sustainable solutions that minimize negative impacts on both the environment and human health.

11.6 STRESS MITIGATION IN SOYBEAN UNDER ABIOTIC STRESS

Drought, cold, heat, salinity, flooding, and heavy metals are only a few examples of abiotic stresses that have a negative impact on plant physiology (Raza et al., 2022). Depending on the species and developmental stage of the plant, several coping mechanisms are used by plants to tackle abiotic stress (Mickelbart et al., 2015). Stress tolerance mechanisms for battling abiotic stresses in crop plants are crucial to be explored to ensure healthy crop growth and maximum output. (Raza et al., 2022). Senescence and leaf abscission are two processes that plants use as an escape mechanism to reduce the size of their canopy under stress (Kooyers, 2015). Senescence provides an ease for nutrients to be transferred from senescent cells to other areas of the plant that are still growing, and increasing plant development (Sachdev et al., 2023). The transport of proteins from chloroplasts to their digesting location in the central vacuole is regulated by a number of vesicular trafficking pathways (Xie et al., 2015), and autophagy is well explored among these pathways during senescence (Xie et al., 2015). The increased activity of autophagy during abiotic stress has a positive impact, while reduced activity has a negative impact on plants (Wang et al., 2017). Similar to this, apoptosis occurs to get rid of unhealthy, diseased, or undesirable cells, keeping cellular equilibrium (Sachdev et al., 2023). Abiotic stresses have harmful impacts on plants‘ physiology, however; plants have developed several strategies to cope with these stresses that include senescence, autophagy, and apoptosis.

A complex network of cellular, physiological, and morphological defenses is involved in a plant's reaction to harmful abiotic circumstances (de Melo et al., 2022). The cuticle, modification of membrane lipids to remodel the membrane, stimulation of non-enzymatic and enzymatic antioxidant systems to protect from reactive-oxygen species, aggregation of osmolytes, and installation of molecular chaperones in response to cell requirement are all factors that contribute towards plants reaction at the cellular level (He et al., 2020). Abiotic stresses usually cause the malfunction of proteins which could be analyzed via various techniques such as protein microarray technology (Gupta et al., 2016) and ribosome profiling (Ingolia, 2014). Reactive-oxygen species are molecules that are actively produced by plants and are exploited as signaling molecules, which are primarily produced at the apoplast of NADPH oxidases and some oxidases and peroxidases, as well as at the chloroplast, mitochondria, peroxisome, and possibly other cellular compartments, via various pathways (Mignolet-Spruyt et al., 2016). Decoding reactive oxygen species signals occur via different redox reactions by variations in the structure and function of proteins by modifying the binding of transcription factors to DNA and impacting the process of transcription (Dietz et al., 2016). Gene regulation factors, such as alternative splicing factors, are active at various levels after receiving the transduced stress trigger (Laloum et al., 2018). Plants have several mechanisms to cope with adverse abiotic conditions, including defenses at the cellular, physiological, and morphological levels which include activation of defense systems, malfunction of proteins, and activation of gene regulation factors (Figure 11.4).

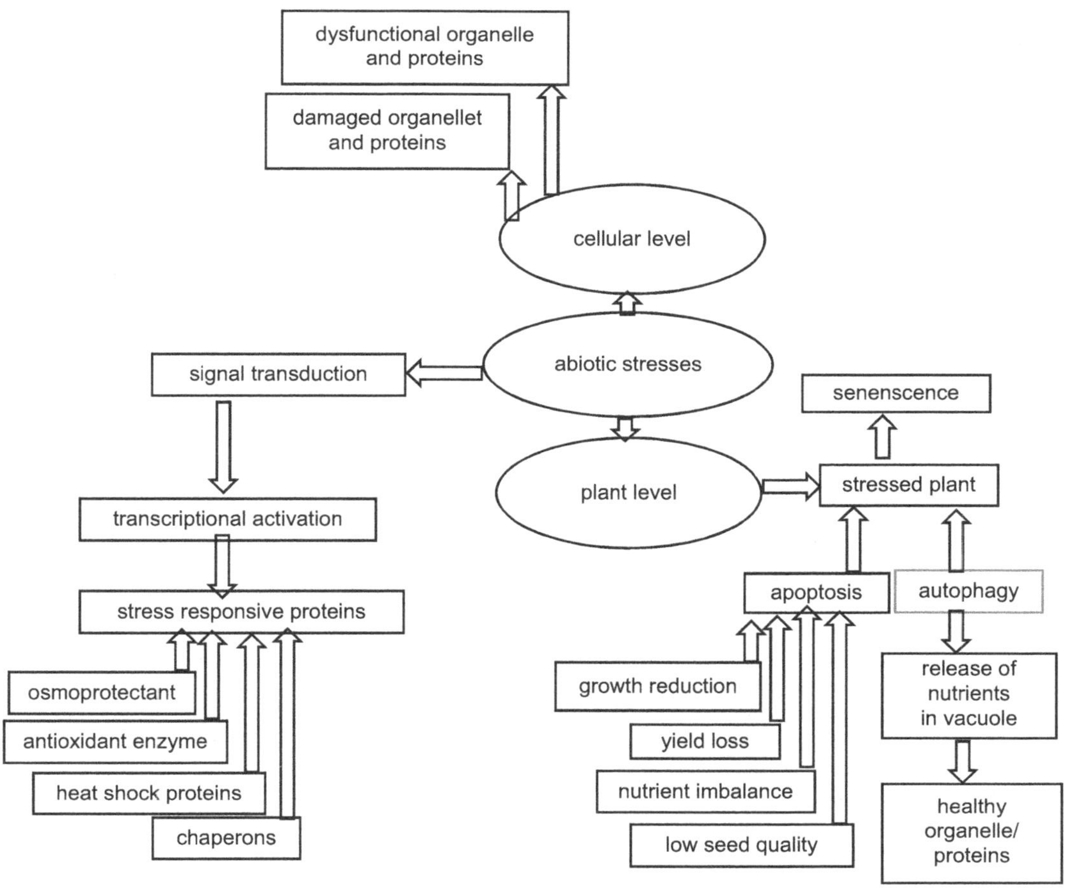

FIGURE 11.4 Mechanism of stress mitigation in soybean against abiotic stress. Abiotic stresses in soybean affects both at cellular and plant levels and mitigate the stress by adopting the ways of senescence, autophagy, or signal transduction for the formation of specific responsive proteins.

11.7 CONCLUSION AND PROSPECTS

The primary source of dietary protein and oil is soybean. Consumption of plant-based whole foods and plant-based protein products is rising quickly on the global market, and this trend is predicted to continue as more studies document their positive benefits on environmental sustainability and human health. According to dietary recommendations that promote using soybean oil as part of a balanced diet for overall health, it is the most extensively used oil. Various techniques for extracting oil in this review were outlined, along with any potential drawbacks. The advantages and risks of employing soybean byproducts have been thoroughly covered. A number of external environmental factors have a detrimental effect on both soybean growth and oil quality. Moreover, proteomics research showed that diverse abiotic stressors had a deleterious effect on the proteins involved in protein metabolism, photosynthesis, and tricarboxylic acid. Plants adopt strategies such as senescence, autophagy, or apoptosis to mitigate the stress response. At the same time, plants produce signaling molecules to cope with stresses for the improvement in crop growth and yield.

AUTHOR CONTRIBUTIONS

Conception, S.K; writing-original draft preparation, F.Y. and R.P.; writing – review and editing. S.K. All authors have read and agreed to the published version of the manuscript.

FUNDING

This research received no external funding.

The authors declare no conflict of interest.

REFERENCES

Abdelghany, A. M., Zhang, S., Azam, M., Shaibu, A. S., Feng, Y., Li, Y., Tian, Y., Hong, H., Li, B., & Sun, J. (2020). Profiling of seed fatty acid composition in 1025 Chinese soybean accessions from diverse ecoregions. *The Crop Journal*, *8*(4), 635–644.

Abulfaraj, A. A., & Jalal, R. S. (2021). Use of plant growth-promoting bacteria to enhance salinity stress in soybean (Glycine Max L.) plants. *Saudi Journal of Biological Sciences*, *28*, 3823–3834.

Adegoye, G. A., Olorunwa, O. J., Alsajri, F. A., Walne, C. H., Wijewandana, C., Kethireddy, S. R., Reddy, K. N., & Reddy, K. R. (2023). Waterlogging effects on soybean physiology and hyperspectral reflectance during the reproductive stage, *Agriculture*, *13*, 844.

Afrin, S., Akhtar, N., Khanam, T., & Hossain, F. (2021). Alleviative effects of zinc on biomass yield and antioxidative enzymes activity in leaves of soybean (Glycine max L.) under salt stress. *American Journal of Agriculture and Forestry*, *9*, 147–155.

Agarwal, N., Singh, A., & Kumar, A. (2015). Salinity effects on growth and productivity of two soybean (Glycine max L.) genotypes. *Indian Journal of Scientific Research*, *6*(2), 59–69.

Ainsworth, E. A., Yendrek, C. R., Sitch, S., Collins, W. J., & Emberson, L. D. (2012). The effects of tropospheric ozone on net primary productivity and implications for climate change. *Annual Review of Plant Biology*, *63*, 637–661.

Akram, S., Siddiqui, M. N., Hussain, B. M. N., Al Bari, M. A., Mostofa, M. G., Hossain, M. A., & Tran, L. S. P. (2017). Exogenous glutathione modulates salinity tolerance of soybean [Glycine max (l.) Merrill] at reproductive stage. *Journal of Plant Growth Regulation*, *36*(4), 877–888.

Al-Zahrani, H. S., Nahar, K., Alharby, H. F., Alsamadany, H., Hakeem, K. R., & Hasanuzzaman, M. (2022). Zinc supplementation enhances glutathione-mediated antioxidant defense and glyoxalase systems to conferring salt tolerance in soybean (*Glycine max* L.). *Agronomy*, *12*, 1032.

Amirjani, M. R. (2010). Effect of salinity stress on growth, mineral composition, proline content, antioxidant enzymes of soybean. *American Journal of Plant Physiology*, *5*(6), 350–360.

Anjum, N. A., Sofo A., Scopa A., Roychoudhury A., Gill S. S., & Iqbal M. (2015). Lipids and proteins – major targets of oxidative modifications in abiotic stressed plants. *Environmental Science and Pollution Research*, *22*, 4099–4121.

Arnett, D. K., Blumenthal, R. S., Albert, M. A., Buroker, A. B., Goldberger, Z. D., Hahn, E. J., Himmelfarb, C. D., Khera, A., Lloyd-Jones, D., McEvoy, J. W., Michos, E. D., Miedema, M. D., Muñoz, D., Smith, S. C., Virani, S. S., Williams, K. A., Yeboah, J., & Ziaeian, B. (2019). 2019 ACC/AHA guideline on the primary prevention of cardiovascular disease: A report of the American College of Cardiology/American Heart Association Task Force on clinical practice guidelines. Circulation, *140*. 10.1161/cir.0000000000000678.

Arrutia, F., Binner, E., Williams, P., & Waldron, K. W. (2020). Oilseeds beyond oil: Press cakes and meals supplying global protein requirements. *Trends in Food Science and Technology*, *100*, 88–102.

Ashlock, L., Rodibaugh, R., Hettiarachchy, N., & Proctor, A. (2000). *Arkansas soybean production handbook*. Little Rock, AR, USA: University of Arkansas Division of Agriculture, Vol. 18.

Assefa, Y., Purcell, L. C., Salmeron, M., Naeve, S., Casteel, S. N., Kovács, P., Archontoulis, S., Licht, M., Below, F., Kandel, H., Lindsey, L. E., Gaska, J., Conley, S., Shapiro, C., Orlowski, J. M., Golden, B. R., Kaur, G., Singh, M., Thelen, K., Laurenz, R., Davidson, D., & Ciampitti, I. A. (2019). Assessing variation in US soybean seed composition (protein and oil). *Frontiers in Plant Science*, *10*, 298.

Ault, T. R. (2020) On the essentials of drought in a changing climate. *Science*, *368*, 256–260.

Baghel, L., Kataria, S., & Guruprasad, K. N. (2016). Static magnetic field treatment of seeds improves carbon and nitrogen metabolism under salinity stress in soybean: Static magnetic field ameliorates salinity stress. *Bioelectromagnetics*, *37*(7), 455–470.

Bailey-Serres, J., Parker, J. E., Ainsworth, E. A., Oldroyd, G. E., & Schroeder, J. I. (2019). Genetic strategies for improving crop yields. *Nature*, *575*, 109–118.

Balfagón, D., Sengupta, S., Gómez-Cadenas, A., Fritschi, F. B., Azad, R. K., & Mittler, R. (2019). Jasmonic acid is required for plant acclimation to a combination of high light and heat stress. *Plant Physiology*, *181*, 1668–1682.

Battisti, I., Ebinezer, L. B., Lomolino, G., Masi, A., & Arrigoni, G. (2021). Protein profile of commercial soybean milks analyzed by label-free quantitative proteomics. *Food Chemistry*, *352*, 129299.

Battistini, C., Gullón, B., Ichimura, E. S., Gomes, A. M. P., Ribeiro, E. P., Kunigk, L., Moreira, J. U. V., & Jurkiewicz, C. (2018). Development and characterization of an innovative synbiotic fermented beverage based on vegetable soybean. *Brazilian Journal of Microbiology*, *49*, 303–309. 10.1016/j.bjm.2017.08.006.

Begum, N., Hasanuzzaman, M., Li, Y., Akhtar, K., Zhang, C., & Zhao, T. (2022). Seed germination behavior, growth, physiology and antioxidant metabolism of four contrasting cultivars under combined drought and salinity in soybean. *Antioxidants*, *11*(3), 498.

Bellaloui, N., Smith, J. R., Mengistu, A., Ray, J. D., & Gillen, A. M. (2017). Evaluation of exotically-derived soybean breeding lines for seed yield, germination, damage, and composition under dryland production in the midsouthern USA. *Frontiers in Plant Science*, *8*, 176.

Bhat, P., Honson, D., & Guttman, M. (2021). Nuclear compartmentalization as a mechanism of quantitative control of gene expression. *Nature Reviews Molecular Cell Biology*, *22*, 653–670. 10.1038/s41580-021-00387-1.

Bhuiyan, M. S., Malek, M. A., Emon, R. M., Khatun, M. K., Khandaker, M. M., & Alam, M. (2022). Increased yield performance of mutation induced Soybean genotypes at varied agro-ecological conditions. *Brazilian Journal of Biology*, *5*, 84.

Bilal, S., Shahzad, R., Khan, A. L., Al-Harrasi, A., Kim, C. K., & Lee, I. J. (2019). Phytohormones enabled endophytic Penicillium funiculosum LHL06 protects Glycine max L. from synergistic toxicity of heavy metals by hormonal and stress-responsive proteins modulation. *Journal of Hazardous Materials*, *379*, 120824.

Blanco, A., Högy, P., Zikeli, S., Pignata, M. L., & Rodriguez, J. H. (2022) Assessment of elevated Coxygen concentrations and heat stress episodes in soybean cultivars growing in heavy metal polluted soils: Crop nutritional quality and food safety. *Environmental Pollution*, *303*, 119123.

Blasbalg, T. L., Hibbeln, J. R., Ramsden, C. E., Majchrzak, S. F., & Rawlings, R. R. (2011). Changes in consumption of omega-3 and omega-6 fatty acids in the United States during the 20th century. *American Journal of Clinical Nutrition*, *93*(5), 950–962.

Bu, T., Lu, S., Wang, K., Dong, L., Li, S., Xie, Q., Xu, X., Cheng, Q., Chen, L., Fang, C., Li, H., Liu, B., Weller, J. L., & Kong, F. (2021). A critical role of the soybean evening complex in the control of photoperiod sensitivity and adaptation. *Proceedings of the National Academy of Sciences of the United States of America*, *118*(8), e2010241118.

Buwjoom, T., Maneewan, B., & Yamauchi, K. (2016). The using of fermented vegetable soybean waste and banana stem in black-bone chicken diets. In Proceedings of the 17th Asian; Australasian Association of Animal Production Societies Animal Science Congress, Fukuoka, Japan, *24*, 1689–1692.

Cao, X., Zhong, C., Zhu, C., Zhu, L., Zhang, J., Wu, L., & Jin, Q. (2018). Ammonium uptake and metabolism alleviate peg-induced water stress in rice seedlings. *Plant Physiology and Biochemistry*, *132*, 128–137.

Chatterjee, C., Gleddie, S., & Xiao, C. W. (2018). Soybean bioactive peptides and their functional properties. *Nutrients*, *10*, 1211.

Chebrolu, K. K., Fritschi, F. B., Ye, S., Krishnan, H. B., Smith, J. R., & Gillman, J. D. (2016). Impact of heat stress during seed development on soybean seed metabolome. *Metabolomics*, *12*, 28.

Chen, C. T., Lu, C. T., Tzen, J. T. C., & Yang, C. Y. (2021). Physiological properties and molecular regulation in different edamame cultivars under drought stress. *Agronomy*, *11*(5), 939.

Chen, H. T., Liu, X. Q., Zhang, H. M., Yuan, X. X., Gu, H. P., Cui, X. Y., & Chen, X. (2018). Advances in salinity tolerance of soybean: Genetic diversity, heredity, and gene identification contribute to improving salinity tolerance. *Journal of Integrative Agriculture*, *17*(10), 2215–2221.

Chen, K., Wang, Y., Zhang, R., Zhang, H., & Gao, C. (2019, 29). CRISPR/Cas genome editing and precision plant breeding in agriculture. *Annual Review of Plant Biology*, *70*, 667–697.

Chu, S., Li, H., Zhang, X., Yu, K., Chao, M., Han, S., & Zhang, D. (2018). Physiological and proteomics analyses reveal low-phosphorus stress affected the regulation of photosynthesis in soybean. *International Journal of Molecular Sciences*, *19*(6), 1688.

Çirka, M., Kaya, A. R., & Eryiğit, T. (2021). Influence of temperature and salinity stress on seed germination and seedling growth of soybean (Glycine Max L.). *Legume Research: An International Research*, *1*.

Clemente, T. E., & Cahoon, E. B. (2009). Soybean oil: Genetic approaches for modification of functionality and total content. *Plant Physiology*, *151*(3), 1030–1040.

Cohen, I., Zandalinas, S. I., Fritschi, F. B., Sengupta, S., Fichman, Y., & Azad, R. K. (2021). The impact of water deficit and heat stress combination on the molecular response, physiology, and seed production of soybean. *Physiologia Plantarum*, *172*, 41–52.

Coward, L., Smith, M., Kirk, M., & Barnes, S. (1998). Chemical modification of isoflavones in soyfoods during cooking and processing. *American Journal of Clinical Nutrition*, *68*(6), 1486S–1491S.

Damanik, R., Marbun, P., & Sihombing, L. (2016). Antioxidant activity of seedling growth in selected soybean genotypes (Glycine max (L.) Merrill) responses of submergence, in IOP Conference Series: *Earth Environmental Science*, *41*, 012003.

Das, A., Rushton, P. J., & Rohila, J. S. (2017). Metabolomic Profiling of soybeans (Glycine max L.) reveals the importance of sugar and nitrogen metabolism under drought and heat stress. *Plants (Basel)*, *6*(2), 21.

De Franciscis, P., Colacurci, N., Riemma, G., Conte, A., Pittana, E., Guida, M., & Schiattarella, A. (2019). A nutraceutical approach to menopausal complaints. *Medicina (Kaunas)*, *55*(9), 544.

de Melo, B. P., Carpinetti, P. D. A., Fraga, O. T., Rodrigues-Silva, P. L., Fioresi, V. S., de Camargos, L. F., & Ferreira, M. F. D. S. (2022). Abiotic stresses in plants and their markers: A practice view of plant stress responses and programmed cell death mechanisms. *Plants*, *11*, 1100.

De Oliveira, M. R. (2021) *Screening and breeding soybean for flood tolerance*. Fayetteville, United States: University of Arkansas. pp. 1–5.

Deol, P., Fahrmann, J., Yang, J., Evans, J. R., Rizo, A., Grapov, D., Salemi, M., Wanichthanarak, K., Fiehn, O., Phinney, B., Hammock, B. D., & Sladek, F. M. (2017). Omega-6 and omega-3 oxylipins are implicated in soybean oil-induced obesity in mice. *Scientific Reports*, *7*(1), 12488.

Dietz, K. J., Turkan, I., & Krieger-Liszkay, A. (2016). Redox- and reactive oxygen species-dependent signaling in and from the photosynthesizing chloroplast. *Plant Physiology*, *171*, 1541–1550.

Djanta, M. K. A., Agoyi, E. E., Agbahoungba, S., Quenum, F. J. B., Chadare, F. J., Assogbadjo, A. E., Agbangla, C., & Sinsin, B. (2020). Vegetable soybean, edamame: Research, production, utilization and analysis of its adoption in sub-Saharan Africa. *Journal of Horticulture and Forestry*, *12*(1), 1–12.

Dong, D., Fu, X., Yuan, F., Chen, P., Zhu, S., Li, B., Yang, Q., Yu, X., & Zhu, D. (2014). Genetic diversity and population structure of vegetable soybean (Glycine max (l.) Merr.) In China as revealed by SSR markers. *Genetic Resources and Crop Evolution*, *66* (9), 173–183.

Du, Y., Zhao, Q., Chen, L., Yao, X., Zhang, W., Zhang, B., & Xie, F. (2020). Effect of drought stress on sugar metabolism in leaves and roots of soybean seedlings. *Plant Physiology and Biochemistry*, *146*, 1–12.

Dubey, A., Saiyam, D., Kumar, A., Hashem, A., Abd_Allah, E. F., & Khan, M. L. (2021). Bacterial Root Endophytes: Characterization of their competence and plant growth promotion in soybean (*Glycine max* (l.) Merr.) under drought stress. *International Journal of Environmental Research and Public Health*, *18*, 931.

Egamberdieva, D., Wirth, S., Jabborova, D., Räsänen, L. A., & Liao, H. (2017). Coordination between Bradyrhizobium and Pseudomonas alleviates salt stress in soybean through altering root system architecture. *Journal of Plant Interactions*, *12*(1), 100–107.

El Sabagh, A., Islam, M. S., Ueda, A., Saneoka, H., & Barutçular, C. (2015). Increasing reproductive stage tolerance to salinity stress in soybean. *International Journal of Agriculture and Crop Sciences*, *8*(5), 738–745.

El-Esawi, M. A., Alaraidh, I. A., Alsahli, A. A., Alamri, S. A., Ali, H. M., & Alayafi, A. A. (2018). Bacillus firmus (sw5) augments salt tolerance in soybean (Glycine max l.) by modulating root system architecture, antioxidant defense systems and stress-responsive genes expression. *Plant Physiology and Biochemistry*, *132*, 375–384.

Ergo, V. V., Lascano, R., Vega, C. R., Parola, R., & Carrera, C. S. (2018). Heat and water stressed field-grown soybean: A multivariate study on the relationship between physiological-biochemical traits and yield. *Environmental and Experimental Botany*, *148*, 1–11.

Eze, O. F., Chatzifragkou, A., & Charalampopoulos, D. (2022). Properties of protein isolates extracted by ultrasonication from soybean residue (okara). *Food Chemistry*, *368*, 130837.

Fang, Y., Cao, D., Yang, H., Guo, W., Ouyang, W., Chen, H., Shan, Z., Yang, Z., Chen, S., Li, X., Chen, L., & Zhou, X. (2021). Genome-Wide identification and characterization of soybean *GmLOR* gene family and expression analysis in response to abiotic stresses. *International Journal of Molecular Sciences*, *22*(22), 12515.

Farhangi-Abriz, S., Ghassemi-Golezani, K. (2018). How can salicylic acid and jasmonic acid mitigate salt toxicity in soybean plants? *Ecotoxicology and Environmental Safety*, *147*, 1010–1016.

Farhoudi, R., Modhej, A., & Afrous, A. (2015). Effect of salt stress on seedlings growth and ions homeostasis of soybean (Glysin max) cultivars. *Journal of Scientific Research and Development*, *5*(8), 118–121.

Farooq, M., Gogoi, N., Barthakur, S., Baroowa, B., Bharadwaj, N., Alghamdi, S. S., & Siddique, K. H. M. (2017). Drought stress in grain legumes during reproduction and grain filling. *Journal of Agronomy and Crop Science*, *203*(2), 81–102.

Feng, Z., Ding, C., Li, W., Wang, D., & Cui, D. (2020). Applications of metabolomics in the research of soybean plant under abiotic stress. *Food Chemistry*, *310*, 125914.

Fenta, B., Beebe, S., Kunert, K., Burridge, J., Barlow, K., Lynch, J., & Foyer, C. (2014). Field phenotyping of soybean roots for drought stress tolerance. *Agronomy (Basel)*, *4*(3), 418–435.

Fercha, A., Capriotti, A. L., Caruso, G., Cavaliere, C., Stampachiacchiere, S., Zenezini Chiozzi, R., & Laganà, A. (2016). Shotgun proteomic analysis of soybean embryonic axes during germination under salt stress. *Proteomics*, *16*(10), 1537–1546.

Flowers, T. J., & Colmer, T. D. (2015). Plant salt tolerance: Adaptations in halophytes. *Annals of Botany*, *115*, 327–331.

Gao, J., Yang, S., Cheng, W., Fu, Y., Leng, J., Yuan, X., Jiang, N., Ma, J., & Feng, X. (2017). GmILPA1, encoding an anaphase-promoting complex-like protein, affects leaf petiole angle. *Plant Physiology*, *1174*(2), 1167–1176.

Ghassemi-Golezani, K., Farhangi-Abriz, S., & Bandehagh, A. (2018). Salicylic acid and jasmonic acid alter physiological performance, assimilate mobilization and seed filling of soybean under salt stress. *Acta Agriculturae Slovenica*, *111*(3), 597–607.

Gillman J., Biever J. J., Ye S., Spollen W. G., Givan S. A., Lyu Z., Joshi T., Smith J. R., & Fritschi F. B. (2019). A seed germination transcriptomic study contrasting two soybean genotypes that differ in terms of their tolerance to the deleterious impacts of elevated temperatures during seed fill. *BMC Research Notes*, *12*, 522.

Gharechahi, J., Vahidi, M. F., Ding, X. Z., Han, J. L., & Salekdeh, G. H. (2020). Temporal changes in microbial communities attached to forages with different lignocellulosic compositions in cattle rumen. *FEMS Microbiology Ecology*, *96*. 10.1093/femsec/fiaa069.

Guo, W., Chen, L., Chen, H., Yang, H., You, Q., Bao, A., Chen, S., Hao, Q., Huang, Y., Qiu, D., Shan, Z., Yang, Z., Yuan, S., Zhang, C., Zhang, X., Jiao, Y., Tran, L. S. P., Zhou, X., & Cao, D. (2020). Overexpression of GMWRI1B in soybean stably improves plant architecture and associated yield parameters, and increases total seed oil production under field conditions. *Plant Biotechnology Journal*, *18*(8), 1639–1641.

Gupta, S., Manubhai, K. P., Kulkarni, V., & Srivastava, S. (2016). An overview of innovations and industrial solutions in Protein Microarray Technology. *Proteomics*, *16*(8), 1297–1308.

Hamayun, M., Sohn, E. Y., Khan, S. A., Shinwari, Z. K., Khan, A. L., & Lee, I. J. (2021). Silicon alleviates the adverse effects of salinity and drought stress on growth and endogenous plant growth hormones of soybean (Glycine max l.). *Pakistan Journal of Botany*, *42*(3), 1713–1722.

Hamner, K. C. (1940) Interrelation of light and darkness in photoperiodic induction. *Botanical Gazette*, *101*(3), 658–687.

Han, K. H. (2012). Omega-3-fatty acid and triglyceride. *Korean Journal of Internal Medicine*, *83*, 724–727.

Hao, Y., Fan, X., Guo, H., Yao, Y., Ren, G., Lv, X., & Yang, X. (2020). Overexpression of the bioactive lunasin peptide in soybean and evaluation of its anti-inflammatory and anti-cancer activities in vitro. *Journal of Bioscience and Bioengineering*, *129*, 395–404.

Hasanuzzaman, M., Nahar, K., Rahman, A., Mahmud, J. A., Hossain, M. S., & Fujita, M. (2016). Soybean production and environmental stresses. In M. Mira (ed), *Environmental stresses in soybean production*. pp. 61–102. Academic Press.

Hashem, A., Abd_Allah, E. F., Alqarawi, A. A., Wirth, S., & Egamberdieva, D. (2019). Comparing symbiotic performance and physiological responses of two soybean cultivars to arbuscular mycorrhizal fungi under salt stress. *Saudi Journal of Biological Sciences*, *26*(1), 38–48.

Hashimoto, T., Mustafa, G., Nishiuchi, T., & Komatsu, S. (2020). Comparative analysis of the effect of inorganic and organic chemicals with silver nanoparticles on soybean under flooding stress. *International Journal of Molecular Sciences*, *21*, 1300.

Haun, W., Coffman, A., Clasen, B. M., Demorest, Z. L., Lowy, A., Ray, E., Retterath, A., Stoddard, T., Juillerat, A., Cedrone, F., Mathis, L., Voytas, D. F., & Zhang, F. (2014). Improved soybean oil quality by targeted mutagenesis of the fatty acid desaturase 2 gene family. *Plant Biotechnology Journal*, *12*(7), 934–940.

He, F., & Chen, J. (2013). Consumption of soybean, soy foods, soy isoflavones and breast cancer incidence: Differences between Chinese women and women in western countries and possible mechanisms. *Food Science and Human Wellness*, *2*, 146–161.

He, L., Jin, N., & Yu, Q. (2020). Impacts of climate change and crop management practices on soybean phenology changes in China. *Science of the Total Environment*, *707*, 135638.

He, Y., Chen, Y., Yu, C. L., Lu, K. X., Jiang, Q. S., Fu, J. L., Wang, G. M., & Jiang, D. A. (2016). Photosynthesis and yield traits in different soybean lines in response to salt stress. *Photosynthetica*, *54*(4), 630–635.

Hossain, M. M., Liu, X., Qi, X., Lam, H. M., & Zhang, J. (2014). Differences between soybean genotypes in physiological response to sequential soil drying and rewetting. *Crop Journal*, *2*(6), 366–380.

Hossain, M. N., Sarker, U., Raihan, M. S., Al-Huqail, A. A., Siddiqui, M. H., & Oba, S. (2022). Influence of salinity stress on color parameters, leaf pigmentation, polyphenol and flavonoid contents, and antioxidant activity of *amaranthus lividus* leafy vegetables. *Molecules*, *27*, 1821.

Hu, W., Strom, N., Haarith, D., Chen, S., & Bushley, K. E. (2018). Mycobiome of cysts of the soybean cyst nematode under long term crop rotation. *Frontiers in Microbiology*, *9*, 386.

Ingolia, N. T. (2014). Ribosome profiling: New views of translation, from single codons to genome scale. *Nature Reviews Genetics*, *15*, 205–213.

Jan, S. A., Tabassum, R., & Bashir, H. (2022). Speed breeding methods for soybean improvement: Recent advances. *Journal of Nutritional Health & Food Engineering*, *12*, 41–42. 10.15406/jnhfe.2022.12.00354.

Jayachandran, M., & Xu, B. (2019). An insight into the health benefits of fermented soy products.*Food Chemistry*, *271*, 362–371.

Ji, W., Cong, R., Li, S., Li, R., Qin, Z., Li, Y., Zhou, X., Chen, S., & Li, J. (2016). Comparative proteomic analysis of soybean leaves and roots by iTRAQ provides insights into response mechanisms to short-term salt stress. *Frontiers in Plant Science*, *7*, 573.

Jia, W., Ma, M., Chen, J., & Wu, S. (2021). Plant morphological, physiological and anatomical adaption to flooding stress and the underlying molecular mechanisms. *International Journal of Molecular Sciences*, *22*, 1088.

Jianing, G., Yuhong, G., Yijun, G., Rasheed, A., Qian, Z., Zhiming, X., Mahmood, A., Shuheng, Z., Zhuo, Z., Xiaoxue, W., & Jian, W. (2022). Improvement of heat stress tolerance in soybean (Glycine max L), by using conventional and molecular tools. Frontiers in Plant Science, *13*, 993189.

Jin, T., Sun, Y., Zhao, R., Shan, Z., Gai, J., & Li, Y. (2019). Overexpression of peroxidase gene GsPRX9 confers salt tolerance in soybean. *International Journal of Molecular Sciences*, *20*, 3745.

Johnson, L. A., & White, P. J. (2008) Soybeans: Chemistry, production, processing, and utilization. Urbana, IL, USA: AOCS Press, *120*(3), 234–239.

Jumrani, K., & Bhatia, V. S. (2019). Interactive effect of temperature and water stress on physiological and biochemical processes in soybean. *Physiology and Molecular Biology of Plants*, *25*(3), 667–681.

Kachare, S., Tiwari, S., Tripathi, N., & Thakur, V. V. (2019). Assessment of genetic diversity of soybean [*Glycine max* (L.) Merr.] genotypes using qualitative traits and microsatellite makers. *Agricultural Research*, *9*(2), 1–5.

Kairam, R., & Sran, H. S. (2022). Advancements in biotechnology to improve plant abiotic stress tolerance. *Agronomy*, *12*, 265.

Kamshybayeva, A., Omarova, Z., Bulegenova, A., Iskakov, B., Abugalieva, S., Jatayeva, K., Shumenov, B., Shintemirov, A., & Kozhakhmetov, S. (2017). Phenotypic and molecular responses of soybean (Glycine max (L.) Merrill) to salinity stress. *Plants*, *6*, 14.

Kang, S. M., Adhikari, A., Khan, M. A., Kwon, E. H., Park, Y. S., & Lee, I. J. (2021). Influence of the rhizobacterium rhodobacter sphaeroides KE149 and biochar on waterlogging stress tolerance in Glycine max l. *Environments*, *8*, 94.

Kargar, S. M. A., & Kareh, S. J. (2016). The evaluation of seed germination on 12 soybean cultivars under different salinity stress levels. *Plant Breeding and Seed Science*, *73*(1), 39.

Kataria, S., Baghel, L., Jain, M., & Guruprasad, K. N. (2019). Magnetopriming regulates antioxidant defense system in soybean against salt stress. *Biocatalysis and Agricultural Biotechnology*, *18*(101090), 101090.

Kaur, G., Zurweller, B. A., Nelson, K. A., Motavalli, P. P., & Dudenhoeffer, C. J. (2017). Soil waterlogging and nitrogen fertilizer management effects on corn and soybean yields. *Agronomy, Soils & Environmental Quality*, *109*, 97–106.

Ke, D., He, Y., Fan, L., Niu, R., Cheng, L., Wang, L., & Zhang, Z. (2022). The Soybean TGA transcription factor gmtga13 plays important roles in the response to salinity stress. *Plant Biology (Stuttgart)*, *24*(2), 313–322.

Khan, M. S. A. (2014). Effects of salt and water stress on leaf production, sodium and potassium ion accumulation in soybean. *Journal of Plant Sciences (Science Publishing Group)*, *2*, 209. 10.11648/j.jps.20140205.19.

Khalid, M., Rehman, H. M., Ahmed, N., Nawaz, S., Saleem, F., Ahmad, S., Uzair, M., Rana, I. A., Atif, R. M., Zaman, Q. U., & Lam, H. M. (2022). Using exogenous melatonin, glutathione, proline, and glycine betaine treatments to combat abiotic stresses in crops. *International Journal of Molecular Sciences*, *23*(21), 12913.

Khan, M. A., Asaf, S., Khan, A. L., Adhikari, A., Jan, R., Ali, S., Imran, M., Kim, K. M., & Lee, I. J. (2019). Halotolerant rhizobacterial strains mitigate the adverse effects of NaCl stress in soybean seedlings. *BioMed Research International*, 1–15.

Khan, M. A., Asaf, S., Khan, A. L., Jan,R., Kang, S. M., Kim, K. M., & Lee, I. J. (2020). Thermotolerance effect of plant growth-promoting Bacillus cereus SA1 on soybean during heat stress. *BMC Microbiology*, *20*(1), 1–4.

Khan, M. N., Ahmed, I., Ud Din, I., Noureldeen, A., Darwish, H., & Khan, M. (2022). Proteomic insight into soybean response to flooding stress reveals changes in energy metabolism and cell wall modifications. *PloS One*, *17*, e0264453.

Khan, M. N., & Komatsu, S. (2016). Proteomic analysis of soybean root including hypocotyl during recovery from drought stress. *Journal of Proteomics*, *144*, 39–50.

Khan, S., Khan, R. U., Alam, W., & Sultan, A. (2018). Evaluating the nutritive profile of three insect meals and their effects to replace soya bean in broiler diet. *Journal of Animal Physiology and Animal Nutrition*, *102*(2), e662–e668.

Kidokoro, S., Watanabe, K., Ohori, T., Moriwaki, T., Maruyama, K., & Mizoi, J. (2015). Soybean DREB 1/CBF-type transcription factors function in heat and drought as well as cold stress-responsive gene expression. *Plant Journal*, *81*, 505–518.

Kim, I. S., Kim, C. H., & Yang, W. S. (2021). Physiologically active molecules and functional properties of soybeans in human health – a current perspective. *International Journal of Molecular Sciences*, *22*, 4054.

Kim, Y., Seo, C. W., Khan, A. L., Mun, B. G., Shahzad, R., Ko, J. W., Yun, B. W., Park, S. K., & Lee, I. J. (2018). Exo-ethylene application mitigates waterlogging stress in soybean (*Glycine max* L.). *BMC Plant Biology*, *18*, 254.

Komatsu, S., Hiraga, S., & Nouri, M. Z. (2014). Analysis of flooding-responsive proteins localized in the nucleus of soybean root tips. *Molecular Biology Reports*, *41*, 1127–1139.

Komatsu, S., Sakata, K., & Nanjo, Y. (2015). 'Omics' techniques and their use to identify how soybean responds to flooding. *Journal of Analytical Science and Technology*, *6*, 9.

Komatsu, S., Yamaguchi, H., Hitachi, K., Tsuchida, K., Kono, Y., & Nishimura, M. (2021). Proteomic and biochemical analyses of the mechanism of tolerance in mutant soybean responding to flooding stress. *International Journal of Molecular Sciences*, *22*, 9046.

Kooyers, N. J. (2015). The evolution of drought escape and avoidance in natural herbaceous populations. *Plant Science*, *234*, 155–162.

Kosová, K., Vítámvás, P., Prášil, I. T., & Renaut, J. (2011). Plant proteome changes under abiotic stress-contribution of proteomics studies to understanding plant stress response. *Journal of Proteomics*, *74*(8), 1301–1322.

Krishnan, H. B., Kim, W. S., Oehrle, N. W., Smith, J. R., & Gillman, J. D. (2020). Effect of heat stress on seed protein composition and ultrastructure of protein storage vacuoles in the cotyledonary parenchyma cells of soybean genotypes that are either tolerant or sensitive to elevated temperatures. *International Journal of Molecular Sciences*, *21*, 4775.

Kuczyaski, J., Tomas, Z., Nawracaa, J., Graczabernaciak, J., & Tyczewska, A. (2020). Chilling stress tolerance of two soya bean cultivars: Phenotypic and molecular responses. *Journal of Agronomy and Crop Science*, *206*(6), 759–772.

Kudełka, W., Kowalska, M., & Popis, M. (2021). Quality of soybean products in terms of essential amino acids composition. *Molecules*, *26*, 5071.

Kumar, A. (2017). Germination behaviour of soybean varieties under different salinity stress. *International Journal of Applied Agricultural Research*, *12*, 69–76.

Laloum, T., Martín, G., & Duque, P. (2018). Alternative splicing control of abiotic stress responses. *Trends in Plant Science*, *23*, 140–150.

Lamichhane, J. R., Debaeke, P., Steinberg, C., You, M. P., Barbetti, M. J., & Aubertot, J. N. (2018). Abiotic and biotic factors affecting crop seed germination and seedling emergence: A conceptual framework. *Plant and Soil*, *432*, 1–28.

Li, H., Fan, W., Zhu, C., Zhang, T., Jiang, J., He, L., Lu, H., Wang, H., Poovaiah, B. W., & Du, L. (2019). Enhanced salt tolerance of rhizobia-inoculated soybean correlates with decreased phosphorylation of the transcription factor GmMYB183 and altered flavonoid biosynthesis. *Molecular & Cellular Proteomics*, *18*(11), 2225–2243.

Li, H., Wang, X. H., Li, Q., Xu, P., Liu, Z. N., Xu, M., & Cui, X. Y. (2022). GmCIPK21, a CBL-interacting protein kinase confers salt tolerance in soybean (Glycine max. L). *Plant Physiology and Biochemistry*, *184*, 47–55.

Li, H., Yuan, Y., Zhu, J., Wang, T., Wang, D., & Xu, Y. (2020). Zein/soluble soybean polysaccharide composite nanoparticles for encapsulation and oral delivery of lutein. *Food Hydrocolloids*, *103*, 105715.

Li, K. P., Wong, C. H., Cheng, C. C., Cheng, S. S., Li, M. W., Mansveld, S., & Bergsma, A. (2021). Gmdnj1, a type-i heat shock protein 40 (hsp40), is responsible for both growth and heat tolerance in soybean. *Plant Direct*, *5*(1), 181.

Li, Y., Chen, Q., Nan, H., Li, X., Lu, S., Zhao, X., Liu, B., Guo, C., Kong, F., & Cao, D. (2017). Overexpression of GmFDL19 enhances tolerance to drought and salt stresses in soybean. *PLoS One*, *12*(6), e0179554.

Lin, P., Wang, K., Zhou, C., Xie, Y., Yao, X., & Yin, H. (2018). Seed transcriptomics analysis in camellia oleifera uncovers genes associated with oil content and fatty acid composition. *International Journal of Molecular Sciences*, *19*(1), 118.

Lin, X., Liu, B., Weller, J. L., Abe, J., & Kong, F. (2021). Molecular mechanisms for the photoperiodic regulation of flowering in soybean. *Journal of Integrative Plant Biology*, *63*(6), 981–994.

Liu, A. (2019). Transcriptomic reprogramming in soybean seedlings under salt stress: Soybean transcriptome under salt stress. *Plant, Cell & Environment*, *42*(1), 98–114.

Liu, A., Xiao, Z., Wang, Z., Lam, H. M., & Chye, M. L. (2021). Galactolipid and phospholipid profile and proteome alterations in soybean leaves at the onset of salt stress. *Frontiers in Plant Science*, *12*, 644408.

Liu, H. S., Mahfuz, S. U., Wu, D., Shang, Q. H., & Piao, X. S. (2020). Effect of chestnut wood extract on performance, meat quality, antioxidant status, immune function, and cholesterol metabolism in broilers. *Poultry Science*, *99*, 4488–4495.

Liu, R., & Lal, R. (2014 14). Synthetic apatite nanoparticles as a phosphorus fertilizer for soybean (Glycine max). *Scientific Reports*, *4*(1), 5686.

Liu, X., Pi, B., Du, Z., Yang, T., Gu, M., Sun, S., & Yu, B. (2022). The transcription factor gmbhlh3 confers cl−/salt tolerance to soybean by upregulating GmCLC1 expression for maintenance of anion homeostasis. *Environmental and Experimental Botany*, *194*(104755), 104755.

Liu,Y., Wu, J., Sun, N., Tu, C., Shi, X., Cheng, H., Liu, S., Li, S., Wang, Y., Zheng, Y., & Uversky, V. N. (2017). Intrinsically disordered proteins as important players during desiccation stress of soybean radicles. *Journal of Proteome Research*, *16*(7), 2393–2409.

Lu, X., Xiong, Q., Cheng, T., Li, Q. T., Liu, X. L., Bi, Y. D., Li, W., Zhang, W. K., Ma, B., Lai, Y. C., Du, W. G., Man, W. Q., Chen, S. Y., & Zhang, J. S. (2017). A PP2C-1 allele underlying a quantitative trait locus enhances soybean 100-seed weight. *Molecular Plant*, *9*, 670–684.

Ma, C., Borgatta, J., Hudson, B. G., Tamijani, A. A., De La Torre-Roche, R., Zuverza-Mena, N., Shen, Y., Elmer, W., Xing, B., Mason, S. E., & Hamers, R. J. (2020). Advanced material modulation of nutritional and phytohormone status alleviates damage from soybean sudden death syndrome. *Nature Nanotechnology*, *15*(12), 1033–1042.

Ma, X., Zhu, Q., Chen, Y., & Liu, Y. G. (2016). CRISPR/Cas9 platforms for genome editing in plants: Developments and applications. *Molecular Plant*, *9*, 961–974. 10.1016/j.molp.2016.04.009.

Mak, M., Babla, M., Xu, S. C., O'Carrigan, A., Liu, X. H., Gong, Y. M., Holford, P., & Chen, Z. H. (2014). Leaf mesophyll K+, H+ and Ca2+ fluxes are involved in drought-induced decrease in photosynthesis and stomatal closure in soybean. *Environmental and Experimental Botany*, *98*, 1–12.

Mangena, P. (2020). Role of benzyladenine seed priming on growth and physiological and biochemical response of soybean plants grown under high salinity stress condition. *International Journal of Agronomy*, *2020*, 1–5.

Maurel, C., Boursiac, Y., Luu, D. T., Santoni, V., Shahzad, Z., & Verdoucq, L. (2015). Aquaporins in plants. *Physiological Reviews*, *95*, 1321–1358.

Messina, M., Shearer, G., & Petersen, K. (2021). Soybean oil lowers circulating cholesterol levels and coronary heart disease risk, and has no effect on markers of inflammation and oxidation. *Nutrition*, *89*, 111343.

Miao, L., Yang, S., Zhang, K., He, J., Wu, C., Ren, Y., Gai, J., & Li, Y. (2020). Natural variation and selection in gmsweet39 affect soybean seed oil content. *New Phytologist*, *225*, 1651–1666.

Michaletti, A., Naghavi, M. R., Toorchi, M., Zolla, L., & Rinalducci, S. (2018). Metabolomics and proteomics reveal drought-stress responses of leaf tissues from spring-wheat. *Scientific Reports*, *8*(1), 5710.

Mickelbart, M. V., Hasegawa, P. M., & Bailey-Serres, J. (2015). Genetic mechanisms of abiotic stress tolerance that translate to crop yield stability. *Nature Reviews Genetics*, *16*(4), 237–251.

Mignolet-Spruyt, L., Xu, E., Idanheimo, N., Hoeberichts, F. A., Muhlenbock, P., Brosche, M., Van Breusegem, F., & Kangasjarvi, J. (2016). Spreading the news: Subcellular and organellar reactive oxygen species production and signaling. *Journal of Experimental Botany*, *67*, 3831–3844.

Min, C. W., Kim, Y. J., Gupta, R., Kim, S. W., Han, W. Y., Ko, J. M., Kang, H. W., Yoon, W. B., Choung, M. G., Kim, Y. C., & Kim, S. T. (2016). High-throughput proteome analysis reveals changes of primary metabolism and energy production under artificial aging treatment in glycine max seeds. *Applied Biological Chemistry*, *59*, 841–853.

Min, C. W., Lee, S. H., Cheon, Y. E., Han, W. Y., Ko, J. M., Kang, H. W., Kim, Y. C., Agrawal, G. K., Rakwal, R., Gupta, R., & Kim, S. T. (2017). In-depth proteomic analysis of Glycine max seeds during controlled deterioration treatment reveals a shift in seed metabolism. *Journal of Proteomics*, *169*, 125–135.

Miransari, M. (2016) Soybean, protein, and oil production under stress. In M. Miransari (ed), *Environmental stresses in soybean production* pp. 157–176.

Mishra, N., Tripathi, M. K., Tiwari, S., Tripathi, N., Ahuja, A., Sapre, S., & Tiwari, S. (2021). Cell suspension culture and *in vitro* screening for drought tolerance in soybean using poly-ethylene glycol. *Plants*, *10*(3), 517–536.

Moradi, A., Dai, S., Wong, E. O. Y., Zhu, G., Yu, F., Lam, H. M., Wang, Z., Burlingame, A., Lin, C., Afsharifar, A., Yu, W., Wang, T., & Li, N. (2021). Isotopically dimethyl labeling-based quantitative proteomic analysis of phosphoproteomes of soybean cultivars. *Biomolecules*, *11*(8), 1218.

Moustafa-Farag, M., Mahmoud, A., Arnao, M. B., Sheteiwy, M. S., Dafea, M., Soltan M., Elkelish, A., Hasanuzzaman, M., & Ai, S. (2020) |Melatonin-induced water stress tolerance in plants: Recent advances. *Antioxidants*, *9*, 809.

Munns, R., & Gilliham, M. (2015). Salinity tolerance of crops – what is the cost? *New Phytologist*, *208*, 668–673.

Mustafa, G., Hasan, M., Yamaguchi, H., Hitachi, K., Tsuchida, K., & Komatsu, S. (2020). A comparative proteomic analysis of engineered and biosynthesized silver nanoparticles on soybean seedlings. *Journal of Proteome Research*, *19*(8), 3354–3363.

Mustafa, G., & Komatsu, S. (2016). Insights into the response of soybean mitochondrial proteins to various sizes of aluminum oxide nanoparticles under flooding stress. *Journal of Proteome Research*, *15*, 4464–4475.

Mustafa, G., & Komatsu, S. (2014). Quantitative proteomics reveals the effect of protein glycosylation in soybean root under flooding stress. *Frontiers in Plant Science*, *5*, 627.

Mutava, R. N., Prince, S. J., Syed, N. H., Song, L., Valliyodan, B., Chen,W., & Nguyen, H. T. (2015). Understanding abiotic stress tolerance mechanisms in soybean: A comparative evaluation of soybean response to drought and flooding stress. *Plant Physiology and Biochemistry*, *86*, 109–120.

Nanjo, Y., Jang, H. Y., Kim, H. S., Hiraga, S., Woo, S. H., & Komatsu, S. (2014). Analyses of flooding tolerance of soybean varieties at emergence and varietal differences in their proteomes. *Phytochemistry*, *106*, 25–36.

Nguyen, Q. H., Vu, L. T. K., Nguyen, L. T. N., Pham, N. T. T., Nguyen, Y. T. H., Van Le, S., & Chu, M. H. (2019). Overexpression of the GmDREB6 gene enhances proline accumulation and salt tolerance in genetically modified soybean plants. *Scientific Reports*, *9*(1), 19663.

Nigam, B., Dubey, R. S., & Rathore, D. (2022). Protective role of exogenously supplied salicylic acid and PGPB (Stenotrophomonas Sp.) on spinach and soybean cultivars grown under salt stress. *Scientia Horticulturae (Amsterdam)*, *293*(110654), 110654.

Nile, S. H., Venkidasamy, B., Samynathan, R., Nile, A., Shao, K., Chen, T., Sun, M., Khan, M. U., Dutta, N., Thiruvengadam, M., Shariati, M. A., Rebezov, M., & Kai, G. (2022). Soybean processing wastes: Novel insights on their production, extraction of isoflavones, and their therapeutic properties. *Journal of Agricultural and Food Chemistry*, *70*(23), 6849–6863.

Ning, L. H., Kan, G. Z., Shao, H. B., & Yu, D. Y. (2018). Physiological and transcriptional responses to salt stress in salt-tolerant and salt-sensitive soybean (Glycine max [L.] Merr.) seedlings. *Land Degradation & Development*, *29*, 2707–2719.

Oh, M., & Komatsu, S. (2015). Characterization of proteins in soybean roots under flooding and drought stresses. *Journal of Proteomics*, *114*, 161–181.

Oh, M., Nanjo, Y., Wang, X., & Komatsu, S. (2014). Gel-free proteomic analysis of calcium effects on soybean root tip under flooding and drought stresses, *Plant and Cell Physiology*, 1405–1420.

Ohama, N., Sato, H., Shinozaki, K., & Yamaguchi-Shinozaki, K. (2017). Transcriptional regulatory network of plant heat stress response. *Trends in Plant Science*, *22*, 53–65.

Ondrasek, G., Rathod, S., Manohara, K. K., Gireesh, C., Anantha, M. S., Sakhare, A. S., Parmar, B., Yadav, B. K., Bandumula, N., Raihan, F., Zielińska-Chmielewska, A., Meriño-Gergichevich, C., Reyes-Díaz, M., Khan, A., Panfilova, O., Seguel Fuentealba, A., Romero, S. M., Nabil, B., Wan, C., Shepherd, J., & Horvatinec, J. (2022). Salt stress in plants and mitigation approaches. *Plants*, *11*, 717.

Ort, N. W. W., Morrison, M. J., Cober, E. R., Samanfar, B., & Lawley, Y. E. (2022). Photoperiod affects node appearance rate and flowering in early maturing soybean. *Plants*, *11*, 871.

Ortiz, A. C., De Smet, I., Sozzani, R., & Locke, A. M. (2022). Field-grown soybean shows genotypic variation in physiological and seed composition responses to heat stress during seed development. *Environmental and Experimental Botany*, *195*, 104768.

Osman, M. S., Badawy, A. A., Osman, A. I., & Abdel Latef, A. A. H. (2021). Ameliorative impact of an extract of the halophyte Arthrocnemum macrostachyum on growth and biochemical parameters of soybean under salinity stress. *Journal of Plant Growth Regulation*, *40*(3), 1245–1256.

Otie, V., Udo, I., Shao, Y., Itam, M. O., Okamoto, H., An, P., & Eneji, E. A. (2021). Salinity effects on morpho-physiological and yield traits of soybean (Glycine max L.) as mediated by foliar spray with brassinolide. *Plants*, *10*(3), 541.

Ozga, J. A., Kaur, H., Savada, R. P., & Reinecke, D. M. (2017). Hormonal regulation of reproductive growth under normal and heat-stress conditions in legume and other model crop species. *Journal of Experimental Botany*, *68*, 1885–1894.

Pan, J., Sharif, R., Xu, X., & Chen, X. (2021). Mechanisms of waterlogging tolerance in plants: Research progress and prospects. *Frontiers in Plant Science*, *11*, 627331.

Parcell, J. (2018). Global edible vegetable oil market trends. *Biomedical Journal of Scientific & Technical Research*, *2*, 10.26717/bjstr.2018.02.000680.

Parveen, A. U. H. M., Akhtar, J., & Basra, S. M. (2016). Interactive effect of salinity and potassium on growth, biochemical parameters, protein and oil quality of soybean genotypes. *Pakistan Journal of Agricultural Sciences*, *53*, 69–78.

Pathanet, M. S., Lee, J. D., Shannon, J. G., & Nguyen, H. T. (2007). Recent advances in breeding for drought and salt stress tolerance in soybean. Advances in molecular breeding toward drought and salt tolerant crops. *International Journal of Molecular Sciences*, 739–773.

Peighambardoust, S. H., Karami, Z., Pateiro, M., & Lorenzo, J. M. (2021). A review on health-promoting, biological, and functional aspects of bioactive peptides in food applications. *Biomolecules*, *11*, 631.

Pereira, Y. C., Silva, F. R. D., Silva, B. R. S. D., Cruz, F. J. R., Marques, D. J., & Lobato, A. K. D. S. (2020). 24-epibrassinolide induces protection against waterlogging and alleviates impacts on the root structures, photosynthetic machinery and biomass in soybean. *Plant Signaling & Behavior*, *15*, 1805885.

Pi, E., Xu, J., Li, H., Fan, W., Zhu, C., Zhang, T., Jiang, J., He, L., Lu, H., Wang, H., Poovaiah, B. W., & Du, L. (2019). Enhanced salt tolerance of rhizobia-inoculated soybean correlates with decreased phosphorylation of the transcription factor gmmyb183 and altered flavonoid biosynthesis. *Molecular & Cellular Proteomics*, *18*(11), 2225–2243.

Pi, E., Zhu, C., Fan, W., Huang, Y., Qu, L., Li, Y., Zhao, Q., Ding, F., Qiu, L., Wang, H., & Poovaiah, B. W. (2018). Quantitative phosphoproteomic and metabolomic analyses reveal GmMYB173 optimizes flavonoid metabolism in soybean under salt stress. *Molecular & Cellular Proteomics*, *17*(6), 1209–1224.

Ploschuk, R., Miralles, D., & Striker, G. (2022). A quantitative review of soybean responses to waterlogging: Agronomical, morpho-physiological and anatomical traits of tolerance. *Plant and Soil*, *475*, 237–252.

Putri, P. H., Susanto, G. W. A., & Artari, R. (2017). Response of soybean genotypes to salinity in germination stage. *Nusantara Bioscience*, *9*(2), 133–137.

Qin, J., Gu, F., Liu, D., Yin, C., Zhao, S., & Chen, H. (2013). Proteomic analysis of elite soybean Jidou17 and its parents using iTRAQ-based quantitative approaches. *Proteome Science*, *11*, 12.

Rafiee-Yarandi, H., Ghorbani, G. R., Alikhani, M., Sadeghi-Sefidmazgi, A., & Drackley, J. K. (2016). A comparison of the effect of soybeans roasted at different temperatures versus calcium salts of fatty acids on performance and milk fatty acid composition of mid-lactation Holstein cows. *Journal of Dairy Science*, *99*(7), 5422–5435.

Rasheed, A., Mahmood, A., Maqbool, R., Albaqami, M., Sher, A., & Sattar, A. (2022). Key insights to develop drought-resilient soybean: A review. *Journal of King Saud University – Science*, *34*, 102089.

Ratnaparkhe, M. B., Satput, G. K., Kumawat, G., Chandra,S., Kamble, V., Kavishwar, R., Singh,V., Singh, J., Singh, A. K., & Kumar, V. (2022). Genomic designing for abiotic stress tolerant soybean. In C. Kole (eds), *Genomic designing for abiotic stress resistant oilseed crops*. pp. 1–73. Cham: Springer.

Rayaprolu, S. J., Hettiarachchy, N. S., Chen, P., Kannan, A., & Mauromostakos, A. (2013). Peptides derived from high oleic acid soybean meals inhibit colon, liver and lung cancer cell growth. *Food Research International*, *50*(1), 282–288.

Raza, A., Charagh, S., Garcıa-Caparro,-S.P., Rahman, M. A., Ogwugwa, V. H., Saeed, F., & Jin, W. (2022). Melatonin-mediated temperature stress tolerance in plants. *GM Crops & Food*, *13*(1), 196–217.

Rehman, S. U., Qanmber, G., Tahir, M. H. N., Irshad, A., Fiaz, S., Ahmad, F., Ali, Z., Sajjad, M., Shees, M., Usman, M., & Geng, Z. (2021). Characterization of vascular plant one-zinc finger (VOZ) in soybean (Glycine max and Glycine soja) and their expression analyses under drought condition. *PLoS One*, *16*(7), e0253836.

Ribichich, K. F., Chiozza, M., Ávalos-Britez, S., Cabello, J. V., Arce, A. L., Watson, G., Arias, C., Portapila, M., Trucco, F., Otegui, M. E., & Chan, R. L. (2020). Successful field performance in warm and dry environments of soybean expressing the sunflower transcription factor HB4. *Journal of Experimental Botany*, *71*(10), 3142–3156.

Rizzo, G., & Baroni, L. (2018). Soy, Soy foods and their role in vegetarian diets, *Nutrients*, *10*(1), 43.

Roychoudhury, A., & Basu, S. (2021). Abscisic acid control of plant macro element membrane transport systems in response to water deficit and high salinity. In A. Roychoudhury, D. K. Tripathi, & R. Deshmukh (eds), *Transporters and plant osmotic stress*. India: Academic Press, pp. 89–99.

Sabagh, A. E., Hossain, A., Islam, M. S., Iqbal, M. A., Fahad, S., Ratnasekera, D., & Llanes A. (2020). Consequences and mitigation strategies of heat stress for sustainability of soybean (Glycine max L. Merr.) production under the changing climate. In A. Hossain (ed), *Plant stress physiology*. London, United Kingdom, pp. 1–20.

Sachdev, S., Ansari, S. A., & Ansari, M. I. (2023). Oxidative stress triggered damage to cellular biomolecules. In S. Sachdev (ed), *Reactive oxygen species in plants: The right balance*. Singapore: Springer Nature Singapore, pp. 45–59.

Sadak, M. S., Abd El-Hameid, A. R., Zaki, F. S., Dawood, M. G., & El-Awadi, M. E. (2020). Physiological and biochemical responses of soybean (Glycine max L.) to cysteine application under sea salt stress. *Bulletin of the National Research Centre*, *44*, 1–10.

Sahu, V. K., Tiwari, S., Gupta, N., Tripathi, M. K., & Yasin, M. (2020). Evaluation of physiological and biochemical contents in *desi* and *kabuli* chickpea. *Legume Research: An International Journal*, *10*, 1197–1208.

Sakazono, S., Nagata, T., Matsuo, R., Kajihara, S., Watanabe, M., & Ishimoto, M. (2014). Variation in root development response to flooding among 92 soybean lines during early growth stages. *Plant Production Science*, *17*, 228–236.

Sarker, U., & Oba, S. (2018). Catalase, superoxide dismutase and ascorbate-glutathione cycle enzymes confer drought tolerance of *Amaranthus tricolor*. *Scientific Reports*, *8*, 16496.

Sarker, U., & Oba, S. (2020). The response of salinity stress-induced *A. tricolor* to growth, anatomy, physiology, non-enzymatic and enzymatic antioxidants. *Frontiers in Plant Science*, *11*, 559876.

Sasidharan, R., Hartman, S., Liu, Z., Martopawiro, S., Sajeev, N., van Veen, H., Yeung, E., & Voesenek, L. A. C. J. (2018). Signal dynamics and interactions during flooding stress. *Plant Physiology*, *176*, 1106–1117.

Sathi, K. S., Masud, A., Falguni, M. R., Ahmed, N., Rahman, K., & Hasanuzzaman, M. (2022). Screening of soybean genotypes for waterlogging stress tolerance and understanding the physiological mechanisms. *Advances in Agriculture*, *2022*, 554–665.

Schwalbert, R., Stefanello, L. O., Schwalbert, R. A., Tarouco, C. P., Drescher, G. L., Trentin, E., Tassinari, A., da Silva, I. B., Brunetto, G., & Nicoloso, F. T. (2021). Soil tillage affects soybean growth and promotes heavy metal accumulation in seeds. *Ecotoxicology and Environmental Safety*, *216*, 112191.

Shahbaz, M., & Ashraf, M. (2013). Improving salinity tolerance in cereals. *Critical Reviews in Plant Sciences*, *32*, 2.

Shahid, M. A., Sarkhosh, A., Khan, N., Balal, R. M., Ali, S., Rosi, L., Gomez, C., Mattson, N., Nasim, W., & Garcia-Sanchez, F. (2020). Insights into the physiological and biochemical impacts of salt stress on plant growth and development. *Agronomy*, *10*, 938.

Sharifi, P., Amirnia, R., & Shirani Bidabadi, S. (2022). Role of silicon in mediating heat shock tolerance in soybean. *Gesunde Pflanzen*, *74*(2), 397–411.

Sheteiwy, M. S., Shao, H., Qi, W., Daly, P., Sharma, A., Shaghaleh, H., Hamoud, Y. A., El-Esawi, M. A., Pan, R., Wan, Q., & Lu, H. (2021). Seed priming and foliar application with jasmonic acid enhance salinity stress tolerance of soybean (Glycine Max L.) seedlings. *Journal of the Science of Food and Agriculture*, *101*(5), 2027–2041.

Shi, K., Qi, Y., Chen, F., Meng, Y., Luo, X., Shuai, H., Zhou, W., Ding, J., Du, J., Liu, J., Yang, F., Wang, Q., Liu, W., Yong, T., Wang, X., Feng, Y., & Yang, W. (2017). Salt stress represses soybean seed germination by negatively regulating ga biosynthesis while positively mediating ABA biosynthesis. *Frontiers in Plant Science*, *8*, 1372.

Shimamura, S., Yoshioka, T., Yamamoto, R., Hiraga, S., Nakamura, T., Shimada, S., & Komatsu, S. (2014). Role of abscisic acid in flood-induced secondary aerenchyma formation in soybean (*Glycine max*) hypocotyls. *Plant Production Science*, *17*, 131–137.

Shu, K., Qi, Y., Chen, F., Meng, Y., Luo, X., Shuai, H., Zhou, W., Ding, J., Du, J., Liu, J., Yang, F., Wang, Q., Liu, W., Yong, T., Wang, X., Feng, Y., & Yang, W. (2017). Salt stress represses soybean seed germination by negatively regulating ga biosynthesis while positively mediating ABA biosynthesis. *Frontiers in Plant Science*, *8*, 1372.

Shu, Y., Zhou, Y., Mu, K., Hu, H., Chen, M., He, Q., & Huang, S. (2020). A transcriptomic analysis reveals soybean seed pre-harvest deterioration resistance pathways under high temperature and humidity stress. *Genome*, *63*, 115–124.

Shuai, H., Meng, Y., Luo, X., Chen, F., Zhou, W., & Dai, Y. (2017). Exogenous auxin represses soybean seed germination through decreasing the gibberellin/abscisic acid (GA/ABA) ratio. *Scientific Reports*, *7*, 1–11.

Silva, B. R. S., Batista, B. L., & Lobato, A. K. S. (2021). Anatomical changes in stem and root of soybean plants submitted to salt stress. *Plant Biology (Stuttgart)*, *23*(1), 57–65.

Singh, A., & Meena, R. S. (2020). Response of bioregulators and irrigation on plant height of indian mustard (*Brassica juncea* L.). *Journal of Oilseed Brassica*, *1*, 9–14.

Smith, J. R., Gillman, J. D., Bellaloui, N., Gillen, A., & Ray, J. D. (2017). *Soybean* germplasm line ds25-1 with heat tolerance and competitive yield under heat str*ess.*USDA-ARS: Columbia, MO, USA, pp. 1–7.

Soliman, M. H., Abdulmajeed, A. M., Alhaithloul, H., Alharbi, B. M., El-Esawi, M. A., Hasanuzzaman, M., & Elkelish, A. (2020). Saponin biopriming positively stimulates antioxidants defense, osmolytes metabolism and ionic status to confer salt stress tolerance in soybean. *Acta Physiologiae Plantarum*, *42*(7), 1–5.

Song, H., Taylor, D. C., & Zhang, M. (2023). Bioengineering of soybean oil and its impact on agronomic traits. *International Journal of Molecular Sciences*, *24*(3), 2256.

Song, K., Yim, W. C., & Lee, B. M. (2017). Expression of heat shock proteins by heat stress in soybean. *Plant Breeding and. Biotechnology*, *5*(4), 344–353.

Staniak, M., Szpunar-Krok, E., & Kocira, A. (2023). Responses of soybean to selected abiotic stresses-photoperiod, temperature and water. *Agriculture*, *13*(1), 146.

Sukit, K., Buaream, M., Tonglian, B., Nattakarn, K., Kohsyo, Y., Sontaya, N., Chamnian, Y., & Koh-En, Y. (2017). Fermented green vegetable soybean meal increases dressed carcass meat and bone weight in black-boned chickens (fah luang chicken, gallus gallus). *American Journal of Animal and Veterinary Sciences*, *12*(3), 176–181.

Sun, Z., Su, C., Yun, J., Jiang, Q., Wang, L., Wang, Y., Cao, D., Zhao, F., Zhao, Q., Zhang, M., Zhou, B., Zhang, L., Kong, F., Liu, B., Tong, Y., & Li, X. (2019). Genetic improvement of the shoot architecture and yield in soya bean plants via the manipulation of gmmir156b. *Plant Biotechnology Journal*, *17*(1), 50–62.

Syed, N. H., Prince, S. J., Mutava, R. N., Patil, G., Li, S., Chen, W., Babu, V., Joshi, T., Khan, S., & Nguyen, H. T. (2015). Core clock, SUB1, and ABAR genes mediate flooding and drought responses via alternative splicing in soybean. *Journal of Experimental Botany*, *66*, 7129–7149.

Taheri, S., Gantait, S., Azizi, P., & Mazumdar, P. (2022). Drought tolerance improvement in Solanum lycopersicum: An insight into "OMICS" approaches and genome editing. *3 Biotech*, *12*, 63.

Taketani, T., Oyama, C., Mihara, A., Tanabe, Y., Abe, M., Hirade, T., Yamamoto, S., Bo, R., Kanai, R., Tadenuma, T., Michibata, Y., Yamamoto, S., Hattori, M., Katsube, Y., Ohnishi, H., Sasao, M., Oda, Y., Hattori, K., Yuba, S., Ohgushi, H., & Yamaguchi, S. (2015). Ex vivo expanded allogeneic mesenchymal stem cells with bone marrow transplantation improved osteogenesis in infants with severe hypophosphatasia. Cell Transplantation, *112*(08), 1931–1943.

Tamang, B. G., Li, S., Rajasundaram, D., Lamichhane, S., & Fukao, T. (2021). Overlapping and stress-specific transcriptomic and hormonal responses to flooding and drought in soybean. *Plant Journal*, *107*, 100–117.

Tamang, B. G., Magliozzi, J. O., Maroof, M. S., & Fukao, T. (2014). Physiological and transcriptomic characterization of submergence and reoxygenation responses in soybean seedlings. *Plant, Cell & Environment*, *37*, 2350–2365.

Thangavel, P., Puga-Olguín, A., Rodríguez-Landa, J. F., & Zepeda, R. C. (2019). Genistein as potential therapeutic candidate for menopausal symptoms and other related diseases. *Molecules*, *24*, 3892.

Thomey, M. L., Slattery, R. A., Köhler, I. H., Bernacchi, C. J., & Ort, D. R. (2019). Yield response of field-grown soybean exposed to heat waves under current and elevated [Coxygen]. *Global Change Biology*, *25*(12), 4352–4368.

Thu, N. B. A., Nguyen, Q. T., Hoang, X. L. T., Thao, N. P., & Tran, L. S. P. (2014). Evaluation of drought tolerance of the Vietnamese soybean cultivars provides potential resources for soybean production and genetic engineering. *BioMed Research International*, *2014*, 736–809.

Tidke, S. A., Ramakrishn, D., Kiran, S., Kosturkova, G., & Ravishanka, G. A. (2015). Nutraceutical potential of soybean: Review. *Asian Journal of Clinical Nutrition*, *7*(2), 22–32.

Tiwari, M., Kumar, R., Min, D., & Jagadish, S. V. K. (2022). Genetic and molecular mechanisms underlying root architecture and function under heat stress-a hidden story. *Plant, Cell & Environment*, *45*(3), 771–788.

Toscano, S., Ferrante, A., & Romano, D. (2019). Response of Mediterranean ornamental plants to drought stress. *Horticulturae*, *5*, 6.

Tripathi, P., Rabara, R. C., Reese, R. N., Miller, M. A., Rohila, J. S., Subramanian, S., Shen, Q. J., Morandi, D., Bücking, H., Shulaev, V., & Rushton, P. J. (2016). A toolbox of genes, proteins, metabolites and promoters for improving drought tolerance in soybean includes the metabolite cholesterol and stomatal development genes. *BMC Genomics*, *17*, 1–22.

Utami, M. M., & Hertamawati, R. T. (2020). Dietary edamame soybean isoflavon concentrate on improving carcass quality of broilers. In *IOP Conference Series: Earth and Environmental Science*, *411*, 012035.

Valdés-López, O., Batek, J., Gomez-Hernandez, N., Nguyen, C. T., Isidra-Arellano, M. C., Zhang, N., Joshi, T., Xu, D., Hixson, K. K., Weitz, K. K., Aldrich, J. T., Paša-Tolić, L., & Stacey, G. (2016). Soybean roots grown under heat stress show global changes in their transcriptional and proteomic profiles. *Frontiers in Plant Science*, *7*, 517.

Veas, R. E., Ergo, V. V., Vega, C. R., Lascano, R. H., Rondanini, D. P., & Carrera, C. S. (2021). Soybean seed growth dynamics exposed to heat and water stress during the filling period under field conditions. *Journal of Agronomy and Crop Science*, *208*, 472–485.

Wang, H., Wang, H., Shao, H., & Tang, X. (2016a). Recent advances in utilizing transcription factors to improve plant abiotic stress tolerance by transgenic technology. *Frontiers in Plant Science*, *7*, 67.

Wang, N., Zhong, X., Cong, Y., Wang, T., Yang, S., Li, Y., & Gai, J. (2016b). Genome-wide analysis of phosphoenolpyruvate carboxylase gene family and their response to abiotic stresses in soybean. *Scientific Reports*, *6*. 10.1038/srep38448.

Wang, L., Liu, L., Ma, Y., Li, S., Dong, S., & Zu, W. (2018a). Transcriptome profiling analysis characterized the gene expression patterns responded to combined drought and heat stresses in soybean. *Computational Biology and Chemistry*, *77*, 413–429.

Wang, G., Chang, L., Zhang, R., Wang, S., Wei, X., Rickert, E., Krost, P., Xiao, L., & Weinberger, F. (2018b). Can targeted defense elicitation improve seaweed aquaculture? *Journal of Applied Phycology*, *31*, 1845–1854. 10.1007/s10811-018-1709-6.

Wang, W. Y., Yong, G. W. J., & Jaitrong, W. (2018c). The ant genus Rhopalomastix (Hymenoptera: Formicidae: Myrmicinae) in Southeast Asia, with descriptions of four new species from Singapore based on morphology and DNA barcoding. Zootaxa, 4532. 10.11646/zootaxa.4532.3.1.

Wang, X., & Komatsu, S. (2020). Proteomic techniques for the development of flood-tolerant soybean. *International Journal of Molecular Sciences*, *21*(20), 7497.

Wang, X., & Komatsu, S. (2022). The role of phytohormones in plant response to flooding. *International Journal of Molecular Sciences*, *23*, 6383.

Wang, X., Li, F., Chen, Z., Yang, B., Komatsu, S., & Zhou, S. (2021). Proteomic analysis reveals the effects of melatonin on soybean root tips under flooding stress. *Journal of Proteomics*, *232*, 104064.

Wang, X., Li, Y., Zhang, H., Sun, G., Zhang, W., & Qiu, L. (2015). Evolution and association analysis of gmcyp78a10 gene with seed size/weight and pod number in soybean. *Molecular Biology Reports*, *42*, 489–496.

Wang, X., Sakata, K., & Komatsu, S. (2018). An integrated approach of proteomics and computational genetic modification effectiveness analysis to uncover the mechanisms of flood tolerance in soybeans. *International Journal of Molecular Sciences*, *19*, 1301.

Wang, X., Xu, C., Cai, X., Wang, Q., & Dai, S. (2017). Heat-responsive photosynthetic and signaling pathways in plants: Insight from Proteomics. *International Journal of Molecular Sciences*, *18*(10), 2191.

Wang, Y., Jiang, L., Chen, J., Tao, L., An, Y., Cai, H., & Guo, C. (2018). Overexpression of the alfalfa wrky11 gene enhances salt tolerance in soybean. *PLoS One*, *13*(2), e0192382.

Waśkiewicz, A., Szcześniewska, D., Szostak-Węgierek, D., Kwaśniewska, M., Pająk, A., Stepaniak, U., Kozakiewicz, K., Tykarski, A., Zdrojewski, T., Zujko, M. E., & Drygas, W. (2016). Are dietary habits of the Polish population consistent with the recommendations for prevention of cardiovascular disease? — WOBASZ II project. *Kardiologia Polska*, *74*, 969–977. 10.5603/kp.a2016.0003.

Wei, J., Liu, X., Li, L., Zhao, H., Liu, S., Yu, X., Shen, Y., Zhou, Y., Zhu, Y., Shu, Y., & Ma, H. (2020 26). Quantitative proteomic, physiological and biochemical analysis of cotyledon, embryo, leaf and pod reveals the effects of high temperature and humidity stress on seed vigor formation in soybean. *BMC Plant Biology, 20*(1), 127.

Weisany, W., Sohrabi, Y., Heidari, G., Siosemardeh, A., & Ghassemi-Golezani, K. (2012). Changes in antioxidant enzymes activity and plant performance by salinity stress and zinc application in soybean (*'Glycine max'*L.). *Plant Omics*, *5*, 60–67.

Whitham, S. A., Qi, M., Innes, R. W., Ma, W., Lopes-Caitar, V., & Hewezi, T. (2016). Molecular soybean-pathogen interactions. *Annual Review of Phytopathology*, *54*, 443–468.

Widyasari, K., Tran, P. T., Shin, J., Son, H., & Kim, K. H. (2022). Overexpression of purple acid phosphatase gmpap2.1 confers resistance to soybean mosaic virus in a susceptible soybean cultivar. *Journal of Experimental Botany*, *73*(5), 1623–1642.

Wijewardana, C., Henry, W. B., & Reddy, K. R. (2017). Evaluation of drought tolerant maize germplasm to induced drought stress. *Journal of the Mississippi Academy of Sciences*, *62*, 316–329.

William, H., Coffman, A., Clasen, B. M., Demorest, Z. L., Lowy, A., Ray, E., Retterath, A., Stoddard, T., Juillerat, A., Cedrone, F., Mathis, L., Voytas, D. F., & Zhang, F. (2014). Improved soybean oil quality by targeted mutagenesis of the fatty acid desaturase 2 gene family. *Plant Biotechnology Journal*, *12*(7), 934–940.

Wünsche, J. N. IV (2021) International symposium on horticulture in europe–SHE. Stuttgart (Germany), 8-12 March 2021, https://www.ishs.org/symposium/507

Xie, H., Yang, D. H., Yao, H., Bai, G., Zhang, Y. H., & Xiao, B. G. (2016). Itraq-based quantitative proteomic analysis reveals proteomic changes in leaves of cultivated tobacco (Nicotiana tabacum) in response to drought stress. *Biochemical and Biophysical Research Communications*, *9*(1), 768–775.

Xie, Q., Michaeli, S., Peled-Zehavi, H., & Galili, G. (2015). Chloroplast degradation: One organelle, multiple degradation pathways. *Trends in Plant Science*, *20*, 264–265.

Xin, H., Jing-Pei, C., & Xiao-Yan, Z. (2015). Pyrolysis kinetics of soybean straw using thermos gravimetric analysis. *Fuel*, *169*, 93–98.

Xu, Z., Li, C., Zhang, C., Yu, Y., van der Werf, W., & Zhang, F. (2020). Intercropping maize and soybean increases efficiency of land and fertilizer nitrogen use; A meta-analysis. *Field Crops Research*, *246*, 107661.

Yang, Z., Chi, Y., Cui, Y., Wang, Z., Hu, D., Yang, H., Bhat, J. A., Wang, H., Kan, G., Yu, D., & Huang, F. (2022). Ectopic expression of gmrnf1a encoding a soybean e3 ubiquitin ligase affects Arabidopsis silique development and dehiscence. *Planta*, *255*(3), 55.

Yasmeen, F., Raja, N. I., Mustafa, G., Sakata, K., & Komatsu, S. (2016). Quantitative proteomic analysis of post-flooding recovery in soybean root exposed to aluminum oxide nanoparticles. *Journal of Proteomics*, *143*, 136–150.

Yi, J. Y., Kim, H. J., & Chung, M. S. (2017). Manufacture of low-benzo(a) pyrene sesame seed (Sesamum indicum l.) Oil using a self-designed apparatus. *PLoS One*, *12*(3), 573–585.

Yijun, G., Zhiming, X., Jianing, G., Qian, Z., Rasheed, A., Hussain, M. I., Ali, I., Shuheng, Z., Hassan, M. U., Hashem, M., Mostafa, Y. S., Wang, Y., Chen, L., Xiaoxue, W., & Jian, W. (2022). The intervention of classical and molecular breeding approaches to enhance flooding stress tolerance in soybean – an review. *Frontiers in Plant Science*, *13*, 1085368.

Yin, X., & Komatsu, S. (2015). Quantitative proteomics of nuclear phosphoproteins in the root tip of soybean during the initial stages of flooding stress. *Journal of Proteomics*, *119*, 183–195.

Yin, X., & Komatsu, S. (2016). Nuclear proteomics reveals the role of protein synthesis and chromatin structure in root tip of soybean during the initial stage of flooding stress. *Journal of Proteome Research*, *15*, 2283–2298.

Yin, X., Nishimura, M., Hajika, M., & Komatsu, S. (2016). Quantitative proteomics reveals the flooding-tolerance mechanism in mutant and abscisic acid-treated soybean. *Journal of Proteome Research*, *15*, 2008–2025.

Yin, X., Sakata, K., & Komatsu, S. (2014). Phosphoproteomics reveals the effect of ethylene in soybean root under flooding stress. *Journal of Proteome Research*, *13*(12), 5618–5634.

Yin, Y., Tian, X., He, X., Yang, J., Yang, Z., & Fang, W. (2022 15). Exogenous melatonin stimulated isoflavone biosynthesis in NaCl-stressed germinating soybean (Glycine max L.). *Plant Physiology and Biochemistry*, *185*, 123–131.

Yin, Y., Yang, R., Han, Y., & Gu, Z. (2015). Comparative proteomic and physiological analyses reveal the protective effect of exogenous calcium on the germinating soybean response to salt stress. *Journal of Proteomics*, *113*, 110–126.

Yu, X., Yang, A., & James, A. T. (2017). Comparative proteomic analysis of drought response in roots of two soybean genotypes. *Crop and Pasture Science*, *68*, 609–619.

Yue, Y., Liu, N., Jiang, B., Li, M., Wang, H., Jiang, Z., Pan, H., Xia, Q., Ma, Q., & Han, T. (2017). A single nucleotide deletion in J encoding GmELF3 confers long juvenility and is associated with adaption of tropic soybean. *Molecular Plant*, *10*, 656–658.

Yuhong, G., Rasheed, A., Zhao, Z., Gardiner, J. J., Ilyas, M., & Wang, P. (2021). Role of conventional and molecular techniques in soybean yield and quality improvement: A critical review. *Notulae Botanicae Horti Agrobotanici Cluj-Napoca*, *49*, 1–22.

Zafar, S. A., Zaidi, S. S., Gaba, Y., Singla-Pareek, S. L., Dhankher, O. P., Li, O., Mansoor, S., & Pareek, A. (2020). Engineering abiotic stress tolerance via CRISPR/ Cas-mediated genome editing. *Journal of Experimental Botany*, *71*(20), 470–479.

Zahra, N., Hafeez, M. B., Ghaffar, A., Kausar, A., Al-Zeidi, M., Siddique, K. H. M., & Farooq, M. (2023). Plant photosynthesis under heat stress: Effects and management. *Environmental and Experimental Botany*, *206*, 105178.

Zaki, S. S., Belal, E. E. E., & Rady, M. M. (2019). Cyanobacteria and glutathione applications improve productivity, nutrient contents, and antioxidant systems of salt-stressed soybean plant. *International Letters of Natural Sciences*, *76*, 72–85.

Zandalinas, S. I., Fichman, Y., Devireddy, A. R., Sengupta, S., Azad, R. K., & Mittler, R. (2020). Systemic signaling during abiotic stress combination in plants. *Proceedings of the National Academy of Sciences of the United States of America*, *117*(24), 13810–13820.

Zhang, L. X., Liu, W., Tsegaw, M., Xu, X., Qi, Y. P., Sapey, E., Liu, L. P., Wu, T. T., Sun, S., & Han, T. F. (2020). Principles and practices of the photo-thermal adaptability improvement in soybean. *Journal of Integrative Agriculture*, *19*(2), 295–310.

Zhang, S., Zhang, Z., Bales, C., Gu, C., DiFonzo, C., Li, M., Song, Q., Cregan, P., Yang, Z., & Wang, D. (2017). Mapping novel aphid resistance QTL from wild soybean, glycine Soja 85-32. Züchter Genet. *Breeding Research*, *130*, 1941–1952.

Zhang, X. Z., Zheng, W. J., Cao, X. Y., Cui, X. Y., Zhao, S. P., Yu, T. F., Chen, J., Zhou, Y. B., Chen, M., Chai, S. C., Xu, Z. S., & Ma, Y. Z. (2019). Genomic analysis of stress associated proteins in soybean and the role of *gmsap16* in abiotic stress responses in *arabidopsis* and soybean. *Frontiers in Plant Science*, *10*, 145.

Zhang, Y., Fang, Q., Zheng, J., Li, Z., Li, Y., Feng, Y., Han, Y., & Li, Y. (2022). GmLecRlk, a Lectin receptor-like protein kinase, contributes to salt stress tolerance by regulating salt-responsive genes in soybean. *International Journal of Molecular Sciences*, *23*(3), 1030.

Zhao, B., Dai, A., Wei, H., Yang, S., Wang, B., Jiang, N., & Feng, X. (2016). Arabidopsis klu homologue gmcyp78a72 regulates seed size in soybean. *Plant Molecular Biology*, *90*, 33–47.

Zhong, Z., Furuya, T., Ueno, K., Yamaguchi, H., Hitachi, K., Tsuchida, K., Tani, M., Tian, J., & Komatsu, S. (2020). Proteomic analysis of irradiation with millimeter waves on soybean growth under flooding conditions. *International Journal of Molecular Sciences*, *21*(2), 486.

Zhou, Z., Jiang, Y., Wang, Z., Gou, Z., Lyu, J., Li, W., Yu, Y., Shu, L., Zhao, Y., Ma, Y., & Fang, C. (2015). Resequencing 302 wild and cultivated accessions identifies genes related to domestication and improvement in soybean. *Nature Biotechnology*, *33*(4), 408–414.

Zhu, L., Feng, L., Yu, X., Fu, X., Yang, Q., Jin, H., & Yuan, F. (2022). Development and application of an in vitro method to evaluate anthracnose resistance in soybean Germplasm. *Research Square*, *11*(5), 657.

Zorb, C., Geilfus, C. M., & Dietz, K. J. (2019). Salinity and crop yield. *Plant Biology*, *21*, 31–38.

Zuffo, A., Mario, A., Fã¡bio, A., Gonzã¡lez, A. J., Teodoro, P. E., Teodoro., Pereira Ribeiro, L., & Busch (2020). Multi trait stability index: A tool for simultaneous selection of soya bean genotypes in drought and saline stress. *Journal of Agronomy and Crop Science*, *206*(7), 1–8.

12 Genome Editing Systems for Speed Breeding of Tomato (*Solanum lycopersicon* L.)

Sonupriya Sahu, Madhusmita Barik, Jatindra Nath Mohanty, Animesh Pattnaik, and Rukmini Mishra
Department of Botany, School of Applied Sciences, Centurion University of Technology and Management, Jatni, Bhubaneswar, Odisha, India

Raj Kumar Joshi
Department of Biotechnology, Rama Devi Women's University, Bhubaneswar, Odisha, India

12.1 INTRODUCTION

Solanum lycopersicum L., or tomatoes, are among the most significant vegetable crops in terms of economic value, with an annual production of 186.82 million metric tonnes. China is the world's top producer of tomatoes, followed by Turkey and India together accounting for 30% of the overall output (FAO, 2017). The tomato is a rich source of many important macromolecules such as vitamins (alpha-tocopherol), carotenoids (lycopene), phytochemicals (flavonoids), proteins, and polysaccharides (pectin) (Kumar et al., 2021). Tomato seeds are a major source of proteins, chemicals, food supplements, and oil (Rajan et al., 2022). In other words, the nutritional and economic values of tomato have made it a model vegetable for study (Li et al., 2018a). However, many biotic and abiotic variables currently hinder tomato production, and the agricultural and breeding community faces a significant problem in the creation of stress-tolerant crops (Fahad et al., 2017; Zhou et al., 2019).

To fulfill the demands of the world's population, contemporary agriculture must instead turn to new techniques as outdated ones reach their limits. Speed breeding (SB) is a term used to describe different techniques used to enhance the cycle turnover (Ghosh et al., 2018). One of the foundational elements of SB is rapid generation advancement. By significant reduction in the generation period, thereby shortening the duration of breeding programs, speeds up the development of novel varieties that better meet consumer needs, and effectively tackles the urgent challenges of climate change and the imperative for sustainable agriculture (Wanga et al., 2021). Rapid generation turnover (RGT), *in vitro* cultivation of developing embryos, fast generation cycling (FGC), rapid generation cycling (RGC), and accelerated single-seed descent (aSSD) are some of the approaches used for accelerated breeding in plants. During the last few years, these SB techniques have been used to improve annual generation turnover by up to three times over traditional generation advancement systems in a number of economically important cereal crops. One of the approaches for shortening the generation time is the modification of agronomic practices (Samantara et al., 2022). As a typical SB generation takes 65–100 days on average, SB of tomatoes would be useful for both scientific and applied research (Wanga et al., 2021). The process of developing new varieties through traditional breeding takes a long period, and therefore SB techniques could be integrated with the recently developed genome editing approaches towards

 DOI: 10.1201/b23372-12

faster growth and greater resilience to stresses (Wang et al., 2022a, 2022b). Sequence-specific nucleases (SSNs) are utilized in genome editing procedures to facilitate the cleavage of double-stranded DNA at specific sites and their subsequent repair through the cell's endogenous repair mechanism results in precise genetic mutation (Mishra et al., 2019). What's more, the CRISPR/Cas system is widely accepted by the scientific community due to its simplicity, non-requirement of complex protein chemistry, and the ability for simultaneous introduction of double-stranded breaks (DSBs) at multiple sites (Mishra et al., 2019). What's more, the CRISPR/Cas9 system has emerged as the preferred technique for the precise modification of plants, animals, and microorganisms alike. Recent studies have shown that CRISPR/Cas system has been used for editing several traits in tomato (Salava et al., 2021; Xia et al., 2021; Chaudhuri et al., 2022). These edited tomato lines could be integrated into the SB approaches to accelerate tomato breeding for the development of desirable traits precisely and within a short period. Therefore, in this chapter, we focus on the recent development in the SB approaches and their integration into the CRISPR/Cas editing system for the improvement of tomato.

12.2 BREEDING APPROACHES IN TOMATO

Being a model plant, extensive genetic and molecular characterization has been carried out in tomato including the development of high-density genetic maps, complete genome sequence, gene-tagged molecular markers, introgression lines, and genetic mutants (Sato et al., 2012). Re-sequencing hundreds of genomes from cultivars, wild cousins, and landraces of tomatoes has shown a large frequency of chromosomal rearrangements as a result of residual wild introgressions and very little genetic diversity (Schouten et al., 2019). The genetic basis of tomatoes narrowed during the domestication process, hindering intra-populational breeding strategies from achieving significant genetic improvements (Souza et al., 2012). Tomatoes' resistance to inbreeding allows for the creation and upkeep of inbred lines despite their low genetic variety. As a result, the recombination of genetic diversity appears to be a great substitute for achieving better genotypes (Souza et al., 2012; Cappetta et al., 2020). Furthermore, agronomically important features like disease resistance and quality attributes have been introduced by retaining genome fragments from wild cousins, greatly enhancing the genetic heterogeneity inside the tomato gene pool (D'Esposito et al., 2019).

12.3 MOLECULAR MARKER AIDED SELECTION IN TOMATO BREEDING

In the early 1980s, the landscape of plant breeding underwent a dramatic transformation with the advent of various molecular marker systems. Using the molecular markers, a highly saturated tomato reference linkage map was created from the interspecific population of *L. esculentum* LA925 and *L. pennellii* LA716 (Shirasawa et al., 2010). Additionally, the tomato genome sequencing projects also allowed for the development of even more precise genetic markers. Nonetheless, a significant challenge persisted in the form of an insufficient number of DNA markers capable of detecting polymorphism within the cultivated tomato species (Hamilton et al., 2012). As a result, current studies have focused on using high-resolution markers like single nucleotide polymorphisms (SNPs) to discover changes among closely related species variants (Sacco et al., 2015; Esposito et al., 2020). Marker-assisted selection (MAS) has been the main method used to add molecular markers into the conventional phenotypic selection (PS) in tomato. This strategy included chromosomal regions carrying particular genes or quantitative trait loci (QTLs) in an effort to improve the plant selection process (Donato et al., 2017). MAS has played a pivotal role in mapping genes or QTLs associated with stress responses and flower and fruit development and their subsequent introgression into cultivated lines (reviewed in Osei et al., 2018). However, the process is slower and often suffers from selection breakage due to recombination and/or the emergence of new sources of stress.

12.4 ACCELERATED MARKER DEVELOPMENT THROUGH GENOTYPING BY SEQUENCING

To capture a maximum number of informative loci, it is essential to have an appropriate abundance of markers at your disposal (Zhang et al., 2019). One effective approach is to employ genotyping by sequencing (GBS), which allows for the efficient generation of densely packed marker structures. In contrast, the conserved SNPs linked to a potential gene mutation can be found directly using the restriction site-associated RNA sequencing (RAR-seq) approach (Alabady et al., 2015). Single primer enrichment technology (SPET), a customizable technique for targeted genotyping in tomatoes, was recently developed to improve structure design and enhance the capacity to multiplex for genotyping in tomato (Barchi et al., 2019). Leveraging prior genomic selection (GS) data can aid in creating a set of markers that are optimized for the next steps. For instance, Liabeuf et al. (2018) streamlined the "SolCAP array" of 7,700 SNPs (Sim et al., 2012) in order to filter populations with low recombination. Furthermore, it's worth noting that the prediction accuracy can also be influenced by the minor allele frequency threshold (MAF) (Zhang et al., 2019). It is crucial to incorporate proven genomic signatures into tomato breeding selection processes in an efficient manner. Human selection exerts changes in genomic regions that impact agronomically relevant traits (Smykal et al., 2018). Understanding the genetic mechanisms controlling phenotypic diversity among cultivars and the history of the population depends on the identification of selection signatures (Gao et al., 2019). Estimating differences in allele frequencies between or within a population can enhance trait selection. The chromosome landscape of tomato cultivars has changed significantly due to the introgression of prominent genetic fragments from wild *Solanum* species. For instance, 79% of chromosome 9 in tomato accounts for the introgression of genetic fragments from wild *Solanum peruvianum* encoding tomato mosaic virus (ToMV) resistance gene Tm2. To pinpoint homologous recombination breakpoints, it's recommended to employ an average coverage of 8× in tomato (Gonda et al., 2019). Numerous statistical methods have been evaluated for estimating marker effects in tomatoes (Liabeuf, et al., 2018). The best approach should be chosen based on the particular situation, taking into account elements like population size, heritability, genetic architecture, and computing needs (Heslot et al., 2013). The genomic best linear unbiased prediction (GBLUP) and the ridge regression best linear unbiased prediction (RR-BLUP) (Whittaker et al., 2000), are recommended when assessing traits influenced by many small-effect genes within a closely related training population (TRN). Conversely, when traits are governed by major-effect QTLs or when predicting unrelated individuals, Bayesian methods can yield higher prediction accuracy (De los Campos et al., 2013). Empirical research, however, shows that in tomatoes, regression-based and Bayesian approaches do not significantly vary from one another (Liabeuf et al., 2018).

12.5 ACCELERATION OF TOMATO BREEDING THROUGH GENOMIC SELECTION

Plant breeding programs aiming at traits with polygenic inheritance have promising opportunities to increase their efficiency through the use of GS (Voss-Fels et al., 2018; Xu et al., 2020). SNPs and other comprehensive genomic data are used to determine an individual's prospective breeding value. Recent advancements in high-throughput genotyping (HTG) systems enable the generation of thousands of markers, allowing for comprehensive genome-wide scanning. Allelic connections between the markers and phenotypes are used to forecast the phenotypic value of possible selections. A genomic estimated breeding value (GEBV), expressed as an index-dependent linear function, is computed for each breeding candidate. To obtain GEBVs for a genotyped testing set (TST), GS combines genotypic and phenotypic data from a TRN within a training set (TRS) (Figure 12.1). The GS model is then used to predict the breeding values of non-phenotyped individuals in the following round of selection. This pioneering exploration of GS simulations in plants was first reported by Meuwissen and collaborators (Meuwissen et al., 2001). Plant breeders now have appropriate GS

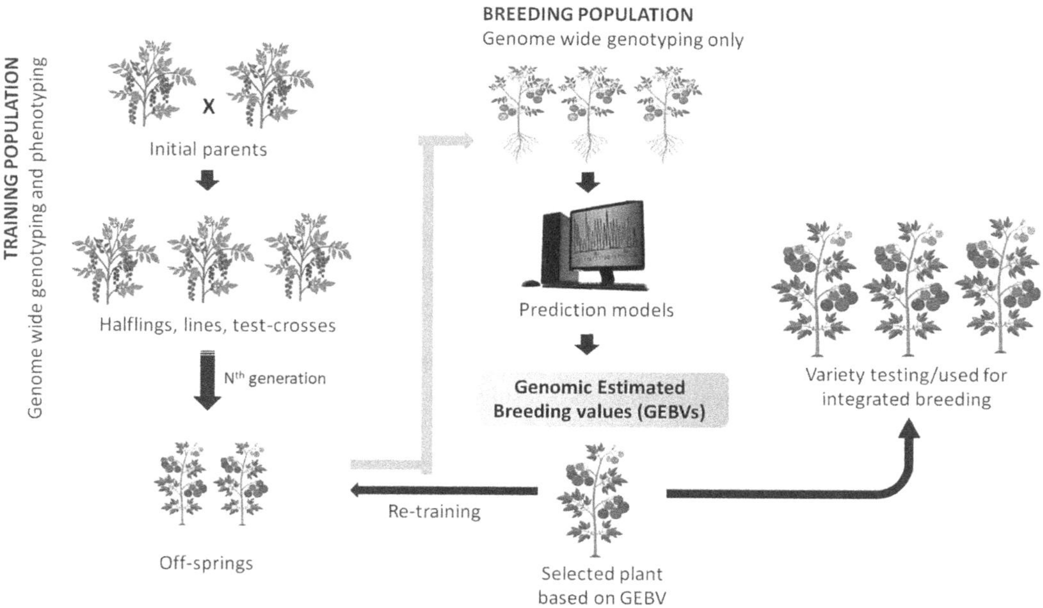

FIGURE 12.1 Steps of genomic selection. The genotypic and phenotypic information from the training sets of plants facilitates the adjustment of the genomic selection model to select for desired variety based on genomic estimated breeding values.

models for evaluating the link between true breeding values (TBVs) and estimated breeding values (EBVs) thanks to a thorough evaluation of linear regression and unbiased Bayesian prediction techniques (Figure 12.1). Genomic-assisted screening of the breeding populations has the potential to expedite genetic progress in each breeding cycle, potentially reducing the time required for selection by up to two-thirds. This acceleration is particularly valuable when selecting traits with low heritability. While individual marker effects are typically quite modest, the cumulative wealth of genome-wide marker data holds the promise of elucidating the entire genetic variance (Voss-Fels et al., 2018). Nevertheless, integrating GS into actual plant breeding schemes poses challenges for breeders. The candidate count in a plant breeding program is often constrained by budgetary and temporal considerations, which inevitably influence the rate of genetic advancement in crop improvement. To date, various GS-based breeding approaches have been conceptualized for a range of crops, including tomato, resulting in increased rates of genetic gain (Duangjit et al., 2016; Voss-Fels et al., 2018). The extensive use of GS in plant breeding calls for a number of important activities: cost reduction in GS, the development of economical genotyping and phenotyping platforms, the establishment of diverse and regularly updated training populations (TRN), the creation of highly efficient and versatile genomic prediction models, the enhancement of agronomic practices to shorten breeding cycle durations, the establishment of robust decision support systems, and the implementation of open-source breeding programs. Additionally, it is imperative to precisely calibrate important GS-influencing elements such as the density of the markers, size of the training set, genetic relatedness between the TRN and TST species, population structure, phenotyping techniques, and statistical model dependability (Heslot et al., 2015).

12.6 TOMATO IMPROVEMENT THROUGH REVERSE BREEDING

Backcrossing represents another widely employed tomato breeding strategy for the purpose of transferring a valued trait from a contributor parent into the genetic background of a recipient parent. To add desirable features, backcrossing procedures employing exotic or elite elements are frequently employed. Yet, the constant emergence of new alleles, the potential for linkage drags, crossbreeding with older varieties or exotic materials, and the prolonged breeding cycles stemming from intricate crossing schemes can collectively diminish the genetic gain. GS can surely improve the response to genetic selection, which is accomplished by choosing genotypes with high breeding value within a diverse population. Researchers have achieved significant rates of genetic gain by a modification of the traditional backcrossing approach, in which the breeding value of the recurrent parent determines the chosen lines for each generation (Shelton et al., 2015). Moreover, genes with significant impacts can be chosen within each generation after crossbreeding with the recurrent line through a combination of GS with gene-specific-marker analysis. Thus, it is expected that the GS method will gradually upsurge the genetic gain at every generation. Through the usage of gene-specific molecular, one can identify candidate genotypes for selection that carry particular alleles, such as resistance characteristics. Furthermore, a subset of the markers used to apply the GS model can be selected in order to identify unwanted segments that come from a wild donor. Notably, resistance gene introgressions on specific chromosomes have resulted in the integration of sizable wild genome portions into cultivated tomatoes (Schouten et al., 2019). Developing this strategy, genome-wide selection in backcross generation using high-throughput markers may result in more recovery of the recipient parent traits from BC_1 to BC_3 without sacrificing the introgression of advantageous features.

The choice of parental lines to create F_1 hybrid tomatoes is based on the lines' breeding value. Yamamoto and group used a series of 96 large-fruited F1 tomato lines to create GS models in order to obtain precise breeding values (Yamamoto et al., 2017). These models were subsequently validated using segregating populations derived from crosses. Based on genotypic and phenotypic data for parameters like fruit weight and soluble solids, the GS models correctly predicted parental combinations that produced superior hybrids. It's important to note that these models' predicted effectiveness differed based on the particular features and parental pairings. Small populations with low genetic variety greatly lower the selection benefit of the GS strategy, even while the necessity to fix favorable alleles in the gene pool may enhance inbreeding. Using GS techniques to manage elite genetic variability and gradually raise the frequency of advantageous alleles can be very beneficial (Falk et al., 2010). Assessing a greater number of selfing progenies may increase the predictive accuracy of the parental cross. Advances in genotyping technologies and our understanding of the tomato genome have made it possible for breeders to quickly identify important alleles among elite germplasm (Ercolano et al., 2014; Sacco et al., 2015). Using a set of approved markers, these important alleles can be carefully paired to create new lines.

12.7 GENOME EDITING APPROACHES FOR SPEED BREEDING IN TOMATO

Introducing tomato varieties with high yield and resistance to several stresses requires a quick and inexpensive process. In recent times, traditional mutagenesis procedures are being quickly replaced by genome engineering technologies. Model plant and crop genomes are frequently subjected to induced targeted insertions, deletions, and particular sequential modifications using customized site-specific nucleases (Voytas et al., 2013). With genome editing, it is possible to precisely manipulate particular genomic sequences for crop improvement, in contrast to random mutagenesis caused by gamma radiation or EMS. In the recent past, several SSNs, such as meganucleases, zinc finger nucleases (ZFNs), and transcription activator-like effector nucleases (TALENs), have been widely employed in plants (Voytas & Gao, 2014). More recently, the 'clustered regularly interspaced short palindromic repeats' and the 'CRISPR associated protein' (CRISPR/Cas)—an effective genome editing system based on bacterial monomeric DNA endonuclease has gained popularity rapidly because it is easier to use and more straightforward than ZFN and TALEN (Zhang et al., 2018) (Figure 12.2). CRISPR/Cas relies on RNA-DNA interaction for sequence-specific genome editing. CRISPR/Cas systems are categorized into five major types, with the well-studied CRISPR/Cas9

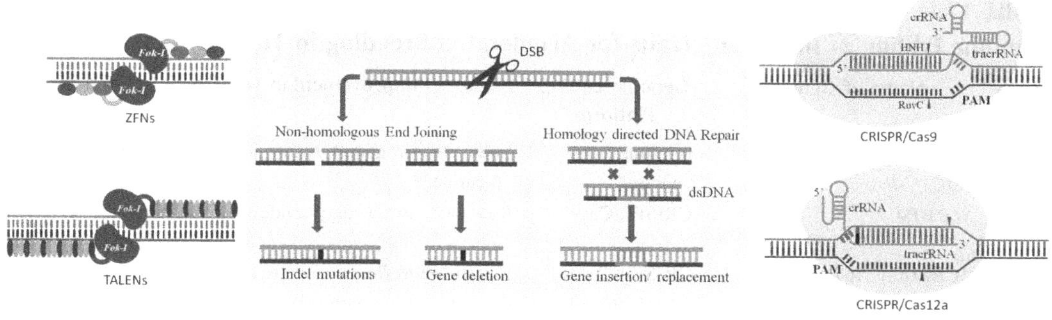

FIGURE 12.2 Double-stranded breaks (DSBs) are produced when transcriptional activator-like effector nucleases (TALENs) and zinc finger nucleases (ZFNs) bring two FokI monomers close to the target site. CRISPR/Cas9 platform identifies the target site using a 100 nt single guide (sgRNA) and creates DSB using Cas9 endonuclease. A 40–45 nt CRISPR RNA is used by the CRISPR/Cas12a system to identify the target site, and Cas12a endonuclease is used to create DSB. DSBs are repaired using homologous recombination (HR) or non-homologous end joining (NHEJ).

belonging to type II (Makarova et al., 2015). The Cas9 protein and synthetic single-guide RNA (sgRNA) are involved in CRISPR/Cas9; sgRNA directs Cas9 to cause site-specific breaks in DNA and thereby activates the endogenous DNA repair pathways (Mishra et al., 2022). What's more, CRISPR/Cas has become a popular method for genome engineering in plants, and since then, a lot of research has been done to address many issues in plant biology and agricultural development (reviewed in Mishra et al., 2019; Wang et al., 2019; Joshi et al., 2020). Additionally, CRISPR/Cpf1 (Cas12a), a novel member of the CRISPR/Cas family, has gradually found applications in several plant species, including the tomato, rice, soybean, and cotton (Mishra et al., 2018; Vu et al., 2020; Zheng et al., 2022). The efficient use of CRISPR/Cpf1 in conjunction with geminiviral multireplicons resulted in a notable boost in tomato genome editing, particularly for homology-directed repair (HDR)-based edits, demonstrating its role in tomato genome editing (Vu et al., 2020).

While CRISPR/Cas9 has been largely embraced by the plant species, numerous technological advancements in the CRISPR/Cas platforms have also facilitated the precise improvement of tomato (Table 12.1). These advancements concentrate on increasing editing efficiency, decreasing the possibility of off-target effects, speeding up the creation of transgene-free and genome-edited plants, and making it easier to induce genetic changes in many cell types, tissues, or organs. For instance, in tomato mutants generated by CRISPR/Cas9, the utilization of multiplex sgRNAs, rather than single sgRNAs, could enhance the likelihood of obtaining substantial deletions (Santillán Martinez et al., 2020). In a recent study, the CRISPR/Cas9 construct carrying the anthocyanin-enhancing gene complex PAP1/MYB75 resulted in transgene-free gene-edited tomato plants with enhanced color pattern development (Hu et al., 2019).

A recently developed CRISPR/Cas9 toolkit, featuring Cas9 expression under the control of a fruit-specific phosphoenolpyruvate carboxylase 2 gene promoter, enabled precise gene editing in tomato fruits (Feder et al., 2020). Additionally, a recent breakthrough in nanoparticle-mediated tomato genome editing was made possible by a straightforward foliar spray of nanoparticles coated with CRISPR/Cas9 complexes (Doyle et al., 2019). More recently, genome editing has also facilitated the *de novo* generation of plant meristem to avoid the complex process of tissue culture (Maher et al., 2020). These strategies provide greater opportunities for functional characterization of genetic traits for stress tolerance and yield improvement in tomato. Likewise, CRISPR/Cas9 mediated inactivation of tomato *Self Prunning (SP)* and *Self Prunning 5 G (SP5G)* has contributed to smaller internodes and accelerated flowering thereby contributing to higher yield (Soyk et al., 2017; Zhang et al., 2018). In yet another study, CRISPR/Cas-mediated

TABLE 12.1
Genome Editing of Important Traits for Accelerated Breeding in Tomato

Sl No	Name of Gene(s)	Genome Editing Platform	Nature of Improvement in Tomato	References
1	*SlDDB1, SlDET1 & SlCYC-B*	Cytidine base editor	Improved carotenoid, lycopene, and β-carotene	Hunziker et al., 2022
2	*SLPMR4*	CRISPR/Cas9	Resistance to powdery mildews	Santillán et al., 2020
3	*ACET1a & ACET1b*	CRISPR/Cas9	Resistance to *Botrytis cinerea*	Jeon et al., 2020
4	*eIF4E1*	CRISPR/Cas9	Resistance to tomato potyvirus	Yoon et al., 2020
5	*SlLBD40*	CRISPR/Cas9	Improved tolerance to drought	Liu et al., 2020
6	*SP, SP5G and SlER*	CRISPR/Cas9	Accelerated flowering, reduced internodes	Rehman et al., 2022
7	*SlJAZ2*	CRISPR/Cas9	Resistance to bacteria	Ortigosa et al., 2019
8	*ALS*	Cytidine base editor	Improved tolerance to herbicide chlorsulfuron	Shimatani et al., 2017
9	*MYB12*	CRISPR/Cas9	Improved carotenoid, lycopene, and β-carotene	Yang et al., 2023
10	*PSY1 or CRTISO*	CRISPR/Cas9	Improved carotenoid, lycopene, and β-carotene	Dahan-Meir et al., 2018
11	*SGR1, LCY-E, Blc, LCY-B1, & LCY-B2*	CRISPR/Cas9	Improved lycopene content	Li et al., 2018b
12	*SSADH, GABA-TP1, GABA-TP2, GABA-TP3 and CAT9*	CRISPR/Cas9	GABA accumulation	Li et al., 2018c
13	*SlARF5 and SlARF7*	CRISPR/Cas9	Parthenocarpy	Xia et al., 2021
14	TYLCV genome	CRISPR/Cas9	Resistance to yellow leaf curl virus	Thiruppathi, 2022
15	*FAS, MULT, FW2.2, CycB*	CRISPR/Cas9	Improved fruit size and higher yield	Zsögön et al., 2018
16	*SP, SlGGP1, CLAVATA3, WUSCHEL and SP5G*	CRISPR/Cas9	Short plant, reduced internodes, accelerated flower, resistance to diseases	Li et al., 2018d
17	*CLV3, S or SP*	CRISPR/Cas9	Improved tomato fruit yield.	Zsögön et al., 2018
18	*SlGAD2 and SlGAD3*	CRISPR/Cas9	GABA accumulation	Nonaka et al., 2017
19	*SlIAA9 or SlAGL6*	CRISPR/Cas9	Parthenocarpy	Minh et al., 2020
20	*SlALC*	CRISPR/Cas9	Extended shelf life of fruit	Erpen-Dalla et al., 2019
21	*SlMlo1*	CRISPR/Cas9	Resistance to powdery mildew	Pramanik et al., 2021

mutagenesis of RECQ4 has resulted in the augmentation of crossover frequency in interspecific tomato hybrids (Maagd et al., 2020). In crop breeding, crossover creation during plant meiosis is essential for appropriate chromosome segregation and allows for the ideal blending of features by combining parental alleles on chromosomes. The process begins with the generation of numerous DSBs, but only a few result in crossovers. In *Arabidopsis thaliana*, a limited number of genes have been identified that influence the conversion of DNA crossover intermediate structures into non-crossovers. To explore the potential for enhancing crossover production in interspecific hybrids, the CRISPR/Cas9 mutagenesis technique was utilized to knock out the RecQ4 gene in an interspecific tomato hybrid, resulting in a biallelic recq4 mutant in the F_1 hybrid of tomato. This will greatly expedite the breeding process by introducing beneficial alleles from wild cousins into cultivated tomato lines.

12.8 CONCLUSION

The gene editing approach for SB in tomato is very novel. This comprehensive analysis has shed light on the promising role of genome editing systems in advancing crop improvement, ultimately paving the way for accelerated breeding. The integration of cutting-edge genomic technologies, including the CRISPR-Cas9 system, offers tomato breeders a powerful arsenal to expedite the development of improved tomato varieties. By precisely targeting and modifying specific genes associated with desired traits, genome editing systems can significantly reduce the time and resources required for traditional breeding approaches. Future prospects in tomato breeding are indeed stirring as genome editing techniques continue to evolve and become more precise; breeders can anticipate even greater efficiencies in trait improvement, disease resistance, and overall crop performance. Moreover, the ongoing expansion of genomic databases and the availability of comprehensive genomic resources will further enhance our understanding of tomato genetics, enabling breeders to make informed decisions and develop superior cultivars more rapidly. Additionally, the synergy between genome editing and other emerging technologies, such as GS, MAS, large-scale phenotyping and genotyping, and reverse breeding holds immense potential for accelerating the tomato breeding process. These integrated approaches will likely lead to the development of tomato varieties that are not only more productive and resilient but also tailored to meet the evolving demands of consumers and the challenges of changing agricultural landscapes.

AUTHOR CONTRIBUTIONS

The manuscript was conceptualized by RM and RKJ. SS, MB, JNM and AP wrote the manuscript. The manuscript was critically revised by RM and RKJ. All the authors have read and approved the final manuscript.

ACKNOWLEDGMENT

This research was supported by grants from the Science & Technology Dept., Govt. of Odisha, India (ST-BT-MISC-0009-2023-4688/ST) (to R K Joshi). RM is supported by a research grant from SERB, Govt. of India (SPG/2021/001604). RKJ acknowledge the infrastructure support developed through the CURIE program (DST/CURIE/01/2019) of DST, Govt. of India.

The authors declare that they have no conflict of interest.

REFERENCES

Alabady, M. S., Rogers, W. L., & Malmberg, R. L. (2015). Development of transcriptomic markers for population analysis using restriction site associated RNA sequencing (RARseq). *PLoS One*, *10*, e0134855. 10.1371/journal.pone.0134855

Barchi, L., Acquadro, A., Alonso, D., Aprea, G., Bassolino, L., Demurtas et al. (2019). Single primer enrichment technology (SPET) for high-throughput genotyping in tomato and eggplant germplasm. *Frontiers in Plant Science*, *10*, 1005. 10.3389/fpls.2019.01005

Cappetta, E., Andolfo, G., Di Matteo, A., & Ercolano, M. R. (2020). Empowering crop resilience to environmental multiple stress through the modulation of key response components. *Journal of Plant Physiology*, *246*, 153134. 10.1016/j.jplph.2020.153134.

Chaudhuri, A., Halder, K., & Datta, A. (2022). Classification of CRISPR/Cas system and its application in tomato breeding. *Theoretical and Applied Genetics*, *135*, 367–387. 10.1007/s00122-021-03984-y

Dahan-Meir, T., Filler-Hayut, S., Melamed-Bessudo, C., Bocobza, S., Czosnek, H., Aharoni, A., & Levy, A. A. (2018). Efficient *in planta* gene targeting in tomato using geminiviral replicons and the CRISPR/Cas9 system. *The Plant Journal*, *1*, 5–16. 10.1111/tpj.13932

De los Campos, G., Hickey, J. M., Pong-Wong, R., Daetwyler, H. D., & Calus, M. P. L. (2013). Whole-genome regression and prediction methods applied to plant and animal breeding. *Genetics*, *193*(2), 327–345.

D'Esposito, D., Cappetta, E., Andolfo, G., Ferriello, F., Borgonuovo, C., et al. (2019). Deciphering the biological processes underlying tomato biomass production and composition. *Plant Physiology and Biochemistry, 143*, 50–60. 10.1016/j.plaphy.2019.08.010.

Donato, A. D., Andolfo, G., Ferrarini, A., Delledonne, M., & Ercolano, M. R. (2017). Investigation of orthologous pathogen recognition gene-rich regions in solanaceous species. *Genome*, *60*, 850–859. 10.1139/gen-2016-0217.

Doyle, C., Higginbottom, K., Swift, T. A., Winfield, M., Bellas, C., Benito- Alifonso, D., Fletcher, T., Galan, C. M., Edwards, K., & Whitney, H. M. (2019). A simple method for spray-on gene editing in planta. *BioRxiv*, 805036. 10.1101/805036

Duangjit, J., Causse, M., & Sauvage, C. (2016). Efficiency of genomic selection for tomato fruit quality. *Molecular Breeding*, *36*, 29. 10.1007/s11032-016-0453-3.

Ercolano, M. R., Sacco, A., Ferriello, F., D'Alessandro, R., Tononi, P. et al. (2014). Patchwork sequencing of tomato San Marzano and Vesuviano varieties highlights genome-wide variations. *BMC Genomics*, *15*, 138. 10.1186/1471-2164-15-138

Erpen-Dalla Corte, L., Mahmoud, M. L., Moraes, S. T., Mou, Z., Grosser, W. J., Dutt, M. et al. (2019). Development of improved fruit, vegetable, and ornamental crops using the CRISPR/Cas9 genome editing technique. *Plants*, *8*(12), 601. 10.3390/plants8120601

Esposito, S., Cardi, T., Campanelli, G., Sestili, S., Díez, M. J., et al. (2020). ddRAD sequencing-based genotyping for population structure analysis in cultivated tomato provides new insights into the genomic diversity of Mediterranean 'da serbo' type long shelf-life germplasm. *Horticulture Research*, *7*, 134. 10.1038/s41438-020-00353-6.

Fahad, S., Bajwa, A. A., Nazir, U., Anjum, S. A., Farooq, A., et al. (2017). Crop production under drought and heat stress: Plant responses and management options. *Frontiers in Plant Science*, *8*, 1147. 10.3389/fpls.2017.01147

Falk, D. E. (2010). Generating and maintaining diversity at the elite level in crop breeding. *Genome*, *53*, 982–991. 10.1139/G10-081

FAO (2017). FAOSTAT. Food and Agriculture Organization of the United Nations, Rome, Italy.

Feder, A., Jensen, S., Wang, A. Q., Courtney, L., Middleton, L. et al. (2020). Tomato fruit as a model for tissue specific gene silencing in crop plants. *Horticulture Research*, *7*, 142. 10.1038/s41438-020-00363-4

Gao, L., Gonda, I., Sun, H., Ma, Q., Bao, K. et al. (2019). The tomato pan-genome uncovers new genes and a rare allele regulating fruit flavor. *Nature Genetics*, *51*, 1044–1051. 10.1038/s41588-019-0410-2

Ghosh, S., Watson, A., Gonzalez-Navarro, O. E., Ramirez-Gonzalez, R. H. et al. (2018). Speed breeding in growth chambers and glasshouses for crop breeding and model plant research. *Nature Protocols*, *13*, 2944–2963. 10.1038/s41596-018-0072-z

Gonda, I., Ashrafi, H., Lyon, D. A., Strickler, S. R., Hulse-Kemp, A. M. et al. (2019). Sequencing-based bin map construction of a tomato mapping population, facilitating high-resolution quantitative trait loci detection. *The Plant Genome*, *12*, 1. 10.3835/plantgenome2018.02.0010

Hamilton, J. P., Sim, S.-C., Stoffel, K., Van Deynze, A., Buell, C. R., & Francis, D. M. (2012). Single nucleotide polymorphism discovery in cultivated tomato via sequencing by synthesis. *The Plant Genome*, *5*, 17–29.

Heslot, N., Jannink, J. L., & Sorrells, M. E. (2015). Perspectives for genomic selection applications and research in plants. *Crop Science*, *55*, 1–12. 10.2135/cropsci2014.03.0249

Heslot, N., Rutkoski, J., Poland, J., Jannink, J. L., & Sorrells, M. E. (2013). Impact of marker ascertainment bias on genomic selection accuracy and estimates of genetic diversity. *PLoS One*, *8*, e74612. 10.1371/journal.pone.0074612

Hu, N., Xian, Z. Q., Li, N., Liu, Y. D., Huang, W. et al. (2019). Rapid and user-friendly open-source CRISPR/Cas9 system for single- or multi-site editing of tomato genome. *Horticulture Research*, *6*, 7. 10.1038/s41438-018-0082-6

Hunziker, J., Nishida, K., Kondo, A., Ariizumi, T., & Ezura, H. (2022). Phenotypic characterization of high carotenoid tomato mutants generated by the target-AID base-editing technology. *Frontiers in Plant Science*, *13*, 848560. 10.3389/fpls.2022.848560

Jeon, J. E., Kim, J. G., Fischer, C. R., Mehta, N., Dufour-Schroif, C. et al. (2020). A pathogen-responsive gene cluster for highly modified fatty acids in tomato. *Cell*, *180*(1), 176–187. 10.1016/j.cell.2019.11.037

Joshi, R. K., Bharat, S. S., & Mishra, R. (2020). Engineering drought tolerance in plants through CRISPR/Cas genome editing. *3Biotech, 10*(9), 400. 10.1007/s13205-020-02390-3.

Kumar, M., Tomar, M., Bhuyan, D. J. et al. (2021). Tomato (Solanum lycopersicum L.) seed: A review on bioactives and biomedical activities. *Biomedicine & Pharmacotherapy*, *142*, 112018. 10.1016/j.biopha.2021.112018

Li, Y., Wang, H., Zhang, Y., & Martin, C. (2018a). Can the world's favorite fruit, tomato, provide an effective biosynthetic chassis for high-value metabolites? *Plant Cell Reports*, *37*, 1443–1450. 10.1007/s00299-018-2283-8

Li, R., Li, R., Li, X., Fu, D., Zhu, B. et al. (2018b). Multiplexed CRISPR/Cas9-mediated metabolic engineering of γ-aminobutyric acid levels in Solanum lycopersicum. *Plant Biotechnol Journal*, *16*(2),415–427. 10.1111/pbi.12781

Li, T., Yang, X., Yu, Y., Si, X., Zhai, X. et al. (2018c) Domestication of wild tomato is accelerated by genome editing. *Nature Biotechnology*, *36*(12), 1160–1163. 10.1038/nbt.4273

Li, X., Wang, Y., Chen, S., Tian, H., Fu, D. et al. (2018d). Lycopene is enriched in tomato fruit by CRISPR/Cas9-mediated multiplex genome editing. *Frontiers in Plant Science*, *9*, 559. 10.3389/fpls.2018.00559

Liabeuf, D., Sim, S. C., & Francis, D. M. (2018). Comparison of marker-based genomic estimated breeding values and phenotypic evaluation for selection of bacterial spot resistance in tomato. *Phytopathology*, *108*, 392–401. 10.1094/PHYTO-12-16-0431-R.

Liu, L., Zhang, J., Xu, J., Li, Y., Guo, L. et al. (2020). CRISPR/Cas9 targeted mutagenesis of SlLBD40, a lateral organ boundaries domain transcription factor, enhances drought tolerance in tomato. *Plant Science*, *301*, 110683. 10.1016/j.plantsci.2020.110683

Maagd, R. A., Loonen, A., Chouaref, J., Pelé, A., Meijer-Dekens, F. et al. (2020). CRISPR/Cas inactivation of RECQ 4 increases homeologous crossovers in an interspecific tomato hybrid. *Plant Biotechnology Journal*, *18*(3), 805–813. 10.1111/pbi.13248

Maher, M. F., Nasti, R. A., Vollbrecht, M., Starker, C. G., Clark, M. D., & Voytas, D. F. (2020). Plant gene editing through de novo induction of meristems. *Nature Biotechnology*, *38*, 84–89. 10.1038/s41587-019-0337-2

Makarova, K. S., Wolf, Y. I., Alkhnbashi, O. S., Costa, F., Shah, S. A. et al. (2015). An updated evolutionary classification of CRISPR-Cas systems. *Nature Reviews Microbiology*, *13*, 722–736. 10.1038/nrmicro3569

Meuwissen, T., Hayes, B. J., & Goddard, M. E. (2001). Prediction of total genetic value using genome-wide dense marker maps. *Genetics*, *157*, 1819–1829. 10.1093/genetics/157.4.1819

Minh, B. M., Hanh, H. H., Hue, H. T. (2020). Construction of CRISPR/Cas9 expression vectors harbouring gRNA targeted on SlIAA9 gene of tomato. *Vietnam Journal of Biotechnology*, *18*(1), 147–156. 10.15625/1811-4989/18/1/15273

Mishra, R., Joshi, R. K., & Zhao, K. (2022). Genome editing is revolutionizing crop improvement. In K. Zhao, R. Mishra, & R. K. Joshi (eds), *Genome editing technologies for crop improvement* (pp. 3–41). Singapore: Springer International Publishing. 10.1007/978-981-19-0600-8_1

Mishra, R., Joshi, R. K., & Zhao, K. (2019). Base editing in crops: Current advances, limitations and future implications. *Plant Biotechnology Journal*, *18*, 20–31. 10.1111/pbi.13225.

Mishra, R., Joshi, R. K., & Zhao, K. (2018). Genome editing in rice: Recent advances, challenges and future implications. *Frontiers in Plant Science*, *9*, 1361. 10.3389/fpls.2018.01361

Nonaka, S., Arai, C., Takayama, M., Matsukura, C., & Ezura, H. (2017). Efficient increase of ɣ-aminobutyric acid (GABA) content in tomato fruits by targeted mutagenesis. *Scientific Reports*, *7*(1), 7057. 10.1038/s41598-017-06400-y

Ortigosa, A., Gimenez-Ibanez, S., Leonhardt, N., & Solano, R. (2019). Design of a bacterial speck resistant tomato by CRISPR/Cas9-mediated editing of SlJAZ2. *Plant Biotechnology Journal*, *17*(3), 665–673. 10.1111/pbi.13006

Osei, M. K., Prempeh, R., Adjebeng, J., Opoku, J., & Danquah A. (2018). Marker-assisted selection (MAS): A fast-track tool in tomato breeding. In *Recent advances in tomato breeding and production*. London, UK: IntechOpen, pp. 93–113. 10.5772/intechopen.70226

Pramanik, D., Shelake, R. M., Park, J., Kim, M. J., Hwang, I. et al. (2021). CRISPR/Cas9-mediated generation of pathogen-resistant tomato against tomato yellow leaf curl virus and powdery mildew. *International Journal of Molecular Sciences*, *22*(4), 1878. 10.3390/ijms22041878

Rajan, A., Kumar, S., Sunil, C. K. et al. (2022). Recent advances in the utilization of industrial byproducts and wastes generated at different stages of tomato processing: Status report. *Journal of Food Processing and Preservation*, *46*, e17063. 10.1111/jfpp.17063.

Rehman, F., Gong, H., Bao, Y., Zeng, S., Huang, H., & Wang, Y. (2022). CRISPR gene editing of major domestication traits accelerating breeding for Solanaceae crops improvement. *Plant Molecular Biology*, *108*(3), 157–173. 10.1007/s11103-021-01229-6

Sacco, A., Ruggieri, V., Parisi, M., Festa, G., & Rigano, M. M. (2015). Exploring a tomato landraces collection for fruit-related traits by the aid of a high-throughput genomic platform. *PLoS One*, *10*, e0137139. 10.1371/journal.pone.0137139.

Salava, H., Thula, S., Mohan, V., Kumar, R., & Maghuly, F. (2021). Application of genome editing in tomato breeding: Mechanisms, advances, and prospects. *International Journal of Molecular Sciences*, *22*(2), 682. 10.3390/ijms22020682

Samantara, M. J., Bohra, A., Mohapatra, S. R., Prihatini, R., & Asibe, F. (2022). Breeding more crops in less time: A perspective on speed breeding. *Biology*, *11*, 275. 10.3390/biology11020275.

Santillán Martinez, M. I., Bracuto, V., Koseoglou, E., Appiano, M., Jacobsen, E. et al. (2020). CRISPR/Cas9-targeted mutagenesis of the tomato susceptibility gene PMR4 for resistance against powdery mildew. *BMC Plant Biology*, *20*, 284. 10.1186/s12870-020-02497-y

Sato, S., Tabata, S., Hirakawa, H., Asamizu, E., & Shirasawa, K. et al. (2012). The tomato genome sequence provides insights into fleshy fruit evolution. *Nature*, *485*, 635–641. 10.1038/nature11119

Schouten, H. J., Tikunov, Y., Verkerke, W., Finkers, R., Bovy, A. et al. (2019). Breeding has increased the diversity of cultivated tomato in the Netherlands. *Frontiers in Plant Science*, *10*, 1606. 10.3389/fpls. 2019.01606.

Shelton, A., Tracy, W., Shelton, A. C., & Tracy, W. F. (2015). Recurrent selection and participatory plant breeding for improvement of two organic open-pollinated sweet corn (Zea mays L.) populations. *Sustainability*, *7*, 5139–5152. 10.3390/su7055139

Shimatani, Z., Kashojiya, S., Takayama, M., Terada, R., Arazoe, T. et al. (2017). Targeted base editing in rice and tomato using a CRISPR-Cas9 cytidine deaminase fusion. *Nature Biotechnology*, *35*, 441–444. 10. 1038/nbt.3833

Shirasawa, K., Asamizu, E., Fukuoka, H., Ohyama, A., Sato, S. et al. (2010) An interspecific linkage map of SSR and intronic polymorphism markers in tomato. *Theoretical and Applied Genetics*, *121*(4), 731–739. 10.1007/s00122-010-1344-3.

Sim, S. C., Van Deynze, A., Stoel, K., Douches, D. S., Zarka, D. et al. (2012). High-density SNP genotyping of tomato (Solanum lycopersicum L.) reveals patterns of genetic variation due to breeding. *PLoS One*, *7*, e45520. 10.1371/journal.pone.0045520

Smykal, P., Nelson, M. N., Berger, J. D., & von Wettberg, E. J. B. (2018). The impact of genetic changes during crop domestication. *Agronomy*, *8*, 119. 10.3390/agronomy8070119

Souza, L. M., Paterniani, M., Melo, P. C. T., & Melo, A. M. T. (2012). Diallel cross among fresh market tomato inbreeding lines. *Horticulture Research*, *30*, 246–251. 10.1590/S0102-05362012000200011.

Soyk, S., Lemmon, Z. H., Oved, M., Fisher, J., Liberatore, K. L. et al. (2017). Bypassing negative epistasis on yield in tomato imposed by a domestication gene. *Cell*, *169*, 1142–1155. 10.1016/j.cell.2017.04.032

Thiruppathi, D. (2022). CRISPR keeps going "wild": A new protocol for DNA-free genome editing of tetraploid wild tomatoes. *Plant Physiology*, *189*(1), 10–11. 10.1093/plphys/kiac061.

Voss-Fels, K. P., Cooper, M., & Hayes, B. J. (2018). Accelerating crop genetic gains with genomic selection. *Theoretical and Applied Genetics*, *132*, 669–686. 10.1007/s00122-018-3270-8

Voytas, D. F., & Gao, C. (2014). Precision genome engineering and agriculture: Opportunities and regulatory challenges. *PLoS Biology*, *12*(6), e1001877. 10.1371/journal.pbio.1001877

Voytas, D. F. (2013). Plant genome engineering with sequence-specific nucleases. *Annual Review of Plant Biology*, *64*, 327–350. 10.1146/annurev-arplant-042811-105552

Vu, T. V., Sivankalyani, V., Kim, E. J., Doan, D. T. H., Tran, M. T. et al. (2020). Highly efficient homology-directed repair using CRISPR/Cpf1-geminiviral replicon in tomato. *Plant Biotechnology Journal*, *18*, 2133–2143. 10.1111/pbi.13373

Wang, L., Wang, Y., Makhmoudova, A., Nitschke, F., Tetlow, I. et al. (2022a). CRISPR-Cas9-mediated editing of starch branching enzymes results in altered starch granules in brassica napus. *Plant Physiology*, *188*, 1866–1886. 10.1093/plphys/kiab535

Wang, Z., Wang, L., Han, L., Cheng, Z., Liu, X. et al. (2022b). HECATE2 acts with GLABROUS3 and tu to boost cytokinin biosynthesis and regulate cucumber fruit wart formation. *Plant Physiology*, *187*, 1619–1635. 10.1093/plphys/kiab377

Wang, T., Zhang, H. Y., & Zhu, H. L. (2019). CRISPR technology is revolutionizing the improvement of tomato and other fruit crops. *Horticulture Research*, *6*, 77. 10.1038/s41438-019-0159-x

Wanga, M. A., Shimelis, H., Mashilo, J., & Laing, M. D. (2021). Opportunities and challenges of speed breeding: A review. *Plant Breeding* *140*(2), 185–194. 10.1111/pbr.12909.

Whittaker, J. C., Thompson, R., & Denham, M. C. (2000). Marker-assisted selection using ridge regression. *Genetic Research* *75*, 249–252. 10.1017/s0016672399004462

Xia, X., Cheng, X., Li, R., Yao, J., Li, Z. et al. (2021). Advances in application of genome editing in tomato and recent development of genome editing technology. *Theoretical and Applied Genetics*, *134*, 2727–2747. 10.1007/s00122-021-03874-3

Xu, Y., Liu, X., Fu, J., Wang, H., Wang, J. et al. (2020). Enhancing genetic gain through genomic selection: From livestock to plants. *Plant Communications*, *1*, 2641–2666. 10.1016/j.xplc.2019.100005

Yamamoto, E., Matsunaga, H., Onogi, A., Ohyama, A., Miyatake, K., et al. (2017). Efficiency of genomic selection for breeding population design and phenotype prediction in tomato. *Heredity*, *118*, 202–209. 10.1038/hdy.2016.84.

Yang, T., Ali, M., Lin, L., Li, P., He, H. et al. (2023). Recoloring tomato fruit by CRISPR/Cas9-mediated multiplex gene editing. *Horticulture Research*, *10*(1), uhac214. 10.1093/hr/uhac214

Yoon, Y. J., Venkatesh, J., Lee, J. H., Kim, J., Lee, H. E. et al. (2020). Genome editing of eIF4E1 in tomato confers resistance to pepper mottle virus. *Frontiers in Plant Science*, *11*, 1098. 10.3389/fpls.2020.01098

Zhang, H., Yin, L., Wang, M., Yuan, X., & Liu, X. (2019). Factors affecting the accuracy of genomic selection for agricultural economic traits in maize, cattle, and pig populations. *Frontiers in Genetics*, *10*, 189. 10.3389/fgene.2019.00189

Zhang, S., Jiao, Z., Liu, L., Wang, K., & Zhong, D. et al. (2018). Enhancer-promoter interaction of SELF PRUNING 5G shapes photoperiod adaptation. *Plant Physiology*, *178*(4), 1631–1642. 10.1104/pp.18.01137

Zheng, W., Mishra, R., Joshi, R. K., & Zhao, K. (2022). Targeted gene replacement in plants using CRISPR-Cas technology. In K. Zhao, R. Mishra, R. K. Joshi (eds), *Genome editing technologies for crop improvement* (pp. 139–160). Singapore: Springer International Publishing. 10.1007/978-981-19-0600-8_7

Zhou, R., Kong, L., Wu, Z., Rosenqvist, E., Wang, Y. et al. (2019). Physiological response of tomatoes at drought, heat and their combination followed by recovery. *Physiologia Plantarum*, *165*(2), 144–154. 10.1111/ppl.12764

Zsögön, A., Čermák, T., Naves, E. R., Notini, M. M., & Edel, K. H. (2018). De novo domestication of wild tomato using genome editing. *Nature Biotechnology*, *36*(12), 1211–1216. 10.1038/nbt.4272

13 Speed Germplasm Generation and Precision-Designed Breeding in Tomato Based on the CRISPR/Cas9 System

Jiao Wang, Shuting Ding, and Kai Shi
Department of Horticulture, Zhejiang University, Hangzhou, China

13.1 INTRODUCTION

Tomato (*Solanum lycopersicum* L.), originating from the Andes mountain region in South America, is known as the world's largest vegetable crop. The total production of tomato in the world was 189 million tons in 2021, with nearly 100 billion US dollars per year of economic value, according to the Food and Agriculture Organization of the United Nations (FAO) (www.fao.org). Tomato is low in calories and rich in nutritional qualities such as antioxidant lycopene, flavonoids, various minerals, and vitamins. In addition to its significant economic and nutrient value, with a short life cycle, high seed production, and easily genetically transformable, tomato has been used as an important model for studying fleshy fruit physiology and development, as well as disease resistance (Zhang et al., 2020; Mauxion et al., 2021).

Global climate change causes frequent occurrences of biotic and abiotic stresses, such as pathogen invasion, drought, and high-temperature stresses, posing a threat to both the yield and quality of vegetable crops. With the improvement in people's living standards, consumers are more concerned about the nutrition and health of vegetable crops. However, tomato lacks excellent germplasm resources to cope with the changing environment and the growing needs of consumers (Marefatzadeh-Khameneh et al., 2021). Thus, tomato breeding is becoming increasingly important.

Genomics has been widely applied in crop breeding with significant achievements based on the rapid advancement of high-throughput sequencing technology and computer science. Genomics and its derivative technologies have become cutting-edge research fields, influencing the development of the global agricultural industry. In 2012, over 300 researchers from 14 countries including China, the United States, the Netherlands, and Israel, completed the whole genome sequencing of the cultivated tomato "Heinz1706" (Sato et al., 2012). They identified approximately 34,727 genes, among which 97.4% (33,840 genes) were precisely located on chromosomes. At the same time, a framework map of the genome of the wild tomato ancestor, the ground cherry tomato, was drawn, which provides a foundation for subsequent mining of functional genes and the study of the evolution from wild to cultivated species, thus promoting the rapid application of genomics in breeding (Sato et al., 2012). In 2014, the genomes of 360 tomato germplasms from around the world, including cultivated and wild tomato germplasm materials, were re-sequenced (Lin et al., 2014). The changes in the tomato genome caused by prolonged domestication and breeding were analyzed. The history of tomato breeding, including four main stages of domestication, improvement, differentiation, and introgression was revealed. The study revealed that the enlargement of tomato fruits went through two evolutionary processes, from the blueberry-sized cultivated tomato close relative (*S. pipmpinellifolum*, SP) to the small-fruited cherry tomato

DOI: 10.1201/b23372-13

(*S. lycopersicum* var. cerasiforme, SLC), and then to the ordinary large-fruited cultivated tomato (*S. lycopersicum* var. lycopersicum, SLL). The Solanaceae Genomics Network (https://solgenomics.net/) integrates a large amount of basic information, identification data, genome sequences, SNPs, and GWAS analysis outcomes for a variety of tomato germplasm. These free and easily accessible resources provide valuable support for tomato genetic research, and quality improvement and are of great significance for tomato breeding (Mata-Nicolás et al., 2020).

Genome sequencing provides variant targets for genome-directed editing and lays the foundation for the rapid development of superior breeding materials. Traditional hybrid breeding is often limited by interspecies reproductive isolation and linkage drag (Giovannoni, 2018; Westram et al., 2022). Genome editing breeding can break through these limitations, rapidly and accurately improving target traits (Chen et al., 2019). Emmanuelle Charpentier and Jennifer A. Doudna were awarded the 2020 Nobel Prize in Chemistry for their development of the genome editing technique, CRISPR/Cas9. It is among the limited number of technologies to win the Nobel Prize within a short period after its introduction. CRISPR/Cas9 gene editing involves the use of a single-stranded guide RNA (sgRNA) to direct a cluster of regularly interspaced short palindromic repeats (CRISPR) RNA to search and anchor target DNA in the genome, followed by the action of the nuclease Cas protein to cause cleavage of the target DNA, stimulating cell's DNA self-repair mechanism (Shan et al., 2013). During DNA repair, target DNA (i.e., gene knockout) can be destroyed, or the missing sequences of target DNA can be corrected, or even a single base (i.e., base editing) can be introduced, and foreign genes can be introduced at specific sites (i.e., gene knock-in), thereby achieving precise editing at specific sites and multiple loci. Previous gene modification technologies before CRISPR/Cas9 mainly include zinc finger nuclease and transcription activator-like effector nuclease technology (Zhang et al., 2010; Chen & Gao, 2013). However, these two technologies have high requirements for experimental conditions, low editing efficiency, and high cost, and are subject to significant limitations in application. In contrast, CRISPR/Cas9 technology has simple operation, high editing efficiency, low cost, and diverse editing functions, enabling gene-targeted knockout, knock-in, base editing, homology repair (HDR), guided editing, etc. Since its introduction in 2012, the CRISPR/Cas9 technology has seen rapid adoption and improvement in molecular biology-related laboratories worldwide, playing an important role in basic and applied research in life sciences (Brooks et al., 2014; Chen et al., 2019; Salava et al., 2021). It quickly enters the application fields of agricultural variety breeding and human disease treatment clinical trials, becoming one of the most widely influential and rapidly applied biotechnologies (Wang et al., 2017; El-Mounadi et al., 2020).

Currently, the successful creation of superior gene mutants in fruit size, fruit nutrition, and plant defense has been accelerated by CRISPR/Cas9 genome editing technology, thus accelerating genetic improvement in tomato breeding materials.

13.2 ENHANCING TOMATO FRUIT WEIGHT AND SIZE THROUGH GENOME EDITING TECHNOLOGY

In the domestication of tomato, fruit weight has increased by hundreds of times. During the two evolution processes from wild-ground cherry tomato to cherry tomato and then to large-fruit cultivated tomato, more than ten fruit-weight genes have been subject to human-directed selection (Lin et al., 2014). Among them, the *cell size regulator* (*CSR*, *fw11.3*) gene was cloned in 2017. Mutations in this gene lead to an increase in tomato peel cell size, fruit weight, and juicy flesh. The artificial domestication and selection process of cherry tomato (the ancestor of tomato) has obtained the mutant genotype CSR-D of the *CSR* gene, which is 1.4 Kb shorter than *CSR* and encodes a protein that is 194 amino acids shorter. Co-expression and gene ontology (GO) clustering analysis shows that the CSR protein is related to cell differentiation in fruit tissues (Mu et al., 2017). Another negative regulator of tomato fruit size, *excessive number of floral organs* (*ENO*) gene, was discovered in 2020. *ENO* is a novel member of the AP2/ERF transcription factor superfamily. It can limit the generation of floral stem cells by regulating *WUSCHEL* gene expression in floral organs, thereby affecting fruit size. During the long-term domestication process of tomato, the ENO promoter was mutated, leading to increased tomato fruit size (Yuste-Lisbona et al., 2020).

The *locule number* (*lc*) and *fasciated* (*fas*) mainly increase tomato fruit size by enhancing the number of locules. The effect of *lc* on the number of locules is relatively small, increasing the number of locules from 2 to 3 or 4. In contrast, *fas* can increase the number of locules from 2 to 6. The epistatic interaction between these two loci can increase the number of locules to more than 8, doubling the fruit size in tomato. *Lc* is located on chromosome 2 of tomato. Two SNPs highly correlated with the number of locules were identified downstream of the *WUSCHEL* (*WUS*) gene using a genetic mapping method, suggesting that the phenotype of *lc* may be related to *WUS* (Muños et al., 2011). The *fas* mutation arises from the inversion of a 294 kb chromosomal segment on chromosome 11. People have attributed the multilocular phenotype of *fas* to the inactivation of a YABBY transcription factor gene near the inversion site for a long time (Cong et al., 2008). Later a study suggested that the phenotype of *fas* is more likely due to the reduced expression of another gene *CLAVATA3* (*CLV3*) near the inversion site (Xu et al., 2015). Two other loci controlling the number of locules, Fasciated and branched (FAB) and Fasciated inflorescence (FIN), were also identified in tomato. FAB encodes the receptor kinase CLV1 of CLV3, while FIN encodes an arabinosyltransferase that can modify CLV3 post-translationally (Xu et al., 2015).

In bustling cities, there is a great demand for fresh fruits and vegetables, which has given rise to the emerging industry of urban agriculture. However, the space in cities is limited, and in order to harvest higher yields in small planting spaces, most crops grown in urban agricultural systems are short in stature and grow quickly. To enrich people's dining tables, developing more new types of "urban crops" has become an important goal for scientists. A study showed that three genes, precocious growth termination (*SP*), condensed shoots, rapid flowering (*SP5G*), and regulator of tomato stem length (*SlER*), were mutated in tomato using the CRISPR genome editing technology. The *SP* gene regulates plant growth and maturation, the *SP5G* gene controls plant flowering time, and the *SlER* gene regulates the length of the plant's stem. By designing a strategy of trait stacking, the three modified mutant genes were combined to ultimately cultivate a new variety of tomato plants with a compact shape, like a bouquet of roses, and small tomatoes clustered together like grapes. Compared to other tomato varieties, the new tomato variety can stop growing earlier and bloom and bear fruit earlier, reaching maturity in just 40 days. The new tomato fruits are small, taste good, are environmentally friendly to grow, do not occupy a large area of land, and are very suitable for urban agricultural systems and even space station planting (Kwon et al., 2020).

Over the last few years, the rapid development of genome editing technology has brought revolutionary changes to crop breeding. Using genome editing technology to domesticate wild plants with natural stress resistance is a brand-new strategy to obtain excellent tomato germplasm. Using genome editing technology, multiple genes that regulate yield and quality, among other traits, have been precisely targeted within coding and regulatory regions without sacrificing their natural resistance to salt and scab disease in wild tomato. The researchers used a multi-target CRISPR/Cas9 vector system to precisely target the coding region of flowering photoperiod sensitivity, plant type and fruit synchronous maturation control genes *SP* and *SP5G*, as well as the cis-regulatory elements of *CLV3* and *WUS* genes controlling fruit size, and upstream open reading frame of the vitamin C synthase gene *GGP1* involved in vitamin C biosynthesis, resulting in the generation of 140 independent gene-edited plants (Li et al., 2018a). The genotype and phenotype identification of the offspring population demonstrated that the photoperiod sensitivity of wild tomato flowering was eliminated through genome editing, breaking the geographical range limitation of cultivation and achieving the first step of wild plant domestication. Meanwhile, the "indeterminate" growth type of currant tomato, which has late flowering and sparse fruiting, was transformed into a "double determinate" compact growth type, increasing the rate of fruit set, synchronization of fruit maturity, and harvest index. The modification of the cis-regulatory elements of the small peptide gene *CLV3* and its downstream signaling pathway gene *WUS* and

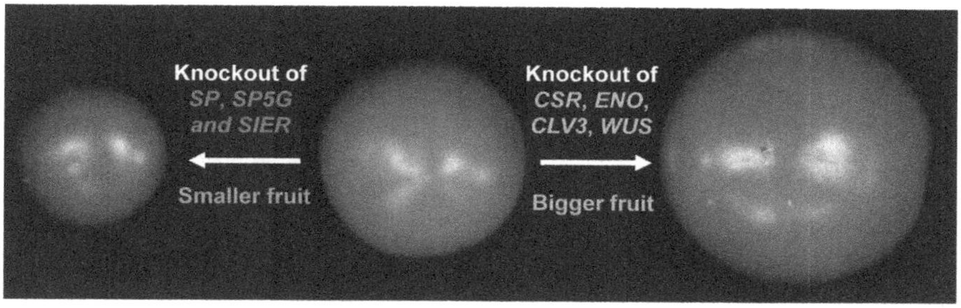

FIGURE 13.1 Key genes regulating fruit size.

GGP1 increased the fruit size and vitamin C content of wild tomato. Experimental treatments involving salt stress and fungus infection confirmed that precisely introducing these crucial agronomic traits did not compromise the innate defense in wild tomato plants (Li et al., 2018a). This study demonstrates for the first time that genome editing technology can achieve rapid domestication of wild tomato plants, providing novel strategies for the precise creation and design of novel tomato germplasm (Figure 13.1).

13.3 BREEDING HIGH-QUALITY TOMATO USING GENOME EDITING SYSTEM

In traditional breeding processes, excessive focus on fruit size and yield has resulted in the loss of certain genetic loci controlling flavor and nutritional quality. A recent study identified glucose, fructose, citric acid, malic acid, and 29 volatile compounds as the primary substances influencing the flavor of tomato. Through the analysis of flavor compound contents in over 400 tomato samples from various regions worldwide, combined with genome sequencing and bioinformatics analysis, more than 250 genetic loci controlling tomato flavor were uncovered, thus providing the first-ever elucidation of the genetic basis of tomato flavor. Among these loci, two control sugar content, and five control acidity, and it was also discovered that some volatile compounds can enhance the fruit's sweetness, while others can impart a floral aroma to the fruit (Tieman et al., 2017). These findings offer a practical roadmap for utilizing gene editing technology to rapidly breed delicious tomato. From hundreds of tomato genotypes, Zhu et al. (2018) assembled and investigated a comprehensive dataset comprising metabolomes, transcriptomes, and genomes. By mGWAS and eQTL analysis, 535 genes, 970 SNPs (single nucleotide polymorphisms), and 371 metabolites associated with tomato flavor were identified. They also discovered genetic loci influencing the content of various nutrients, including vitamins and amino acids. This study provides valuable knowledge regarding the genetic regulation of flavor and nutritional content, offering a foundation for whole-genome design breeding strategies in tomato.

Lycopene is recognized as a bioactive compound in the treatment of chronic illnesses and lowering the risk of cardiovascular diseases and cancer (Clinton et al., 2009). Enhancing the lycopene levels in tomato fruit is a crucial goal for improving tomato fruit quality. Li et al. (2018b) utilized the CRISPR/Cas9 multiplex genome editing approach to elevate lycopene levels within the fruit by stimulating lycopene synthesis while concurrently preventing its conversion into α and β-carotenes. They simultaneously knocked out five genes involved in carotenoid metabolic, resulting in a remarkable 5.1-fold amplification of lycopene level. Zsögön et al. (2018) developed a genome engineering using CRISPR-Cas9 to integrate favorable agricultural traits with beneficial traits found in wild *S. pimpinellifolium*, facilitating the successful de novo domestication of this wild species. Through the modification of lycopene β-cyclase, responsible for the conversion of a significant portion of lycopene into β-carotene, a 500% increase in fruit lycopene levels has been achieved compared with commercial cherry tomato. Significantly, the enhancement in lycopene content did not have an adverse impact on the levels of β-carotene or lutein. Hunziker et al. (2020)

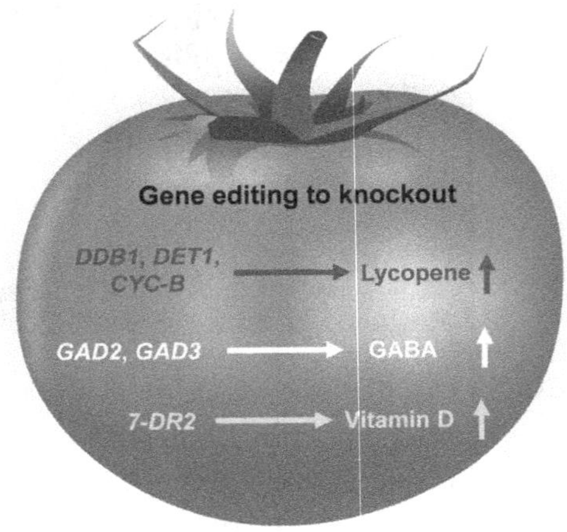

FIGURE 13.2 Using gene editing technology to generate high-quality tomato.

utilized target-AID base-editing technology, which led to the replacement of cytidine with thymine, to introduce substitution mutations in *SlDDB1*, *SlDET1*, and *SlCYC-B* genes that regulate the accumulation of carotenoid, inducing an improvement in lycopene accumulation.

γ-Aminobutyric acid (GABA), an amino acid, does not participate in protein synthesis but possesses health-promoting functions. Elevating GABA content can further augment the ability of tomato fruit to reduce blood pressure. GAD serves as a crucial enzyme in GABA biosynthesis and includes a C-terminal autoinhibitory domain, the removal of which can increase GAD activity. Among the five GAD genes in tomato, SlGAD2 and SlGAD3 are specifically active during the process of fruit development. The researchers used CRISPR-Cas9 technology to remove the autoinhibitory domains of SlGAD2 and SlGAD3, resulting in the insertion of a stop codon directly preceding the autoinhibitory domains. Remarkably, this modification increased the GABA content by 7–15 times without compromising plant growth, fruit size, or yield (Nonaka et al., 2017). And the CRISPR-edited tomato with high GABA content is already available on the market.

Vitamin D is an essential nutrient for the human body, playing a crucial role in maintaining the calcium-phosphorus balance and supporting the health of bones, teeth, and muscles. The amount of vitamin D in most foods is insufficient, and plants are considered to be highly deficient sources of this nutrient. 7-DHC which is a precursor in the biosynthesis of vitamin D3 is found in the leaves and unripe green fruits, and does not normally accumulate in mature fruit. In an attempt to facilitate the accumulation of 7-DHC in mature tomato fruits, *Sl7-DR2* was knocked out by CRISPR-Cas9 gene editing technology. The inactivation of Sl7-DR2 did not have any impact on tomato growth, development, or yield. On the contrary, the loss of Sl7-DR2 activity resulted in a substantial elevation of 7-DHC levels in tomato. The conversion of 7-DHC to vitamin D3 in these tomatoes resulted in a content equivalent to the vitamin D3 found in 28 grams of tuna or 2 medium-sized eggs. Gene *7-DR2* knockout can also be applied to other *Solanaceae* plants such as pepper, eggplant, and potato (Li et al., 2022) (Figure 13.2).

13.4 GENOME EDITING SYSTEM FOR GENERATING TOMATO WITH HIGH DISEASE RESISTANCE

Tomato faces numerous threats from viruses, bacteria, and fungi, which can lead to reduced yield and quality, potentially resulting in significant economic losses (Borrelli et al., 2018). To establish resistance in varieties, the utilization of resistance genes (R genes) is a common approach. However, the durability of resistance conferred by a single R gene is often limited, as pathogens can easily lose or mutate their corresponding R gene effectors (Dangl et al., 2013). By employing a gene stacking approach, introducing multiple R genes can effectively enhance a plant's resistance durability (Fuchs, 2017). Simultaneously, this can be achieved through the inactivation of susceptibility genes (S genes), which promote the infectivity and proliferation capabilities of pathogens (Zaidi et al., 2018). Compared to stacking multiple R genes, disrupting the S gene presents a simpler strategy for achieving broad-spectrum and long-lasting resistance in plants. S genes are typically conserved within plant species (Dong & Ronald, 2019). Once the S gene has been characterized in a plant species, it can serve as a reference for identifying its homologous genes in other species.

DMR6 (downy mildew resistant 6) is previously reported in Arabidopsis as a factor related to susceptibility to disease. In tomato, there are two homologous genes: *SlDMR6-1* and *SlDMR6-2*. Pathogen infection upregulates the expression of *SlDMR6-1* but not *SlDMR6-2*. Consistently, *Sldmr6-1* mutants exhibit stronger defense against various types of pathogens such as fungi, bacteria, and oomycetes. The defense is associated with elevated defense hormone salicylic acid (SA) content and upregulation of immune response-related gene expression in plants. Experiments in field conditions also indicate that the resistance acquired by *Sldmr6-1* mutants is repeatable (Thomazella et al., 2021).

Necrotrophic fungi, such as *Alternaria solani* and *Botrytis cinerea*, pose a significant danger to the cultivation of tomato. However, it has not been documented that any single R gene confers a strong defense against necrotrophic fungi. Recently, Ramirez Gaona et al. (2023) conducted EMS mutagenesis on tomato Micro-Tom (MT). They found that compared to the wild-type MT, the lesion diameter of the mutant strain M2042 decreased by 20–30% with *B. cinerea* infection by a detached leaf assay (DLA). Researchers constructed an F2 population by crossing M2042 with Moneymaker and used a BSA-seq analysis to identify the candidate gene PUB17. Subsequent fine mapping revealed that SNP within PUB17 exhibited co-segregation with the immune trait. Using gene editing technology, *pub17* mutants were generated and exhibited significantly increased immunity to *A. solani* and *B. cinerea*. Researchers also introduced EMS mutant alleles into various tomato breeding lines and successfully obtained F1 hybrids with homozygous *pub17* alleles, which showed no difference in agronomic traits and fruiting compared to F1 generations without pub17 mutations.

Many candidate immune-related genes have been identified in tomato, but before the emergence of CRISPR/Cas technology, validating the functional importance of these genes was technically challenging (Liao et al., 2015; Ahammed et al., 2018; Zhang et al., 2018; Ding et al., 2021). While CRISPR/Cas technology has been applied in editing genes that are related to defense responses in plants (Nekrasov et al., 2017; Xu et al., 2019; Ortigosa et al., 2019; Wang et al., 2022; Ding et al., 2023), all documented researches have focused on one or a few genes, providing limited insights into the specificity, mutation types, and heritability of CRISPR/Cas9 genome editing in tomato. Zheng et al. (2019) established the Plant Genome Editing Database (http://plantcrispr.org/cgi-bin/crispr/index.cgi), offering valuable information on tomato lines with mutations in immune-related genes induced by CRISPR/Cas9 technology. Zhang et al. (2020) assessed the efficiency of gRNA through transient expression, and then tomato transformation was performed using *Agrobacterium* pools which contained 2–4 Cas9/gRNA vectors to mutate 63 candidate genes associated with resistance. The generation of numerous tomato mutants, produced through CRISPR/Cas9 targeting of resistance-related genes, highlights the effectiveness and robustness of this mutagenesis method for gene editing in tomato. It is noteworthy that the gene modifications introduced were inherited stably across successive generations. Additionally, off-target mutations can pose a potential issue in functional genomics research. The researchers identified 18 potential off-target sites for 12 gRNAs through PCR and Sanger sequencing in 12 T0 plants, 68 T1 plants, and 44 T2 plants and examined off-target mutations. No off-target modifications were detected in both Cas9 and non-Cas9 control plants, indicating that gRNA and CRISPR-mediated mutations exhibit high specificity. Therefore, the CRISPR/Cas system can serve as a crucial tool for enhancing tomato resistance to disease (Figure 13.3).

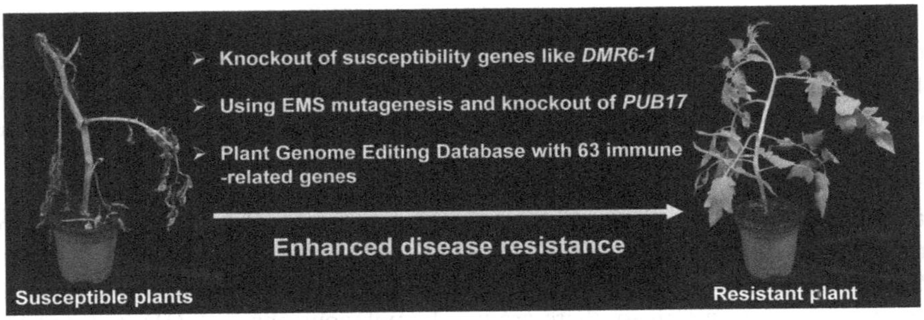

FIGURE 13.3 Approaches to generate tomato with high disease resistance.

13.5 CONCLUSIONS AND PERSPECTIVES

The CRISPR/Cas9 system has proven successful in plant breeding, including not only model plants such as tomato and Arabidopsis but also many crops such as rice, wheat, corn, soybeans, tobacco, sorghum, potatoes, poplars, apples, and bananas (Chen et al., 2019). It plays a crucial role in increasing crop yield, enriching nutritional components, and enhancing disease resistance and stress resilience. Compared with traditional hybrid breeding and genetic engineering breeding technologies, gene editing breeding technology has the following advantages.

One major advantage of CRISPR/Cas9 genome editing technology is that it can simultaneously edit multiple genes, achieving rapid improvement of multiple traits in crops, significantly reducing breeding time, and lowering costs. Researchers from Brazil, Germany, and the United States collaborated to use the CRISPR/Cas9 system to perform a one-time reverse genetic engineering operation on wild tomato, modifying six genes related to yield and quality in modern tomato varieties, achieving rapid domestication of wild tomato (Zsögön et al., 2018). This gene editing allowed the fruit of the wild tomato, which was originally the size of peas, to increase 3-fold, and the number of fruits per plant to increase 10-fold. The lycopene content in the fruit increased by 1-fold compared to wild tomato, and increased by 5-fold compared to the widely cultivated tomato varieties. Because traditional breeding methods cannot simultaneously consider traits such as yield, nutrition content, and disease resistance during domestication and variety breeding, most modern tomato varieties have gradually lost important nutrients such as lycopene. This study undoubtedly provides a new way of crop breeding.

Another advantage of CRISPR/Cas9 genome editing technology is that it can avoid the safety risks of genetic engineering breeding technology. In genetically modified organisms (GMOs) crop breeding, the selection of marker genes and the integration of some non-target DNA increase the safety risks and regulatory costs of GMOs. Since the CRISPR/Cas9 system has been applied to crop breeding, scientists have been working to develop genome editing systems without foreign DNA insertion to completely avoid the safety risks of GMOs (El-Mounadi et al., 2020). Furthermore, due to the continuous improvement of genome editing tools, scientists have developed a variety of new genome editing systems such as single-base substitution editing, guide editing, and circular editing (Li et al., 2020), which will break through the technical bottlenecks of traditional breeding and achieve breeding goals that traditional breeding cannot achieve. In summary, CRISPR/Cas9 has emerged as the most prominent genome editing technology in tomato breeding due to its higher editing efficiency, simpler operation, lower cost, and ability to edit multiple targets simultaneously. As described above, this technology exhibits significant potential for improving crucial agronomic traits in tomato.

From a technical perspective, traditional breeding is the foundation of molecular breeding. The establishment of the molecular breeding technology system largely depends on comprehending the molecular mechanisms underlying the formation of agronomic traits in crops. The localization, cloning, and functional analysis of key genes or QTL controlling important agronomic traits can provide target genes for genome editing breeding. In the future, using the genetic effects of QTL, the interaction between QTL, and the interaction between QTL and the environment, information simulation and prediction of various possible gene type combinations can be performed. These traditional breeding technologies can combine with genome editing breeding technology, ultimately achieving the target-oriented design breeding in tomato.

Genome editing, particularly CRISPR-Cas9, has great potential in improving crop yield, nutritional value, and stress resilience. To ensure the effective and sustainable utilization of this technology for crop improvement, the scientific community must address the biosafety and societal concerns about gene editing, including target gene site selection, guide RNA design, off-target effects, and the delivery method. Furthermore, it is crucial to educate the general public about this technology's properties and to reevaluate the regulations on genome-edited plants in various countries. Since the transgene carrying the CRISPR-Cas9 cassette has the potential to be removed by gene segregation, the genome-modified plants may be transgene-free which exhibit similarities to naturally occurring varieties that possess genetic variations. Therefore, commercialization of genome-edited plants or products may potentially circumvent the stringent biosafety regulations required for transgenic plants.

REFERENCES

Ahammed, G. J., Li, X., Zhang, G., Zhang, H., Shi, J., Pan, C., Yu, J., & Shi, K. (2018). Tomato photorespiratory glycolate-oxidase-derived H_2O_2 production contributes to basal defence against *Pseudomonas syringae*. *Plant, Cell & Environment*, *415*, 1126–1138.

Borrelli, V. M. G., Brambilla, V., Rogowsky, P., Marocco, A., & Lanubile, A. (2018). The enhancement of plant disease resistance using CRISPR/Cas9 technology. *Frontiers in Plant Science*, *9*, 1245.

Brooks, C., Nekrasov, V., Lippman, Z. B., & Van Eck, J. (2014). Efficient gene editing in tomato in the first generation using the clustered regularly interspaced short palindromic repeats/CRISPR-associated9 system. *Plant Physiology*, *166*, 1292–1297.

Chen, K., Wang, Y., Zhang, R., Zhang, H., & Gao, C. (2019). CRISPR/Cas genome editing and precision plant breeding in agriculture. *Annual Review of Plant Biology*, *70*, 667–697.

Chen, K., & Gao, C. (2013). TALENS: Customizable molecular DNA scissors for genome engineering of plants. *Journal of Genetics and Genomics*, *40*, 271–279.

Clinton, S. K. (2009). Lycopene: Chemistry, biology, and implications for human health and disease. *Nutrition Reviews*, *56*, 35–51.

Cong, B., Barrero, L. S., & Tanksley, S. D. (2008). Regulatory change in YABBY-like transcription factor led to evolution of extreme fruit size during tomato domestication. *Nature Genetics*, *40*, 800–804.

Dangl, J. L., Horvath, D. M., & Staskawicz, B. J. (2013). Pivoting the plant immune system from dissection to deployment. *Science*, *341*, 746–751.

Ding, S., Lv, J., Hu, Z., Wang, J., Wang, P., Yu, J., Foyer, C. H., & Shi, K. (2023). Phytosulfokine peptide optimizes plant growth and defense via glutamine synthetase GS2 phosphorylation in tomato. *EMBO Journal, 42*, e111858.

Ding, S., Shao, X., Li, J., Ahammed, G. J., Yao, Y., Ding, J., Hu, Z., Yu, J., & Shi, K. (2021). Nitrogen forms and metabolism affect plant defence to foliar and root pathogens in tomato. *Plant, Cell & Environment*, *44*, 1596–1610.

Dong, O. X., & Ronald, P. C. (2019). Genetic engineering for disease resistance in plants: Recent progress and future perspectives. *Plant Physiology*, *180*, 26–38.

El-Mounadi, K., Morales-Floriano, M. L., & Garcia-Ruiz, H. (2020). Principles, applications, and biosafety of plant genome editing using CRISPR-Cas9. *Frontiers in Plant Science*, *11*, 56.

Fuchs, M. (2017). Pyramiding resistance-conferring gene sequences in crops. *Current Opinion in Virology*, *26*, 36–42.

Giovannoni, J. (2018). Tomato multiomics reveals consequences of crop domestication and improvement. *Cell*, *172*, 6–8.

Hunziker, J., Nishida, K., Kondo, A., Kishimoto, S., Ariizumi, T., & Ezura, H. (2020). Multiple gene substitution by target-AID base-editing technology in tomato. *Scientific Reports*, *10*, 20471.

Kwon, C., Heo, J., Lemmon, Z. H., Capua, Y., Hutton, S. F., Van Eck, J., Park, S. J., & Lippman, Z. B. (2020). Rapid customization of Solanaceae fruit crops for urban agriculture. *Nature Biotechnology*, *38*, 182–188.

Li, C., Zhang, R., Meng, X., Chen, S., Zong, Y., Lu, C., Qiu, J., Chen, Y., Li, J., & Gao, C. (2020). Targeted, random mutagenesis of plant genes with dual cytosine and adenine base editors. *Nature Biotechnology*, *38*, 875–882.

Li, J., Scarano, A., Gonzalez, N. M., D'Orso, F., Yue, Y., Nemeth, K., Saalbach, G., Hill, L., de Oliveira Martins, C., Moran, R., Santino, A., & Martin, C. (2022). Biofortified tomatoes provide a new route to vitamin D sufficiency. *Nature Plants*, *8*, 611–616.

Li, T., Yang, X., Yu, Y., Si, X., Zhai, X., Zhang, H., Dong, W., Gao, C., & Xu, C. (2018a). Domestication of wild tomato is accelerated by genome editing. *Nature Biotechnology*, *36*, 1160–1163.

Li, X., Wang, Y., Chen, S., Tian, H., Fu, D., Zhu, B., Luo, Y., & Zhu, H. (2018b). Lycopene is enriched in tomato fruit by CRISPR/Cas9-mediated multiplex genome editing. *Frontiers in Plant Science*, *9*, 559.

Liao, Y., Tian, M., Zhang, H., Li, X., Wang, Y., Xia, X., Zhou, J., Zhou, Y., Yu, J., Shi, K., & Klessig, D. F. (2015). Salicylic acid binding of mitochondrial alpha-ketoglutarate dehydrogenase E2 affects mitochondrial oxidative phosphorylation and electron transport chain components and plays a role in basal defense against tobacco mosaic virus in tomato. *New Phytologist*, *2053*, 1296–1307.

Lin, T., Zhu, G., Zhang, J., Xu, X., Yu, Q., Zheng, Z., Zhang, Z., Lun, Y., Li, S., & Wang, X., et al. (2014). Genomic analyses provide insights into the history of tomato breeding. *Nature Genetics*, *46*, 1220–1226.

Marefatzadeh-Khameneh, M., Fabriki-Ourang, S., Sorkhilalehloo, B., Abbasi-Kohpalekani, J., & Ahmadi, J. (2021). Genetic diversity in tomato (*Solanum lycopersicum* L.) germplasm using fruit variation implemented by tomato analyzer software based on high throughput phenotyping. *Genetic Resources and Crop Evolution*, *68*, 2611–2625.

Mata-Nicolás, E., Montero-Pau, J., Gimeno-Paez, E., Garcia-Carpintero, V., Ziarsolo, P., Menda, N., Mueller, L. A., Blanca, J., Cañizares, J., & van der Knaap, E., et al. (2020). Exploiting the diversity of tomato: The development of a phenotypically and genetically detailed germplasm collection. *Horticulture Research*, *7*, 66.

Mauxion, J., Chevalier, C., & Gonzalez, N. (2021). Complex cellular and molecular events determining fruit size.*Trends in Plant Science*, *26*, 1023–1038.

Mu, Q., Huang, Z., Chakrabarti, M., Illa-Berenguer, E., Liu, X., Wang, Y., Ramos, A., & van der Knaap, E. (2017). Fruit weight is controlled by *cell size regulator* encoding a novel protein that is expressed in maturing tomato fruits. *PLoS Genetics*, *13*, e1006930.

Muños, S., Ranc, N., Botton, E., Bérard, A., Rolland, S., Duffé, P., Carretero, Y., Le Paslier, M., Delalande, C., & Bouzayen, M., et al.. (2011). Increase in tomato locule number is controlled by two single-nucleotide polymorphisms located near WUSCHEL. *Plant Physiology*, *156*, 2244–2254.

Nekrasov, V., Wang, C., Win, J., Lanz, C., Weigel, D., & Kamoun, S. (2017). Rapid generation of a transgene-free powdery mildew resistant tomato by genome deletion. *Scientific Reports*, *7*, 482.

Nonaka, S., Arai, C., Takayama, M., Matsukura, C., & Ezura, H. (2017). Efficient increase of ɣ-aminobutyric acid (GABA) content in tomato fruits by targeted mutagenesis. *Scientific Reports*, *7*, 7057.

Ortigosa, A., Gimenez-Ibanez, S., Leonhardt, N., & Solano, R. (2019). Design of abacterial speck resistant tomato by CRISPR/Cas9-mediated editing of SlJAZ2. *Plant Biotechnology Journal*, *17*, 665–673.

Ramirez Gaona, M., van Tuinen, A., Schipper, D., Kano, A., Wolters, P. J., Visser, R. G. F., van Kan, J. A. L., Wolters, A. A., & Bai, Y. (2023). Mutation of PUB17 in tomato leads to reduced susceptibility to necrotrophic fungi. *Plant Biotechnology Journal*, *21*, 2157–2159.

Salava, H., Thula, S., Mohan, V., Kumar, R., & Maghuly, F. (2021). Application of genome editing in tomato breeding: Mechanisms, advances, and prospects. *International Journal of Molecular Sciences*, *22*, 682.

Sato, S., Tabata, S., Hirakawa, H. et al. (2012). The tomato genome sequence provides insights into fleshy fruit evolution. *Nature*, *485*, 635–641.

Shan, Q., Wang, Y., Li, J. et al. (2013). Targeted genome modification of crop plants using a CRISPR-Cas system. *Nature Biotechnology*, *31*, 686–688.

Thomazella, D. P. T., Seong, K., Mackelprang, R., Dahlbeck, D., Geng, Y., Gill, U. S., Qi, T., Pham, J., Giuseppe, P., Lee, C. Y., Ortega, A., Cho, M. J., Hutton, S. F., & Staskawicz, B. (2021). Loss of function of a DMR6 ortholog in tomato confers broad-spectrum disease resistance. *Proceedings of the National Academy of Sciences of the United States*, *118*, e2026152118.

Tieman, D., Zhu, G., Resende, M. F., Jr, Lin, T., Nguyen, C., Bies, D., Rambla, J. L., Beltran, K. S., Taylor, M., Zhang, B., Ikeda, H., Liu, Z., Fisher, J., Zemach, I., Monforte, A., Zamir, D., Granell, A., Kirst, M., Huang, S., & Klee, H. (2017). A chemical genetic roadmap to improved tomato flavor. *Science*, *355*, 391–394.

Wang, H., Li, M., Lee, C. M., Chakraborty, S., Kim, H., Bao, G., & Leong, K. W. (2017). CRISPR/Cas9-based genome editing for disease modeling and therapy: Challenges and opportunities for nonviral delivery. *Chemical Reviews*, *117*, 9874–9906.

Wang, J., Wang, A., Luo, Q., Hu, Z., Ma, Q., Li, Y., Lin, T., Liang, X., Yu, J., Foyer, C. H., & Shi, K. (2022). Glucose sensing by regulator of G protein signaling 1 RGS1 plays a crucial role in coordinating defense in response to environmental variation in tomato. *New Phytologist*, *236*, 561–575.

Westram, A. M., Stankowski, S., Surendranadh, P., & Barton, N. H. (2022). Reproductive isolation, speciation, and the value of disagreement: A reply to the commentaries on 'What is reproductive isolation?'. *Journal of Evolutionary Biology*, *35*, 1200–1205.

Xu, C., Liberatore, K. L., MacAlister, C. A., Huang, Z., Chu, Y., Jiang, K., Brooks, C., Ogawa-Ohnishi, M., Xiong, G., & Pauly, M., et al. (2015). A cascade of arabinosyltransferases controls shoot meristem size in tomato. *Nature Genetics*, *47*, 784–792.

Xu, J., Hua, K., & Lang, Z. (2019). Genome editing for horticultural crop improvement. *Horticulture Research*, *6*, 113.

Yuste-Lisbona, F. J., Fernández-Lozano, A., Pineda, B., Bretones, S., Ortíz-Atienz, A., García-Sogo, B., Müller, N. A., Angosto, T., Capel, J., & Moreno, V., et al. (2020). *ENO* regulates tomato fruit size through the floral meristem development network. *PNAS*, *117*, 8187–8195.

Zaidi, S. S., Mukhtar, M. S., & Mansoor, S. (2018). Genome editing: Targeting susceptibility genes for plant disease resistance. *Trends in Biotechnology*, *36*, 898–906.

Zhang, F., Maeder, M. L., Unger-Wallace, E., Hoshaw, J. P., Reyon, D., Christian, M., Li, X., Pierick, C. J., Dobbs, D., & Peterson, T., et al. (2010). High frequency targeted mutagenesis in *Arabidopsis thaliana* using zinc finger nucleases. *PNAS*, *107*, 12028–12033.

Zhang, H., Hu, Z., Lei, C., Zheng, C., Wang, J., Shao, S., Li, X., Xia, X., Cai, X., Zhou, J., Zhou, Y., Yu, J., Foyer, C. H., & Shi, K. (2018). A plant phytosulfokine peptide initiates auxin-dependent immunity through cytosolic Ca^{2+} signaling in tomato. *The Plant Cell*, *30*, 652–667.

Zhang, N., Roberts, H. M., Van Eck, J., & Martin, G. B. (2020). Generation and molecular characterization of CRISPR/Cas9-induced mutations in 63 immunity-associated genes in tomato reveals specificity and a range of gene modifications. *Frontiers in Plant Science*, *11*, 10.

Zheng, Y., Zhang, N., Martin, G. B., & Fei, Z. (2019). Plant Genome Editing Database PGED: A call for submission of information about genome edited plant mutants. *Molecular Plant*, *12*, 127–129.

Zhu, G., Wang, S., Huang, Z., Zhang, S., Liao, Q., Zhang, C., Lin, T., Qin, M., Peng, M., Yang, C., Cao, X., Han, X., Wang, X., van der Knaap, E., Zhang, Z., Cui, X., Klee, H., Fernie, A. R., Luo, J., & Huang, S. (2018). Rewiring of the fruit metabolome in tomato breeding. *Cell*, *172*, 249–261.

Zsögön, A., Čermák, T., Naves, E. R., Notini, M. M., Edel, K. H., Weinl, S., Freschi, L., Voytas, D. F., Kudla, J., & Peres, L. E. P. (2018). De novo domestication of wild tomato using genome editing. *Nature Biotechnology*, *36*, 1211–1216.

14 Speed Breeding in Fruit Trees

Current Achievements and Future Perspectives

Vadim Lebedev
Branch of the Shemyakin-Ovchinnikov Institute of Bioorganic Chemistry of the Russian Academy of Sciences, Prospekt Nauki 6, Pushchino, Moscow Region, Russia

14.1 INTRODUCTION

The world's population is continuously increasing; by 2050, it is projected to increase by almost 2 billion to reach 9.7 billion people (UN, 2022). Due to the population growth, the demand for food supply will also grow. The population increases mainly due to the urban population growth, and the ratio of food producers to food consumers is constantly decreasing. Over the past 30 years, the number of people employed in agriculture has fallen 1.6 times and now only 2–4% of the population is involved in food production in developed countries (World Bank, 2021). There are concerns that the yield improvement is starting to stabilize: for the main crops, it increases by 0.8–1.2% annually, but this indicator needs to be doubled to meet the increasing demand for food of plant origin (Li et al., 2018). One of the main ways to increase crop productivity is to breed new varieties.

Climate change also poses a serious threat to agriculture and food security. Changes in various parameters such as CO_2 concentration in the atmosphere, temperature, and precipitation affect plant growth. Their effect is different: a higher concentration of CO_2 usually increases photosynthesis, but a lack of precipitation and high temperature reduce yields. For example, every °C increase in temperature reduces global wheat production by 6% (Asseng et al., 2015). Overall, according to the global agricultural economic model, these biophysical shocks of climate change will reduce major crop yields by 17%, increase market prices by 20%, and reduce consumption by 3% by 2050 (Nelson et al., 2014).

Besides, climate change affects crop yields through increased biotic stresses. Rising temperatures directly affect the reproduction, development, survival, and spread of pests, and indirectly affect the relationships between pests, their habitat, and other insect species (Skendžić et al., 2021). Climate change can contribute to plant infection by changing the evolution of pathogens and the emergence of new strains, changing the host–pathogen interaction, as well as shifting the pathogen area (Singh et al., 2023). It is assumed that any potential increase in yields over the next five decades will be offset by climate change causing changes in diseases from known and new pathogens (Chaloner et al., 2021).

As a result, to ensure food security, it is extremely important to improve the characteristics of agriculture crops through breeding (Zhang et al., 2023). The improvement rates of the genetic potential of yields should be higher than the rates in current breeding programs to compensate for an increasing world population and global climate change (Voss-Fels et al., 2019). The traditional breeding techniques consist of creating genetic diversity, usually by crossing, followed by the

DOI: 10.1201/b23372-14

identification of the best individuals from large populations, as a rule, based on phenotype analysis. This is a very long process that can take 10–15 years due to the many steps of hybridization, selection, and evaluation required (Cazzola et al., 2021). New approaches are required to speed up the crop breeding process.

14.2 FRUIT AND NUT TREE BREEDING

Fruit and nut trees are an important component of the global sustainable food production system and play an important role in food and nutrition security. Fruits possess attractive colour, taste, and aroma, representing a valuable source of many biologically active compounds (Sabbadini et al., 2021). Besides, fruits are promoted as functional foods as they are a rich source of biologically active compounds such as antioxidants, polyphenols, and vitamins (Nayak et al., 2021). With the growth of the world's population and the diversity of food habits, it is necessary to constantly increase the production of fruits and nuts to ensure food security in the future (Savadi et al., 2021).

The fruit tree industry is also facing challenges caused by global climate change. Forecasts indicate a decrease in water availability in many important fruit-growing areas (Laurens et al., 2018). An increase in global average temperature will increase the risk of the spread of existing – and the emergence of invasive – pathogens and pests in orchards (Endo et al., 2020). One of the main ways to protect the global fruit industry from these challenges is by breeding varieties, resistant to water or biotic stresses. Besides these concerns, also standard for other crops, fruit trees have problems unique to them. Climate change can significantly reduce fruit production due to reduced accumulation of chilling temperatures needed for budbreak, and decreased abundance of insect pollinators, which are necessary for fruit development (Herath et al., 2023).

The growing demand for horticultural products, the changing climatic conditions, as well as a change in consumer preferences require the breeding of new varieties of fruit trees (Ahmad et al., 2021). Traditional breeding of fruit crops goes slow, and its results are unpredictable due to the duration of juvenile periods and heterozygosity and the inability to extrapolate seedling traits to fruiting adult trees (Savadi et al., 2021). Consequently, new methods of fruit tree breeding are required for the rapid creation of cultivars with increased productivity, resistance to abiotic and biotic stress factors, as well as improved nutritional quality.

The main cause of the slow breeding of fruit trees is the long juvenile phase. The juvenile period is the stage of vegetative growth that continues until the first flowering. During the juvenile phase, the tree does not bloom even in favorable environmental conditions (Hackett, 1985). While some traits, e.g., disease resistance, can be assessed on young plants, reproductive traits, such as size, taste, and shelf life, can only be assessed on fruits and a long juvenile phase delays the assessment. Long generation cycles, in turn, lead to an increase in the physical size of trees, which requires significant areas for planting hybrid populations in the field plus associated large labor and financial means (Flachowsky et al., 2011). As a result, the selection of just one elite hybrid requires significant efforts to maintain thousands of plantlets.

The juvenile period of fruit trees takes many years. In apple trees, it lasts approximately 5–12 years depending on the cultivar (Yamagishi et al., 2014), in pears can reach 9–14 years (Freiman et al., 2012), stone fruits (plum, cherry, peach) take 3–7 years (Srinivasan et al., 2012; McGarry et al., 2017), in walnut, 8–10 years (Jin et al., 2022), in citrus ranges from 6 to 20 years depending on the species and genotype (Endo et al., 2005). After completion of the juvenile period, flowering is dependent upon sufficient winter chilling and spring warming. Moreover, flowering and fruiting are highly dependent on various environmental factors such as severe winter frosts, late spring frosts, and rainy and windy weather during the pollination period (Petri et al., 2018). Finally, some fruit trees have a biennial bearing that hampers the fruit trait evaluation. As a result, the breeding of new pome fruit cultivars usually takes 15–25 years if well-known varieties with commercial fruit quality are used for crossing (Lespinasse, 2009). A similar breeding duration was reported for citrus – an average of 23 years is required to breed one promising hybrid (Shimizu, 2022) and for nuts, the development and release of a pecan cultivar took about 27 years (Thompson & Grauke, 2003). Significantly more time is required when a new trait (e.g., resistance to diseases) is to be introduced into elite lines or commercial cultivars from a wild species. Such introgression breeding programs require a number of backcrossing or pseudo-backcrossing with high-quality cultivars (Kumar et al., 2020). At least five generation cycles are required to eliminate the majority of undesired wild apple alleles carried along with the trait of interest (Joshi et al., 2009). In the end, the creation of a novel apple cultivar, combining a trait from a wild apple and the fruit quality of the best cultivars, takes over 50 years (Schouten et al., 2006). More than 50 years are needed to produce a new apple cultivar that combines a trait of a wild apple tree and the fruit quality at the level of the best cultivars (Schouten et al., 2006). For example, it took 56 years and five hybrid generations to produce a cultivar with the gene of resistance to apple scab from *Malus* × *floribunda* (Le Roux et al., 2012).

Flowering and fruiting are energy-consuming processes, and their too-early onset can weaken the tree. In this regard, the juvenile phase is a natural defense mechanism: the tree does not flower until it reaches a competitive size in the environment (Hackett, 1985). In total, the transition from juvenile to reproductive state is a very important stage in the life of a perennial woody plant (Khan et al., 2014). As the time of the onset of flowering is crucial for reproductive success, plants have developed mechanisms that incorporate endogenous and exogenous cues to precisely regulate periods of vegetative and reproductive growth (Varkonyi-Gasic et al., 2013). Unlike annual plants, cycles of coordinated vegetative and reproductive growth in woody perennial species are interrupted by dormancy periods in temperate regions, which allow survival under unfavorable winter conditions. Fruit trees are generally challenging subjects for studying the molecular genetic mechanisms of flowering due to heterozygous genomes, long juvenile periods, poorly understood gene functions, and a lack of genomic databases (Song et al., 2013). Nevertheless, by affecting these mechanisms, the juvenility period can be shortened. Over the past decades, a number of methods have been developed that can accelerate plant breeding.

14.3 APPROACHES FOR BREEDING ACCELERATION

The history of plant breeding can be divided into four major stages, which successively replace one another, integrating existing techniques with new technologies to increase breeding efficiency (Wallace et al., 2018). Breeding 1.0 lasted about 10,000 years and Breeding 2.0 about 100 years (since the end of the 19th century). These stages relate to conventional plant breeding based only on phenotypic selection, but Breeding 2.0 is distinguished by the scientific basis, which established the discoveries of Darwin and Mendel and using theory and methods of quantitative genetics. Since the 1990s, Breeding 3.0 has come, when breeders started to switch to the use of molecular markers and genomic data, which provided efficient alternatives to traditional breeding methods. Now a new level, Breeding 4.0, begins to combine technological advances in genetics and information systems such as genome resequencing, genome editing, high-throughput phenotyping, and machine learning approaches (Wallace et al., 2018). These latest biotechnological innovations in plant breeding are called new plant breeding techniques (NPBTs) (Lusser et al., 2011). These techniques include (1) methods of conventional genetic transformation including cisgenesis, intragenesis, grafting on transgenic rootstocks, and agro-infiltration; (2) genome editing via oligonucleotide-directed mutagenesis and zinc finger nuclease (ZFN) technology; (3) epigenetic modification of gene expression using RNA-dependent DNA methylation; (4) reverse breeding; (5) synthetic genomics. Unlike traditional breeding methods, NPBTs are able to make targeted modifications in the genome and do not alter the genetic background, which is particularly important for elite fruit cultivars, which are highly recognizable and valued by consumers (Salonia et al., 2020).

Following “the breeder’s equation”, first formulated back in 1937, genetic gains (response to selection) can be improved by increasing selection intensity, selection accuracy, and genetic variation, or by decreasing the length of the breeding cycle (Moose & Mumm, 2008). Considering the complexity of the genetic parameters in this equation, it is easier and faster to manipulate the cycle of selection to increase genetic gain (Cobb et al., 2019). Further on, we give examples of various methods of reducing the duration of the breeding cycle of fruit trees, both traditional and biotechnological.

14.4 TRADITIONAL METHODS

14.4.1 NATURAL EARLY FLOWERING GENOTYPES

Natural forms had been a source of genetic diversity for breeders for many hundreds of years before hybridization started to be used in breeding. Natural early flowering genotypes are good candidates for use in breeding since their survival under natural selection demonstrated their viability. However, as in the case of natural mutant woody forms aberrant in lignin biosynthesis (Lebedev & Shestibratov, 2021), very few examples of such trees are known. The early flowering genotypes are almost unknown for most fruit trees. We are aware of early flowering reports for seedlings of several citrus species – trifoliate orange (*Poncirus trifoliata*), grapefruit (*Citrus paradisi*), pummelo (*C. grandis*), yamamikan (*C. intermedia*), and ponkan-tangerine (*C. reticulata*) (Flachowsky et al., 2011). However, this phenotype was unstable, which made it difficult to use it to shorten the juvenile period (Endo et al., 2020). Dewi et al. (2013) studied precocious flowering in various crosses between Citrus species and in progenies of 50 Citrus accessions (pummelo and relatives, mandarin and relatives) pollinated with grapefruit pollen, 1–16% of seedlings flowered within 3 years after seed germination (Dewi et al., 2013).

It has been reported that seedlings of temperate fruit trees (apple, pear, apricot and cherry) can vary in the duration of the juvenile stage, but flowering at the age of several months was not (Flachowsky et al., 2009). At the same time, early flowering walnut (*Juglans regia* L.) seedlings ("Early Mature" phenotype) from trees originating from Central Asia and Iran are known. These plants flowered sometimes in the year of sowing and had dwarf growth habits, but some genotypes due to early bud break were susceptible to late spring frosts and blight disease (Germain et al., 1997). Progeny from these trees was propagated *in vitro* and microshoots flowered after three to six subcultures (Breton et al., 2004). At the end of the first growing season in the greenhouse, the mean height of flowering was two times lower compared to the height of nonflowering trees. Precocious and dwarf walnut genotypes from Iran showed genetic stability both under field and *in vitro* conditions (Vahdati et al., 2014). Early flowering walnut accessions and typical plants from Iran were compared using SSR markers (Ebrahimi et al., 2017). The two types of studied genotypes had different alleles and "Early Mature" trees were more distinct and diverse from "Normal Growth" plants.

Early flowering varieties of walnut are also known in China. The typical juvenile period for walnut is 8–10 years, but the precocious walnut cultivars found in Xinjiang province flowered within 3 years and sometimes within a single year (Jin et al., 2023). In the offspring obtained from crossing early flowering cultivar "Xinwen 81" and later flowering local walnut, seedings were selected that flowered one or two years after planting (Jin et al., 2022). In these seedlings, the flowering-related genes *FLOWERING LOCUS T* (FT), *SUPPRESSOR OF OVEREXPRESSION OF CONSTANS1* (SOC1), and *LEAFY* (LFY) are highly expressed compared to others and a set of miRNAs associated with the floral transition was additionally identified. Identification of the genetic mechanisms of early flowering may contribute to the acceleration of walnut breeding (Jin et al., 2022). Studies demonstrated that the *JrFLC* gene regulates the transition to the reproductive phase in early flowering walnuts from Xinjiang province and may be used for shortening of juvenile phase in walnut (Jin et al., 2023).

In fruit trees, somatic mutations ("bud sports") are another source of genetic diversity (Ban et al., 2022). A favorable mutation in a commercial variety is a great advantage as compared to hybridization in traditional breeding (Ruiz et al., 2019). Dozens of commercially available cultivars of peaches, nectarines, and apples originating from bud sport mutations are known

(Ban & Jung, 2023), but no early flowering mutations for these species were reported. At the same time, two reports are known about spontaneous mutants with a short juvenile phase in citrus. A spontaneous early flowering mutant of trifoliate orange (*P. trifoliata*) was found in China in 1976 (Zhang et al., 2009). Seedling of these mutants bloomed first in the one year after germination (the wild-type plants produced flowers after 5–8 years) and flowered 2–3 times every year. The study showed that three genes (*BAM*, *FT* and *TF1*) may play a critical role in the precocious phenotype of mutant trifoliate orange (Zhang et al., 2009). The spontaneous sport mutant "x11" was selected from trees of the sweet orange "Tobias" in Brazil in 2006 (Pinheiro et al., 2014). "x11" differs from "Tobias" more compact habits and increased seed number in fruits. Seedlings of "x11" produced fertile flowers after one to two years and flowered repeatedly in all seasons. Transgenic "x11" lines expressing a reporter gene *uidA* exhibited the first flowers within 3–5 months after acclimatization in the greenhouse. This mutant has an advantage, as it is of great commercial interest as the sweet orange compared to trifoliate orange (Pinheiro et al., 2014). Probably, these citrus mutants have a common mechanism of early flowering, which makes flower initiation independent of the season.

14.4.2 Agrotechnical Techniques

Agrotechnical methods aimed at accelerating plant flowering are widely used in breeding. In particular, for annual crops, Lee Hickey and his colleagues developed "speed breeding" to accelerate the flowering and seed set in order to move to the next breeding cycle as quickly as possible. The essence of this technique is the cultivation of plants in a controlled-environment room with optimal temperature and humidity regimes and extended photoperiod (22 hours of light instead of natural variable photoperiod (10–16 hours), which enables achieving 4–6 generations per year for spring and durum wheat, barley, chickpea, pea, and canola instead of 2–3 under normal greenhouse conditions (Watson et al., 2018). Other techniques to enhance the cycle turnover include the development of homozygous lines, using cultivation under stress conditions and *in vitro* culture of immature embryos, using immature seed harvest (Samantara et al., 2022).

Agrotechnical methods for reducing the length of the juvenile period have also long been used in fruit tree breeding. The photoperiod and temperature conditions are the most important environmental factors that regulate the flowering of trees (Hanke et al., 2007). Besides, since in trees the length of the juvenile phase is inversely correlated with vigor, it can be shortened using environmental conditions favorable for intensive growth (van Nocker & Gardiner, 2014). Using optimal conditions for fast early seedling growth (high temperature and humidity, high irradiance, long photoperiod, high red:far-red ratio, elevated concentration of CO_2, and non-limiting fertigation) and applications of a gibberellin biosynthesis inhibitor, it was possible to reduce the juvenile period of the apple to 18 months (Volz et al., 2009). Some apple hybrids between carriers of resistance to scab and fire blight formed first flowers within 14 months after germination (Baumgartner et al., 2011).

A direct relationship between the induction of flowering and plant size was shown on transgenic apple plants expressing the reporter *uidA* gene (Yao et al., 1999). Accelerated flowering of the transgenic apple cultivar "Royal Gala" was achieved by optimizing growing conditions in the greenhouse, which allowed extended growing seasons. Apple plants produced stems 2 m high and 100 nodes on average at the end of the first year and flowered in the second year. Plants that formed less than 80 nodes in the first year did not bloom in the second year. Probably, flowering requires sufficient node numbers to be produced (Yao et al., 1999). Aldwinckle (1975) previously obtained apple flowering 16–20 months after germination and plants produced flowers on 90–100 nodes.

On the whole, the flowering of apple seedlings could be shortened to 10 months under optimum growing conditions in the greenhouse, but plants that grow continuously can grow very tall (up to 3 m) and become difficult to manage (van Nocker & Gardiner, 2014). The large size of woody plants by the time of flowering increases costs and makes their management difficult.

This does not allow to widely use their cultivation in controlled environment facilities, unlike for annual crops.

Other agrotechnical techniques, besides control of environmental factors for flowering induction and growth acceleration, are also known. In his review, Zimmerman (1972) lists grafting (on dwarfing rootstocks or into old trees), girdling and scoring, shoot and root pruning, fertilization, and chemical treatments. The use of dwarf rootstocks made it possible to accelerate the flowering of seedling scions by 2–4 years compared to ungrafted seedlings (Hackett 1985). Water stress or defoliation induced the second flowering of potted apple trees in any one year, suggesting that the loss of leaves is the primary reason for reflowering (Jones, 1987). Growth can also be controlled by plant growth regulators. However, this method is not widespread for reducing juvenile period due to high variability among genotypes, species, and experiments (van Nocker & Gardiner, 2014). Seeds of most temperate fruit trees require a chilling period (stratification) for germination. Overcoming seed dormancy is another way of shortening the breeding cycle. Embryo rescue or treatment of bioactive gibberellins is used for this purpose (van Nocker & Gardiner, 2014).

Less is known about studies with other, besides apple, species of fruit trees. A significant correlation has been found between pecan tree growth and precocity levels (Thompson & Grauke, 2003). Under high-intensity light in growth chambers, some chestnut seedlings form catkins in their second or third year (Baier et al., 2012). Using a forcing growth protocol (drip fertirrigation, controlled temperature, continuous light, and pinching lateral shoots) for olive seedlings, the first flowering occurred 28 months after germination and almost all seedlings flowered within five years, while a juvenile phase of the olive tree lasts usually over 10 years (Santos-Antunes et al., 2005). Evaluation of the relationship between architectural parameters and the duration of the juvenile period in olive seedlings showed that "Plant height" and "Secondary shoots insertion angle" are the most correlated with the time of flowering (Hammami et al., 2021). A new olive selection strategy, combining quantitative (height) and qualitative (architectural types) criteria, makes it possible to exclude seedlings with a long juvenile period at the early stage of the breeding cycle and increase the efficiency of breeding programs. However, the strong growth-precocity relationship is undesirable as the common objective in fruit tree breeding is to create cultivars that combine precocity and compact crown size during fruiting (Thompson & Grauke, 2003).

On the whole, it is hardly possible to induce the flowering of fruit trees through agrotechnical approaches earlier than in the second year after germination (Flachowsky et al., 2009). A more radical reduction of the juvenile period of fruit trees can be achieved only by genetic and genomic modification.

The speed breeding technology combines well with biotechnological methods such as marker-assisted selection (MAS), genome editing, genome selection (GS), and high-throughput phenotyping (Samantara et al., 2022), which will be considered further. For example, the speed-breeding technique was successfully integrated with MAS based on single-nucleotide polymorphism (SNP) markers for the development of salinity-tolerant rice (Rana et al., 2019). It took only 17 months to achieve six generations of backcrossing, i.e., less than three months per generation. Whole genome sequencing showed that 93.5% of the hybrid genome was from a high-yielding elite variety and only 2.7% from a donor of salt resistance. A combination of genomic selection (GS) and speed breeding was used to reduce the breeding cycle for quantitative traits in wheat (Watson et al., 2019). A combination of speed breeding and high-throughput phenotyping creates opportunities for multiple trait selection (Wanga et al., 2021).

14.5 BIOTECHNOLOGICAL METHODS

The effectiveness of NPBTs requires knowledge about the genetic control of agronomically important traits; they remain limited in fruit trees compared to major crops. Studies on model grassy objects cannot always be extrapolated to woody species. For many decades since the 1930s, when Chailakhan hypothesized the existence of the flower-inducing hormone "florigen" (Putterill & Varkonyi-Gasic, 2016), efforts have been made to clarify the mechanisms of flowering initiation in plants. The molecular nature of florigen was identified as the protein encoded by the *FT* genes (Corbesier et al., 2007). *FT* genes have been isolated and characterized in many plants including perennial woody species. The products of *FT* genes are small (about 175 amino acids), globular proteins from the phosphatidylethanolamine binding protein (PEBP) family that induce plant flowering (Putterill & Varkonyi-Gasic, 2016). FT and other floral integrator genes, such as SOC1 and TERMINAL FLOWER1 (TFL1) genes, regulate the expression of flower meristem-determining genes, LFY and APETALA1 (AP1), which ultimately induces flowering (Savadi et al., 2021). Another gene, TERMINAL FLOWER1/CENTRORADIALIUS (TFL1/CEN), is closely related to FT (amino acid sequences are 98% similar), but has an opposite function, as it represses flowering by competing with FT for the binding of transcription factors (Bennett & Dixon, 2021). Studies have shown that changes in the expression of flowering genes affect the transition to the reproductive phase in plants. Thus, with the help of biotechnological methods, it is possible to shorten the juvenile phase of fruit trees, affecting the expression of key genes controlling the initiation of flowering. The most promising approaches are the upregulation of flowering promoters FT genes or the downregulation of flowering repressors *FTL1* genes.

14.5.1 Genetic Engineering

A promising area is transgenic manipulation by the duration of the juvenile phase. Genetic engineering offers various advantages over classical breeding. First, it enables the insertion of genes without changing the genetic background of plants, which is especially valuable for fruit trees, because fruit production in the world is based on a limited number of cultivars that both consumers and growers use. Second, it enables the use of the genes of valuable traits from any source. The plant blooms when the flowering genes are activated under the influence of certain inner and outer signals. Overexpression of flowering-related genes can lead to acceleration of flowering (Table 14.1).

As the classical genetic engineering of plants (transgenesis) causes public concern about transgenic crops, two alternative transformation concepts have been developed – intragenesis and cisgenesis (Holme et al., 2013). Intragenesis is a genetic transformation of plants by genes from cross-compatible species, but recombination of genetic elements is admissible *in vitro* (Rommens, 2004). The requirements for cisgenesis are more stringent: it also introduces genes from sexually compatible species, but these genes must be identical to natural genes, including all regulatory elements (Schouten et al., 2006). A number of intragenic early flowering fruit trees are known, where promoters of native genes were replaced, but cisgenic plants were not reported. Nevertheless, the commercialization of both cisgenic and intragenic plants currently is subject to the same regulatory rules as other transgenic crops. Stringent regulation makes it difficult for transgenic fruit trees to enter the market and to date, only virus-resistant plum HoneySweet and non-browning Arctic® apples have been approved by regulatory agencies (Lobato-Gomez et al., 2021).

Accelerated breeding technology (fast track breeding, Wenzel et al., 2013; Endo et al., 2020; "FasTrack" breeding, Petri et al., 2018) based on the utilization of transgenic early flowering fruit trees allows to not only shorten the breeding cycle but also obtain transgene-free genotypes. Its concept is to cross the line with single-copy early flowering transgene and closely related wild species with a monogenic trait of interest (Flachowsky et al., 2009 According to Mendel's laws, 25% of F1 hybrids contain both valuable traits and they are used in backcrosses with high-quality cultivars to reduce the proportion of the wild species' genome. Ultimately, null segregants containing only the valuable trait, and not the transgene, are selected but these plants can be obtained much more quickly than traditional breeding methods (Patocchi et al., 2021). In the United States, these null segregants are not considered to be genetically modified plants and can be released into the environment without any restrictions (Callahan et al., 2016).

TABLE 14.1
Application of Genetic Transformation Techniques for Early Flowering Induction in Fruit Trees

Species	Gene	Promoter*	Strategy	Early Flowering	Habit	References
Actinidia eriantha	AcFT (kiwifruit)	CaMV 35S	overexpression	no (AcFT)	reduced growth (AcFT)	Varkonyi-Gasic et al., 2013
	AdCEN (kiwifruit)			delayed flowering (AdCEN)	normal (AdCEN)	
Apple (*Malus × domestica*)	MdTFL1 (apple)	CaMV 35S	antisense	8–25 months	weak vigor	Kotoda et al., 2006
	BpMADS4 (birch)	CaMV 35S	overexpression	*in vitro*	reduced growth	Flachowsky et al., 2007
	AP1 (Arabidopsis)	CaMV 35S	overexpression	no	normal	Zhu et al., 2009
	LFY (Arabidopsis)	CaMV 35S	overexpression	no	dwarfism	Flachowsky et al., 2010
	MdFT1 (apple)	CaMV 35S	overexpression	*in vitro*	dwarfism	Kotoda et al., 2010
	MdFT1 (apple)	CaMV 35S	overexpression	*in vitro* (CaMV 35 S)	? (CaMV35S)	Trankner et al., 2010
		AtSUC2		no (AtSUC2)	normal (AtSUC2)	
	MdTFL1 (apple)	–	virus induction	1.5–2 months	?	Sasaki et al., 2011
	AtFT (Arabidopsis)	–	virus induction	1.5–2 months (FT)	?	Yamagishi et al., 2011
	MdFT1 (apple)			no (MdFT1)		
	MdTFL1-1 (apple)	CaMV 35S	RNAi	*in vitro*	reduced growth	Flachowsky et al., 2012
	PtFT1 (poplar)	Gmhsp17.5-E	overexpression	after heat treatment	normal	Wenzel et al., 2013
	PtFT2 (poplar)					
	AtFT (Arabidopsis)	CaMV 35S	overexpression	*in vitro* (CaMV 35 S)	death	Tanaka et al., 2014
		rolC		*in vitro* (rolC)	? (rolC)	
	AtFT + MdTFL1	–	virus induction	1.5–3 months	?	Yamagishi et al., 2014
	PcFT2 (pear)	CaMV 35S	overexpression	no	prolonged growth	Freiman et al., 2015
	MdTFL1-1 (apple)	Gmhsp17.5-E	RNAi	no	normal	Weigl et al., 2015a
	BpMADS4 (birch)	CaMV 35S	overexpression	*in vitro*	spindly habit	Tomes et al., 2023
Blueberry (*Vaccinium corymbosum*)	VcFT (blueberry)	CaMV 35S	overexpression	*in vitro*	dwarfism	Song et al., 2013
Citrange (*Citrus sinensis ×*	LFY (Arabidopsis)	CaMV 35S	overexpression	12–20 months (LFY)	weeping habit (LFY)	Pena et al., 2001
	AP1 (Arabidopsis)			13–15 months (AP1)	normal (AP1)	

(Continued)

TABLE 14.1 *(Continued)*
Application of Genetic Transformation Techniques for Early Flowering Induction in Fruit Trees

Species	Gene	Promoter*	Strategy	Early Flowering	Habit	References
Poncirus trifoliata)	FT (citrus)	CaMV 35S	overexpression	*in vitro* (FT citrus)	death (FT citrus)	Moore et al., 2016
	FT (poplar)			no (FT poplar)	? (FT poplar)	
	CcFT1 (clementine)	CaMV 35S	overexpression	no (CcFT1)	death (35 S:CcFT3)	Soares et al., 2020
	CcFT3 (clementine)	NOS		*in vitro* (35 S:CcFT3)	normal (AtSUC2:CcFT3)	
		AtHSP		no (NOS:CcFT3)		
		AtSUC2		no (AtHSP:CcFT3)		
				1 year (AtSUC2:CcFT3)		
Citrus excelsa	AtFT (Arabidopsis)	–	virus induction	1 year	normal	Velazquez et al., 2016
	CiFT (sweet orange)					
Grapefruit	FT (citrus)	CaMV 35S	overexpression	*in vitro*	death	Moore et al., 2016
(*Citrus paradisi*)	CsAP1 (sweet orange)	AtRD29A	overexpression	4.5 years (CsAP1)	normal	Orbović et al., 2021
	CsLFY (sweet orange)			8 years (CsLFY)		
	PtFT1-scFv (trifoliate orange)	CaMV 35S	overexpression	3–6 months	dwarfism	Sinn et al., 2021
Kiwiberry (*Actinidia arguta*)	AcFT1-6xHA (kiwifruit)	CaMV 35S	overexpression	yes	?	Herath et al., 2023
Kiwifruit (*Actinidia chinensis*)	AcFT (kiwifruit)	CaMV 35S	overexpression	*in vitro*	death	Moss et al., 2018
	AcFT*					
	AcFT1-6xHA (kiwifruit)	CaMV 35S	overexpression	yes	?	Herath et al., 2023
Kumquat (*Fortunella crassifolia*)	AP1 (Arabidopsis)	CaMV 35S	overexpression	11–23 months	normal	Duan et al., 2010
Kumquat (*Fortunella hindsii*)	miR3954 (citrus)	CaMV 35S	overexpression	7 months	normal	Liu et al., 2017
Olive (*Olea europaea*)	MtFTa1 (alfalfa)	CaMV 35S	overexpression	*in vitro*	reduced growth	Haberman et al., 2017
Pear (*Pyrus communis*)	CiFT (mandarin)	CaMV 35S	overexpression	*in vitro*	?	Matsuda et al., 2009
	MdTFL1 (apple)	CaMV 35S	RNAi	*in vitro*	reduced vigor	Freiman et al., 2012
	BpMADS4 (birch)	CaMV 35S	overexpression	6–18 months	normal	Tomes et al., 2023

Plum (*Prunus domestica*)	PtFT1 (poplar)	CaMV 35S	overexpression	1–10 months	shrub-type habit	Srinivasan et al., 2012
Sweet orange	CsFT (sweet orange)	CaMV 35S	overexpression	1 year	stunted growth	Pons et al., 2014
(*Citrus chinensis*)	FT (citrus)	CaMV 35S	overexpression	*in vitro*	death	Moore et al., 2016
	CsAP1 (sweet orange)	AtRD29A	overexpression	no	normal	Orbović et al., 2021
	CsLFY (sweet orange)					
Trifoliate orange	CiFT (mandarin)	CaMV 35S	overexpression	12–33 weeks	dwarfism	Endo et al., 2005
(*Poncirus trifoliata*)	CiFT (mandarin)	CaMV 35S	overexpression	a few months	dwarfism	Endo et al., 2009
	ToFT (trifoliate orange)	CaMV 35S	overexpression	*in vitro*	?	Wu et al., 2022
	AtFT (Arabidopsis)					

Notes

* Promoters: strong constitutive (CaMV 35 S), weak constitutive (NOS), phloem-specific (AtSUC2, rolC), heat inducible (AtHSP, Gmhsp17.5-E), stress inducible (AtRD29A)

14.5.1.1 Citrus

Citrus is the most important fruit crop in subtropical and tropical regions, with global production exceeding 160 million tons in 2021 (FAOSTAT, 2021). The great economic importance of citrus fruits stimulates the development of new cultivars, but conventional citrus breeding is a long-term and expensive process due to the long juvenile period of trees. Acceleration of flowering of fruit trees using genetic engineering was first demonstrated namely on citrus when the transfer of the *Arabidopsis AP1* or *LFY* genes under CaMV 35S promoter into citrange (*Citrus sinensis* Osb. × *Poncirus trifoliata* L. Raf.) induced flowering at an age of 12–20 months, this trait being transmitted to the offspring (Pena et al., 2001). The effects of the two genes were different: *LFY* caused abnormalities in the vegetative organs, whereas *AP1* did not. The contradictory effects of the genes can be due to some specific interactions between them and host plants (Pena et al., 2001). Transgenic plants of "Meiwa" kumquat (*Fortunella crassifolia* Swingle) with the *35S:AP1* construct also exhibited normal growth (Duan et al., 2010).

Expression of the *FT* gene (*CiFT*) from the satsuma mandarin (*Citrus unshiu* Marc.) under the CaMV 35S promoter caused an even greater effect: the transgenic trifoliate orange (*P. trifoliata*) began flowering within 12 weeks of being transferred to a greenhouse (Endo et al., 2005). However, transgenic plants differed from the control by dwarf growth and enhanced branching. The *P35S:CiFT* gene was later used in gene construct for co-introduction and co-expression with a target gene (Endo et al., 2009). The antisense sequence of d-limonene synthase (*CitMTSE1*) gene was used for the metabolic engineering of aroma components in trifoliate orange (*P. trifoliata*). Some transgenic lines started to flower 30 weeks after transformation and to bear fruit one year after transformation. This allowed estimating the limonene levels in flowers and fruits. Some lines demonstrated dwarf habit and small leaves as previously reported (Endo et al., 2005). In a similar work, Pons et al. (2014) simultaneously blocked the expression of a carotenoid biosynthesis gene by RNAi and overexpressed the *FT* gene from sweet orange (*CsFT*) in "Pineapple" sweet orange plants. The transgenic lines produced fruits with a "golden" phenotype and significantly increased β-carotene content within one year after grafting in the greenhouse. Besides, they bore fruit twice a year instead of once. Transgenic plants had stunted growth in comparison to wild-type seedlings (Pons et al., 2014). The Endo et al. (2009) and Pons et al. (2014) studies are examples of the use of early flowering plants for the evaluation of metabolically engineered fruits.

Thus, ectopic expression of flowering genes in citrus, along with accelerated flowering, led to abnormal phenotypes, such as dwarf or weeping growth form, and curled leaves (Pena et al., 2001; Endo et al., 2005). Possibly, these effects were due to the expression of flowering genes under the strong constitutive CaMV 35S promoter. The effect of this promoter was also manifested during the transformation of citrus ("Carrizo" citrange, "Hamlin" sweet orange, and "Duncan" grapefruit) by the citrus FT gene, when, instead of shoots, explants produced flower buds that eventually died (Moore et al., 2016). Transformation of trifoliate orange by *FT* genes from trifoliate orange (*ToFT*) or *Arabidopsis* (*AtFT*) driven by the CaMV 35S promoter also induced flower formation in some lined *in vitro*, which did not form roots and died (Wu et al., 2022). The other transgenic lines formed flowers earlier than 10 months after planting in the greenhouse. Moore et al. (2016) suggested that a decrease in FT activity in transgenic citrus using nonconstitutive or inducible promoters was necessary. This approach was implemented in Soares et al. (2020), where the citrus homologue of FT, cloned from *Citrus clementina* (*CcFT1* or *CcFT3*), was transferred to the "Carrizo" citrange rootstock (*C. sinensis* × *P. trifoliata*), driven by the phloem specific *AtSUC2* promoter from *A. thaliana*. No morphological deviations occurred, but of the two constructs, only the *AtSUC2-CcFT3* vector induced early flowering (16 months after transformation). Also, the use of a poplar FT gene did not induce flowering in transgenic citrus plants for two years (Moore et al., 2016). *A. thaliana* stress-inducible promoter AtRD29A was used to transfer sweet orange *CsAP1* or *CsLFY* genes to commercially important "Duncan" grapefruit and "Valencia" sweet orange (Orbović et al., 2021). The most precocious line with the *CsAP1* gene (grapefruit line) flowered after just 4.5 years under non-inductive greenhouse conditions and its T1 siblings flowered after 18 months. Transgenic lines with the *CsLFY* gene flowered after 8 years and the T1 siblings flowered at the age 3–8 months but all flowers of these seedlings fell off. All transgenic plants were morphologically normal and did not differ from non-transgenic control (Orbović et al., 2021).

The dependence of phenotypic deviations on the expression level of the flowering genes was demonstrated by Sinn et al. (2021). A chimeric protein construct including a translational fusion of *P. trifoliata* FT1 (PtFT1) with a single-chain variable fragment antibody (scFv) under the CaMV 35S promoter was transferred to grapefruit (*C. paradisi*). This caused flowering after 3–6 months and a number of different effects – from moderate leaf curling to dwarfed growth and a complete absence of thorns. Moderate expression of the transgene resulted in early flowering largely without morphological aberrations (Sinn et al., 2021). Clonal propagation of transgenic lines via rooted cuttings preserved the early flowering phenotype. However, there was no obvious correlation between the early or delay flowering response and the expression level of *A. thaliana AP1*, *CiAP1,* and *CiLFY* genes in all transgenic kumquat lines (Duan et al., 2010). This was previously reported by Pena et al. (2001) for transgenic citrange plants. At the same time, the *CiFT* and *CiTFL1* genes were expressed differently in early flowering and delay flowering transgenic kumquat lines, and the expression pattern was opposite. Acceleration or delay of flowering in different transgenic lines may be caused by the interaction of foreign *AP1* and endogenous flowering genes (Duan et al., 2010). Wu et al. (2022) showed the effect of FT genes on the expression of other flowering genes. Overexpression of *ToFT* or *AtFT* genes in transgenic trifoliate orange clearly increased expression levels of endogenous *ToLFY* and *ToAP1* genes.

The introduction of multiple genes ("stacking" or "pyramiding") in one plant allows for a combination of several new traits or the improvement of complex traits, but there are very few studies on fruit trees (Lebedev, 2023a). An early flowering citrange line with the *AP1* gene was used for retransformation to establish a gene-stacking system for citrus trees (Cervera et al., 2009). The line AP1-13.20 (Pena et al., 2001) was selected because it contained a single copy of the T-DNA and demonstrated early and regular flowering each spring. For repeated transformation, antibiotic resistance (*hpt*) and herbicide resistance (*bar*) genes were used as selectable markers. Most of the double transformants began flowering a year after being transferred to the greenhouse, just like the original transgenic citrange plants with the *AP1* gene. Retransformed rapid-cycling plants flowered and bore fruit every spring in the greenhouse for more than 3 years, the expression of the *AP1* transgene was stable and uniform which confirms the suitability of this system (Cervera et al., 2009).

Besides overexpression of flowering genes, other genes were used to accelerate the flowering of citrus plants. Studies on kumquat (*F. hindsii*) showed that the overexpression of miR3954 significantly accelerated flowering in transgenic plants (Liu et al., 2017). Transgenic plants had no phenotypic differences, but their offspring flowered at the stage of 18–19 leaves, while the offspring of the control did it at the stage of 35 leaves. The miR3954 has a specific expression in the flowers of various citrus species and the early flowering phenotype of kumquat may be caused by the interaction between miR3954 and a NAC transcription factor (Liu et al., 2017).

The fast-track breeding system of citrus was created based on precocious trifoliate orange plants with the *35S:CiFT* gene to impart disease resistance (Endo et al., 2020). The transgenic line of trifoliate orange 2–11 (Endo et al., 2005) was used as a donor of citrus tristeza virus (CTV) resistance and the early flowering phenotype. This line was crossed with CTV susceptible "Hyuganatsu" (*C. tamurana*). The resulting F1 generation was subjected to three backcrossings with "Hyuganatsu" and clementine mandarin (*C. clementina*) to remove the undesirable features of trifoliate orange. To increase the efficiency of the selection, MAS was used: using CAPS markers in the BC progenies, the genetic loci of virus resistance and transferred T-DNA were identified (Endo et al., 2020). The segregation ratio among the BC progenies corresponded to the Mendelian fashion and about 25% of plants had CTV resistance but contained no T-DNA. As they have no T-DNA segments, they cannot be considered GM plants, and no regulatory restrictions for food safety and security are applicable to them. Using the fast-track breeding system, the backcrossing breeding cycle was reduced from five to one year (Endo et al., 2020).

An alternative to stable transformation for the induction of flowering is the use of viral vectors. Both FT proteins and viruses use the phloem as a transport system, and coupling an FT ortholog with a virus-based vector amplifying the inserted sequence can promote flowering (McGarry et al., 2017). The use of viral vectors is a fast system for delivering proteins to plants, requiring no transformation and regeneration techniques *in vitro* and causing no risk of somaclonal variation. To prevent phenotypic changes, vectors based on citrus leaf blotch virus (CLBV) containing the *A. thaliana* or citrus *FT* genes were also used (Velazquez et al., 2016). Citrus plants inoculated with any of these vectors flowered after 4–6 months, and showed no changes in the morphology of the plant, leaf, flower, or fruit. The analysis demonstrated the vector to be not integrated into the genome and is not transmitted with pollen (Velazquez et al., 2016). Authors suggest that this viral vector should not be considered as a risky genetically modified organism and this approach is capable of accelerating the breeding of new cultivars.

14.5.1.2 Apple

Apple (*Malus* × *domestica* Borkh.) is the major temperate fruit tree and its global production was about 93 million tons in 2021 (FAOSTAT, 2021). The reduction of the juvenile period in apple was first reported by Japanese researchers and the suppression strategy was used. The transfer of the *MdTFL1* antisense gene to the "Orin" apple tree led to the flowering of transgenic lines in a greenhouse after 8–25 months, while the control did not start to flower after 6 years (Kotoda et al., 2006). The expression level of the *MdTFL1* gene correlated with a decrease in the juvenile phase. Flowers were normal and, after pollination, fruits and seeds were obtained. Later, the *MdTFL1-1* gene was silenced in apple plants using an RNAi-based approach (Flachowsky et al., 2012). The observed effects are much stronger than that of Kotoda et al. (2006). Transgenic lines of the cultivar "Pinova" flowered since the sixth month in tissue culture and within a few weeks after transfer to the greenhouse, and their growth was reduced. Most of the transgenic flowers had abnormalities indicating that *MdTFL1* genes are also involved in floral organ development (Flachowsky et al., 2012). However, flowers were fertile and most fruits had normal fruit size. All transgenic seedlings developed flowers within a few weeks but their height was only 4–40% of the height of non-transgenic seedlings after two years of cultivation in the greenhouse (Flachowsky et al., 2012). To exclude the undesirable effects of constitutive suppression of the *MdTFL1* gene on the growth of apple plants, the heat-inducible promoter Gmhsp17.5-E from soybean was used (Weigl et al., 2015a). RNAi silencing construct was transferred in "Pinova" and "Gala" apples but transgenic lines did not induce early flowering after heat shock treatment. The authors suggested the reason for the lack of early flowering in the ineffectiveness of gene construct or the low expression level.

Overexpression of apple FT also accelerated flowering. Transgenic apples carrying the *35S:MdFT1* produced flowers *in vitro* 2–6 months after regeneration (Kotoda et al., 2010). *In vitro* flowering lines showed an extremely weak growth and some lines senesced, but the expression level of *MdFT1* was high. At the same time, attempts to shorten the flowering time in apple trees with the help of *A. thaliana* genes were unsuccessful, probably due to differences in the mechanisms of flower bud formation. No flowering was observed for two years in a greenhouse of the apple rootstock M26 with the *Arabidopsis AP1* gene (Zhu et al., 2009). The transfer of the *Arabidopsis LFY* gene to the "Pinova" apple tree also failed to cause flowering for two years in a greenhouse but resulted in a columnar phenotype with shorter internodes (Flachowsky et al., 2010). It should be noted that the same *LFY* and *AP1* genes from *A. thaliana* induced early flowering in citrus (Pena et al. (2001). This can be indicative of a difference in the formation of flowers in evergreen and deciduous fruit trees. Transfer of FT-like gene from pear (*Pyrus communis*), *PcFT2*, does not induce flowering in the transgenic apple cultivar "Galaxy" after 4 years of growth in the greenhouse (Freiman et al., 2015). However, transgenic apple plants overexpressing this gene showed a delay in bud dormancy and leaf senescence.

A more successful approach for the induction of early flowering was the use of the *BpMADS4* gene of the silver birch (*Betula pendula* Roth.) from the AP1/SQUA/FUL-related genes (Flachowsky et al., 2007). After its transfer to the seedlings of the "Pinova" apple cultivar, in some lines flowers formed on shoots *in vitro*. In the greenhouse, flowering was observed after 3–4 months, but the level of transcription did not correlate with flowering. The flowers were fertile and bore fruits, containing the usual number of seeds (Flachowsky et al., 2007). The transgenic plants obtained in this work were later used for accelerated apple tree breeding. The most promising genotype of the apple tree proved to be a clone T1190 with the *BpMADS4* gene, which contained one copy of T-DNA and quickly started to flower in a greenhouse (Flachowsky et al., 2009). For instance, hybrids of this clone with the wild apple tree *M. fusca* started to flower after a few weeks of growth in a greenhouse, after reaching a height of 40 cm. These transgenic F1 seedlings when pollinated with "Topaz" cultivar yielded seeds of the first pseudo-backcross (BC01), confirming the achievability of shortening the breeding cycle to one year (Flachowsky et al., 2009). The prospects of using this early flowering apple tree in rapid breeding programs were confirmed in subsequent investigations.

Currently, apple production is dominated by several cultivars that were obtained quite a long time ago and are susceptible to a number of diseases, and enhancing resistance is the second major objective in apple breeding after increasing the marketability of the fruits (Gardiner et al., 2007). New cultivars will reduce the application of pesticides in apple production, but new races of pathogens appear faster than classical breeding can respond to them (Flachowsky et al., 2011). In addition, donors of resistance are usually wild *Malus* species and several generations of backcrosses are required to remove unwanted alleles, which further lengthens the breeding process. In this regard, the fast-track breeding system for apple was used specifically to obtain disease-resistant genotypes.

In Germany, hybrids of clone T1190 with *M. fusca* were used to pyramid several genes of disease resistance (Flachowsky et al., 2011). Transgenic early flowering F1 seedlings were pollinated by "Regia" cultivar (donor of fire blight resistance and scab resistance genes) or pollen of clone 98/6–10 (donor of powdery mildew resistance genes). The inheritance of disease-resistance genes was assessed using SSR and SCAR markers. In the future, plants that carry disease-resistance genes will be backcrossed with apple elite cultivars (for example, "Golden Delicious"), to consistently remove all undesirable traits of wild apples. This is the first study on the use of transgenic early flowering plants in combination with MAS to accelerate the breeding cycles of the classical fruit-tree breeding program (Flachowsky et al., 2011).

A Swiss research group crossed T1190 with the ornamental apple cultivar "Evereste" carrying the fire blight resistance locus (Le Roux et al., 2012). The seedlings, containing the early flowering transgene and the disease resistance locus, were then crossed with the apple cultivar "Topaz", "Maloni Sally", or "Milwa" (pseudo-backcrosses, BC01). The BC01 seedlings containing the transgene and locus estimated using SSRs were then pollinated with pollen from the "Royal Gala" cultivar. The crossings were carried out at intervals of about a year, which confirms the prospects of using early flowering plants for accelerated breeding. Although such seedlings contained an average of 24.3% of the "Evereste" genome, disease resistance in hybrids increased significantly (Le Roux et al., 2012). This work had been successfully continued, and the fifth generation null segregants (BC'4) carrying the fire blight Fb_E resistance gene were reported to be produced for 7 years (Schlatholter et al., 2018). The genotypes maintained a high level of resistance to fire blight and only 4% of the "Evereste" genome was preserved in some seedlings. Some of the null segregants carried both the fire blight resistance locus and the apple scab resistance gene, the presence of which was determined using SCAR and SSR markers. Using the early flowering genotype, seedlings of the fifth generation were produced 3–4 times faster than when using classical breeding (7 years instead of 20–25 years) (Schlatholter et al., 2018).

Finally, investigators from the United States recently used the clone T1190 to select apple trees for the resistance to blue mold, the most economically important postharvest disease (Luo et al., 2020). Some wild apple trees, such as *M. sieversii*, are natural sources of blue mold resistance. The transgenic clone was crossed with F1 "Gala" × *M. sieversii* hybrids. In MBC1 offspring, seedlings with blue mold resistance-associated 193-bp allele were identified. The use of MAS based on SNP markers for background selection showed that the portion of the wild apple genome in hybrids was reduced to 16.1% (Luo et al., 2020).

Recent comparative whole genome sequencing of the transgenic early flowering apple clone T1190 showed that it contains only one transgene copy, located on chromosome 4, and partial DNA sections of the transformation vector are absent (Patocchi et al., 2021). Although the T1190 line was used in breeding the most, it also had disadvantages. The slender phenotype of T1190 could make it difficult to evaluate resistance to *Erwinia amylovora* and rather drooping plant habit negatively affected the fruiting of hybrids (Flachowsky et al., 2011; Le Roux et al., 2012). Characterization of additional *BpMADS4* overexpressing lines obtained based on Pinova seedlings or an *in vitro* culture of four different apple cultivars ("Pinova", "Gala", "Mitchgla Gala", and "Santana') showed that each line had a unique molecular character (Weigl et al., 2015b). However, transgenic lines did not differ in flower organ development and pollen viability compared with wild-type plants. The introgression of the early flowering gene into different linkage groups will improve the integration of genes encoding valuable traits during hybridization (Weigl et al., 2015b).

In addition to the deformed growth habit of early flowering apples, plants with the *35S:BpMADS4* construct have another problem: the continuous formation of new flowers due to a constitutive promoter (Wenzel et al., 2013). The *PtFT1* and *PtFT2* genes from *Populus trichocarpa* under heat-inducible promoter were used to solve it. These genes were transferred to the Pinova apple tree, and the heat treatment at 42°C in a greenhouse induced the formation of flowers. Pollination with *Malus* × *robusta* 5 pollen led to the formation of fruits (Wenzel et al., 2013). The importance of choosing a promoter was also shown in other works. Overexpression of the *MdFT1* gene in the "Pinova" apple tree under the CaMV 35S promoter led to flowering *in vitro*, but the lines with the gene under the *Suc2* promoter did not flower during the first year in a greenhouse (Trankner et al., 2010). In Tanaka (2014), the apple semi-dwarfing rootstock "JM2" transformed the *FT* gene from *A. thaliana* under the CaMV 35S or rolC promoter. Constitutive promoter induced flower formation directly from the callus, but there was no vegetative growth and the explants died. With the rolC promoter, flowering was observed *in vitro* for 4–8 months after transformation, as well as in a greenhouse (Tanaka et al., 2014).

Recently, the *35S:BpMADS4* construct was transferred to the apple cultivar "Royal Gala" (Tomes et al., 2023). Transgenic apple plants flowered *in vitro* or several weeks after grafting on the rootstock. Despite a spindly appearance of stems, fruits were produced and early flowering seedlings were observed in the offspring.

As in citrus, acceleration of apple flowering has been successfully achieved using viral vectors. About 10% of infected seedlings flowered within 1.5–2 months after inoculation of the apple latent spherical virus (ALSV) vector containing apple *MdTFL1* gene (Sasaki et al., 2011). Flowers were normal and after pollination, fruits contained fertile seeds. Yamagishi et al. (2011) reported that the inoculation of the ALSV vector expressing an FT gene of *A. thaliana* increased the proportion of flowering seedlings to 30%, but in this work, MdFT1-ALSV did not induce early flowering. The flowers were pollinated and produced fruits containing seeds. The examination indicated that ALSV was not transmitted by seeds and the next-generation seedlings were all virus-free (Yamagishi et al., 2011). To evaluate the flowering-inductive effect of *FT* genes, Yamagishi et al. (2014) tested 18 orthologues of FT from the genomes of 13 plant species including perennial (apple, Japanese apricot, poplar, and grapevine). All FT orthologues accelerated flowering in *A. thaliana* after inoculation of ALSV vectors, but only four accelerated flowering in apples. About 30% of apple seedlings flowered after inoculation with AtTSF or AtFT from *A. thaliana*, while inoculation with GmFT2a (*Glycine max*) or GtFT1 (*Gentiana trifloral*) yielded a flowering rate of approximately 10%. The authors concluded that the expression of an FT gene alone is not sufficient to induce flowering. Simultaneous expression of AtFT and suppression of *MdTFL1-1* in apples through an ALSV vector induced flowering of more than 90% of seedlings within 1.5–3 months after inoculation (Yamagishi et al., 2014). In almost all early flowering apple seedlings, the formation of flower buds occurred continuously. These plants produced fruits after pollination indicating that apples can complete one life cycle (from the seed to the seed) within one year. The presence of ALSV was tested in seed embryos and seedlings and the rate of seed transmission of ALSV was extremely low. Thus, the progeny of inoculated plants does not contain transgenes and is not subject to restrictions for genetically modified plants (Yamagishi et al., 2014). The authors believe that this is the first report of a successful reduction of the generation cycle in apples to one year without the use of stable genome modification.

14.5.1.3 Pear

Shortly after early flowering transgenic apple plants were produced, a similar work with pear was carried out (Matsuda et al., 2009). Japanese researchers transferred the Citrus FT (*CiFT*) gene to the European pear (*P. communis* L.) cultivars "La France" and "Ballade" (Endo et al., 2005). Most of the transgenic lines of both varieties flowered during *in vitro* cultivation. The expression level correlated with floral bud differentiation *in vitro*. All the transgenic plants with the CiFT gene demonstrated a changed number of floral organs but pollen was viable (Matsuda et al., 2009). The early flowering phenotype was passed on to the offspring – transgenic seedlings formed flowers within 10 months after transfer to a greenhouse.

Another strategy (RNAi suppression) was used by Freiman et al. (2012). Suppression of the *MdTFL1* gene using RNAi reduced the expression of the native pear genes *PcTFL1-1* and *PcTFL1-2* in pear "Spadona". Transgenic lines formed flowers after several weeks in tissue culture. Plants in the greenhouse flowered after 1–8 months, whereas non-transgenic "Spadona" plants failed to flower even after 8–10 years. The pollen of the transgenic line was fertile, and fruits were produced. The transgenic F1 seedlings formed flowers starting from 4 months after germination. The authors believe that, using this genotype, the breeding cycle of pear could be reduced to two years and it is possible to obtain four hybrid generations in just 9 years compared to the 50 years required for traditional breeding. This fast breeding method can be used to transfer disease-resistance traits from wild pear species (Freiman et al., 2012).

Recently, the *BpMADS4* gene under the CaMV 35S promoter was used to accelerate pear breeding. After the transformation of the "Conference" cultivar, some transgenic pear lines flowered *in vitro*, but none of them survived (Tomes et al., 2023). After agrobacterial inoculation, the transgenic lines flowered within 6–18 months, unlike the control, but the habit was similar to non-transgenic plants. The flowers were pollinated, and fruits were borne. About half of the seedlings formed flowers 6–8 weeks after their transfer to a greenhouse. The transfer of the same gene to the apple tree "Royal Gala" revealed differences between two fruit tree species (Tomes et al., 2023). Transgenic pear plants began to bloom later than apple, transgenic lines of which flowered 6 months after grafting. In addition, the *35S:BpMADS4* transgene caused a spindly phenotype in apples but transgenic pears showed an upright, tree-like habit. The phenotype of early flowering pear and apple seedlings also differed, suggesting different effects of *BpMADS4* on plant development processes. These results indicate that the *BpMADS4* gene can shorten the pear breeding cycles to less than 2 years and accelerate the development of new pear cultivars (Tomes et al., 2023).

14.5.1.4 Plum

The work to accelerate the breeding of stone fruits began later than with other types of fruit trees. In a report by Srinivasan et al. (2012), the *FT1* gene from *P. trichocarpa* driven by the CaMV 35S promoter was transferred to the "BlueByrd" plum. Several plantlets flowered *in vitro* but were unable to form roots. Some lines flowered within 1 to 10 months after transfer to the greenhouse, other lines formed flowers several months later. The flowers were pollinated, and fruits were borne. Seedlings also showed early flowering. Thus, these lines can be used to speed up plum breeding. Besides early flowering, transgenic plum lines demonstrated pleiotropic phenotypes: instead of a typical upright tree growth habit they demonstrated increased lateral branching in a shrub-type growth habit. Similar tree growth architecture was also reported for citrus trees over-expressing FT (Endo et al., 2005). Unlike control plums, FT plums demonstrated continuous flowering with altered dormancy requirements. Evaluation of *PtFT1* expression in leaves showed that levels of transcripts were higher in flowering lines compared to non-flowering ones. However, increased formation of lateral branches was observed both at high and low expression. To change the habit, low expression of the *PtFT1* gene is sufficient, in contrast to flowering.

Subsequently, these FT plums were used in hybridization programs called "FasTrack" breeding (Petri et al., 2018). The use of "FasTrack" breeding in a greenhouse provides for year-round flowering and fruit production. This technology was used to produce new stoneless plum cultivars (Callahan et al., 2015). The "FasTrack" system was also used to transfer the PPV resistance trait from transgenic plum 'Honeysweet' into 'French' type plums (Petri et al., 2018). The selection of seedlings in the offspring that do not contain the early flowering transgene will enable excluding the genetically modified status for them, and they could be grown as cultivars produced by traditional breeding.

14.5.1.5 Olive

Olive juvenility was dramatically shortened through overexpression of the *MtFTa1* gene from *Medicago truncatula* driven by the CAMV 35S promoter (Haberman et al., 2017). This gene was transferred into embryogenic cultures of the olive cultivar "Picual" and flowering in transgenic lines *in vitro* was observed 2–20 weeks after somatic embryo germination. Most of the vegetative buds developed into flower buds and micropropagation of transgenic olive shoots was difficult. In the greenhouse these plants produced new flowers all year-round, that is, they were no longer dependent on winter chilling, and fruiting after pollination (Haberman et al., 2017).

14.5.1.6 Kiwifruit

The genus *Actinidia* includes fruit perennial vines with features of development specific to woody species (Varkonyi-Gasic et al., 2013). The most known among them is kiwifruit (*A. chinensis*), which typically takes 5 years to first flowering under glasshouse conditions. Attempts to transfer constitutively expressed *AcFT* gene from *A. chinensis* into two species (*A. chinensis* and *A. deliciosa*) were unsuccessful but it was possible to obtain transgenic plants for *A. eriantha* known as a rich source of vitamin C (Varkonyi-Gasic et al., 2013). The *35S:AcFT* construct did not induce early flowering, but transgenic lines demonstrated reduced growth and survival rates. The levels of transgene expression in surviving lines were relatively low. Authors suggested that high expression of FT may disrupt processes of regeneration and transformation, or cause growth retardation, as has already been observed in other species (Endo et al., 2005; Kotoda et al., 2010). By contrast, the transfer of the *35S:AdCEN* construct into *A. eriantha* plants did not affect growth, but delayed flowering, that is, the effect of a *CEN* gene was opposite from that of an *FT* gene (Varkonyi-Gasic et al., 2013).

Later, *A. chinensis* "Hort16A" was transformed by two vectors, containing AcFT coding sequences under the control of the CaMV 35S promoter: the full-length and modified, in which the unusual region was removed (Moss et al., 2018). After the transformation of the AcFT construct, regeneration of flower-like initials was observed, which died within 2 months, whereas the modified AcFT construct (*pro35S:AcFT**) induced flowers *in vitro* but plantlets were also not viable.

To obtain viable precocious *Actinidia* plants, constitutive expression of a translational fusion of the *AcFT1* gene with six copies of the hemagglutinin (HA) tag sequence (Herath et al., 2023) was used recently. It has been hypothesized that chimeric FT proteins may promote flowering to a lesser extent compared to native FT (Corbesier et al. 2007). Moderate phenotypic change was observed in early flowering citrus with chimeric FT proteins (Sinn et al., 2021). Besides flowering *in vitro*, viable transgenic plants of *A. chinensis* were obtained that failed to flower in tissue culture (Herath et al., 2023). These plants developed flowers after the formation of more than 30 leaves, and this trait was inherited; in some seedlings, the flowering was observed 3–14 weeks after germination. A similar phenotype was observed when the *35S:AcFT1*-6xHA construct was transferred to *A. arguta*, which produces nutritionally rich small fruit with a thin edible skin (Latocha, 2017), whereas the transformation of *AcFT1* without the HA label led to non-viable *in vitro* flower and fruit development (Herath et al., 2023). Thus, expression of tagged *FT* gene can significantly accelerate flowering of Actinidia species while producing viable plants and makes possible a rapid-cycle breeding strategy.

14.5.1.7 Blueberry

Blueberry is a woody perennial fruit shrub that is valued for its excellent nutritional quality due to the high content of biologically active compounds. The overexpression of the blueberry *FT* gene (*VcFT*) significantly accelerates flowering in the highbush blueberry (*Vaccinium corymbosum* L.) cultivar "Aurora" (Song et al., 2013). Control blueberry plants did not bloom *in vitro*, but all transgenic lines with *35S:VcFT* produced flowers with normal morphology after culturing *in vitro* for 6 weeks. Transgenic plants flowered after 6–10 weeks in the greenhouse without any chilling treatment, which usual plants require for flowering induction (Song et al., 2013). In addition, the expression of the *VcFT* gene under the CaMV 35S promoter resulted in significantly lower height compared to control plants. A similar dwarf phenotype has been reported in FT-overexpressing fruit trees such as trifoliate orange (Endo et al., 2005) and plum (Srinivasan et al., 2012). It is possible that overexpression of the *FT* gene disturbed the control of the major upstream genes, which respond to inductive photoperiod (CO-like gene) and regulated by autonomous control and vernalization (FLC-like gene) (Song et al., 2013).

It was later shown that the frequency of adventitious shoot regeneration from leaves of the transgenic line was two times lower compared to the control, suggesting that *VcFT* overexpression in blueberry has a negative impact on shoot regeneration (Gao et al., 2016). One-year-old transgenic blueberry flowered regardless of vernalization and showed continuous flowering, while the non-transgenic plants did not have floral buds until they were three years old. Transgenic plants were two times shorter and had fewer branches than non-transgenic plants. After four years of growing without any chilling, transgenic plants died, indicating that overexpression of the *VcFT* gene is not sufficient to completely replace the vernalization in blueberry plants (Gao et al., 2016).

14.5.2 Transgrafting

Grafting is an ancient horticultural technique practiced for over 2,000 years (Baron et al., 2019). In orchards, grafting is used for control of tree size, adaptation to biotic and abiotic stresses, and improving fruit quality. Besides, fruit rootstocks can induce early flowering in scions. This effect in grafted plants is achieved through the transfer of flowering proteins from the rootstock to the scion. The apple dwarfing rootstock M9 (a spontaneous mutant overexpressing *FT* genes) induced early flowering in apple scion via upregulation of key flowering time genes including *MdFT* genes in the vasculature (Foster et al., 2014). Bennici et al. (2021) found that the flowering intensity in "Tarocco Scirè" sweet orange grafted onto "C35" citrange was associated with the upregulation of the *CiFT2* gene. An interesting method of reducing the juvenile period of fruit trees is transgrafting technology that combines transgenic rootstock and non-transgenic scion (Habibi et al., 2022). In this case, the transgenic early flowering rootstock overexpressing flowering genes is able to stimulate flowering in the scion via the movement of FT proteins across a graft junction (van Nocker & Gardiner, 2014). Rootstocks are used in horticulture significantly less than scions (cultivars), and a selected transgenic rootstock with high expression can be used as a universal donor of FT proteins for a wide range of cultivars that would not need to be genetically modified in each case. Another advantage is that the scions and their products (fruits) will remain transgene-free and overcome the regulatory restrictions for transgenic crops (Lobato-Gomez et al., 2021). The disadvantage of this method is that the transport of flowering proteins from the rootstock to the graft is not completely effective.

Studies of flowering induction in fruit trees via graft-transmitted FT proteins or RNAs from the transgenic rootstock are rather rare and have yielded contradictory results. Attempts to induce early flowering in transgrafted apple plants were unsuccessful. Grafting of the wild-type plants on an apple line with the *MdFT1* gene, flowering *in vitro*, did not induce flowering for 6 months, although the non-grafted line flowered continuously (Trankner et al., 2010). Plants grafted on the T1187 or 1190 lines with the *BpMADS4* gene did not flower for two years (Flachowsky et al., 2011). The effect of early flowering from RNAi-suppressed *MdTFL1* in transgenic apple rootstocks was not graft-transmissible in scions within 14 months after grafting (Flachowsky et al., 2012). Finally, grafting a non-transgenic apple tree onto a rootstock with the FT genes from poplar did not lead to flowering and the *PtFT1* and *PtFT2* transcripts were not transferred to the receptor scion (Wenzel et al., 2013). However, Freiman et al. (2015) showed that the chimeric PcFT2–YFP protein moved through graft junction to wild-type apple scion and potentially may function as a mobile signal for flowering.

However, grafting of sweet orange scions onto transgenic "Carrizo" citrange rootstocks containing the *CcFT3* gene driven by the phloem-specific *AtSUC2* promoter induced the formation of flower buds after 30–40 days (Soares et al., 2020). FT transcript levels in flowering scions were two-fold higher compared to shoots grafted onto the non-transgenic rootstocks. Thus, the flowering signal was transmissible through the graft junction resulting in early blooming of sweet orange scions (Soares et al., 2020). Song et al. (2019) also reported that transgenic blueberry rootstocks overexpressing *35S:VcFT* construct induced flowering of non-transgenic scions within 2 weeks after receiving full chilling. However, under non-chilling conditions, non-transgenic scions on transgenic rootstocks did not flower.

Recently, Wu et al. (2022) showed that FT proteins can be transmitted over long distances in wild-type trifoliate orange and sweet orange grafted onto transgenic trifoliate orange rootstocks expressing fusion constructs (*35S:ToFT-GFP* and *35S:AtFT-GFP*). GFP signal could be clearly detected in non-transgenic scions up to the top, 1.5 m above the graft junction. The movement of FT fusion proteins through graft unions was also detected but not the transfer of FT mRNAs. However, transgenic rootstocks did not accelerate flowering of any grafted scions within 2 years after grafting (Wu et al., 2022). The authors suggest several possible reasons for the lack of early flowering in scions. First, the reduction of FT protein amount during long-distance transport through phloem is possible. Second, the transition to the reproductive phase in trees is a very complicated process in which multiple genes are involved. For example, transported FT protein from rootstock may suppress the expression of endogenous FT proteins in scion. Third, to accelerate flowering in grafted citrus, coordinated transport of several FT proteins may be required. For example, two (*CiFT2* and *CiFT3*) of the three known FT homologs in Satsuma mandarin are involved in floral induction (Wu et al., 2022). Thus, long-distance transport of florigenic signals in woody perennials requires further investigation.

In addition, grafting can be used for studying the physiological and genetic mechanisms of flowering (Mudge et al., 2009). Of interest is the early flowering of trees, which is not directly related to flowering genes. It is known that overexpression of nitrogen metabolism genes can accelerate flowering and a transgenic line of downy birch (*B. pubescens*) with a pine glutamine synthetase gene was identified that formed inflorescences at the age of 3 years instead of 10 years in wild-type plants (Lebedev et al., 2021). Transgenic rootstocks with such genes can not only improve the nitrogen use efficiency of fruit trees in orchards but also potentially accelerate the flowering of grafted scions.

14.5.3 Genome Editing

Genome editing involves the production of precise, stable, and inheritable modifications at specific DNA locations using site-specific nucleases without leaving exogenous DNA (Salonia et al., 2020). This technique is more precise than traditional breeding or standard genetic engineering methods (Abdallah et al., 2015). Initially, ZFNs and transcription activator-like effector nucleases were used for targeting mutagenesis, but these techniques require two different artificial proteins, are expensive and time-consuming (Arencibia et al., 2019). The third nuclease, clustered regularly interspaced short palindromic repeat-associated protein 9 (CRISPR/Cas9) system, involving single-guide RNA and a nuclease protein, revolutionized genome editing due to its simplicity, versatility, design flexibility, and efficiency (Kumar & Jain, 2015). Although the CRISPR/Cas9 system is widely used for genome editing, the integration of this cassette can lead to undesirable changes (off-target effects) and plant lethality (Son & Park, 2022). In recent years, several new CRISPR-based systems have been created to improve the specificity and overcome the limitations of the CRISPR/Cas9 system including Cas12a (Cpf1), Cas12b (C2c1), and Cas13 (C2c2) (Biswas et al., 2021).

Regarding the regulatory aspects of genome-edited crops, there are two approaches: the "product-based" approach and "process-based" (Nadakuduti et al., 2018). In the United States, genome editing is considered to be a novel breeding tool that does not require regulation, as edited plants (products) do not contain any DNA sequences from other organisms and are indistinguishable from plants obtained by traditional breeding methods. In contrast, the European Union chose a "process-based" approach, the argument is that the process of genome editing is not natural, and edited plants should be regulated similarly to transgenic plants.

Genome editing has many advantages over conventional and transgenic breeding, but some features of fruit trees make the application of this technology difficult (Savadi et al., 2021). These features include heterozygosity and polyploidization of fruit tree genomes, long juvenile period, low efficiency of transformation and regeneration, and lack of well-assembled genome sequences. Nevertheless, the first genome editing of fruit trees was carried out on sweet oranges about 10 years ago (Jia & Wang, 2014), only a year later than reports on the first CRISPR-edited plants in general (Li et al., 2013; Shan et al., 2013). Since then, genome editing has been successfully applied to such trees as apple, pear, citrus species, and walnut (Savadi et al., 2021; Rukavtsova et al., 2023). The breeding cycle when using genome editing is shorter than when using genetic engineering, since biosafety evaluation of transgenic plants is not required, which for transgenic fruit trees can take several years (Savadi et al., 2021). The main editing targets were genes associated with disease resistance and carotenoid biosynthesis. However, several studies in recent years focused on the induction of early flowering.

Conditions for the application of the CRISPR/Cas9 system were optimized to knock out the *TFL1* gene on the apple "Gala" and the pear "Conference" (Charrier et al., 2019). Targeting *MdTFL1.1* gene in apple induced early flowering (1–6 months after the beginning of their micropropagation) in 93% of transgenic lines. On the contrary, the flowering of the pear transgenic lines targeted in the *PcTFL1.1* gene appeared after 4–12 months of their micropropagation, and the rate of edition was only 9%. The lower rate of mutated phenotype in pears may be due to the need to suppress both native *TFL1* genes (*PcTFL1.1* and *PcTFL1.2*) (Charrier et al., 2019).

Two kiwifruit *CEN*-like genes, *AcCEN4* and *AcCEN*, were simultaneously targeted in "Hort16A" (*A. chinensis*) plants using the CRISPR-Cas9 system (Varkonyi-Gasic et al., 2019). Unlike the transformation with the *35S:AcFT1* construct, when non-differentiated *in vitro* flowers were produced, all CEN4 lines demonstrated normal morphology *in vitro* and developed flowers 8–9 months after transformation. CEN4 lines formed flowers at the stage of six or more leaves and had a compact growth habit. The number of targeted genes and alleles generally correlated with changes in the flowering time and plant habits. Pollination enabled the production of fruits with viable seeds (Varkonyi-Gasic et al., 2019).

Genome editing was also performed on two taxonomically distant *Actinidia* species, the diploid *A. eriantha* and tetraploid *A. arguta* (kiwiberry). Bi-allelic edits in *CEN* or *CEN4* gene using a construct previously used for *A. chinensis* (Varkonyi-Gasic et al., 2019), also yielded the same effect for *A. arguta* (Herath et al., 2023). Early flowering dwarf lines were produced, in which flowers and fruits displayed normal morphology and fertility. To edit the *A. arguta* tetraploid, a new construct was developed targeting all four alleles of *AaCEN4* and *AaCEN* genes. Two different phenotypes were observed: dwarf lines with very rapid flowering at the ends of shoots, or lines with longer growth and continuous abundant flowering on lateral shoots (Herath et al., 2023). Thus, the CRISPR-Cas9-mediated editing of the *CEN* genes successfully accelerated the flowering time in various *Actinidia* species.

14.5.4 Genomics-Assisted Breeding

Genomics-assisted breeding (GAB) technology is a powerful tool in plant breeding using molecular markers for the prediction of phenotype from genotype. Novel genomics-based approaches, MAS and GS, can significantly improve breeding efficiency by the rapid selection of promising genotypes with valuable traits (Iwata et al., 2016). The concept of MAS is the existence of associations between DNA markers and genes of agronomically important traits and the use of the presence/absence of these markers as replacement/supplement of phenotype-based selection (Collard et al., 2005). The MAS is most effective for traits controlled by a limited number of quantitative trait loci with large effects and is applied for tagging of these traits in fruit and nut trees (Manzoor et al., 2023). However, flowering time is a complex trait and the role of molecular markers in fast-track breeding systems is limited to confirming the presence of disease-resistance genes and assessing the proportion of the genome of wild species in backcrosses (Schlatholter et al., 2018; Luo et al., 2020; Endo et al., 2020).

GS technology was developed to evaluate complex traits, such as yield, that are regulated by a large number of genes. Unlike MAS, which tracks limited numbers of genes with large effects, GS analyses large numbers of DNA markers to predict genomic breeding values (GEBV) of individual genotypes (Meuwissen et al., 2001). For this, genomic prediction models based on genotyping and phenotyping of a "training population" are first developed; then they are applied to a genotyped "test population" for selection parents with the highest ranked GEBV. One of the advantages of GS over conventional breeding methods is that parents for crossing can be chosen at the early stages of cultivar development (Hickey et al., 2019). GS can be combined with speed breeding (Watson et al., 2019).

GS is capable of significantly accelerating the breeding process and improving genetic gain for woody plants, mainly by reducing the evaluation under field conditions, as it allows evaluation of the genetic value of plants at a very young age (up to seedlings) (Lebedev et al., 2020). The gain in time becomes maximum for traits that are usually evaluated in mature age, e.g., fruit characteristics. The application of GS can be very useful for the selection of the best seedlings in the final generation of the breeding program (Schlatholter et al., 2018). However, compared with annual crops (Budhlakot et al., 2022; Sandhu et al., 2022) and forest trees (Lebedev et al., 2020), few GS studies have been reported in fruit tree species (Muranty et al., 2015; Imai et al.; 2019; Sun et al., 2023). Owing to the use of GS, the breeding cycle length for apple in New Zealand conditions could be reduced from 7 to 4 years (Kumar et al., 2012). Later, Kumar et al. (2020) suggested a two-stage selection of apple: foreground selection for monogenic traits using MAS and background selection of parents using GS. Propagation of prospective parents in fast-growth chambers to shorten the juvenile phase could reduce the breeding cycle of apples to two years.

14.5.5 Big Data and Artificial Intelligence

The integration of classical breeding approaches and "omics" technologies such as genomics, transcriptomics, proteomics, metabolomics, and phenomics has significantly increased the production of complex datasets (Rai, 2022). In recent years, the number of sequenced plant genomes has also grown rapidly and not only major crops but also orphan ones are already available. Processing these data, and predicting and selecting the best lines for their use in breeding programs has become a serious problem. These plant breeding problems can be solved using artificial intelligence (AI) approaches such as machine learning and deep learning (Xu et al., 2022). Machine learning algorithms are a powerful tool for analyzing and interpreting heterogeneous, large, and unstructured datasets (omics data), allowing the development of predictive models that can speed up the breeding process (Chao et al., 2023). With the help of these models, it is possible to identify genetic factors associated with the desirable traits and predict the productivity of different varieties in specific environmental conditions, increasing the efficiency of choosing parents for crossing. For example, two different algorithms of ML were used to develop models for detecting different stresses of tomato plants in response to temperature changes, low content of water and nutrients in the greenhouse (Elvanidi & Katsoulas, 2022). This is one of the newest approaches in modern plant breeding; its application on fruit trees has not yet been reported.

The use of AI is of particular interest in the processing of digital phenotyping data. Phenotyping is the measurement of any parameter of growth, development, physiology, and biochemistry of plants; in breeding, it is the basis for selecting the desired genotype from the entire set of hybrids when creating new varieties. However, one of the significant restrictions of classical breeding is its inability to establish correlations between the genotype and the phenotype (Rai, 2022). Image-based automated high-throughput phenotyping allows to replace the laborious and inconsistent method of traditional manual phenotyping, providing rapid, accurate, and non-destructive measurement of large plant populations (Kim, 2020). Despite its importance, its use in the breeding of fruit trees is rather limited. For example, it is only recently that digital phenotyping methods have been used to evaluate the fruit characteristics in apple (Keller et al., 2023) and pear (Lebedev, 2023b).

At the same time, the GS method is based on establishing a correlation between genotype and phenotype data obtained by genotyping and phenotyping the so-called "training populations" (Bhat et al., 2016). Unfortunately, many images of agricultural plants are characterized by poor visibility of interesting objects and backgrounds, which leads to an increased probability of errors during their analysis (Nabwire et al., 2021). Besides, traditional linear methods are not capable of detecting nonlinear effects and epistatic interactions that control complex quantitative phenotypes in GS (Bhat et al., 2023). Neural networks combining various AI-based algorithms were developed to improve the accuracy of these GS models (Niazian & Niedbała, 2020).

14.6 CONCLUSION

The increasing world population and global climate change demand the development of new crop varieties with high yields, enhanced resistance to stresses, and improved nutritional quality. The use of modern genomics-based approaches can accelerate plant breeding by reducing the amount of time, space, and resources, but requires preliminary studies at the molecular level. Unfortunately, fruit and nut trees are not model species and their research is limited compared to major crops due to the long juvenile period, heterozygous genome, and large size of plants. However, only new biotechnological techniques can improve fruit and nut trees, since they allow rapid, precise, and effective genetic modifications without changing the characteristics of elite cultivars.

Manipulating juvenility through transgenesis is a promising approach, but it sometimes results in abnormal growth and development. The use of inducible promoters or regulatory genes could stimulate early flowering without deleterious effects associated with constitutive overexpression of structural flowering genes. Agrobacterium-mediated transformation is unable to control the site of transgene insertion into the genome, but this problem of unintended effects can be solved using targeted genome editing approaches. In addition, genome-edited crops do not require such strict biosafety regulation as transgenic plants, and multiplex genome editing allows the pyramiding and modifications of several genes at the same time, which is especially important for perennial plants. The combination of high-throughput genotyping and phenotyping in GS programs will accelerate the identification and selection of new parents, or new candidate cultivars in hybrid populations. Finally, the use of AI is capable of processing big omics data with great accuracy, allowing us to improve our understanding of the genetics, physiology, and biochemistry of fruit trees under different environmental conditions. Using these techniques, fruit tree breeders will be able to accelerate the development, testing, and commercialization of new cultivars better adapted to changing climate, which will satisfy the ever-increasing demand for fresh and processed fruits and improve global food security.

REFERENCES

Abdallah, N. A., Prakash, C. S., & McHughen, A. G. (2015). Genome editing for crop improvement: Challenges and opportunities. *GM Crops & Food*, *6*(4), 183–205. 10.1080/21645698.2015.1129937.

Ahmad, S., Makhmale, S., Bosamia, T. C., Sangh, C., & Nawade, B. (2021). Speed breeding: A potential tool for mitigating abiotic stresses. In A.C. Rai, A. Rai , K.K. Rai , V.P. Rai , & A. Kumar (Eds.), *Stress tolerance in horticultural crops* (pp. 51–61). Woodhead Publishing. 10.1016/B978-0-12-822849-4.00012-7

Aldwinckle, H. S. (1975). Flowering of apple seedlings 16 to 20 months after germination. *Scientia Horticulturae*, *10*, 124–126.

Arencibia, A., D'Afonseca, V., Chakravarthi, M., *et al.* (2019). Learning from transgenics: Advanced gene editing technologies should also bridge the gap with traditional genetic selection. *Electronic Journal of Biotechnology*, *41*, 22–29. 10.1016/j.ejbt.2019.06.001

Asseng, S., Ewert. F., Martre, P., Rötter, R. P., Lobell, D. B., Cammaran, D., *et al.* (2015). Rising temperatures reduce global wheat production. *Nature Climate Change*, *5*, 143–147. 10.1038/NCLIMATE2470.

Baier, K., Maynard, C., & Powell, W. A. (2012). Early flowering in chestnut species induced under high intensity, high dose light in growth chambers. *Journal of The American Chestnut Foundation*, *26*, 8–10.

Ban, S., & Jung, J. H. (2023). Somatic mutations in fruit trees: Causes, detection methods, and molecular mechanisms. *Plants*, *12*, 1316. 10.3390/plants12061316

Ban, S., El-Sharkawy, I., Zhao, J., Fei, Z., & Xu, K. (2022). An apple somatic mutation of delayed fruit maturation date is primarily caused by a retrotransposon insertion-associated large deletion. *The Plant Journal*, *111*, 1609–1625. 10.1111/tpj.15911

Baron, D., Amaro, A. C. E., Pinac, A., & Ferreira, G. (2019). An overview of grafting re-establishment in woody fruit species. *Scientia Horticulturae*, *243*, 84–91. 10.1016/j.scienta.2018.08.012

Baumgartner, I. O., Patocchi, A., Franck, L., Kellerhals, M., & Broggini, G. A. L. (2011). Fire blight resistance from 'Evereste' and *Malus sieversii* used in breeding for new high quality apple cultivars: Strategies and results. *Acta Horticulturae*, *896*, 391–397. 10.17660/ActaHortic.2011.896.56

Bennett, T., & Dixon, L. E. (2021). Asymmetric expansions of FT and TFL1 lineages characterize differential evolution of the EuPEBP family in the major angiosperm lineages. *BMC Biology*, *19*, 181. 10.1186/s12915-021-01128-8

Bennici, S., Las Casas, G., Distefano, G., Gentile, A., Lana, G., Di Guardo, M., *et al.* (2021). Roostock affects floral induction in citrus engaging the expression of the FLOWERING LOCUTS T (CiFT). *Agriculturae*, *11*, 140. 10.3390/agriculture11020140

Bhat, J. A., Ali, S., Salgotra, R. K., Mir, Z. A., Dutta, S., Jadon, V., Tyagi, A., Mushtaq, M., Jain, N., Singh, P. K., Singh, G. P., & Prabhu, K. V. (2016). Genomic selection in the era of next generation sequencing for complex traits in plant breeding. *Frontiers in Genetics*, *7*, 221. 10.3389/fgene.2016.00221

Bhat, J. A., Feng, X., Mir, Z.,A., Raina, A., & Kadambot, H. M. (2023). Recent advances in artificial intelligence, mechanistic models, and speed breeding offer exciting opportunities for precise and accelerated genomics-assisted breeding. *Physiologia Plantarum*, *175*, e13969. 10.1111/ppl.13969

Biswas, S., Zhang, D., & Shi, J. (2021). CRISPR/Cas systems: Opportunities and challenges for crop breeding. *Plant Cell Reports*, *40*, 1979–1998. 10.1007/s00299-021-02708-2

Breton, C., Cornu, D., Chriqui, D., Sauvanet, A., Capelli, P., Germain, E., & Jay-Allemand, H. (2004). Somatic embryogenesis, micropropagation and plant regeneration of "Early Mature" walnut trees (*Juglans regia*) that flower in vitro. *Tree Physiology*, *24*, 425–435. 10.1093/treephys/24.4.425.

Budhlakot, I. N., Kushwaha, A. K., Rai, A., Chaturvedi, K. K., Kumar, A., Pradhan, A. K., Kumar, U., Kumar, R. R., Juliana, P., Mishra, D. C., & Kumar, S. (2022). Genomic selection: A tool for accelerating the efficiency of molecular breeding for development of climate-resilient crops. *Frontiers in Genetics*, *13*, 832153. 10.3389/fgene.2022.832153

Callahan, A. M., Srinivasan, C., Dardick, C., Scorza, R., Goldman, I. L., & Ortiz, R. (2016). Rapid cycle breeding: application of transgenic early flowering for perennial trees. *Plant Breeding Reviews*, *40*, 299–334. 10.1002/9781119279723.ch7

Callahan, A., Dardick, C., Tosetti, R., Lalli, D., & Scorza, R. (2015). 21st century approach to improving burbank's 'Stoneless' plum. *HortScience*, *50*(2), 189–194. 10.21273/HORTSCI.50.2.195

Cazzola, F., Bermejo, C. J., Gatti, I., & Cointry, E. (2021). Speed breeding in pulses: An opportunity to improve the efficiency of breeding programs. *Crop and Pasture Science*, *72*, 165–172. 10.1071/CP20462

Cervera, M., Navarro, L., & Pena, L. (2009). Gene stacking in 1-year cycling APETALA1 citrus plants for a rapid evaluation of transgenic traits in reproductive tissues. *Journal of Biotechnology*, *140*, 278–282. 10.1016/j.jbiotec.2009.01.024

Chaloner, T. M., Gurr, S. J., & Bebber, D. P. (2021). Plant pathogen infection risk tracks global crop yields under climate change. *Nature Climate Change*, *11*, 710–715. 10.1038/s41558-021-01104-8

Chao, H., Zhang, S., Hu, Y., Ni, Q., Xin, S., Zhao, L., Ivanisenko, V. A., Orlov, Y. L., & Ming, C. (2023). Integrating omics databases for enhanced crop breeding. *Journal of Integrative Bioinformatics*, *20*, 20230012. 10.1515/jib-2023-0012

Charrier, A., Vergne, E., Dousset, N., Richer, A., Petiteau, A., & Chevreau, E. (2019). Efficient targeted mutagenesis in apple and first time edition of pear using the CRISPR-Cas9 system. *Frontiers in Plant Science*, *10*, 40. 10.3389/fpls.2019.00040

Cobb, J. N., Juma, R. U., Biswas, P. S., Arbelaez, J. D., Rutkoski, J., Atlin, G., Hagen, T., Quinn, M., & Ng, E. H. (2019). Enhancing the rate of genetic gain in public-sector plant breeding programs: Lessons from the breeder's equation. *Theoretical and Applied Genetics*, *132*, 627–645. 10.1007/s00122-019-03317-0

Collard, B. C. Y., Jahufer, M. Z. Z., Brouwer, J. B., & Pang, E. C. K. (2005). An introduction to markers, quantitative trait loci (QTL) mapping and marker-assisted selection for crop improvement: The basic concepts. *Euphytica*, *142*, 169–196. 10.1007/s10681-005-1681-5

Corbesier, L., Vincent, C., Jang, S., Fornara, F., Fan, Q., Searle, I., Giakountis, A., Farrona, S., Gissot, L., Turnbull, C., & Coupland, G. (2007). FT protein movement contributes to long-distance signaling in floral induction of Arabidopsis. *Science*, *316*, 1030–1033. 10.1126/science.1141752

Dewi, P. S., Wakana, A., Tanimoto, Y., Sakai, K., & Kajiwara, K. (2013). Precocious flowering of Citrus seedlings and its use for determination of cultivars generating male sterile progenies. *Scientia Horticulturae*, *160*, 1–11. 10.1016/j.scienta.2013.05.028.

Duan, Y. X., Fan, J., & Guo, W. W. (2010). Regeneration and characterization of transgenic kumquat plants containing the *Arabidopsis APETALA1* gene. *Plant Cell, Tissue and Organ Culture*, *100*, 273–281. 10.1 007/s11240-009-9646-3

Ebrahimi, A., Zarei, A., Fardadonbeh, M. Z., & Lawson, S. (2017). Evaluation of genetic variability among "Early Mature" Juglans regia using microsatellite markers and morphological traits. *PeerJ*, *5*, e3834. 10.7717/peerj.3834

Elvanidi, A., & Katsoulas, N. (2022). Machine learning-based crop stress detection in greenhouses. *Plants*, *12*, 52–71. 10.3390/plants12010052

Endo, T., Fujii, H., Omura, M., & Takehiko Shimada, T. (2020). Fast-track breeding system to introduce CTV resistance of trifoliate orange into citrus germplasm, by integrating early flowering transgenic plants with marker-assisted selection. *BMC Plant Biology*, *20*, 224. 10.1186/s12870-020-02399-z

Endo, T., Shimada, T., Fujii, H., Kobayashi, Y., Araki, T., & Omura, M. (2005). Ectopic expression of an FT homolog from citrus confers an early flowering phenotype on trifoliate orange (*Poncirus trifoliata* L. Raf.). *Transgenic Research*, *14*, 703–712. 10.1007/s11248-005-6632-3

Endo, T., Shimada, T., Fujii, H., Nishikawa, F., Sugiyama, A., Nakano, M., Shimizu, T., Kobayashi, Y., Araki, T., Peña, L., & Omura, M. (2009). Development of a CiFT coexpression system for functional analysis of genes in citrus flowers and fruit. *Journal Japanese Society for Horticultural Science*, 7874–7883. 10.2503/jjshs1.78.74

FAOSTAT (2021) http://www.fao.org/faostat/en/. Accessed 25

Flachowsky, H., Hanke, M-V., Peil, A., Strauss, S. H., & Fladung, M. (2009). A review on transgenic approaches to accelerate breeding of woody plants. *Plant Breeding*, *128*, 217–226. 10.1111/j.1439-0523.2008.01591.x

Flachowsky, H., Hattasch, C., Hofer, M., Peil, A., & Hanke, M.-V. (2010). Overexpression of LEAFY in apple leads to a columnar phenotype with shorter internodes. *Planta*, *231*, 251–263. 10.1007/s00425-009-1041-0

Flachowsky, H., Le Roux, P.-M., Peil, A., Patocchi, A., Richter, K., & Hanke, M.-V. (2011). Applying a high-speed breeding technology to apple (*Malus domestica*) based on transgenic early flowering plants and marker-assisted selection. *New Phytologist*, *192*, 364–377. 10.1111/j.1469-8137.2011.03813.x

Flachowsky, H., Peil, A., Sopanen, T., Elo, A., & Hanke, M-V. (2007). Overexpression of BpMADS4 from silver birch (*Betula pendula* Roth.) induces early flowering in apple (*Malus domestica* Borkh.). *Plant Breeding*, *126*, 137–145. 10.1111/j.1439-0523.2007.01344.x

Flachowsky, H., Szankowski, I., Waidmann, S., Peil, A., Trankner, C., & Hanke, M. V. (2012). The MdTFL1 gene of apple (Malus × domestica Borkh.) reduces vegetative growth and generation time. *Tree Physiology*, *32*(10), 1288–1301. 10.1093/treephys/tps080

Foster, T. M., Watson, A. E., van Hooijdonk, B. M., & Schaffer, R. J. (2014). Key flowering genes including FT-like genes are upregulated in the vasculature of apple dwarfing rootstocks. *Tree Genetics & Genomes*, *10*, 189–202. 10.1007/s11295-013-0675-z

Freiman, A., Golobovitch, S., Yablovitz, Z., Belausov, E., Dahan, Y., Peer, R., Avraham, L. *et al.* (2015). Expression of FLOWERING LOCUS T2 transgene from *Pyrus communis* L. delays dormancy and leaf senescence in *Malus x domestica* Borkh, and causes early flowering in tobacco. *Plant Science*, *241*, 164–176. 10.1016/j.plantsci.2015.09.012

Freiman, A., Shlizerman, L., Golobovitch, S., Yablovits, Z., Korchinsky, R., Cohen, Y., Samach, A., Chevreau, E., Le Roux, P.-M., Patocchi, A., & Flaishman, M. A. (2012). Development of a transgenic early flowering pear (*Pyrus communis* L.) genotype by RNAi silencing of PcTFL1-1 and PcTFL1-2. *Planta*, *235*, 1239–1251. 10.1007/s00425-011-1571-0

Gao, X., Walworth, A. E., Mackie, C., & Song, G. Q. (2016). Overexpression of blueberry FLOWERING LOCUS T is associated with changes in the expression of phytohormone-related genes in blueberry plants. *Horticulture Research*, *3*, 16053. 10.1038/hortres.2016.53

Gardiner, S. E., Bus, V. G. M., Rusholme, R. L., Chagne, D., Rikkerink, E. H. A. (2007) Apple. In C. Kole (Ed.), *Genome mapping and molecular breeding in plants* (pp. 1–62). Berlin: Springer.

Germain, E., Delort, F., & Kanivets, V. (1997). Precocious maturing walnut population originating from central Asia: Their behavior in France. *Acta Horticulturae, 442*, 83–90. 10.17660/ActaHortic.1997.442.10.

Haberman, A., Bakhshian, O., Cerezo-Medina, S., *et al.* (2017). A possible role for flowering locus T-encoding genes in interpreting environmental and internal cues affecting olive (*Olea europaea* L.) flower induction. *Plant Cell Environment*, *40*, 1263–1280. 10.1111/pce.12922

Habibi, F., Liu, T., Folta, K., & Sarkhosh, A. (2022). Physiological, biochemical, and molecular aspects of grafting in fruit trees. *Horticulture Research*, *9*, uhac032. 10.1093/hortre/uhac032

Hackett, W. P. (1985). Juvenility, maturity, and rejuvenation in woody plants. *Horticultural Reviews*, *7*, 109–155 Ed. J. Janick (Hoboken, NJ, USA: John Wiley & Sons, Inc). 10.1002/9781118060735.ch3

Hammami, S. B. M., Lorenzo León, L., Rapoport, H. F., & de la Rosa, R. (2021). A new approach for early selection of short juvenile period in olive progenies. *Scientia Horticulturae*, *281*, 109993. 10.1016/j.scienta.2021.109993.

Hanke, M-V., Flachowsky, H., Peil, A., & Hättasch, C. (2007). No flower no fruit: Genetic potentials to trigger flowering in fruit trees. *Genes Genomes Genomics*, *1*, 1–20.

Herath, D., Wang, T., Voogd, C., Peng, Y., Douglas, M., Putterill, J., *et al.* (2023). Strategies for fast breeding and improvement of Actinidia species. *Horticultural Research*, *10*(3), uhad016. 10.1093/hr/uhad016

Hickey, L. T., Hafeez, A. N., Robinson, H., Jackson, S. A., Leal-Bertioli, S. C. M., Tester, M., Gao, C., Godwin, I. D., Hayes, B. J., & Wulff, B. B. H. (2019). H. Breeding crops to feed 10 billion. *Nature Biotechnology*, *37*, 744–754. 10.1038/s41587-019-0152-9

Holme, I. B., Wendt, T., & Holm, P. B. (2013). Intragenesis and cisgenesis as alternatives to transgenic crop development. *Plant Biotechnology Journal*, *11*, 395–407. 10.1111/pbi.12055

Imai, A., Kuniga, T., Yoshioka, T., Nonaka, K., Mitani, N., Fukamachi, H. *et al.* (2019) Single-step genomic prediction of fruit-quality traits using phenotypic records of non-genotyped relatives in citrus. *PLoS One*, *14*(8), e0221880. 10.1371/journal.pone.0221880

Iwata, H., Minamikawa, M. F., Kajiya-Kanegae, H., Ishimori, M., & Hayashi, T. (2016). Genomics-assisted breeding in fruit trees.*Breeding Science*, *66*, 100–115. 10.1270/jsbbs.66.100

Jia, H., & Wang, N. (2014). Targeted genome editing of sweet orange using Cas9/sgRNA. *Public Library of Science One, 9*, e93806. 10.1371/journal.pone.0093806

Jin, Q., Mo, R., Chen, W., Zhang, Q., Sheng, F., Wu, C., Zhang, R., & Luo, Z. (2022). Identification and comparative analysis of genes and microRNAs involved in the floral transition of the xinjiang early-flowering walnut (*Juglans regia* L.). *Horticulturae*, *8*, 136. 10.3390/horticulturae8020136

Jin, Q., Zhang, R., Chen, L., & Luo, Z. (2023). Isolation of *FLOWERING LOCUS C* and preliminary characterization in the floral transition of Xinjiang precocious walnut. *Horticulturae*, *9*(5), 582. 10.3390/horticulturae9050582

Jones, H. G. (1987). Repeat flowering in apple caused by water stress or defoliation. *Trees*, *1*, 135–138. 10.1007/BF00193555

Joshi, S., Soriano, J. M., Schaart, J., Broggini, G. A. L., Szankowski, I., Jacobsen, E., Krens, F., & Schouten, H. J. (2009). Approaches for development of cisgenic apples. *Transgenic Plant Journal*, *3*, 40–46.

Keller, B., Jung, M., Bühlmann-Schütz, S., Hodel, M., Studer, B., Broggini, G. A. L., & Patocchi, A. (2024). The genetic basis of apple shape and size unraveled by digital phenotyping. *G3 Genes|Genomes|Genetics*, jkae045. 10.1101/2023.08.25.554767

Khan, M. R. G., Ai, X-Y., & Zhang, J-Z. (2014). Genetic regulation of flowering time in annual and perennial plants. *Wiley Interdisciplinary Reviews: RNA*, *5*, 347–359. 10.1002/wrna.1215

Kim, J. Y. (2020). Roadmap to high throughput phenotyping for plant breeding. *Journal of Biosystems Engineering*, *45*, 43–55. 10.1007/s42853-020-00043-0

Kotoda, N., Hayashi, H., Suzuki, M., Igarashi, M., Hatsuyama, Y., Kidou, S., *et al.* (2010). Molecular characterization of FLOWERING LOCUS T-Like genes of apple (*Malus* × *domestica* Borkh.). *Plant Cell Physiology*, *51*, 561–575. 10.1093/pcp/pcq021

Kotoda, N., Iwanami, H., Takahashi, S., & Abe, K. (2006). Antisense expression of MdTFL1, a TFL1-like gene, reduces the juvenile phase in apple. *Journal of the American Society for Horticultural Science*, *131*, 74–81. 10.21273/JASHS.131.1.74

Kumar, S., Chagne, D., Bink, M. C. A. M., Volz, R. K., Whitworth, C. *et al.* (2012). Genomic Selection for Fruit Quality Traits in Apple (Malus6domestica Borkh.). *Public Library of Science One*, *7*(5), e36674. doi:10.1371/journal.pone.0036674

Kumar, S., Hilario, E., Deng, C. H., & Molloy, C. (2020). Turbocharging introgression breeding of perennial fruit crops: A case study on apple. *Horticulture Research*, *7*, 47. 10.1038/s41438-020-0270-z

Kumar, V., & Jain, M. (2015). The CRISPR–Cas system for plant genome editing: Advances and opportunities. *Journal of Experimental Botany*, *66*(1), 47–57. 10.1093/jxb/eru429

Latocha, P. (2017). The nutritional and health benefits of Kiwiberry (*Actinidia arguta*) – a review. *Plant Foods for Human Nutrition*, *72*, 325–334. doi: 10.1007/s11130-017-0637-y

Laurens, F., Aranzana, M. J., Arus, P., Bassi, D., Bink, M., Bonany, J., *et al.* (2018). An integrated approach for increasing breeding efficiency in apple and peach in Europe. *Horticultural Research*, *5*, 11. 10.1038/s41438-018-0016-3

Le Roux, P.-M., Flachowsky, H., Hanke, M.-V., Gessler, C., & Patocchi, A. (2012). Use of a transgenic early flowering approach in apple (*Malus domestica* Borkh.) to introgress fire blight resistance from 'Evereste'. *Molecular Breeding*, *30*, 857–874. 10.1007/s11032-011-9669-4

Lebedev, V. (2023a). Impact of intron and retransformation on transgene expression in leaf and fruit tissues of field-grown pear trees. *International Journal of Molecular Sciences*, *24*(16), 12883. 10.3390/ijms241612883

Lebedev, V. (2023b). Fruit characteristics of transgenic pear (*Pyrus communis* L.) trees during long-term field trials. *Plant Foods for Human Nutrition*, *78*(2), 445–451. 10.1007/s11130-023-01076-4

Lebedev, V. G., Lebedeva, T. N., Chernodubov, A. I., & Shestibratov, K. A. (2020). Genomic selection for forest tree improvement: Methods, achievements and perspectives. *Forests*, *11*(11), 1190. 10.3390/f11111190

Lebedev, V. G., Popova, A. A., & Shestibratov, K. A. (2021). Genetic engineering and genome editing for improving nitrogen use efficiency in plants. *Cells*, *10*(12), 3303. 10.3390/cells10123303

Lebedev, V. G., & Shestibratov, K. A. (2021). Genetic engineering of lignin biosynthesis in trees: Compromise between wood properties and plant viability. *Russian Journal of Plant Physiology*, *68*(4), 596–612. 10.1134/S1021443721030109

Lespinasse, Y. (2009). Review of pome fruit breeding in Europe: Which strategies for the near future? *Acta Horticulturae*, *814*, 865–872. 10.17660/ActaHortic.2009.814.147

Li, H., Rasheed, A., Hickey, L. T., & He, Z. (2018). Fast-forwarding genetic gain. *Trends in Plant Science*, *23*, 184–186. 10.1016/j.tplants.2018.01.007

Li, J. F., Norville, J. E., Aach, J., McCormack, M., Zhang, D., Bush, J., *et al.* (2013). Multiplex and homologous recombination-mediated genome editing in *Arabidopsis* and *Nicotiana benthamiana* using guide RNA and Cas9. *Nature Biotechnology*, *31*, 688–691. 10.1038/nbt.2654

Liu, Y., Ke, L., Wu, G., Xu, Y., Wu, X., Xia, R., Deng, X., & Xu, Q. (2017). miR3954 is a trigger of phasi RNA s that affects flowering time in citrus. *The Plant Journal*, *92*, 263–275. 10.1111/tpj.13650

Lobato-Gomez, M., Hewitt, S., Capell, T., Christou, P., Dhingra, A., & Giron-Calva, P. S. (2021). Transgenic and genome-edited fruits: Background, constraints, benefits, and commercial opportunities. *Horticulture Research*, *8*, 1–16. 10.1038/s41438-021-00601-3

Luo, F., Norelli, J. L., Howard, N. P., Wisniewski, M., Flachowsky, H., Hanke, M.-V., *et al.* (2020). Introgressing blue mold resistance into elite apple germplasm by rapid cycle breeding and foreground and background DNA-informed selection. *Tree Genetics and Genomes*, *16*(2), 28. 10.1007/s11295-020-1419-5

Lusser, M., Parisi, C., Plan, D., & Cerezo, E. R. (2011). New plant breeding techniques. In *State-of-the-art and prospects for commercial development, JRC scientific and technical reports* (pp. 1–220). Luxembourg: Publications Office of the European Union. 10.2791/54761

Manzoor, I., Samantara, K., Bhat, M. S., Farooq, I., Bhat, K. M., Mir, M. A., & Wani, S. H. (2023). Advances in genomics for diversity studies and trait improvement in temperate fruit and nut crops under changing climatic scenarios. *Frontiers in Plant Science*, *13*, 1048217. 10.3389/fpls.2022.1048217

Matsuda, N., Ikeda, K., Kurosaka, M., Takashina, T., Isuzugawa, K., Endo, T., & Omura, M. (2009). Early flowering phenotype in transgenic pears (*Pyrus communis* L.) expressing the CiFT gene. *Journal of the Japanese Society for Horticultural Science*, *78*, 410–416. 10.2503/jjshs1.78.410

McGarry, R. C., Klocko, A. L., Pang, M., Strauss, S. H., & Ayre, B. G. (2017) Virus-induced flowering: An application of reproductive biology to benefit plant research and breeding. *Plant Physiology*, *173*, 47–55. 10.1104/pp.16.01336

Meuwissen, T. H., Hayes, B. J., & Goddard, M. E. (2001). Prediction of total genetic value using genome-wide dense marker maps. *Genetics*, *157*(4), 1819–1829. 10.1093/genetics/157.4.1819

Moore, G. A., Rogers, K. L., Kamps, T. L., Pajon, M., Stover, J., Febres, V. J., & Stover, E. (2016). Rapid cycling plant breeding in citrus.*Citrograph*, *7*, 80–85.

Moose, S. P., & Mumm, R. H. (2008). Molecular plant breeding as the foundation for 21st century crop improvement. *Plant Physiology*, *147*, 969–977. 10.1104/pp.108.118232

Moss, S. M., Wang, T., Voogd, C., Brian, L. A., Wu, R., Hellens, R. P., & Varkonyi-Gasic, E. (2018). AcFT promotes kiwifruit in vitro flowering when overexpressed and Arabidopsis flowering when expressed in the vasculature under its own promoter. *Plant Direct*, *2*, e00068. 10.1002/pld3.68

Mudge, K., Goldschmidt, E., Scofield, S. R., & Janick, J. (2009). A history of grafting. *Horticultural Reviews*, *35*, 437–487. 10.1002/9780470593776.ch9

Muranty, H., Troggio, M., Sadok, I. *et al.* (2015). Accuracy and responses of genomic selection on key traits in apple breeding.*Horticulture Research*, *2*, 15060. 10.1038/hortres.2015.60

Nabwire, S., Suh, H. K., Kim, M. S., Baek, I., & Cho, B. K. (2021). Application of artificial intelligence in phenomics. *Sensors*, *21*(13), 4363. 10.3390/s21134363

Nadakuduti, S. S., Buell, C. R., Voytas, D. F., Starker, C. G., & Douches, D. S. (2018). Genome editing for crop improvement – applications in clonally propagated polyploids with a focus on potato (*Solanum tuberosum* L.). *Frontiers in Plant Science*, *9*, 1607. 10.3389/fpls.2018.01607

Nayak, S. N., Aravind, B., Malavalli, S. S., Sukanth, B. S., Poornima, R., Bharati, P., Hefferon, K., Kole, C., & Puppala, N. (2021) Omics technologies to enhance plant based functional foods: An overview. *Frontiers in Genetics*, *12*, 742095. 10.3389/fgene.2021.742095

Nelson, G. C., Valin, H., Sands, R. D., *et al.* (2014). Climate change effects on agriculture: Economic responses to biophysical shocks. *Proceedings of the National Academy of Sciences of the United States of America*, *111*, 3274–3279. 10.1073/pnas.1222465110

Niazian, M., & Niedbała, G. (2020). Machine learning for plant breeding and biotechnology. *Agriculture*, *10*(10), 436. 10.3390/agriculture10100436

Orbović, V., Ravanfar, S. A., Acanda, Y., Narvaez, J., Merritt, B. A., Levy, A., & Lovatt, C. J. (2021). Stress-inducible Arabidopsis thaliana RD29A promoter constitutively drives Citrus sinensis APETALA1 and LEAFY expression and precocious flowering in transgenic citrus spp. *Transgenic Research. 30*, 687–699. 10.1007/s11248-021-00260-z

Patocchi, A., Keilwagen, J., Berner, T., Wenzel, S., Broggini, G. A. L., Altschmied, L., Hanke, M-V., & Flachowsky, H. (2021). No evidence of unexpected transgenic insertions in T1190 – A transgenic apple used in rapid cycle breeding – Following whole genome sequencing. *Frontiers in Plant Science*, *12*, 715737. 10.3389/fpls.2021.715737

Peña, L., Martín-Trillo, M., Juárez, J., Pina, J. A., Navarro, L., & Martínez-Zapater, J. M. (2001). Constitutive expression of Arabidopsis LEAFY or APETALA1 genes in citrus reduces their generation time. *Nature Biotechnology*, *19*,263–267. 10.1038/85719

Petri, C., Alburquerque, N., Faize, M., Scorza, R., & Dardick, C. (2018). Current achievements and future directions in genetic engineering of European plum (*Prunus domestica* L.). *Transgenic Research*, *27*, 225–240. 10.1007/s11248-018-0072-3

Pinheiro, T. T., Figueira, A., & Latado, R. R. (2014). Early-flowering sweet orange mutant 'x11' as a model for functional genomic studies of *Citrus*. *BMC Research Notes*, *7*, 511. http://www.biomedcentral.com/1756-0500/7/511

Pons, E., Alquezar, B., Rodrıguez, A., Martorell, P., Genoves, S., Ramon, D., Rodrigo, M. J., Zacarıas, L., & Pena, L. (2014). Metabolic engineering of β-carotene in orange fruit increases its in vivo antioxidant properties. *Plant Biotechnology Journal*, *12*, 17–27. 10.1111/pbi.12112

Putterill, J., & Varkonyi-Gasic, E. (2016). FT and florigen long-distance flowering control in plants. *Current Opinion in Plant Biology*, *33*, 77–82. 10.1016/j.pbi.2016.06.008

Rai, K. K. (2022). Integrating speed breeding with artificial intelligence for developing climate-smart crops. *Molecular Biology Reports*, *49*, 11385–11402. 10.1007/s11033-022-07769-4

Rana, M. M., Takamatsu, T., Baslam, M., Kaneko, K., Itoh, K., Harada, N., Sugiyama, T., Ohnishi, T., Kinoshita, T., Takagi, H. *et al.* (2019). Salt tolerance improvement in rice through efficient SNP marker-assisted selection coupled with speed-breeding. *International Journal of Molecular Sciences*, *20*, 2585. 10.3390/ijms20102585

Rommens, C. M. (2004). All-native DNA transformation: A new approach to plant genetic engineering. *Trends in Plant Science*, *9*, 457–464. 10.1016/j.tplants.2004.07.001

Ruiz, D., García-Gómez, B. E., Egea, J., Molina, A., Martínez-Gómez, P., & Campoy, J. A. (2019). Phenotypical characterization and molecular fingerprinting of natural early-flowering mutants in apricot (*Prunus armeniaca* L.) and Japanese plum (*P. salicina* Lindl.). *Scientia Horticulturae*, *254*, 187–192. 10.1016/j.scienta.2019.05.002

Rukavtsova, E. B., Zakharchenko, N. S., Lebedev, V. G., & Shestibratov, K. A. (2023). CRISPR-Cas genome editing for horticultural crops improvement: Advantages and prospects. *Horticulturae*, *9*(1), 38. 10.3390/horticulturae9010038

Sabbadini, S., Capocasa, F., Battino, M., Mazzoni, L., & Mezzetti, B. (2021). Improved nutritional quality in fruit tree species through traditional and biotechnological approaches. *Trends in Food Science & Technology*, *117*, 125–138. 10.1016/j.tifs.2021.01.083

Salonia, F., Ciacciulli, A., Poles, L., Pappalardo, H. D., La Malfa, S., & Licciardello, C. (2020) New plant breeding techniques in Citrus for the improvement of important agronomic traits. A review. *Frontiers in Plant Science*, *11*, 1234. 10.3389/fpls.2020.01234

Samantara, K., Bohra, A., Mohapatra, S. R., Prihatini, R., Asibe, F., Singh, L., Reyes, V. P., Tiwari, A., Maurya, A. K., Croser, J. S., Wani, S. H., Siddique, K. H. M., & Varshney, R. K. (2022). Breeding more crops in less time: A perspective on speed breeding. *Biology*, *11*, 275. 10.3390/biology11020275

Sandhu, K. S., Merrick, L. F., Sankaran, S., Zhang, Z., & Carter, A. H. (2022). Prospectus of genomic selection and phenomics in cereal, legume and oilseed breeding programs. *Frontiers in Genetics*, *12*, 829131. 10.3389/fgene.2021.829131

Santos-Antunes, F., Leon, L., de la Rosa, R., Alvarado, J., Mohedo, A., Trujillo, I., & Rallo, L. (2005). The length of the juvenile period in olive as influenced by vigor of the seedlings and precocity of the parents. *Horticultural Science, 40*, 1213–1215. 10.21273/HORTSCI.40.5.1213

Sasaki, S., Yamagishi, N., & Yoshikawa, N. (2011). Efficient virus-induced gene silencing in apple, pear, and Japanese pear using Apple latent spherical virus vectors. *Plant Methods*, *7*, 15. 10.1186/1746-4811-7-15

Savadi, S., Mangalassery, S., & Sandesh, M. S. (2021). Advances in genomics and genome editing for breeding next generation of fruit and nut crops. *Genomics*, *113*, 3718–3734. 10.1016/j.ygeno.2021.09.001

Schlathölter, I., Jänsch, M., Flachowsky, H., Broggini, G. A. L., Hanke, M.-V., & Patocchi, A. (2018). Generation of advanced fire blight resistant apple (*Malus domestica*) selections of the 5th generation within seven years applying the early flowering approach. *Planta*, *247*, 1475–1488. 10.1007/s00425-018-2876-z

Schouten, H. J., Krens, F. A., & Jacobsen, E. (2006). Cisgenic plants are similar to traditionally bred plants: International regulations for genetically modified organisms should be altered to exempt cisgenesis. *EMBO Reports, 7*, 750–753. 10.1038/sj.embor.7400769

Shan, Q., Wang, Y., Li, J., Zhang, Y., Chen, K., Liang, Z., *et al.* (2013). Targeted genome modification of crop plants using a CRISPR-Cas system. *Nature Biotechnology*, *31*, 686–688. 10.1038/nbt.2650

Shimizu, T. (2022). Breeding new premium quality cultivars by citrus breeding 2.0 in Japan: An integrative approach suggested by genealogy. *Horticulturae*, *8*, 559. 10.3390/horticulturae8060559

Singh, B. K., Delgado-Baquerizo, M., Egidi, E., Guirado, E., Leach, J. E., Liu, H., & Trivedi, P. (2023). Climate change impacts on plant pathogens, food security and paths forward. *Nature Reviews Microbiology*, *21*, 640–656. 10.1038/s41579-023-00900-7

Sinn, J. P., Held, J. B., Vosburg, C., Klee, S. M., Orbovic, V., Taylor, E. L., Gottwald, T. R., Stover, E., Moore, G. A., & McNellis, T. W. (2021). Flowering locus T chimeric protein induces floral precocity in edible citrus. *Plant Biotechnology Journal*, *19*, 215–217. 10.1111/pbi.13463

Skendžić, S., Zovko, M., Živković, I. P., Lešić, V., & Lemić, D. (2021). The impact of climate change on agricultural insect pests. *Insects*, *12*, 440. 10.3390/insects12050440

Soares, J. M., Weber, K. C., Qiu, W., *et al.* (2020). The vascular targeted citrus FLOWERING LOCUS T3 gene promotes non-inductive early f lowering in transgenic Carrizo rootstocks and grafted juvenile scions. *Scientific Reports*, *10*, 21404. 10.1038/s41598-020-78417-9

Son, S., & Park, S. R. (2022). Challenges facing CRISPR/Cas9-Based genome editing in plants. *Frontiers in Plant Science*, *13*, 902413. 10.3389/fpls.2022.902413

Song, G. Q., Walworth, A., Lin, T., Chen, Q., Han, X., Zaharia, I., & Zhong, G. (2019). VcFT-induced mobile florigenic signals in transgenic and transgrafted blueberries. *Horticultural Research*, *6*, 105. 10.1038/s41438-019-0188-5

Song, G. Q., Walworth, A., Zhao, D. Y., Jiang, N., & Hancock, J. F. (2013). The Vaccinium corymbosum FLOWERING LOCUS T-like gene (VcFT): A flowering activator reverses photoperiodic and chilling requirements in blueberry. *Plant Cell Reports*, *32*, 1759–1769. 10.1007/s00299-013-1489-z

Srinivasan, C., Dardick, C., Callahan, A., & Scorza R. (2012). Plum (*Prunus domestica*) trees transformed with poplar FT1 result in altered architecture, dormancy requirement, and continuous flowering. *Public Library of Science One*, *7*, e40715. 10.1371/journal.pone.0040715

Sun, M., Zhang, M., Kumar, S., Qin, M., Liu, Y., Wang, R., Qi, K., Zhang, S., Chang, W., Li, J., & Wu, J. (2024). Genomic selection of eight fruit traits in pear. *Horticultural Plant Journal*, *10*, 318–326. 10.1016/j.hpj.2023.04.008

Tanaka, N., Ureshino, A., Shigeta, N., Mimida, N., Komori, S., Takahashi, S., Tanaka-Moriya, Y., & Wada, M. (2014). Overexpression of Arabidopsis FT gene in apple leads to perpetual flowering. *Plant Biotechnology*, *31*, 11–20. 10.5511/plantbiotechnology.13.0912a

Thompson, T. E., & Grauke, L. J. (2003). Pecan tree growth and precocity. *Journal of the American Society for Horticultural Science*, *128*, 63–66. 10.21273/JASHS.128.1.0063

Tomes, S., Gunaseelan, K., Dragulescu, M., Wang, Y-Y., Guo, L., Schaffer, R. J., & Varkonyi-Gasic, E. (2023). A MADS-box gene-induced early flowering pear (*Pyrus communis* L.) for accelerated pear breeding. *Frontiers in Plant Science*, *14*, 1235963. 10.3389/fpls.2023.1235963

Tränkner, C., Lehmann, S., Hoenicka, H., Hanke, M. V., Fladung, M., Lenhardt, D., Dunemann, F., Gau, A., Schlangen, K., Malnoy, M., *et al.* (2010). Overexpression of an FT-homologous gene of apple induces early flowering in annual and perennial plants. *Planta*, *232*, 1309–1324. 10.1007/s00425-010-1254-2

UN (2022). World Population Prospects 2022, Standard Projections, Compact File, Estimates tab, Total Population as of 1 January (thousands) column". United Nations Department of Economic and Social Affairs, Population Division. 2022.

Vahdati, K., Hassani, D., & Rezaee, R. (2014). Behavior of some early mature and dwarf Persian walnut trees in Iran. *Acta Horticulturae*, *1050*, 189–196. 10.17660/ActaHortic.2014.1050.24.

van Nocker, S., & Gardiner, S. E. (2014). Breeding better cultivars, faster: Applications of new technologies for the rapid deployment of superior horticultural tree crops. *Horticultural Research*, *1*, 14022. 10.1038/hortres.2014.22

Varkonyi-Gasic, E., Moss, S. M. A., Voogd, C., Wang, T., Putterill, J., & Hellens, R. P. (2013). Homologs of FT, CEN and FD respond to developmental and environmental signals affecting growth and flowering in the perennial vine kiwifruit. *New Phytologist*, *198*, 732–746. 10.1111/nph.12162

Varkonyi-Gasic, E., Wang, T., Voogd, C., Jeon, S., Drummond, R. S. M., Gleave, A. P., *et al.* (2019). Mutagenesis of kiwifruit CENTRORADIALIS-like genes transforms a climbing woody perennial with long juvenility and axillary flowering into a compact plant with rapid terminal flowering. *Plant Biotechnology Journal*, *17*, 869–880. 10.1111/pbi.13021

Velazquez, K., Agüero, J., Vives, M. C., Aleza, P., Pina, J. A., Moreno, P., *et al.* (2016). Precocious flowering of juvenile citrus induced by a viral vector based on Citrus leaf blotch virus: A new tool for genetics and breeding. *Plant Biotechnology Journal*, *14*, 1976–1985. 10.1111/pbi.12555

Volz, R. K., Rikkerink, E., Austin, P., Lawrence, T., & Bus, V. G. M. (2009). "Fast-breeding" in apple: A strategy to accelerate introgression of new traits into elite germplasm. *Acta Horticulturae*, *814*, 163–168. 10.17660/ActaHortic.2009.814.20

Voss-Fels, K. P., Stahl, A., & Hickey, L. T. (2019). Q&A: Modern crop breeding for future food security. *BMC Biology*, *17*, 18. 10.1186/s12915-019-0638-4

Wallace, J. G., Rodgers-Melnick, E., & Buckler, E. S. (2018). On the road to breeding 4.0: Unraveling the good, the bad, and the boring of crop quantitative genomics. *Annual Review of Genetics*, *52*, 26.1–26.24. 10.1146/annurev-genet-120116-024846

Wanga, M. A., Shimelis, H., Mashilo, J., & Laing, M. D. (2021). Opportunities and challenges of speed breeding: A review. *Plant Breeding*, *140*, 185–194. 10.1111/pbr.12909

Watson, A., Ghosh, S., Williams, M. J., Cuddy, W. S., Simmonds, J., Rey, M.-D., Hatta, M. A. M., Hinchliffe, A., Steed, A., Reynolds, D., *et al.* (2018). Speed breeding is a powerful tool to accelerate crop research and breeding. *Nature Plants*, *4*, 23–29. 10.1038/s41477-017-0083-8

Watson, A., Hickey, L. T., Christopher, J., Rutkoski, J., Poland, J., & Hayes, B. J. (2019). Multivariate genomic selection and potential of rapid indirect selection with speed breeding in spring wheat. *Crop Breeding and Genetics*, *59*, 1945–1959. 10.2135/cropsci2018.12.0757

Weigl, K., Flachowsky, H., Peil, A., & Hanke, M.-V. (2015a). Heat mediated silencing of *MdTFL1* genes in apple (*Malus × domestica*). *Plant Cell Tissue and Organ Culture*, *123*, 511–521. 10.1007/s11240-015-0855-7

Weigl, K., Wenzel, S., Flachowsky, H., Peil, A., & Hanke, M. V. (2015b). Integration of BpMADS4 on various linkage groups improves the utilization of the rapid cycle breeding system in apple. *Plant Biotechnology Journal*, *13*, 246–258. 10.1111/pbi.12267

Wenzel, S., Flachowsky, H., & Hanke, M.-V. (2013). The Fast-track breeding approach can be improved by heat-induced expression of the FLOWERING LOCUS T genes from poplar (*Populus trichocarpa*) in apple (*Malus x domestica* Borkh.). *Plant Cell, Tissue and Organ Culture*, *115*, 127–137. 10.1007/s11240-013-0346-7

World Bank (2021). World development indicators 2021. https://databank.worldbank.org/source/world-development-indicators

Wu, Y-M., Ma, Y-J., Wang, M., Zhou, H., Gan, Z-M., Zeng, R-F., Ye, L-X., Zhou, J-J., Zhang, J-Z., & Hu, C-G. (2022). Mobility of FLOWERING LOCUS T protein as a systemic signal in trifoliate orange and its low accumulation in grafted juvenile scions. *Horticulture Research*, *9*, uhac056. 10.1093/hr/uhac056

Xu, Y., Zhang, X., Li, H., Zheng, H., Zhang, J., Olsen, M. S., Varshney, R. K., Prasanna, B. M., & Qian Q. (2022). Smart breeding driven by big data, artificial intelligence, and integrated genomic-enviromic prediction. *Molecular Plant*, *15*, 1664–1695. 10.1016/j.molp.2022.09.001

Yamagishi, N., Kishigami, R., & Yoshikawa, N. (2014). Reduced generation time of apple seedlings to within a year by means of a plant virus vector: A new plant-breeding technique with no transmission of genetic modification to the next generation. *Plant Biotechnology Journal*, *12*, 60–68. 10.1111/pbi.12116

Yamagishi, N., Sasaki, S., Yamagata, K., Komori, S., Nagase, M., Wada, M., Yamamoto, T., & Yoshikawa, N. (2011). Promotion of flowering and reduction of a generation time in apple seedlings by ectopical expression of the *Arabidopsis thaliana FT* gene using the *Apple latent spherical virus* vector. *Plant Molecular Biology*, *75*, 193–204. 10.1007/s11103-010-9718-0

Yao, J.-L., Cohen, D., Van den Brink, R., & Morris, B. (1999). Assessment of expression and inheritance patterns of three transgenes with the aid of techniques for promoting rapid flowering of transgenic apple trees. *Plant Cell Reports*, *18*, 727–732. 10.1007/s002990050650

Zhang, C., Jiang, S., Tian, Y., Dong, X., Xiao, J., Lu, Y., Liang, T., Zhou, H., Xu, D., Zhang, H., Luo, M., & Xia, Z. (2023). Smart breeding driven by advances in sequencing technology. *Modern Agriculture*, *1*, 43–56. 10.1002/moda.8

Zhang, J.-Z., Li, Z.-M., Yao, J.-L., & Hu, C.-G. (2009). Identification of flowering-related genes between early flowering trifoliate orange mutant and wild-type trifoliate orange (*Poncirus trifoliata* L. Raf.) by suppression subtraction hybridization (SSH) and microarray. *Gene*, *430*, 95–104. 10.1016/j.gene.2008.09.023.

Zhu, L. H., Li, X. Y., & Welander, M. (2009). Can Arabidopsis AP1 gene shorten the juvenility of apple? *Acta Horticulturae*, *829*, 259–264. 10.17660/ActaHortic.2009.829.37

Zimmerman, R. H. (1972). Juvenility and flowering in woody plants: A review. *Horticultural Science*, *7*, 447–455.

15 Molecular, Morpho-Functional, and Agronomic Phenotyping Tools for Grapevine Selection and Breeding

J. Miguel Costa and Luísa Carvalho
LEAF – Linking Landscape, Environment, Agriculture and Food – Research Center, Associated Laboratory TERRA, Instituto Superior de Agronomia, Universidade de Lisboa, Tapada da Ajuda, Lisboa, Portugal

Rui Figueira
CIBIO/InBio, Centro de Investigação em Biodiversidade e Recursos Genéticos, Laboratório Associado, Instituto Superior de Agronomia, Universidade de Lisboa, Tapada da Ajuda, Lisboa, Portugal

CIBIO/InBio, Centro de Investigação em Biodiversidade e Recursos Genéticos, Laboratório Associado, Universidade do Porto, Campus Agrário de Vairão, Vairão, Portugal

Hiroyuki Shimono
Crop Science Laboratory, Faculty of Agriculture, Iwate University, Agri - Innovation Center, Iwate University, 3-18-8, Ueda, Morioka, Iwate, Japan

Jorge Marques da Silva
BioISI – Biosystems and Integrative Sciences Institute, Universidade de Lisboa, Faculdade de Ciências, Lisboa, Portugal

15.1 CHALLENGES OF THE WINE SECTOR: CLIMATE CHANGE AND PLANTING MATERIAL

Wine production is a global and highly competitive industry. Wine production areas span a wide geographical and climatic range, but the most important ones are located in mid-latitude regions with a typical Mediterranean-type climate, characterized by seasonal variability and stressful environments during summer, such as the south of Europe (Fraga et al., 2012). France, Italy, Spain, and Portugal are the leading EU wine producers representing altogether 53.4% of the world's production (OIV, 2023). However, viticulture in Mediterranean areas is increasingly exposed to higher air and soil temperatures and more severe water deficits coupled with more pronounced inter-annual and intra-annual climate variability (Costa et al., 2016; Fonseca et al., 2023). Climate change is adverse to berry and leaf/canopy physiology, yield, and vines' longevity and represents a

 DOI: 10.1201/b23372-15

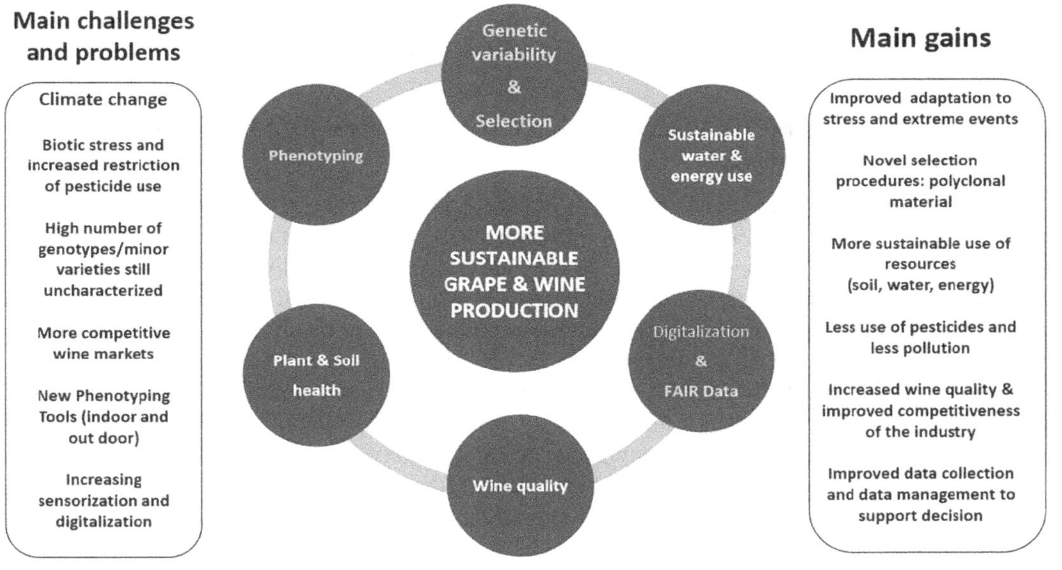

FIGURE 15.1 Diagram summarizing the main challenges and problems for the wine industry and factors involved in the sustainability and competitiveness of grape and wine production in Portugal. Plant selection and breeding have a prominent role in the possible solutions for the sector, namely in terms of climate change adaptation. Digitalization and data management are additional tools to support more precise crop and soil management as well as grapevine phenotyping and breeding (Adapted from Costa et al., 2022 and Clones4Climate Change project, LEAF – ISA, U. Lisboa).

major threat to the sustainability of wine production in the region (Carvalho et al., 2015; Dinis et al., 2018; Costa et al., 2022).

Irrigation emerged as a short-term strategy to minimize the negative impacts of climate change in Mediterranean conditions, but increased water scarcity and the need for more sustainable water use emerged as major issues for Mediterranean viticulture (Costa et al., 2016; OIV, 2021) (Figure 15.1). Therefore, complementary and/or alternative strategies to mitigate the effects of climate change are in demand in the wine industry. They encompass improved plant and soil management strategies, more precise irrigation, and very importantly, the use of better adapted plant material (varieties and rootstocks) to heat and drought (Figure 15.1).

Selection of better-adapted varieties, clones, and rootstocks, in combination with improved quality of nursery planting materials, are basic factors in optimizing the performance of vineyards (Ollat et al., 2015). The large diversity that characterizes grapevine germplasm collections is a valuable asset for breeding and selection programs, namely for resistance to biotic and abiotic stresses (Shecori et al., 2022). In addition, minority varieties are being studied and rediscovered and some may provide higher adaptation potential, namely due to their long-span ripening period and/or high berry acidity levels, which counter balance the observed decreases in acidity in wine blends due to climate change (Santesban, 2020). As a consequence, a detailed characterization of the agronomic and physiological responses of these minority varieties to heat and drought or to irrigation is paramount.

Digitalization of the agriculture sector is a rapidly evolving reality (EU, 2023; McFadden et al., 2023), and it has been driven by the convergence of agronomic, biological, informatics, and engineering research fields. Most of the remote sensing devices used in agronomical applications can be applied to plant phenotyping and selection (Costa et al., 2019; Mir et al., 2019) (Figure 15.1).

In this chapter, we review conservation and selection strategies in grapevine and the available phenotyping tools for viticulture and selection. The relevance of data management for phenotyping is discussed. Novel digital tools permit improved monitoring and the development of digital proxies of traits or improved modelling approaches for grapevine phenotyping data are addressed.

15.2 GRAPEVINE BIODIVERSITY: RELEVANCE, CONSERVATION, AND SELECTION ISSUES

There are thousands of grapevine varieties worldwide (Lacombe et al., 2013; OIV, 2017), and there is a considerable number of ancient varieties with high levels of intra-varietal diversity that can be used for breeding (Gonçalves and Martins, 2022). This large genetic diversity favors phenotypic variability of molecular, morpho-functional, and agronomic traits that are relevant for breeding and selection. For example, berry weight and berry volume can vary by up to a 23-fold factor between varieties (Houel et al., 2013), and regarding grapevine phenology, the range of the time of budburst can vary up to 39 days between early and late varieties (Boursiquot et al., 1995; Tello and Ibanez, 2023). Literature reports variation in other traits such as pollen viability (Tello et al., 2018), or photosynthesis and water use efficiency (Torregrosa et al., 2017; Bota et al. 2022; Mairata et al., 2022). However, there is still the need to better comprehend the morpho-physiological and genetic behavior to increase our understanding of the stress adaptation mechanisms limiting the development of non-empirical breeding programs for improved resistance to stress (Torregrosa et al., 2017).

A major aim for new and better-adapted varieties and clones to warmer and dryer Mediterranean climates is a longer phenological cycle with a later maturation that avoids the acute summer heat stress, protecting berry aromas and acidity. Other breeding aims include tolerance to water deficits, a balanced wine profile, satisfactory acidity, and a reasonable alcoholic degree (Ibanez et al., 2015). Resistance to biotic stress is also on demand to reduce pesticide use (Onofre, 2022; Töpfer and Trapp, 2022).

Sources of genetic variability for vine growers can range from varietal creation, such as new hybrid genotypes, to the introduction of varieties from other regions. However, one of the most straightforward strategies is to explore and use clonal variability within the varieties as it requires no regulatory modifications (Duchêne et al., 2016). Clonal differences in grapevine are due to multiple factors, including somatic mutations, epigenetic changes, and biotic determinants such as viruses (Franks et al., 2002). Clone collections are available worldwide and are used to explore phenotypic variation that can improve adaptation to climate change such as late-ripening clones or clones with a high ability to maintain berry characteristics stable under warmer conditions (Duchêne et al., 2016). In addition, grapevine diversity is a very important asset regarding resistance to pests and diseases, especially in the context of more restrictive legislation on the use of pesticides (e.g., the Green Deal in the EU) (Onofre, 2022). Grapevine diversity is crucial as well to respond to the needs of the modern wine industry concerning fast-changing consumer profiles while helping to sustain the quality, identity, and differentiation of grapes and wines (Onofre, 2022).

However, natural selection due to environmental pressures in the past, artificial selection due to human selection, and the commercial use of few varieties caused loss of genes and alleles (Grassi and Lorenzis, 2021), resulting in genetic erosion. Therefore, *Vitis vinifera* diversity must be protected, preserved, and better characterized. Part of the strategy to protect diversity relies on optimized strategies to preserve germplasm and to carry out phenotyping that can favor conservation and selection. Such monitoring methodologies can also be useful in grapevine stress monitoring and support quality control of the new planting material.

15.2.1 Conservation of Grapevine Genetic Resources: *In Situ* and *Ex Situ*

The main aim of conservation is to maintain the inter- and intra-varietal diversity in *Vitis* (Gonçalves and Martins, 2022). Conservation of *Vitis* germplasm can be achieved *in situ* (e.g., protected areas) or *ex situ* (e.g., field collections, tissue culture, or by cryopreservation) (Shecori et al., 2022). Field plantings (Figure 15.2) are commonly used as a strategy to preserve *Vitis*

FIGURE 15.2 Image of the experimental fields of PORVID (Portuguese Association for Grapevine Diversity) in Pegões (38°38'54.9"N 8°38'38.2"W; Setúbal district, Southern Portugal). PORVID has about 15 ha of vineyards devoted to clonal selection under field conditions (irrigated and non-irrigated, spur pruning, 4,000 plants/ha (1 m × 2.5 m) and 3 plants per clone) with rows with a Northeast–Southwest orientation (Carvalho et al., 2018). The experimental fields are laid out according to row-column designs with 4–6 replicates and 3–4 plants per experimental unit. Climate and soil information is available "in farm". (A) Experimental field overview by a satellite image (source Google maps); (B) UAV image showing the rows; and (C) two side-by-side rows of clones grown in field conditions.

germplasm in Portugal and other grapevine-growing countries, such as Israel, United States, Spain, Italy, and France (Shecori et al., 2022). Field plantings permit the readily available material to be used in selection and breeding programs and for further vegetative propagation (Figure 15.2). However, field collections are costly and depend on environmental conditions and hazards (such as fires or floods). Moreover, different varieties react differently to cultivation, maintenance, and environmental conditions that oblige to have fields in different regions (terroirs). This is the case of Portuguese grapevine collections managed by the Portuguese Association for Grapevine Diversity (PORVID) (Figure 15.2).

Improved preservation strategies must be considered for gene banks to keep genetic resources for future breeding programs. *In vitro* technologies are an alternative strategy, based on slow growth storage (SGS) and cryopreservation (Shecori et al., 2022). The SGS technique is an *in vitro* approach to preserving vegetatively propagated species by controlling the growth of plantlets and economizing storage space, labor, and costs (Benelli et al., 2022). Moreover, SGS prolongs the interval between subcultures, thus lowering the risk of losing germplasm through handling errors (e.g., contamination) and decreasing the risk of genetic instability by reducing the number of subcultures (Benelli et al., 2022). Cryopreservation (at −196°C) in liquid nitrogen is another possibility for long-term preservation of *Vitis* germplasm, which can work as a backup for field collections for indigenous cultivars (Bi et al., 2017). Conservation of grapevine biodiversity must involve not only investments in human resources and technologies but also close cooperation with grape growers and field technicians because many local varieties are residually cultivated in marginal agricultural areas or in a rural context endangered by different factors of biodiversity erosion (Biasi and Brunori, 2015).

15.2.2 Breeding and Selection

Breeding and selection programs in grapevine have been largely focused on agronomic traits and in obtaining new cultivars with higher yield, improved berry composition traits, and preferably more resistant to biotic and abiotic stress (Duchâne et al., 2016; Carvalho et al., 2021). To achieve this aim, two strategies are often implemented: (i) screening of existing *Vitis* diversity (Gonçalves and Martins, 2022); and (ii) creation of individuals with favorable traits (Töpfer and Trapp, 2022). The first is well accepted by the sector but the breeding of new grapevine varieties is a slow and expensive process. In grapevine, the breeding cycle length can reach 20–25 years, when accounting for phenotyping new varieties (Töpfer and Trapp, 2022). In addition, breeding involves a complex administrative process and acceptance by regulatory boards (Gonçalves and Martins, 2022), as well as by consumers, namely in what concerns the case of novel fungus-resistant varieties (Kiefer and Szolnoki, 2023). These limitations do not apply to clonal selection within an authorized variety that is well established within the sector (regulators and consumers). Consequently, grapevine selection programs in recent decades have been largely based on the clonal selection of commercial varieties (Ibanez et al., 2015; Roach et al., 2021; Mairata et al., 2022).

Early clonal selections aimed to provide virus-free material with maximum productivity but currently many new clones exist, offering a wider selection of traits, namely for sensory characteristics (e.g., Chardonay) (Roach et al., 2021). In the case of Portuguese viticulture and of autochthonous varieties, the selection of groups of clones enabled higher yields, making varieties such as the Portuguese Touriga Nacional relevant for commercial vineyards. Currently, Portuguese researchers are selecting clones of varieties more tolerant to abiotic stress to face climate change. This is the case of Aragonez (Syn. Tempranillo) (Carvalho et al., 2020), and of the Portuguese variety Arinto, which has high acidity and tolerance to abiotic stress (Carvalho et al., 2023). Studies on disease-resistant clones are also being implemented to respond to the impending need to decrease the use of pesticides in viticulture.

15.3 SCREENING AND SELECTION METHODOLOGIES IN GRAPEVINE

Selection and breeding are linked to plant phenotyping. Phenotyping can be implemented either in the field or indoor conditions (greenhouse, lab) depending on the characteristics of the plant material, the selection methods and related technologies, or, very importantly, the budget available (low cost vs high-cost approaches) (Fiorani and Schurr, 2013; Costa et al., 2019; Poorter et al., 2023). Plant and crop phenotyping involves multiple protocols and tools aiming at retrieving data on multiple plant traits, preferably in a simple, fast, and robust way. Measurements can have different resolutions in space and time and can range from quantification of molecular traits to plant population traits (Gago et al., 2015; Costa et al., 2019; Poorter et al., 2023).

Grapevine has large genetic variability that must be reflected in selection trials that, therefore, must be large, and use adequate experimental designs (Gonçalves et al., 2010). Adequate phenotyping methods must enable robust screening of plant traits while permitting monitoring for a large set of individuals, in a short period and at controlled costs (CYMMIT, 2013; Costa et al., 2019; Reynolds et al., 2019; Carvalho et al. 2021). Contact and destructive measurements are used not only in phenotyping and selection of cereal crops (CYMMYT, 2013) but also in woody species such as grapevine (Carvalho et al., 2021). However, contact and destructive techniques are unable to provide the high-throughput phenotyping (HTP) required to screen large populations. Therefore, remote sensing approaches based on imaging techniques expanded in plant science and the agro-industry as they became cheaper and more user-friendly. RGB infrared and multispectral imaging and LiDAR (light detection and ranging) can support crop monitoring and phenotyping (Costa et al., 2013; Matese et al., 2022; Moreno and Andújar, 2023; Poorter et al., 2023). In viticulture, commonly used approaches include hyper and multispectral, thermal, and RGB sensors which can be hand-held or transported in mobile ground-based platforms (Gutierrez et al., 2015; Milella et al., 2019; Matese et al., 2022; Palacios et al., 2023) or even in aerial platforms (e.g., UAVs) (Gago et al., 2015; Matese et al., 2022).

Plant phenotyping to support selection and breeding often combines several sensors, resulting in massive, complex, and non-homogeneous datasets (Carvalho et al., 2021). This emphasizes the need for multidisciplinary teams and cross-sectoral expertise, which can be a limitation for small research groups (Carvalho et al., 2021; Ferro and Catania, 2023) or small/medium-sized enterprises. Fortunately, international phenotyping networks (e.g., the International Plant Phenotyping Network – IPPN or the EMPHASIS project) permit the sharing of international phenotyping facilities and know-how. However, getting access to them is not always easy, and often they are not adequate for mature woody Mediterranean crops such as grapevine (Costa et al., 2019).

15.3.1 Controlled and Semi-Controlled Conditions

Plant physiologists and molecular biologists usually use indoor, strictly controlled, or semi-controlled environments to analyze in more detail the influence of selected variables on above- and below-ground growth and other plant responses (Watt et al., 2020). Phenotyping done under controlled or semi-controlled environmental conditions (e.g., in a phytotron or greenhouse) improves precision and repeatability as they minimize disturbances due to environmental variability (Figure 15.3). Selection or phenotyping trials under indoor platforms are often based on a "plant-to-sensor" system in which sensors remain fixed, and plants are transported manually or using a conveyor belt. This approach normally applies to small size potted plants and plants grown under uniform environmental conditions (Langstroff et al., 2022). However, the "plant-to-sensor" system may modify a plant's microclimate and influence its morpho-physiological response (e.g., leaf morphology, plant architecture, leaf gas exchange), resulting in uncontrolled errors in the characterization of the phenotype, especially at more mature phenotypical stages.

Other phenotyping strategies are based on a "sensor-to-plant" system in which sensors move near plants to be observed, minimizing the impact on plant growth and morpho-physiology and giving flexibility to the sensor's movement. Their major limitations relate to the need for hardware and software integration when collecting real-time population data and extracting phenotypic parameters during sensor movement (Bian et al., 2022), which results in particularly complex when several sensors are used. Phenotyping under controlled growing conditions enables good cultivation of plants in substrate or in hydroponics which can facilitate the use of rhizotrons coupled to imaging systems for root phenotyping (Krzyzaniak et al., 2021).

15.3.2 Field Phenotyping

Growing plants in the field are the most common in plant breeding, as a means to better quantify genotype × environment × management (G × E × M) interactions, to model growth, and to predict yield based on crop modeling (Watt et al., 2020). Field phenotyping enables work with crops/plants grown under natural environmental conditions. Important traits such as phenology, growth habits, and yield

FIGURE 15.3 Low-cost experimental setup under semi-controlled climate conditions (greenhouse) based on manual irrigation and manual plant transportation. Grapevine plants grown from cuttings of pruned wood for two months in the greenhouse were subjected to drought stress for seven days. (A) RGB image of plants in the trial; (B) thermal image of a well-watered plant; (C) thermal image of a water-stressed plant.

can only be accurately evaluated in the field (Klodt et al., 2015; Bian et al., 2022). In the case of grapevine, the common phenotyping traits include growth and vigor, yield, and berry composition.

Field phenotyping also involves important challenges that are shared with commercial vineyard production (see Gatti et al., 2022), namely in terms of soil heterogeneity (depth and fertility) and microclimatic conditions (e.g., radiation, air, and soil temperature). This requires using adequate experimental designs and optimizing the statistical setup (Gonçalves et al., 2010; Carvalho et al, 2021; Coupel-Ledru et al., 2022). Therefore, the existence of multiple field trials, located in different sites, and covering a range of natural conditions (pedoclimatological regions) would provide the most robust setup to cover the needs of crop science, phenotyping, and breeding to tackle future production challenges (Morisse et al., 2022). Moreover, variations in weather conditions occurring during field phenotyping measurements can influence the robustness of phenotype analyses. Data/image retrieval and the performance of imaging tools impact image quality (blurred images, unfocused images) and plant behavior. For example, thermal imaging measurements are not reproducible under windy or cloudy days as they modify stomatal behavior, transpiration, and leaf evaporative cooling (Costa et al., 2013). The same applies to fluctuating radiation or wind speed conditions that disturb passive-type spectroscopic measurements or leaf temperature (Costa et al., 2013; Ge et al., 2016). Another challenge for field phenotyping with imaging sensors is to distinguish between plants in the foreground and the field and other plants in the background, namely when using RGB (Klodt et al., 2015) or thermal imaging (Costa et al., 2013) although the use of artificial backgrounds or nighttime measurements helped to minimize the problem (Victorino et al., 2022; Palacios et al., 2023).

15.3.2.1 Ground-Based Measurements

Ground-based field phenotyping in grapevine must consider aspects such as canopy architecture and training systems. In grapevine, the vertical shoot position (VSP) system is dominant because it is more suitable for mechanization. As a consequence, grapevine field phenotyping must adapt to the

FIGURE 15.4 Ground-based phenotyping in an experimental vineyard of the *Vitis vinifera* variety Arinto at the Portuguese Association for Grapevine Diversity (PORVID) in Pegões. (A) Field overview; (B) surface leaf temperature measurement with an infrared thermometer; (C) measurement of reflectance indexes at the sunlit side of the canopy.

monitoring of "wall crops" (Planas et al., 2022) (see Figures 15.2 and 15.4). The VSP training system favors the development of ground-based mobile field phenotyping platforms ("phenomobiles") (Deery et al., 2014) but there is still the need to optimize the operational speed of mobile platforms, especially under irregular soil surface conditions (Anthony et al., 2014; Zhang et al., 2020), and optimize image retrieval under variable distance to the target leaves/plants, or variable radiation exposure. This applies to the different robots being developed for viticulture or grapevine phenotyping applications such as the PHENOBOT (Kicherer et al., 2015) or the VINBOT (Lopes et al., 2016).

Ground-based phenotyping can involve occasional contact and destructive measurements, as well as imaging measurements. Examples of the former are leaf gas exchange (using infrared gas analyzers or parameters), water potential, or point measurements of NDVI or leaf Chl content. Point measurement demands time and labor due to the large number of samples/measuring representative points (leaves, vines) in a vineyard. Imaging sensors can also be used in ground-based measurements, supported by either mobile or fixed platforms that can be handheld or secured by tripods or towers using different sensors (RGB, multispectral, hyperspectral, infrared, etc.) (Kicherer et al., 2015; Gutierrez et al., 2018; Milella et al., 2019; Tardaguila et al., 2021).

15.3.3 Aerial and Satellite Platforms

Aerial and satellite platforms enable data collection of the Earth's surface, atmosphere, and oceans using RGB, infrared imaging, or spectral imaging. Aerial and satellite phenotyping platforms are used as an alternative or additional option to overcome the limitations posed by ground-based platforms and to speed up the screening of a large number of plants. The major limitations to overcome include the resolution and the quality of the data retrieved (Xu et al., 2021) (Table 15.1 and Figures 15.5 and 15.6).

TABLE 15.1
Pros and Cons of Using Ground-Based, UAV, and Satellite Platforms at the Level of Acquisition and Data Processing and for a Potential Application to Crop Monitoring and Crop/Plant Phenotyping (++ Excellent, + Good, o Fair/medium, - Low) (Adapted from La Fata 2021)

		Ground-based	UAV	Satellite
Data acquisition	Areal coverage/speed	–	+	++
	Costs	++	+	+/–
	Flexibility	+	++	–
	Autonomy	**Not applied**	–	++
	Dependence on cloud cover	++	++	–
	Frequency of data retrieval	++	++	+
Raw data & data processing	Payload	**Not applied**	–	++
	Resolution	++	++	o
	Processing speed	++	–	+

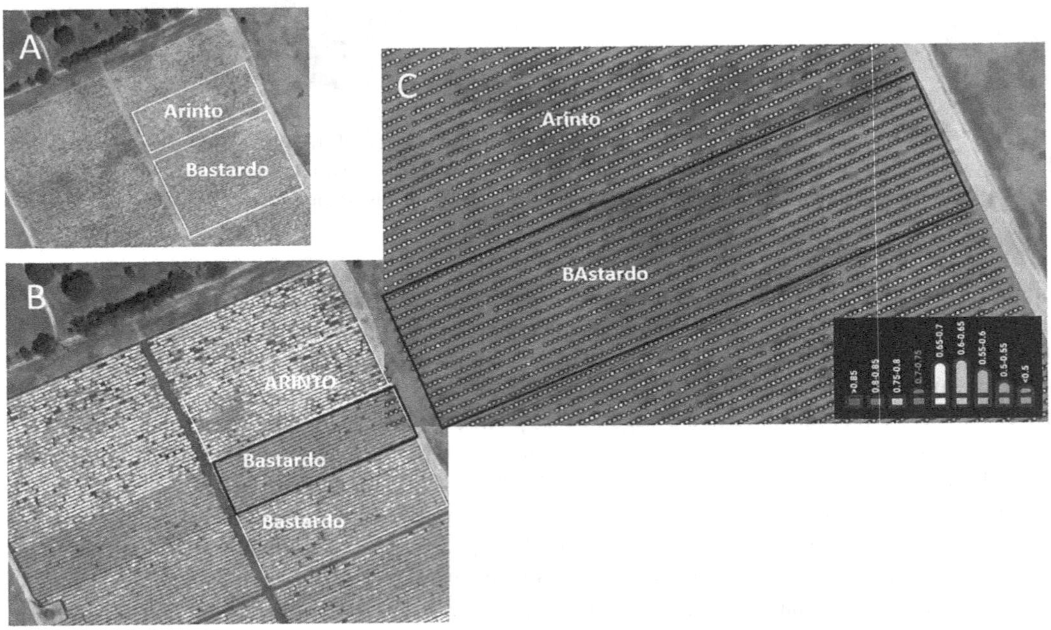

FIGURE 15.5 Aerial images taken on 14 August 2023 (fight time; 12–13 h, Tair max=32.4°C, average wind speed= 2.7 ms^{-1}, RH min= 27%). From the PORVID collection, Pegões, South Portugal. (A) RGB images showing the area occupied by *Vitis vinifera* Bastardo and Arinto clones; (B) NDVI images from different varieties and clones of grapevine and (C) processed NDVI imaging showing the individuals in the field (Ceres Imaging, USA). Images can distinguish different varieties (B, C) but they are not able to distinguish different groups of clones (three plants/group) with a basic image processing approach (C). The lowest values of NDVI observed in Bastardo are due to a green leafhopper (*Empoasca vitis* (Goethe)) attack.

Indeed, when comparing genotypes in field conditions, differences can be rather subtle, below the discrimination capacity and the efficiency of these tools for plant phenotyping. This is the case of phenotyping for drought tolerance, described for rice paddies (Madhav et al., 2013) but the same problem can be envisaged for vineyards (Figures 15.5 and 15.6). Also, the precise separation

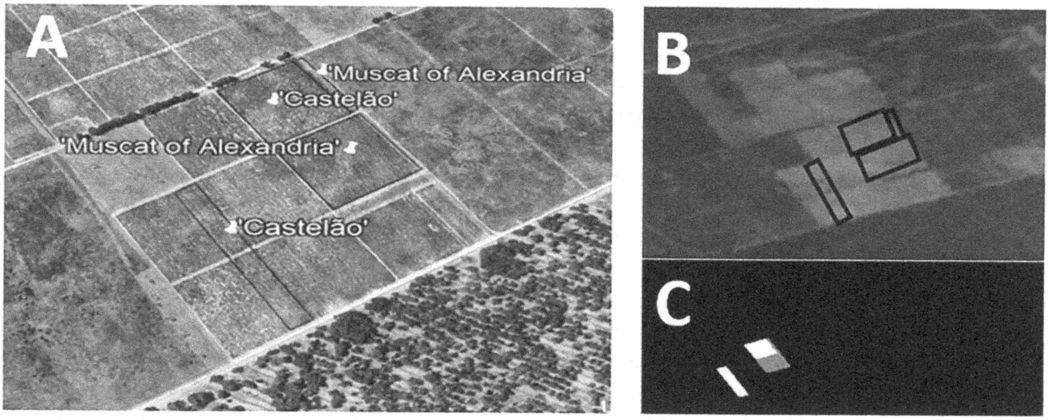

FIGURE 15.6 (A) Aerial view (Google Earth) of a vineyard located in Pegões, composed of two *Vitis vinifera* varieties with different phenology traits, "Muscat of Alexandria" and "Castelão" and (B) their original (RGB) and (C) processed Sentinel-2 imagery used to compute the index NDVI, which differed between the two varieties between July and October, but especially in August and September.

between plants and soil can pose problems, namely for aerial NDVI and thermal images. This would also require not only improved detector performance but also additional image processing for a clear separation between soil and plant data (Xu et al., 2021)

There is still the need to optimize the use of aerial and satellite imaging applied to phenotyping purposes although major progress has been already achieved, namely for stress monitoring (Araus et al., 2022). In the case of satellite imagery, data can be retrieved on crop's or plant's performance for traits such as photosynthetic activity, growth and leaf area expansion, and water status. The use of satellites has advantages and disadvantages over other aerial platforms (e.g., UAVs), and the issue of image resolution still being a major limitation (Table 15.1). The basic principle is that photosynthetic plant tissues interact with electromagnetic radiation in the ultraviolet (UV) and, more significantly, the visible and infrared range, resulting in reflectance spectra that provide information on the physiological and phenological condition of plants (Barradas et al., 2021; Cavaco et al., 2022; Matese et al., 2022). Reflectance spectroscopy is extensively used for short-range measurements, namely at the leaf level, resulting in many reflectance indices. New hyperspectral cameras permit to obtaining of spatial representations of these indices. Some of them are broadband and can be calculated from data gathered with multispectral optical devices installed on satellites. More than 100 vegetation indices (VI) based on remote sensing methods have been used in viticulture, and three major groups of VI can be applied in precision viticulture and grapevine phenotyping: (a) *plant biomass indices*; (b) *photosynthetic activity indices*; and (c) *water stress indices* (Giovos et al., 2021). Some examples of VI used in precision viticulture and potential application to HTP are the following:

a. **Plant biomass indices:** The most frequently calculated index from satellite imagery is NDVI (Normalized Difference Vegetation Index). This index indicates the presence of chlorophyll and it is used to measure the density of the vegetation cover. Sentinel-2 imagery is frequently used to map NDVI in vineyards. For example, NDVI was used to monitor the effects of stress (e.g., heatwaves). Reluy et al. (2022) used Sentinel-2 imagery to calculate NDVI in a vineyard in southern France, immediately before and after a heatwave (intra-annual comparison) and in years where heatwaves occurred or were absent (inter-annual comparison), concluding that only severe leaf damage was detected, and a combination of inter-annual and intra-annual data provided the best results. Satellite imagery based on NDVI can also be used to monitor grapevine phenology (Cunha et al., 2010) and to estimate grapevine's leaf area. Johnston et al. (2003) showed that the Leaf

Area Index measured by satellite imagery is highly correlated with the same parameter measured with ground-based techniques.

Vineyard spatiotemporal variability was accessed by Pastonchi et al. (2020) and Sozzi et al. (2019) comparing NDVI computed from Sentinel-2 and unmanned aerial vehicle imagery and concluding that they led to similar results. Also, Sams et al. (2022) concluded that NDVI computed from satellite imagery (Landsat 8 and Sentinel-2) and a manned aircraft led to similar yield estimation. Vélez et al. (2021) reported that they were able to map the "Appellation of Origin" areas and terroir characteristics with NDVI computed from Landsat 8 imagery. On the contrary, Khaliq et al. (2019) compared NDVI computed from Sentinel-2 and from an unmanned aerial vehicle imagery, reaching opposite conclusions, as they concluded that grapevine NDVI obtained from Sentinel-2 is confounded by the NDVI retrieved from the inter-row vegetation. Other studies, in table grape concluded that proximal sensing was more accurate in terms of yield and quality characteristics as compared to Landsat 8 sensing (Anastasiou et al., 2018; Kasimati et al., 2021). Furthermore, the authors observed that an alternative vegetation index, Green Normalized Difference Vegetation Index (GNDVI), was more accurate than NDVI. NDVI strongly correlates with vines' vigor and can be a relevant trait for phenotyping and selection. However, the open access/free satellite imagery of Sentinel-2 has very limited resolution (maximal pixel resolution 10 m) to phenotype individual plants, as it is required in HTP for breeding programs (Table 15.1). Such image resolution may be provided by using unmanned aerial vehicles (UAVs) (Matese et al., 2022) or by commercial high-resolution satellites, such as GeoEye-1, capable of acquiring panchromatic images with 0.41 m resolution and multispectral images with 1.65 m resolution (Pengra et al., 2015).

b. **Photosynthetic activity indices:** The Photosynthetic Reflectance Index (PRI) developed by Gamon (1992) is the most used photosynthetic activity index. Short-term (daily) PRI variations reflect the xanthophyll cycle, a photoprotection process based on the de-epoxidation of violaxanthin into zeaxanthin under excess light, and the reverse epoxidation reaction under limiting light. Violaxanthin is a photochemically active xanthophyll, while zeaxanthin dissipates absorbed light energy in the form of heat. This dissipation protects the photosynthetic apparatus from photooxidation but decreases photosynthetic efficiency (Chen and Cheng, 2003). Long-term changes in chlorophyll and carotenoid content during senescence processes or in response to environmental stresses are documented in different plant groups (e.g., Araújo et al. 2013; Cruz de Carvalho et al., 2019). Long-term (seasonal) PRI fluctuations reflect variations in the ratio of carotenoids to chlorophylls. Carotenoids have a photoprotective function and consequently, PRI is interpreted as an index of stress or vegetation health. PRI has been measured in grapevines but only with proximal imaging systems (González-Flor et al., 2019), where it adequately predicted diurnal dynamics of stomatal aperture and transpiration, paving the way to be used in Vitis HTP-based satellite imagery, when images with enough resolution became available.

c. **Water stress indices:** An index based on the R970/R900 ratio (WI, water index) was first proposed by Penuelas et al. (1997), and three years later a related broadband index, possibly the most used today, the Normalized Difference Water Index (NDWI), was proposed by Gao (1996). In precision viticulture, several attempts were made to estimate vineyards water stress using VI, measured by proximal methods, namely the WI and the PRI (González-Flor et al. 2019). NDWI has been used, together with NDVI, to determine the optimal date of application of N fertilizer to a vineyard in Sicilia, showing the usefulness of these indices in decision support systems (Comparetti and Marques da Silva, 2022). However, satellite imagery-derived water indices, so far, haven't been used for grapevine HTP, but they might prove useful when higher-resolution satellite optics become available and/or cheaper.

d. **Other image processing approaches:** The steep improvement in computing power and the development of artificial intelligence paved the way to alternative uses, non-based on VI, of satellite imagery in precision agriculture and precision viticulture (Romero et al., 2023). Karakizi et al. (2016), for instance, successfully used a technique of object-based image analysis to detect vineyards on the landscape and identify the grapevine variety. Recent commercial applications include water stress risk estimation, carbon balance potentials, and ecosystem management with nature and biodiversity conservation indicators (Araújo et al., 2023). At present, VI derived from satellite imagery is not used in HTP in grapevine but these approaches might be extended, in the future, for *Vitis* HTP, after being further studied and optimized.

15.4 MOLECULAR AND GENETIC TOOLS APPLIED TO GRAPEVINE SELECTION AND BREEDING

Molecular tools are available in *Vitis* at two levels with different rates of success and development. The first, well-established and widespread is the use of DNA markers to identify genotypes. Simple sequence repeats (SSRs, or microsatellites) are used to identify grapevine varieties, with several international databases available for the effect (Pl@ntGrape, *le catalogue des vignes cultivées en France*, http://plantgrape.plantnet-project.org/, Swiss *Vitis* microsatellite database). The most widespread database is the International Variety Catalogue (VIVC, https://www.vivc.de/), which offers a set of nine SSR *loci* that are used as a reference to identify a large set of varieties that can be combined with four or six more to identify a larger number of varieties (respectively, Dong et al., 2018; Cretazzo et al., 2022). This method also enables to identify previously undescribed varieties and new crosses between varieties (Cretazzo et al., 2022; Gonçalves et al, 2023). The nine VIVC Genres081/GrapeGen06 nuclear SSRs are used in National Variety Catalogues in European Countries to describe their local varieties. Recent breakthroughs in grapevine genotyping enabled to identify 46 grapevine clones of four major varieties (Cabernet Sauvignon, Chardonnay, Merlot, and Sauvignon blanc) using short-read sequencing technology (Urra et al., 2023).

Vitis vinifera varieties are highly susceptible to diseases, such as powdery or downy mildew, and the use of back crossings between these varieties and resistant genotypes of other species of the genus *Vitis* can lead to the production of new genotypes that ideally would maintain the traditional variety's characteristics while displaying resistance to biotic stress, thus enabling decreases in the use of pharmaceuticals for plant protection (Wagner et al., 2023). This method is, however, time-consuming and the expected results are often difficult to achieve.

From the beginning of the 21st century, marker-assisted selection (MAS) enabled to improve and speed up the selection process for some traits, namely in breeding toward resistance against powdery and downy mildew (Dalbó et al., 2001; Pauquet et al., 2001). This tool enables fast and efficient selection of resistance *loci* and although at the start it was mostly used for biotic stress resistance, it is now common for several other traits. Traditionally, one common problem of the use of resistant inter-species hybrids is a loss of aromas and flavors when compared to the *V. vinifera* counterparts. Regarding this, MAS also offered a solution, by developing markers for specific aromas, that can be used to screen resistant hybrids with favorable aromas such as the muscat-flavour-related marker targeting the *VvDXS* gene (Emanuelli et al., 2014).

One major trait in dried and table grapes is seedlessness. Traditionally, this trait was obtained through parthenocarpy or cross-breeding using one seedless parent. However, the latter is time-consuming and the former gives rise to small berries that are not highly valued in the markets. Therefore, MAS emerged as an option, and Linkage groups coupled to berry size and seedlessness were identified from quantitative trait loci (QTL) maps deriving from many seedless parental combinations (reviewed by Akkurt et al., 2019).

A cheaper, easier to use, and enabling high-throughput detection single-nucleotide polymorphism (SNP) marker based on those QTL results was developed for seedless selection, the marker p3_*VvAGL11* (Mejía et al., 2011). Recently, however, this marker has been described as causing too many false positives (Ocarez et al., 2020; Fu-Qiang et al., 2023), which may be due to the intrinsic defects of the marker itself, a yet undescribed and complex genetic architecture, or recombination events between the marker and the causal SNP in the coding region (Ocarez et al., 2020), and better ones have been developed such as the SNP site Chr18: 26889437 (A/C) (Fu-Qiang et al., 2023), or the e7_VviAGL11 SNP-based marker (Ocarez et al., 2020), both in the *VviAGL11* gene.

Significant improvements were made upon the development of efficient gene transfer protocols that enable the overexpression of plant genes that can limit pathogen growth (e.g., the overexpression of serotonin N-acetyltransferase, *VvSNAT2,* Yu et al., 2019), the silencing of plant susceptibility genes (e.g., the knockdown of the downy mildew susceptibility gene *VviLBDIf7* in Pinot Noir using RNAi, Marcianò et al., 2021), or the downregulation of pathogen virulence effector genes (these were reviewed in detail by Capriotti et al., 2020). The development and application of such methods have been aided by the use of SSR (simple sequence repeats), CAPS (cleaved amplified polymorphic sequences), and SCAR (sequence-characterized amplified region) in species such as *V. riparia*, *V. rupestris*, and *V. berlandieri* that harbor many resistance genes. These markers enable the screening of large germplasm collections for those traits (Prazzoli et al., 2019).

The elusive, hard-to-define, and quantified definition of "tolerance to abiotic stress" involves a plethora of genes and pathways that make adequate markers of tolerance hard to find and use. Nevertheless, several individual genes were targeted to enhance tolerance to several abiotic stresses, such as the NAC transcription factor (*VvNAC17*), whose overexpression increased tolerance to salinity, cold, and drought stresses in transgenic plants of *Arabidopsis* (Ju et al., 2020), or N-acetylserotonin methyltransferase (*ASMT*), whose overexpression in tobacco plants increased tolerance to salt and osmotic stresses (Yu et al., 2022), or of the Aux/IAA transcription factor gene *VvIAA18,* that was also overexpressed in tobacco plants, increasing their levels of tolerance to salt stress (Li et al., 2020). It is noteworthy to emphasize that, so far, none of these constructs was done in grapevine, and therefore its response to the manipulation of those genes is still unknown.

Changing berry characteristics within a variety to accommodate wine makers' needs and consumers' demands, when achieved through conventional selection, is time-consuming and laborious, but it is a well-proven method, with hundreds of commercially available clones of many varieties (Gonçalves and Martins, 2022). For faster and more targeted changes in aroma profiles, several attempts were made, to change patterns of gene expression as was the case of the three similar chalcone synthase genes catalyzing the first step of the flavonoid pathway (*VvCHS1,2,3*) (Harris et al., 2013). When their expression patterns in the biosynthesis of different flavonoid compounds were established, the authors proposed that their differential regulation by MYB transcription factors may enable to alteration of the flavonoid profile in grapes (Harris et al., 2013). Indeed, this was achieved, but via manipulation of the expression of the genes encoding flavonoid 3'-hydroxylase (F3' H) and flavonoid 3' 5'-hydroxylase (F3' 5' H), which provided the potential to manipulate flavonoid composition in Shiraz grapevines (Robinson et al., 2019). The authors also altered the expression of the tannin synthesis genes anthocyanidin reductase (ANR) and leucoanthocyanidin reductase (LAR) in the same variety, further enabling the ability to manipulate wine color and tannin properties (Robinson et al., 2021). Recently, metabolomic approaches were used to profile wine quality traits and to identify their genetic basis in inter-specific hybrids of *V. aestivalis*-derived cv. "Norton" × *V. vinifera* cv. "C. Sauvignon" to enable the full characterization of aromas of the whole hybrid population (Awale et al., 2022).

15.5 DATA MANAGEMENT AND FAIR DATA IN GRAPEVINE PHENOTYPING

15.5.1 Challenges in Data Collection and Analysis

Plant and crop phenotyping and selection and breeding trials involve the acquisition of a high amount of data. Data can be retrieved from multiple experimental controlled or field conditions, using specific protocols and multiple experimental parameters, methodologies, and equipment, and either by hand measurements or high-throughput means (Costa et al., 2019; Pommier et al., 2019). All data components of a particular experimental assay or field survey often result in heterogeneous and complicated datasets, which need to be adequately recorded in tables or other data structures, with well-defined schemas and sufficient documentation as metadata. This can be managed through planning, before the data capture stage: defining table structure, data types, data precision, definition of identifiers, table relations, and data description. All these components are critical to ensure data quality, fit for use, and proper documentation, potentiating any future use and reuse of data.

Data, unlike other assets, do not lose value when reused. In consequence, data is often used and reused, not only for the initial purpose it was created but also in different situations and fields of application. Furthermore, different platforms create datasets for different scopes (taxonomic, genetic, trait, morphological, chemical, meteorological, pedological), which need to be aggregated to support applications that reflect biological processes. This data aggregation is, however, not trivial and can be a big challenge, because often data terms and concepts are defined with the scope of a domain. When changing of domain, the concept associated with a term might change, even if the term is the same. To avoid this, the development or use of controlled vocabularies and ontologies (e.g., Crop Ontology, AGROVOC) is essential, although this is seen many times as an unnecessary level of complexity, expertise, and resource intensiveness.

In addition, the completeness of data is a critical data quality component that can limit data aggregation. For example, when depositing genetic sequences in GenBank, metadata should be filled to include sampling and specimen data of the living organism that provided that sequence. However, this metadata is filled in depending on the willingness of the authors. Not providing that metadata will restrict possibilities in including that genetic diversity in further analyses and data streams involving taxonomy, agroecology, or other applications at the population level. It will hamper, in addition, further studies (e.g., morphological or trait analysis) on the biological material, if the link between the sequence and the physical object is lost due to poor metadata documentation.

The "discoverability" is essential to enable the aggregation of relevant data to analyze a specific research problem. For example, it is important that all of the relevant data related to phenotyping observations of a plant species can be identified and collected for analysis, regardless of that data being scattered through assay outputs of different levels (biophysical, morphological, physiological, biochemical) and scales (cell, organ, plant, ecosystem). The adoption and implementation of the FAIR data principles (Wilkinson et al., 2016) in data management would maximize the potential of observations and records being used when relevant.

The FAIR acronym stands for Findable, Accessible, Interoperable, and Reusable, a set of principles that establish technical and data governance best practices that ensure discoverability and reuse of data. Questions related to persistent identifiers, data description through metadata, and data and metadata indexing in searchable resources all relate to the *findable* principle. To ensure *access*, data should be retrievable by its identifier through open and free protocols implemented universally that manage authentication and authorization. Even if data becomes not available anymore, metadata should be always accessible. A critical aspect is the *interoperable* property that data should enable, to allow aggregation of data from different domains or sources. This requires data and metadata to use vocabularies and ontologies, in support of their data schemas, and also the identification of references to other data based on links between ontologies. All of these previous principles contribute to promoting the final *reuse* goal, which adds principles related to the rich description of attributes through metadata, the necessity of identifying a data usage license, and the provenance and being attached to data standards relevant to the community. The implementation of these principles is not trivial and requires high levels of coordination, planning, and collaboration, not only at organizations but also and mainly at the level of the communities.

Data management and analysis can generate heavy computational tasks. Luckily, in the last decades, we have observed a rapid development of computational capacity, not only at the desktop level but also in servers, clusters, and cloud services. Distributed computing and high-performance computing are also becoming available through funding from national or international research funds. This means that the solution to complex computational problems is now within the reach of researchers and small organizations, which can use shared resources made available by commercial or research infrastructure providers, at accessible cost levels. In this context, Research Infrastructures (RI) can play a more central role, promoting coordination and best principles within the community, access to shared data platforms and resources, and fostering innovation. One of these RI at the European level is EMPHASIS (https://emphasis.plant-phenotyping.eu/), which results from the investment of several European countries with the goal of enabling facilities, resources, and services for plant phenotyping across Europe. This level of coordination and focus can facilitate links to other thematics of computing infrastructures – genetic sequence data, biodiversity and ecosystem data, food and agriculture – making possible to find solutions based on phenotyping to help solve current challenges of viticulture.

15.5.2 Modelling and Phenotyping Applied to Crops and Grapevine

Crop phenotyping and modeling are closely related. Genetic improvement of plant performance will be an important solution to increase the productivity and quality of agricultural products, particularly in the context of ongoing climate change. Recent advances in genotyping technologies have improved our ability to screen candidate high-yielding cultivars faster and more efficiently using methods such as genome-wide association studies and genomic prediction (Millet et al., 2019; de los Campos et al., 2020; Keller et al., 2020; Robert et al., 2020). Although genotyping a given cultivar is now easier and less expensive than ever before using genome-wide DNA sequencing, phenotypic analyses based on plant yield must be repeatedly evaluated based on data from season-long field trials across a range of environments. The yield of a given cultivar is not stable across environments since a plant's yield results from the interactions among many physiological processes that respond to environmental fluctuations throughout the growing season, and yield is also affected by management practices (M). Thus, strong genotype by environment interactions (G × E), genotype by management interactions (G × M), and three-way (G × E × M) interactions make it difficult to predict plant yields across a range of environments and management regimes. Compared to annual crop species such as wheat (*Triticum aestivum*), soybean (*Glycine max*), maize (*Zea mays*), and rice (*Oryza sativa*), perennial plant species such as grapevine (*Vitis vinifera*) are more challenging, since they can be affected by carryover effects of the environment and management regime in previous years in addition to current-year effects.

To accelerate plant breeding, weather-driven growth models would be powerful tools. Such models can predict plant yield by integrating mathematical descriptions of plant physiological processes in response to changes in their environments and management practices throughout the cropping cycle. In the models, cultivar characteristics are expressed by using cultivar-specific parameters in the mathematical equations that describe a plant's physiological processes. The parameters can be used for genomic analysis of yield (Chapman et al., 2003; Chenu et al., 2009; Jones et al., 2019; Kadam et al., 2019; Jighly et al., 2023; Shimono et al., 2023), leaf expansion (Reymond et al., 2003), and flowering time (Nakagawa et al., 2005; Uptmoor et al., 2008). Furthermore, a growth model that incorporates genomic information such as DNA markers and SNPs can predict the performance of new cultivars in environments where they have not been previously tested (Kadam et al., 2019; Jighly et al., 2023; Shimono et al., 2023). This can provide opportunities to virtually (*in silico*) design the genome of plants so they are better adapted to target environments such as different regions and also to future climate scenarios.

Relatively few mathematical models are currently available for grapevine, an economically important perennial plant species. Several models of grapevine have been developed to evaluate the effects of environments (Bindi, 1997; Schmidt et al., 2019), water management (Nogueira Junior et al., 2018; Zhu et al., 2018; Zhu et al., 2021; Kang et al., 2023), sink-source balance (Cola et al., 2014), cultivar difference (Zhu et al., 2021), and disease (Brischetto et al., 2021). These models can be used with the abovementioned genetic techniques to support the use of genomic data, and this will help to accelerate grapevine breeding and mitigate its response to ongoing climate change, which threatens future plant growth, yield, and quality.

During previous grapevine breeding programs, large phenotyping datasets for yield and quality have been accumulated for crops grown under a diverse set of growing conditions at different locations, in different years, and with different cultivars and management practices. These datasets can be reused (upcycled) to increase breeding efficiency by calculating the genotypic coefficients for yield and its plasticity (Shimono et al., 2023). The reuse of these data for yield phenotyping has great potential to help breeders identify promising varieties/genotypes that will permit them to boost grapevine yield and quality and maintain these improvements despite climate change.

15.6 CONCLUSIONS AND PROSPECTS

Grapevine is a worldwide economically relevant crop, and the wine industry is largely dependent on its capacity for adaptation to climate change, and on new tools to support higher efficiency and sustainability of the supply chain (Costa et al., 2022). The use of superior plant material is a suitable approach to increase the resilience and competitiveness of the global wine business. Improved agronomic and morpho-physiological characterization of novel varieties, clones, or ancient autochthonous varieties and the increased availability of varietal-specific or clone-specific data will support researchers, breeders, and viticulturists in making the wine sector better prepared to face extreme climate conditions and to satisfy more demanding and diverse consumers.

Grapevine selection and breeding are major strategies to improve grapevine plant material and more effective grapevine phenotyping is a step toward that improvement. This implies optimizing field, greenhouse, and phytotron phenotyping approaches and technologies while developing novel solutions regarding data analysis and processing. New developments will encompass increased digitalization and automatization of procedures in both controlled and field conditions. In the case of field phenotyping, more robust remote sensing technologies (e.g., RGB, thermal or multispectral imaging) for both ground-based and aerial imaging systems are expected. However, there are still challenges posed by accuracy, interoperability, data storage, computation power, and financial costs that hinder the digital transformation in grapevine phenotyping and breeding. Development of novel phenotyping platforms must also take into account the costs of the equipment and its maintenance, and up-gradings (e.g., software, detectors) (Poorter et al., 2023). At present, we are still studying the best-performing sensing approaches, but the excessive number of sensors should be avoided as redundancy.

Aerial and satellite imagery for HTP in grapevine should also be considered and optimized for practical application by the academy and the industry. The future of HTP depends on increasing spatial-temporal resolution, turnaround time in data analysis, sensor integration, human-machine interaction, throughput, operational stability, operability, automation, and accessibility (Gill et al., 2022) Detectors and related image retrieval and processing must be optimized (see Romero et al., 2023).

The use of image-based optical devices to phenotype plants had a very positive impact on plant breeding as well as on precision agriculture/viticulture. However, these devices, especially when based on mobile platforms for field phenotyping, have limited spatial resolution and work speed. This can be overcome by using plant wearables that can permanently capture individual plant data (Zhang et al., 2023). The publisher Frontiers and the World Economic Forum have co-published the "Top 10 Emerging Technologies for 2023" report, where wearable plant sensors were included. Given their positioning on a plant's surface, plant wearables can provide information about the state of their internal condition/environment and about the plant's microclimate. Wearable sensors were used to assess not only the elongation rate, leaf temperature, hydration status, bioelectric potential, and stress response but also air temperature, humidity, and light in the plant's microenvironment as well as the presence of pesticides and toxic gases (Zhang et al., 2023).

Radio fequency identification (RFID) technology has been used in various commercial and industrial sectors for a long, including the logistics chain for agricultural products, guaranteeing their traceability (Luvisi et al., 2010). RFID tagging of living organisms began with pets and livestock and was only partially extended to plants. The first electronic systems to label individual plants were external, usually plastic strips, which, although useful, presented the risk of getting damaged or lost (Kumagai and Miller, 2006). For this reason, systems are being developed for the internal placement of RFID devices in plants. In species that support grafting well, the only ones studied so far, the internal placement of RFID microchips does not seem to cause measurable physiological problems. This technology was successfully tested in vineyards in Italy. After the microchips were implanted, only 0.2% turned out to be inoperable due to damage in the process. After two years of commercial viticulture, 100% of the successfully implanted microchips remained operational (Luvisi et al., 2010).

Finally, we should also envisage the development of artificial intelligence applications for plant and crop phenotyping together with modeling approaches or the development of digital twins (Purcell et al., 2023). The future of grapevine phenotyping and breeding will encompass an increased capacity to use available phenotyping data to carry out large meta-analyses to help decipher the genetic architecture of traits across environments. In this context, sustainable long-term data management following the FAIR principles will be one of the major challenges of modern phenotyping and breeding applied to grapevine and other crops.

REFERENCES

Akkurt, M., Tahmaz, H., & Veziroğlu, S. (2019). Recent developments in seedless grapevine breeding. *South African Journal of Enology and Viticulture*, *40*(2). 10.21548/42-2-3342

Anastasiou, E., Balafoutis, A., Darra, N., Psiroukis, V., Biniari, A., Xanthopoulos, G., & Fountas, S. (2018). Satellite and proximal sensing to estimate the yield and quality of table grapes. *Agriculture*, *8*(6), 94. 10.3390/agriculture8070094

Anthony, D., Elbaum, S., Lorenz, A., & Detweiler, C. (2014). On crop height estimation with UAVs. *IEEE/RSJ International Conference on Intelligent Robots and Systems*, Chicago, IL, USA, pp. 4805–4812. 10.1109/IROS.2014.6943245

Araújo, J., Pimenta, V., Campos, J., Pinheiro, P., Porto, J. V., Manso, J., Fontes, N., & Graça, A. (2023). Innovation co-development for viticulture and enology: Novel tele-detection web-service fuses vineyard data. 43rd World Congress of Vine and Wine, BIO Web of Conferences, *56*, 01006. 10.1051/bioconf/20235601006

Araújo, S., Duque, A., Marques da Silva, J., Santos, D., Bernardes da Silva, A., & Fevereiro, P. (2013). Water deficit and recovery response of *Medicago truncatula* plants expressing the ELIP-like DSP22. *Biologia Plantarum*, *57*, 159–163. 10.1007/s10535-012-0235-7

Awale, M., Liu, C., & Kwasniewski, M. T. (2022). Generating novel aroma phenotypes using commercial wine samples to characterize an F1 population. *Frontiers in Plant Science*, *13*, 894492. 10.3389/fpls.2022.894492

Araus, Jose Luis, Kefauver, Shawn Carlisle, Vergara-Díaz, Omar, Gracia-Romero, Adrian, Rezzouk, Fatima Zahra, Segarra, Joel, Buchaillot, Maria Luisa, Chang-Espino, Melissa, Vatter, Thomas, Sanchez-Bragado, Rut, Fernandez-Gallego, José Armando, Serret, Maria Dolores, & Bort, Jordi (2022). Crop phenotyping in a context of global change: What to measure and how to do it. Journal of Integrative Plant Biology, 64, 592–618. 10.1111/jipb.13191.

Barradas, A., Correia, P. M. P., Silva, S., Mariano, P., Pires, M. C., Matos, A. R., Bernardes da Silva, A. B., & Marques da Silva, J. (2021). Comparing machine learning methods for classifying plant drought stress from leaf reflectance spectra in *Arabidopsis thaliana*. *Applied Sciences*, *11*, 6392. 10.3390/app11146392

Benelli, C., Tarraf, W., Izgu, T., & de Carlo, A. (2022). *In Vitro* conservation through slow growth storage technique of fruit species: An overview of the last 10 years. *Plants*, *11*(23), 3188. 10.3390/plants11233188

Bi, W.-L., Pan, C., Hao, X.-Y., Cui, Z.-H., Kher, M. M., Marković, Z., Wang, Q.-C., & Teixeira da Silva, J. A. (2017). Cryopreservation of grapevine (*Vitis* Spp.) – a review. *In Vitro Cellular & Developmental Biology - Plant*, *53*, 449–460.

Bian, L., Zhang, H., Ge, Y., Čepl, J., Stejskal J., & EL-Kassaby, Y. A. (2022). Closing the gap between phenotyping and genotyping: Review of advanced, image-based phenotyping technologies in forestry. *Annals of Forest Science*, *79*, 22. 10.1186/s13595-022-01143-x

Biasi, R., & Brunori, E. (2015). The on-farm conservation of grapevine (*Vitis vinifera L.*) landraces assures the habitat diversity in the viticultural agro-ecosystem. *Vitis*, *54*, 265–269.

Bindi, M., Migulietta, F., Gozzinia, B., Orlandini, S., & Seghi, L. (1997). A simple model for simulation of growth and development in grapevine (*Vitis vinifera* L). I. Model description. *Vitis*, *36*, 67–71.

Bota, J., Arroyo-Garcia, R., Tortosa, I., & Medrano, H. (2022). Exploiting genetic diversity to improve environmental sustainability of Mediterranean vineyards. In J. M. Costa, S. Catarino, J. M. Escalona, & P. Comuzzo (eds), *Improving sustainable viticulture and winemaking practices* (pp. 25–44). London, United Kingdom: Academic Press. 10.1016/B978-0-323-85150-3.00010-4

Boursiquot, J. M., Dessup, M., & Rennes, C. (1995). Distribution des principaux caracteres phenologiques, agronomiques et technologiques chez *Vitis vinifera* L. *Vitis*, *34*, 31–35.

Brischetto, C., Bove, F., Fedele, G., & Rossi, V. (2021). A weather-driven model for predicting infections of grapevines by Sporangia of *Plasmopara viticola*. *Frontiers in Plant Science, 12*, 636607. 10.3389/fpls.2021.636607

Capriotti, L., Baraldi, E., Mezzetti, B., Limera, C., & Sabbadini, S. (2020). Biotechnological approaches: Gene overexpression, gene silencing, and genome editing to control fungal and oomycete diseases in grapevine. *International Journal of Molecular Sciences*, *21*(16), 5701. 10.3390/ijms21165701

Carvalho, L. C., Coito, J. L., Colaço, S., Sangiogo, M., & Amâncio, S. (2015). Heat stress in grapevine: The pros and cons of acclimation. *Plant, Cell & Environment*, *38*, 777–789. 10.1111/pce.12445.

Carvalho, L. C., Coito, J. L., Gonçalves, E. F., Lopes, C., & Amâncio, S. (2018). Physiological and agronomical responses to environmental fluctuations of two Portuguese grapevine varieties during three field seasons. *Ciência e Técnica Vitivinicola*, *33*, 1–14. 10.1051/ctv/20183301001

Carvalho, L. C., Gonçalves, E., Amâncio, S., & Martins, A. (2020). Selecting Aragonez genotypes able to outplay climate change driven abiotic stress. *Frontiers in Plant Science*, *11*, 599230. 10.3389/fpls.2020.599230

Carvalho, L. C., Gonçalves, E. F., Marques da Silva, J., & Costa, J. M. (2021). Potential phenotyping methodologies to assess inter- and intravarietal variability and to select grapevine genotypes tolerant to abiotic stress. *Frontiers in Plant Science*, *12*, 718202. 10.3389/fpls.2021.718202

Carvalho, L. C., Pinto, T., Cammisano, A., Cid, J., Faísca-Silva, D., Costa, J. M., Amâncio, S., Martins, A., & Gonçalves, E. F. (2023). Polyclonal selection for abiotic stress tolerance in Arinto: Implications in yield and quality of the must. Proceedings of the 44th OIV Congress, Cadis, Spain.

Cavaco, A. M., Utkin, A. B., Marques da Silva, J., & Guerra, J. (2022). Making sense of light: The use of optical spectroscopy techniques in plant sciences and agriculture. *Applied Sciences*, *12*, 997. 10.3390/app12030997

Chapman, S., Cooper, M., Podlich, D., & Hammer, G. (2003). Evaluating plant breeding strategies by simulating gene action and dryland environment effects. *Agronomy Journal*, *95*(1), 99–113.

Chen, L.-S., & Cheng, L. (2003). Both xanthophyll cycle-dependent thermal dissipation and the antioxidant system are up-regulated in grape (Vitis labrusca L. cv. Concord) leaves in response to N limitation. *Journal of Experimental Botany*, *54*(390), 2165–2175. 10.1093/jxb/erg220

Chenu, K., Chapman, S. C., Tardieu, F., McLean, G., Welcker, C., & Hammer, G. L. (2009). Simulating the yield impacts of organ-level quantitative trait loci associated with drought response in maize: A "gene-to-phenotype" modelling approach. *Genetics*, *183*(4), 1507–1523.

Cola, G., Mariani, L., Salinari, F., Civardi, S., Bernizzoni, F., Gatti, M., & Poni, S. (2014). Description and testing of a weather-based model for predicting phenology, canopy development and source–sink balance in *Vitis vinifera* L. cv. Barbera. *Agricultural and Forest Meteorology*, *184*, 117–136.

Comparetti, A., & Marques da Silva, J. (2022). Use of sentinel-2 satellite for spatially variable rate fertiliser management in a Sicilian vineyard. *Sustainability*, *14*, 1688. 10.3390/su14031688

Costa, J. M., Grant, O. M., & Chaves, M. M. (2013). Thermography to explore plant–environment interactions. *Journal of Experimental Botany*, *64*(13), 3937–3949.

Costa, J. M., Vaz, M., Escalona, J., Egipto, R., Lopes, C., Medrano, H., & Chaves, M. M. (2016). Modern viticulture in southern Europe: Vulnerabilities and strategies for adaptation to water scarcity. *Agricultural Water Management*, *164*, 5–18.

Costa, J. M., Marques da Silva, J., Pinheiro, C., Barón, M., Mylona, P., Centritto, M., et al. (2019). Opportunities and limitations of crop phenotyping in southern European countries. *Frontiers in Plant Science*, *10*, 1125. 10.3389/fpls.2019.01125

Costa, J. M., Catarino, S., Escalona, J. M., & Comuzzo, P. (2022). Achieving a more sustainable wine supply chain – environmental and socioeconomic issues of the industry. In J. M. Costa, S. Catarino, J. Escalona, P. Comuzzo, P. (eds), *Improving sustainable viticulture and winemaking practices* (pp. 1–24). London, United Kingdom: Academic Press, Elsevier, ISBN: 9780323851503. doi: 10.1016/B978-0-323-85150-3.00009-8

Coupel-Ledru, A., Lebon, E., Goutouly, J. P., Christophe, A., Gago, P., Brault, C., This, P., Doligez, A., & Simonneau, T. (2022). Phenotyping for drought tolerance in grapevine populations: The challenge of heterogeneous field conditions. In J. M. Costa, S. Catarino, J. Escalona, P. Comuzzo, P. (eds), *Improving sustainable viticulture and winemaking practices*(pp. 65–83). London, United Kingdom: Academic Press, Elsevier.

Cretazzo, E., Moreno Sanz, P., Lorenzi, S., Benítez, M. L., Velasco, L., & Emanuelli, F. (2022). Genetic characterization by SSR markers of a comprehensive wine grape collection conserved at Rancho de la Merced (Andalusia, Spain). *Plants*, *11*, 1088. 10.3390/plants11081088

Cruz de Carvalho, R., Branquinho, C., & Marques da Silva, J. (2019). Desiccation rate affects chlorophyll and carotenoid content and the recovery of the aquatic moss *Fontinalis antipyretica* (Fontinalaceae). *Hattoria*, *10*, 53–60. 10.18968/hattoria.10.0_53

Cunha, M., Marçal, A. R. S., & Rodrigues, A. (2010). A comparative study of satellite and ground-based vineyard phenology. In I. Manakos, & C. Kalaitzidis (eds), *Imaging Europe* (pp. 68–77), IOS Press. 10.3233/978-1-60750-494-8-68

CYMMYT, (2013). Technical Report of the Research Program on Climate Change, Agriculture and Food Security. CIMMYT_2013_Technical_Report.pdf Assessed on 20 September 2023 https://repository.cimmyt.org/xmlui/bitstream/handle/10883/1288/96144.pdf

Dalbó, M., Ye, G., Weeden, N., Wilcox, W., & Reisch, B. (2001). Marker-assisted selection for powdery mildew resistance in grapes. *Journal of the American Society for Horticultural Science*, *126*(1), 83–89.

de los Campos, G., Perez-Rodriguez, P., Bogard, M., Gouache, D., & Crossa, J. (2020). A data-driven simulation platform to predict cultivars' performances under uncertain weather conditions. *Nature Communications*, *11*(1), 4876.

Deery, D., Jimenez-Berni, J., Jones, H., Sirault, X., & Furbank, R. (2014). Proximal remote sensing buggies and potential applications for field-based phenotyping. *Agronomy*, *4*, 349–379. 10.3390/agronomy4030349

Dinis, L.-T., Malheiro, A., Luzio, A., Fraga, H., Ferreira, H., Gonçalves, I., Pinto, G., Correia, C. M., & Moutinho-Pereira, J. (2018). Improvement of grapevine physiology and yield under summer stress by kaolin-foliar application: Water relations, photosynthesis and oxidative damage. *Photosynthetica*, *56*, 641–665.

Dong, Z., Liu, W., Li, X., Tan, W., Zhao, Q., Wang, M., Ren, R., Ma, X., & Tang, X. (2018) Genetic relationships of 34 grapevine varieties and construction of molecular fingerprints by SSR markers. *Biotechnology & Biotechnological Equipment*, *32*(4), 942–950. 10.1080/13102818.2018.1450162

Duchêne, E. (2016). How can grapevine genetics contribute to the adaptation to climate change? *OENO One*, *50*(3), 113–124.

Emanuelli, F., Sordo, M., Lorenzi, S., Battilana, J., & Grando, M. S. (2014). Development of user-friendly functional molecular markers for *VvDXS* gene conferring muscat flavor in grapevine. *Molecular Breeding*, *33*, 235–241.

EU (2023). The future of farming is here. Shaping Europe's digital future. https://digital-strategy.ec.europa.eu/en/policies/future-farming. Accessed on 19 November 2023.

Fiorani, F., Schurr, U. (2013) Future scenarios for plant phenotyping. *Annual Review of Plant Biology*, *64*, 267–291.

Fonseca, A., Fraga, H., & Santos, J. (2023). Exposure of Portuguese viticulture to weather extremes under climate change. *Climate Services*, *30*. 10.1016/j.cliser.2023.100357

Fraga, H., Malheiro, A. C., Moutinho-Pereira, J., & Santos, J. A. (2012). An overview of climate change impacts on European viticulture. *Food and Energy Security*, *1*, 94–110. 10.1002/fes3.14

Franks, T., Botta, R., & Thomas M. R. (2002). Chimerism in grapevines: implications for cultivar identity, ancestry and genetic improvement. *Theoretical and Applied Genetics*, *104*(2), 192–199. 10.1007/s001220100683

Ferro, M. V., & Catania, P. (2023). Technologies and innovative methods for precision viticulture: A comprehensive review. *Horticulturae*, *9*. 399. 10.3390/horticulturae9030399

Fu-Qiang, W., Lu, B., Peng-Peng, Q., Shuo, G., Jing-Han, G., Chen-Shuo, G., Jian-Fu, J., Chong-Huai, L., Yong, W., Guo-Tian, L., Yue-Jin, W., Yan, X. (2023). Development and application of KASP marker for high throughput detection of the seedless trait in grapevine. *Journal of Integrative Agriculture*, *22*(11), 3269–3283. 10.1016/j.jia.2023.10.014

Gago, J., Douthe, C., Coopman, R. E., Gallego, P. P., Ribas-Carbo, M., Flexas, J., Escalona, J., & Medrano, H. (2015). UAVs challenge to assess water stress for sustainable agriculture. *Agricultural Water Management*, *153*, 9–19. 10.1016/j.agwat.2015.01.020

Galvan, F. E. R., Pavlick R., Trolley, G., Aggarwal, S., Sousa, D., Starr, C., Forrestel, E., Bolton, S., Alsina, M. M., Dokoozlian, N., & Gold, K. M. (2023). Scalable early detection of grapevine viral Infection with airborne imaging spectroscopy. *Phytopathology*, *113*, 1369–1614. 10.1094/PHYTO-01-23-0030-R

Gamon, J., Penuelas, J., & Field, C. (1992). A narrow-waveband spectral index that tracks diurnal changes in photosynthetic efficiency. *Remote Sensing of Environment*, *41*, 35–44.

Gao, B.-C. (1996). NDWI – A normalized difference water index for remote sensing of vegetation liquid water from space. *Remote Sensing of Environment*, *58*, 257–266.

Gatti, M., Garavani, A., Squeri, C. Diti, I., De Monte, A., Scotti, C., & Poni, S. (2022). Effects of intra-vineyard variability and soil heterogeneity on vine performance, dry matter and nutrient partitioning. *Precision Agriculture*, *23*, 150–177. 10.1007/s11119-021-09831-w

Ge, Y., Bai, G., Stoerger, V., & Schnable, J. C. (2016). Temporal dynamics of maize plant growth, water use, and leaf water content using automated high throughput RGB and hyperspectral imaging. *Computers and Electronics in Agriculture*, *127*, 625–632. 10.1016/j.compag.2016.07.028.

Gill, T., Gill, S. K., Saini, D. K., Chopra, Y., de Koff, J. P., & Sandhu, K. S. (2022). A comprehensive review of high throughput phenotyping and machine learning for plant stress phenotyping. *Phenomics*, *2*(3), 156–183. 10.1007/s43657-022-00048-z

Giovos, R., Tassopoulos, D., Kalivas, D., Lougkos, N., & Priovolou, A. (2021). Remote sensing vegetation indices in viticulture: A critical review. *Agriculture*, *11*, 457. 10.3390/agriculture11050457

Gonçalves, E., St Aubyn, A., & Martins, A. (2010). Experimental designs for evaluation of genetic variability and selection of ancient grapevine varieties: A simulation study. *Heredity*, *104*, 552–562. 10.1038/hdy.2009

Gonçalves, E., & Martins A. (2022). Efficient assessment and large-scale conservation of intra-varietal diversity of ancient grapevine varieties: Case study Portugal. *Plants*, *11*, 1917. 10.3390/plants11151917

Gonçalves, E., Faísca-Silva, D., Rocheta, M., Pinto, T., Carvalho, L. C., Pereira, L., Barrias, S., Rocha, S., Lopes, A., Guimarães, J., et al. (2023). Large-scale screening of unknown varieties in a grapevine intra-varietal variability collection. Proceedings of the 44th OIV Congress, Cádis, Spain.

González-Flor, C., Serrano, L., & Gorchs, G. (2019). Use of reflectance indices to assess vine water status under mild to moderate water deficits. *Agronomy*, *9*, 346. 10.3390/agronomy9070346

Grassi, F., & De Lorenzis, G. (2021). Back to the origins: Background and perspectives of grapevine domestication. *International Journal of Molecular Sciences*, *22*, 4518. 10.3390/ijms22094518

Gutiérrez, S., Tardaguila, J., Fernández-Novales, J., & Diago M. P. (2015). Data mining and non-invasive proximal sensing for precision viticulture. Conference Proceedings Paper – Sensors and Applications. 10.3390/ecsa-2-S2003

Gutiérrez, S., Fernández-Novales, J., Diago, M. P., & Tardaguila, J. (2018). On-The-Go hyperspectral imaging under field conditions and machine learning for the classification of grapevine varieties. *Frontiers in Plant Science*, *9*, 1102. 10.3389/fpls.2018.01102

Harris, N. N., Luczo, J. M., Robinson, S. P., & Walker, A. R. (2013). Transcriptional regulation of the three grapevine chalcone synthase genes and their role in flavonoid synthesis in Shiraz. *Australian Journal of Grape and Wine Research*, *19*(2), 221–229. 10.1111/ajgw.12026

Houel, C., Martin-Magniette, M.-L., Nicolas, S. D., Lacombe, T., Le Cunff, L., Franck, D., Torregrosa, L., Conéjéro, G., Lalet, S., This, P., & Adam-Blondon, A-F. (2013). Genetic variability of berry size in the grapevine (*Vitis vinifera* L.). *Australian Journal of Grape and Wine Research*, *19*, 208–220. 10.1111/ajgw.12021

Ibáñez, J., Carreño, J., Yuste, J., & Martínez-Zapater, J. M. (2015). Grapevine breeding and clonal selection programmes in Spain. In A. Reynolds (ed.), *Grapevine breeding programs for the wine industry* (pp. 183–209). Amsterdam, The Netherlands: Elsevier.

Jighly, A., Thayalakumaran, T., O'Leary, G. J., Kant, S., Panozzo, J., Aggarwal, R., Hessel, D., Forrest, K. L., Technow, F., Tibbits, J. F. G., et al. (2023). Using genomic prediction with crop growth models enables the prediction of associated traits in wheat. *Journal of Experimental Botany*, *74*, 1389–1402.

Johnston, L. F., Roczen, D. E., Youkhana, S. K., Nemani, R. R., & Bosch, D. F. (2003). Mapping vineyard leaf area with multispectral satellite imagery. *Computers and Electronics in Agriculture*, *38*, 33–44.

Jones, M. R., Singels, A., Chinorumba, S., Patton, A., Poser, C., Singh, M., Martine, J. F., Christina, M., Shine, J., Annandale, J., et al. (2019). Exploring process-level genotypic and environmental effects on sugarcane yield using an international experimental dataset. *Field Crops Research*, *244*, 107622.

Ju, Y.-L., Yue, X.-F., Min, Z., Wang, X.-H., Fang, Y.-L., & Zhang, J.-X. (2020) *VvNAC17*, a novel stress-responsive grapevine (*Vitis vinifera* L.) NAC transcription factor, increases sensitivity to abscisic acid and enhances salinity, freezing, and drought tolerance in transgenic *Arabidopsis*. *Plant Physiology and Biochemistry*, *146*, 98–111. 10.1016/j.plaphy.2019.11.002

Kadam, N. N., Jagadish, S. V. K., Struik, P. C., van der Linden, C. G., & Yin, X. (2019). Incorporating genome-wide association into eco-physiological simulation to identify markers for improving rice yields.*Journal of Experimental Botany*, *70*(9), 2575–2586.

Kang, C., Diverres, G., Karkee, M., Zhang, Q., & Keller, M. (2023). Decision-support system for precision regulated deficit irrigation management for wine grapes. *Computers and Electronics in Agriculture*, *208*, 107777.

Karakizi, C., Oikonomou, M., & Karantzalos, K. (2016). Vineyard detection and vine variety discrimination from very high resolution satellite data. *Remote Sensing*, *8*, 235. 10.3390/rs8030235

Kasimati, A., Espejo-Garcia, B., Vali, E., Malounas, I., & Fountas, S. (2021). Investigating a selection of methods for the prediction of total soluble solids among wine grape quality characteristics using Normalized Difference Vegetation Index data from proximal and remote sensing. *Frontiers in Plant Science*, *12*, 683078. 10.3389/fpls.2021.683078

Keller, B., Ariza-Suarez, D., de la Hoz, J., Aparicio, J. S., Portilla-Benavides, A. E., Buendia, H. F., Mayor, V. M., Studer, B., & Raatz, B. (2020). Genomic prediction of agronomic traits in common bean (*Phaseolus vulgaris* L.) under environmental stress. *Frontiers in Plant Science*, *11*, 01001. 10.3389/fpls.2020.01001

Khaliq, A., Comba, L., Biglia, A., Aimonino, D. R., Chiaberge, M., & Gay, P. (2019). Comparison of satellite and UAV-based multispectral imagery for vineyard variability assessment. *Remote Sensing*, *11*, 436. 10.3390/rs11040436

Kiefer, C., & Szolnoki, G. (2023). Consumer acceptance of fungus-resistant grape varieties—an exploratory study using sensory evaluation tests among consumers in Germany. *Sustainability*, *15*, 10664. 10.3390/su151310664

Kicherer, A., Herzog, K., Pflanz, M., Wieland, M., Rüger, P., Kecke, S., Kuhlmann, H., & Töpfer, R. (2015). An automated field phenotyping pipeline for application in grapevine research. *Sensors*, *15*, 4823–4836. 10.3390/s150304823

Klodt, M., Herzog, K., Töpfer, R., & Cremers, D. (2015). Field phenotyping of grapevine growth using dense stereo reconstruction. *BMC Bioinformatics*, *16*, 143. 10.1186/s12859-015-0560-x

Krzyzaniak, Y., Cointault, F., Loupiac, C., Bernaud, E., Ott, F., Salon, C., Laybros, A., Han, S., Héloir, M.-C., Adrian, M., & Trouvelot, S. (2021). *In situ* phenotyping of grapevine root system architecture by 2D or 3D imaging: Advantages and limits of three cultivation methods. *Frontiers in Plant Science*, *12*, 638688. 10.3389/fpls.2021.638688

Kumagai, M. H., & Miller, P. (2006). Development of electronic barcode for use in plant pathology and functional genomics. *Plant Molecular Biology*, *61*, 515–523.

Lacombe, T., Boursiquot, J. M., Laucou, V., di Vecchi-Staraz, M., Péros, J.-P., & This, P. (2013). Large-scale parentage analysis in an extended set of grapevine cultivars (*Vitis vinifera* L.). *Theoretical and Applied Genetics*, *126*, 401–414. 10.1007/s00122-012-1988-2

La Fata, A. (2021). Thermography to assess grapevine status and traits opportunities and limitations in crop monitoring and phenotyping – a review. MSc Thesis, Engenharia de Viticultura e Enologia – Double Degree, U. Lisboa, 63p.

Langstroff, A., Heuermann, M. C., Stahl, A., & Junker, A. (2022). Opportunities and limits of controlled-environment plant phenotyping for climate response traits. *Theoretical and Applied Genetics*, *135*(1), 1–16. 10.1007/s00122-021-03892-1

Li, W., Dang, C., Ye, Y., Wang, Z., Hu, B., Zhang, F., Zhang, Y., Qian, X., Shi, J., Guo, Y., Zhou, Q., Wang, T., Chen, X., & Wang, F. (2020). Overexpression of grapevine *vviaa18* gene enhanced salt tolerance in tobacco. *International Journal of Molecular Sciences*, *21*(4). 10.3390/ijms21041323

Lopes, C., Graça, J., Sastre, J., Reyes, M., Guzman, R., Braga, R., Monteiro, A., & Pinto, P. (2016). Vineyard yield estimation by vinbot robot—Preliminary results with the white variety Viosinho. In Proceedings of the 11th International Terroir Congress, McMinnville, OR, USA, 10–14 July 2016, p. 516.

Luvisi, A., Triolo, E., Rinaldelli, E., Bandinelli, R., Pagano, M., & Ginic, B. (2010). Radiofrequency applications in grapevine: From vineyard to web. *Computers and Electronics in Agriculture*, *70*, 256–259.

Madhav, M. S., Laha, G. S., Padmakumari, A. P., Somasekhar, N., Mangrauthia, S. K., & Viraktamath, B. C. (2013). Phenotyping rice for molecular plant breeding. In S. Panguluri, & A. Kumar (eds), *Phenotyping for plant breeding*. (pp. 1–40). New York, NY: Springer. 10.1007/978-1-4614-8320-5_1

Mairata, A., Tortosa, I., Douthe, C., Escalona, J. M., Pou, A., & Medrano, H. (2022). Comparing selection criteria to select grapevine clones by water use efficiency. *Agronomy*, *12*, 1963. 10.3390/agronomy12081963

Marcianò, D., Ricciardi, V., Marone Fassolo, E., Passera, A., Bianco, P. A., Failla, O., Casati, P., Maddalena, G., De Lorenzis, G., & Toffolatti, S. L. (2021). RNAi of a putative grapevine susceptibility gene as a possible downy mildew control strategy. *Frontiers in Plant Science*, *12*, 667319. 10.3389/fpls.2021.667319

Matese, A., Di Gennaro, S. F., Orlandi, G., Gatti, M., & Poni, S. (2022). Assessing grapevine biophysical parameters: From unmanned aerial vehicles hyperspectral imagery. *Frontiers in Plant Science*, *13*, 898722. 10.3389/fpls.2022.898722

McFadden, J., Njuki, E., & Griffin, T. (2023). Precision agriculture in the digital era: Recent adoption on U.S. farms. Economic Information Bulletin No. EIB-248, p. 52. Available at: https://www.ers.usda.gov/webdocs/publications/105894/eib-248.pdf. Accessed on 15 November 2023.

Mejía, N., Soto, B., Guerrero, M., Casanueva, X., Houel, C., Miccono, M.Á., Ramos, R., Le Cunff, L., Boursiquot, J.-M., Hinrichsen, P., & Adam-Blondon, A.-F. (2011). Molecular, genetic and transcriptional evidence for a role of *VvAGL11* in stenospermocarpic seedlessness in grapevine. *BMC Plant Biology*, *11*, 57. 10.1186/1471-2229-11-57

Milella, A., Marani, R., Petitti, A., & Reina, G. (2019). In-field high throughput grapevine phenotyping with a consumer-grade depth camera. *Computers and Electronics in Agriculture*, *156*, 293–306. 10.1016/j.compag.2018.11.026

Millet, E. J., Kruijer, W., Coupel-Ledru, A., Alvarez-Prado, S., Cabrera-Bosquet, L., Lacube, S., Charcosset, A., Welcker, C., van Eeuwijk F., & Tardieu F. (2019). Genomic prediction of maize yield across European environmental conditions. *Nature Genetics*, *51*(6), 952–956.

Mir, R. R., Reynolds, M., Pinto, F., Khan, M. A., & Bhat, M. A. (2019). High-throughput phenotyping for crop improvement in the genomics era. *Plant Science*, *282*, 60–72. 10.1016/j.plantsci.2019.01.007

Moreno, H., & Andújar, D. (2023). Proximal sensing for geometric characterization of vines: A review of the latest advances. *Computers and Electronics in Agriculture*, *210*, 107901. 10.1016/j.compag.2023.107901

Morisse, M., Wells, D. M., Millet, E. J., Lillemo, M., Fahrner, S., Cellini, F., Lootens, P., Muller, O., Herrera, J. M., Bentley, A. R., & Janni, M. (2022). A European perspective on opportunities and demands for field-based crop phenotyping. *Field Crops Research*, *276*, 108371. 10.1016/j.fcr.2021.108371

Nakagawa, H., Yamagishi, J., Miyamoto, N., Motoyama, M., Yano, M., & Nemoto, K. (2005). Flowering response of rice to photoperiod and temperature: a QTL analysis using a phenological model. *Theoretical and Applied Genetics*, *110*(4), 778–786.

Nogueira Júnior, A. F., Amorim, L., Savary, S., & Willocquet, L. (2018). Modelling the dynamics of grapevine growth over years. *Ecological Modelling*, *369*, 77–87.

Ocarez, N., Jiménez, N., Núñez, R., Perniola, R., Marsico, A. D., Cardone, M. F., Bergamini, C., & Mejía, N. (2020). Unraveling the deep genetic architecture for seedlessness in grapevine and the development and validation of a new set of markers for *VviAGL11*-based gene-assisted selection. Genes, *11*, 151. 10.3390/genes11020151

OIV (2017). Focus OIV 2017. Distribution of the world's grapevine varieties. ISBN: 979-10-91799-89-8. Available online at: https://www.oiv.int/public/medias/5888/en-distribution-of-the-worlds-grapevine-varieties.pdf Accessed on 20 September 2023.

OIV (2021). Sustainable use of water in winegrape vineyards. Available online at: https://www.oiv.int/public/medias/7949/2021-oiv-collective-expertise-document-sustainable-use-of-wa.pdf. Accessed on 20 September 2023.

OIV (2023). State of the world vine and wine sector in 2022. Available online at: OIV_State_of_the_world_Vine_and_Wine_sector_in_2022_2.pdf. Accessed on 29 September 2023.

Ollat, N., Peccoux, A., Papura, D., Esmenjaud, D., Marguerit, E., Tandonnet, J.-P., Bordenave, L., Cookson, S. J., Barrieu, F., Rossdeutsch, L., Lecourt, J., Lauvergeat, V., Vivin, P., Bert, P.-F., & Delrot, S. (2015). Rootstocks as a component of adaptation to environment. In H. Gerós, M. M. Chaves, H. Medrano, & S. Delrot (eds), *Grapevine: Climate and environmental stress, biotic and abiotic factors* (pp. 60–108). Oxford: Wliey Blackwell.

Onofre, J. (2022). European wine policy framework – the path toward sustainability. In J. M. Costa, S. Catarino, J. Escalona, & P. Commuzzo (eds), *Improving sustainable viticulture and winemaking practices* (pp. 485–499). London, United Kingdom: Elsevier Academic Press.

Palacios, F., Diago, M. P., Melo-Pinto, P. et al. (2023). Early yield prediction in different grapevine varieties using computer vision and machine learning. *Precision Agriculture*, *24*, 407–435. 10.1007/s11119-022-09950-y

Pastonchi, L., Di Gennaro, S. F., Toscano, P., & Matese, A. (2020). Comparison between satellite and ground data with UAV-based information to analyse vineyard spatio-temporal variability. *OENO One*, *54*(4), 919–934.

Pauquet, J., Bouquet, A., This, P., & Adam-Blondon, A. F. (2001). Establishment of a local map of AFLP markers around the powdery mildew resistance gene run1 in grapevine and assessment of their usefulness for marker assisted selection. *Theoretical and Applied Genetics*, *103*, 1201–1210.

Pengra, B., Long, J., Dahal, D., Stehman, S. V., & Loveland, T. R. (2015). A global reference database from very high resolution commercial satellite data and methodology for application to Landsat derived 30 m continuous field tree cover data. *Remote Sensing of Environment*, *165*, 234–248.

Penuelas, J., Pinol, J., Ogaya, R., & Filella, I. (1997). Estimation of plant water concentration by the reflectance Water Index WI (R900/R970). *International Journal of Remote Sensing*, *18*, 2869–2875.

Planas, S., Román, C., Sanz, R., & Rosell-Polo, J. R. (2022). Bases for pesticide dose expression and adjustment in 3D crops and comparison of decision support systems. *Science of The Total Environment*, *806*, 150357. 10.1016/j.scitotenv.2021.150357.

Pommier, C., Michotey, C., Cornut, G., Roumet, P., Duchêne, E., Flores, R., Lebreton, A., Alaux, M., Durand, S., Kimmel, E. et al. (2019). Applying FAIR principles to plant phenotypic data management in GnpIS. *Plant Phenomics*, *2019*, 1671403. 10.34133/2019/1671403

Poorter, H., Hummel, G. M., Nagel, K. A., Fiorani, F., von Gillhaussen, P., Virnich, O., Schurr, U., Postma, J. A., van de Zedde, R., & Wiese-Klinkenberg, A. (2023). Pitfalls and potential of high-throughput plant phenotyping platforms. *Frontiers in Plant Science*, *14*, 1233794. 10.3389/fpls.2023.1233794

Prazzoli, M. L., Lorenzi, S., Perazzolli, M., Toffolatti, S., Failla, O., & Grando, M. S. (2019). Identification of disease resistance-linked alleles in *Vitis vinifera* germplasm. *BIO Web of Conferences*, *13*, 01004. 10.1051/bioconf/20191301004

Purcell, W., Neubauer, T., & Mallinger, K. (2023). Digital Twins in agriculture: Challenges and opportunities for environmental sustainability. *Current Opinion in Environmental Sustainability*, *61*, 101252. 10.1016/j.cosust.2022.101252

Reluy, N. P., Baghdadi, N., Simonneau, T., Bazzi, H., Hajj, M., Pret, V., Amin, G., & Daret, E. (2022). Can we detect the damage of a heatwave on vineyards using Sentinel-2 optical remote sensing data? *Oeno One*, *56*(1), 145–159. 10.20870/oeno-one.2022.56.1.4632

Reymond, M., Muller, B., Leonardi, A., Charcosset, A., & Tardieu, F. (2003). Combining quantitative trait loci analysis and an ecophysiological model to analyze the genetic variability of the responses of maize leaf growth to temperature and water deficit. *Plant Physiology*, *131*(2), 664–675.

Reynolds, D., Baret, F., Welcker, C., Bostrom, A., Ball, J., Cellini, F., Lorence, A., Chawade, A., Khafif, M., Noshita, K., Mueller-Linow, M., Zhou, J., & Tardieu, F. (2019). What is cost-efficient phenotyping? Optimizing costs for different scenarios. *Plant Science*, *282*, 14–22. 10.1016/j.plantsci.2018.06.015

Roach, M., Borneman, A., Schmidt, S., & Krstic, M. (2021). Applying genomics to grapevine clones. *Wine & Viticulture Journal*, *36*(2), 39–41.

Robert, P., Le Gouis, J., Consortium, T. B., & Rincent, R. (2020). Combining crop growth modeling with trait-assisted prediction improved the prediction of genotype by environment Interactions.*Frontiers in Plant Science*, *11*, 827. 10.3389/fpls.2020.00827

Robinson, S. P., Pezhmanmehr, M., Speirs, J., McDavid, D. A. J., Hooper, L. C., Rinaldo, A. R., Bogs, J., Ebadi, A., & Walker, A. R. (2019). Grape and wine flavonoid composition in transgenic grapevines with altered expression of flavonoid hydroxylase genes. *Australian Journal of Grape and Wine Research*, *25*(3), 293–306. 10.1111/ajgw.12393

Robinson, S. P., Bogs, J., McDavid, D. A. J., Hooper, L. C., Speirs, J., & Walker, A. R. (2021). Transgenic grapevines with decreased expression of tannin synthesis genes have altered grape and wine flavonoid composition. *Australian Journal of Grape and Wine Research*, *27*(1), 106–117. 10.1111/ajgw.12468

Sams, B., Bramley, R. G. V., Sanchez, L., Dokoozlian, N., Ford, C., & Pagay, V. (2022). Remote sensing, yield, physical characteristics, and fruit composition variability in cabernet sauvignon vineyards. *American Journal of Enology and Viticulture*, *73*(2), 93–105.

Santesban, L. G. (2020). Strategies for adapting vineyards to a changing climate. (Re)-Learning from Mediterranean viticulture, IVES Technical Reviews, 10.20870/ives-tr.2020.3788

Schmidt, D., Bahr, C., Friedel, M., & Kahlen, K. (2019). Modelling approach for predicting the impact of changing temperature conditions on grapevine canopy architectures. *Agronomy*, *9*, 426. 10.3390/agronomy9080426

Shimono, H., Abe, A., Kim, C. H., Sato, C., & Iwata, H. (2023). Upcycling rice yield trial data using a weather-driven crop growth model. *Communications Biology*, *6*, 764. 10.1038/s42003-023-05145-x

Shecori, S., Kher, M. M., Tyagi, K., Lerno, L., Netzer, Y., Lichter, A., Ebeler, S. E., & Drori, E. (2022) A field collection of indigenous grapevines as a valuable repository for applied research. *Plants*, *11*, 2563. 10.3390/plants11192563

Sozzi, M., Kayad, A., Marinello, F., Taylor, J. A., & Tisseyre, B. (2019). Comparing vineyard imagery acquired from Sentinel-2 and unmanned aerial vehicle (UAV) platform. *OENO One*, *2*, 189–197.

Tardaguila, J., Stoll, M., Gutiérrez, S., Proffitt, T., & Diago, M. P. (2021). Smart applications and digital technologies in viticulture: A review. *Smart Agricultural Technology*, *1*, 100005. 10.1016/j.atech.2021.100005

Tello, J., Montemayor, M. I., Forneck, A., & Ibáñez, J. (2018). A new image-based tool for the high throughput phenotyping of pollen viability: Evaluation of inter- and intra-cultivar diversity in grapevine. *Plant Methods*, *14*, 3. 10.1186/s13007-017-0267-2

Tello, J., & Ibáñez, J. (2023). Review: Status and prospects of association mapping in grapevine. *Plant Science*, *327*, 111539. 10.1016/j.plantsci.2022.111539

Töpfer, R., & Trapp, O. (2022). A cool climate perspective on grapevine breeding: Climate change and sustainability are driving forces for changing varieties in a traditional market. *Theoretical and Applied Genetics*, *135*, 3947–3960. 10.1007/s00122-022-04077-0

Torregrosa, L., Bigard, A., Doligez, A., Lecourieux, D., Rienth, M., Luchaire, N., Pieri, P., Chatbanyong, R., Shahood, R., Farnos, M., Roux, C., Adiveze, A., Pillet, J., Sire, Y., Zumstein, E., Veyret, M., Le Cunff, L., Lecourieux, F., Saurin, N., Muller, B., Ojeda, H., Houel, C., Péros, J.-P., This, P., Pellegrino, A., & Romieu, C. (2017). Developmental, molecular and genetic studies on grapevine response to temperature open breeding strategies for adaptation to warming. *OENO One*, *51*(2), 155–165. 10.20870/oeno-one.2017.51.2.1587

Uptmoor, R., Schrag, T., Stutzel, H., & Esch, E. (2008). Crop model based QTL analysis across environments and QTL based estimation of time to floral induction and flowering in Brassica oleracea. *Molecular Breeding*, *21*(2), 205–216.

Urra, C., Sanhueza, D., Pavez, C., Tapia, P., Núñez-Lillo, G., Minio, A., Miossec, M., Blanco-Herrera, F., Gainza, F., Castro, A., Cantu, D., & Meneses, C. (2023). Identification of grapevine clones via high-throughput amplicon sequencing: A proof-of-concept study. *Genes Genomes Genetics*, *13*(9), 145. 10.1093/g3journal/jkad145

Vélez, S., Barajas, E., Blanco, P., Rubio, J. A., & Castrillo, D. (2021). Spatio-temporal analysis of satellite imagery (NDVI) to identify terroir and vineyard yeast differences according to appellation of origin (AOP) and biogeographic origin. *J*, *4*, 244–256. 10.3390/j4030020

Victorino, G., Braga, R. P., Santos-Victor, J., & Lopes, C. M. (2022). Overcoming the challenge of bunch occlusion by leaves for vineyard yield estimation using image analysis. *OENO One*, *56*(1), 117–131. 10.20870/oeno-one.2022.56.1.4863

Wagner, M., Stanbury, P., Dietrich, T., Döring, J., Ewert, J., Foerster, C., Freund, M., Friedel, M., Kammann, C., Koch, M., Owtram, T., Schultz, H. R., Voss-Fels, K., & Hanf, J. (2023). Developing a sustainability vision for the global wine industry. *Sustainability*, *15*(13), 10487. 10.3390/su151310487

Watt, M., Fiorani, F., Usadel, B., Rascher, U., Muller, O., & Schurr, U. (2020). Phenotyping: New windows into the plant for breeders. *Annual Review of Plant Biology*, *71*(1), 689–712.

Wilkinson, M. D., Dumontier, M., Aalbersberg, I. J., Appleton, G., Axton, M., Baak, A., Blomberg, N., Boiten, J.-W., Santos, L. B. S., Bourne, P. E., Bouwman, J., Brookes, A. J., Clark, T., Crosas, M., Dillo, I., Dumon, O., Edmunds, S., Evelo, C. T., Finkers, R., Gonzalez-Beltran, A., Gray, A. J. G., Groth, P., Goble, C., & Grethe, J. S. (2016). The FAIR guiding principles for scientific data management and stewardship. *Scientific Data*, *3*, 160018.

Xu, R., Li, C., & Bernardes, S. (2021). Development and testing of a UAV-based multi-sensor system for plant phenotyping and precision agriculture. *Remote Sensing*, *13*, 3517. 10.3390/rs13173517

Yu, Y., Bian, L., Jiao, Z., Yu, K., Wan, Y., Zhang, G., & Guo, D. (2019). Molecular cloning and characterization of a grapevine (*Vitis vinifera* L.) serotonin N-acetyltransferase (*VvSNAT2*) gene involved in plant defense. *BMC Genomics*, *20*(1), 880. 10.1186/s12864-019-6085-3

Yu, Y., Ni, Y., Qiao, T., Ji, X., Xu, J., Li, B., & Sun, Q. (2022). Overexpression of *VvASMT1* from grapevine enhanced salt and osmotic stress tolerance in *Nicotiana benthamiana*. *PLoS ONE*, *17*(6), e0269028. 10.1371/journal.pone.0269028

Zhang, Z., Kayacan, E., Thompson, B., & Chowdhary G. (2020). High precision control and deep learning-based corn stand counting algorithms for agricultural robot. *Autonomous Robots*, *44*, 1289–1302. 10.1007/s10514-020-09915-y

Zhang, C., Kong, J., Wu, D., Guan, Z., Ding, B., & Chen, F. (2023). Wearable sensor: An emerging data collection tool for plant phenotyping. *Plant Phenomics*, *5*, 0051. 10.34133/plantphenomics.0051

Zhu, J., Génard, M., Poni, S., Gambetta, G. A., Vivin, P., Vercambre, G., Trought, M. C. T., Ollat, N., Delrot, S., & Dai, Z. (2018). Modelling grape growth in relation to whole-plant carbon and water fluxes. *Journal of Experimental Botany*, *70*, 2502–2521.

Zhu, J., Parker, A., Gou, F., Agnew, R., Yang, L., Greven, M., Raw, V., Neal, S., Martin, D., Trought, M. C. T., Huth, N., & Brown, H. D. (2021). Developing perennial fruit crop models in APSIM next generation using grapevine as an example. *in silico Plants*, *3*(2), diab021. 10.1093/insilicoplants/diab021

16 Rice Submergence Tolerance

Emerging Molecular Breeding Strategies in the Post-Genomic Era

Ashish Gautam
Department of Life Science, School of Life Sciences, Central University of Karnataka, Kadaganchi, Kalaburagi, Karnataka

Department of Plant Breeding and Genetics, Assam Agricultural University, Jorhat, Assam, India

Anamika Thakur
Department of Biotechnology, Dr. Y.S. Parmar University of Horticulture and Forestry, Nauni, Solan, Himachal Pradesh, India

Charupriya Chauhan
Department of Genetics and Plant Breeding, School of Agriculture, Lovely Professional University, Phagwara, Punjab, India

Vinay Sharma
International Crops Research Institute for the Semi-Arid Tropics (ICRISAT), Hyderabad, Telangana, India

Department of Genetics and Plant Breeding, Ch. Charan Singh University, Meerut, Uttar Pradesh, India

Karthick Babu Sivakumar
Department of Plant Breeding and Genetics, Punjab Agricultural University, Ludhiana, Punjab, India

Rakesh Kumar
Department of Life Science, School of Life Sciences, Central University of Karnataka, Kadaganchi, Kalaburagi, Karnataka

Sanjay Kumar Chetia
Department of Plant Breeding and Genetics, Assam Agricultural University, Jorhat, Assam, India

Regional Agricultural Research Station, Assam Agricultural University, Titabar, Assam, India

DOI: 10.1201/b23372-16

16.1 INTRODUCTION

Climate change is indeed an escalating global crisis driven by the current surge in greenhouse gases from both human and natural activities. The Intergovernmental Panel on Climate Change (IPCC) highlighted in its 2007 and 2014 reports that average global temperatures exceeded the preindustrial era, marking a period of intensified global warming. An increase in the average global temperatures and sea level gives rise to global warming, which is influencing weather patterns, causing more frequent and severe heat waves, melting glaciers and ice caps, rising sea levels, altered precipitation patterns, and more intense weather events. Climate change triggers a cascade of risks, from its physical impacts on ecosystems to agroecosystems, agricultural production, food chains, incomes, and trade. These risks have profound economic and social consequences, impacting global livelihoods, food security, and nutrition. Particularly in developing countries where a majority of the workforce is employed in agriculture, climate change poses a significant threat. Shifts in weather patterns, extreme weather events, and environmental disruptions are making food production more challenging and uncertain in these regions, exacerbating food and nutritional security issues. The effects of climate change on precipitation and rising sea levels have led to imbalances in water availability in regions that may suffer from both flooding and droughts (White, 2010). Frequent occurrences of rainfall-triggered flooding and subsequent waterlogging are agricultural catastrophes seen in many regions of the globe, leading to severe production losses (Schmidhuber et al., 2007). In addition to the immediate effect on plant growth, submergence imposes complex abiotic stresses on crops, including reduced light availability (Vervuren et al., 2003), oxygen depletion (Youssef et al., 2016), and an alteration in soil chemistry (Sun et al., 2007). The alterations in both physical and chemical environments resulting from excessive rainfall can significantly affect crop health, development, and productivity. In the current environmental conditions, the decline in crop productivity caused by heavy precipitation has had a notable adverse effect on the grain supply and overall food security (Urban et al., 2015). According to the Food and Agriculture Organization (FAO), floods ranked as the second most severe agricultural catastrophe, surpassed only by droughts. Over the decade spanning from 2008 to 2018, these floods accounted for a staggering $21 billion in losses, impacting both crops and livestock in the developing world (FAO, 2021). Currently, rice (*Oryza sativa* L.) stands out as one of the crops most vulnerable to flooding. Rice holds significant importance as a staple food, nourishing the diets of more than 3.5 billion people worldwide and playing a vital role in ensuring food security for numerous nations in Africa and Asia (Samal et al., 2018; Chukwu et al., 2019; Gautam et al., 2022b). Floods may hamper rice crops at any developmental stage, regardless of whether they manifest as brief and sudden flash floods or prolonged periods of stagnant inundation. Rice can withstand anaerobic conditions to some extent, but severe inundation, whether total or partial submergence, can bring about various environmental stresses. Consequently, rice cultivars exhibit varying responses to changes in floodwater patterns, making it challenging to transfer flood-resistant strains from one location to another (Dar et al., 2017). Within rice agroecosystems, three distinct types of inundation occur at different stages of development and for varying durations. The first kind of flood is submergence during anaerobic germination. In rainfed lowland ecosystems, intense and abrupt rainfall often follows direct seeding cultural practices. Under this condition, rice seeds are submerged completely and suffer anoxia or hypoxia, which results in low or poor germination, poor crop establishment, and seedling death. The second type involves complete submergence during the vegetative growth stage, either at the seedling or tillering stage, which can last for days or even weeks. This issue is common in river basins and often leads to yield losses due to flash floods. The third kind of flooding occurs in lowland rice cultivating areas, where the water level ranges from 20 to 50 cm, depending on whether it is a rainfed or irrigated system. However, seasonal fluctuations and the duration of rain can cause water levels to exceed 50 cm, reaching heights of up to 4 m, depending on the crop's growth stage. This flooding can persist for an extended period, potentially lasting for several months, depending on the specific location.

The extent of crop losses depends on the frequency and intensity of the flooding, ranging from yield reduction due to moderate stagnant flooding to complete crop loss when subjected to complete submergence. Every year, floods inflict varying levels of damage in South and Southeast Asian nations like India, Bangladesh, Nepal, Myanmar, Cambodia, Laos, Philippines, Vietnam, and Thailand (Ahern et al., 2005; Acuin et al., 2011). Flooding impacts approximately 22 million hectares of rainfed lowland areas in South and Southeast Asia, encompassing 6.2 million hectares of rice fields in India (Azarin et al., 2017; Dar et al., 2017). Among these 22 million hectares, 15 million hectares of rainfed lowlands are specifically vulnerable to short-term flash floods (Singh et al., 2017), leading to an estimated economic loss of one billion US dollars (Mackill et al., 2012). These flood-prone environments constitute roughly 7% of the world's total rice acreage and contribute 4% to global rice production (Yang et al., 2017).

Although many high-yielding rice varieties experience significant flood-related losses, Asia has seen the development of various landraces that exhibit tolerance to different types of flooding. For instance, deepwater rice varieties possess the ability to elongate quickly to endure flooding, whereas most of the high-yielding modern varieties cannot elongate sufficiently for survival. In cases where floods persist for over a few days, rice plants exert significant energy in futile attempts to escape, leaving them unable to recover once the flooding subsides. The search for submergence-tolerant rice genotypes began at the International Rice Research Institute (IRRI) in 1970 (Vergara & Mazaredo, 1975). Among various rice varieties, FR13A emerged as the most tolerant cultivar, having been developed through pure-line selection from the landrace "Dhalputtia", which traces its origins to Orissa, India (HilleRisLambers & Vergara, 1982). After that, it was frequently exploited as a valuable donor in breeding initiatives to enhance tolerance. Several breeding lines exhibiting submergence tolerance have been developed. Among these, IR49830-7-1-2-1 stands out as a remarkably robust and high-yielding indica-type line (Mackill et al., 1993), officially introduced as "Popoul" in Cambodia back in 1999.

Nevertheless, despite submergence tolerance, this variety did not gain widespread acceptance among farmers due to the absence of several essential characteristics, notably the preferred local grain quality (Neeraja et al., 2007). A significant milestone in the evolution of submergence-tolerance breeding occurred with the discovery of the primary quantitative trait locus (QTL) named Submergence 1 (SUB1), which governs this trait (Xu & Mackill, 1996). This initial research ultimately culminated in the mapping of the SUB1 region in FR13A and subsequently in the recognition of the vital ethylene-responsive factor (ERF) gene, SUB1A-1, which has been recognized as essential and adequate for submergence tolerance (Xu et al., 2006). Utilizing the sequence data, specialized molecular markers for SUB1 were developed, enabling a precise marker-assisted backcrossing (MABC) system. This system has proven effective in the targeted incorporation of the SUB1 QTL into extensively cultivated "mega-varieties" across South and Southeast Asia, as well as in Africa (Neeraja et al., 2007; Septiningsih et al., 2009; Bailey-Serres et al., 2010; Mackill et al., 2012). Through successive backcrossing with the recipient parent, improved varieties with enhanced submergence tolerance carry the FR13A SUB1 locus while retaining all other attributes of the original variety. These new varieties maintain their original grain quality and other traits preferred by local communities, thereby increasing their acceptance among farmers and facilitating their official release in the specified countries. These revolutionary cultivars are known as "Sub1" or "Scuba" rice. Numerous high-yielding rice varieties possessing submergence tolerance through the SUB1 gene have been successfully developed. These varieties have demonstrated a yield advantage of 1–3 tons even after enduring floods lasting 10–15 days. Notable flood-tolerant varieties that have been developed and are currently being grown include Swarna Sub1 in India, Samba Mahsuri in Bangladesh, IR64-Sub1 in the Philippines, and Ciherang Sub1 in Indonesia and Nepal. Currently, efforts are underway to combine submergence tolerance with tolerance to stagnant flooding, aiming to create rice varieties with enhanced resilience and adaptability to various types of flooding. In this chapter, we will review the advancements in deciphering the molecular and physiological underpinnings of various forms of submergence tolerance. We will also explore endeavors to apply this knowledge to create improved rice varieties that effectively address future challenges.

16.2 PHYSIOLOGICAL, MORPHOLOGICAL, AND MOLECULAR RESPONSES OF PLANTS UNDER SUBMERGENCE

Submergence impairs severe crop losses, especially rice cultivation at flood-prone low land conditions is one of the major concerns, as submergence conditions may directly affect the plant's growth and development, which can lead to irreversible damage to the plant tissues. Often, submergence tolerance has been a challenging trait of interest for researchers, as the effect of submergence is modestly affected by several environmental factors like duration and degree of submergence (Jackson & Ram, 2003). Extensive genetic manipulations and submergence tolerance utilization will remain elusive until the understanding of physiological, biochemical, and molecular responses is uncovered.

The major physiological stress plants encounter under submergence is hypoxia. Hypoxia is the condition where a prevalence of low oxygen levels and eventually oxygen deprivation occurs. Hypoxic conditions can result in impaired cellular function as they hinder the electron transport cycles in mitochondria, which directly affects the synthesis of adenosine triphosphates (ATPs). Anoxic condition is a much more severe form of hypoxia, where the plant is completely deprived of oxygen. Under prolonged flooding, most plant tissues, often entire plants may get submerged, resulting in severe hypoxia. Responses to the submergence are fairly interrelated as several molecular responses eventually incur considerable effects on the physiology and morphology of plants to offer better tolerance. The plant response to low oxygen levels (hypoxia) under submergence originated from the expression of several hypoxia-responsive genes that sense the oxygen-deprived condition and modify the cellular mechanisms like energy/carbohydrate metabolism (Das et al., 2005; Loreti & Perata, 2020). Hypoxic signal sensing by several hypoxia-responsive genes, such as group VII ethylene response factors (ERF-VII) and SNF1-related kinase, was identified in rice (Loreti et al., 2018). Hypoxic and anoxic conditions can affect the seed germination and seedling growth. Yemelianov et al. (2020) showed the change in cellular pH due to potassium efflux in rice during hypoxic stress and canvassing the cellular ion buffering mechanism under submergence stress. Regulation of cellular CO_2 concentration and photosynthetic efficiency were interestingly controlled by the concentration of cellular silica level (Pan et al., 2021). Trait discovery studies on anaerobic seed germination in rice uncover several anaerobic respiration-related factors and enzymes viz RAMY3D. Conversion from aerobic to anaerobic respiration through pyruvate dehydrogenase bypass was a key biochemical pathway in water lodging conditions (Miro & Ismail, 2013). Seedling growth and plant height have been enhanced under flooding stress as an adaptive response. The mechanisms that rice undergoes to maintain submergence tolerance are illustrated in Figure 16.1.

The sucrose synthase gene plays a notable role in seedling growth during the flood adaptation in japonica rice (Li et al., 2020). Hypoxic conditions in turn induce the plant system to develop specialized tissues. Photosynthetic efficiency and photosystem II activity depleted considerably under water lodging and flood conditions. Genotypes like FR-13A had higher photosynthetic activity under submergence as an adaptive response (Das & Sarkar, 2001; Sarkar et al., 2006; Alpuerto et al., 2016; Barik et al., 2019). Submergence-tolerant cultivars were found to maintain more carbohydrates and chlorophyll levels than the non-tolerant ones (Sarkar et al., 1996). Hypoxic conditions are the main causative factor in the generation of reactive oxygen species (ROS) under oxidative conditions. Interestingly, ROS concentration and accumulation can increase in plants under submergence stress, as anaerobic respiration, disruption of canonical respiration pathways, and hindrance in the energy metabolism, especially the electron transport chains in mitochondrial membranes due to hyponic/anoxic conditions are apparent effects of submergence stress (Wu & Yang, 2016).

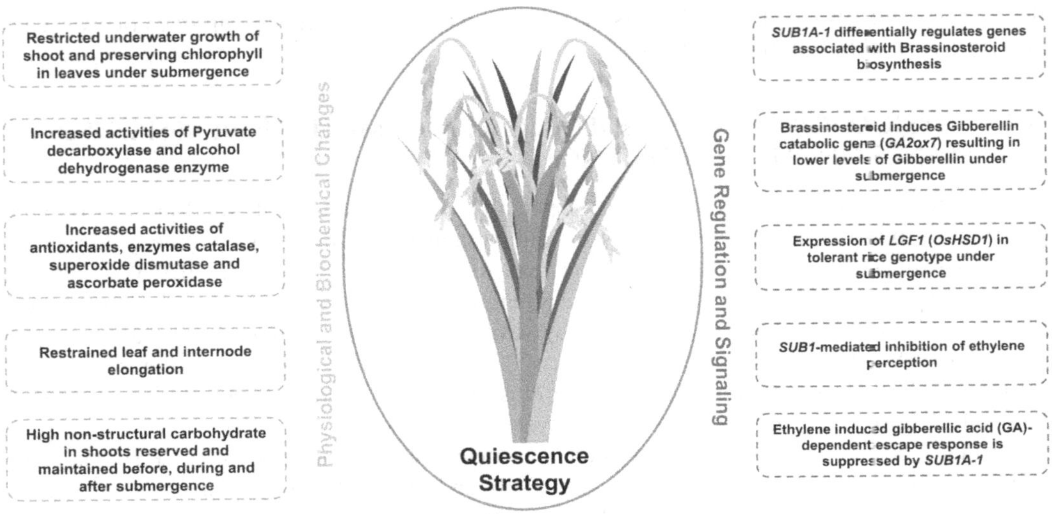

FIGURE 16.1 Mechanisms in rice to cope with submergence.

Sarkar et al. (2001), Das et al. (2004), and Panda and Sarkar (2013) reported the adaptive response of different genotypes of rice on antioxidant defense against ROS and leaf gaseous exchange under submergence conditions. Unlike other plant species, oxidative stress itself is not a potential harm to a non-aerobic rice crop. ROS provides a considerable level of oxidation under semi-aerobic conditions, yet prolonged and waterlogging stress could hamper the cellular machinery, senescence, and growth of rice. Genotypes with oxidative and deoxidative buffering capacity could serve as a donor for submergence tolerance (Upadhyay, 2018). Enzymes like GR, DHAR, and APX play a crucial role in leveraging stress tolerance under flood.

A major morphological adaptation that plants often generate under hypoxic conditions is the development of an adventitious root system and aerenchyma cells. Development of aerenchyma cells and the root system structure like changes in frequency and size of root cortex cells confers potential adaptability to flooding stress in plants (Lin and Sauter, 2020; Yamauchi et al., 2021). Shoot elongation is an empirically analyzed trait in rice for submergence tolerance (Vriezen et al., 2003). Enhanced shoot growth and plant height were found in several flood-tolerant lines, which have been identified to improve the standability of the crop under flood. Thus, a large extent of the physiological and morphological mechanisms result from interaction through phytohormone-induced cellular responses. Elongation of leaf lamina, leaf sheath, and development of aerenchymatic roots are traits of interest for breeding submergence-tolerant rice cultivars.

Molecular mechanisms that counteract the cellular damages incurred by submergence could be phyto-hormonal-driven cellular adaptive mechanisms. Ethylene response factor genes have been mapped and validated in the SUB1 QTL region from submergence tolerant line FR-13A (Xu & Mackill, 1996; Xu et al., 2006; Septiningsih et al., 2009). Deepwater rice cultivars have a similar mechanism to grow under higher water levels for a prolonged time by internode elongation (Hattori et al., 2007, 2008, 2009). The ethylene response factor-related genes SNORKEL 1 & 2 have a similar function to SUB1. However, several research works were conducted to identify the appropriate criteria for designing submergence tolerance rice cultivars (Nagai et al., 2010). Molecular mechanisms related to ethylene response factors in submergence tolerance were also reported in studies by Fukao et al. (2006), Jackson (2008), and Jung et al. (2010). Bailey-Serres et al.'s (2012) group VII ERFs are involved in sensing the low oxygen conditions and are considered a major factor in submergence tolerance. Plant hormones like gibberellins, auxins, brassinosteroids, and ethylene have enhanced the shoot growth in rice (Schmitz et al., 2013; Nghi et al., 2021).

Interestingly, Fukao and Bailey-Serres (2008) reported that submergence tolerance in rice is induced by SUB1A gene which considerably reduces the gibberellins response by SLR1 and SLR2 proteins. SUB1A was found to coordinate the reduced levels of auxin expression but improved the ABA responsiveness as reported by Alpuerto et al. (2022). With this overwhelming evidence, we could consider SUB1-related genes are primarily involved in increased responsiveness of stress-related hormones like ethylene and ABA. Eventually, they lower the level of auxins and gibberellins as they hinder the cellular defense mechanisms. However, SUB1 QTL was not only found to control ethylene response factors but also involved in major signal transduction cascades, involving Mitogen-activated protein kinase. MPK3 gene governs the major regulatory pathway that offers submergence tolerance (Singh & Sinha, 2016).

Proteomic studies on rice submergence regimes lead to an understanding of the molecular mechanism of flood tolerance. Barding et al. (2012, 2013) and Locke et al. (2018) reported differential metabolic regulations by SUB1 through NMR and proteomic studies. Enzymes with hydrolase activity were found to be a major player in regulating cellular stress. Hydrolyzing O-glycosyl enzymes were found to be differentially expressed proteins (DEPs) in the submergence and standard conditions (Xiong et al., 2019).

16.3 BREEDING STRATEGIES FOR SUBMERGENCE TOLERANCE

16.3.1 Breeding for Submergence Tolerance in the Pre-Genomic Era

Submergence tolerance in rice is uncommon, and most of the landraces and cultivars don't tolerate the flood stand. International deepwater rice observational nurseries in IRRI started screening for donors for deepwater rice cultivation and submergence tolerance in the early 1970s (Hillerislambers & Seshu, 1982). Goda Heenati, Thavalu, and Kurkaruppan from Sri Lanka and FR13A and FR43A from India are long-known donors for submergence tolerance. Several other donors used for breeding flood tolerance are given in Tables 16.1 and 16.2. These lines were identified as donors to get used in breeding programs from the 1970s (Vergara & Mazaredo, 1975). In a large-scale screening experiment, done by CRRI, India, less than 1% of the lines were found to survive submergence (Sarkar & Mohanty, 2010). Few submergence-tolerant lines have been identified in rice and they have been extensively used in breeding programmes. Goda Heenati, FR13A, and Kurkaruppan are notable donors of this trait. FR13A is the most common donor for breeding submergence tolerance around the world as it is bestowed with a higher tolerance level for two weeks under submergence. However, FR13A doesn't have the traits of commercial interest and agronomic value, restricting it from being used directly as a variety. Still, as a landrace, this line has been cultivated by farmers in a few pockets of the Indian state of Orissa. It has several undesirable traits like photoperiod sensitivity, tallness, and poor grain yield. Though trait discovery-led breeding provides great opportunities for breeding the crops, very few studies used to uncover submergence-responding traits. One such trait is the aquatic tillering ability under submergence. Gomosta et al. (1982) studied the difference in the aquatic tillering ability among deepwater rice genotypes Habiganj Aman and the cultivated types.

Early inheritance studies by Mohanty and Khush (1985) and Haque et al. (1989) showed that submergence tolerance has significant dominant and additive effects. With 0.7 narrow sense heritability, this trait is highly heritable with high genotypic variance. Genetic studies in the F2 and backcross populations from the crosses between tolerant genotypes FR13A and BKNFR with susceptible variety IR42 were carried out by Mishra et al. (1996). The segregation pattern was found to be 3:1, as this trait was reported to be governed by a single major gene. All of these donors were agronomically poor, with low yield potential and

TABLE 16.1
Donors Identified/Available and Released for Breeding Flood Tolerance in Rice

S. No.	Donors for Submergence Tolerance	Country	Progress Made in Breeding for Submergence Tolerance	Parentage	References
1	IC516149	India	Multiple stress tolerant lines. Potential donor for drought, submergence, and anaerobic germination	Derived line from FR13A	Chakraborty et al. (2023)
2	Madhuka	India	Flash flood tolerant genotype with high internode elongation capacity during submergence	Cultivar	Singh (1981)
3	Bhutahuti	India	Early maturing line with high internode elongation	Cultivar	Datta (1982)
4.	FR 13 A	India	Donor for the majority of breeding programs for submergence tolerance in rice.	Landrace	Bailey-Serres et al. (2010)
5	IR40931-33-1-3-2	IRRI (Philippines)	Derived line from FR13A	FR13A	Septiningsih et al. (2009)
6	Habiganj Aman	India	Line with high aquatic tillering ability	Landrace	Gomosta et al. (1982)
7	Samudrabali	India	Submergence-tolerant landrace from India	Landrace	Barik et al. (2020)
8	Basnamundi	India	Submergent tolerant landrace from India	Landrace	Barik et al. (2020)
9	Godaba	India	Submergent tolerant landrace from India	Landrace	Barik et al. (2020)
10	Surudaka	India	Submergent tolerant landrace from India	Landrace	Barik et al. (2020)
11	Dokarakuji	India	Submergent tolerant landrace from India	Landrace	Barik et al. (2020)
12	FR43A	India	Submergence-tolerant landrace from India	Landrace	Bailey-Serres et al. (2010)
13	BR24	Bangladesh	Widely grown cultivar with moderate submergence tolerance	Cultivar	Zhang et al. (2023)
14	Goda Heenati	Sri Lanka	Potential donor line for breeding submergence tolerance	Landrace	Panda et al. (2021)
15	Kurkaruppan	Sri Lanka	Potential donor line for breeding submergence tolerance	Landrace	Panda et al. (2021)
16	Thavalu15314	Sri Lanka	Potential donor line for breeding submergence tolerance	Landrace	Panda et al. (2021)
17	Thavalu15325	Sri Lanka	Potential donor line for breeding submergence tolerance	Landrace	Panda et al. (2021)
18	BKNFR	India	Improved variety with semi-dwarf stature	Derived line from FR13A	Mishra et al. (1996)
19	IR49830-7-1-2-2	IRRI (Philippines)	A derived line from FR13A, the donor parent for the famous Swarna sub1 Variety	FR13A	Septiningsih et al. (2009)
21	Shwe-Thwe-Yin-Hye	Myanmar	Multiple stress-resistant donor, moderately tolerant to submergence stress	Landrace	Zhang et al. (2023)
23	Cisanggarun	Indonesia	Multiple stress-resistant donor	Landrace	Zhang et al. (2023)

TABLE 16.2
Varieties Identified/Available and Released for Breeding Flood Tolerance in Rice

S. No.	Varieties for Submergence Tolerance	Varietal Specifications	Parentage	References
1	CR1009 Sub1	Variety released through marker-assisted integration of sub1 QTL in CR1009, an existing long-duration variety	CR 1009 x FR13 A	Robin et al. (2019)
2	Samba Mahsuri-Sub1	Improved line-by-marker aided selection with SUB1 QTL in the background of the famous cultivar Samba Mashuri	Samba Mashuri x IR49830	Septiningsih et al. (2009)
3	Ranjit sub1	Improved line-by-marker aided selection with SUB1 QTL	Ranjit x Swarna sub1	Chetia et al. (2018), Gautam et al. (2022b)
4	Bahadur sub1	Improved line-by-marker aided selection with SUB1 QTL	Bahadur x Swarna sub1	Gautam et al. (2018, 2022a, 2022b)
5	IR64-Sub1	Improved line-by-marker aided selection with SUB1 QTL	IR64 x IR49830	Septiningsih et al. (2009)
6	Swarna Sub1	Improved line-by-marker aided selection with SUB1 QTL	Swarna x IR49830	Septiningsih et al. (2009)

tall stature; the need to develop short semi-dwarf, submergence-tolerant varieties was the major thrust of research at IRRI. Several breeding programs were launched to introduce/utilize the submergence tolerance in erstwhile elite high-yielding semi-dwarf cultivars (Mohanty & Chaudhary, 1986; Mackill et al., 1993). Numerous cultivars were identified in India during the late 1970s and 1980s for deep, semi-deep, and flash flood tolerance. Few among them are FR13A & FR43B as flash flood tolerant types and Marsh 1& 2 as semi-deepwater rice in Orissa, Madhukar in Uttar Pradesh, BR13, BR14 in Bihar, and Habiganj 1 & 2 in Assam. Several other types of varieties from Tamil Nadu and Kerala (Co14, PTB7, and PTB15) also served as a source for submergence tolerance (Singh, 1981). An interesting study (Datta, 1982) showed that the earlier growing types are most suited for flood tolerance. The study canvasses the internode elongation as a trait to breed, as the genotypes that show elongated internodes can outplay the deepwater condition. Some early maturing types like Madhukar, GS 9, Bhutahuti, and TC30 have significantly higher internode elongation rates, which escapes the ill effects of submergence tolerance. Figure 16.2 depicts the methodology used in the development of submergence-tolerant lines.

Breeding for submergence tolerance is tedious not because of the trait complexity but the inaccuracy and difficulty in phenotyping. Phenotyping for submergence tolerance needs controlled water-lodged conditions for a specific period, say 12–14 days. Selecting the breeding materials and segregating generations by phenotyping is an uphill task. Thus, most breeding programs worldwide use specific donors for breeding for submergence tolerance via marker-assisted selection (MAS). Most importantly, a well-known potential donor to serve this purpose has a tolerant allele at the SUB1 locus of a landrace FR13A that has been used (Bailey-Serres et al., 2010; Breseghello & Coelho, 2013). There is no report of any induced mutant with submergence tolerance in rice. However, a study (Datta, 1982) showed that a promising mutant (Mutant 8R1) was subjected to several screening procedures for flood tolerance.

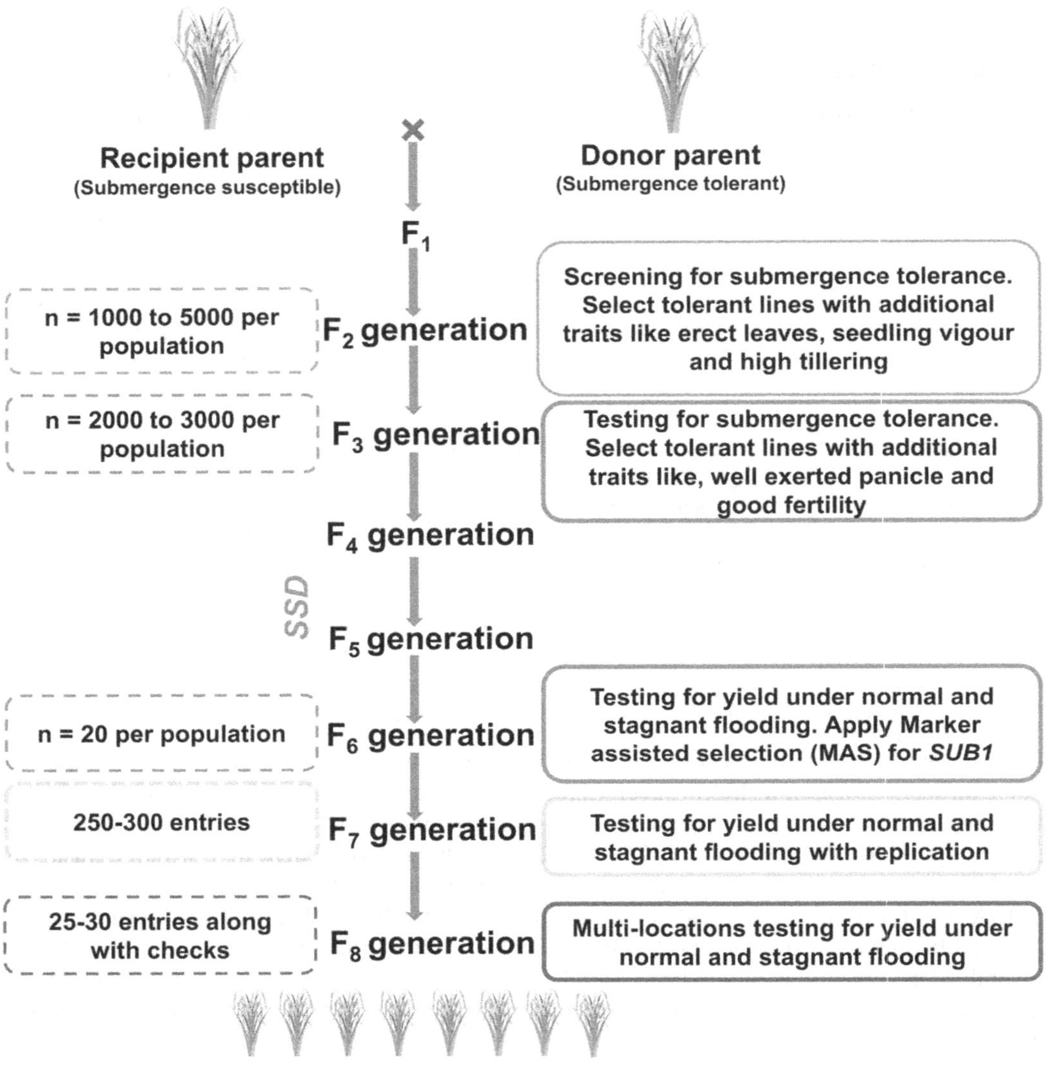

FIGURE 16.2 Methodology for developing submergence-tolerant lines.

16.4 GENOMICS ERA (HIGH-THROUGHPUT GENOTYPING USING THE NGS PLATFORM)

The Genomics Era has overwhelmingly shifted our understanding and approach toward enhancing submergence tolerance in plants, predominantly leveraging the potent capabilities of next-generation sequencing (NGS) platforms. NGS, with its capacity to execute high-throughput genotyping, has been pivotal in advancing our genomic insights, especially in understanding the genetic architecture underpinning the ability of plants to endure submergence stress. A landmark revelation in this domain was identifying and comprehensively characterizing the SUB1A gene in rice, a notable determinant for submergence tolerance (Xu et al., 2006). SUB1A governs a cascade of ethylene-mediated physiological responses under inundation conditions, strategically restraining growth and conserving energy to enhance the plant's survivability during and after submergence. Additionally, NGS has enabled researchers to explore genome-wide association studies (GWAS) to uncover various QTLs related to submergence tolerance. This comprehensive genetic mapping has enabled the elucidation of complex gene networks and metabolic pathways that plants leverage to modulate their physiological and metabolic states during submergence (Septiningsih et al., 2009).

Consequently, the bountiful genomic data produced by NGS platforms, mapping genotypic variability among assorted cultivars and wild relatives, are synergistically amalgamated with traditional breeding methodologies. This convergence aims to fabricate novel cultivars that inherently possess augmented resilience against submergence stress, ensuring their survival and maintaining yield stability under fluctuating environmental conditions. As we plunge deeper into the genomics era, the intertwined pathways of classical breeding and contemporary genomic science beckon the prospect of achieving sustainable agricultural paradigms aptly tailored to navigate the challenges imposed by a dynamically altering global climate.

16.4.1 Utilising Transgenic Methodologies to Shield Rice from Submergence

The endeavor of securing rice against prolonged submergence has significantly advanced through transgenic technology, especially in an epoch where global climate shifts pose stark challenges. Classical breeding has faced its share of constraints. Yet, transgenics have ushered in a novel paradigm, infusing beneficial genes into rice genomes, thereby providing an additional defense against floods. The prominent Sub1 gene, by modulating shoot elongation during underwater phases, enables the rice to utilize its energy judiciously, ensuring not only a superior survival rate during inundation but also safeguarding other pivotal agronomic traits and maintaining yield (Hattori et al., 2009).

16.4.2 Harnessing the CRISPR-Cas9 Mechanism for Enhanced Rice Resilience

The CRISPR-Cas9 genome editing technology has bestowed a precise and dependable mechanism for modifying genomes, primarily exemplified in the quest for developing submergence-tolerant rice. This technology equips researchers with the means to alter designated genes precisely, accentuating certain traits such as enhanced submergence endurance. A detailed study by Xu et al. (2021) exemplifies how mutants were formulated by intercepting the OsTPP7 gene, which is instrumental in regulating shoot growth in the initial stages, signaling a leap forward in developing robust rice strains.

16.4.3 The Dawn of Base Editing and Its Implications for Rice Cultivation

The advent of innovative CRISPR technologies, particularly base editing, has emerged as a beacon for precision-driven genome editing, offering a mechanism that circumvents double-strand breaks, thus reducing unintended mutations and bolstering specificity. Li et al. (2022) illuminate the potential inherent in this technology, depicting the successful utilization of an optimized adenine base editor to instigate specific point mutations within the rice genome. This feat underscores the capacity of base editing to meticulously modulate traits crucial for agricultural stability, enhancing crop robustness. Diversifying and expanding upon conventional CRISPR-Cas9 mechanisms, myriad CRISPR modifications have surfaced, each endowed with distinctive utilities that enhance its applicability in genome editing. Incorporating varied Cas proteins and calibrating the CRISPR system to accommodate diverse genetic engineering needs have remarkably expanded its applicability. This encompasses the capacity to target an expanded range of genomic locations and engineer various editing outcomes, providing researchers with a sophisticated toolkit for nuanced genetic modifications essential for delving into intricate traits like submergence tolerance in rice.

As the global climate undergoes palpable shifts, threatening agricultural equilibriums, the concerted and advanced exploration of transgenics, CRISPR technologies, and pioneering CRISPR modifications emerge as a pivotal ally in sustaining rice production. Developing rice varieties that showcase superior submergence tolerance without compromising on other vital agronomic features is crucial. This amalgamation of genetic engineering and genome editing technologies is poised to serve as a crucial ally in navigating the multifaceted challenges that loom over the future of rice cultivation, ensuring the sustainability and productivity of this essential crop amidst fluctuating environmental scenarios.

16.4.4 Submergence QTL

Rice is seriously at risk from flooding. Flooding is becoming more frequent due to climate change (Rao et al., 2016). Rice farming is in jeopardy on a global scale. Plants are sequentially subjected to various challenges due to floods (Septiningsih & Mackill, 2018). Growing global climate change worsens this problem because enhanced, high-yielding mega rice cultivars are vulnerable to flooding (Afrin et al., 2018). Over 100 million people's livelihoods are impacted yearly by flash floods threatening over 22 million hectares of rice paddy. The population of the globe is growing at an unprecedented rate; it is anticipated to reach 9 billion people by the year 2050 (Leridon, 2020). So, to solve difficulties with food scarcity, the current situation calls for increased food production soon (Wang et al., 2016). Only a few native rice landraces are cultivated by underprivileged farmers due to numerous flooding stress types common in Asian countries (Barik et al., 2020).

A QTL is a genomic region that controls quantitative traits and is controlled by several genes with minor effects. Researchers have made significant progress in the past 20 years in determining the genetic basis for increasing rice flooding tolerance. By utilizing molecular biology and molecular markers, technological advances have identified several QTL/genes of flooding tolerance through the cross between parent of a flooding-sensitive and a flooding-tolerant rice variety. Research on flooding tolerance advanced significantly by identifying the primary QTL *Sub1* on chromosome 9, which is the key factor to flooding tolerance. In addition, several other QTLs and genes, including *AG1, AG2, SK1,* and *SK2,* have been identified and can be used as genetic marker for selecting cultivars that are tolerant to flood by utilizing the MABC method.

Various studies have identified the *Sub1* gene as a common QTL. This QTL was first identified and mapped on chromosome 9 within the local rice landrace FR13A (Ram et al., 2002). Xu et al. (2006) further delineated QTLs responsible for 70% of the phenotypic variance in flooding tolerance, with the remaining 30% attributed to other minor QTLs situated on chromosomes 1, 2, 5, 7, and 10, respectively. The *Sub1* QTL encodes putative ethylene response components, including *Sub1A*, *Sub1B*, and *Sub1C*. During submergence, the expression of these genes is upregulated. Notably, genes *Sub1B* and *Sub1C* are present across all types of rice. However, in local rice landraces like *FR13A* and other tolerant rice genotypes, the *Sub1A* gene has been identified as the key player within QTL Sub1, solely responsible for conferring flash flooding tolerance (Bailey-Serres et al., 2010). Aligning with these findings, the *Sub1* gene is localized within a 0.16 cM genomic region on chromosome 9. Xu et al. (2006) utilized a population of 3,000 F_2 plants in their study. Two allele variants of the *Sub1A* gene, known as *Sub1A-1* and *Sub1A-2*, have been documented. Of these, the *Sub1A-2* gene variant is specifically present in tolerant rice genotypes (Xu et al., 2006; Septiningsih et al., 2012).

Both alleles (*Sub1A-1* and *Sub1A-2*) encode the same proteins except for Ser186 and Pro186, respectively (Bailey-Serres et al., 2010). The *Sub1* gene controls the lengthening of shoots, stores carbohydrates, and protects chlorophyll levels. Therefore, as flooding decreases, the *Sub1* gene controls plant survival and regenerates the growth of the tolerant rice genotype (Pucciariello & Perata, 2013). Flooding caused the *Sub1A* gene in some rice cultivars to limit the production of ethylene, which allowed the gibberellic acid signaling repressors to accumulate mRNA and protein (Fukao et al., 2019). In order to prevent prolonged growth and maintain essential carbs for survival and recovery during flooding conditions, the *Sub1A*-mediated control of gibberellic acid response represses genes required for sucrose and starch catabolism. This quiescence method is frequently used in rice breeding to create rice varieties that can withstand flooding (Septiningsih et al., 2009).

The anaerobic germination capacity of rice BC_2F_2 populations studied by Angaji et al. (2010) crossed a rice variety sensitive to flooding with the tolerant donor, Khao Hlan On, originated from Myanmar. Five different QTLs were discovered in this study, *qAG1.2*, *qAG7.2*, *qAG9.1*, *qAG9.2*, and *qAG3.1*, explaining 17.9–33.5% of the phenotypic variance. The QTL (*qAG9.2*), with a phenotypic variance of 33.5% and 20.3 LOD score, located on the long arm of chromosome 9. It was determined that the influence of this important QTL was restricted to a 50 KB chromosomal area. The *Trehalose-6-phosphate* gene (*qAG9.2*) underlies QTL and was subsequently cloned and functionalized (Loreti 2018). The gene associated with trehalose-6-phosphate metabolism is OsTPP, under anaerobic germination, which results in starch mobilization and rapidly elongates coleoptiles, increasing germination resistance to flooding stress (Singh et al., 2017). This gene *qAG9.2* evolved to enhance CihehrangSub1 for anaerobic germination tolerance; the QTL contributes to high-yielding rice genotypes (Toledo et al., 2015). Rice AG potential has been the subject of additional QTL-targeted studies (Baltazar et al., 2014). Three pairs of epistatic loci were detected on chromosomes 2, 3, 5, and 11, accounting for 0.5–19.6% phenotypic variance. Two QTL on chromosomes 5 and 11 were identified by Jiang et al. (2006) in a different study using a population (USSR5/ N22) with 11–15.5% PVE. In addition to two minor QTLs, major QTLs originated from the Chinese floods tolerant rice variety Ma-Zhan Red which has also been found on chromosome 7. In their investigation of the Nanhi QTLs (from a landrace), Baltazar et al. (2019) found that chromosome 7 has a rich genomic region to the major QTL. Four QTLs, named as *qAG3*, *qAG7.1*, *qAG7.2,* and *qAG7.3*, were detected in the $F_{2:3}$ populations with 8.1–12.6% PVE and LOD 5.7–7.7. Additionally, Zhang et al.'s (2017) study used 5,291 SNPs for the GWAS to assess expression patterns in 432 Indica cultivars. The researchers detected 15 loci and concluded gene *LOC Os06g03520* confers flooding resistance on rice at the time of germination. In anaerobic rice, the *hexokinase stretch-carrying gene HXK6* was identified and found to be involved in coleoptile elongation. Biparental QTL mapping and a GWAS analysis identified 88 loci with AG potential. The researchers examined the six different rice cultivars' RNA sequences and found that the rate at which coleoptiles elongate in anoxic conditions varied (Hsu & Tung, 2017). Yang et al. (2019) conducted a study using recombinant inbred lines (RIL) derived from hybridization between japonica and an indica cultivar. They employed a high-density bin map comprising 2,711 bin loci and successfully identified 25 loci. Concurrently, Nishimura et al. (2020) utilized a population of chromosome-segment replacement lines within the IR64 genetic background under anaerobic conditions, discovering a novel QTL termed *qACE3.1.* This QTL confers anaerobic germination tolerance, specifically through coleoptile elongation. Additionally, Jeong et al. (2020) employed RIL derived from a cross between a photoblastic rice and the Korean Japanese rice cultivar Nampyeong, revealing three QTLs, i.e., *qAG1*, *qAG3*, and *qAG11*. Among these, the *AG1* gene was identified as the candidate gene for the anaerobic germination QTLs (*qAG.9.2*), influencing coleoptile elongation during anaerobic germination. Further molecular investigations are required to fully elucidate the role of the *HXK6* gene in anaerobic germination tolerance. Furthermore, GWAS studies have been instrumental in identifying novel genes associated with anaerobic germination (Hsu & Tung, 2015). GWAS has identified a strong candidate gene for a *DUF* domain protein, although its significance for AG tolerance and functional characterization is unclear (Zhang et al., 2017). For a better understanding of the mechanism underlying anaerobic germination tolerance in rice, more research is needed to clarify the specifics of this crucial QTL's biological process response (Kuya et al., 2019). This QTL accounted for 69% PVE for submergence tolerance with an LOD score of 36 (Dixit et al., 2017). On chromosome 9, a 0.16 cM area containing 3,000 F_2 individual plants was used to map and identify the *Sub1* QTL (Xu et al., 2000). The locus was subsequently narrowed down to 0.075 cM by Xu et al. (2006) utilizing 4,022 F_2 individual plants and a fine-scale physical map approach. The putative ERF is encoded by a group of three genes at the *Sub-1* locus: *Sub1A*, *Sub1B*, and *SUB1C*. Table 16.3 provides information on the QTLs and multiple trait associations (MTAs) found for traits associated with submergence tolerance.

TABLE 16.3
Quantitative Trait Loci and MTAs Identified for Submergence Tolerance-Related Traits

Population Name	QTLs	Chromosome No	Function of QTLs	Linked Primers	PVE (%)	References
IR4093 1-26 × PI543851 (F2)	*Sub1*	9	A major locus for submergence tolerance	RFLP markers	69	Xu and Mackill (1996)
Kinmaze × DV85 (RIL)	*qAG1, qAG2, qAG5a, qAG5b, qAG7*	1, 2, 5, 5, 7	Increases tolerance to anaerobic germination	–	10.5–19.6	Jiang et al. (2004)
USSR5 × N22 (F2)	*qAG-5* and *qAG-11*	5 and 11	Increases tolerance to anaerobic germination	RM169 and RM254	15.5 11.0	Jiang et al. (2006)
	LSD1, EDS1, and *PAD4*		*LSD1* enhances disease susceptibility, *EDS1* and *PAD* regulates the development of lysigenous aerenchyma	–	–	Muhlenbock et al. (2007)
W0120 × C9285 (NILs)	*qTIL12, qNE112*, and *qLE112*	12	Internode elongation in response to water rise	RM3739, M10-3	14–36	Hattori et al. (2007)
	SK1 and *SK2*		*SKs* genes encoding ethylene-responsive factors, regulate the internode elongation in deepwater rice.	–	–	Hattori et al. (2009)
Khao Hlan On × IR64 (BC2F2)	*qAG1-2, qAG 3-1, qAG-7-2 7, qAG-9-1, qAG-9-2*	1, 3, 7 and 9	All the QTL increases tolerance of flooding during germination	RM149,RM327, RM478, RM520, RM582, RM5485, RM7571,RM10694, RM23877	17.9–33.5	Angaji et al. (2010)
Madabaru × IR72 (F2:3)	*qSUB1.1, qSUB2.1, qSUB9.1, qSUB12.1*	1,2,9 and 12	Regulating of ethylene-dependent pathway mediated by the *Sub1A* gene	RM 511, RM 463	52.3, 36.4, 18.6, 16.3	Septiningsih et al. (2012)
IR42 × Ma-Zhan Red (BC2F2)	Six QTLs	2,5,6,7	Increases tolerance of flooding during germination	RM263, RM5378, RM204, RM402, RM313, RM28766	31.7	Septiningsih et al. (2013)
Nanhi × IR64	*qAG7*	7	Increases tolerance to anaerobic germination		22.3	Baltazar et al. (2014)
Nipponbare and IR64 (RIL)	*HXK6*	1	Encoded a hexokinase	SNP	–	Hsu and Tung (2015)
IR64 × Khao Hlan On (NILs)	*OsTPP7*	9	Improved germination and elongation of coleoptiles.	RM3769, RM24141, RM24161, RM105		Kretzschmar et al. (2015)
–	*AUX1*	1	Initiation of rice collateral root control	DFR-F2, DFR- R2 and DFR-LB2	–	Zhao et al. (2015)

IR42 × FR13A (RIL)	*qSUB1.1, qSUB4.1, qSUB8.1, qSUB9.1, qSUB10.1*	1, 4 ,8, 9 and 10	Enhancing submergence tolerance		20.20, 12.40, 13, 69, 15.80	Gonzaga et al. (2015)
–	*LOC_Os06g03520*		Encoding DUF domain containing protein and enhance coleoptile elongation	–	–	Zhang et al. (2017)
Tai Nguyen × Anda	*qAG1a, qAG1b*, and *qAG8*	1 and 8	Increases tolerance to anaerobic germination		5.5, 14.1, and 5.9	Kim et al. (2019)
–	*SD1*		*SEMIDAWRF* gene representing a gibberellin biosynthesis gene, promoted elongation of internode in rice	–	–	Cheng et al. (2022)
IR64 × Kharsu 80 A (BC2F2)	*qAG7.1, qAG7.2, qAG7.3* on 7 and *qAG3* o	3	Increases tolerance to anaerobic germination	id3002377-id3004190	8.1–12.6	Baltazar et al. (2019)
–	*TIR1*	5	Encoding a F-Box auxin receptor protein	–	–	Nghi et al. (2019)
PBR × Nampyeong (BC2F2)	*qAG1, qAG3, qAG11*	1, 3, 11	Confirm tolerance for anaerobic germination	–	13.44, 6.71, and 14.52	Jeong et al. (2020)
T65 × C9285 (NILs)	*ACE1 and DEC1*	3 and 12	*ACE1* imparts divisional capability on intercalary meristematic cells and hence controls internode elongation along with GA; *DEC1* limits the elongation of the internodes	–	–	Nagai et al. (2020)
IR64/Koshihikari (CSSLs)	*qACE3.1*	3	Elongation of coleoptiles under anaerobic germination	P1676-3		Nishimura et al. (2020)
–	*LOC_Os03g31550*	3	Represents a xanthine dehydrogenase enzyme	SNP chip	–	Rohilla et al. (2020)

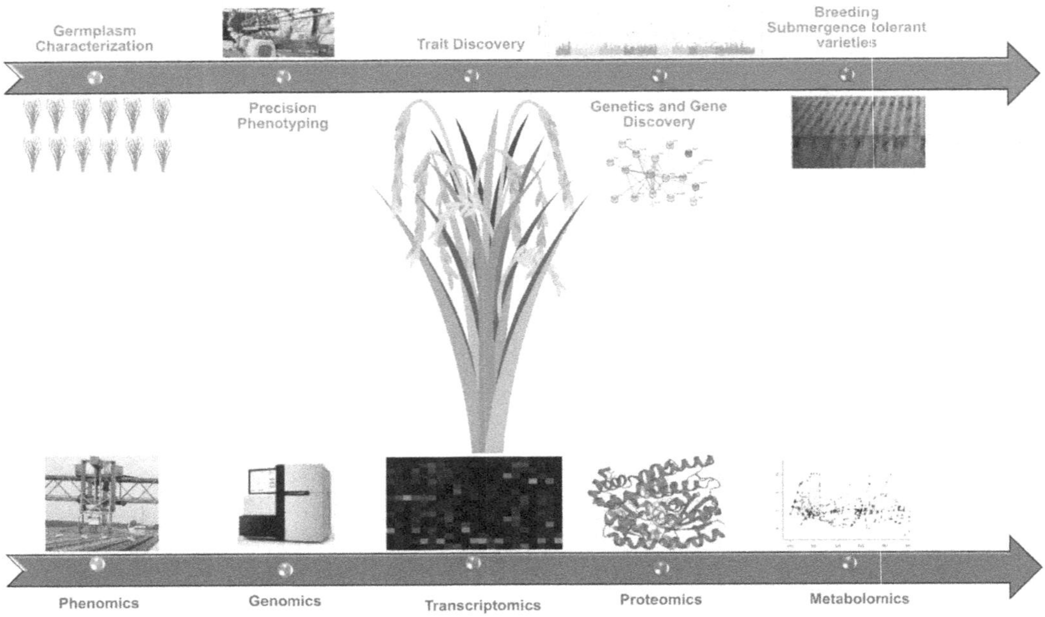

FIGURE 16.3 Integration of new omics and high-throughput phonemics approaches to breed new submergence-tolerant varieties.

Researchers have mostly used the FR13A landrace as the best donor for submergence tolerance. Therefore, it has been demonstrated that *Sub1A-1* is the main indicator of rice flooding resistance (Septiningsih et al., 2009). Kurokawa et al. (2018) reported gene *Leaf Gas Film1* (LGF1/OsHDS1), which is identified as predominantly significant for surface hydrophobicity and leaf gas film retention in leaves of submerged rice. Increased flood tolerance in rice (Septiningsih et al., 2009) significantly contributes to the underwater gas exchange. The discovery of such a gene also provides a chance to investigate differences in the ability of different types of rice to retain a gas film, as well as the potential to pinpoint alleles that have higher expression of *LGF1/OsHSD1* gene (Winkel et al., 2014). The integration of new omics and high throughput phonemics methods, which was done to breed new submergence-tolerant types, is depicted in Figure 16.3.

Hamamura and Kupkanchanakul (1979) used di-allele crosses and an investigation of the progeny for submersion resistance to report partial dominance and contribution of 5–6 genes governing the floating ability of rice species. Additionally, researchers typically utilize the lowest-length internode location and evaluate the internodes' overall length to assess the rice features in deepwater Inouye (1983).

Tripathi and Rao (1985) state that a single dominant gene is responsible for early nodal differentiation. Suge (1987) did characterize the impact of complementary genes controlling internode elongation, and this elongation in deepwater rice genotypes is caused by a single recessive gene (*dw3*), according to Eiguchi et al. (1993). In deepwater rice, another study found associated genomic regions in a RIL (IR74 and Jalmagna). Total 26 QTLs that control plant elongation and flooding tolerance have been reported (Sripongpangkul et al., 2000). Two QTLs for the least amount of internode elongation were located on chromosomes 3 and 12 (Nemoto et al. 2004). Table 16.4 displays the improved varieties that were released using recurrent parents and donors, and the markers associated with QTL.

Kawano et al. (2008) conducted a mapping study utilizing internode elongation rate and the least elongated internode as indicators. They successfully identified two QTLs/genes associated with each indicator, located at chromosome loci 3 and 12, and 1 and 12, respectively. Meanwhile,

TABLE 16.4
List of the Improved Varieties Released Using the Molecular Breeding Method

Improved Variety	Parent (Donor)	Parents (Recurrent)	QTLs	Population	Markers Linked	Country	Reference
Swarna-Sub1	IR49830-7	Swarna	*Sub1*	BC3F2	RM219, RM464A, RM316	Philippine	Neeraja et al. (2007)
AS996-Sub1	IR64 Sub1	AS996	*Sub1*	BC3F1	ART5, SC3	Vietnam	Huyen et al. (2012)
BR11-Sub1	IR40931-33-1-3-2	BR11	*Sub1*	BC2F2	RM833	Bangladesh	Iftekharuddaula et al. (2015)
NSIC Rc222- Sub1 + AG1	Ciherang- Sub1 + AG1	NSIC Rc222	*Sub1* + *AG1*	BC1F1	RM8300 for Sub1, RM3769, RM24141 for AG1	Philippine	Rafael et al. (2015)
Ciherang-Sub1, PSB Rc18-Sub1	IR64-Sub1	Ciherang, PSB Rc18	*Sub1*	BC1F1, BC3F3	RM495, RM220 RM243, RM10793 RM1008, RM543 RM71, RM208	Philippine	Septiningsih et al. (2015)
Ciherang-Sub1 + AG1	IR64-AG1	Ciherang-Sub1	*Sub1* + *AG1*	BC2F2	RM231, RM561, RM316, RM23833	Philippine	Toledo et al. (2015)
BRRI dhan33-Sub1	BR11	BRRI dhan33	*Sub1*	BC4F3	RM8300, Sub1C173, RM296, RM23915	Bangladesh	Iftekharuddaula et al. (2016)
MR219-Sub1	Swarna-Sub1	MR219	*Sub1*	BC2F2	RM8300, RM219, RM23805	Malaysia	Ahmed et al. (2016) and Mojulat et al. (2017)
Jyothi-Sub1	Swarna-Sub1	Jyothi	*Sub1*	BC2F3	Sub1BC1, Sub1 BC2, Sub1BC3, ART5, IYT1	India	John et al. (2017)
Jaya-Sub1	Swarna-Sub1	Jaya	*Sub1*	BC2F3	RM23668, RM23679, RM316, RM8303	India	Arya and Shylaraj (2018)
CO 43-Sub1	FR13A	CO 43	*Sub1*	BC3F3	ART5, RM23877, RM219, RM3769	India	Hifzur et al. (2018)
PSBRc82-AG1, PSBRc82- AG2, PSBRc82-AG1-AG2	Kho Hlan On for AG1, Ma-Zhan Red for AG2	PSBRc82	*AG1*, *AG2*, *AG1* + *AG2*	–	RM3769, RM24141 for AG1,	Philippine	Mondal et al. (2020)
Ciherang-Sub1-AG1, Ciherang-Sub1-AG2, Ciherang-Sub1-AG1-AG2	Kho Hlan On for AG1, Ma-Zhan Red for AG2	Ciherang-Sub1	*Sub1* + *AG1*, *Sub1* + *AG2*, *Sub1* + *AG1* + *AG2*	–	RM8300 for Sub1, RM3769, RM24141 for AG1	Philippine	Mondal et al. (2020)
Aiswarya-Sub1	Swarna Sub1	Aiswarya	*Sub1* + *Saltol*	BC2F2	ART3, ART5, Sub1C173, Sub1AB1, Sub1BC1	India	Nair and Shylaraj (2021)
Maudamani-Sub1	Swarna-Sub1	Maudamani	*Sub1*	BC3F2	Sub1- A203, Sub1-BC2, RM8300	India	Pandit et al. (2021)

Hattori et al. (2007) employed three indicators-internode length, lowest elongation of internodes, and the total number of elongated internodes to analyze QTL mapping. Their study revealed genomic regions on chromosome locations 1, 3, and 12 associated with stem elongation in deepwater rice. In a related study, Hattori et al. (2008) identified QTLs on chromosome 12 that influence internode elongation in deepwater rice cultivars, using near-isogenic lines within the genetic background of Taichung65. Furthermore, through positional cloning and functional investigation, Hattori et al. (2009) detected two putative ERFs, *SK1* and *SK2* genes, corresponding to *Snorkel1* and *Snorkel2* on chromosome 12, respectively. These genes were found to be absent or dysfunctional in all Japanese non-deepwater rice types. The expression of these genes was significant in various parts of the plant, including the leaf sheath, blade, and basal stem area, involving nodes and internodes. The expansion of deepwater rice was associated with cell elongation and was regulated by developmental, hormonal, and environmental cues (Cho & Kende, 1997). The increased expression during internode lengthening was linked to softer cell walls and a change in cellulose microfibril orientation (Sauter et al., 1993). Additionally, ethylene was found to stimulate aerenchyma formation and elongation in the internode (Steffens et al., 2011). Nagai et al. (2012) investigated and identified two QTLs, *qTIL2* and *qTIL4*, controlling early internode elongation in deepwater rice during the initial seedling phase. Furthermore, a genome-wide eQTL study mapping expression QTLs based on RIL populations (Taichung 65 and Bhadua) found 10,047 *eQTLs*, including 2,902 cis-eQTLs and 7,145 trans-eQTLs, with a majority located on chromosomes 1, 4, and 12 (Kuroha et al., 2017). Using 148 inbred lines from a cross between IR10F365 and Ciherang-Sub1, Singh et al. (2017) investigated QTLs associated with stationary floods. They identified major QTLs for yield components and important agronomic traits at the IRRI. Further validation of these QTLs through targeted molecular genetics and advanced molecular breeding is essential for their utilization. A detailed overview of breeding lines utilized for varietal improvement via the molecular breeding approach is presented in Table 16.5.

Despite substantial advances in using the MABC approach to develop rice varieties that can withstand flooding, there are still some serious problems with flood tolerance in molecular rice

TABLE 16.5
A List of Breeding Lines Used for Varietal Improvement Using Molecular Breeding Approach

Breeding Lines	Name	Year	Country
IR05F102 (Swarna)	Improved Suwarna	2009	India
	INPARA-5	2009	Indonesia
	BRRI dhan-51	2010	Bangladesh
	Swarna -Sub1	2011	Nepal
	Yemyoke Khan	2011	Myanmar
IR07F102 (IR64)	NSIC Rc194	2009	Philippines
	INPARA-4	2009	Indonesia
IR07F290 (BR11)	BRRI dhan-52	2010	Bangladesh
IR09F436 (Ciherang)	INPARA-5	2011	Indonesia
	BINA dhan11	2013	Bangladesh
IR07F101 (S. Mahsuri)	S. Mahsuri-Sub1	2011	Nepal
	S. Mahsuri-Sub1	2012	India
	BINA dhan 12	2013	Bangladesh
IR49830-7-1-2-1-3	Popoul	1999	Cambodia
IRRI 119	PSB Rc 68	1997	Philippines

breeding. The severity and duration of floods have gotten harsher and longer recently due to extreme weather conditions made worse by climatic factors. The *Sub1* introgressed mega-type varieties lack yield stability under different flood regimes and are vulnerable to stationary flooding. Therefore, a superior allele that can offer even higher tolerance and stability for flood stress rather than the *Sub1A* gene or unique genes must be identified. There currently needs to be more understanding of all plant stress-response mechanisms. To reduce flood stress, MABC is a widely used method for developing flood-tolerant rice varieties. Additionally, methods for editing the genome, particularly the CRISPR/Cas9 system, mostly based on designed site-specific nucleases, have recently seen a significant improvement. These novel methods produce local point mutations, deletions, or insertions that enhance the functionality of genes linked to biotic and abiotic stress.

16.5 CONCLUSION AND FUTURE PROSPECTS

The incorporation of the SUB1 region via marker-assisted breeding has notably improved submergence tolerance in numerous high-yield crop varieties, while not impeding their growth, productivity, or grain quality. Although there have been notable breakthroughs in utilizing the MABC technique to create flood-tolerant rice varieties, certain critical challenges remain to be addressed in the context of enhancing flood tolerance through molecular rice breeding. As a result of exacerbated climatic factors, extreme weather conditions have led to an escalation in both the intensity and duration of floods in recent years. Mega varieties introgressed with Sub1 exhibit vulnerability to stagnant flooding and fail to deliver consistent yields across various flood regimes. Hence, it is imperative to discover a superior allele, distinct from the Sub1A gene, capable of delivering even greater tolerance and stability under flood stress. Currently, there is a lack of comprehensive understanding regarding the complete mechanisms governing plant stress responses. To advance further, it is crucial to investigate and establish connections between stress response data in specific cells, tissues, and organisms. Furthermore, endeavors should be made to enhance the development of superior, flood-resistant varieties with high yields by combining multiple genes such as Sub1, AG1, AG2, SK1, and SK2 into a single elite genetic background capable of thriving in diverse flood conditions and enhancing grain yields within rainfed agroecosystems. In light of the abiotic and biotic stresses prevalent in rainfed lowland areas, the cultivation of rice varieties resilient to diverse stresses can be realized by combining the Sub1 gene with other traits or genes associated with tolerance, ultimately enhancing adaptability to the challenging rice habitat. The effectiveness and outcomes of gene pyramiding can be influenced by various factors. The effectiveness of pyramiding is significantly enhanced when the characteristics of the gene of interest are well understood, and when the most reliable markers are utilized during the selection process. Conversely, pyramiding is less likely to succeed when QTLs have minor effects. This is due to the possibility that these loci may yield a higher rate of false positives, and imprecise localization of loci may necessitate selecting a larger number of markers to cover extensive genomic regions. The effectiveness of selection and breeding is also dependent on the breeder's ability to recognize the target gene and the desired traits. Moreover, the availability of funding resources might affect the number of populations and crosses that can be incorporated into the pyramiding strategy within the breeding program. In addition to MABC, genome editing techniques, particularly the CRISPR/Cas9 system, have seen significant advancements in recent years. These techniques rely on engineered site-specific nucleases and have been widely adopted in the development of flood-tolerant rice varieties. These emerging techniques enhance the functionality of genes associated with biotic and abiotic stress by introducing precise local point mutations, deletions, or insertions in the target genes. To bolster flooding tolerance in rice, genes like Sub1A, SK1, SK2, OsTPP7, SD1-DW, AG1, and AG2 can be modified using various genome editing tools. The progression of new technologies in molecular genetics, genome editing tools, and contemporary breeding methods, including the MAS, can be collectively employed to develop flood-tolerant rice varieties tailored for various types of flooding ecosystems.

REFERENCES

Acuin J., Firestone R., Htay T. T., Khor G. L., Thabrany H., Saphonn V., et al. (2011). Southeast Asia: An emerging focus for global health. *Lancet*, *377*, 534–535. doi: 10.1016/S0140-6736(10)61426-2.

Afrin, W., Nafis, M. H., Hossain, M. A., Islam, M. M., & Hossain, M. A. (2018). Responses of rice (*Oryza sativa* L.) genotypes to different levels of submergence. *Comptes Rendus Biologies*, *341*, 85–96.

Ahern M., Kovats R. S., Wilkinson P., Few R., & Matthies F. (2005). Global health impacts of floods: Epidemiological evidence. *Epidemiologic Reviews*, *27*:36–46. doi: 10.1093/epirev/mxi004.

Ahmed, A. U., Hernandez, R., & Naher, F. (2016). Adoption of stress-tolerant rice varieties in Bangladesh. In Franz W. Gatzweiler & von Braun Joachim (Eds.), *Technological and institutional innovations for marginalized smallholders in agricultural development* (pp. 241–255). Cham: Springer International Publishing.

Alpuerto J. B., Fukuda M., Li S., Hussain R. M., Sakane K., & Fukao T. (2022). The submergence tolerance regulator SUB1A differentially coordinates molecular adaptation to submergence in mature and growing leaves of rice (*Oryza sativa* L.). *The Plant Journal*, *110*(1), 71–87.

Alpuerto J. B., Hussain R. M. F., & Fukao T. (2016). The key regulator of submergence tolerance, SUB1A, promotes photosynthetic and metabolic recovery from submergence damage in rice leaves. *Plant, Cell and Environment*, *39*, 672–684.

Angaji, S. A., Septiningsih, E. M., Mackill, D. J., & Ismail, A. M. (2010). QTLs associated with tolerance of flooding during germination in rice (*Oryza sativa* L.). *Euphytica*, *172*, 159–168.

Arya, K. V., & Shylaraj, K. S. (2018). Introgression of *Sub1* QTL (Submergence tolerant QTL) into the elite rice variety Jaya by marker-assisted backcross breeding. *Journal of Tropical Agriculture*, *56*, 114–128.

Azarin, K. V., Usatov, A. V., & Kostylev, P. I. (2017). Molecular breeding of submergence-tolerant rice. *Annual Research & Review in Biology*, *18*(1), 1–10

Bailey-Serres J., Fukao T., Gibbs D. J., Holdsworth M. J., Lee S. C., & Licausi F. (2012). Making sense of low oxygen sensing. *Trends in Plant Science*, *17*, 129–138.

Bailey-Serres J., Fukao T., Ronald P., Ismail A., Heuer S., & Mackill D. (2010). Submergence tolerant rice: SUB1's journey from landrace to modern cultivar. *Rice*, *3*(2), 138–147.

Baltazar, M. D., Ignacio, J. C. I., Thomson, M. J., Ismail, A. M., Mendioro, M. S., & Septiningsih, E. M. (2014). QTL mapping for tolerance of anaerobic germination from IR64 and the aus landrace Nanhi using SNP genotyping. *Euphytica*, *197*, 251–260.

Baltazar, M. D., Ignacio, J. C. I., Thomson, M. J., Ismail, A. M., Mendioro, M. S., & Septiningsih, E. M. (2019). QTL mapping for tolerance to anaerobic germination in rice from IR64 and the *aus* landrace Kharsu 80A. *Breeding Science*, *69*(2), 227–233.

Barding, G. A., Béni, S., Fukao, T., BaileySerres, J., & Larive C. K. (2013). Comparison of GC-MS and NMR for metabolite profiling of rice subjected to submergence stress. *Journal of Proteome Research*, *12*, 898–909.

Barding, G. A., Fukao, T., Béni,S., Bailey-Serres, J., & Larive, C. K. (2012). Differential metabolic regulation governed by the rice SUB1A gene during submergence stress and identification of alanylglycine by 1H NMR spectroscopy. *Journal of Proteome Research*, *11*, 320–330.

Barik, J., Kumar, V., Lenka, S. K., & Panda, D. (2020). Assessment of variation in morpho-physiological traits and genetic diversity in relation to submergence tolerance of five indigenous lowland rice landraces. *Rice Science*, *27*(1), 32–43.

Barik, J., Panda, D., Mohanty, S. K., & Lenka, S. K. (2019). Leaf photosynthesis and antioxidant response in selected traditional rice landraces of Jeypore tract of Odisha, India to submergence. *Physiology and Molecular Biology of Plants*, *25*, 847–863.

Breseghello, F., & Coelho, A. S. G. (2013). Traditional and modern plant breeding methods with examples in rice (*Oryza sativa* L.). *Journal of Agricultural and Food Chemistry*, *61*(35), 8277–8286.

Chakraborty, K., Roy, S., Jena, P., Dash, G. K., Barik, M., Behera, M., … & Swain, P. (2023). Identification of multiple abiotic stress tolerant donors for climate-resilient rice (*Oryza sativa*) development. *Indian Journal of Agricultural Sciences*, *93*(3), 258–262.

Cheng X., Huang, Y., Tan, Y. et al. (2022). Potentially useful dwarfing or semi-dwarfing genes in rice breeding in addition to the *sd1* gene. *Rice*, *15*, 66. https://doi.org/10.1186/s12284-022-00615-y

Chetia, S. K., Kalita, M., Verma, R. K., Barua, B., Ahmed, T., Modi, M. K., & Singh, N. K. (2018). Flood proofing of Ranjit, a popular variety of North-Eastern India through transfer of Sub1 rice QTL by modified marker-assisted backcross breeding. *Indian Journal of Genetics and Plant Breeding*, *78*(2), 166–173.

Cho, H. T., & Kende, H. (1997). Expression of expansin genes is correlated with growth in deepwater rice. *Plant Cell*, *9*, 1661–1671.

Chukwu, S. C., Rafii, M. Y., Ramlee, S. I., Ismail, S. I., Oladosu, Y., Okporie, E., & Jalloh, M. (2019). Marker-assisted selection and gene pyramiding for resistance to bacterial leaf blight disease of rice (*Oryza sativa* L.). *Biotechnology & Biotechnological Equipment 33*, 440–455.

Dar, M. H., Chakravorty, R., Waza, S. A., Sharma, M., Zaidi, N. W., Singh, A. N., Singh, U. S., & Ismail, A. M. (2017). Transforming rice cultivation in flood prone coastal Odisha to ensure food and economic security *Food Security*, *9*(4), 711–722.

Das, K. K., Panda, D., Nagaraju, M., Sharma, S. G., & Sarkar, R. K. (2004). Antioxidant enzymes and aldehyde releasing capacity of rice cultivars (*Oryza sativa* L.) as determinants of anaerobic seedling establishment capacity. Bulgarian *Journal of Plant Physiology*, *30* (1–2), 34–44.

Das, K. K., & Sarkar, R. K. (2001). Post flood changes on the status of chlorophyll, carbohydrate and nitrogen content and its association with submergence tolerance in rice. *Plant Archives*, *1*, 15–19.

Das, K. K., Sarkar, R. K., & Ismail, A. M. (2005). Elongation ability and non-structural carbohydrate levels in relation to submergence tolerance in rice. *Plant Science*, *168*, 131–136.

Datta, S. K. (1982). Elongation of rice in deepwater. In Proceedings of the 1981 International Deepwater Rice Workshop (pp. 261–269). Los Baños, Philippines: International Rice Research Institute.

Dixit, S., Singh, A., Sandhu, N., Bhandari, A., Vikram, P., & Kumar, A. (2017). Combining drought and submergence tolerance in rice: marker-assisted breeding and QTL combination effects. *Molecular Breeding*, *37*, 1–12.

Eiguchi, M., Hirano, H. Y., Sano, Y., & Morishima, H. (1993). Effects of water depth on internodal elongation and floral induction in a deepwater-tolerant rice line carrying the dw3 gene. *Japanese Journal of Breeding*, *43*, 135–139.

FAO. (2021). The impact of disasters and crises on Agriculture and Food Security. Food and Agriculture Organization of the United Nations Rome, Italy.

Fukao, T., & Bailey-Serres, J. (2008). Submergence tolerance conferred by Sub1A is mediated by SLR1 and SLRL1 restriction of gibberellins responses in rice. *Proceedings of the National Academy of Sciences of the United States of America*, *105*, 16814–16819.

Fukao, T., Barrera-Figueroa, B. E., Juntawong, P., & Pena-Castro, J. M. (2019). Submergence and waterlogging stress in plants: A review highlighting research opportunities and understudied aspects. *Frontiers in Plant Science*, *10*, 340.

Fukao, T., Xu, K., Ronald, P. C., & Bailey-Serres, J. (2006). A variable cluster of ethylene response factor-like genes regulates metabolic and developmental acclimation responses to submergence in rice. *Plant Cell*, *18*, 2021–2034.

Gautam, A., Chetia, S. K., Modi, M. K., & Ahmed, T. (2018). Phenotypic screening and evaluation of Sub1 introgressed lines in popular rice varieties Ranjit and Bahadur of Assam, India. *International Journal of Current Microbiology and Applied Sciences*, *7*(9), 1744–1755.

Gautam, A., Chetia, S. K., Sharma, V., Phukon, M., Modi, M. K., & Ahmed, T. (2022b). Morpho-molecular investigation of recurrent parent genome recovery of Sub1 introgressed advance lines of rice (*Oryza sativa* L.). *Indian Journal of Genetics and Plant Breeding*, *82*(4), 411–418. doi: 10.31742/ISGPB.82.4.4

Gautam, A., Chetia, S. K., Sharma, V., Verma R. K., Phukon, M., Kalita, M., Modi, M. K., & Ahmed, T. (2022a). SSR marker-based DNA fingerprinting of Sub1 introgressed lines in the background of traditional rice varieties of Assam India. *Indian Journal of Biochemistry and Biophysics*, *59*, 350–356. http://nopr.niscair.res.in/handle/123456789/59381

Gomosta, A. R., Sarkar, M. H., & Haque M. Z. (1982). Yield compensating ability by aquatic tillers in deepwater rice. In Proceedings of the 1981 International Deepwater Rice Workshop (pp. 293–298). The Philippines: International Rice Research Institute Los Baños.

Gonzaga, Z. J. C., Carandang, J., Sanchez, D. L. et al. (2015). Mapping additional QTLs from FR13A to increase submergence tolerance in rice beyond SUB1. *Euphytica*, *209*, 627–636. https://doi.org/10.1007/s10681-016-1636-z

Hamamura, K., & Kupkanchanakul, T. (1979). Inheritance of floating ability in rice. *Japanese Journal of Breeding*, *29*, 211–216.

Haque, Q. A., Hille Ris Lambers, D., Tepora, N. M., & Dela Cruz, Q. D. (1989). Inheritance of submergence tolerance in rice. *Euphytica*, *41*, 247–251.

Hattori, Y., Miura, K., Asano, K., Yamamoto, E., Mori, H., Kitano, H., Matsuoka, M., & Ashikari, M. (2007). A major QTL confers rapid internode elongation in response to water rise in deepwater rice.*Breeding Science*, *57*, 305–314.

Hattori, Y., Nagai, K., Furukawa, S., Song, X. J., Kawano, R., Sakakibara, H., Wu, J., Matsumoto, T., Yoshimura, A., Kitano, H., Matsuoka, M., Mori, H., & Ashikari, M. (2009). The ethylene response factors SNORKEL1 and SNORKEL2 allow rice to adapt to deep water. *Nature*, *460*(7258), 1026–1030.

Hattori, Y., Nagai, K., Mori, H., Kitano, H., Matsuoka, M., & Ashikari, M. (2008). Mapping of three QTL that regulate internode elongation in deepwater rice. *Breeding Science*, *58*, 39–46.

Hifzur, R., Vijayalakshmi, D., Sasikala, R., Sudha, M., Kaliyaperumal, A. K., Raha, S., Panneerselvam, N., Ramanathan, V., Nallatha J., Sabariappan, R., & Raveendran, M. (2018). Introgression of submergence tolerance into CO 43, a popular rice variety of India, through marker-assisted backcross breeding. *Czech Journal of Genetics and Plant Breeding*, *54*, 101–108.

HilleRisLambers, D., & Seshu, D. V. (1982). Some ideas on breeding procedures and requirements for deepwater rice improvement. In Proceedings of the 1981 International Deepwater Rice Workshop, IRRI, Manila, Philippines (pp. 29–44).

HilleRisLambers, D., & Vergara, B. S. (1982). Summary results of an international collaboration on screening methods for flood tolerance. In Proceedings of the 1981 International Deepwater Rice Workshop. International Rice Research Institute, Los Banos, pp. 347–353.

Hsu, S. K., & Tung, C. W. (2015). Genetic mapping of anaerobic germination-associated QTLs controlling coleoptile elongation in rice. *Rice*, *8*, 38.

Hsu, S. K., & Tung, C. W. (2017). RNA-Seq analysis of diverse rice genotypes to identify the genes controlling coleoptile growth during submerged germination. *Frontiers in Plant Science*, *8*, 762.

Huyen, L. T. N., Cuc, L. M., Ismail, A. M., & Ham, L. H. (2012). Introgression of the salinity tolerance QTLs into AS996, the elite rice variety of Vietnam. *American Journal of Plant Sciences*, *3*, 981–987.

Iftekharuddaula, K. M., Ahmed, H. U., Ghosal, S., Amin, A., Moni, Z. R., Ray, B. P., Barman, H. N., Siddique, M. A., Collard, B. C. Y., & Septiningsih, E. M. (2016). Development of early maturing submergence-tolerant rice varieties for Bangladesh. *Field Crops Research*, *190*, 44–53.

Iftekharuddaula, K. M., Ahmed, H. U., Ghosal, S., Moni, Z. R., Amin, A., & Ali, S. (2015). Development of new submergence tolerant rice variety for Bangladesh using marker-assisted backcrossing. *Rice Science*, *22*, 16–26.

Inouye, J. (1983). Relation between elongation ability and internode elongation of floating rice under rising water conditions. *Journal of Tropical Agriculture*, *27*, 181–186.

Jackson, M. B. (2008). Ethylene-promoted elongation: An adaptation to submergence stress. *Annals of Botany*, *101*, 229–248.

Jackson, M. B., & Ram, P. C. (2003). Physiological and molecular basis of susceptibility and tolerance of rice plants to complete submergence. *Annals of Botany*, *91*(2), 227–241.

Jeong, J. M., Cho, Y. C., Jeong, J. U., Mo, Y. J., Kim, C. S., Kim, W. J., Baek, M. K., & Kim, S. M. (2020). QTL mapping and effect confirmation for anaerobic germination tolerance derived from the japonica weedy rice landrace PBR. *Plant Breeding*, *139*, 83–92.

Jiang, L., Hou, M., Wang, C., & Wan, J. (2004). Quantitative trait loci and epistatic analysis of seed anoxia germinability in rice (*Oryza sativa*). *Rice Science*, *11*, 238–244.

Jiang, L., Liu, S., Hou, M., Tang, J., Chen, L., Zhai, H., & Wan, J. (2006). Analysis of QTLs for seed low temperature germinability and anoxia germinability in rice (*Oryza sativa* L.). *Field Crops Research*, *98*, 68–75.

John, D., & Shylaraj, K. S. (2017). Introgression of *sub1* QTL into an elite rice (*Oryza sativa* L.) variety Jyothi through marker-assisted backcross breeding. *Journal of Tropical Agriculture*, *55*, 1–11.

Jung, K. H., Seo, Y. S., Walia, H., Cao, P., Fukao, T., & Canlas, P. E. (2010). The submergence tolerance regulator *Sub1A* mediates stress-responsive expression of AP2/ERF transcription factors. *Plant Physiology*, *152*, 1674–1692.

Kawano, R., Doi, K., Yasui, H., Mochizuki, T., & Yoshimura, A. (2008). Mapping of QTLs for floating ability in rice. *Breeding Science*, *58*, 47–53.

Kim, S. M., Kim, C. S., Jeong, J. U., Reinke, R. F. , & Jeong, J. M. (2019). Marker-assisted breeding for improvement of anaerobic germination in japonica rice (*Oryza sativa*). *Plant Breeding*, *138*(6), 810–819.

Kretzschmar, T., Pelayo, M. A., Trijatmiko, K. R., Gabunada, L. F., Alam, R., Jimenez, R., Mendioro, M. S., Slamet-Loedin, I. H., Sreenivasulu, N., Bailey-Serres, J., Ismail, A. M., Mackill, D. J., & Septiningsih, E. M. (2015). A trehalose-6-phosphate phosphatase enhances anaerobic germination tolerance in rice. *Nature Plants*, *1*, 15124.

Kuroha, T., Nagai, K., Kurokawa, Y., Nagamura, Y., Kusano, M., Yasui, H., Ashikari, M., & Fukushima, A. (2017). eQTLs regulating transcript variations associated with rapid internode elongation in deepwater rice. *Frontiers in Plant Science*, *8*, 1753.

Kurokawa, Y., Nagai, K., Huan, P. D., Shimazaki, K., Qu, H., Mori, Y., Toda, Y., Kuroha, T., Hayashi, N., Aiga, S., & Itoh, J. I. (2018). Rice leaf hydrophobicity and gas films are conferred by a wax synthesis gene (*LGF 1*) and contribute to flood tolerance. *New Phytologist*, *218*, 1558–1569.

Kuya, N., Sun, J., Iijima, K., Venuprasad, R., & Yamamoto, T. (2019). Novel method for evaluation of anaerobic germination in rice and its application to diverse genetic collections. *Breeding Science*, *69*, 633–639.

Leridon, H. (2020). World population outlook: Explosion or implosion?. *Population and Society*, *573*, 1–4.

Li, C., Zong, Y., Wang, Y., Jin, S., Zhang, D., Song, Q., … & Gao, C. (2022). Expanded base editing in rice and wheat using a Cas9-adenosine deaminase fusion. *Genome Biology*, *19*(1), 59. doi: 10.1186/s13059-018-1443-z.

Li, Y. S., Ou, S. L., & Yang, C. Y. (2020). The seedlings of different japonica rice varieties exhibit differ physiological properties to modulate plant survival rates under submergence stress. *Plants*, *9*(8), 982.

Lin, C., & Sauter, M. (2020). Control of root system architecture by phytohormones and environmental signals in rice. *Israel Journal of Plant Sciences*, *67*(1-2), 98–109.

Locke, A. M., Barding, G. A., Sathnur, S., Larive, C. K., & Bailey-Serres J. (2018). Rice SUB1A constrains remodelling of the transcriptome and metabolome during submergence to facilitate post-submergence recovery. *Plant, Cell and Environment*, *41*, 721–736.

Loreti, E., & Perata, P. (2020). The many facets of hypoxia in plants. *Plants*, *9*(6), 745.

Loreti, E., Valeri, M. C., Novi, G., & Perata P. (2018). Gene regulation and survival under hypoxia requires starch availability and metabolism. *Plant Physiology*, *176*(2), 1286–1298.

Mackill, D. J., Amante, M. M., Vergara B. S., & Sarkarung S. (1993). Improved semidwarf rice lines with tolerance to submergence of seedlings. *Crop Science*, *33*(4), 749–753.

Mackill, D. J., Ismail, A. M., Singh, U. S., Labios, R. V., & Paris, T. R. (2012a). Development and rapid adoption of submergence-tolerant (Sub1) rice varieties. *Advances in Agronomy*, *115*, 299–352.

Miro, B., & Ismail, A. M. (2013). Tolerance of anaerobic conditions caused by flooding during germination and early growth in rice (*Oryza sativa* L.). *Frontiers in Plant Science*, *4*, 269.

Mishra, S. B., Senadhira, D., & Manigbas, N. L. (1996). Genetics of submergence tolerance in rice (*Oryza sativa* L.). Field Crops Research, *46*(1–3), 177–181.

Mohanty, H. K., & Chaudhary, R. C. (1986). Progress in rainfed lowland rice. *Breeding for submergence tolerance in rice in India* (pp. 191–200). Manila, Philippines: IRRI.

Mohanty, H. K., & Khush, G. S. (1985). Diallel analysis of submergence tolerance in rice, *Oryza sativa* L. *Theoretical and Applied Genetics*, *70*, 467–473.

Mojulat, W. C., Yusop, R., Ismail, R., Juraimi, A. S., Harun, A. R., Ahmed, F., Tanweer, F. A., & Latif, M. (2017). Analysis of simple sequence repeat markers linked to submergence tolerance on newly developed rice lines derived from MR263 × Swarna-Sub1. *Sains Malaysiana*, *46*, 521–528.

Mondal, S., Khan, M. I. R., Dixit, S., Pompe, P. C., Septiningsih, E. M., & Ismail, A. M. (2020). Growth, productivity and grain quality of *AG1* and *AG2* QTLs introgression lines under flooding in direct-seeded rice system. *Field Crops Research*, *248*, 107713.

Muhlenbock, P., Plaszczyca, M., Plaszczyca, M., Mellerowicz, E., & Karpinski, S. (2007). Lysigenous aerenchyma formation in Arabidopsis is controlled by lesion-simulating disease1. *Plant Cell*, *19*, 3819–3830.

Nagai, K., Hattori, Y., & Ashikari, M. (2010). Stunt or elongate? Two opposite strategies by which rice adapts to floods. *Journal of Plant Research*, *123*, 303–309.

Nagai, K., Kuroha, T., Ayano, M., Kurokawa, Y., Angeles-Shim, R. B., Shim, J. H., Yasui, H., Yoshimura, A., & Ashikari, M. (2012). Two novel QTLs regulate internode elongation in deepwater rice during the early vegetative stage. *Breeding Science*, *62*, 178–185.

Nagai, K., Mori, Y., Ishikawa, S., Furuta, T., Gamuyao, R., Niimi, Y., Hobo, T., Fukuda, M., Kojima, M., Takebayashi, Y., Fukushima, A., Himuro, Y., Kobayashi, M., Ackley, W., Hisano, H., Sato, K., Yoshida, A., Wu, J., Sakakibara, H., Sato, Y., & Ashikari, M. (2020). Antagonistic regulation of the gibberellic acid response during stem growth in rice. *Nature*, *584*(7819), 109–114.

Nair, M. M., & Shylaraj, K. S. (2021). Introgression of dual abiotic stress tolerance QTLs (*Saltol* QTL and *Sub1* gene) into rice (*Oryza sativa* L.) variety Aiswarya through marker-assisted backcross breeding. *Physiology and Molecular Biology of Plants*, *27*, 497–514.

Neeraja, C. N., Maghirang-Rodriguez, R., Pamplona, A., Heuer, S., Collard, B. C., Septiningsih, E. M., Vergara, G., Sanchez, D., Xu, K., Ismail, A. M., & Mackill, D. J. (2007). A marker-assisted backcross approach for developing submergence-tolerant rice cultivars. *Theoretical and Applied Genetics*, *115*(6), 767–776.

Nemoto, K., Ukai, Y., Tang, D. Q., Kasai, Y., & Morita, M. (2004). Inheritance of early elongation ability in floating rice revealed by diallel and QTL analyses. *Theoretical and Applied Genetics*, *109*, 42–47.

Nghi, K. N., Tondelli, A., Vale, G., Tagliani, A., Mare, C., Perata, P., & Pucciariello, C. (2019). Dissection of coleoptile elongation in japonica rice under submergence through integrated genome-wide association mapping and transcriptional analyses. *Plant, Cell & Environment*, *42*(6), 1832–1846.

Nghi, K. N., Tagliani, A., Mariotti, L., Weits, D. A., Perata, P., & Pucciariello, C. (2021). Auxin is required for the long coleoptile trait in japonica rice under submergence. *New Phytologist*, *229*, 85–93.

Nishimura, T., Sasaki, K., Yamaguchi, T., Takahashi, H., Yamagishi, J., & Kato, Y. (2020). Detection and characterization of quantitative trait loci for coleoptile elongation under anaerobic conditions in rice. *Plant Production Science*, *23*, 374–383.

Pan, T., Zhang, J., He, L., Hafeez, A., Ning, C., & Cai, K. (2021). Silicon enhances plant resistance of rice against submergence stress. *Plants*, *10*(4), 767.

Panda, D., Barik, J., & Sarkar, R. K. (2021). Recent advances of genetic resources, genes and genetic approaches for flooding tolerance in rice. *Current Genomics*, *22*(1), 41–58.

Panda, D., & Sarkar, R. K. (2013). Characterization of leaf gas exchange and anti-oxidant defense of rice (*Oryza sativa* L.) cultivars differing in submergence tolerance owing to complete submergence and consequent re-aeration. *Agricultural Research*, *2*, 301–308.

Pandit, E., Pawar, S., Barik, S. R., Mohanty, S. P., Meher, J., & Pradhan, S. K. (2021). Marker-assisted backcross breeding for improvement of submergence tolerance and grain yield in the popular rice variety 'Maudamani'. *Agronomy*, *11*, 1263.

Pucciariello, C., & Perata, P. (2013). Quiescence in rice submergence tolerance: an evolutionary hypothesis. *Trends in Plant Science*, *18*, 377–381.

Rafael, A. B., Barroga, W. V., Marcelo, P. N. M., Septiningsih, E. M., & Manangkil, O. E. (2015). Introgression of *Sub1* and *AG1* tolerance genes to NSIC Rc222 through marker-assisted backcrossing. *The Philippine Journal of Crop Science*, *40*(1). https://agris.fao.org/search/en/providers/122430/records/6473a35313d110e4e7a77884

Ram, P. C., Singh, B. B., Singh, A. K., Ram, P., Singh, P. N., Singh, H. P., Boamfa, I., Harren, F., Santosa, E., Jackson, M. B., & Setter, T. L. (2002). Submergence tolerance in rainfed lowland rice: physiological basis and prospects for cultivar improvement through marker-aided breeding. *Field Crops Research*, *76*, 131–152.

Rao, C. S., Gopinath, K. A., Prasad, J. V. N. S., & Singh, A. K. (2016). Climate resilient villages for sustainable food security in tropical India: concept, process, technologies, institutions, and impacts. *Advances in Agronomy*, *140*, 101–214.

Robin, S., Jeyaprakash, P., Amudha, K., Pushpam, R., Rajeswari, S., Manonmani, S., & Ganesamurthy, K. (2019). Rice CR1009 Sub 1 (IET 22187)-A new flood tolerant rice variety. *Electronic Journal of Plant Breeding*, *10*(3), 995–1004.

Rohilla, M., Singh, N., Mazumder, A., Sen, P., Roy, P., Chowdhury, D., Singh, N. K., & Mondal, T. K. (2020). Genome-wide association studies using 50 K rice genic SNP chip unveil genetic architecture for anaerobic germination of deep-water rice population of Assam, India. *Molecular Genetics and Genomics*, *295*(5), 1211–1226.

Samal, R., Roy, P. S., Sahoo, A., Kar, M. K., Patra, B. C., Marndi, B. C., & Gundimeda, J. N. R. (2018). Morphological and molecular dissection of wild rices from eastern India suggests distinct speciation between O. rufipogon and O. nivara populations. *Scientific Reports*, *8*(1), 2773.

Sarkar, R. K., & Mohanty, P. (2010). An overview on submergence tolerance in rice: farmers' wisdom and amazing science. *Journal of Plant Biology*, *37*, 191–199.

Sarkar, R. K., Das, S., & Ravi, I. (2001). Changes in certain antioxidative enzymes and growth parameters as a result of complete submergence & subsequent re-aeration of rice cultivars differing in submergence tolerance. *Journal of Agronomy and Crop Science*, *187*, 69–74.

Sarkar, R. K., De, R. N., Reddy, J. N., & Ramakrishnayya, G. (1996). Studies on the submergence tolerance mechanism in relation to carbohydrate, chlorophyll and specific leaf weight in rice (*Oryza sativa* L.). *Journal of Plant Physiology*, *149*, 623–625.

Sarkar, R. K., Reddy, J. N., Sharma, S. G., & Ismail, A. M. (2006). Physiological basis of submergence tolerance in rice and implications for crop improvement. *Current Science*, *91*, 899- 906.

Sauter, M., Seagull, R. W., & Kende, H. (1993). Internodal elongation and orientation of cellulose microfibrils and microtubules in deepwater rice. *Planta*, *190*, 354–362.

Schmidhuber, J., & Tubiello, F. N. (2007) Global food security under climate change. *Proceedings of the National Academy of Sciences of the United States of America*, *104*, 19703–19708.

Schmitz, A. J., Folsom, J. J., Jikamaru, Y., Ronald, P., & Walia H. (2013). SUB1A-mediated submergence tolerance response in rice involves differential regulation of the brassinosteroid pathway. *New Phytologist*, *198*, 1060–1070.

Septiningsih, E. M., Hidayatun, N., Sanchez, D. L., Nugraha, J. C., Pamplona, A. M., Collard, B. C. Y., Ismail, A. M., & Mackill, D. J. (2015). Accelerating the development of new submergence tolerant rice varieties: the case of Ciherang-Sub1 and PSB Rc18-Sub1. *Euphytica*, *202*, 259–268.

Septiningsih, E. M., Ignacio, J. C., Sendon, P. M., Sanchez, D. L., Ismail, A. M., & Mackill, D. J. (2013). QTL mapping and confirmation for tolerance of anaerobic conditions during germination derived from the rice landrace Ma-Zhan Red. *TAG. Theoretical and Applied Genetics*, *126*(5), 1357–1366.

Septiningsih, E. M., & Mackill, D. J. (2018). Genetics and breeding of flooding tolerance in rice. In Sasaki Takuji & Motoyuki Ashikari (Eds.), *Rice genomics, genetics and breeding* (pp. 275–295). Singapore: Springer.

Septiningsih, E. M., Pamplona, A. M., Sanchez, D. L., Neeraja, C. N., Vergara, G. V., Heuer, S., Ismail, A. M., & Mackill, D. J. (2009). Development of submergence tolerant rice cultivars: the Sub1 locus and beyond. *Annals of Botany*, *103*, 151–160.

Septiningsih, E. M., Sanchez, D. L., Singh, N., Sendon, P. M., Pamplona, A. M., Heuer, S., & Mackill, D. J. (2012). Identifying novel QTLs for submergence tolerance in rice cultivars IR72 and Madabaru. *TAG. Theoretical and Applied Genetics*, *124*(5), 867–874.

Singh, A., Septiningsih, E. M., Balyan, H. S., Singh, N. K., & Rai, V. (2017). Genetics, physiological mechanisms and breeding of flood-tolerant rice (*Oryza sativa* L.). *Plant Cell Physiology, 58*(2), 185–197.

Singh, P., & Sinha, A. K. (2016). A positive feedback loop governed by SUB1A1 interaction with MITOGEN-ACTIVATED PROTEIN KINASE3 imparts submergence tolerance in rice. *The Plant Cell*, *28*(5), 1127–1143.

Singh, U. S. (1981). Deep water rice in India: Problems and prospects. In Proceedings of International Deep Water Rice Workshop (pp. 355–362).

Sripongpangkul, K., Posa, G. B. T., Senadhira, D. W., Brar, D., Huang, N., Khush, G. S., & Li, Z. K. (2000). Genes/QTLs affecting flood tolerance in rice. *Theoretical and Applied Genetics*, *101*, 1074–1081.

Steffens, B., Geske, T., & Sauter, M. (2011). Aerenchyma formation in the rice stem and its promotion by H_2O_2. *New Phytologist*, *190*, 369–378.

Suge, H. (1987). Physiological genetics of internodal elongation under submergence in floating rice. *Japanese Journal of Genetics*, *62*, 69–80.

Sun, L., Chen, S., Chao, L., & Sun, T. (2007). Effects of flooding on changes in Eh, pH and speciation of cadmium and lead in contaminated soil. *Bulletin of Environmental Contamination and Toxicology*, *79*, 514–518.

Toledo, A. M. U., Ignacio, J. C. I., Casal, C., Gonzaga, Z. J., Mendioro, M. S., & Septiningsih, E. M. (2015). Development of improved Ciherang-Sub1 having tolerance to anaerobic germination conditions. *Plant Breeding and Biotechnology*, *3*, 77–87.

Tripathi, R. S., & Balakrishna Rao, M. J. (1985). Inheritance studies of characters associated with floating habit and their linkage relationship in rice. *Euphytica*, *34*, 875–881.

Upadhyay, R. K. (2018). Oxidative injury and its detoxification in rice plants after submergence stress. *Proceedings of the National Academy of Sciences, India Section B: Biological Sciences*, *88*, 15–21.

Urban, D. W., Roberts, M. J., Schlenker, W., & Lobell, D. B. (2015). The effects of extremely wet planting conditions on maize and soybean yields. *Climate Change*, *130*, 247–260.

Vergara, B. S., & Mazaredo, A. (1975). Screening for resistance to submergence under greenhouse conditions. In Proceedings of the International Seminar on Deepwater Rice, August 1974. Bangladesh Rice Research Institute, Dhaka, pp. 67–70

Vervuren, P., Blom, C. W. P. M., & De-Kroon, H. (2003). Extreme flooding events on the Rhine and the survival and distribution of riparian plant species. *Journal of Ecology*, *91*, 135–146.

Vriezen, W. H., Zhou, Z., & Van Der Straeten, D. (2003). Regulation of Submergence-induced enhanced shoot elongation in *Oryza sativa* L. *Annals of Botany*, *91*(2), 263–270.

Wang, H., Wang, H., Shao, H., & Tang, X. (2016). Recent advances in utilizing transcription factors to improve plant abiotic stress tolerance by transgenic technology. *Frontiers in Plant Science*, *7*, 67.

White, I. (2010). *Water and the city risk, resilience and planning for a sustainable future.* New York, NY: Routledge Taylor and Francis Group.

Winkel, A., Pedersen, O., Ella, E., Ismail, A. M., & Colmer, T. D. (2014). Gas film retention and underwater photosynthesis during field submergence of four contrasting rice genotypes. *Journal of Experimental Botany*, *65*, 3225–3233.

Wu, Y. S., & Yang, C. Y. (2016). Physiological responses and expression profile of NADPH oxidase in rice (*Oryza sativa*) seedlings under different levels of submergence. *Rice*, *9*, 1–10.

Xiong, Q., Cao, C., Shen, T., Zhong, L., He, H., & Chen, X. (2019). Comprehensive metabolomic and proteomic analysis in biochemical metabolic pathways of rice spikes under drought and submergence stress. *Biochimica et Biophysica Acta (BBA)-Proteins and Proteomics, 1867*(3), 237–247.

Xu, C., Park, S. J., Van Eck, J., & Lippman, Z. B. (2021). Control of inflorescence architecture in tomato by BTB/POZ transcriptional regulators. *Genes & Development*, *35*(1–2), 122–136. doi: 10.1101/gad.340798.120.

Xu, K. N., & Mackill, D. J. (1996). A major locus for submergence tolerance mapped on rice chromosome 9. *Molecular Breeding*, *2*(3), 219–224.

Xu, K., Xu, X., Fukao, T., Canlas, P., Maghirang-Rodriguez, R., Heuer, S., Ismail, A. M., Bailey-Serres, J., Ronald, P. C., & Mackill, D. J. (2006). *Sub1A* is an ethylene-response-factor-like gene that confers submergence tolerance to rice. *Nature*, *442*, 705–708.

Xu, K., Xu, X., Ronald, P. C., & Mackill, D. J. (2000). A high-resolution linkage map of the vicinity of the rice submergence tolerance locus Sub1. *Molecular Genetics and Genomics*, *263*, 681–689.

Yamauchi, T., Noshita, K., & Tsutsumi, N. (2021). Climate-smart crops: key root anatomical traits that confer flooding tolerance. *Breeding Science*, *71*(1), 51–61.

Yang, J., Sun, K., Li, D., Luo, L., Liu, Y., Huang, M., Yang, G., Liu, H., Wang, H., Chen, Z., & Guo, T. (2019). Identification of stable QTLs and candidate genes involved in anaerobic germination tolerance in rice via high-density genetic mapping and RNA-Seq. *BMC Genomics*, *20*, 1–15.

Yang, S. Y., Wu, Y. S., Chen, C. T., Lai, M. H., & Yen, H. M., & Yang, C. Y. (2017). Physiological and molecular responses of seedlings of an upland rice ('Tung Lu 3') to total submergence compared to those of a submergence-tolerant lowland rice ('FR13A'). *Rice*, *10* , p. 42

Yemelyanov, V.V., Chirkova, T.V., Shishova, M.F., & Lindberg, S.M. (2020). Potassium Efflux and Cytosol Acidification as Primary Anoxia-Induced Events in Wheat and Rice Seedlings. *Plants*, *9*, 1216. https://doi.org/10.3390/plants9091216

Youssef, M. S., Mira, M. M., Renault, S., Hill, R. D., & Stasolla, C. (2016). Phytoglobin expression influences soil flooding response of corn plants. *Annals of Botany*, *118*, 919–931.

Zhang, C., Li, M., Rey, J. D., Feng, T., Lafitte, R., Zheng, T., & Li, Z. (2023). Simultaneous improvement and genetic dissection of drought and submergence tolerances in rice (*Oryza sativa* L.) by selective introgression. *Frontiers in Plant Science, 14*, 1134450.

Zhang, M., Lu, Q., Wu, W., Niu, X., Wang, C., Feng, Y., Xu, Q., Wang, S., Yuan, X., Yu, H., & Wang, Y. (2017). Association mapping reveals novel genetic loci contributing to flooding tolerance during germination in indica rice. *Frontiers in Plant Science*, *8*, 678.

Zhao, H., Ma, T., Wang, X., Deng, Y., Ma, H., Zhang, R. , & Zhao, J (2015). OsAUX1 controls lateral root initiation in rice (*Oryza sativa* L.). *Plant Cell and Environment*, *38*, 220–222.

17 Speed Breeding in Vegetable Crops

Implications and Challenges

Ruma Devi and Shivani Chauhan
Department of Vegetable Science, Punjab Agricultural University, Ludhiana, India

17.1 INTRODUCTION

The present global human population is 7.8 billion and is projected to increase by roughly 9.9 billion by the year 2050. There is a need to increase food production by 60–80% more to sustain around 10 billion population in 2050 (Hickey et al., 2019). It is anticipated that climatic variations characterized by increasing temperatures, the higher occurrence of floods, and prolonged periods of drought would result in the emergence of previously unknown diseases and more regular occurrences of pest outbreaks. Consequently, a flexible and adaptive approach to plant breeding will be necessary to effectively address these challenges. This emphasized the necessity of augmenting the existing rate of genetic improvement in essential food crops to protect global food security. Enhancing the rate of genetic advancement entails the implementation of hastened crop breeding pipelines to facilitate the prompt dissemination of enhanced crop cultivars (Lin et al., 2016). The usual method for developing a new crop variety involves the careful selection of parental genotypes that possess desired qualities, followed by mating between them and subsequent selections in advanced generations to develop superior cultivars to meet the market demand (Shimelis & Laing, 2012). The primary breeding objectives within a crop cultivar development program encompass increasing the level of productivity potential, improvement of dietary value, and improved resistance to both environmental stress factors (Breseghello & Coelho, 2013). The fundamental crop improvement program involves choosing favorable parents followed by their mating and selections in advanced generations in multiple environmental conditions to identify the best cultivars and multiplication of seeds and distribution (Shimelis & Laing, 2012). However, in the absence of a pre-breeding program, these traditional methods can take a prolonged duration of ten years to develop an improved variety (Ahmar et al., 2020). Field selection will take a season followed by subsequent progression from one generation to the next due to the fundamental principles of the crop. The duration of legume crops typically varies from 3 to 6 months, and crops like cassava complete their lifecycle within 15–18 months. Furthermore, due to adverse climatic conditions, such as temperature fluctuations, uneven rainfall, photoperiod, etc., only one crop cycle is allowed within a year (Laux et al., 2010). Newer techniques like double haploid breeding (Dwivedi et al., 2015) and speed breeding (SB) (Hickey et al., 2019) have the potential to reduce the duration of the crop. The concept of SB was highlighted when NASA decided to grow crops in space. SB refers to a set of methodologies that aim to expedite the reproductive processes of crop genotypes by manipulating the surrounding environmental variables. This manipulation is done to accelerate the vegetative and seed production stages, hence facilitating the progression to subsequent breeding generations in the most efficient manner possible. Multiple selection methods, such as single seed descent (SSD), single pod descent (SPD), single plant selection (SPS), clonal selection, and marker-assisted selection (MAS), can be integrated with SB to

DOI: 10.1201/b23372-17

speed up the breeding cycle and maximize resource utilization (Hickey et al., 2017; Samineni et al., 2019; Watson et al., 2018). SB enables more generations per year (three to nine) than the conventional method (Ghosh et al., 2018). Consequently, the utilization of SB techniques enables the expeditious formation of stable homozygous genotypes followed by generation advancement, which accelerates the development and introduction of new cultivars (Watson et al., 2018). Moreover, the application of SB technology aligns with the selection using markers and phenotyping methods enabling the simultaneous assessment and selection of numerous traits. In this chapter, we will discuss various components and methods of SB in vegetable crops and the challenges associated.

17.2 APPROACHES USED IN SB

SB techniques involve creating a controlled environment to provide all the necessary conditions for plant growth and exert influence on its development at every stage (Ghosh et al., 2018). SB involves the utilization of environment-controlled chambers equipped with supplementary LED lighting to extend the duration of photoperiods, as well as the regulation of temperature and humidity levels. In order to improve plant growth, multiple environmental variables must be controlled to promote and enhance photosynthesis as well as vegetative and reproductive growth. In the context of horticultural crops, various methodologies are used to cultivate plants in highly regulated environments where significant abiotic growth factors can be optimized during their development. In this section, we will discuss various components and their strategies involved in SB.

17.3 EXTENDED LIGHT PERIOD (PHOTOPERIOD)

The most popular method to speed up the breeding cycle for crop plants is to lengthen the photoperiod (Ghosh et al., 2018). It has previously been used to increase day length in winter to encourage plant growth. The two main lighting sources being used are sodium vapor lamps and light-emitting diodes (LEDs) (Ghosh et al., 2018). Therefore, it is crucial to ascertain the optimal light quality, intensity, and photoperiod that induce flower initiation in different crops and genotypes. Light serves as the primary energy source for the process of plant photosynthesis and subsequent growth. Artificial lighting is utilized as a photoperiodic stimulus to regulate the blossoming process as well as a supplementary light source to expedite plant growth and enhance the overall quality and yield of crops. Recently, there has been an increase in attention toward utilizing LEDs as a viable lighting solution in horticulture systems. This is mostly due to their notable advantages, such as their excellent radiant efficiency, extended lifespan, minimal heat generation, narrow spectral output, and ability to cater to the specific light intensity and wavelength needs of various plant species. Any light source that emits a spectrum encompassing photosynthetic active radiation within the range of 400–700 nm, including blue, red, and far-red wavelengths, is considered appropriate for speed breeding (Ghosh et al., 2018). The utilization of LEDs, either individually or in conjunction, presents a viable approach for achieving a suitable variety of wavelengths. When undertaking indoor SB operations in areas with erratic availability of electricity, solar power solutions offer a practical and sustainable approach.

17.4 TEMPERATURE TREATMENT

The rate of plant development is significantly influenced by temperature. The temperature reactions vary based on the specific type of plant and their respective phenological stage. Modifications to soil and air temperatures exert an influence on the processes of germination and growth, hence resulting in accelerated growth, flowering, seed production, and maturation. Plant development is influenced by both low and high temperatures, leading to a transition from the vegetative to the flowering stages (McClung et al., 2016). The temperature range required for germination in the majority of crops is generally seen to be between 12 and 30°C. In contrast, the optimal temperature

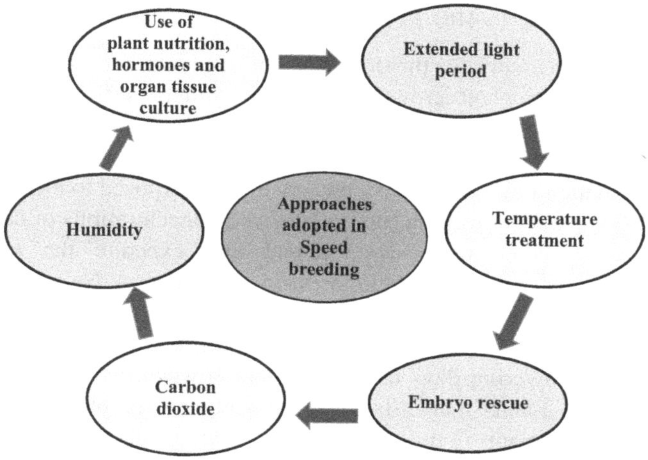

FIGURE 17.1 Different approaches adopted in speed breeding.

range for development, flowering, and seed generation in most crops normally falls within the range of 25–30°C. The rate of vegetative development in plants is positively correlated with an increase in temperature, as long as it remains within the optimal range. Typically, across a wide range of plant species, it is observed that the vegetative phase tends to need a higher optimal temperature compared to the development stage (Hatfield & Prueger, 2015). The relationship between temperature and photoperiod has a significant impact on the process of flowering in numerous species. The ideal temperature and photoperiod conditions were employed to expedite the process of flowering within key limitations (Adams et al., 1999). The manipulation of light and temperature as a means of regulating flowering in plants has been a long-standing practice in the field of environmental management (Ghosh et al., 2018) (Figure 17.1).

17.5 EMBRYO CULTURE

Embryo culture is a technique utilized to promote the development of plant embryos and the cultivation of several crop species. The methodology entails the acquisition and nurturing of underdeveloped seeds within a controlled environment, utilizing a growth medium that may or may not contain a plant growth regulator (PGR). The utilization of PGRs has proven to be effective in the cultivation of faba bean, resulting in the attainment of six to eight generations within a single year (Mobini et al., 2020). The utilization of this methodology has also been observed in the context of pea (Ochatt et al., 2002) and soybean, at a frequency of seven and five generations, respectively (Roumet & Morin, 1997). Embryo rescue is employed in conjunction with stress treatments to reduce the generation time in pea plants (Ribalta et al., 2014). The use of PGR flurprimidol often inhibits plant growth and is commonly employed in conjunction with embryo rescue techniques to accelerate the maturation of peas, hence reducing the generation time (Ribalta et al., 2017). In each of these methodologies, the selection of young embryos is employed to optimize the generation time of various crop species, along with the manipulation of photoperiod variation, changing of light quality, adjustments in temperature, and soil managing strategies. The size of the embryo has a significant impact on the rate of germination and should be taken into account while performing embryo rescue. This rapid breeding method's use is limited because of species-specific protocols, recalcitrant seeds, the need for specialized handling experience, and the necessity for infrastructure construction (Wang et al., 2011). Nevertheless, embryo rescue can supplement methods, such as prolonged photoperiod, and aid in the development of efficient species-specific protocols.

17.6 CARBON DIOXIDE (>400 PPM)

Carbon dioxide (CO_2) is necessary for the development of plants. Increasing CO_2 concentrations, on the other hand, enhances plant growth and photosynthetic efficiency, resulting in earlier flowering and greater biomass. The concentration of carbon dioxide is essential for modulating the opening of stomatal pores, through which plants exchange gases with their surroundings. The average leaf photosynthetic rates experience a 40% rise when exposed to elevated amounts of CO_2 ranging from 475 to 600 parts per million (ppm). Elevated concentrations of carbon dioxide (CO_2) have been found to stimulate plant development and expedite the transformation from photosynthetic growth to reproductive development in angiosperms. Nevertheless, it is important to note that diverse crop species have distinct reactions when exposed to elevated amounts of carbon dioxide (CO_2). For example, increased concentrations of CO_2 of 400/700 and 350/650ppm resulted in a decrease in flowering days in soybean and cowpea by 2 and 12 days, respectively (Springer & Ward, 2007). Plants that utilize the C4 pathway for photosynthesis exhibit lower responses compared to other plants in the presence of elevated levels of carbon dioxide (CO_2). This reduced responsiveness is attributed to the decrease in stomatal conductance caused by elevated CO_2 levels, which indirectly enhances photosynthesis by mitigating water stress during periods of drought. Moreover, the utilization of CO_2-supplemented growth chambers has been found to expedite the process of soybean breeding (Nagatoshi & Fujita, 2019). However, the manipulation of CO_2 levels necessitates the use of growth chambers, CO_2 cylinders, and regulators, hence resulting in supplementary operational expenses.

17.7 HUMIDITY

Humidity has direct effects on plant water relations and indirect effects on photosynthetic growth and fertilization. The stomatal movement of plants is determined by the vapor pressure deficit and air humidity. At low humidity, the transpiration rate is extremely elevated; consequently, plants respond by closing stomata to mitigate the evaporation loss. Maintaining humidity in controlled environment chambers poses a significant challenge, with the optimal range of 60–70% being considered desirable. It may be advantageous to maintain a lower humidity level for crops that exhibit greater adaptability to arid environments (Ghosh et al., 2018).

17.8 NUTRIENTS, HORMONES, AND TISSUE CULTURE

The utilization of nutrients and growth regulators has been utilized to expedite the process of growth as well as stimulate the occurrence of blooming and seed formation. Additionally, these techniques have been utilized to facilitate the germination of underdeveloped seeds under controlled laboratory conditions (Bermejo et al., 2016). Various reactions to the adoption of PGRs can be noticed in controlled environments, such as playhouses and controlled chambers, where factors like photoperiod and temperatures are closely monitored and manipulated. In this particular case, the synergistic application of auxin and cytokinin hormones, specifically flurprimidol (0.3 μM), indole-3-acetic acid (5.7 μM), and zeatin (2.3 μM), resulted in the stimulation of flowering – 100% and subsequent seed production – 90% in faba bean under in vitro conditions (Mobini et al., 2020). In addition, a greater number of seeds was achieved with the external application of 6-benzylaminopurine (5–10 M BAP) on the fourth day following the initiation of flowering in the faba bean (Mobini et al., 2020). Double haploidy is a highly advanced approach in the field of in vitro tissue culture, which can be effectively employed in the process of SB for crops. This technique involves the rescue of haploid embryos, which are subsequently subjected to chromosome doubling to obtain homozygous lines. This approach has demonstrated the ability to achieve homozygosity within two generations, as opposed to the usual breeding methods that typically need six or more generations (Fuente et al., 2013).

17.9 IMPORTANT FACTORS FOR SB

The constituents of SB are deliberately modified according to the particular crop to minimize the duration of each generation. Additionally, crop-specific procedures are established to further enhance the efficiency of the process. SB procedures consist of three separate methodologies: the utilization of a climate-controlled chamber, the application of glasshouse conditions, and the construction of a tabletop growth cabinet (Watson et al., 2018). Watson et al. (2018) utilized longer photoperiod, additional lighting, and temperature control in a glasshouse and concluded that this way of growing crops can produce up to six generations of long-day or day-neutral crops each year, whereas normal methods can only produce one to two generations. Currently, SB has been effectively employed for long-day and day-neutral plant species that do not require vernalization (Table 17.1). The optimum settings facilitate the process of germination in immature seeds, accelerate plant development, and stimulate early flowering and maturity, thus resulting in a reduction in generation time (Ghosh et al., 2018; Watson et al., 2018).

TABLE 17.1
Some Examples of Vegetable Crops

Crop	Type	Controlled Environment Condition	Field Condition (Generation Time in Days)	References
Tomato	Day-neutral	Light tolerance gene CAB13 was introgressed which increased productivity under continuous light	80	Velez-Rameriz et al. (2014)
Capsicum	Day-neutral	12–20 h; 26/20°C day/night; Fluorescent and LED lamps	75	Choi et al. (2023)
Chilli	Day-neutral	Photoperiod, Light intensity Experiment I: 240, 300, 360, and 420 $\mu mol \cdot m^{-2} \cdot s^{-1}$ PPFD, respectively, with a 12 h photoperiod Experiment II: a 14, 16, 18, and 20 h photoperiod, respectively, with 420 $\mu mol \cdot m^{-2} \cdot s^{-1}$ PPFD Experiment III: under PPFD of 420 $\mu mol \cdot m^{-2} \cdot s^{-1}$ and a 12 h photoperiod	38–40 for flowering	Liu et al. (2022)
Radish		Rapid generation advance Extending light exposure through supplemental light with SSD		Ghosh et al. (2018)
Lettuce		12 h light/12 h dark cycle		Mao et al. (2019)
Amaranthus	Short day	Photoperiod and temperature	28 days for flowering	Jahne et al. (2020)
		LD-16 h in the initial stage; 35/30°C day/night; 150mmol SD-8 h later stage; 30/25°C	60	Stetter et al. (2016)
Faba bean	Day-neutral or long day	18 h – 22/18 °C light/dark; 178 μmol	54	Mobini et al. (2020)
Pea	Long day	22 h – 22/17 °C light/dark; 360/500 μmol	51	Watson et al. (2018)

PPFD: photosynthetic photon flux density (Table 17.1).

17.10 ADVANTAGES

SB is an innovative technique that exhibits considerable potential for expediting breeding cycles and enhancing selection efficacy, irrespective of variations in day length, temperature, and precipitation associated with seasonal changes. SB is estimated to be around two to three times more rapid compared to traditional breeding methods. SB is a cost-effective and efficient method that minimizes labor requirements, making it a viable option for widespread adoption, including in economically weak areas. Numerous germplasms could be cultivated densely in a small area, which is particularly advantageous for large-scale population screening. SB can be conducted throughout the year, irrespective of external environmental conditions and fluctuations in climate.

17.11 CHALLENGES

The utilization of accelerated breeding approaches is a valuable strategy for expediting the process of conventional breeding. Nevertheless, the successful implementation of this technology necessitates a proficient understanding, efficient phenotyping facilities, suitable infrastructure, and consistent resources for advancement and research (Shimelis et al., 2019). One significant obstacle that may restrict the implementation of SB within the public sector pertains to the scarcity of proficient and enthusiastic plant breeders, as well as plant breeding technicians, particularly in developing nations (Shimelis et al., 2019). The number of scientists with expertise in plant breeding is limited due to the availability of postgraduate programs in this field being restricted to a select few universities in underdeveloped nations. In many nations, the legislative and administrative infrastructure for the management of the current state of plant breeders' rights and seed regulation is insufficiently established, impeding the advancement of crop breeding initiatives that have the potential to generate advantages along the entire value chain, encompassing farmers and consumers (Tripp et al., 2007). Therefore, it is crucial for developing nations to make necessary adjustments to their policies and practices regarding investments in crop genetic research and education. The implementation of this modification is vital to ensure the long-term viability of crop enhancement efforts and the integration of scientific progressions such as SB.

Sophisticated architecture is necessary for accelerated breeding to effectively control parameters such as soil moisture, temperature, and photoperiod. However, public plant breeding projects in many underdeveloped countries face limited institutional support. This restricted the utilization of advanced breeding techniques, such as SB and biotechnological techniques (Byerlee & Fischer, 2002). Moreover, there exists a restricted accessibility of specialist equipment that is necessary for the initial identification of features during the process of generation advancement (Ribaut et al., 2010). In addition, significant reliance on donor agencies, sometimes known as the 'donor mindset', coupled with a lack of cooperation in provincial breeding projects results in redundant efforts and inefficient allocation of resources. Hence, it is imperative to foster proactive collaboration across national and regional agencies to facilitate the establishment of infrastructure as well as to facilitate the exchange of resources and expertise after its implementation. There is a potential to mitigate the expenses associated with the establishment of new infrastructure through the implementation of novel, localized technology. For instance, the utilization of modified transport containers equipped with solar energy temperature and light control mechanisms is an avenue for cost reduction (Chiurugwi et al., 2018).

The effective control of environmental variables such as moisture levels, temperature, and photoperiod within indoor cultivation facilities necessitates the presence of dependable water and energy provisions. In order to effectively operate indoor SB facilities, it is imperative to have access to energy sources that are both cost-effective and environmentally friendly, while also ensuring consistent and dependable power supply for cooling, heating, and lighting purposes (O'Connor et al., 2013). The successful manipulation of temperature and day length in breeding efforts for SB is considerably impeded by the problem of irregular electrical sources. The cultivation of crops in agricultural fields necessitates certain essential procedures such as soil preparation, fertilizer, water management, and other conventional farming practices. These practices incur significant expenses and necessitate major expenditures in infrastructure. In the context of developing nations, the implementation of SB techniques necessitates the exploration of inventive approaches to address the challenges associated with water and electric power provision, including the use of environment-friendly solar energy solutions. It is conceivable to design a compact indoor breeding kit that has LED lights and temperature controls, which may be powered by solar energy utilizing currently available technologies. One possible strategy entails the utilization of SB concepts inside partially regulated field-based methods. This method entails using high-density planting techniques while effectively managing moisture and nutrient stress. By adopting this strategy, it becomes feasible to cultivate large populations at a comparatively reduced expense.

The strategy employed by SB involves the stimulation of early flowering in photoperiod-responsive crops. Nevertheless, several plant species exhibit distinct responses in terms of their photoperiod requirements and sensitivity to light intensity. The successful use of SB in a plant species that exhibits short-day and day-neutral characteristics necessitates the establishment of standardized protocols tailored to the specific species and varieties involved. The variations in photoperiodic conditions between short-day and long-day durations can span from minutes to hours, posing significant challenges for achieving standardization in scientific studies (Jackson & Jackson, 2009). The aforementioned limitation of SB can be effectively addressed with the implementation of distinct photoperiods for the vegetative and reproductive stages, as illustrated in the case of amaranth (Stetter et al., 2016).

17.12 COMBINING SB WITH MARKER-ASSISTED SELECTION, GENOMIC SELECTION, TRANSGENICS, AND GENOME EDITING

The combination of various current approaches with the SB platform has promise for enhancing the efficiency of the breeding program. The molecular-assisted selection (MAS) technique utilizes markers at the DNA level for the indirect selection of desired traits. The concept of marker-assisted selection involves the identification of a DNA marker that is tightly linked with the gene of interest. The process of molecular-assisted selection (MAS) involves the utilization of three distinct types of selection: foreground selection, background selection, and recombinant selection. The extensive utilization of MAS in expediting the generation time of various crops has been documented in previous studies (Das et al., 2017; Prabhu et al., 2009). The marker associated with the trait can be utilized to selectively identify the target plant. Moreover, the plant possessing the required characteristics can undergo rapid generation (SB) to establish stability. The application of 'seed chipping' in conjunction with barcoding has the potential to enhance the efficiency of high-throughput marker-assisted selection (MAS) when performed at the level of individual plants. SB can carry out breeding activities, including processes such as crosses between them, producing mapping populations, and field evaluation for certain features (Watson et al., 2018). Additionally, SB can be combined with transgenic, pyramiding, and backcrossing techniques to speed up these procedures (Watson et al., 2018). As a result, combining MAS and SB will speed up plant breeding and lower project costs (Figure 17.2).

In a similar vein, the process of genomic selection (GS) determines the genomic estimated breeding values (GEBVs) for the sum of the effects of all quantitative trait loci by using compact markers at the whole genome level. These values explain all the genetic differences that can exist for desired traits (Hayes et al., 2009). Plants exhibiting elevated GEBV have been selected for progression to subsequent generations. Genome sequencing (GS) is primarily designed to investigate complex genetic features, such as yield, which are influenced by the interaction of several genes. Genomic selection (GS) possesses a notable advantage over alternative breeding approaches in terms of its ability to expedite the process of variety development, as well as its

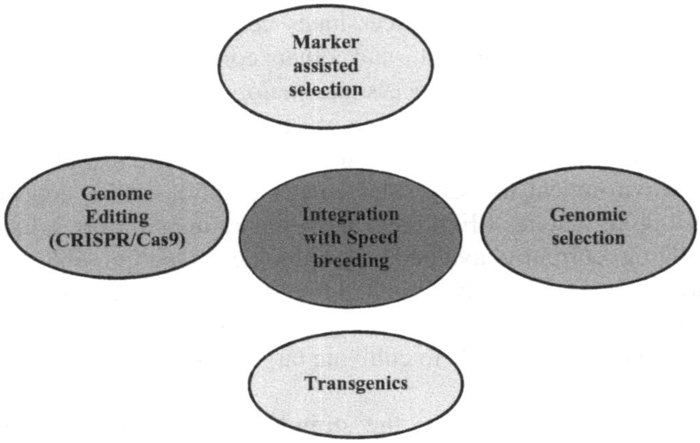

FIGURE 17.2 Schematic representation of integrated breeding methods with speed breeding.

efficient utilization of time and resources. The biggest concern associated with genome sequencing (GS) at present is the cost involved. The cost associated with genomic selection (GS) can be reduced by employing it in a subsequent generation or by limiting its application to specific traits that potentially transcend the threshold value using alternative strategies like selection-based procedures (Riaz et al., 2016).

Therefore, combining selection breeding and genomic selection techniques holds promise for improving selection precision and achieving substantial annual genetic advancements (Gorjanc et al., 2018; Hickey et al., 2017). In this method, the parental individuals were chosen based on their GEBVs, resulting in a more efficient selection process. Subsequently, progeny produced by this selected group was further selected. The procedure mentioned above is iterated multiple times to facilitate an accelerated reproductive cycle. The combination of SB and genomic selection techniques can effectively enhance the genetic advantage achieved by parent selection in successive generations within a breeding program. The utilization of the combination approach demonstrates a reduction in the issue of inbreeding when compared to the implementation of either phenotypic selection or the genomic selection scheme. The integration of this technique, in conjunction with various characteristics, has the potential to yield more benefits, such as the normalization of the differential vegetation index and the measurement of canopy temperature. These advancements can be facilitated by the utilization of high-throughput phenotyping platforms, hence enabling the early-stage identification and selection of desirable qualities (Crain et al., 2018).

By employing techniques such as genome editing and mutagenesis, the process of SB can be utilized to cultivate biofortified foods. This involves the reduction of anti-nutritional elements found in Brassica species, as well as the enhancement of vitamin content in staple crops like rice. Variability holds utmost significance in a varietal developmental scheme. Nevertheless, the protracted process of domestication has resulted in a reduction in the pool of genetic variation that is accessible for selective breeding purposes. The CRISPR/Cas system currently facilitates the rapid generation of novel genomic diversity (Shen et al., 2017). The CRISPR/Cas system possesses the capability of multiplexing, enabling the simultaneous alteration of multiple targets and the rapid accumulation of various phenotypes within a single generation (Zhou et al., 2019). The enhancement of inflorescence design and flower output was achieved by the inhibition of regulatory components using CRISPR/Cas technology (Liu et al., 2013; Soyk et al., 2017). In the context of polyploid crops, such as potatoes, which possess several copies of genes, the task of modifying or eliminating a specific target gene that negatively impacts potato quality by conventional breeding methods is challenging. In this scenario, the genome editing platform, with CRISPR-Cas9 being the most prevalent, appears to exhibit greater efficiency, particularly when applied to the genome of polyploid plants. The potential manipulation involves the addition, deletion, or substitution of the gene of interest. The process of CRISPR-Cas9-mediated editing utilizes single-stranded guide RNA (sgRNA) as a means to direct the Cas9 enzyme toward a particular target sequence (Doudna & Charpentier, 2014; Haroon et al., 2019). Nevertheless, the cultivation of genome-edited plants necessitates the implementation of in vitro tissue culture manipulation, and the regulatory framework governing these plants falls under the purview of transgenic plant regulation. This integration and tissue culture-related technological manipulations and advancements might bypass the regulatory consequences associated with genetically engineered products.

17.13 CONCLUSION

Typically, the process of breeding requires approximately four to six generations in order to establish a genetically stable lineage with regard to a particular feature. The growing need for enhanced agricultural crop output has prompted the need for expedited crop development, wherein the use of fast-growing varieties with reduced generation time holds potential advantages for crop research. SB offers researchers worldwide the opportunity to examine the combined potential of genome editing technologies and selective breeding within the context of SB, thereby transforming the field of plant study. The implementation of SB techniques has the potential to expedite the progress of developing cultivars that exhibit desirable qualities in the market. This is achieved by minimizing the temporal, spatial, and resource requirements associated with selecting and genetically enhancing the superior crop types. This technology enables plant breeders to expedite the delivery of enhanced crop varieties. Efficient integration of SB into a crop development project relies on the implementation of streamlined procedures that minimize labor requirements and utilize cost-effective facilities. In addition, the incorporation of SB into conventional breeding methods, marker-assisted selection, and genome editing techniques has the potential to improve the efficiency of selecting superior genotypes and lines that possess desirable traits, such as increased yield and improved nutritional characteristics, as well as enhanced resistance to both biotic and abiotic stresses. The selection approaches that are best suitable for the practice of accelerated breeding encompass SSD, single pod development, and SPS methods. However, the extensive adoption of SB techniques in numerous developing nations, particularly within public plant breeding initiatives, is constrained by the scarcity of adequately educated plant breeders and technicians, as well as insufficient infrastructure and unreliable access to water and energy. Presently, there exists a deficiency in governmental assistance, both in terms of policy and financial resources, to facilitate and maintain the practice of SB within public plant breeding initiatives.

REFERENCES

Ahmar, S., Gill, R. A., Jung, K. H., Faheem, A., Qasim, M. U., Mubeen, M., & Zhou, W. (2020). Conventional and molecular techniques from simple breeding to speed breeding in crop plants: recent advances and future outlook. *International Journal of Molecular Sciences, 21,* 2590.

Adams, S., Pearson, S., Hadley, P., & Patefield, W. M. (1999). The effects of temperature and light integral on the phases of photoperiod sensitivity in *Petunia×hybrida. Annals of Botany*, *83*, 263–269. 10.1006/anbo.1998.0817.

Bermejo, C., Gatti, I., & Cointry, E. (2016). In vitro embryo culture to shorten the breeding cycle in Lentil (*Lens Culinaris* Medik). *Plant Cell, Tissue and Organ Culture, 127,* 585–590.

Breseghello, F., & Coelho, A. S. G. (2013). Traditional and modern plant breeding methods with examples in rice (*Oryza Sativa* L.). *Journal of Agricultural and Food Chemistry, 61,* 8277–86.

Byerlee, D., & Fischer, K. (2002). Accessing modern science, policy and institutional options for agricultural biotechnology in developing countries. *World Development, 61,* 8277–86.

Chiurugwi, T., Kemp, S., Powell, W., & Hickey, L. T. (2018). Speed breeding orphan crops. *Theoretical and Applied Genetics, 132,* 607–616.

Choi, H., Back, S., Kim, G. W., Lee, K., Venkatesh, J., Lee, H. B., Kwon, J., & Kang, B. (2023). Development of a speed breeding protocol with flowering gene investigation in pepper (Capsicum annuum). *Frontiers in Plant Science, 14.* 10.3389/fpls.2023.1151765.

Crain, J., Mondal, S., Rutkoski, J., Singh, R. P., & Poland, J. (2018). Combining high-throughput phenotyping and genomic information to increase prediction and selection accuracy in wheat breeding. *The Plant Genome, 11,* 1–14.

Das, G., Patra, J. K., & Baek, K. (2017). Insight into MAS: a molecular tool for development of stress resistant and quality of rice through gene stacking. *Frontiers in Plant Science, 8,* 985.

Doudna, J. A. & Charpentier, E. (2014). The new frontier of genome engineering with Crispr-Cas9. *Science, 346*, 1258096.

Dwivedi, S. L., Britt, A. B., Tripathi, L., Sharma, S., Upadhyaya, H. D., & Ortiz, R. (2015). Haploids: constraints and opportunities in plant breeding. *Biotechnology Advances, 33,* 812–29.

De La Fuente, G. N., Frei, U. K., & Lübberstedt, T. (2013). Accelerating plant breeding. *Trends in Plant Science*, *18*, 667–672. 10.1016/j.tplants.2013.09.001.

Ghosh, S., Watson, A., Gonzalez-Navarro, O. E. *et al.* (2018). Speed breeding in growth chambers and glasshouses for crop breeding and model plant research. *Nature Protocols, 13,* 2944–2963.

Gorjanc, G., Gaynor, R. C., & Hickey, J. M. (2018). Optimal cross selection for long-term genetic gain in two-part programs with rapid recurrent genomic selection. *Theoretical and Applied Genetics, 131,* 1953–1966.

Haroon, M., Afzal, R., Idrees, F., Sunny, A., & Khan, A. S. (2019). Genome editing and speed breeding; game changers to boost the crop production. *The International Journal of Biological Research, 2,* 295–300.

Hatfield, J. L., & Prueger, J. H. (2015). Temperature extremes: effect on plant growth and development. *Weather and Climate Extremes*, 1–7.

Hayes, B. J., Bowman, P. J., Chamberlain, A. J., & Goddard, M. E. (2009). Invited review: genomic selection in dairy cattle: progress and challenges. *Journal of Dairy Science, 92,* 433–43.

Hickey, L. T., Hafeez, A., Robinson, H. *et al.* (2019). Breeding crops to feed 10 billion. *Nature Biotechnology, 37,* 744–754.

Hickey, L. T. Germa, S. E., Diaz, J. E., Ziems, L. *et al.* (2017). Speed breeding for multiple disease resistance in barley. *Euphytica*, 213, 64.

Jackson, S. D., & Jackson, S. D. (2009). Plant responses to photoperiod. *New Phytologist, 181,* 517–31.

Jahne, F., Hahn, V., Wurschum, T. *et al.* (2020). Speed breeding short-day crops by LED-controlled light schemes. *Theoretical and Applied Genetics, 133,* 2335–2342.

Laux, P., Jackel, G., Tingem, R. M., & Kunstmann, H. (2010). Impact of climate change on agricultural productivity under rainfed conditions in Cameroon—a method to improve attainable crop yields by planting date adaptations. *Agricultural and Forest Meteorology, 150,* 1258–1271.

Lin, Z., Cogan, N. O. I., Pembleton, L. W. *et al.* (2016). Genetic gain and inbreeding from genomic selection in a simulated commercial breeding program for perennial ryegrass. *Plant Genome, 9,* 1–12.

Liu, C., Teo, Z. W., Bi, Y. *et al.* (2013). A conserved genetic pathway determines inflorescence architecture in arabidopsis and rice. *Developmental Cell, 24,* 612–22.

Liu, K., He, R., He, X. *et al.* (2022). Speed breeding scheme of hot pepper through light environment modification. *Sustainability, 14,* 12225.

Mao, H., Hang, T., Zhang, X., & Lu, N. (2019). Both multi-segment light intensity and extended photoperiod lighting strategies, with the same daily light integral, promoted *Lactuca sativa* L. growth and photosynthesis. *Agronomy, 9,* 857.

Mcclung, C. R., Lou, P., Hermand, V., & Kim, J. A. (2016). The importance of ambient temperature to growth and the induction of flowering. *Frontiers in Plant Science, 7,* 1266.

Mobini, S., Khazaei, H., Warkentin, T. D., & Vandenberg, A. (2020). Shortening the generation cycle in faba bean (vicia faba) by application of cytokinin and cold stress to assist speed breeding. *Plant Breeding, 139,* 1181–1189.

Nagatoshi, Y., & Fujita, Y. (2019). Accelerating soybean breeding in a CO2-supplemented growth chamber. *Plant and Cell Physiology*, *60*, 77–84. 10.1093/pcp/pcy189.

O'Connor, D. J., Wright, G. C., Dieters, M. J. *et al.* (2013). Development and application of speed breeding technologies in a commercial peanut breeding program. *Peanut Science, 40,* 107–114.

Ochatt, S. J., Sangwan, R. S., Marget, P., Ndong, A., Rancilliac, M., & Perney, P. (2002). New approaches towards the shortening of generation cycles for faster breeding of protein legumes. *Plant Breeding, 121* 436–440.

Prabhu, A. S., Filippi, M. C., Silva, G. B., Lobo, L. S., & Morais, O. P. (2009). An unprecedented outbreak of rice blast on a newly released cultivar BRS Colosso in Brazil. In *Advances in genetics, genomics and control of rice blast disease*. Springer Science+Business Media B.V.

Riaz, A., Periyannan, S., Aitken, E., & Hickey, L. (2016). A rapid phenotyping method for adult plant resistance to leaf rust in wheat. *Plant Methods, 2,* 12–17.

Ribalta, F. M., Croser, J. S., Erskine, W., Finnegan, P. M., & Lulsdorf, M. M. (2014). Antigibberellin-induced reduction of internode length favors in vitro flowering and seed-set in different pea genotypes. *Biologia Plantarum, 58,* 39–46.

Ribalta, F. M., Karen, M. P., Kylie, N. *et al.* (2017). Precocious floral initiation and identification of exact timing of embryo physiological maturity facilitate germination of immature seeds to truncate the lifecycle of pea. *Plant Growth Regulation, 81,* 345–353.

Ribaut, J. M., De Vicente, M. C., & Delannay, X. (2010). Molecular breeding in developing countries: challenges and perspectives. *Current Opinion in Plant Biology, 13,* 213–218.

Roumet P., & Morin, F. (1997). Germination of immature soybean seeds to shorten reproductive cycle duration. *Crop Science, 37,* 521–525.

Samineni, S., Sen, M., Sajja, S. B., & Gaur, P. M. (2019). Rapid generation advance (RGA) in chickpea to produce up to seven generations per year and enable speed breeding. *Crop Journal, 8,* 1–6.

Shen, L., Hua, Y., Fu, Y., *et al.* (2017). Rapid generation of genetic diversity by multiplex crispr/cas9 genome editing in rice. *Science China Life Sciences, 60,* 506–515.

Shimelis, H, Gwata, E. T., & Laing, M. D. (2019). Crop improvement for agricultural transformation in Southern Africa, *Transforming Agriculture in Southern Africa* (p. 346). 10.4324/9780429401701-13.

Shimelis, H., & Laing, M. (2012). Timelines in conventional crop improvement: pre-breeding and breeding procedures. *Australian Journal of Crop Science, 6,* 1542–1549.

Soyk, S., Lemmon, Z. H., Oved, M. *et al.* (2017). Bypassing negative epistasis on yield in tomato imposed by a domestication gene. *Cell, 169,* 1142–1155.

Springer, C. J., & Ward, J. K. (2007). Flowering time and elevated atmospheric CO_2. *New Phytologist, 176,* 243–255.

Stetter, M. G., Zeitler, L., Steinhaus, A. *et al.* (2016). Crossing methods and cultivation conditions for rapid production of segregating populations in three grain amaranth species. *Frontiers in Plant Science, 7,* 816.

Tripp, R., Louwaars, N., & Eaton, D. (2007). Plant variety protection in developing countries. A report from the field. *Food Policy, 32,* 354–371.

Velez-Ramirez, A. I., van Ieperen, W., Vreugdenhil, D., van Poppel, Pieter M. J. A., Heuvelink, Ep, & Millenaar, F. F. (2014). A single locus confers tolerance to continuous light and allows substantial yield increase in tomato. *Nature Communications*, *5*. 10.1038/ncomms5549.

Wang, X., Wang, Y., Zhang, G., & Ma, Z. (2011). An integrated breeding technology for accelerating generation advancement and trait introgression in cotton. *Plant Breeding, 130,* 569–573.

Watson, A., Ghosh, S., Williams, M. J., Cuddy, W. S. *et al.* (2018). Speed breeding is a powerful tool to accelerate crop research and breeding. *Nature Plants, 4,* 23–29.

Zhou, J., Xin, X., He, Y., Chen, H., Li, Q., Tang, X., Zhong, Z., Deng, K., Zheng, X., Akher, S. A., Cai, G., Qi, Y., & Zhang, Y. (2019). Multiplex QTL editing of grain-related genes improves yield in elite rice varieties. *Plant Cell Reports, 38,* 475–485.

18 Strategies for Speed Breeding in Crops to Accelerate Plant Improvement

Naglaa A. Abdallah
Department of Genetics Faculty of Agriculture, Cairo University, Cairo, Egypt

National Biotechnology Network of Expertise, ASRT, Egypt

Aladdin Hamwieh
Department of Biotechnology, International Centre for Agricultural Research in the Dry Areas (ICARDA), Giza, Egypt

Khaled Radwan
National Biotechnology Network of Expertise, ASRT, Egypt

Department of Biotechnology, Agricultural Genetic Engineering Research Institute (AGERI), ARC, Giza, Egypt

Michael Baum
Department of Biotechnology, International Centre for Agricultural Research in the Dry Areas (ICARDA), Giza, Egypt

18.1 INTRODUCTION

Hunger is currently rising globally and the proportion of starvation worldwide has increased to 11% in 2016 (UN, 2018). This alarming trend highlights the urgent need for innovative agricultural research and development to address food security challenges. In response, numerous research institutes around the world have been established to tackle these issues and find sustainable solutions. In 2017, approximately 821 million individuals worldwide faced hunger, as reported by the Food and Agriculture Organization (FAO, 2018). These numbers underscore the critical importance of investing in agricultural advancements and sustainable farming practices. Efforts must be made to not only increase food production but also ensure equitable distribution and access to nutritious food for all. Despite efforts to achieve the Sustainable Development Goal of ending hunger worldwide by 2030 (FAO, 2018), food insecurity is on the rise. Plant breeders face additional difficulties in creating tolerant/resistant varieties due to climate change-related phenomena such as floods, droughts, carbon dioxide increases, a 2°C rise in average global temperatures, an increase in the salinity of the soil, and the proliferation of pests and diseases. These challenges make it increasingly difficult for plant breeders to develop crops that can withstand these adverse conditions and still provide adequate nutrition. In addition, limited funding and resources hinder the research and development of new varieties that could potentially combat food insecurity.

Agriculture serves as the primary source of sustenance for the growing population, projected to reach 10 billion by 2057 (www.theworldcounts.com), indicating a 25% growth in population over the

 DOI: 10.1201/b23372-18

next three decades (Hickey et al., 2019). The confluence of food, energy, and water shortages will be exacerbated, resulting in a significant increase in the demand for food by 60%. Historically, traditional breeding techniques have proven effective in the creation of high-yielding cultivars. The widespread adoption of these cultivars, together with the implementation of sound agronomic practices, played a pivotal role in the emergence of the green revolution during the 1960s. The breeding cycles required to generate genetically pure types are characterized by a considerable duration, typically spanning seven to ten years. The decrease in the duration required for the breeding process of a particular variety can significantly impact the annual rate of genetic improvement. Novel technologies are required to generate rapid precise and targeted manipulation in crops. Various technologies have been developed to facilitate rapid generation advancement (RGA). These involve plant transformation, speed breeding, genome editing, doubled haploid (DH) production, marker-assisted selection (MAS), high-throughput genotyping, genomic selection (GS), plant transformation, and shuttle breeding. Opting for the combination of advanced methods such as speed cloning and genome editing is the optimal choice for efficiently achieving the desired generation in a short timeframe, paving the way for speed breeding. By using these methods, it would be easier to quickly develop climate-resilient cultivars that are more productive and better able to handle both living and nonliving stresses.

The initial strategy for obtaining accelerated generational growth entailed cultivating many generations in different geographic locations by employing off-season seeding techniques. The utilization of shuttle-breeding techniques in specific crops to expedite the production of two generations within a single year was found to be viable. The strategy indicated above exhibited satisfactory effectiveness in decreasing the duration of breeding by 50% to introduce a new variety. A technique called DH has gotten a lot of attention as a very promising way to do recurrent GS and make a lot of homozygous lines in just one year. MAS has demonstrated significant potential due to its ability to enhance selection accuracy and enable breeders to concentrate on a smaller number of promising plants, thereby improving the efficiency and precision of selection (Collard & Mackill, 2008; Humphreys & Knox, 2015; Wessel & Botes, 2014; Wiśniewska et al., 2019). The utilization of MAS can also effectively decrease the duration of the selection process for breeders, particularly when employed in the context of the backcrossing technique. The aforementioned approach is referred to as the marker-assisted backcrossing strategy, which was employed by breeders in India to cultivate two drought-tolerant varieties (Geleltu and Pusa 10216), as well as one variation resistant to Fusarium wilt (Mannur et al., 2019), within the chickpea species. The utilization of high-throughput genotyping and sequencing technology has significantly accelerated the rate of genetic gain within the desired conditions. The efficacy of association mapping, high-density genetic mapping, and quantitative trait loci (QTL) has experienced notable enhancements to effectively discern appropriate markers for selection. These novel methods will facilitate breeders and geneticists in acquiring a more comprehensive comprehension of quantitative features, which are regulated by a combination of main and minor genes. However, despite the potential of MAS to expedite the process of variety release, its application becomes more intricate when dealing with quantitative traits that are governed by numerous genes with minor impacts (Desta et al., 2014). The provided text is insufficient in length to undergo academic rephrasing.

The adoption of GS has become widely acknowledged as a prevalent approach within the breeding domain, as indicated by Meuwissen et al. (2001). The utilization of early selection before phenotypic measurements has the potential to significantly expedite breeding progress in a highly cost-effective manner. In certain instances, such as in the case of chickpeas, it is possible to extract DNA from the seed coats of the population during the summer season. This extraction process facilitates early selection before the winter planting period. One of the benefits of GS is that it does not rely on QTL that bestow specific qualities, as is the case with MAS. The underlying principle of this approach involves the utilization of molecular markers and phenotypic data obtained from a specific population, referred to as the training population, to forecast the phenotypic values within the breeding population. This prediction is made without the need for field screening and the subsequent improvement of the resistance gene analogs.

The utilization of DH technology as an expedited method for enhancing crop traits has become a common practice in numerous research institutions, resulting in the development and marketing of over 300 varieties (Sood & Dwivedi, 2015). In recent scholarly works, researchers have employed haploid induction editing technology (HI-Edit) to perform genetic modifications on inbred maize lines, as demonstrated by Kelliher et al. (2019). In the study conducted by Kelliher et al. (2019), the maize pollen parent, which had transformed with Cas9 was employed for pollinating wheat spikes. This process ultimately led to the production of altered wheat embryos.

Recent years have witnessed notable advancements in the creation of sensor modules designed to improve high-throughput phenotyping. These modules encompass several technologies, including imaging cameras, drones, light-emitting diodes (LED) lighting, and portable devices. Their primary purpose is to facilitate the screening of a diverse array of features (Fiorani & Schurr, 2013). The utilization of high-throughput phenotyping technology has resulted in the screening of larger populations within shorter timeframes compared to traditional screening methods, while also achieving enhanced accuracy. Several high-throughput phenotyping platforms have been employed to enhance genetic gain in various crops. For instance, Hu et al. (2020) utilized unmanned aerial vehicles (UAV) to assess grain yield in wheat. Wang et al. (2019) employed a mobile field vehicle to evaluate flowering time. Additionally, Deery et al. (2019) utilized the ArduCrop wireless infrared thermometer and airborne thermography with a manned helicopter to measure canopy temperature. The application of LED light innovation in plant growth was first observed in 1990. This development followed numerous previous attempts to cultivate plants using artificial light to enable plant growth outside of their natural seasons. Speed breeding techniques have been successfully implemented on many crops such as wheat, barley, chickpea, pea, and canola. This method involves subjecting the plants to a photoperiod of 22 hours per day, achieved through the use of LED lights. The utilization of LEDs not only facilitates cost reduction in terms of lighting and cooling but also enables the acceleration of plant growth and development. The findings suggest that there is significant potential for incorporating speed breeding alongside other innovative plant breeding techniques to expedite the process of crop enhancement (Watson et al., 2018). Additionally, it has been observed that the use of speed breeding can reduce the life cycle of crops by as much as 50% (Samineni et al., 2020).

Integration of cutting-edge technologies such as MAS and genome editing into breeding programs accelerates breeding cycles for increasing the rate of genetic gain over time and crop improvement (Figure 18.1).

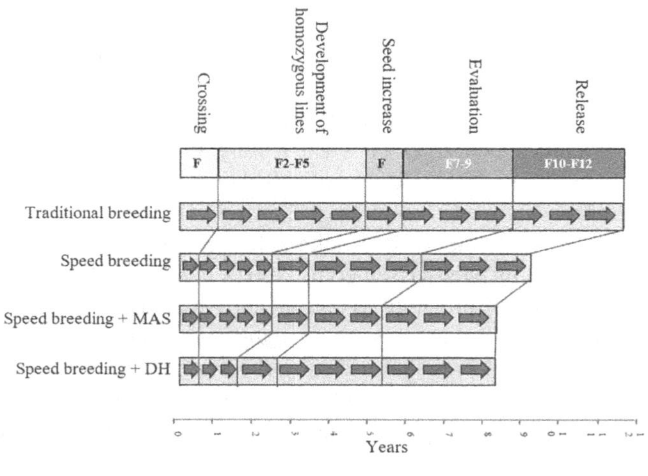

FIGURE 18.1 Diagram illustrating traditional breeding versus developed breeding strategies (speed breeding, speed-double haploid DH, and speed-marker-assisted selection) and their impact on the length of the breeding cycle.

18.2 SHUTTLE BREEDING

The utilization of shuttle-breeding technology enables crop breeders to expedite the development of segregating populations across several geographical regions (including latitude, altitude, and rainfall) within a single year, hence reducing the duration of the breeding cycle. The utilization of the off-season period has long been employed to facilitate the development of breeding populations for various crops and vegetables. The acceleration of the breeding process is highly advantageous; yet, it is subject to a maximum of two generations per year, with occasional instances of three generations. Nevertheless, the task necessitates the identification of a suitable area that is accessible to the breeders during the off-season. One additional challenge encountered during the off-season is the transportation of seeds across governmental boundaries. The process of clearing the seeds from customs is a time-consuming endeavor.

The shuttle-breeding initiative at CIMMYT was established during the early 1960s by Norman Borlaug, who implemented the practice of cultivating two wheat crops annually under different growing circumstances. International Crops Research Institute for the Semi-Arid Tropics (ICARDA) has implemented off-season planting techniques at two distinct locations, namely Tal Hadya North Aleppo, Syria, and Terbol station, Terbol, Lebanon, to cultivate cereal and legume crops. The utilization of shuttle breeding in wheat has proven to be highly efficacious in the identification and selection of photoperiod-insensitive genes (Ppd1 and Ppd2), as well as dwarfing genes (*Rht1* and *Rht2*). The outcomes of shuttle breeding have led to the development of semidwarf wheat varieties that exhibit broad adaptability, high yield potential, tolerance to lodging, and responsiveness to inputs. At ICARDA, the modified pedigree-bulk selection approach is employed for the advancement of F2 to F6 generations in spring bread wheat. This method involves the individual selection of approximately 2000 F2 plants. The shuttle-breeding process involves multiple cycles at different stations located in Lebanon, Morocco, Egypt, and Ethiopia. In Lebanon, the breeding occurs in two cycles, specifically during the winter and summer seasons, at Terbol station (33°,48′34″, 35°59′24, elev. 894 m). In Morocco, there is one breeding cycle during the winter season at Marchouch station. In Egypt, the breeding takes place during the early winter season at Sids station (28°54′21″, 30°57′01″ elev. 32 m). Last, in Ethiopia, there is one breeding cycle during the summer season at Kulumsa station (8°01′09″, 39°09′12″ elev. 2175 m).

18.3 DOUBLED HAPLOID

The DH technique is an efficient method for expediting the formation of homozygous genotypes. This approach involves the application of chemical agents, such as colchicine, to induce chromosome doubling in haploid cells, hence accelerating the process of generating homozygous genotypes. The successful development and application of this technology has been observed in several crops, including barley, rice, wheat, and maize, resulting in a reduction of the time required to release new varieties to a span of three to four years (Brennan & Martin, 2007). Nevertheless, the application of DH in crops such as legumes has been limited due to the intricate tissue culture techniques necessary for embryo maturation. One further drawback of DH is the lack of responsiveness exhibited by certain species.

DH technology has been employed in wheat through the utilization of two techniques. The first technique involves intergeneric hybridization, where wheat pollen is used. Following hybridization, chromosome elimination occurs for two to three days. The phenomenon of chromosome elimination after fertilization has been postulated on the grounds of the impact of genome equilibrium in reciprocal crosses involving diploid and autotetraploid states. One further technological aspect within the field of digital humanities involves the cultivation of anthers or microspores. In Sudan, the emergence of two heat-tolerant DH varieties was documented by Ali et al. (2006). Similarly, in China, the cultivation of over 20 varieties resulting from tissue culture efforts has been observed on a land area exceeding 1 million hectares, as reported by Hu (1997).

The initial cultivar, known as Jinghua 1, was introduced in China in 1983 through the utilization of other cultural techniques. Subsequently, other nations such as France, Hungary, and Sweden have also developed wheat cultivars utilizing other culture methods, as documented by De Buyser et al. (1987), Pauk et al. (1995), and Tuvesson et al. (2003). Between 2005 and 2011, Romania successfully developed four wheat cultivars through the application of intergeneric hybridization techniques, specifically by employing maize pollination (Giura, 2007; Săulescu et al., 2012). The initial DH wheat variety, 'Him Pratham', was developed in India through the application of the chromosomal elimination technique as described by Chaudhary et al. (2015). In the year 2018, the United States of America introduced two wheat types, namely Avery and Langin, through the process of maize hybridization, as documented by Haley et al. (2018).

The techniques of Bulbosum, anther culture, and microspore culture have been utilized in the context of barley breeding to generate DH individuals. Several DH populations have been established to investigate molecular markers in barley, facilitating the development of detailed chromosome-based maps. These maps play a crucial role in pinpointing significant genes, as demonstrated by Rollins et al. (2013). These genes have the potential to be associated with various biochemical pathways, proteins, and enzymes. In their study, Honsdorf et al. (2017) utilized the S42IL DH population, derived from a breeding cross between the German malting variety 'Scarlett' of Hordeum vulgare ssp. vulgare and the accession 'ISR42-8' of Hordeum vulgare ssp. spontaneum. This population was employed to investigate the QTL responsible for regulating grain filling in the presence of terminal drought stress.

Speed breeding: Plant breeders engage in ongoing endeavors to create contemporary cultivars that possess both a substantial yield and significant nutritional content, all while optimizing efficiency, to adequately nourish the expanding global population. Nevertheless, the standard methods employed for crop enhancement are inadequate in meeting the ever-increasing need for food, mostly because of the restricted number of generations that can be produced during a single year during the breeding process. Speed breeding (SB) is an effective technique utilized to expedite the crop breeding process through the manipulation of diverse environmental factors and the integration of artificial intelligence (AI) systems. These factors include photoperiod, light intensity, temperature, soil moisture, soil nutrition, and high-density planting. The utilization of SB greatly expedites the crop breeding process by stimulating early blooming and seed production, hence diminishing the duration required for the creation of each successive breeding generation. According to Wanga et al. (2021), the utilization of single seed descent (SSD) in the plant breeding program offers advantages in terms of time efficiency and cost-effectiveness when compared to traditional breeding methods. Moreover, SSD is considered the most effective approach for implementing speed breeding in controlled environments.

The utilization of LED lights has been employed to expedite the crop breeding cycle through light manipulation, specifically in terms of altering the duration of light exposure per day. The application of LED lighting in plant growth was first recorded in 1990 (Bula et al., 1991; Nakamura & Fasol, 1997). A study on the use of LED technology for cultivating wheat in space was conducted by the National Aeronautics and Space Administration (NASA) of the United States in collaboration with Utah State University. Speed breeding was used to ensure that the generation time of major crops like wheat, chickpea, canola, and barley could be considerably reduced by adaptation of this technology (Table 18.1). The findings of Bugbee and Koerner (1997) revealed a significant reduction in the life cycle of plants when exposed to LED lighting. Nevertheless, advancements have been made in the utilization of this technology in the field of plant breeding, resulting in improved efficiency and reduced costs. In comparison to older light systems, a significant portion (40–60%) of energy is now transformed into light rather than heat, rendering it a more energy-efficient and cooler light system (Stutte et al., 2015). The aforementioned properties render this technology a very suitable approach for implementation in crop breeding endeavors. In addition to the regulation of photoperiod, two additional aspects play a crucial role in establishing an effective speed breeding system. The first factor is the

TABLE 18.1
Comparison of Speed Breeding Control Condition to Accelerate the Generation Time of 4–6 Generations of Wheat, Barley, Chickpea, and Canola Compared to the Greenhouse in Natural Conditions Giving 2–3 Generations

Crop	Breeding method	Generation / 12 months
 Spring & durum wheat	Greenhouse	
	Speed breeding	
 Barley	Greenhouse	
	Speed breeding	
 chickpea	Greenhouse	
	Speed breeding	
 Pea	Greenhouse	
	Speed breeding	
 Canola	Greenhouse	
	Speed breeding	

utilization of an air conditioning system to effectively manage and control the ambient temperature. The second factor involves the optimization of humidity levels within the range of 70–75%. According to Watson et al. (2018), it is recommended to maintain a photoperiod of 22 hours per day and a temperature range of 22–25 °C. During the remaining two hours of darkness, the temperature can be dropped to 17 °C.

Speed breeding techniques have been implemented in a range of crops, particularly in key grain and legume commodities, resulting in the ability to achieve four to eight breeding generations within a single year. Furthermore, the utilization of SB has been widely employed to phenotype agronomic traits in wheat through the implementation of various selection approaches (Li et al., 2018a,b). Additionally, SB has been utilized for the multi-trait phenotyping of quantitative traits in barley (Hickey et al., 2017). The technique of speed breeding has been utilized in the assessment of root characteristics, disease reactivity, and plant height in durum wheat, as demonstrated by Alahmad et al. (2018). This methodology has facilitated breeders in efficiently identifying and integrating disease resistance traits within a limited timeframe. Speed breeding has been utilized in the assessment of disease resistance in wheat, specifically for stripe rust disease (Hickey et al., 2011), leaf rust (Riaz et al., 2016), and Fusarium head blight (Watson et al., 2018). Furthermore, in Australia, the initial spring wheat variety developed through the implementation of the speed breeding technique, which was produced in 2017 (Hickey et al., 2019). The utilization of speed breeding facilities has been employed within a hydroponics system to assess the resistance of pea plants to boron (Bennett et al., 2017). Additionally, these facilities have been utilized to evaluate the resistance of chickpea plants to Fusarium wilt as well as the tolerance of chickpea and wheat plants to high salt levels (Hamwieh et al., unpublished data). The utilization of speed breeding facilities offers a potential benefit in enhancing the genetic gain of strategic crops through the screening of various features (Li et al., 2018b).

Speed breeding is a logical progression in facilitating the large-scale evaluation of diverse plant genotypes to align with the rapid advancements in plant science research. The integration of speed breeding techniques in conjunction with contemporary breeding methods and genetic engineering approaches holds promise for the development of customized crops characterized by enhanced nutritional content, increased productivity, and improved resilience against both abiotic and biotic challenges. Various speed breeding methods have been established for several pulse crops, such as lentil, chickpea, faba bean, grass pea, field pea, and pigeonpea, to achieve a rapid generation turnover of four to six generations per year (Ghosh et al., 2018; Lulsdorf & Banniza, 2018; Samineni et al., 2019; Saxena et al., 2019). In summary, the speed breeding protocol has demonstrated the ability to boost flower induction in lentils by approximately 25–28% of the field time (Mobini et al., 2016). Moreover, the use of speed breeding has resulted in a significant increase in the number of generations per year, with a reported enhancement of 3–4 times (Idrissi, 2020). The application of synthetic brassinosteroids (SB) in faba bean has shown the capacity to decrease the duration of flower initiation and enhance the rates of grain settling. As a result, it yields a maximum of seven generations per year (Mobini et al., 2015). According to several scholarly sources (Mobini & Warkentin, 2016; Ribalta et al., 2017; Samantara et al., 2022), it has been observed that the pea plant can undergo optimization of up to six generations within a single year.

The application of speed breeding techniques in India has facilitated the advancement of three generations per year in chickpea, both in field and greenhouse settings (Gaur et al., 2007). In a study conducted by Watson et al. (2018), it was shown that the number of generations in chickpea has subsequently risen to four generations each year when cultivated in modified glasshouses with the use of sodium vapor lamps. However, it is important to note that this approach is associated with significant energy consumption. In 2019, Samineni et al. implemented a highly cost-efficient approach to chickpea cultivation. This method involved utilizing long photoperiods facilitated by LED lights, which resulted in the possibility of achieving up to five harvests per year. Samineni et al. (2019) conducted a two-year experiment demonstrating that the utilization of six farmed chickpea genotypes, categorized as either early or late maturity, enables the successful attainment of seven generations when employing the early maturity genotypes, and six generations when employing the late maturity genotypes. According to the findings of Samineni et al. (2020), the duration of one life cycle, encompassing the period from seed germination to seed production, was seen to be approximately 50–55 days when employing the speed breeding technique.

The initial application of the SSD method in cereal crops, specifically wheat (*Triticum aestivum*) and barley (*Hordeum vulgare*), facilitated expedited selection in breeding projects (Ghosh et al., 2018; Rai & Rai, 2022). The duration required for the development of recombinant inbred lines (RILs) through SSD was effectively improved to achieve a range of six to eight

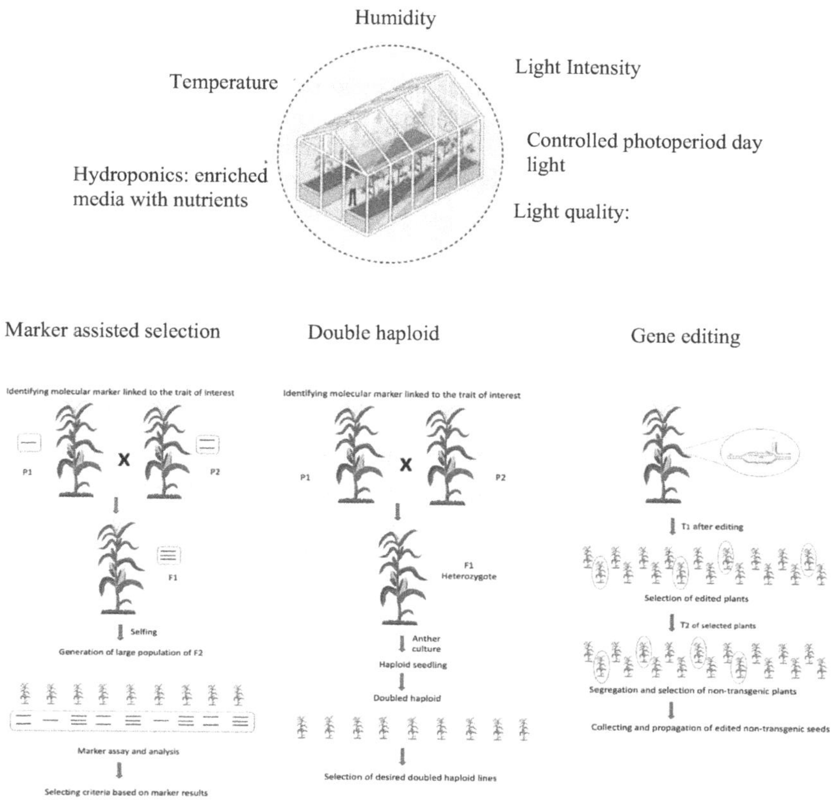

FIGURE 18.2 Diagram speed breeding technology for acceleration double haploid, marker-assisted selection, and gene editing.

generations each year, contingent upon the specific genotype (Cha et al., 2020; Yao et al., 2017). The breeders can expedite the process of early selection by efficiently incorporating the necessary genes into the elite lines using the speed breeding method.

A combination of speed breeding technology and several different technologies (Figure 18.2) such as MAS, double haploid, and genome editing has accelerated the development of improved cultivars (Croser et al., 2021; Lulsdorf et al., 2018; Sood, & Dwivedi, 2015; Xu et al., 2016). Traditional breeding could take 12 years.

18.4 IMPLEMENTING SPEED BREEDING IN CGIAR

Following the successful development of the SB protocol for many crops, it is imperative to expand the implementation of this technology on a broader scale to facilitate international breeding projects, particularly within international centers. ICRISAT has effectively implemented a speed breeding (SB) technique in chickpea cultivation, resulting in the potential for up to six generations per year. This achievement was accomplished inside a controlled greenhouse environment with enough exposure to natural daylight, as documented by Watson et al. (2018) and Gaur et al. (2019). In 2017, the International Center for Agricultural Research in the Dry Areas (ICARDA) initiated the process of testing and implementing the technique known as speed breeding. This endeavor involved the creation of a pilot speed breeding platform located in Rabat, Morocco. The first platform has a maximum capacity of 3000 entries and was utilized to formulate protocols pertaining to speed breeding techniques for both cereals and legumes as mandated. In the year

2021, ICARDA successfully implemented the speed breeding platform in Rabat. This endeavor was part of a larger project aimed at enhancing the crop breeding programs in Arab nations. The funding for this initiative was provided by the Arab Fund for Economic and Social Development (AFESD) and the Crop Trust Project. Before its implementation, the facilities and methodology of the speed breeding platform were thoroughly validated. The initiative undertaken by the Arab Fund for Economic and Social Development (AFESD) centered around three primary elements: (1) the implementation of speed breeding techniques, (2) the utilization of high-throughput precision data collection methods, and (3) the integration of Big Data into breeding programs. The project outputs will assist both ICARDA and national research programs in the development of enhanced varieties, not only for the Central, West Asia, and North Africa (CWANA) region but also on a worldwide scale. This will contribute to the improvement of crop production in light of the problems posed by climate change. The newly developed speed breeding platform has an automated generation advancement facility, which can accommodate 50,000 entries and occupies a working and storage space of 150 m^2. It is anticipated that the complete system will be completely operational by the commencement of 2022, with plans to expand it to encompass 100,000 entries by 2023. The methods of speed breeding have been thoroughly improved and specifically tailored for all crops under the purview of ICARDA, including but not limited to wheat, barley, chickpea, lentil, grass pea, and more recently, faba bean. The shortening of the crop cycle for each crop will enable the cultivation of up to five generations per annum. Additionally, ICARDA has effectively conducted trials on other approaches for trait selection in combination with speed breeding. Among these approaches, disease screening and non-destructive end-use quality have shown the most potential.

The implementation of an optimized protocol for expediting the breeding process has been employed in the breeding cycle of chickpea and lentil at ICARDA. The experiment involved cultivating seeds from five different genotypes in both a growth chamber and a greenhouse. The plants were subjected to a high density of supplemental lighting, specifically using far-red enhanced LED and blue LED lights. The lighting conditions were maintained for a prolonged photoperiod of 22 hours of light at a temperature of 25.5 °C, followed by 2 hours of darkness at a temperature of 15 °C. The experimental procedure was conducted at the Agricultural Research Center in Egypt, utilizing the controlled growth house facilities. The plants received supplementation from blue and far-red LED lights. Chickpea and lentil seeds were sown in 0.5 L containers containing a soil mixture composed of equal parts peat moss and dirt. The onset of flowering occurred within a range of 23–32 days following the sowing of lentil and chickpea seeds, whereas the initial formation of pods took place between 31 and 42 days. The fully developed pods were collected within a timeframe of 51–65 days after the first planting. The seeds that were collected were subjected to a drying process at a temperature range of 30–35°C for 48 hours. Subsequently, the aforementioned seeds were sown utilizing the identical methodology to obtain the subsequent generation. The use of this speed breeding protocol is anticipated to enable the attainment of four to five generations each year, as opposed to the one or two generations achievable by traditional greenhouse or field techniques. This strategy has the potential to reduce both field and labor expenses. Additionally, it facilitates the integration of high-throughput genotyping and GS in the context of chickpea and lentil.

18.5 GENOME EDITING TO ACCELERATE CROP IMPROVEMENT

Targeted mutation refers to a technological approach that facilitates the generation of specific mutations within the genome. This technique is commonly referred to as genome editing (GnEd). The GnEd approach utilizing the CRISPR/Cas system represents a precise, swift, and versatile tool for targeting and modifying any segment of an organism's genome through complementary gRNA to the target sequence. GnEd exhibits the capability to alter specific DNA sequences, involving processes such as deletion, addition, or substitution of genetic bases. This approach proves valuable for annotating extensive genomic data, subsequently facilitating the exploration of genomic diversity in plant populations through conventional breeding methods. Consequently, the integration of speed cloning and the CRISPR/Cas system into crop improvement programs serves as an accessible, secure, and accurate technique for developing enhanced crop varieties to meet the demands of a growing future population (Rehman et al., 2020).

To design the position of the targeted mutation, GnEd necessitates sufficient knowledge regarding the gene in question and its associated function. Hence, the utilization of GnEd is seen as a form of precision breeding that has the potential to enhance the efficacy of plant genetic resources, thereby expediting the development and introduction of novel crop varieties. A comprehensive understanding of the genome of the desired species is imperative to facilitate the advancement of a novel agricultural product possessing the desired characteristic through the utilization of cutting-edge breeding techniques. The objective of targeted gene editing (GE) is to implement specific modifications in genomes by employing programmable site-directed nucleases (SDN) to induce double-strand breaks (DSBs) in the DNA (Abdallah et al., 2015, 2021; Zhang et al., 2019). This approach leverages the cell's DNA repair mechanisms to introduce desired changes. GnEd represents a viable approach for translating gene biology into practical breeding techniques. Notably, various CRISPR variant technologies have already been developed and expeditiously used for both fundamental and applied research, encompassing the domain of plant breeding (Abdallah et al., 2021).

GnEd, which was initiated in the previous decade, has emerged as a rapidly expanding approach that has found use in numerous crop species due to its widespread acceptance, cost efficiency, expedited process, and improved precision targeting (Xu et al., 2016). However, the CRISPR/Cas technique relies on a short guide sequence RNA (sgRNA) to serve as a guide, directing the nuclease to the desired site within the genome (Abdallah et al., 2015; Bortesi et al., 2016). The utilization of CRISPR/Cas techniques has facilitated the manipulation of genes of considerable significance.

The CRISPR/Cpf1 technology offers notable advantages in terms of its efficiency and precision when it comes to genome editing. The Cpf1 endonuclease has a reduced size in comparison to Cas9, thereby necessitating a shorter CRISPR RNA (crRNA). The base editing technique enables the conversion of nucleotides in target DNA without the occurrence of DSBs. The technique of cytosine base editing enables the conversion of cytosine (C) to thymine (T) through the utilization of guide RNA (gRNA) that binds to a modified version of the Cas9 protein known as dead Cas9 (dCas9), which is coupled to C deaminase and uracil DNA glycosylase inhibitor. The technique of adenine base editing has been devised to facilitate the conversion of adenine (A) to guanine (G). The primary editing (PE) technique facilitates the effective substitution of genomic DNA segments with edits dictated by RNA. The efficacy of the process relies on the integration of Cas9 nickase (nCas9) with a reverse transcriptase (RT) and an extended guide RNA (gRNA), specifically the prime editing gRNA (pegRNA), which carries the necessary genetic information for the correction of the target sequence. The pegRNA's 5' end harbors the gRNA responsible for facilitating the binding of nCas9 to the target genomic sequence and inducing a nick in the complementary strand. The 3' region of the pegRNA harbors a sequence that functions as a template for the reverse transcription (RT) system. The resulting single-stranded DNA (ssDNA) derived from this reverse transcription process operates autonomously, independent of the homology-directed repair machinery within the host cell. Epigenome editing (EE) involves the modification of the epigenome by the removal or inactivation of nucleases, as well as the addition of an effector domain. This process enables the introduction of a novel function that can induce alterations in gene expression in an inheritable manner without modifying the underlying DNA sequence. The attainment of epigenetic modifications, such as gene promoter or histone targeting, leading to methylation or demethylation processes, has been proposed as a means to regulate gene expression (Abdallah et al., 2021).

Each of these methods aims to introduce targeted modifications in the genome, generating desired novel alleles to expedite the development and dissemination of new kinds, thereby expanding the existing limited genetic diversity with unique genes. The successful integration of editing components into the genome necessitates two fundamental prerequisites: access to the genome sequences and the utilization of efficient transformation techniques. SDN have been instrumental in the development of three distinct types of genetically engineered organisms. The first form, known as SDN1 or site-directed nuclease 1, operates by inducing DSB in the target gene. This disruption can result in various outcomes, including gene silencing, gene knockout, or alterations in gene activity. The utilization of site-directed nuclease 2 (SDN2) involves the requirement of a short single-stranded DNA as a template-guided mechanism for repairing the targeted DSB. This repair process facilitates the deliberate insertion of one or more mutations at the specific target location. The success of SDN3 relies on the utilization of a substantial donor sequence consisting of double-stranded DNA. This donor sequence facilitates the repair of specific double-stranded breaks (DSBs), hence enabling the incorporation of desired genes or genetic elements at the designated target location (Das et al., 2021; Menz et al., 2020). Several recent publications have focused on the significance of GE in different plant species.

The integration of speed breeding, SNP-MAS, together with gene transformation or genome editing techniques would usher in a novel phase of scientific investigation aimed at augmenting the breeding efficiency for the development of plants that exhibit tolerance to both abiotic and biotic challenges within a reduced timeframe. An illustration of the application of the SB approach can be observed in a rice breeding experiment, wherein transgenic lines resistant to salinity were developed over 1.4 years, encompassing six sequential generations. This process involved three backcrosses and two instances of self-pollination (Rana et al., 2019). To effectively address the emerging issues presented by climate change, it is imperative to implement a transformative revolution.

Genome-edited lines are preselected at the T_1 generation followed by further evaluation at the T_2 generation for elimination of the transgenes and off-target genotypes. Combining speed breeding application with genome editing has been demonstrated in Brassica and soybean (Bao et al., 2020; Murovec et al., 2018; Yang et al., 2017). Uniting genome editing and speed breeding holds the promise of surmounting this crisis, enabling the generation of multiple generations within a single year.

18.6 MARKER-ASSISTED SELECTION AND GENOMIC SELECTION

In contemporary plant breeding projects, the predominant method for plant selection involves the utilization of molecular markers. This approach offers notable advantages in terms of time efficiency and precision compared to traditional breeding methods (Heffner et al., 2009). Molecular markers, such as simple sequence repeat markers, are frequently employed in conventional approaches for MAS within localized breeding programs. Combining speed breeding with MAS has been applied to accelerate genetic gain such as herbicide-tolerant chickpea, disease resistance in barely, introducing valuable characteristics from wild lentil and rapid phenotyping in wheat (Atieno et al., 2017; Croser et al., 2021; Hickey et al., 2017; Lulsdorf et al., 2018).

According to Panjabi et al. (2019), the implementation of speed breeding necessitates the utilization of molecular techniques that are more suited to adaptability. These techniques should offer enhanced control over the breeding process while concurrently advancing our comprehension of the intricate connections between plants and their surrounding environment. The advancement of high-density molecular markers, such as single nucleotide polymorphisms (SNPs), has facilitated the implementation of GS as a potentially valuable method in crop breeding (Meuwissen et al., 2001).

Speed breeding plays a crucial role in comprehending genomic architecture and constructing GS models that connect phenomic and genomic data for specific crop species. The effectiveness of GS models relies on estimating marker-trait associations through genetic markers and employing them for comprehensive phenotyping of test populations. This process aids in predicting significant genomic regions or loci within a particular genome by generating GS models tailored to the specific crop species. These models are prediction-based and are developed by estimating marker-trait associations through genetic markers, followed by extensive phenotyping of test populations (Niazian & Niedbała, 2020).

OMICS techniques provide important data with high quality and quantity of data that has helped in breeding programs (Rajasundaram et al., 2016; Sartor et al., 2019). The GS approach involves the development of computational models using a collection of genotypes, referred to as the training population. These models are designed to learn from the markers that are related to specific qualities of interest. By utilizing the genetic value, the GS method enables the prediction of the most optimal selection. This methodology was introduced by Jannink et al. in 2010. Further development of GS is necessary to enhance the accuracy of trait selection, particularly for complicated quantitative traits such as yield. According to Roorkiwal et al. (2016), the prediction accuracies in chickpea varied from 0.138 for seed yield to 0.912 for 100-seed weight, as demonstrated by the GS. The feasibility and potential of GS in rice have been demonstrated in various studies. For instance, Onogi et al. (2015), Isidro et al. (2015), and Spindel et al. (2015) reported average predictive values of up to 0.85, 0.6, and 0.63, respectively, for flowering date in rice using GS. Similarly, Grenier et al. (2015), Cui et al. (2020), Isidro et al. (2015), and Wang et al. (2015) found predictive values of up to 0.54, 0.58, 0.7, and 0.86, respectively, for plant height in rice using GS. Nevertheless, the predictive capacity for grain yield remains relatively modest, as indicated by an average value of 0.35 derived from six distinct research investigations. The highest prediction achieved thus far stands at 0.54, accomplished through the utilization of 102,795 SNP markers and the GBLUP statistical model (Cui et al., 2020; Grenier et al., 2015; Spindel et al., 2015; Wang et al., 2015; Xu et al., 2016). In the study conducted by Poland et al. (2012), the predictive value for grain yield in wheat using 1726 DArT markers and GBLUP ranged from 0.2 to 0.4. However, other studies (Crossa et al., 2010; Perez-Rodriguez et al., 2012; Zhao et al., 2013) reported higher values for the predictive accuracy of grain yield using the ridge regression best linear unbiased prediction (rrBLUP) statistical model, ranging from 0.3 to 0.6, 0.48 to 0.61, and 0.5 to 0.6, respectively. The enhancement of predictive computational models and the refinement of phenotyping data accuracy have the potential to enhance these values. There are multiple software packages available within the R environment that offer GS assessment. These packages include Bayesian linear regression (BLR) as described by Pérez et al. (2010), genomic best linear unbiased prediction (GBLUP) as proposed by Endelman (2011), genome-wide regression and prediction BGLR as presented by Pérez et al. (2014), genomic estimates of breeding values (BWGS) as discussed by Charmet et al. (2020), genomic prediction of hybrid performance (Predhy) as outlined by Xu et al. (2021), and Solving Mixed Model Equations in R (SOMMER) as introduced by Covarrubias-Pazaran et al. (2021).

18.7 INTEGRATING SPEED BREEDING WITH AI

AI has emerged as a key focus for expediting the enhancement of crop breeding processes. The intelligent and efficient extraction of information from breeding datasets is imperative for the advancement of Next-Gen AI, utilizing models and well-defined algorithms. AI technologies, such as neural deep learning and networks, have been devised to enhance the effectiveness and precision of multi-omics data through the application of relevant models and definitive algorithms (Harfouche et al., 2019). AI holds the potential to accelerate breeding programs by predicting factors associated with disease resistance and agronomic traits and analyzing breeding values in the context of climate changes (Niazian & Niedbała, 2020; Parmley et al., 2019). Notably, breeding programs focused on enhancing crop abiotic stress tolerance aim to bridge the genotype-phenotype gap resulting from alterations in phenotype plasticity. The interplay between environmental shifts and genotypic variations contributes to a reduction in the phenotypic plasticity of economically significant traits (van Dijk et al., 2021).

Plant phenomics utilizes diverse approaches, primarily focused on identifying target traits that govern unique stress responses. This approach ensures the acquisition of precise, reproducible, accurate, and rapid data. Cutting-edge computer vision systems play a pivotal role in estimating and analyzing data through image analysis, while computational analysis is employed to predict biological responses (Esposito et al., 2020). AI plays a key role in phenotyping even the slightest variations among complex traits at various stages of plant growth and development, thereby expediting breeding programs. AI achieves this by providing high-resolution images that facilitate the discrimination of better-performing genotypes within large populations at different plant growth and development stages (Godwin et al., 2019; Harfouche et al., 2019). The data associated with genotype diversity in response to environmental changes are intricate and governed by multiple genes. Addressing this complexity requires an AI-based physiological gravimetric system to accurately record small changes occurring in plants within the soil-plant-atmosphere continuum (van Dijk et al., 2021).

The Next-Gen AI approach, derived from the study and analysis of phenotypic data, is employed to pinpoint QTLs associated with stress responsiveness and crucial agronomic traits (Niazian & Niedbała, 2020; Parmley et al., 2019). Machine learning techniques, including UAV and ground-based equipment, play a vital role in capturing high-throughput phenotypic data relevant to breeding programs. UAVs encompass satellites, drones, phenotyping towers, environmental sensor networks, phenomobiles, autonomous ground vehicles, aircraft, helicopters, zeppelins, and field scanning platforms, commonly equipped with high-resolution cameras and sensors to revolutionize crop phenomics in the field. Additionally, a robotic field-based phenotyping system with side-view stereo imaging has been developed to monitor plant height at regular intervals (Bao et al., 2019). AI and specific software are then utilized to analyze the collected data, identifying genotypes that exhibit superior agronomic or disease-resistant traits (Esposito et al., 2020).

The Next-Gen AI approach resulted from studying and analyzing phenotypic data is used to identify QTLs related to stress-responsive to important agronomic traits (Niazian & Niedbała, 2020; Parmley et al., 2019). Machine learning approaches, such as UAV and ground-based equipment, are used to capture high-throughput phenotypic data relevant to breeding programs. UAVs comprise satellites, drones, phenotyping towers, environmental sensor networks, phenomobiles, autonomous ground vehicles, aircraft, helicopters, zeppelins, and field scanning platforms are commonly used. Equipments are combined with high-resolution cameras and sensors to revolutionize crop phenomics from the field. In addition, a robotic field-based phenotyping system with side-view stereo imaging was developed to capture the plant height at regular intervals of time (Bao et al., 2019). AI or specific software is used to analyze the collected data to identify the genotypes exhibiting the superlative agronomic/disease-resistant traits (Esposito et al., 2020).

18.8 CONCLUSION

For several decades, plant breeders have struggled to develop breed cultivars with high yields that can withstand climate changes and abiotic and biotic stresses. The breeding cycle for important crops such as wheat, rice, and maize is approximately for about 8–12 years. Novel technologies for breeding rapid generation are required for sustainably producing and supplying essential crops that can meet the demands of the ever-increasing world population. The emergence of speed breeding enabled the reduction of time, space, and cost to develop and release improved cultivars. Speed breeding technically mimics the natural environment artificially by creating and alternating light duration, intensity, and temperature-controlled zones to accelerate plant growth and improve crop productivity. Speed breeding is used in early generations to select desirable allelic combinations for multiple traits with improved accuracy and predictability. Speed breeding combined with different breeding strategies such as double haploid, genome editing, and molecular breeding such as MAS and GS can bring more genetic gain to achieve greater yield increase, adapt to climate change and emergency stress problems, and may play a major role for future food security. In addition, implementing Next-Gen AI for speed breeding will facilitate handling the huge OMICS data fast and precisely to understand plant function under severe stress conditions.

REFERENCES

Abdallah, N. A., Hamwieh, A., Radwan, K., Fouad, N., & Prakash, C. (2021). Genome editing techniques in plants: a comprehensive review and future prospects toward zero hunger. *GM Crops & Food*, *12*(2), 601–615. 10.1080/21645698.2021.2021724. 2021 Dec 31; Epub 2022 Feb 9. PMID: 35135438.

Abdallah, N. A., Prakash, C. S., & McHughen, A. G. (2015). Genome editing for crop improvement: challenges and opportunities. *GM Crops & Food*, *6*(4), 183–205.

Alahmad, S., Dinglasan, E., Leung, K. M., Riaz, A., Derbal, N., Voss-Fels, K. P., Able, J. A., Bassi, F. M., Christopher, J., & Hickey, L. T. (2018). Speed breeding for multiple quantitative traits in durum wheat. *Plant Methods*, *14*(1), 1–15.

Ali, A. M., Mustafa, H. M., Tahir, I. S., Elahmadi, A. B., Mohamed, M. S., Ali, M. A., Suliman, A. M., Baum, M., & Ibrahim, A. E. S. (2006). Two doubled haploid bread wheat cultivars for irrigated heat-stressed environments. *Sudan Journal of Agricultural Research*, *6*, 35–42.

Atieno, J., Li, Y., Langridge, P., Dowling, K., Brien, C., Berger, B., Varshney, R., & Sutton, T. (2017). Exploring genetic variation for salinity tolerance in chickpea using image-based phenotyping. *Scientific Reports*, *7*, 1300.

Bao, A., Zhang, C., Huang, Y., Chen, H., Zhou, X., & Cao, D. (2020). Genome editing technology and application in soybean improvement. *Oil Crop Science*, *5*, 31–40. [CrossRef]

Bao, Y., Tang, L., Breitzman, M. W., Salas Fernandez, M. G., & Schnable, P. S. (2019). Field-based robotic phenotyping of sorghum plant architecture using stereo vision. *Journal of Field Robotics*, *36*, 397–415. 10.1002/rob.21830

Bennett, R. G., Ribalta, F. M., Pazos-Navarro, M., Leonforte, A., & Croser, J. S. (2017). Discrimination of boron tolerance in Pisum sativum L. genotypes using a rapid, high-throughput hydroponic screen and precociously germinated seed grown under far-red enriched light. *Plant Methods*, *13*(1), 70.

Bortesi, L., Zhu, C., Zischewski, J., Perez, L., Bassié, L., Nadi, R., Forni, G., Lade, S. B., Soto, E., Jin, X., & Medina, V. (2016). Patterns of CRISPR/Cas9 activity in plants, animals and microbes. *Plant Biotechnology Journal*, *14*(12), 2203–2216.

Brennan, J. P., & Martin, P. J. (2007). Returns to investment in new breeding technologies. *Euphytica*, *157*(3), 337–349.

Bugbee, B., & Koerner, G. (1997). Yield comparisons and unique characteristics of the dwarf wheat cultivar 'USU-Apogee'. *Advances in Space Research*, *20*(10), 1891–1894.

Bula, R. J., Morrow, R. C., Tibbitts, T. W., Barta, D. J., Ignatius, R. W., & Martin, T. S. (1991). Light-emitting diodes as a radiation source for plants. *HortScience*, *26*(2), 203–205.

Cha, J. K., Lee, J. H., Lee, S. M., Ko, J. M., & Shin, D. (2020). Heading date and growth character of Korean wheat cultivars by controlling photoperiod for rapid generation advancement. *Korean Society of Breeding Science*, *52*(1), 20–24.

Charmet, G., Tran, L. G., Auzanneau, J., Rincent, R., & Bouchet, S. (2020). BWGS: AR package for genomic selection and its application to a wheat breeding programme. *PLoS One*, *15*(4), e0222733.

Chaudhary, H. K., Badiyala, A., & Jamwal, N. S. (2015). New frontiers in doubled haploidy breeding in wheat. *Agricultural Research Journal*, *52*(4), 1–12.

Collard, B. C., & Mackill, D. J. (2008). Marker-assisted selection: an approach for precision plant breeding in the twenty-first century. *Philosophical Transactions of the Royal Society B: Biological Sciences*, *363*(1491), 557–572.

Covarrubias-Pazaran, G., Martini, J. W., Quinn, M., & Atlin, G. (2021). Strengthening public breeding pipelines by emphasizing quantitative genetics principles and open-source data management. *Frontiers in Plant Science*, *12*. 10.3389/fpls.2021.681624

Croser, J., Mao, D., Dron, N., Michelmore, S., McMurray, L., Preston, C., Bruce, D., Ogbonnaya, F. C., Ribalta, F. M., Hayes, J., et al. (2021). Evidence for the application of emerging technologies to accelerate crop improvement—A collaborative pipeline to introgress herbicide tolerance into chickpea. *Frontiers in Plant Science*, *12*, 779122.

Crossa, J., de los Campos, G., Perez, P., Gianola, D., Burgueño, J., et al. (2010). Prediction of genetic values of quantitative traits in plant breeding using pedigree and molecular markers. *Genetics*, *186*, 713–724.

Cui, Y., Li, R., Li, G., Zhang, F., Zhu, T., Zhang, Q., Ali, J., Li, Z., & Xu, S. (2020). Hybrid breeding of rice via genomic selection. *Plant Biotechnology Journal*, *18*(1), 57–67.

Das, A., Ghana, P., Rudrappa, B., Gandhi, R., Tavva, V. S., and Mohanty, A. (2021). Genome editing of rice by CRISPR-Cas: end-to-end pipeline for crop improvement. In *Rice genome engineering and gene editing* (pp. 115–134). New York, NY: Humana.

De Buyser, J., Henry, Y., Lonnet, P., Hertzog, R., & Hespel, A. (1987). 'Florin': a doubled haploid wheat variety developed by the anther culture method. *Plant Breeding*, *98*(1), 53–56.

Deery, D. M., Rebetzke, G. J., Jimenez-Berni, J. A., Bovill, W. D., James, R. A., Condon, A. G., Furbank, R. T., Chapman, S. C., & Fischer, R. A. (2019). Evaluation of the phenotypic repeatability of canopy temperature in wheat using continuous-terrestrial and airborne measurements. *Frontiers in Plant Science*, *10*, 875.

Desta, Z. A., & Ortiz, R. (2014). Genomic selection: genome-wide prediction in plant improvement. *Trends in Plant Science*, *19*(9), 592–601.

Endelman, J. B. (2011). Ridge regression and other kernels for genomic selection with R package rrBLUP. *Plant Genome*, *4*(3), 250–255.

Esposito S., Carputo D., Cardi T., & Tripodi P. (2020). Applications and trends of machine learning in genomics and phenomics for next-generation breeding. *Plants*, *9*(1), 34. 10.3390/plants9010034

FAO (2018). *Food outlook – biannual report on the global food markets, Nov 2018*. Rome. p104. License CC BY-NC-SA 3.0 IGO

Fiorani, F., & Schurr, U. (2013). Future scenarios for plant phenotyping. *Annual Review of Plant Biology*, *64*, 267–291.

Gaur, P. M., Samineni, S., Gowda, C. L. L., & Rao, B. V. (2007). Rapid generation advancement in chickpea. *Journal of SAT Agricultural Research*, *3*(1). 10.1016/j.cj.2019.08.003

Gaur, P. M., Samineni, S., Thudi, M., Tripathi, S., Sajja, S. B., Jayalakshmi, V., Mannur, D. M., Vijayakumar, A. G., Ganga Rao, N. V., Ojiewo, C., & Fikre, A. (2019). Integrated breeding approaches for improving drought and heat adaptation in chickpea (*Cicer arietinum* L.). *Plant Breeding*, *138*(4), 389–400.

Ghosh, S., Watson, A., Gonzalez-Navarro, O. E., Ramirez-Gonzalez, R. H., Yanes, L., Mendoza-Suárez, M., Simmonds, J., Wells, R., Rayner, T., Green, P., & Hafeez, A. (2018). Speed breeding in growth chambers and glasshouses for crop breeding and model plant research. *Nature Protocols*, *13*(12), 2944–2963.

Giura, A. (2007). Haploids and doubled haploid lines production by Zea system in Triticum durum and Triticale. In *Cercetări ştiinţifice, Horticultură, Inginerie Genetică, vol XI*. Ed. Agroprint, USAMVB, Timişoara, pp. 32–37.

Godwin, I. D., Rutkoski, J., Varshney, R. K., & Hickey, L. T. (2019). Technological perspectives for plant breeding. *Theoretical and Applied Genetics*, *132*(3), 555–557. 10.1007/s00122-019-03321-4

Grenier, C., Cao, T. V., Ospina, Y., Quintero, C., Châtel, M. H., Tohme, J., Courtois, B., & Ahmadi, N. (2015). Accuracy of genomic selection in a rice synthetic population developed for recurrent selection breeding. *PLoS One*, *10*(8), e0136594.

Haley, S. D., Johnson, J. J., Peairs, F. B., Stromberger, J. A., Hudson-Arns, E. E., Seifert, S. A., Anderson, V. A., Bai, G., Chen, X., Bowden, R. L., & Jin, Y. (2018). Registration of 'Avery'hard red winter wheat. *Journal of Plant Registrations*, *12*(3), 362–366.

Harfouche, A. L., Jacobson, D. A., Kainer, D., Romero, J. C., Harfouche, A. H., Mugnozza, G. S., Moshelion, M., Tuskan, G. A., Keurentjes, J. J., & Altman, A. (2019). Accelerating climate resilient plant breeding by applying next-generation artificial intelligence. *Trends in Biotechnology*, *37*(11), 1217–1235. 10.1016/j.tibtech.2019.05.007

Heffner, E. L., Sorrells, M. E., & Jannink, J. L. (2009). Genomic selection for crop improvement. *Crop Science*, *49*, 1–12. 10.2135/cropsci2008.08.0512

Hickey, L. T., Germán, S. E., Pereyra, S. A., Diaz, J. E., Ziems, L. A., Fowler, R. A., Platz, G. J., Franckowiak, J. D., & Dieters, M. J. (2017). Speed breeding for multiple disease resistance in barley. *Euphytica*, *213*(3), 64.

Hickey, L. T., Lawson, W., Platz, G. J., Dieters, M., Arief, V. N., German, S., Fletcher, S., Park, R. F., Singh, D., Pereyra, S., & Franckowiak, J. (2011). Mapping Rph20: a gene conferring adult plant resistance to *Puccinia hordei* in barley. *Theoretical and Applied Genetics*, *123*(1), 55–68.

Hickey, L. T., Hafeez, A. N., Robinson, H., Jackson, S. A., Leal-Bertioli, S. C. M., Tester, M., Gao, C., Godwin, I. D., Hayes, B. J., & Wulff, B. B. H. (2019). Breeding crops to feed 10 billion. *Nature Biotechnology*, *37*(7), 744–754.

Honsdorf, N., March, T. J., & Pillen, K. (2017). QTL controlling grain filling under terminal drought stress in a set of wild barley introgression lines. *PLoS One*, *12*(10), e0185983.

Hu, H. (1997). In vitro induced haploids in wheat. In *In vitro haploid production in higher plants* (pp. 73–97). Dordrecht: Springer.

Hu, Y., Knapp, S., & Schmidhalter, U. (2020). Advancing high-throughput phenotyping of wheat in early selection cycles. *Remote Sensing*, *12*(3), 574.

Humphreys, D. G., & Knox, R. E. (2015). Doubled haploid breeding in cereals. In J. M. Al-Khayri et al. (Eds.), *Advances in plant breeding strategies: breeding, biotechnology and molecular tools*. New York: Springer International Publishing. 10.1007/978-3-319-22521-0_9

Idrissi, O. (2020). Application of extended photoperiod in lentil: towards accelerated genetic gain in breeding for rapid improved variety development. *Moroccan Journal of Agricultural Sciences*, *1*(1). https://techagro.org/index.php/MJAS/article/view/823

Isidro, J., Jannink, J. L., Akdemir, D., Poland, J., Heslot, N., & Sorrells, M. E. (2015). Training set optimization under population structure in genomic selection. *Theoretical and Applied Genetics*, *128*(1), 145–158.

Jannink, J. L., Lorenz, A. J., & Iwata, H. (2010). Genomic selection in plant breeding: from theory to practice. *Briefings in Functional Genomics*, *9*(2), 166–177.

Kelliher, T., Starr, D., Su, X., Tang, G., Chen, Z., Carter, J., Wittich, P. E., Dong, S., Green, J., Burch, E., & McCuiston, J. (2019). One-step genome editing of elite crop germplasm during haploid induction. *Nature Biotechnology*, *37*(3), 287–292.

Li, A., Jia, S., Yobi, A., Ge, Z., Sato, S. J., Zhang, C., Angelovici, R., Clemente, T. E., & Holding, D. R., (2018a). Editing of an alpha-kafirin gene family increases, digestibility and protein quality in sorghum. *Plant Physiology*, *177*(4), 1425–1438.

Li, H., Rasheed, A., Hickey, L. T., & He, Z. (2018b). Fast-forwarding genetic gain. *Trends in Plant Science*, *23*(3), 184–186.

Lulsdorf, M. M., & Banniza, S. (2018). Rapid generation cycling of an F2 population derived from a cross between *Lens culinaris* Medik. and *Lens ervoides* (Brign.) Grande after aphanomyces root rot selection. *Plant Breeding*, *137*(4), 486–491.

Mannur, D. M., Babbar, A., Thudi, M., Sabbavarapu, M. M., Roorkiwal, M., Sharanabasappa, B. Y., Bansal, V. P., Jayalakshmi, S. K., Yadav, S. S., Rathore, A., & Chamarthi, S. K. (2019). Super Annigeri 1 and improved JG 74: two Fusarium wilt-resistant introgression lines developed using marker-assisted backcrossing approach in chickpea (*Cicer arietinum* L.). *Molecular Breeding*, *39*(1), 1–13.

Menz, J., Modrzejewski, D., Hartung, F., Wilhelm, R., & Sprink, T. (2020). Genome edited crops touch the market: a view on the global development and regulatory environment. *Frontiers in Plant Science*, *11*. 10.3389/fpls.2020.586027

Meuwissen, T. H., Hayes, B. J., & Goddard, M. E. (2001). Prediction of total genetic value using genome-wide dense marker maps. *Genetics*, *157*(4), 1819–1829.

Mobini, S. H., & Warkentin, T. D. (2016). A simple and efficient method of in vivo rapid generation technology in pea (*Pisum sativum* L.). *In Vitro Cellular & Developmental Biology-Plant*, *52*(5), 530–536.

Mobini, S. H., Lulsdorf, M., Warkentin, T. D., & Vandenberg, A. (2016). Low red: far-red light ratio causes faster in vitro flowering in lentil. *Canadian Journal of Plant Science*, *96*(5), 908–918.

Mobini, S. H., Lulsdorf, M., Warkentin, T. D., & Vandenberg, A. (2015). Plant growth regulators improve in vitro flowering and rapid generation advancement in lentil and faba bean. *In Vitro Cellular & Developmental Biology-Plant*, *51*(1), 71–79.

Murovec, J., Guˇcek, K., Bohanec, B., Avbelj, M., & Jerala, R. (2018). DNA-Free Genome Editing of Brassica oleracea and B. rapa protoplasts using CRISPR-Cas9 ribonucleoprotein complexes. *Frontiers in Plant Science*, *9*, 1594.

Nakamura, S., & Fasol, G. (1997). *The blue laser diode: GaN based light emitters and lasers.* Spring, Heidelberg.

Niazian M., & Niedbała G. (2020). Machine learning for plant breeding and biotechnology. *Agriculture*, *10*(10), 436. 10.3390/agriculture10100436

Onogi, A., Ideta, O., Inoshita, Y., Ebana, K., Yoshioka, T., Yamasaki, M., & Iwata, H. (2015). Exploring the areas of applicability of whole-genome prediction methods for Asian rice (*Oryza sativa* L.). *Theoretical and Applied Genetics*, *128*(1), 41–53.

Panjabi, P., Yadava, S. K., Kumar, N., Bangkim, R., & Ramchiary, N. (2019). Breeding Brassica juncea and B. rapa for sustainable oilseed production in the changing climate: progress and prospects. In *Genomic designing of climate-smart oilseed crops* (pp. 275–369). Cham: Springer.

Parmley, K. A., Higgins, R. H., Ganapathysubramanian, B., Sarkar, S., & Singh, A. K. (2019). Machine learning approach for prescriptive plant breeding. *Scientific Reports*, *9*(1), 1–12. 10.1038/ s41598-019-53451-4

Pauk, J., Kertész, Z., Beke, B., Bóna, L., Csösz, M., & Matuz, J. (1995). New winter wheat variety: 'GK Délibáb' developed via combining conventional breeding and in vitro androgenesis. *Cereal Research Communications*, *23*, 251–256.

Pérez-Rodríguez, D. G., González-Camacho, J. M., Crossa, J., Manès, Y., & Dreisigacker, S. (1 December 2012). Comparison between linear and non-parametric regression models for genome-enabled prediction in wheat. *G3 Genes|Genomes|Genetics*, *2*(12), 1595–1605. 10.1534/g3.112.003665

Pérez, L., Soto, E., Villorbina, G., Bassie, L., Medina, V., Muñoz, P., Capell, T., Zhu, C., Christou, P., & Farré, G., (2018). CRISPR/Cas9-induced monoallelic mutations in the cytosolic AGPase large subunit gene APL2 induce the ectopic expression of APL2 and the corresponding small subunit gene APS2b in rice leaves. *Transgenic Research*, *27*(5), 423–439.

Pérez, P., & de Los Campos, G. (2014). Genome-wide regression and prediction with the BGLR statistical package. *Genetics*, *198*(2), 483–495.

Pérez, P., de Los Campos, G., Crossa, J., & Gianola, D. (2010). Genomic-enabled prediction based on molecular markers and pedigree using the Bayesian linear regression package in R. *The Plant Genome*, *3*(2), 106.

Poland, J. A., Endelman, J., Dawson, J., Rutkoski, J., Wu, S., Manes, Y., Dreisigacker, S., Crossa, J., Sánchez-Villeda, H., Sorrells, M., & Jannink, J. L. (2012). Genomic selection in wheat breeding using genotyping-by-sequencing. *Plant Genome*, *5*(3), 103–113.

Rai, A. C., & Rai, K. K. (2022). Speed breeding for rapid cycling of crops for stress management and global food security. In *Microbial biocontrol: Food security and post harvest management* (pp. 23–37). Cham: Springer.

Rajasundaram, D., & Selbig, J. (2016). More effort—more results: recent advances in integrative 'omics' data analysis. *Current Opinion in Plant Biology*, *30*, 57–61. 10.1016/j.pbi.2015.12.010

Rana, M. M., Takamatsu, T., Baslam, M., Kaneko, K., Itoh, K., Harada, N., Sugiyama, T., Ohnishi, T., Kinoshita, T., Takagi, H., & Mitsui, T. (2019). Salt tolerance improvement in rice through efficient SNP marker-assisted selection coupled with speed-breeding. *International Journal of Molecular Sciences*, *20*(10), 2585.

Rehman, N., Inam, S., Shahid, M. M., Ali, G. M., & Khan, M. R. (2020). Speed cloning leading to speed breeding through genome editing. *Advances in Crop Science and Technology*, *8*, 443.

Riaz, A., Periyannan, S., Aitken, E., & Hickey, L. (2016). A rapid phenotyping method for adult plant resistance to leaf rust in wheat. *Plant Methods*, *12*(1), 1–10.

Ribalta, F. M., Pazos-Navarro, M., Nelson, K., Edwards, K., Ross, J. J., Bennett, R. G., Munday, C., Erskine, W., Ochatt, S. J., & Croser, J. S. (2017). Precocious floral initiation and identification of exact timing of embryo physiological maturity facilitate germination of immature seeds to truncate the lifecycle of pea. *Plant Growth Regulation*, *81*(2), 345–353.

Rollins, J. A., Drosse, B., Mulki, M. A., Grando, S., Baum, M., Singh, M., Ceccarelli, S., & Von Korff, M. (2013). Variation at the vernalisation genes Vrn-H1 and Vrn-H2 determines growth and yield stability in barley (*Hordeum vulgare*) grown under dryland conditions in Syria. *Theoretical and Applied Genetics*, *126*(11), 2803–2824.

Roorkiwal, M., Rathore, A., Das, R. R., Singh, M. K., Jain, A., Srinivasan, S., Gaur, P. M., Chellapilla, B., Tripathi, S., Li, Y., Hickey, J. M., Lorenz, A., Sutton, T., Crossa, J., Jannink, J.-L., & Varshney, R. K. (2016). Genome-enabled prediction models for yield related traits in chickpea. *Frontiers in Plant Science*, *7*, 1666. 10.3389/fpls.2016.01666

Samantara, K., Bohra, A., Mohapatra, S. R., Prihatini, R., Asibe, F., Singh, L., Reyes, V. P., Tiwari, A., Maurya, A. K., Croser, J. S., & Wani, S. H. (2022). Breeding more crops in less time: a perspective on speed breeding. *Biology*, *11*(2), 275.

Samineni, S., Sen, M., Sajja, S. B., & Gaur, P. M. (2020). Rapid generation advance (RGA) in chickpea to produce up to seven generations per year and enable speed breeding. *The Crop Journal*, *8*(1), 164–169.

Samineni, S., Sen, M., Sajja, S., & Gaur, P. (2019). Rapid generation advance (RGA) in chickpea to produce up to seven generations per year and enable speed breeding. *The Crop Journal*, *8*, 164–169.

Sartor, R. C., Noshay, J., Springer, N. M., & Briggs, S. P. (2019). Identification of the expressome by machine learning on omics data. *Proceedings of the National Academy of Sciences USA*, *116*(36), 18119–18125. 10.1073/pnas.1813645116

Săulescu, N. N., Ittu, G., Giura, A., Mustătea, P., & Ittu, M. (2012). Results of using Zea method for doubled haploid production in wheat breeding at Nardi Fundulea – Romania. *Romanian Agricultural Research*, *29*, 3–8.

Saxena, K. B., Saxena, R. K., Hickey, L. T., & Varshney, R. K. (2019). Can a speed breeding approach accelerate genetic gain in pigeonpea?. *Euphytica*, *215*(12), 1–7.

Sood, S., & Dwivedi, S. (2015). Doubled haploid platform: an accelerated breeding approach for crop improvement. In *Plant biology and biotechnology* (pp. 89–111). New Delhi: Springer.

Spindel, J., Be Spindel, J., Begum, H., Akdemir, D., Virk, P., Collard, B., Redona, E., Atlin, G., Jannink, J. L., & McCouch, S. R. (2015). Genomic selection and association mapping in rice (*Oryza sativa*): effect of trait genetic architecture, training population composition, marker number and statistical model on accuracy of rice genomic selection in elite, tropical rice breeding lines. *PLoS Genetics*, *11*(2), e1004982.

Stutte, G. W. (2015). Commercial transition to LEDs: a pathway to high-value products. *HortScience*, *50*(9), 1297–1300.

Tuvesson, S., Von Post, R., & Ljungberg, A. (2003). Wheat anther culture. In *Doubled haploid production in crop plants* (pp. 71–76). Dordrecht: Springer.

UN (2018). *The sustainable development goals report*. New York: United Nations. 10.11260/kenkokyoiku.19.7

van Dijk, A. D. J., Kootstra, G., Kruijer, W., & de Ridder, D. (2021). Machine learning in plant science and plant breeding. *iScience*, *24*(1), 101890. 10.1016/j.isci.2020.101890

Wang, X., Xuan, H., Evers, B., Shrestha, S., Pless, R., & Poland, J. (2019). High-throughput phenotyping with deep learning gives insight into the genetic architecture of flowering time in wheat. *GigaScience*, *8*(11), giz120.

Wang, Y., Du, Y., Yang, Z., Chen, L., Condon, A. G., & Hu, Y. G. (2015). Comparing the effects of GA-responsive dwarfing genes Rht13 and Rht8 on plant height and some agronomic traits in common wheat. *Field Crops Research*, *179*, 35–43.

Wanga, M. A., Shimelis, H., Mashilo, J., & Laing, M. D. (2021). Opportunities and challenges of speed breeding: a review. *Plant Breeding*, *140*, 185–194.

Watson, A., Ghosh, S., Williams, M. J., Cuddy, W. S., Simmonds, J., Rey, M. D., Hatta, M. A. M., Hinchliffe, A., Steed, A., Reynolds, D., & Adamski, N. M. (2018). Speed breeding is a powerful tool to accelerate crop research and breeding. *Nature Plants*, *4*(1), 23–29.

Wessel, E., & Botes, W. C. (2014). Accelerating resistance breeding in wheat by integrating marker assisted selection and doubled haploid technology. *South African Journal of Plant and Soil*, *31*, 35–43.

Wiśniewska, H., Majka, M., Kwiatek, M., Gawlowska, M., Surma, M., Adamski, T., Kaczmarek, Z., Drzazga, T., Lugowska, B., Korbas, M., & Belter, J. (2019). Production of wheat doubled haploids resistant to eyespot supported by marker-assisted selection. *Electronic Journal of Biotechnology*, *37*, 11–17.

Xu, R., Yang, Y., Qin, R., Li, H., Qiu, C., Li, L., Wei, P., & Yang, J. (2016). Rapid improvement of grain weight via highly efficient CRISPR/Cas9-mediated multiplex genome editing in rice. *Journal of Genetics* and *Genomics*, *43*(8), 529.

Xu, Y., Ma, K., Zhao, Y., Wang, X., Zhou, K., Yu, G., Li, C., Li, P., Yang, Z., Xu, C., & Xu, S. (2021). Genomic selection: a breakthrough technology in rice breeding. *The Crop Journal*, *9*, 669–677.

Yang, H., Wu, J.-J., Tang, T., Liu, K.-D., Dai, C. (2017). CRISPR/Cas9-mediated genome editing efficiently creates specific mutations at multiple loci using one sgRNA in Brassica napus. *Scientific Reports*, *7*, 7489.

Yao, Y., Zhang, P., Liu, H., Lu, Z., & Yan, G. (2017). A fully in vitro protocol towards large scale production of recombinant inbred lines in wheat (*Triticum aestivum* L.). *Plant Cell, Tissue and Organ Culture (PCTOC)*, *128*(3), 655–661.

Zhang, H. X., Zhang, Y., & Yin, H. (2019). Genome editing with mRNA encoding ZFN, TALEN, and Cas9. *Molecular Therapy*, *27*(4), 735–746.

Zhao, K., Lu, Z. X., Park, J. W., Zhou, Q., & Xing, Y. (2013). GLiMMPS: robust statistical model for regulatory variation of alternative splicing using RNA-seq data. *Genome Biology*, *14*(7), 1–15.

Index

NOTE: Page numbers followed by *f* and *t* indicate figures and tables respectively.

ABE. *See* Adenine base editor (ABE)
Abiotic stress, 131, 180, 208, 212–218, 213–215*t*
  drought, 214–215*t*, 216
  flooding, 212, 213*t*, 216
  salinity, 213–214*t*, 216–218
  stress mitigation in soybean under, 226–227, 227*f*
  temperature, 217
  tolerance to, 44, 306
Accelerated breeding of crops
  CRISPR-Cas serves, 69–72
  genome editing emerging, 68, 68*f*
  GWAS and DNA markers, 58–62
  high-throughput phenotyping for phenomics data collection, 55, 58
  integration of novel breeding strategies, 74
  next-generation sequencing, genomic selection and breeding by design, 65–67
  speed breeding, 73–74
  speed cloning to develop climate-resilient crops, 62–64
Accelerated marker development, 244
Accelerated single-seed descent (aSSD), 18, 242
Acceleration of crop breeding
  advantages and limitations of, 46–47
  applications of speed breeding, 43–46
  crop improvement, 38
  principles of speed breeding, 41–43
  speed breeding techniques, 38–41
acquisition_management.py program, 172
Adenine base editor (ABE), 124, 125
Adenosine triphosphates (ATPs), 321
Advanced analytical techniques, 197*t*
Aerenchyma cells, 322
Aerial and satellite platforms, 301
Aeroponics, 40–41
AFLPs. *See* Amplified fragment length polymorphisms (AFLPs)
AgRenSeq, 64
Agricultural/agriculture, 83
  production, interconnected challenges of, 83
  productivity, 122
Agrobacterium-mediated transformation, 126
Agronomic abiotic stress breeding program, 95
Agrotechnical techniques, 268–269
Airborne thermography, 356
Altered phenotype, 110
American Heart Association, 212
Amplified fragment length polymorphisms (AFLPs), 60
Anaerobic germination, 319
Anopheles gambia, 111
Anthocyanidin reductase (ANR), 306
Apple, 276–279
APX. *See* Ascorbate peroxidase (APX)
Arab Fund for Economic and Social Development (AFESD), 362
Arabidopsis, 131–132
Arabidopsis thaliana, 248
ArduCrop wireless infrared thermometer, 356
Artificial intelligence, 284–285
Artificial lighting, 23, 344
Ascorbate peroxidase (APX), 115
Automated phenotyping, 47
Automatic control processes, 178–179, 179*f*

B+LAN architecture, 173, 174*f*
Backcrossing, 19
Bait library, 62–64, 63*f*
Barley, 44
Bayesian linear regression (BLR), 365
Big data, 284–285
Big data-driven breeding, 85
Biodiversity, 296–298
Biotechnological methods, 269–285, 271–273*t*
Biotechnological research, 7*f*
Biotic stresses, 180
Blueberry, 281
BR response. *See* Brassinosteroid (BR) response
Brassica, 44
Brassinosteroid (BR) response, 130
Breeding. *See also* Speed breeding
  acceleration approaches, 266–267
  by design, 65–67
  programs, 161
  and selection programs in grapevine, 298
Breeding strategies, for submergence tolerance in the pre-genomic era, 323–325
Broad-spectrum multi-pathogen resistance, 130
Bulbosum, 358

Canola, 44
Capture-based target enrichment, 62–64, 63*f*
Carbon dioxide (>400 ppm), 346
  concentration, 42
  levels, 25
Cas9 enzyme, 351
Central, West Asia, and North Africa (CWANA), 362
Cereal crop breeding techniques, 189*f*
  based on controlled mating, 190–193
  based on observable variations, 188–190
  genome editing techniques, 193*t*
  hybridization breeding, 190, 192*t*
  idiotype plant breeding, 190–193, 191*f*
  instinctive selection by farmers for building landraces, 189
  mass selection and pure-line selection, 189–190, 192*t*

origin of crops and plant domestication, 188–189
pedigree analysis, 190, 192*t*
population breeding, 191, 192*t*
rapid generation advancement, 194–202
Cereals
crops, 27–28
fast-forwarding genetic gain in, 60–62
CGIAR, implementing speed breeding in, 361–362
Chickpea, 44
Chilling, 214*t*
Chromatin immunoprecipitation sequencing (ChIP-Seq), 6
Chromosomal doubling, 74
Chromosomes, 8–9
CIMMYT, 357
Circadian cycles, 182
related genes, 3
Circadian rhythms, 182
Citrus, 274–276
Climate change, 83, 149, 215*t*, 319
and planting material, 294–296
wine sector challenges, 294–296, 295*f*
Climate-smart agriculture
advancing agriculture through plant phenomics, 85–87
data-driven approaches for phenotyping, 87–88
data processing, handling, and harmonizing, 92
emergence of new plant breeding technologies, 83
multi-omics, AI, and next generation sequencing, 95
omics-driven breeding, 84
plant phenomics, 84–85
plant phenotyping with sensors, 88–92
speed breeding and NPBTS, 93–95
Clonal selection, 343
Colchicine, 357
Color imaging modality, 181, 183
Colored tomatoes genotypes, 129
Commercial sensor choices, 88
Controlled environment agriculture, 47
Controlled-environment chamber conditions, 25
Conventional breeding, 18–19, 152
methods, 23*f*
Conventional conveyor systems, 86
Conventional plant breeding methods, 38
Conventional transformation methods, 126
Core Adaptive Genomic Constitutions (CAGCs), 67
Cotton, 4, 6
Covering, 25
CRISPR
-associated nucleases, 201
employment of, 129
mediated directed evolution of crops, 115
screen, gene stacking in crops, and gene drive, 110–111
technology aids, 105
CRISPR associated protein (CRISPR/Cas), 246–247
CRISPR/Cas system for plant speed breeding, 67, 69–70, 350
in agriculture, 70
base transitions and transversions, 124–125
deployment of genome editing tools in plants, 125–127, 127*f*
in different crops, 70–72
limitation in plants, 138
machinery for crop improvement, 127–137
variants, 133*f*
CRISPR/Cas9 System, 127–128
genome editing system, high-quality tomato using, 257–258, 258*f*
high disease resistance, genome editing system for, 258–259, 260*f*
and precision-designed breeding, 254–261
speed germplasm generation and precision-designed breeding, 254–261
systems for crop improvement, 133–137*t*
weight enhancement, 255–257, 257*f*
working mechanism of, 124*f*
CRISPR/Cpf1 technology, 363
Crop improvement, 38, 149
advanced technical breakthroughs to enhance, 74*f*
multiplex genome editing technologies for, 115
program, 343
speed breeding for, 42*f*
transformative revolution in, 38
Crop speed breeding. *See also* Speed breeding
challenges and opportunities in global scenario, 9–10
collaboration in speed breeding research of, 10
emerging trends and technological innovation, 3–9, 7*f*
investments in speed breeding technologies, 11
speed breeding and regulatory challenges in crop varieties release, 10–11
Crop Trust Project, 362
Crops
agronomic performances of, 107
characteristics, 90*t*
enhancement, 83
plants, speed breeding of, 107
varieties, genetic advancement of, 151
wild relatives of, 110
Cryopreservation, 298
Cultivars, 40
Cutting-edge computer vision systems, 366
Cutting-edge technologies such, 356
Cytoplasmic male sterility (CMS), 111
Cytosine base editor (CBE), 124, 125

Data management
flow from acquisition to local storage, 177
flow from local to remote, 178
Data processing techniques, 197*t*
De novo domestication, 110
wild crops for, 109*t*
Dead Cas9 (dCas9), 363
Deaminases, 124
Deepwater rice cultivars, 322
Desirable traits, 43
Desired genetic variations, 19
Dhalputtia, 320
Differentially expressed proteins (DEPs), 323
Digital image process, 85
Discoverability, 307
Disease resistance, 44, 129–130
DIY sensors, 89
DMR6 (downy mildew resistant 6), 259
DNA markers, 58–59
advancement in marker platforms, 60
fast-forwarding genetic gain in cereals, 60–62
DNA methylation, 5, 6
DNA-recognition domain, 105

Domestication syndrome, 108*t*, 189
of crops, 107
Donor mindset, 348
Double haploidy (DH), 200, 346, 355, 357–361
technology, 8
Double-stranded breaks (DSBs), 123, 243, 247*f*, 363
Double-stranded RNAs (dsRNAs), 8, 63
Drip irrigation, 25
Drones, 55
Drought
abiotic stresses, 214–215*t*, 216
protein variations in soybean, 221*t*, 224
tolerance, 44

Early generation selection, 43
Electromagnetic spectrum, 85–86
Elite cultivars, 122
Embryo culture, 345
EMPHASIS, 308
Envirotyping, 160
Epigenetic regulation, 5
Epigenome editing (EE), 363
EPSPS gene, 132
Error management
automatic control processes, 178–179, 179*f*
user intervention, 180
ESTs (expressed sequence tags), 60
Ethylene response factors (ERF-VII), 321
Ethylene-responsive factor (ERF) gene, 320
Evaporative cooling, 25
Extended light period (photoperiod), 344

FAIR data principles, 307, 310
Fasciated (fas), 256
Fast-forwarding genetic gain in cereals, 60–62
Fast generation cycling (FGC), 18, 242
FasTrack breeding, 280
Fatty acid desaturase *(FAD2)* gene, 107
Field phenotyping, 299–301, 301*f*
Field selection methods, 37
Flood-prone low land conditions, 321
Flooding
abiotic stresses, 212, 213*t*, 216
protein variations in soybean, 218, 219–220*t*
Floods, 319–320
Fogging, 25
Food
production systems, 54
security, 83
Food and Agriculture Organization (FAO), 319
4-Hydroxyphenyl pyruvate dioxygenase (HPPD), 132
FR13A, 320, 323
Fruit and nut tree breeding, 265–266
Fruit crops, 28
Fruit trees speed breeding, 264–265
agrotechnical techniques, 268–269
apple, 276–279
big data and artificial intelligence, 284–285
biotechnological methods, 269–285, 271–273*t*
blueberry, 281
breeding acceleration approaches, 266–267
citrus, 274–276
fruit and nut tree breeding, 265–266
genetic engineering, 270
genome editing, 282–284
genomics-assisted breeding, 284
kiwifruit, 280–281
natural early flowering genotypes, 267–268
olive, 280
pear, 279
plum, 279–280
transgrafting, 281–282

γ-aminobutyric acid (GABA), 258
GE. *See* Genome editing (GE)
Gene editing, 41
Genes pyramiding, KASP markers for, 62
Genetic. *See also* Genome editing (GE) technologies
diversity, 43
engineering, 41, 106
transformation, 125
Genetic engineering, 270
Genic male sterility (GMS), 113, 114
Genome-based breeding, 66–67
Genome editing (GE) technologies, 43, 47, 123, 193*t*
CRISPR-mediated directed evolution of crops, 115
CRISPR screen, gene stacking in crops, and gene drive, 110–111
domestication of, 107–108
domestication of wild cultivars through genome editing, 107–108
genome editing accelerated re-domestication (GEARED), 108–110
groundcherry, 110
haploid inducer lines using, 111–113
male sterile lines, 113–114
multiplex genome editing technologies for crop improvement, 115
rice, 110
for speed breeding of crop plants, 107
techniques, 21, 157
varietal developments with yield optimization, 106–107
Genome editing (GnEd), to accelerate crop improvement, 362–364
Genome editing accelerated re-domestication (GEARED), 108–110
Genome editing system, 269, 282–284, 362–364
high disease resistance, 258–259, 260*f*
high-quality tomato using, 257–258, 258*f*
speed breeding of tomato. *See* Tomato breeding approaches
Genome selection (GS), 269
Genome sequencing (GS), 188, 195*t*, 349–350
Genome-wide association study (GWAS), 4–5, 58–59, 95, 192, 326
advancement in marker platforms, 60
fast-forwarding genetic gain in cereals, 60–62
KASP markers for genes pyramiding, 62
mapping, 61
Genomes, 85
sequencing, 4
Genomic-assisted breeding, 21
Genomic best linear unbiased prediction (GBLUP), 244, 365
Genomic estimated breeding values (GEBVs), 157, 161–162, 244, 349

Genomic selection, 65–67, 65*f*, 156–157
  application of, 152–156
  cycle, 161–162
  description of, 149–152
  genotyping in, 159–160
  implementation of, 159, 161
  models for, 157–159, 160
  phenotyping in, 160
  through speed breeding cycle, 162*f*
Genomics, 22
Genomics-assisted breeding (GAB), 284
Genomics Era
  CRISPR-CAS9 mechanism, 327
  dawn of base editing and its implications for, 327–328
  Quantitative Trait Loci (QTLs), 328–335
  transgenic methodologies to shield rice, 327
  using the NGS platform, 326–327
Genotyped testing set (TST), 244
Genotyping-by-sequencing (GBS), 244
  accelerated marker development through, 244
Genotyping in genomic selection, 159*f*
Glasshouse speed breeding conditions, 26
Global food security challenges, 44–46
Global network structure, 173–174, 174*f*
Global proteome profiling, 195–196*t*
Grapevine
  aerial and satellite platforms, 301–305
  biodiversity, 296–298
  breeding and selection, 298
  conservation of genetic resources, 296–298, 297*f*
  controlled and semi-controlled conditions, 299
  data collection and analysis, challenges in, 306–308
  data management and fair data in phenotyping, 306–309
  field phenotyping, 299–301, 301*f*
  modelling and phenotyping applied to crops and, 308–309
  molecular and genetic tools, 305–306
  relevance, conservation, and selection issues, 296–298
  screening and selection methodologies in, 298–305
Green Normalized Difference Vegetation Index (GNDVI), 304
Greenhouse environments, 39
  growth chambers, 39
Greenhouse phenotyping, 93*f*
GRF-interacting factor 1 (GIF1), 107
Ground-based equipment, 366
Ground-based field phenotyping, 300–301, 301*f*
Groundcherry, 110
Growth chamber, 39
Growth-regulating factor 4 (GRF4), 107
Guide RNA (gRNA), 105
GWAS. *See* Genome-wide association study (GWAS)

Haploid inducer lines, 111–112
  legumes, 113
  maize, 112–113
  millets, 113
  rice, 112
  wheat, 112
Haploid induction, 111
Heat stress, 215*t*, 217
Heat waves, 215*t*, 217
Herbicide resistance, 72, 132–137
Heterosis, 190
Hexokinase stretch-carrying gene HXK6, 329
High disease resistance, genome editing system for, 258–259, 260*f*
High-Throughout Phenotypic Platforms (HT3P), 188
High-throughput genotyping (HTG), 244
High-throughput phenotyping (HTP), 20, 55, 86, 89*f*, 92, 156, 159, 160, 201–202, 269, 299, 309
  drones, 55
  platforms, 20
  sensors and tools, 58
  with speed breeding, 93*f*
  technologies, 87*f*
  unmanned aerial vehicles (UAVs), 55
High-throughput screening, 201–202, 202*f*
High-throughput technologies, 29–30
Him Pratham, 358
Homemade growth room design, 26
Homozygous lines, production of, 105–106
Hordeum vulgare, 360
Hormones, 346
HTP. *See* High-throughput phenotyping (HTP)
Humidity, 27, 346
  regulation of, 25
Hybrid seed production, 113
Hybridization, 43
  breeding, 190, 192*t*
Hydroponics, 40
Hyperspectral sensors, 58
Hypoxia, 321

ICRISAT, 361
Idiotype plant breeding, 190–193, 191*f*
Imaging system, 170*f*
  environmental adaptations, 175–177
  global network structure, 173–174, 174*f*
  in humid environment, 175*f*
  mini-computer and sensor interactions, 169–173
  RGB-depth-infrared acquisition, 172*f*
  under transparent plastic sheeting, 176*f*
Infrastructure, 48
Integrating speed breeding with AI, 365–366
Intel RealSense D435 cameras, 169, 183–184
Intellectual property, 48
Intergeneric hybridization, 357
Intergovernmental Panel on Climate Change (IPCC), 319
International Center for Agricultural Research in the Dry Areas (ICARDA), 361
International Crops Research Institute for the Semi-Arid Tropics (ICARDA), 357
International Rice Research Institute (IRRI), 320
International Variety Catalogue (VIVC), 305
Intranet, 173
Isoflavonoids, 208, 211, 218
ISR42-8 of Hordeum vulgare ssp., 358

K-mer-based association mapping, 64
KASP markers for genes pyramiding, 62
Kiwifruit, 280–281

Landraces, building, 189
Leaf Gas Film1 (LGF1), 332
LED lighting systems, 22

Legumes, 28, 113
Leucoanthocyanidin reductase (LAR), 306
Lidar, 58
Light, 26–27
  duration, 24, 42
  intensity, 24, 42
  quality, 24
Light Detection and Ranging (LiDAR), 184, 299
Liquid level sensors, 58
LOB1 gene, 72
Local area network (LAN), 173
Locule number (lc), 256
Lycopene, 257

Machine learning techniques (ML), 6, 47, 366
main.py program, role of, 172
Maize, 3, 6, 9, 112–113
  cultivation context, 9
Male sterile (MS)
  lines in plants, 114
  mediated development of, 113–114, 114*t*
Map-based cloning, 108
Marker aided selection (MAS), 243
Marker-assisted backcrossing (MABC) system, 320, 355
Marker-assisted breeding and genomic editing, 200–201, 201*f*
Marker-assisted selection (MAS), 21, 43, 59, 95, 269, 305, 325, 343, 355, 364–365
  and genomic selection, 362–364
Marker platforms, advancement in, 60
Mass selection, 189–190
Meganucleases, 246
Metabolomics, 22, 83
Microsatellites. *See* Simple sequence repeats
Mildew locus (MLO) gene in, 72
Millets, 113
Mini-computer and sensor interactions, 169–173
Modern cultivars, narrow genetic base of, 59*f*
Modern phenotyping, 86
Modern plant biotechnology, 9
Molecular-assisted selection (MAS) technique, 349
Molecular biology, 21
Molecular breeding, 21–22
Monoploids, 111
Morphogenic regulators (MRs), 127
MS. *See* Male sterile (MS)
MTL gene, 112
Multiple breeding cycles, 67
Multispectral sensors, 58
Mutation breeding, 19–20, 122

N-acetylserotonin methyltransferase (*ASMT*), 306
National Aeronautics and Space Administration (NASA), 358
National Lipid Association, 212
Natural diverse populations, 61
Natural early flowering genotypes, 267–268
Natural genetic diversity, 105
Necrotrophic fungi, 259
Network time protocol (NTP), 173
New breeding techniques (NBTs), 194
New plant breeding technologies (NPBTs), 83, 266, 269
  with speed breeding, 93
Next-Gen AI approach, 366
Next-generation sequencing (NGS), 21, 54, 65–67, 326
Normalized Difference Vegetation Index (NDVI), 303–304
Normalized Difference Water Index (NDWI), 304
Novel breeding strategies, integration of, 74
Novel breeding technologies (NBTs), 53–54, 54*f*
Novel crop variety, 151
Novel traits of crops, 156
NPBTs. *See* New plant breeding technologies (NPBTs)
Nutrients, 346
  availability, 27, 42
  content, 44
Nutritional quality, 156

Off-season seeding techniques, 355
Off-target effects, 126
Oilseed crops, 28
Olive, 280
OMICS, 85
  based strategies, for crop improvement, 195–197*t*
  integration of, 22
  technologies, 6, 95, 365
  tools, 5
Optical sensors, 58
Optogenetics, 47
Origin of crops and plant domestication, 188–189
Ornamental herbaceous plants, 86
Oxidative stress, 322

Passive remote sensing, 88–89
Pea, 44
Pear, 279
Pedigree analysis, 190, 192*t*
Phenomics, 83, 84, 85
  in plant biology, 87*f*
  technologies, 85
  transition, 54–55
Phenotyping, 87, 92
  data collection and processing for, 87–88
  of field crop lines, 201–202
  platforms, 87–88
Phosphatidylethanolamine binding protein (PEBP), 270
Photomorphogenesis, 152
Photoperiod regime, manipulation of, 23–24
Photoperiodic flowering, 218
Photosynthetic activity indices, 304
Photosynthetic process, 152
Photosynthetic Reflectance Index (PRI), 304
Phyto-hormonal-driven cellular adaptive mechanisms, 322
Phytohormones, 218
Plant biomass indices, 303–304
Plant growth regulator (PGR), 345
Plant phenotyping
  hardware platforms, 91–92*t*
  procedure, 89*f*
  software platforms, 91*t*
Plant stress response, 168–169
  modalities and, 180–183, 180*t*, 181–183*f*. *See also* RGB-Depth cameras
Plant-to-sensor system, 299
Plants
  biotechnology, 9–10
  breeding techniques, 18

genotypes, selection of, 43
physiology, 40
transformation, 126–127
Plum, 279–280
Polyploidy crops, 111
Population breeding, 191, 192*t*
Population growth, 53
Portuguese Association for Grapevine Diversity (PORVID), 297
Power over ethernet (PoE), 175
Pre-breeding, 54
Pre-processing and data assessment, 197*t*
Precision-designed breeding
genome editing system, high-quality tomato using, 257–258, 258*f*
high disease resistance, genome editing system for, 258–259, 260*f*
and precision-designed breeding, 254–261
speed germplasm generation and, 254–261
weight enhancement, 255–257, 257*f*
Presence-absence variations (PAVs), 192
Primary editing (PE) technique, 72, 363
Prime editing gRNA (pegRNA), 363
Prime editor (PE), 124
Principal genome-editing technologies, 68*f*
Protein variations in soybean, 218–226, 219–223*t*
drought, 221*t*, 224
flooding, 218, 219–220*t*
salinity, 221*t*, 224–225
temperature, 225–226
Proteomics, 22, 83, 197*t*, 207–208. *See also* Soybean
Public acceptance, 48
Pure-line selection, 189–190, 192*t*

QTL. *See* Quantitative trait loci (QTL)
Quality enhancement, 128–129
Quantitative trait loci (QTLs), 63, 95, 243, 244, 260–261, 305, 320, 328–335, 355
alleles, 163
mapping, 59, 60, 61

Radio frequency identification (RFID), 310
Rapid breeding method's, 345
Rapid generation advancement (RGA), 9, 43, 194–202, 355
advantages and disadvantages, 198*f*
double haploidy, 200
evolution, 194*f*
high-throughput screening, 201–202, 202*f*
marker-assisted breeding and genomic editing, 200–201, 201*f*
omics-based strategies, 195–197*t*
phenotyping of field crop lines, 201–202
shuttle breeding, 200, 200*f*
single-seed descent method, 198, 199*f*
speed breeding, 199–200, 199*f*
Rapid generation cycling (RGC), 242
Rapid generation turnover (RGT), 242
Reactive oxygen species (ROS), 321
Recombinant inbred lines (RIL), 329, 360
Regulatory barriers, 47
Remote Desktop Protocol (RDP), 180
Research Infrastructures (RI), 308
Resistance (R) genes, 63
resistance to blast 1(RBL1) gene, 130
Resource deficiency, 162–163
Restriction fragment length polymorphism (RFLP), 114
Reverse breeding, 245–246
Reverse transcription (RT) system, 124, 363
Revolutionized breeding techniques, 72
RGB-Depth cameras, 169
data management, 177–178
error management, 178–180
imaging system, 169–177
mail template for alert, 179
Ribonucleoproteins (RNPs), 127*f*
Rice, 110, 112
Rice submergence tolerance
breeding strategies for, 323–326
Genomics Era, 326–335
molecular breeding strategies, 319–321
Ridge regression (RR), 158
Ridge regression best linear unbiased prediction (RR-BLUP), 244
statistical model, 365
RNA interference (RNAi), 41–42
Robotic field-based phenotyping system, 366

Salinity
abiotic stresses, 213–214*t*, 216–218
protein variations in soybean, 221–222*t*, 224–225
Sanger sequencing, 189
SCAR (sequence characterized amplified region), 114
Secure Socket Shell (SSH), 180
Seed chipping, 349
Seed harvesting, 1
Self Prunning (SP) and *Self Prunning 5 G (SP5G)*, 247
Sensor interactions, mini-computer and, 169–173
Sensor-to-plant system, 299
Sensors, and tools, 58
Sequence-specific nucleases (SSNs), 243
Sequencing technologies, 123
1726 DArT markers, 365
Short hairpin RNAs (shRNAs), 8
Shuttle breeding, 200, 200*f*, 356
Simple mail transfer protocol (SMTP), 179
Simple sequence repeats (SSRs), 305
Single-cross hybrids (SCHs), 111
Single guide RNA (sgRNA), 127*f*
Single nucleotide polymorphism (SNP), 4, 60, 114, 128, 161, 243, 305–306, 364
platforms utilized in crops, 60*t*
Single-nucleotide variants (SNVs), 138
Single plant selection (SPS), 343
Single pod descent (SPD), 343
Single primer enrichment technology (SPET), 244
Single-seed descent (SSD) method, 18, 198, 199*f*, 343, 358
Single sequence repeats (SSRs), 60, 114
Single-stranded DNA (ssDNA), 63
Single-stranded guide RNA (sgRNA), 351
Site-directed nuclease, 1, 364
Site-directed nucleases (SDN), 106
Slow growth storage (SGS), 297–298
SNF1-related kinase, 321
SNP. *See* Single nucleotide polymorphism (SNP)
Soil-plant atmosphere continuum (SPAC), 95
Solanum lycopersicon L., 242–248

Solving Mixed Model Equations in R (SOMMER), 365
Soybean, 3, 5, 6, 44, 207–208
  abiotic stresses and, 212–218, 213–215*t*
  benefits and hazards associated with, 211–212, 211*f*
  oil and human health, 209–212
  oils and oil quality in, 209–210
  processing of, 210*f*
  protein variations in, 218–226, 219–223*t*, 225*f*
  stress mitigation under abiotic stress, 226–227, 227*f*
Speed breeding (SB), 73–74, 199–200, 199*f*
  advantages, 348
  with AI, 365–366
  applications of, 94*t*
  approaches used in, 344
  carbon dioxide (>400 ppm), 346
  cereal crops, 27–28
  in CGIAR, 361–362
  challenges, 30, 47–48, 348–349
  controlled-environment chamber conditions, 25
  conventional breeding, 18–19
  crops, 27, 43–44
  in crops to accelerate plant improvement, 354–356
  doubled haploid, 357–361
  embryo culture, 345
  emerging technologies and innovations in, 47
  extended light period (photoperiod), 344
  fruit crops, 28
  genome editing, 246–248, 362–364
  glasshouse speed breeding conditions, 26
  high-throughput technologies augmenting, 29–30
  homemade growth room design for, 26
  humidity, 27, 346
  increased crop generation through, 153–155
  legumes crops, 28
  light, 26–27
  manipulation of photoperiod regime, 23–24
  marker-assisted selection and, 364–365
  modifying carbon dioxide levels, 25
  molecular breeding, 19–20, 21–22
  nutrient availability, 27
  nutrients, hormones, and tissue culture, 346
  oilseed crops, 28
  parents for, 162
  physical factors affecting plant responses, 26
  process of, 2–3
  programs, 149
  regulation of humidity, 25
  regulation of temperature, 24
  revolutionizing, 22–23, 23*f*
  schematic representation of, 73*f*
  selection methods used in, 156*f*
  shuttle breeding, 356
  techniques, 6, 10, 11
  technologies, 2, 23
  temperature, 27
  temperature treatment, 344–345
  with time, 152*f*
  tools for crop improvement, 150–151*t*
  unit, 2, 2*f*
  value of, 4
  in vegetable crops, 29, 343–344
Speed cloning, 62
  bait library and capture-based target enrichment, 62–64
  k-mer-based association mapping, 64
Speed germplasm generation
  genome editing system, high-quality tomato using, 257–258, 258*f*
  high disease resistance, genome editing system for, 258–259, 260*f*
  and precision-designed breeding, 254–261
  weight enhancement, 255–257, 257*f*
Statistical design, development of, 157–158
Sub1 rice, 320
SUB1A gene, 326
Sucrose synthase gene, 321
Suitable statistical analysis, 158
Sustainable agricultural production, 53

Target environments (TEs), 67
Target Ideal Genomic Constitutions (TIGCs), 67
Targeted breeding, 83
Targeted gene, 71*t*, 110
Targeted mutation, 362
Targeted proteome profiling, 196*t*
Technical expertise, 48
Temperature, 27
  abiotic stresses, 217
  and humidity, 42
  protein variations in soybean, 225–226
  regulation of, 24
  treatment, 344–345
Thermal cameras, 58
Tissue culture, 41, 346
Tolerance against abiotic factors, 130–132
Tomato, 44, 109
Tomato breeding approaches, 243
  accelerated marker development through genotyping by sequencing, 244
  acceleration through genomic selection, 244–245, 245*f*
  double-stranded breaks, 247*f*
  molecular marker aided selection in, 243
  reverse breeding, 245–246
  speed breeding, genome editing approaches for, 246–248
tomato yellow leaf curl virus, 130
Traditional breeding, 1
  methods, 46
  practices, 122
Training population (TP), 161, 285, 365
Transcription activator-like effector nucleases (TALENs), 67, 123, 246, 247*f*
Transcriptomics, 5, 22, 83
Transformation methods, 126
Transgenic technology, 41
Transgrafting, 281–282
Trehalose-6-phosphate gene (qAG9.2), 329
Triticum aestivum, 360
Troubleshooting. *See* Error management

Uninterruptible power supply (UPS) solution, 178
Unmanned aerial vehicles (UAVs), 55, 56–57*t*, 356, 366
User intervention, 180

Vegetable crops, 29
Vegetation indices (VI), 181, 183, 303–304
Versatile crops, 151

Vertical shoot position (VSP) system, 300
Virtual LAN (VLAN), 173, 174*f*
Virtual Networking Computing (VNC), 180
Visual (RGB) sensors, 58
Vitamin D, 258
Vitis germplasm, 296
Vitis vinifera varieties, 305

Water index (WI), 304
Water stress indices, 304
Weight enhancement, 255–257, 257*f*
Wheat, 43–44, 112
  progenitor genomes, 60
Whole genome sequencing (WGS), 138
Wild crops for de novo Domestication, 109*t*
Wild cultivars, domestication of, 107–108
Wine production, 294–296, 295*f*

Year-round production, 40
Yield, 44
  improvement, 127–128
  optimization, varietal developments with, 106–107

Zea mays, 130
Zinc finger nucleases (ZFNs), 21, 67, 123, 246, 247*f*, 283
ZmPLA1 gene, 112–113